Mahesh J. Shah

ELECTRICAL NETWORK THEORY

ELECTRICAL NETWORK THEORY

NORMAN BALABANIAN
THEODORE A. BICKART

Syracuse University

with contributions from the work of the late

SUNDARAM SESHU
University of Illinois

JOHN WILEY & SONS, INC.
NEW YORK · LONDON · SYDNEY · TORONTO

Library of Congress Catalog Card Number: 69-16122
SBN 471 04576 4
Printed in the United States of America

PREFACE

This book was initially conceived as a revision of *Linear Network Analysis* by Sundaram Seshu and Norman Balabanian in the summer of 1965. Before the work of revision had actually started, however, Seshu died tragically in an automobile accident. Since then the conceived revision evolved and was modified to such a great extent that it took on the character of a new book and is being presented as such. We (especially Norman Balabanian) wish, nevertheless, to acknowledge a debt to Seshu for his direct and indirect contributions to this book.

The set of notes from which this book has grown has been used in a beginning graduate course at Syracuse University and at the Berkeley campus of the University of California. Its level would also permit the use of selected parts of the book in a senior course.

In the study of electrical systems it is sometimes appropriate to deal with the internal structure and composition of the system. In such cases topology becomes an important tool in the analysis. At other times only the external characteristics are of interest. Then " systems " considerations come into play. In this book we are concerned with both internal composition and system, or port, characteristics.

The mathematical tools of most importance are matrix analysis, linear graphs, functions of a complex variable, and Laplace transforms. The first two are developed within the text, whereas the last two are treated in appendices. Also treated in an appendix, to undergird the use of impulse functions in Chapter 5, is the subject of generalized functions. Each of the appendices constitutes a relatively detailed and careful development of the subject treated.

In this book we have attempted a careful development of the fundamentals of network theory. Frequency and time response are considered, as are analysis and synthesis. Active and nonreciprocal components (such as controlled sources, gyrators, and negative converters) are treated side-by-side with passive, reciprocal components. Although most of the book is limited to linear, time-invariant networks, there is an extensive chapter concerned with time-varying and nonlinear networks.

v

Matrix analysis is not treated all in one place but some of it is introduced at the time it is required. Thus introductory considerations are discussed in Chapter 1 but functions of a matrix are introduced in Chapter 4 in which a solution of the vector state equation is sought. Similarly, equivalence, canonic forms of a matrix, and quadratic forms are discussed in Chapter 7, preparatory to the development of analytic properties of network functions.

The analysis of networks starts in Chapter 2 with a precise formulation of the fundamental relationships of Kirchhoff, developed through the application of graph theory. The classical methods of loop, node, node-pair, and mixed-variable equations are presented on a topological base.

In Chapter 3 the port description and the terminal description of multiterminal networks are discussed. The usual two-port parameters are introduced, but also discussed are multiport networks. The indefinite admittance and indefinite impedance matrices and their properties make their appearance here. The chapter ends with a discussion of formulas for the calculation of network functions by topological concepts.

The state formulation of network equations is introduced in Chapter 4. Procedures for writing the state equations for passive and active and reciprocal and nonreciprocal networks include an approach that requires calculation of multiport parameters of only a resistive network (which may be active and nonreciprocal). An extensive discussion of the time-domain solution of the vector state equation is provided.

Chapter 5 deals with integral methods of solution, which include the convolution integral and superposition integrals. Numerical methods of evaluating the transition matrix, as well as the problem of errors in numerical solutions, are discussed.

Chapters 6 and 7 provide a transition from analysis to synthesis. The sufficiency of the real part, magnitude, or angle as specifications of a network function are taken up and procedures for determining a function from any of its parts are developed. These include algebraic methods as well as integral relationships given by the Bode formulas. Integral formulas relating the real and imaginary parts of a network function to the impulse or step response are also developed. For passive networks the energy functions provide a basis for establishing analytic properties of network functions. Positive real functions are introduced and the properties of reactance functions and RC impedance and admittance functions are derived from them in depth. Synthesis procedures discussed for these and other network functions include the Darlington procedure and active RC synthesis techniques.

Chapter 8 presents a thorough treatment of scattering parameters and the description of multiport networks by scattering matrices. Both real and complex normalization are treated, the latter including single-frequency and frequency-independent normalization. Reflection and transmission properties of multiport networks, both active and passive, reciprocal and non-reciprocal, are developed in terms of scattering parameters. Applications to filter design and negative resistance amplifiers are discussed.

Concepts of feedback and stability are discussed in Chapter 9. Here the signal flow-graph is introduced as a tool. The Routh-Hurwitz, Liénard-Chipart, and Nyquist criteria are presented.

The final chapter is devoted to time-varying and nonlinear networks. Emphasis is on general properties of both types of network as developed through their state equations. Questions of existence and uniqueness of solutions are discussed, as are numerical methods for obtaining a solution. Attention is also devoted to Liapunov stability theory.

A rich variety of problems has been presented at the end of each chapter. There is a total of 460, some of which are routine applications of results derived in the text. Many, however, require considerable extension of the text material or proof of collateral results which, but for the lack of space, could easily have been included in the text. In a number of the chapters a specific class of problems has been included. Each of these problems, denoted by an asterisk, requires the preparation of a computer program for some specific problem treated in the text. Even though writing computer programs has not been covered and only a minimal discussion of numerical procedures is included, we feel that readers of this book may have sufficient background to permit completion of those problems.

A bibliography is presented which serves the purpose of listing some authors to whom we are indebted for some of our ideas. Furthermore, it provides additional references which may be consulted for specific topics.

We have benefited from the comments and criticisms of many colleagues and students who suggested improvements for which we express our thanks.

Syracuse, New York   N. Balabanian
November, 1968   T. A. Bickart

CONTENTS

ELECTRICAL NETWORK THEORY

.1.

FUNDAMENTAL CONCEPTS

1.1 INTRODUCTION

Electric network theory, like many branches of science, attempts to describe the phenomena that occur in a portion of the physical world by setting up a mathematical model. This model, of course, is based on observations in the physical world, but it also utilizes other mathematical models that have stood the test of time so well that they have come to be regarded as physical reality themselves. As an example, the picture of electrons flowing in conductors and thus constituting an electric current is so vivid that we lose sight of the fact that this is just a theoretical model of a portion of the physical world.

The purpose of a model is to permit us to understand natural phenomena; but, more than this, we expect that the logical consequences to which we are led will enable us to predict the behavior of the model under conditions we establish. If we can duplicate in the physical world the conditions that prevail in the model, our predictions can be experimentally checked. If our predictions are verified, we gain confidence that the model is a good one. If there is a difference between the predicted and experimental values that cannot be ascribed to experimental error, and we are reasonably sure that the experimental analogue of the theoretical model duplicates the conditions of the model, we must conclude that the model is not " adequate " for the purpose of understanding the physical world and must be overhauled.*

---

* An example of such an overhaul occurred after the celebrated Michelson-Morley experiment, where calculations based on Newtonian mechanics did not agree with experimental results. The revised model is relativistic mechanics.

In the case of electric network theory, the model has had great success in predicting experimental results. As a matter of fact, the model has become so real that it is difficult for students to distinguish between the model and the physical world.

The first step in establishing a model is to make detailed observations of the physical world. Experiments are performed in an attempt to establish universal relationships among the measurable quantities. From these experiments general conclusions are drawn concerning the behavior of the quantities involved. These conclusions are regarded as "laws," and are usually stated in terms of the variables of the mathematical model.

Needless to say, we shall not be concerned with this step in the process. The model has by now been well established. We shall, instead, introduce the elements of the model without justification or empirical verification. The process of abstracting an appropriate interconnection of the hypothetical elements of the model in order to describe adequately a given physical situation is an important consideration, but outside the scope of this book.

This book is concerned with the theory of *linear electric networks*. By an electric *network* is meant an interconnection of electrical devices forming a structure with accessible points at which signals can be observed. It is assumed that the electrical devices making up the network are represented by *models*, or hypothetical elements whose voltage-current equations are linear equations—algebraic equations, difference equations, ordinary differential equations, or partial differential equations. In this book we shall be concerned only with lumped networks; hence, we shall not deal with partial differential equations or difference equations.

The properties of networks can be classified under two general headings. First, there are those properties of a network that are consequences of its structure—the *topological properties*. These properties do not depend on the specific elements that constitute the branches of the network but only on how the branches are interconnected; for example, it may be deduced that the transfer function zeros of a ladder network (a specific topological structure) lie in the left half-plane regardless of what passive elements constitute the branches. Second, there are the properties of networks as signal processors. Signals are applied at the accessible points of the network, and these signals are modified or processed in certain ways by the network. These *signal-processing properties* depend on the elements of which the network is composed and also on the topological structure of the network. Thus if the network elements are lossless, signals are modified in certain ways no matter what the structure of the network; further limitations are imposed on these properties by the structure. The properties of lossless ladders, for example, differ from those of lossless lattices. We shall be concerned with both the topological and the signal-processing properties of networks.

1.2 ELEMENTARY MATRIX ALGEBRA

In the analysis of electric networks, as in many other fields of science and engineering, there arise systems of linear equations, either algebraic or differential. If the systems contain many individual equations, the mere process of writing and visualizing them all becomes difficult. Matrix notation is a convenient method for writing such equations. Furthermore, matrix notation simplifies the operations to be performed on the equations and their solution. Just as one learns to think of a space vector with three components as a single entity, so also one can think of a system of equations as one matrix equation. In this section, we shall review some elementary properties of matrices and matrix algebra without great elaboration. In subsequent chapters, as the need for additional topics arises, we shall briefly digress from the discussion at hand for the purpose of introducing these topics.

A *matrix* is a rectangular *array* of quantities arranged in *rows* and *columns*, each quantity being called an *entry*, or *element*, of the matrix. The quantities involved may be real or complex numbers, functions of time, functions of frequency, derivative operators, etc. We shall assume that the entries are chosen from a "field"; that is, they obey an algebra similar to the algebra of real numbers. The following are examples of matrices:

$$\mathbf{M}=\begin{bmatrix}1 & 2\\ 5 & -1\\ 0 & 6\end{bmatrix}, \quad \mathbf{Z}=\begin{bmatrix}\dfrac{s}{s+2} & 2s\\ 2s & 3s^2\end{bmatrix}, \quad \mathbf{V}=\begin{bmatrix}V_1\\ V_2\\ V_3\\ V_4\end{bmatrix}.$$

Square brackets are placed around the entries to enclose the whole matrix. It is not necessary to write the whole matrix in order to refer to it. It is possible to give it a "name" by assigning it a single symbol, such as **M** or **V** in the above examples. We shall consistently use boldface letters, either capital or lower case, to represent matrices.

The *order* of a matrix is an ordered pair of numbers specifying the number of rows and number of columns, as follows: (m, n) or $m \times n$. In the examples above, the orders are $(3, 2)$, $(2, 2)$, and $(4, 1)$, respectively. When the alternate notation is used, the matrices are of order 3×2, 2×2, and 4×1, respectively. The latter is a special kind of matrix called a *column* matrix, for obvious reasons. It is also possible for a matrix to be of order $1 \times n$; such a matrix has a single row and is called a *row* matrix. A matrix in which the number of rows is equal to the number of columns is called a *square* matrix. In the above examples, **Z** is square. For the special cases in which the type of matrix is a column matrix, row matrix, or square matrix the order is determined unambiguously by a single number, which is the number of rows, columns, or either, respectively; for example, if **M** is a square matrix with n rows and n columns, it is of order n.

In order to refer to the elements of a matrix in general terms, we use the notation

$$\mathbf{A} = [a_{ij}]_{m,n}.$$

If the order (m, n) is not of interest, it need not be shown in this expression. The "typical element" is a_{ij}. The above simple expression stands for the same thing as

$$\mathbf{A} = \begin{bmatrix} a_{11} & a_{12} & a_{13} \cdots a_{1n} \\ a_{21} & a_{22} & a_{23} \cdots a_{2n} \\ \cdot & \cdot & \cdot \quad \cdot \\ \cdot & \cdot & \cdot \quad \cdot \\ \cdot & \cdot & \cdot \quad \cdot \\ a_{m1} & a_{m2} & a_{m3} \cdots a_{mn} \end{bmatrix}. \tag{1}$$

BASIC OPERATIONS

Equality. Two matrices $\mathbf{A} = [a_{ij}]$ and $\mathbf{B} = [b_{ij}]$ are said to be equal if they are of the same order and if corresponding elements of the two matrices are identical; that is, $\mathbf{A} = \mathbf{B}$ if $a_{ij} = b_{ij}$ for all i and j.

Multiplication by a Scalar. To multiply a matrix $\mathbf{A} = [a_{ij}]$ by a scalar (i.e., an ordinary number) k, we multiply each element of the matrix by the scalar; that is, $k\mathbf{A}$ is a matrix whose typical element is ka_{ij}.

Addition of Matrices. Addition is defined only for matrices of the same order. To add two matrices we add corresponding elements. Thus if

$\mathbf{A} = [a_{ij}]$ and $\mathbf{B} = [b_{ij}]$, then

$$\mathbf{A} + \mathbf{B} = [a_{ij} + b_{ij}]. \qquad (2)$$

Clearly, addition is commutative and associative; that is,

$$\mathbf{A} + \mathbf{B} = \mathbf{B} + \mathbf{A}$$
$$\mathbf{A} + (\mathbf{B} + \mathbf{C}) = (\mathbf{A} + \mathbf{B}) + \mathbf{C}. \qquad (3)$$

Multiplication of Matrices. If $\mathbf{A} = [a_{ij}]_{m,n}$ and $\mathbf{B} = [b_{ij}]_{n,p}$, then the product of $\mathbf{A}$ and $\mathbf{B}$ is defined as

$$\mathbf{A}\mathbf{B} = \mathbf{C} = [c_{ij}]_{m,p} \qquad (4)$$

where the elements of the product are given by

$$c_{ij} = \sum_{k=1}^{n} a_{ik} b_{kj} = a_{i1}b_{1j} + a_{i2}b_{2j} + \cdots + a_{in}b_{nj}. \qquad (5)$$

That is, the (i, j)th element of the product is obtained by multiplying the elements of the ith row of the first matrix by the corresponding elements in the jth column of the second matrix, then adding these products. This means that multiplication is defined only when the number of columns in the first matrix is equal to the number of rows in the second matrix. Note that the product matrix $\mathbf{C}$ above has the same number of rows as the first matrix and the same number of columns as the second one.

Example:

$$\begin{bmatrix} a_{11} & a_{12} & a_{13} \\ a_{21} & a_{22} & a_{23} \end{bmatrix} \begin{bmatrix} b_{11} & b_{12} & b_{13} \\ b_{21} & b_{22} & b_{23} \\ b_{31} & b_{32} & b_{33} \end{bmatrix}$$

$$= \begin{bmatrix} (a_{11}b_{11} + a_{12}b_{21} + a_{13}b_{31}) & (a_{11}b_{12} + a_{12}b_{22} + a_{13}b_{32}) & (a_{11}b_{13} + a_{12}b_{23} + a_{13}b_{33}) \\ (a_{21}b_{11} + a_{22}b_{21} + a_{23}b_{31}) & (a_{21}b_{12} + a_{22}b_{22} + a_{23}b_{32}) & (a_{21}b_{13} + a_{22}b_{23} + a_{23}b_{33}) \end{bmatrix}.$$

When the product $\mathbf{A}\mathbf{B}$ is defined (i.e., when the number of columns in $\mathbf{A}$ is equal to the number of rows in $\mathbf{B}$), we say that the product $\mathbf{A}\mathbf{B}$ is *conformable*. It should be clear that the product $\mathbf{A}\mathbf{B}$ may be conformable whereas $\mathbf{B}\mathbf{A}$ is not. (Try it out on the above example.) Thus $\mathbf{A}\mathbf{B}$ is not necessarily equal to $\mathbf{B}\mathbf{A}$. Furthermore, this may be the case even if both products are conformable. Thus let

$$\mathbf{A} = \begin{bmatrix} 1 & -1 \\ 1 & 0 \end{bmatrix}, \qquad \mathbf{B} = \begin{bmatrix} 1 & 0 \\ 1 & 1 \end{bmatrix}.$$

Then

$$\mathbf{AB} = \begin{bmatrix} 0 & -1 \\ 1 & 0 \end{bmatrix} \quad \text{and} \quad \mathbf{BA} = \begin{bmatrix} 1 & -1 \\ 2 & -1 \end{bmatrix}$$

which shows that $\mathbf{AB} \neq \mathbf{BA}$ in this case.

We see that matrix multiplication is not commutative as a general rule, although it may be in some cases. Hence, when referring to the product of two matrices $\mathbf{A}$ and $\mathbf{B}$, it must be specified how they are to be multiplied. In the product $\mathbf{AB}$, we say $\mathbf{A}$ is *postmultiplied* by $\mathbf{B}$, and $\mathbf{B}$ is *premultiplied* by $\mathbf{A}$.

Even though matrix multiplication is noncommutative, it is *associative* and *distributive over addition*. Thus if the products $\mathbf{AB}$ and $\mathbf{BC}$ are defined, then

$$(\mathbf{AB})\mathbf{C} = \mathbf{A}(\mathbf{BC})$$

$$\mathbf{A}(\mathbf{B} + \mathbf{C}) = \mathbf{AB} + \mathbf{AC} \tag{6}$$

$$(\mathbf{A} + \mathbf{B})\mathbf{C} = \mathbf{AC} + \mathbf{BC}.$$

Sometimes it is convenient to rewrite a given matrix so that certain *submatrices* are treated as units. Thus, let $\mathbf{A} = [a_{ij}]_{3,5}$. It can be separated or *partitioned* in one of a number of ways, two of which follow.

$$\mathbf{A} = \begin{bmatrix} a_{11} & a_{12} & \vdots & a_{13} & a_{14} & a_{15} \\ \cdots & \cdots & & \cdots & \cdots & \cdots \\ a_{21} & a_{22} & \vdots & a_{23} & a_{24} & a_{25} \\ & & \vdots & & & \\ a_{31} & a_{32} & \vdots & a_{33} & a_{34} & a_{35} \end{bmatrix}$$

$$= \begin{bmatrix} \mathbf{A}_{11} & \vdots & \mathbf{A}_{12} \\ \cdots & \vdots & \cdots \\ \mathbf{A}_{21} & \vdots & \mathbf{A}_{22} \end{bmatrix}$$

where

$$\mathbf{A}_{11} = [a_{11} \quad a_{12}] \quad \mathbf{A}_{12} = [a_{13} \quad a_{14} \quad a_{15}]$$

$$\mathbf{A}_{21} = \begin{bmatrix} a_{21} & a_{22} \\ a_{31} & a_{32} \end{bmatrix} \quad \mathbf{A}_{22} = \begin{bmatrix} a_{23} & a_{24} & a_{25} \\ a_{33} & a_{34} & a_{35} \end{bmatrix}$$

or

$$\mathbf{A} = \begin{bmatrix} a_{11} & a_{12} & \vdots & a_{13} & \vdots & a_{14} & a_{15} \\ a_{21} & a_{22} & \vdots & a_{23} & \vdots & a_{24} & a_{25} \\ \cdots & \cdots & \cdots & \cdots & \cdots & \cdots & \cdots \\ a_{31} & a_{32} & \vdots & a_{33} & \vdots & a_{34} & a_{35} \end{bmatrix}$$

$$= \begin{bmatrix} \mathbf{A}_{11} & \vdots & \mathbf{A}_{12} & \vdots & \mathbf{A}_{13} \\ \cdots & \cdots & \cdots & \cdots & \cdots \\ \mathbf{A}_{21} & \vdots & \mathbf{A}_{22} & \vdots & \mathbf{A}_{23} \end{bmatrix}$$

where

$$\mathbf{A}_{11} = \begin{bmatrix} a_{11} & a_{12} \\ a_{21} & a_{22} \end{bmatrix}, \quad \mathbf{A}_{12} = \begin{bmatrix} a_{13} \\ a_{23} \end{bmatrix}, \quad \mathbf{A}_{13} = \begin{bmatrix} a_{14} & a_{15} \\ a_{24} & a_{25} \end{bmatrix}$$

$$\mathbf{A}_{21} = [a_{31} \quad a_{32}], \quad \mathbf{A}_{22} = [a_{33}], \quad \mathbf{A}_{23} = [a_{34} \quad a_{35}].$$

The submatrices into which $\mathbf{A}$ is partitioned are shown by drawing dotted lines. Each submatrix can be treated as an element of the matrix $\mathbf{A}$ in any further operations that are to be performed on $\mathbf{A}$; for example, the product of two partitioned matrices is given by

$$\mathbf{A}\mathbf{B} = \left[\sum_j \mathbf{A}_{ij} \mathbf{B}_{jk} \right]. \tag{7}$$

Of course, in order for this partitioning to lead to the correct result, it is necessary that each of the submatrix products, $\mathbf{A}_{21}\mathbf{B}_{11}$, etc., be conformable. Matrices partitioned in this fashion are said to be *conformally partitioned*. This is illustrated in the following product of two matrices:

$$\mathbf{A} = \begin{bmatrix} 1 & 0 & \vdots & 1 & -2 \\ 0 & 1 & \vdots & 2 & 0 \\ \cdots & \cdots & \cdots & \cdots & \cdots \\ 4 & 3 & \vdots & -1 & 3 \end{bmatrix} = \begin{bmatrix} \mathbf{U} & \mathbf{A}_{12} \\ \mathbf{A}_{21} & \mathbf{A}_{22} \end{bmatrix}$$

$$\mathbf{B} = \begin{bmatrix} 2 & 1 & \vdots & 1 & 0 \\ -1 & 0 & \vdots & 0 & 1 \\ \cdots & \cdots & \cdots & \cdots & \cdots \\ 0 & 0 & \vdots & 3 & -1 \\ 0 & 0 & \vdots & 1 & -2 \end{bmatrix} = \begin{bmatrix} \mathbf{B}_{11} & \mathbf{U} \\ \mathbf{0} & \mathbf{B}_{22} \end{bmatrix}$$

$$\mathbf{AB} = \begin{bmatrix} \mathbf{B}_{11} & \vdots & \mathbf{U} + \mathbf{A}_{12}\mathbf{B}_{22} \\ \cdots & \vdots & \cdots \\ \mathbf{A}_{21}\mathbf{B}_{11} & \vdots & \mathbf{A}_{21} + \mathbf{A}_{22}\mathbf{B}_{22} \end{bmatrix}$$

$$= \begin{bmatrix} \begin{bmatrix} 2 & 1 \\ -1 & 0 \end{bmatrix} & \vdots & \begin{bmatrix} 1 & 0 \\ 0 & 1 \end{bmatrix} + \begin{bmatrix} 1 & -2 \\ 2 & 0 \end{bmatrix}\begin{bmatrix} 3 & -1 \\ 1 & -2 \end{bmatrix} \\ \cdots & \vdots & \cdots \\ [4 \ 3]\begin{bmatrix} 2 & 1 \\ -1 & 0 \end{bmatrix} & \vdots & [4 \ 3] + [-1 \ 3]\begin{bmatrix} 3 & -1 \\ 1 & -2 \end{bmatrix} \end{bmatrix}$$

$$= \begin{bmatrix} 2 & 1 & 2 & 3 \\ -1 & 0 & 6 & -1 \\ 5 & 4 & 4 & -2 \end{bmatrix}.$$

Differentiation. Let $\mathbf{A}$ be of order $n \times m$. Then, for any point at which $da_{ij}(x)/dx$ exists for $i = 1, 2, \cdots, n$ and $j = 1, 2, \cdots, m$, $d\mathbf{A}(x)/dx$ is defined as

$$\frac{d}{dx}\mathbf{A}(x) = \left[\frac{d}{dx}a_{ij}(x)\right]. \tag{8}$$

Thus the matrix $d\mathbf{A}(x)/dx$ is obtained by replacing each element $a_{ij}(x)$ of $\mathbf{A}(x)$ with its derivative $da_{ij}(x)/dx$. Now it is easy, and left to you as an exercise, to show that

$$\frac{d}{dx}\{\mathbf{A}(x) + \mathbf{B}(x)\} = \frac{d\mathbf{A}}{dx} + \frac{d\mathbf{B}}{dx} \tag{9}$$

$$\frac{d}{dx}\{\mathbf{A}(x)\,\mathbf{B}(x)\} = \frac{d\mathbf{A}}{dx}\mathbf{B} + \mathbf{A}\frac{d\mathbf{B}}{dx} \tag{10}$$

and

$$\frac{d}{dx}\mathbf{A}(f(x)) = \frac{d\mathbf{A}}{df}\frac{df}{dx} = \frac{df}{dx}\frac{d\mathbf{A}}{df}. \tag{11}$$

We see that the familiar rules for differentiation of combinations of functions apply to the differentiation of matrices; the one caution is that the sequence of matrix products must be preserved in (10).

Integration. Let the order of $\mathbf{A}$ be $n \times m$. Then, for any interval on which $\int_{x_1}^{x_2} a_{ij}(y)\, dy$ exists for $i = 1, 2, \cdots, n$ and $j = 1, 2, \cdots, m, \int_{x_1}^{x_2} \mathbf{A}(y)\, dy$ is defined as

$$\int_{x_1}^{x_2} \mathbf{A}(y)\, dy = \left[\int_{x_1}^{x_2} a_{ij}(y)\, dy\right]. \tag{12}$$

Thus the (i, j)th element of the integral of $\mathbf{A}$ is the integral of the (i, j)th element of $\mathbf{A}$.

Trace. The trace of a square matrix $\mathbf{A}$ is a number denoted by tr $\mathbf{A}$ and defined as

$$\mathrm{tr}\ \mathbf{A} = \sum_{i=1}^{n} a_{ii}$$

where n is the order of $\mathbf{A}$. Note that tr $\mathbf{A}$ is simply the sum of the main diagonal elements of $\mathbf{A}$.

Transpose. The operation of interchanging the rows and columns of a matrix is called *transposing*. The result of this operation on a matrix $\mathbf{A}$ is called the *transpose* of $\mathbf{A}$ and is designated $\mathbf{A}'$. If $\mathbf{A} = [a_{ij}]_{m,n}$, then $\mathbf{A}' = [b_{ij}]_{n, m}$, where $b_{ij} = a_{ji}$. The transpose of a column matrix is a row matrix, and vice versa. If, as often happens in analysis, it is necessary to find the transpose of the product of two matrices, it is important to know that

$$(\mathbf{AB})' = \mathbf{B}'\mathbf{A}'; \tag{13}$$

that is, the transpose of a product equals the product of transposes, but in the opposite order. This result can be established simply by writing the typical element of the transpose of the product and showing that it is the same as the typical element of the product of the transposes.

Conjugate. If each of the elements of a matrix $\mathbf{A}$ is replaced by its complex conjugate, the resulting matrix is said to be the *conjugate* of $\mathbf{A}$ and is denoted by $\bar{\mathbf{A}}$. Thus, if $\mathbf{A} = [a_{ij}]_{n, m}$, then $\bar{\mathbf{A}} = [b_{ij}]_{n, m}$, where $b_{ij} = \bar{a}_{ij}$ and $\bar{a}_{ij}$ denotes the complex-conjugate of a_{ij}.

Conjugate Transpose. The matrix that is the conjugate of the transpose of $\mathbf{A}$ or, equivalently, the transpose of the conjugate of $\mathbf{A}$, is called the *conjugate transpose* of $\mathbf{A}$ and is denoted by $\mathbf{A}^*$; that is,

$$\mathbf{A}^* = (\bar{\mathbf{A}}') = (\bar{\mathbf{A}})'. \tag{14}$$

TYPES OF MATRICES

There are two special matrices that have the properties of the scalars 0 and 1. The matrix $\mathbf{0} = [0]$ which has 0 for each entry is called the *zero*, or *null,* matrix. It is square and of any order. Similarly, the *unit* or *identity* matrix **U** is a square matrix of any order having elements on the main diagonal that are all 1, all other elements being zero. Thus

$$
\begin{bmatrix} 1 & 0 \\ 0 & 1 \end{bmatrix}, \quad
\begin{bmatrix} 1 & 0 & 0 \\ 0 & 1 & 0 \\ 0 & 0 & 1 \end{bmatrix}, \quad
\begin{bmatrix} 1 & 0 & 0 & 0 \\ 0 & 1 & 0 & 0 \\ 0 & 0 & 1 & 0 \\ 0 & 0 & 0 & 1 \end{bmatrix}
$$

are unit matrices of order 2, 3, and 4 respectively. It can be readily veri-fied that the unit matrix does have the properties of the number 1; namely, that given a matrix **A**

$$ \mathbf{UA} = \mathbf{AU} = \mathbf{A} \tag{15} $$

where the order of **U** is such as to make the products conformable.

If a square matrix has the same structure as a unit matrix, in that only the elements on its main diagonal are nonzero, it is called a *diagonal* matrix. Thus a diagonal matrix has the form

$$
\mathbf{D} =
\begin{bmatrix}
d_{11} & & & & & \\
 & d_{22} & & & \bigcirc & \\
 & & d_{33} & & & \\
 & & & \cdot & & \\
 & \bigcirc & & & \cdot & \\
 & & & & & \cdot \\
 & & & & & d_{nn}
\end{bmatrix}.
$$

All elements both above the main diagonal and below the main diagonal are zero. A diagonal matrix is its own transpose.

If the elements only below the main diagonal or only above the main diagonal of a square matrix are zero, as in the following,

$$
\mathbf{A} =
\begin{bmatrix}
a_{11} & a_{12} & a_{13} & \cdots & a_{1n} \\
 & a_{22} & a_{23} & \cdots & a_{2n} \\
 & & a_{33} & \cdots & a_{3n} \\
 & & & & \vdots \\
 & \bigcirc & & & \vdots \\
 & & & & a_{nn}
\end{bmatrix}
\quad \text{or} \quad
\mathbf{B} =
\begin{bmatrix}
a_{11} & & & & \\
a_{12} & a_{22} & & \bigcirc & \\
a_{13} & a_{23} & a_{33} & & \\
\vdots & \vdots & \vdots & & \\
\vdots & \vdots & \vdots & & \\
a_{1n} & a_{2n} & a_{3n} & \cdots & a_{nn}
\end{bmatrix},
$$

the matrix is called a *triangular* matrix, for obvious reasons. Also, to be a little more precise, **A** might be called an *upper triangular* matrix and **B** might be called a *lower triangular* matrix.

Symmetric and Skew-Symmetric Matrices. A square matrix is said to be *symmetric* if it is equal to its own transpose: $\mathbf{A} = \mathbf{A}'$ or $a_{ij} = a_{ji}$ for all i and j. On the other hand, if a matrix equals the negative of its transpose, it is called *skew-symmetric*: $\mathbf{A} = -\mathbf{A}'$ or $a_{ij} = -a_{ji}$. When this definition is applied to the elements on the main diagonal, for which $i = j$, it is found that these elements must be zero for a skew-symmetric matrix.

A given square matrix **A** can always be written as the sum of a symmetric matrix and a skew-symmetric one. Thus let

$$\mathbf{A} = [a_{ij}]$$

$$\mathbf{A}_s = [b_{ij}] \quad \text{where} \quad b_{ij} = \frac{a_{ij} + a_{ji}}{2} = b_{ji}$$

$$\mathbf{A}_{ss} = [c_{ij}] \quad \text{where} \quad c_{ij} = \frac{a_{ij} - a_{ji}}{2} = -c_{ji}.$$

Then

$$\mathbf{A} = \mathbf{A}_s + \mathbf{A}_{ss}, \quad \text{since} \quad a_{ij} = b_{ij} + c_{ij}. \tag{16}$$

Hermitian and Skew-Hermitian Matrices. A square matrix is said to be *Hermitian* if it equals its conjugate transpose; that is, **A** is Hermitian if $\mathbf{A} = \mathbf{A}^*$ or, equivalently, $a_{ij} = \bar{a}_{ji}$ for all i and j. As another special case, if a matrix equals the negative of its conjugate transpose, it is called *skew-Hermitian*. Thus **A** is skew-Hermitian if $\mathbf{A} = -\mathbf{A}^*$ or, equivalently, $a_{ij} = -\bar{a}_{ji}$ for all i and j. Observe that a Hermitian matrix having only real elements is symmetric, and a skew-Hermitian matrix having only real elements is skew-symmetric.

DETERMINANTS

With any square matrix **A** there is associated a number called the *determinant* of **A**. Usually the determinant of **A** will be denoted by the symbol det **A** or $|\mathbf{A}|$; however, we will sometimes use the symbol Δ to stand for a determinant when it is not necessary to call attention to the particular matrix for which Δ is the determinant. Note that a matrix and its determinant are two altogether different things. If two matrices have

equal determinants, this does not imply that the matrices are equal; the two may even be of different orders.

The determinant of the $n \times n$ matrix $\mathbf{A}$ is defined as

$$\det \mathbf{A} = \sum \varepsilon\, a_{1\nu_1} a_{2\nu_2} \cdots a_{n\nu_n} \tag{17}$$

or, by the equivalent relation,

$$\det \mathbf{A} = \sum \varepsilon\, a_{\nu_1 1} a_{\nu_2 2} \cdots a_{\nu_n n} \tag{18}$$

where the summation extends over all $n!$ permutations $\nu_1, \nu_2 \cdots, \nu_n$ of the subscripts $1, 2, \cdots, n$ and ε is equal to $+1$ or -1 as the permutation $\nu_1, \nu_2 \cdots, \nu_n$ is even or odd. As a consequence of this definition, the determinant of a 1×1 matrix is equal to its only element, and the determinant of a 2×2 matrix is established as

$$\det \mathbf{A} = \det \begin{vmatrix} a_{11} & a_{12} \\ a_{21} & a_{22} \end{vmatrix} = a_{11}a_{22} - a_{12}a_{21}\,.$$

The product $a_{11}a_{22}$ was multiplied by $\varepsilon = +1$ because $\nu_1, \nu_2 = 1, 2$ is an even permutation of 1, 2; the product $a_{12}a_{21}$ was multiplied by $\varepsilon = -1$ because $\nu_1, \nu_2 = 2, 1$ is an odd permutation of 1, 2. Determinant evaluation for large n by applying the above definition is difficult and not always necessary. Very often the amount of time consumed in performing the arithmetic operations needed to evaluate a determinant can be reduced by applying some of the properties of determinants. A summary of some of the major properties follows:

1. The determinant of a matrix and that of its transpose are equal; that is, $\det \mathbf{A} = \det \mathbf{A}'$.

2. If every element of any row or any column of a determinant is multiplied by a scalar k, the determinant is multiplied by k.

3. Interchanging any two rows or columns changes the sign of a determinant.

4. If any two rows or columns are identical, then the determinant is zero.

5. If every element of any row or any column is zero, then the determinant is zero.

6. The determinant is unchanged if to each element of any row or column is added a scalar multiple of the corresponding element of any other row or column.

Cofactor Expansion. Let $\mathbf{A}$ be a square matrix of order n. If the ith row and jth column of $\mathbf{A}$ are deleted, the determinant of the remaining

matrix of order $n - 1$ is called a *first minor* (or simply a *minor*) of $\mathbf{A}$ or of det $\mathbf{A}$ and is denoted by M_{ij}. The corresponding (first) *cofactor* is defined as

$$\Delta_{ij} = (-1)^{i+j} M_{ij}. \tag{19}$$

It is said that Δ_{ij} is the cofactor of element a_{ij}. If $i = j$, the minor and cofactor are called *principal minor* and *principal cofactor*. More specifically, a principal minor (cofactor) of $\mathbf{A}$ is one whose diagonal elements are also diagonal elements of $\mathbf{A}$.* The value of a determinant can be obtained by multiplying each element of a row or column by its corresponding cofactor and adding the results. Thus

$$\det \mathbf{A} = a_{i1}\Delta_{i1} + a_{i2}\Delta_{i2} + a_{i3}\Delta_{i3} + \cdots + a_{in}\Delta_{in} \tag{20a}$$

$$= a_{1i}\Delta_{1i} + a_{2i}\Delta_{2i} + a_{3i}\Delta_{3i} + \cdots + a_{ni}\Delta_{ni}. \tag{20b}$$

These expressions are called the *cofactor expansions* along a row or column and are established by collecting the terms of (17) or (18) into groups, each corresponding to an element times its cofactor.

What would happen if the elements of a row or column were multiplied by the corresponding cofactors of another row or column? It is left to you as a problem to show that the result would be zero; that is,

$$a_{i1}\Delta_{j1} + a_{i2}\Delta_{j2} + a_{i3}\Delta_{j3} + \cdots + a_{in}\Delta_{jn} = 0 \tag{21a}$$

$$a_{1i}\Delta_{1j} + a_{2i}\Delta_{2j} + a_{3i}\Delta_{3j} + \cdots + a_{ni}\Delta_{nj} = 0. \tag{21b}$$

The *Kronecker delta* is a function denoted by δ_{ij} and is defined as

$$\delta_{ij} = 1 \quad \text{if} \quad i = j$$
$$= 0 \quad \text{if} \quad i \neq j$$

where i and j are integers. Using the Kronecker delta, we can consolidate (20a) and (21a) and write

$$(\det \mathbf{A})\delta_{ij} = \sum_{k=0}^{n} a_{ik} \Delta_{jk}. \tag{22}$$

---

* This definition does not limit the number of rows and columns deleted from Δ to form the minor or cofactor. If one row and column are deleted, we should more properly refer to the *first* principal cofactor. In general, if n rows and columns are deleted, we would refer to the result as the nth principal cofactor.

Similarly, (20b) and (21b) combine to yield

$$(\det \mathbf{A})\delta_{ij} = \sum_{k=0}^{n} a_{ki}\,\Delta_{kj}. \tag{23}$$

Determinant of a Matrix Product. Let $\mathbf{A}$ and $\mathbf{B}$ be square matrices of the same order. The determinant of the product of $\mathbf{A}$ and $\mathbf{B}$ is the product of the determinants; that is,

$$\det(\mathbf{AB}) = (\det \mathbf{A})(\det \mathbf{B}). \tag{24}$$

Derivative of a Determinant. If the elements of the square matrix $\mathbf{A}$ are functions of some variable, say x, then $|\mathbf{A}|$ will be a function of x. It is useful to know that

$$\frac{d|\mathbf{A}|}{dx} = \sum_{i,j=1}^{n} \Delta_{ij}\,\frac{da_{ij}}{dx}. \tag{25}$$

The result follows from the observation that

$$\frac{d|\mathbf{A}|}{dx} = \sum_{i,j=1}^{n} \frac{\partial|\mathbf{A}|}{\partial a_{ij}}\,\frac{da_{ij}}{dx}$$

and from the cofactor expansion for $|\mathbf{A}|$ in (22) or (23) that $\partial|\mathbf{A}|/da_{ij} = \Delta_{ij}$.

Binet-Cauchy Theorem. Consider the determinant of the product $\mathbf{AB}$, assuming the orders are (m, n) and (n, m), with $m < n$. Observe that the product is square of order m. The largest square submatrix of each of the matrices $\mathbf{A}$ and $\mathbf{B}$ is of order m. Let the determinant of each square submatrix of maximum order be called a *major determinant*, or simply a *major*. Then $|\mathbf{AB}|$ is given by the following theorem called the *Binet-Cauchy theorem.*

$$\det \mathbf{AB} = \sum_{\substack{\text{all} \\ \text{majors}}} (\text{products of corresponding majors of } \mathbf{A} \text{ and } \mathbf{B}). \tag{26}$$

The phrase "corresponding majors" means that whatever numbered columns are used for forming a major of $\mathbf{A}$, the same numbered rows are used for forming the major of $\mathbf{B}$.

To illustrate, let

$$\mathbf{A} = \begin{bmatrix} 1 & -1 & 3 \\ 2 & 1 & 0 \end{bmatrix}$$

$$\mathbf{B} = \begin{bmatrix} 2 & 1 \\ -1 & 1 \\ 1 & 0 \end{bmatrix}.$$

In this case $m = 2$ and $n = 3$. By direct multiplication we find that

$$\mathbf{AB} = \begin{bmatrix} 6 & 0 \\ 3 & 3 \end{bmatrix}.$$

The determinant of this matrix is easily seen to be 18. Now let us apply the Binet-Cauchy theorem. We see that there are three determinants of order two to be considered. Applying (26), we get

$$\det \mathbf{AB} = \begin{vmatrix} 1 & -1 \\ 2 & 1 \end{vmatrix} \begin{vmatrix} 2 & 1 \\ -1 & 1 \end{vmatrix} + \begin{vmatrix} 1 & 3 \\ 2 & 0 \end{vmatrix} \begin{vmatrix} 2 & 1 \\ 1 & 0 \end{vmatrix} + \begin{vmatrix} -1 & 3 \\ 1 & 0 \end{vmatrix} \begin{vmatrix} -1 & 1 \\ 1 & 0 \end{vmatrix}$$

$$= (3)(3) + (-6)(-1) + (-3)(-1) = 18.$$

This agrees with the value calculated by direct evaluation of the determinant.

THE INVERSE OF A MATRIX

In the case of scalars, if $a \neq 0$, there is a number b such that $ab = ba = 1$. In the same way, given a square matrix $\mathbf{A}$, we seek a matrix $\mathbf{B}$ such that

$$\mathbf{BA} = \mathbf{AB} = \mathbf{U}.$$

Such a $\mathbf{B}$ may not exist. But if this relationship is satisfied, we say $\mathbf{B}$ is the *inverse* of $\mathbf{A}$ and we write it $\mathbf{B} = \mathbf{A}^{-1}$. The inverse relationship is mutual, so that if $\mathbf{B} = \mathbf{A}^{-1}$, then $\mathbf{A} = \mathbf{B}^{-1}$.

Given a square matrix $\mathbf{A}$, form another matrix as follows:

$$\mathbf{B} = [b_{ij}] \quad \text{where} \quad b_{ij} = \frac{\Delta_{ji}}{\Delta}$$

where $\Delta = \det \mathbf{A}$ and Δ_{ji} is the cofactor of a_{ji}. By direct expansion of $\mathbf{AB}$ and $\mathbf{BA}$ and application of (22) and (23), it can be shown that $\mathbf{B}$ is the inverse of $\mathbf{A}$. (Do it.) In words, to form the inverse of $\mathbf{A}$ we replace each element of $\mathbf{A}$ by its cofactor, then we take the transpose, and finally we divide by the determinant of $\mathbf{A}$.

Since the elements of the inverse of $\mathbf{A}$ have Δ in the denominator, it is clear that the inverse will not exist if det $\mathbf{A} = 0$. A matrix whose determinant equals zero is said to be *singular*. If det $\mathbf{A} \neq 0$, the matrix is *nonsingular*.

The process of forming the inverse is clarified by defining another matrix related to $\mathbf{A}$. Define the *adjoint* of $\mathbf{A}$, written adj $\mathbf{A}$ as

$$\text{adj } \mathbf{A} = [\Delta_{ij}]' = \begin{bmatrix} \Delta_{11} & \Delta_{21} & \Delta_{31} & \cdots & \Delta_{n1} \\ \Delta_{12} & \Delta_{22} & \Delta_{32} & \cdots & \Delta_{n2} \\ \vdots & \vdots & \vdots & & \vdots \\ \Delta_{1n} & \Delta_{2n} & \Delta_{3n} & \cdots & \Delta_{nn} \end{bmatrix}. \tag{27}$$

Note that the elements in the ith row of adj $\mathbf{A}$ are the cofactors of the elements of the ith column of $\mathbf{A}$. The inverse of $\mathbf{A}$ can now be written as

$$\mathbf{A}^{-1} = \frac{1}{\det \mathbf{A}} \text{ adj } \mathbf{A}. \tag{28}$$

Observe, after premultiplying both sides of (28) by $\mathbf{A}$, that

$$\mathbf{A} \cdot [\text{adj } \mathbf{A}] = \mathbf{U} \det \mathbf{A}. \tag{29}$$

Each side of this expression is a matrix, the left side being the product of two matrices and the right side being a diagonal matrix whose diagonal elements each equal det $\mathbf{A}$. Taking the determinant of both sides yields

$$(\det \mathbf{A})(\det \text{ adj } \mathbf{A}) = (\det \mathbf{A})^n$$

or

$$\det \text{ adj } \mathbf{A} = (\det \mathbf{A})^{n-1}. \tag{30}$$

In some of the work that follows in later chapters, the product of two matrices is often encountered. It is desirable, therefore, to evaluate the result of finding the inverse and adjoint of the product of two matrices $\mathbf{A}$ and $\mathbf{B}$. The results are

$$(\mathbf{AB})^{-1} = \mathbf{B}^{-1}\mathbf{A}^{-1} \tag{31}$$

$$\text{adj } (\mathbf{AB}) = (\text{adj } \mathbf{B})(\text{adj } \mathbf{A}). \tag{32}$$

Obviously, the product **AB** must be conformable. Furthermore, both **A** and **B** must be square and nonsingular.

In the case of the first one, note that

$$(\mathbf{AB})(\mathbf{B}^{-1}\mathbf{A}^{-1}) = \mathbf{A}(\mathbf{BB}^{-1})\mathbf{A}^{-1} = \mathbf{AA}^{-1} = \mathbf{U}.$$

Hence **AB** is the inverse of $\mathbf{B}^{-1}\mathbf{A}^{-1}$, whence the result. For the second one, we can form the products $(\mathbf{AB})(\text{adj } \mathbf{AB})$ and $(\mathbf{AB})(\text{adj } \mathbf{B})(\text{adj } \mathbf{A})$ and show by repeated use of the relationship $\mathbf{M}(\text{adj } \mathbf{M}) = \mathbf{U}\,(\det \mathbf{M})$ that both products equal $\mathbf{U}\,(\det \mathbf{AB})$. The result follows.

PIVOTAL CONDENSATION

By repeated application of the cofactor expansion, the evaluation of the determinant of an $n \times n$ array of numbers can be reduced to the evaluation of numerous 2×2 arrays. It is obvious that the number of arithmetic operations grows excessively as n increases. An alternate method for determinant evaluation, which requires significantly fewer arithmetic operations, is called *pivotal condensation*. We will now develop this method.

Let the $n \times n$ matrix **A** be partitioned as

$$\mathbf{A} = \left[\begin{array}{c:c} \mathbf{A}_{11} & \mathbf{A}_{12} \\ \hdashline \mathbf{A}_{21} & \mathbf{A}_{22} \end{array}\right]. \tag{33}$$

where the submatrix $\mathbf{A}_{11}$ is of order $m \times m$ for some $1 \le m < n$. Assume $\mathbf{A}_{11}$ is nonsingular. Then $\mathbf{A}_{11}^{-1}$ exists, and **A** may be factored as follows:

$$\mathbf{A} = \left[\begin{array}{c:c} \mathbf{U} & \mathbf{0} \\ \hdashline \mathbf{A}_{21}\mathbf{A}_{11}^{-1} & \mathbf{U} \end{array}\right]\left[\begin{array}{c:c} \mathbf{A}_{11} & \mathbf{0} \\ \hdashline \mathbf{0} & \mathbf{A}_{22} - \mathbf{A}_{21}\mathbf{A}_{11}^{-1}\mathbf{A}_{12} \end{array}\right]\left[\begin{array}{c:c} \mathbf{U} & \mathbf{A}_{11}^{-1}\mathbf{A}_{12} \\ \hdashline \mathbf{0} & \mathbf{U} \end{array}\right]. \tag{34}$$

The validity of this factorization is established by performing the indicated matrix multiplications and observing that the result is (33).

Now, by repeated application of the cofactor expansion, it may be shown (Problem 35) that the determinant of a triangular matrix is equal to the product of its main diagonal elements. Since

$$\left[\begin{array}{c:c} \mathbf{U} & \mathbf{0} \\ \hdashline \mathbf{A}_{21}\mathbf{A}_{11}^{-1} & \mathbf{U} \end{array}\right] \quad \text{and} \quad \left[\begin{array}{c:c} \mathbf{U} & \mathbf{A}_{11}^{-1}\mathbf{A}_{12} \\ \hdashline \mathbf{0} & \mathbf{U} \end{array}\right]$$

are triangular with "ones" on the main diagonal, their determinants
are unity. So, only the middle matrix in (34) needs attention. This matrix,
in turn, can be factorized as

$$\begin{bmatrix} \mathbf{A}_{11} & \mathbf{0} \\ \mathbf{0} & \mathbf{A}_{22} - \mathbf{A}_{21}\mathbf{A}_{11}^{-1}\mathbf{A}_{12} \end{bmatrix} = \begin{bmatrix} \mathbf{A}_{11} & \mathbf{0} \\ \mathbf{0} & \mathbf{U} \end{bmatrix}\begin{bmatrix} \mathbf{U} & \mathbf{0} \\ \mathbf{0} & \mathbf{A}_{22} - \mathbf{A}_{21}\mathbf{A}_{11}^{-1}\mathbf{A}_{12} \end{bmatrix}. \quad (35)$$

The determinants of the matrices on the right are simply det $\mathbf{A}_{11}$ and
$\det(\mathbf{A}_{22} - \mathbf{A}_{21}\mathbf{A}_{11}^{-1}\mathbf{A}_{12})$, respectively.

Since the determinant of a product of matrices equals the product of
the determinants, then taking the determinant of both sides of (34) and
using (35) leads to

$$\det \mathbf{A} = (\det \mathbf{A}_{11}) \{\det(\mathbf{A}_{22} - \mathbf{A}_{21}\mathbf{A}_{11}^{-1}\mathbf{A}_{12})\}. \quad (36)$$

If $\mathbf{A}_{11}$ is the scalar $a_{11} \neq 0$ (i.e., the order of $\mathbf{A}_{11}$ is 1×1), then the last
equation reduces to

$$\det \mathbf{A} = a_{11} \det \left(\mathbf{A}_{22} - \frac{1}{a_{11}} \mathbf{A}_{21}\mathbf{A}_{12}\right) = a_{11} \det \frac{a_{11}\mathbf{A}_{22} - \mathbf{A}_{21}\mathbf{A}_{12}}{a_{11}}.$$

Now, according to the properties of a determinant, multiplying each row
of a matrix by a constant $1/a_{11}$, will cause the determinant to be multiplied by $1/a_{11}^m$, where m is the order of the matrix. For the matrix on the
right side whose determinant is being found the order is $n-1$, one less
than the order of $\mathbf{A}$. Hence

$$\det \mathbf{A} = \frac{1}{a_{11}^{n-2}} \det (a_{11}\mathbf{A}_{22} - \mathbf{A}_{21}\mathbf{A}_{12}). \quad (37)$$

This is the mathematical relation associated with pivotal condensation.
The requirement that the pivotal element a_{11} be nonzero can always be
met, unless all elements of the first row or column are zero, in which case
det $\mathbf{A} = 0$ by inspection. Barring this, a nonzero element can always be
placed in the (1, 1) position by the interchange of another row with row 1
or another column with column 1. Such an interchange will require a
change of sign, according to property 3 for determinants. It is of primary
significance that repeated application of (37) reduces evaluation of det $\mathbf{A}$
to evaluation of the determinant of just one 2×2 array.

The example that follows illustrates the method of pivotal condensation.

$$\det \begin{bmatrix} 0 & 0 & 1 & 2 \\ -1 & 3 & 0 & -2 \\ 4 & -1 & 3 & 1 \\ 2 & 0 & 2 & 1 \end{bmatrix} = (-1) \det \begin{bmatrix} 1 & 0 & 0 & 2 \\ 0 & 3 & -1 & -2 \\ 3 & -1 & 4 & 1 \\ 2 & 0 & 2 & 1 \end{bmatrix} \begin{matrix} \text{(inter-} \\ \text{change of} \\ \text{columns} \\ \text{1 and 3)} \end{matrix}$$

$$= -\frac{1}{1^{4-2}} \det \left\{ 1 \times \begin{bmatrix} 3 & -1 & -2 \\ -1 & 4 & 1 \\ 0 & 2 & 1 \end{bmatrix} \right.$$

$$\left. - \begin{bmatrix} 0 \\ 3 \\ 2 \end{bmatrix} \begin{bmatrix} 0 & 0 & 2 \end{bmatrix} \right\}$$

$$= -\det \left\{ \begin{bmatrix} 3 & -1 & -2 \\ -1 & 4 & 1 \\ 0 & 2 & 1 \end{bmatrix} - \begin{bmatrix} 0 & 0 & 0 \\ 0 & 0 & 6 \\ 0 & 0 & 4 \end{bmatrix} \right\}$$

$$= -\det \begin{bmatrix} 3 & -1 & -2 \\ -1 & 4 & -5 \\ 0 & 2 & -3 \end{bmatrix}$$

$$= -\frac{1}{3^{3-2}} \det \left\{ 3 \times \begin{bmatrix} 4 & -5 \\ 2 & -3 \end{bmatrix} \right.$$

$$\left. - \begin{bmatrix} -1 \\ 0 \end{bmatrix} \begin{bmatrix} -1 & -2 \end{bmatrix} \right\}$$

$$= -\frac{1}{3} \det \left\{ \begin{bmatrix} 12 & -15 \\ 6 & -9 \end{bmatrix} - \begin{bmatrix} 1 & 2 \\ 0 & 0 \end{bmatrix} \right\}$$

$$= -\frac{1}{3} \det \begin{bmatrix} 11 & -17 \\ 6 & -9 \end{bmatrix}$$

$$= -1.$$

Many of the steps included here for completeness are ordinarily eliminated by someone who has become facile in using pivotal condensation to evaluate a determinant.

LINEAR EQUATIONS

Matrix notation and the concept of matrices originated in the desire to handle sets of linear algebraic equations. Since, in network analysis, we are confronted with such equations and their solution, we shall now turn our attention to them. Consider the following set of linear algebraic equations:

$$a_{11}x_1 + a_{12}x_2 + a_{13}x_3 + \cdots + a_{1n}x_n = y_1$$

$$a_{21}x_1 + a_{22}x_2 + a_{23}x_3 + \cdots + a_{2n}x_n = y_2$$

$$\vdots \qquad \vdots \qquad \vdots \qquad \qquad \vdots \qquad \vdots$$

$$a_{m1}x_1 + a_{m2}x_2 + a_{m3}x_3 + \cdots + a_{mn}x_n = y_m.$$

Such a system of equations may be written in matrix notation as

$$\begin{bmatrix} a_{11} & a_{12} & \cdots & a_{1n} \\ a_{21} & a_{22} & \cdots & a_{2n} \\ \vdots & \vdots & & \vdots \\ a_{m1} & a_{m2} & \cdots & a_{mn} \end{bmatrix} \begin{bmatrix} x_1 \\ x_2 \\ \vdots \\ x_n \end{bmatrix} = \begin{bmatrix} y_1 \\ y_2 \\ \vdots \\ y_m \end{bmatrix}. \tag{39}$$

This fact may be verified by carrying out the multiplication on the left. In fact, the definition of a matrix product, which may have seemed strange when it was introduced earlier, was so contrived precisely in order to permit the writing of a set of linear equations in matrix form.

The expression can be simplified even further by using the matrix symbols $\mathbf{A}$, $\mathbf{x}$, and $\mathbf{y}$, with obvious definitions, to yield

$$\mathbf{Ax} = \mathbf{y}. \tag{40}$$

This single matrix equation can represent any set of any number of linear equations having any number of variables. The great economy of thought and of expression in the use of matrices should now be evident. The remaining problem is that of solving this matrix equation, or the corresponding set of scalar equations, by which we mean finding a set of values for the x's that satisfies the equations simultaneously. If a solution exists, we say the equations are *consistent*.

Each column (or row) of a matrix is identified by its elements. It can be thought of as a *vector*, with the elements playing the role of components of the vector. Although vectors having more than three dimensions cannot be visualized geometrically, nevertheless the terminology of space vectors is useful in the present context and can be extended to n-dimensional space. Thus, in (40), $\mathbf{x}$, $\mathbf{y}$, and each column and each row of $\mathbf{A}$ are vectors. If the vector consists of a column of elements, then it is more precisely called a *column vector*. *Row vector* is the complete name for a vector that is a row of elements. The modifiers "column" and "row" are used only if confusion is otherwise likely. Further, when the word "vector" is used alone, it would most often be interpreted as "column vector."

Now, given a set of vectors, the question arises as to whether there is some relationship among them or whether they are independent. In ordinary two-dimensional space, we know that any two vectors are independent of each other, unless they happen to be collinear. Furthermore, any other vector in the plane can be obtained as some linear combination of these two, and so three vectors cannot be independent in two-dimensional space.

In the more general case, we will say that a set of m vectors, labeled x_i ($i = 1$ to m), is *linearly dependent* if a set of constants k_i can be found such that

$$\sum_{i=1}^{m} k_i \mathbf{x}_i = \mathbf{0} \qquad (k_i \text{ not all zero}). \qquad (41)$$

If no such relationship exists, the vectors are *linearly independent*. Clearly, if the vectors are dependent, then one or more of the vectors can be expressed as a *linear combination* of the remaining ones by solving (41).

With the notion of linear dependence, it is possible to tackle the job of solving linear equations. Let us partition matrix $\mathbf{A}$ by columns and examine the product.

$$\mathbf{Ax} = [\mathbf{a}_1 \quad \mathbf{a}_2 \cdots \mathbf{a}_n] \begin{bmatrix} x_1 \\ x_2 \\ \vdots \\ x_n \end{bmatrix}$$

$$= x_1 \mathbf{a}_1 + x_2 \mathbf{a}_2 + \cdots + x_n \mathbf{a}_n.$$

Expressed in this way, we see that $\mathbf{Ax}$ is a linear combination of the column vectors of $\mathbf{A}$. In fact, there is a vector $\mathbf{x}$ that will give us any desired combination of these column vectors. It is evident, therefore, that $\mathbf{y}$ must be a linear combination of the column vectors of $\mathbf{A}$, if the equation $\mathbf{y} = \mathbf{Ax}$ is to have a solution. An equivalent statement of this condition is the following: The maximum number of linearly independent vectors in the two sets $\mathbf{a}_1, \cdots, \mathbf{a}_n$ and $\mathbf{a}_1, \mathbf{a}_2, \cdots, \mathbf{a}_n, \mathbf{y}$ must be the same if the system of equations $\mathbf{y} = \mathbf{Ax}$ is to be consistent.

A more compact statement of this condition for the existence of a solution, or consistency, of $\mathbf{y} = \mathbf{Ax}$ can be established. Define the *rank* of a matrix as the order of the largest nonsingular square matrix that can be obtained by removing rows and columns of the original matrix. If the rank of a square matrix equals its order, the matrix must be non-singular, so its determinant is nonzero. In fact, it can be established as a theorem that *the determinant of a matrix is zero if and only if the rows and columns of the matrix are linearly dependent.* (Do it.) Thus the rows and columns of a nonsingular matrix must be linearly independent. It follows that the rank of a matrix equals the maximum number of linearly independent rows and columns.

Now consider the two matrices $\mathbf{A}$ and $[\mathbf{A} \quad \mathbf{y}]$, where the second matrix is obtained from $\mathbf{A}$ by appending the column vector $\mathbf{y}$ as an extra column. We have previously seen that the maximum number of linearly independent column vectors in these two matrices must be the same for consistency, so we conclude that the rank of the two matrices must be the same; that is, the system of equations $\mathbf{y} = \mathbf{Ax}$ is consistent if and only if

$$\text{rank } \mathbf{A} = \text{rank } [\mathbf{A} \quad \mathbf{y}]. \tag{42}$$

This is called the *consistency condition*.

Example

Suppose $\mathbf{A}$ is the following matrix of order 3×4:

$$\mathbf{A} = \begin{bmatrix} 2 & 1 & 4 & 5 \\ -1 & 2 & -7 & -5 \\ 3 & 4 & 1 & 5 \end{bmatrix}.$$

By direct calculation, it is found that each of the four square matrices of order 3 obtained by removing one column of $\mathbf{A}$ is singular—has zero determinant. However, the 2×2 matrix

$$\begin{bmatrix} 2 & 1 \\ -1 & 2 \end{bmatrix}$$

obtained by deleting the third row and third and fourth columns is non-singular. Thus rank $\mathbf{A} = 2$. This also tells us that the column vectors

$$\mathbf{a}_1 = \begin{bmatrix} 2 \\ -1 \\ 3 \end{bmatrix} \quad \text{and} \quad \mathbf{a}_2 = \begin{bmatrix} 1 \\ 2 \\ 4 \end{bmatrix}.$$

are linearly independent. The column vectors

$$\mathbf{a}_3 = \begin{bmatrix} 4 \\ -7 \\ 1 \end{bmatrix} \quad \text{and} \quad \mathbf{a}_4 = \begin{bmatrix} 5 \\ -5 \\ 5 \end{bmatrix}.$$

are linear combinations of $\mathbf{a}_1$ and $\mathbf{a}_2$; in particular,

$$\mathbf{a}_3 = 3\mathbf{a}_1 - 2\mathbf{a}_2 \quad \text{and} \quad \mathbf{a}_4 = 3\mathbf{a}_1 - \mathbf{a}_2.$$

If $\mathbf{y} = \mathbf{A}\mathbf{x}$ is to have a solution, then $\mathbf{y}$ must be a linear combination of $\mathbf{a}_1$ and $\mathbf{a}_2$. Suppose $\mathbf{y} = \alpha\mathbf{a}_1 + \beta\mathbf{a}_2$. Then we must solve

$$\alpha\mathbf{a}_1 + \beta\mathbf{a}_2 = \mathbf{A}\mathbf{x} = x_1\mathbf{a}_1 + x_2\mathbf{a}_2 + x_3\mathbf{a}_3 + x_4\mathbf{a}_4$$

or, since

$$\mathbf{a}_3 = 3\mathbf{a}_1 - 2\mathbf{a}_2 \quad \text{and} \quad \mathbf{a}_4 = 3\mathbf{a}_1 - \mathbf{a}_2,$$

then

$$(\alpha - 3x_3 - 3x_4)\mathbf{a}_1 + (\beta + 2x_1 + x_2)a_2 = x_1\mathbf{a}_1 + x_2\mathbf{a}_2.$$

Thus for any x_3 and x_4, $\mathbf{x}$ is a solution of $\mathbf{y} = \mathbf{A}\mathbf{x}$ if $x_1 = \alpha - 3x_3 - 3x_4$ and $x_2 = \beta + 2x_1 + x_2$. The fact that the solution is not unique is a consequence of the fact that the rank of $\mathbf{A}$ is less than the number of columns of $\mathbf{A}$. This is also true of the general solution of $\mathbf{y} = \mathbf{A}\mathbf{x}$, to which we now turn our attention.

GENERAL SOLUTION OF $y = Ax$

Suppose the consistency condition is satisfied and rank $A = r$. Then the equation $y = Ax$ can always be partitioned as follows:

$$
\begin{array}{cc}
r & n - r \\
\text{cols.} & \text{cols.}
\end{array}
$$

$$
\begin{array}{c}
r \\
\text{rows} \\
\\
m - r \\
\text{rows}
\end{array}
\begin{bmatrix}
A_{11} & A_{12} \\
\\
A_{21} & A_{22}
\end{bmatrix}
\begin{bmatrix}
x_1 \\
\\
x_2
\end{bmatrix}
=
\begin{bmatrix}
y_1 \\
\\
y_2
\end{bmatrix}.
\tag{43}
$$

This is done by first determining the rank r by finding the highest order submatrix whose determinant is nonzero. The equations are then rearranged (and the subscripts on the x's and y's modified), so that the first r rows and columns have a nonzero determinant; that is, A_{11} is nonsingular. The equation can now be rewritten as

$$
A_{11}x_1 + A_{12}x_2 = y_1 \tag{44a}
$$

$$
A_{21}x_1 + A_{22}x_2 = y_2. \tag{44b}
$$

The second of these is simply disregarded, because each of the equations in (44b) is a linear combination of the equations in (44a). You can show that this is a result of assuming that the consistency condition is satisfied. In (44a), the second term is transposed to the right and the equation is multiplied through by A_{11}^{-1}, which exists since A_{11} is nonsingular. The result will be

$$
x_1 = A_{11}^{-1}(y_1 - A_{12}x_2). \tag{45}
$$

This constitutes the solution. The vector x_1 contains r of the elements of the original vector x; they are here expressed in terms of the elements of y_1 and the remaining $m - r$ elements of x.

Observe that the solution (45) is not unique if $n > r$. In fact, there are exactly $q = n - r$ variables, the elements of x_2, which may be selected arbitrarily. This number q is an attribute of the matrix A and is called the *nullity*, or *degeneracy*, of A.

For the special case of homogeneous equations, namely the case where $y = 0$, it should be observed from (45) that a nontrivial solution exists only if the nullity is nonzero. For the further special case of $m = n$ (i.e., when A is a square matrix), the nullity is nonzero and a nontrivial solution exists only if A is singular.

To illustrate the preceding, consider the following set of equations:

$$\begin{bmatrix} -1 & 1 & 1 & 0 & 0 & 0 & 0 & 0 \\ 0 & 0 & -1 & 1 & 1 & 0 & 0 & 0 \\ 0 & 0 & 0 & 0 & -1 & 1 & 1 & 0 \\ 0 & 0 & 0 & 0 & 0 & 0 & -1 & 1 \\ 1 & -1 & 0 & -1 & 0 & -1 & 0 & -1 \end{bmatrix} \begin{bmatrix} x_1 \\ x_2 \\ x_3 \\ x_4 \\ x_5 \\ x_6 \\ x_7 \\ x_8 \end{bmatrix} = \begin{bmatrix} 1 \\ 2 \\ 3 \\ 4 \\ -10 \end{bmatrix}.$$

We observe that the first four rows and columns 2, 4, 6, and 8 of A form a unit matrix (which is nonsingular) and so rank $A \geq 4$. In addition, the fifth row is equal to the negative of the sum of the first four rows. Thus the rows of A are not linearly independent and rank $A < 5$. Since $4 \leq$ rank $A < 5$, we have established that rank $A = 4$. For precisely the same reasons it is found that rank $[A \quad y] = 4$. Thus the consistency condition is satisfied. Now the columns can be rearranged and the matrices partitioned as follows:

$$\begin{bmatrix} 1 & 0 & 0 & 0 & \vdots & -1 & 1 & 0 & 0 \\ 0 & 1 & 0 & 0 & \vdots & 0 & -1 & 1 & 0 \\ 0 & 0 & 1 & 0 & \vdots & 0 & 0 & -1 & 1 \\ 0 & 0 & 0 & 1 & \vdots & 0 & 0 & 0 & -1 \\ \hdotsfor{9} \\ -1 & -1 & -1 & -1 & \vdots & 1 & 0 & 0 & 0 \end{bmatrix} \begin{bmatrix} x_2 \\ x_4 \\ x_6 \\ x_8 \\ \cdots \\ x_1 \\ x_3 \\ x_5 \\ x_7 \end{bmatrix} = \begin{bmatrix} 1 \\ 2 \\ 3 \\ 4 \\ \cdots \\ -10 \end{bmatrix}.$$

This has been partitioned in the form of (43) with $\mathbf{A}_{11} = \mathbf{U}$, a unit matrix. The bottom row of the partitioning is discarded, and the remainder is rewritten. Thus

$$\mathbf{U}\begin{bmatrix} x_2 \\ x_4 \\ x_6 \\ x_8 \end{bmatrix} = \begin{bmatrix} 1 \\ 2 \\ 3 \\ 4 \end{bmatrix} - \begin{bmatrix} -1 & 1 & 0 & 0 \\ 0 & -1 & 1 & 0 \\ 0 & 0 & -1 & 1 \\ 0 & 0 & 0 & -1 \end{bmatrix}\begin{bmatrix} x_1 \\ x_3 \\ x_5 \\ x_7 \end{bmatrix}.$$

Since the inverse of $\mathbf{U}$ is itself, this constitutes the solution. In scalar form, it is

$$x_2 = 1 + x_1 - x_3$$
$$x_4 = 2 + x_3 - x_5$$
$$x_6 = 3 - x_5 - x_7$$
$$x_8 = 4 + x_7 .$$

For each set of values for x_1, x_3, x_5, and x_7 there will be a set of values for x_2, x_4, x_6, and x_8. In a physical problem the former set of variables may not be arbitrary (though they are, as far as the mathematics is concerned); they must often be chosen to satisfy other conditions of the problem.

CHARACTERISTIC EQUATION

An algebraic equation that often appears in network analysis is

$$\lambda \mathbf{x} = \mathbf{A}\mathbf{x} \tag{46}$$

where $\mathbf{A}$ is a square matrix of order n. The problem, known as the *eigenvalue problem*, is to find scalars λ and vectors $\mathbf{x}$ that satisfy this equation. A value of λ, for which a nontrivial solution of $\mathbf{x}$ exists, is called an *eigenvalue*, or *characteristic value*, of $\mathbf{A}$. The corresponding vector $\mathbf{x}$ is called an *eigenvector*, or *characteristic vector*, of $\mathbf{A}$.

Let us first rewrite (46) as follows:

$$(\lambda \mathbf{U} - \mathbf{A})\mathbf{x} = \mathbf{0}. \tag{47}$$

This is a homogeneous equation, which we know will have a nontrivial solution only if $\lambda \mathbf{U} - \mathbf{A}$ is singular or, equivalently,

$$\det (\lambda \mathbf{U} - \mathbf{A}) = \mathbf{0}. \tag{48}$$

The determinant on the left-hand side is a polynomial of degree n in λ and is known as the *characteristic polynomial* of $\mathbf{A}$. The equation itself is known as the *characteristic equation* associated with $\mathbf{A}$. For each value of λ that satisfies the characteristic equation, a nontrivial solution of (47) can be found by the methods of the preceding subsection.

To illustrate these ideas, consider the 2×2 matrix

$$\mathbf{A} = \begin{bmatrix} 5 & 1 \\ -2 & 2 \end{bmatrix}.$$

The characteristic polynomial is

$$\det \begin{bmatrix} \lambda - 5 & -1 \\ 2 & \lambda - 2 \end{bmatrix} = \lambda^2 - 7\lambda + 12 = (\lambda - 3)(\lambda - 4).$$

The values 3 and 4 satisfy the characteristic equation $(\lambda - 3)(\lambda - 4) = 0$ and hence are the eigenvalues of $\mathbf{A}$. To obtain the eigenvector corresponding to the eigenvalue $\lambda = 3$, we solve (47) by using the given matrix $\mathbf{A}$ and $\lambda = 3$. Thus

$$\left(\begin{bmatrix} 3 & 0 \\ 0 & 3 \end{bmatrix} - \begin{bmatrix} 5 & 1 \\ -2 & 2 \end{bmatrix} \right) \begin{bmatrix} x_1 \\ x_2 \end{bmatrix} = \begin{bmatrix} -2 & -1 \\ 2 & 1 \end{bmatrix} \begin{bmatrix} x_1 \\ x_2 \end{bmatrix} = \begin{bmatrix} 0 \\ 0 \end{bmatrix}.$$

The result is

$$\begin{bmatrix} x_1 \\ x_2 \end{bmatrix} = \begin{bmatrix} x_1 \\ -2x_1 \end{bmatrix}$$

for any value of x_1. The eigenvector corresponding to the eigenvalue $\lambda = 4$ is obtained similarly.

$$\begin{bmatrix} -1 & -1 \\ 2 & 2 \end{bmatrix} \begin{bmatrix} x_1 \\ x_2 \end{bmatrix} = \begin{bmatrix} 0 \\ 0 \end{bmatrix}$$

from which

$$\begin{bmatrix} x_1 \\ x_2 \end{bmatrix} = \begin{bmatrix} x_1 \\ -x_1 \end{bmatrix}$$

for any value of x_1.

SIMILARITY

Two square matrices **A** and **B** of the same order are said to be *similar* if a nonsingular matrix S exists such that

$$S^{-1}AS = B. \tag{49}$$

The matrix **B** is called the *similarity transform* of **A** by **S**. Furthermore, **A** is the similarity transform of **B** by S^{-1}.

The reason that similarity of matrices is an important concept is the fact that similar matrices have equal determinants, the same characteristic polynomials, and, hence, the same eigenvalues. These facts are easily established. Thus, by applying the rule for the determinant of a product of square matrices, the determinants are equal, because

$$|B| = |S^{-1}AS| = |S^{-1}|\,|A|\,|S| = |S^{-1}S|\,|A| = |A|.$$

The characteristic polynomials are equal because

$$|\lambda U - B| = |\lambda U - S^{-1}AS| = |S^{-1}(\lambda U - A)S|$$
$$= |S^{-1}|\,|\lambda U - A|\,|S| = |S^{-1}S|\,|\lambda U - A|$$
$$= |\lambda U - A|.$$

Since the eigenvalues of a matrix are the zeros of its characteristic polynomial, and since **A** and **B** have the same characteristic polynomials, their eigenvalues must be equal.

An important, special similarity relation is the similarity of **A** to a diagonal matrix

$$\Lambda = \begin{bmatrix} \lambda_1 & & & & & \\ & \lambda_2 & & & \bigcirc & \\ & & \cdot & & & \\ & & & \cdot & & \\ & & & & \cdot & \\ & \bigcirc & & & \lambda_{n-1} & \\ & & & & & \lambda_n \end{bmatrix}$$

Now, if **A** and Λ are similar, then the diagonal elements of Λ are the eigenvalues of **A**. This follows from the fact that **A** and Λ have the same eigenvalues and, as may easily be shown, the eigenvalues of Λ are its diagonal elements.

Next we will show that **A** *is similar to a diagonal matrix* Λ *if and only if* **A** *has* n *linearly independent eigenvectors*. First, suppose that **A** is similar to Λ. This means $\Lambda = S^{-1}AS$ or, equivalently,

$$S\Lambda = AS. \tag{50}$$

Now partition S by columns; that is, set $S = [S_1 S_2 \cdots S_n]$, where the S_i are the column vectors of S. Equating the jth column of AS to the jth column of SΛ, in accordance with (50), we get

$$\lambda_j S_j = AS_j. \tag{51}$$

By comparing with (46), we see that S_j is the eigenvector corresponding to λ_j. Since **A** is similar to Λ, S is nonsingular, and its column vectors (eigenvectors of **A**) are linearly independent. This establishes the necessity.

Now let us suppose that **A** has n linearly independent eigenvectors. By (50), the matrix S satisfies $S\Lambda = AS$. Since the n eigenvectors of **A** (column vectors of S) are linearly independent, S is nonsingular, and $S\Lambda = AS$ implies $\Lambda = S^{-1}AS$. Thus Λ is similar to **A** and, equivalently, **A** is similar to Λ.

We have just shown that, if the square matrix S having the eigenvectors of **A** as its column vectors is nonsingular, then **A** is similar to the diagonal matrix $\Lambda = S^{-1}AS$.

Example

As an illustration take the previously considered matrix

$$A = \begin{bmatrix} 5 & 1 \\ -2 & 2 \end{bmatrix}.$$

Earlier we found that $\lambda_1 = 3$ and $\lambda_2 = 4$ are the eigenvalues and that, for arbitrary, nonzero s_{11} and s_{12},

$$S_1 = \begin{bmatrix} s_{11} \\ -2s_{11} \end{bmatrix} \quad \text{and} \quad S_2 = \begin{bmatrix} s_{12} \\ -s_{12} \end{bmatrix}$$

are the corresponding eigenvectors. Let $s_{11} = s_{12} = 1$; then

$$S = \begin{bmatrix} 1 & 1 \\ -2 & -1 \end{bmatrix}$$

and therefore

$$S^{-1} = \begin{bmatrix} -1 & -1 \\ 2 & 1 \end{bmatrix}.$$

Then, of course,

$$S^{-1}AS = \begin{bmatrix} -1 & -1 \\ 2 & 1 \end{bmatrix} \begin{bmatrix} 5 & 1 \\ -2 & 2 \end{bmatrix} \begin{bmatrix} 1 & 1 \\ -2 & -1 \end{bmatrix} = \begin{bmatrix} 3 & 0 \\ 0 & 4 \end{bmatrix} = \Lambda$$

The procedure so far available to us for ascertaining the existence of S such that A is similar to Λ requires that we construct a trial matrix having the eigenvectors of A as columns. If that trial matrix is nonsingular, then it is S, and S exists. It is often of interest to know that an S exists without first constructing it. The following theorem provides such a criterion: *The n eigenvectors of A are distinct and, hence, S exists, if—*

1. *The eigenvalues of A are distinct.*

2. *A is either symmetric or Hermitian.**

SYLVESTER'S INEQUALITY

Consider the matrix product PQ, where P is a matrix of order $m \times n$ and rank r_P and where Q is a matrix of order $n \times k$ and rank r_Q. Let r_{PQ} denote the rank of the product matrix. Sylvester's inequality is a relation between r_P, r_Q, and r_{PQ} which states that[†]

$$r_P + r_Q - n \leq r_{PQ} \leq \min \{r_P, r_Q\}. \tag{52}$$

Note that n is the number of columns of the first matrix in the product or the number of rows of the second one.

As a special case, suppose P and Q are nonsingular square matrices of order n. Then $r_P = r_Q = n$, and, by Sylvester's inequality, $n \leq r_{PQ} \leq n$ or $r_{PQ} = n$. This we also know to be true by the fact that $|PQ| = |P| \, |Q| \neq 0$, since $|P| \neq 0$ and $|Q| \neq 0$. As another special case, suppose $PQ = 0$. Then r_{PQ} is obviously zero, and, by Sylvester's inequality, $r_P + r_Q \leq n$.

---

* Proofs may be found in R. Bellman, *Introduction to Matrix Analysis*, McGraw-Hill Book Co., Inc., New York, 1960, Chs. 3 and 4.

† A proof of Sylvester's inequality requires an understanding of some basic concepts associated with finite dimensional vector spaces. The topic is outside the scope of this text and no proof will be given. For such a proof see F. R. Ganthmacher, *The Theory of Matrices*, Vol. I, Chelsea Publishing Co., New York, 1959.

NORM OF A VECTOR

One of the properties of a space vector is its length. For an n-vector the notion of length no longer has a geometrical interpretation. Nevertheless, it is a useful concept, which we shall now discuss.

Define the *norm* of an n-vector $\mathbf{x}$ as a non negative number $\|\mathbf{x}\|$ that possesses the following properties:

1. $\|\mathbf{x}\| = 0$ if and only if $\mathbf{x} = \mathbf{0}$.

2. $\|\alpha\mathbf{x}\| = |\alpha| \, \|\mathbf{x}\|$, where α is a real or complex number.

3. $\|\mathbf{x}_1 + \mathbf{x}_2\| \leq \|\mathbf{x}_1\| + \|\mathbf{x}_2\|$, where $\mathbf{x}_1$ and $\mathbf{x}_2$ are two n-vectors.

A vector may have a number of different norms satisfying these properties. The most familiar norm is the Euclidean norm, defined by

$$\|\mathbf{x}\|_2 = (\mathbf{x}^*\mathbf{x})^{1/2} = \left(\sum_{i=1}^{n} |x_i|^2 \right)^{\frac{1}{2}}. \tag{53}$$

This is the square root of the sum of the squares of the components of the vector. The Euclidean norm is the one we are most likely to think about when reference is made to the length of a vector; however, there are other norms that are easier to work with in numerical calculations. One such norm is

$$\|\mathbf{x}\|_1 = \sum_{i=1}^{n} |x_i|; \tag{54}$$

that is, the sum of the magnitudes of the vector components. For want of a better name, we shall call it the *sum-magnitude* norm. Another such norm is

$$\|\mathbf{x}\|_\infty = \overset{\max}{i} |x_i|; \tag{55}$$

that is, the magnitude of the component having the largest magnitude. We shall call it the *max-magnitude* norm. It is a simple matter to show that $\|\mathbf{x}_2\|$, $\|\mathbf{x}\|_1$, and $\|\mathbf{x}\|_\infty$ each satisfy the stated properties of a norm.

That each of these norms is a satisfactory measure of vector length can be established by several observations. If any one of these norms is nonzero, the other two are nonzero. If any one of them tends toward zero as a limit, the other two must do likewise.

A matrix is often thought of as a *transformation*. If $\mathbf{A}$ is a matrix of order $m \times n$ and $\mathbf{x}$ is an n-vector, then we think of $\mathbf{A}$ as a matrix that transforms $\mathbf{x}$ into the m-vector $\mathbf{Ax}$. We will later need to establish bounds on the norm of the vector $\mathbf{Ax}$; to do this we introduce the norm of a matrix.

The matrix $\mathbf{A}$ is said to be bounded if there exists a real, positive constant K such that

$$\|\mathbf{Ax}\| \leq K \|\mathbf{x}\| \tag{56}$$

for all $\mathbf{x}$. The greatest lower bound of all such K is called the norm of $\mathbf{A}$ and is denoted by $\|\mathbf{A}\|$. It is easy to show that the matrix norm has the usual properties of a norm; that is,

1. $\|\mathbf{A}\| = 0$ if and only if $\mathbf{A} = \mathbf{0}$;

2. $\|\alpha\mathbf{A}\| = |\alpha|\|\mathbf{A}\|$, where α is a real or complex number; and

3. $\|\mathbf{A}_1 + \mathbf{A}_2\| \leq \|\mathbf{A}_1\| + \|\mathbf{A}_2\|$.

In addition, it is easily deduced that $\|\mathbf{A}_1\mathbf{A}_2\| \leq \|\mathbf{A}_1\| \|\mathbf{A}_2\|$.

By the definition of the greatest lower bound, it is clear that

$$\|\mathbf{Ax}\| \leq \|\mathbf{A}\| \|\mathbf{x}\|. \tag{57}$$

It is possible to show that a vector exists such that (57) holds as an equality. We will not do so in general, but will take the cases of the sum-magnitude norm in (54), the max-magnitude norm in (55), and the Euclidean norm in (53).

Thus, by using the sum-magnitude norm in (54), we get

$$\|\mathbf{Ax}\|_1 = \sum_{i=1}^{m} \Big| \sum_{j=1}^{n} a_{ij} x_j \Big| \leq \sum_{i=1}^{m} \sum_{j=1}^{n} |a_{ij}| |x_j|$$

$$\leq \Big\{ \max_j \sum_{i=1}^{m} |a_{ij}| \Big\} \sum_{j=1}^{n} |x_j| \leq \Big\{ \max_j \sum_{i=1}^{m} |a_{ij}| \Big\} \|\mathbf{x}\|_1. \tag{58}$$

The first step and the last step follow from the definition of the sum-magnitude norm. The second step is a result of the triangle inequality for complex numbers. Suppose the sum of magnitudes of a_{ij} is the largest for the kth column; that is, suppose $\max_j \sum_{i=1}^{m} |a_{ij}| = \sum_{i=1}^{m} |a_{ik}|$. Then (58) is satisfied as an equality when $x_j = 0$ for $j \neq k$ and $x_k = 1$. Therefore

$$\|\mathbf{A}\|_1 = \max_j \sum_{i=1}^{m} |a_{ij}|. \tag{59}$$

Thus the sum-magnitude norm of a matrix $\mathbf{A}$ is the sum-magnitude norm of the column vector of $\mathbf{A}$ which has the largest sum-magnitude norm.

Next let us use the max-magnitude norm in (55). Then

$$\|\mathbf{A}\mathbf{x}\|_\infty = \overset{\max}{\underset{i}{}} \left| \sum_{j=1}^{n} a_{ij}\,x_j \right| \leq \overset{\max}{\underset{i}{}} \sum_{j=1}^{n} |a_{ij}|\,|x_j|$$

$$\leq \left\{ \overset{\max}{\underset{i}{}} \sum_{j=1}^{n} |a_{ij}| \right\} \overset{\max}{\underset{j}{}} |x_j| \leq \left\{ \overset{\max}{\underset{i}{}} \sum_{j=1}^{n} |a_{ij}| \right\} \|x_j\|_\infty. \quad (60)$$

The pattern of steps here is the same as in the preceding norm except that the max-magnitude norm is used. Again, suppose the sum of magnitudes of a_{ij} is largest for the kth row; that is, suppose $\overset{\max}{\underset{i}{}} \sum_{j=1}^{n}|a_{ij}| = \sum_{j=1}^{n}|a_{kj}|$. Then (60) is satisfied as an equality when $x_j = \operatorname{sgn}(a_{kj})$. (The function $\operatorname{sgn} y$ equals $+1$ when y is positive and -1 when y is negative.) Therefore

$$\|\mathbf{A}\|_\infty = \overset{\max}{\underset{i}{}} \sum_{j=1}^{n} |a_{ij}|. \quad (61)$$

Thus the max-magnitude norm of a matrix $\mathbf{A}$ is the sum-magnitude norm of that row vector of $\mathbf{A}$ which has the largest sum-magnitude norm.

Finally, for the Euclidean norm, although we shall not prove it here, it can be shown* that

$$\|\mathbf{A}\mathbf{x}\|_2 = (\mathbf{x}^*\mathbf{A}^*\mathbf{A}\mathbf{x})^{1/2} \leq |\lambda_m|^{1/2}(\mathbf{x}^*\mathbf{x})^{1/2} = |\lambda_m|^{1/2}\|\mathbf{x}\| \quad (62)$$

* Tools for showing this will be provided in Chapter 7.

where λ_m is the eigenvalue of $\mathbf{A}^*\mathbf{A}$ having the largest magnitude. It can also be shown that a vector $\mathbf{x}$ exists such that (62) holds as an equality. Therefore

$$\|\mathbf{A}\|_2 = |\lambda_m|^{1/2}. \quad (63)$$

Example

As an illustration, suppose $\mathbf{y} = \mathbf{A}\mathbf{x}$, or

$$\begin{bmatrix} y_1 \\ y_2 \\ y_3 \end{bmatrix} = \begin{bmatrix} 1 & -1 \\ -2 & 0 \\ 3 & 4 \end{bmatrix} \begin{bmatrix} x_1 \\ x_2 \end{bmatrix}.$$

From (59) the sum-magnitude norm of **A** is

$$\|\mathbf{A}\|_1 = \max\ \{6,\ 5\} = 6.$$

From (61) the max-magnitude norm of **A** is

$$\|\mathbf{A}\|_\infty = \max\ \{2,\ 2,\ 7\} = 7.$$

As for the Euclidean norm, we first find that

$$\mathbf{A}^*\mathbf{A} = \begin{bmatrix} 1 & -2 & 3 \\ -1 & 0 & 4 \end{bmatrix} \begin{bmatrix} 1 & -1 \\ -2 & 0 \\ 3 & 4 \end{bmatrix} = \begin{bmatrix} 14 & 11 \\ 11 & 17 \end{bmatrix}.$$

The characteristic equation of **A*****A** is

$$|\lambda \mathbf{U} - \mathbf{A}^*\mathbf{A}| = \begin{vmatrix} \lambda - 14 & -11 \\ -11 & \lambda - 17 \end{vmatrix} = \lambda^2 - 31\lambda + 117$$
$$= (\lambda - 26.64)(\lambda - 4.36).$$

Hence $\lambda_m = 26.64$ and

$$\|\mathbf{A}\|_2 = \sqrt{26.64} = 5.18.$$

We also know, by substituting the above matrix norms into (57), that

$$(|y_1| + |y_2| + |y_3|) \leq 6(|x_1| + |x_2|)$$

$$\max\ (|y_1|, |y_2|, |y_3|) \leq 7 \max\ (|x_1|, |x_2|)$$

and

$$(|y_1|^2 + |y_2|^2 + |y_3|^2)^{1/2} \leq 5.18(|x_1|^2 + |x_2|^2)^{1/2}.$$

In this section, we have given a hasty treatment of some topics in matrix theory largely without adequate development and proof. Some of the proofs and corollary results are suggested in the problems.

1.3 NOTATION AND REFERENCES

The signals, or the variables in terms of which the behavior of electric networks is described, are voltage and current. These are functions of time t and will be consistently represented by lower-case symbols $v(t)$ and $i(t)$. Sometimes the functional dependence will not be shown explicitly when there is no possibility of confusion; thus v and i will be used instead of $v(t)$ and $i(t)$.

The Laplace transform of a time function will be represented by the capital letters corresponding to the lower-case letter representing the time function. Thus, $I(s)$ is the Laplace transform of $i(t)$, where s is the complex frequency variable, $s = \sigma + j\omega$. Sometimes the functional dependence on s will not be shown explicitly, and $I(s)$ will be written as plain I.

The fundamental laws on which network theory is founded express relationships among voltages and currents at various places in a network. Before these laws can even be formulated it is necessary to establish a system for correlating the sense of the quantities i and v with the indications of a meter. This is done by establishing a reference for each voltage and current. The functions $i(t)$ and $v(t)$ are real functions of time that can take on negative as well as positive values in the course of time. The system of references adopted in this book is shown in Fig. 1. An arrow

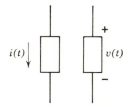

Fig. 1.   Current and voltage references.

indicates the reference for the current in a branch. This arrow does not mean that the current is always in the arrow direction. It means that, whenever the current is in the arrow direction, $i(t)$ will be positive. Similarly, the plus and minus signs at the ends of a branch are the voltage reference for the branch. Whenever the voltage polarity is actually in the sense indicated by the reference, $v(t)$ will be positive. Actually, the symbol for voltage reference has some redundancy, since showing only the plus sign will imply the minus sign also. Whenever there is no possibility of confusion, the minus sign can be omitted from the reference.

For a given branch the direction chosen as the current reference and the polarity chosen as the voltage reference are arbitrary. Either of the two possibilities can be chosen as the current reference, and either of the two possibilities can be chosen as the voltage reference. Furthermore, the reference for current is independent of the reference for voltage. However, it is often convenient to choose these two references in a certain way, as shown in Fig. 2. Thus, with the current-reference arrow drawn along-

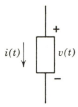

Fig. 2.   Standard references.

side the branch, if the voltage-reference plus is at the tail of the current reference, the result is called the *standard reference*. If it is stated that the standard reference is being used, then only one of the two need be shown; the other will be implied. It must be emphasized that there is no requirement for choosing standard references, only convenience.

1.4   NETWORK CLASSIFICATION

It is possible to arrive at a classification of networks in one of two ways. One possibility is to specify the kinds of elements of which the network is composed and, on the basis of their properties, to arrive at some generalizations regarding the network as a whole. Thus, if the values of all the elements of a network are constant and do not change with time, the network as a whole can be classified as a *time-invariant* network.

Another approach is to focus attention on the points of access to the network and classify the network in terms of the general properties of its responses to excitations applied at these points. In this section we shall examine the second of these approaches.

LINEARITY

Let the excitation applied to a network that has no initial energy storage be labeled $e(t)$ and the response resulting therefrom $w(t)$. A *linear* network is one in which the response is proportional to the excitation and the principle of superposition applies. More precisely, if the response to an excitation $e_1(t)$ is $w_1(t)$ and the response to an excitation $e_2(t)$ is $w_2(t)$, then the network is linear if the response to the excitation $k_1e_1(t) + k_2 e_2(t)$ is $k_1w_1(t) + k_2 w_2(t)$.

This scalar definition can be extended to matrix form for multiple excitations and responses. Excitation and response vectors, $e(t)$ and $w(t)$, are defined as column vectors

$$e(t) = \begin{bmatrix} e_a(t) \\ e_b(t) \\ \vdots \end{bmatrix} \quad \text{and} \quad w(t) = \begin{Bmatrix} w_a(t) \\ w_b(t) \\ \vdots \end{Bmatrix}$$

where e_a, e_b, etc., are excitations at positions a, b, etc.; and w_a, w_b, etc., are the corresponding responses. Then a network is linear if the excitation vector $k_1 e_1(t) + k_2 e_2(t)$ gives rise to a response vector $k_1 w_1(t) + k_2 w_2(t)$, where w_i is the response vector to the excitation vector e_i.

TIME INVARIANCE

A network that will produce the same response to a given excitation no matter when it is applied is *time invariant*. Thus, if the response to an excitation $e(t)$ is $w(t)$, then the response to an excitation $e(t + t_1)$ will be $w(t + t_1)$ in a time-invariant network. This definition implies that the values of the network components remain constant.

PASSIVITY

Some networks have the property of either absorbing or storing energy. They can return their previously stored energy to an external network, but never more than the amount so stored. Such networks are called *passive*. Let $E(t)$ be the energy delivered to a network having one pair of terminals from an external source up to time t. The voltage and current at the terminals, with standard references, are $v(t)$ and $i(t)$. The power delivered to the network will be $p(t) = v(t) i(t)$. We define the network to be passive if

$$E(t) = \int_{-\infty} v(x)\, i(x)\, dx \geq 0 \qquad\qquad (64)$$

or

$$E(t) = \int_{t_0}^{t} v(x) i(x)\, dx + E(t_0) \geq 0.$$

This must be true for any voltage and its resulting current for all t.

Any network that does not satisfy this condition is called an *active network*; that is, $\int_{-\infty}^{t} v(x)\, i(x)\, dx < 0$ for some time t.

If the network has more than one pair of terminals through which energy can be supplied from the outside, let the terminal voltage and current matrices be

$$\mathbf{v}(t) = \begin{bmatrix} v_1(t) \\ v_2(t) \\ \cdot \\ \cdot \\ \cdot \\ v_n(t) \end{bmatrix} \quad \text{and} \quad \mathbf{i}(t) = \begin{bmatrix} i_1(t) \\ i_2(t) \\ \cdot \\ \cdot \\ \cdot \\ i_n(t) \end{bmatrix}$$

with standard references. The instantaneous power supplied to the network from the outside will then be

$$p(t) = \sum_{j=1}^{n} v_j(t)\, i_j(t) = \mathbf{v}'(t)\, \mathbf{i}(t). \tag{65}$$

The network will be passive if, for all t,

$$E(t) = \int_{-\infty}^{t} \mathbf{v}'(x)\, \mathbf{i}(x)\, dx \geq 0 \tag{66}$$

RECIPROCITY

Some networks have the property that the response produced at one point of the network by an excitation at another point is invariant if the positions of excitation and response are interchanged (excitation and response being properly interpreted). Specifically, in Fig. 3a the network

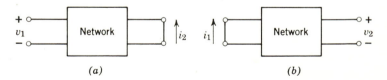

(a)   (b)

Fig. 3.   Reciprocity condition.

is assumed to have no initial energy storage; the excitation is the voltage $v_1(t)$ and the response is the current $i_2(t)$ in the short circuit. In Fig. 3b, the excitation is applied at the previously short-circuited point, and the response is the current in the short-circuit placed at the position of the previous excitation. The references of the two currents are the same relative to those of the voltages. A *reciprocal* network is one in which, for any pair of excitation and response points, here labeled 1 and 2, $i_1 = i_2$ if $v_2 = v_1$. If the network does not satisfy this condition, it is *nonreciprocal*.

Up to the last chapter of this book we shall be concerned with networks that are linear and time invariant. However, the networks will not be limited to passive or reciprocal types. The latter types of network do have special properties, and some procedures we shall discuss are limited to such networks. When we are discussing procedures whose application is limited to passive or reciprocal networks, we shall so specify. When no specification is made, it is assumed that the procedures and properties under discussion are generally applicable to both passive and active, and to both reciprocal and nonreciprocal, networks. The final chapter of the book will be devoted to linear, time-varying and to nonlinear networks.

1.5 NETWORK COMPONENTS

Now let us turn to a classification of networks on the basis of the kinds of elements they include. In the first place, our network can be characterized by the adjective "lumped." We assume that all electrical effects are experienced immediately throughout the network. With this assumption we neglect the influence of spatial dimensions in a physical circuit, and we assume that electrical effects are lumped in space rather than being distributed.

In the network model we postulate the existence of certain elements that are defined by the relationship between their currents and voltages. The three basic elements are the resistor, the inductor, and the capacitor. Their diagrammatic representations and voltage-current relationships are given in Table 1. The resistor is described by the resistance parameter R or the conductance parameter G, where $G = 1/R$.

The inductor is described by the inductance parameter. The reciprocal of L has no name, but the symbol Γ (an inverted L) is sometimes used. Finally, the capacitor is described by the capacitance parameter C. The reciprocal of C is given the name elastance, and the symbol D is sometimes used.

A number of comments are in order concerning these elements. First, the v-i relations ($v = Ri$, $v = L\, di/dt$, and $i = C\, dv/dt$) satisfy the linearity condition, assuming that i and v play the roles of excitation and response, as the case may be. (Demonstrate this to yourself.) Thus networks of R, L, and C elements are linear. Second, the parameters R, L, and C are constant, so networks of R, L, and C elements will be time invariant.

Table 1

		Voltage-Current Relationships		
Element	Parameter	Direct	Inverse	Symbol
Resistor	Resistance R Conductance G	$v = Ri$	$i = \dfrac{1}{R} v = Gv$	
Inductor	Inductance L Inverse Inductance Γ	$v = L \dfrac{di}{dt}$	$i(t) = \dfrac{1}{L} \displaystyle\int_0^t v(x)\, dx + i(0)$	
Capacitor	Capacitance C Elastance D	$i = C \dfrac{dv}{dt}$	$v(t) = \dfrac{1}{C} \displaystyle\int_0^t i(x)\, dx + v(0)$	

In the third place, assuming standard references, the energy delivered to each of the elements starting at a time when the current and voltage were zero will be

$$E_R(t) = \int_{-\infty}^{t} Ri^2(x)\, dx \tag{67}$$

$$E_L(t) = \int_{-\infty}^{t} L \frac{di(x)}{dx} i(x)\, dx = \int_0^{i(t)} Li'\, di' = \tfrac{1}{2} Li^2(t) \tag{68}$$

$$E_C(t) = \int_{-\infty}^{t} C \frac{dv(x)}{dx} v(x)\, dx = \int_0^{v(t)} Cv'\, dv' = \tfrac{1}{2} Cv^2(t). \tag{69}$$

Each of the right-hand sides is non-negative for all t. Hence networks of R, L, and C elements are *passive*. Finally, networks of R, L, and C elements are *reciprocal*, but demonstration of this fact must await later developments.

It should be observed in Table 1 that the inverse v-i relations for the inductance and capacitance element are written as definite integrals. Quite often this inverse relationship is written elsewhere as an indefinite integral (or antiderivative) instead of a definite integral. Such an expression is incomplete unless there is added to it a specification of the initial values $i(0)$ or $v(0)$, and in this sense is misleading. Normally one thinks of the voltage $v(t)$ and the current $i(t)$ as being expressed as explicit functions such as $\varepsilon^{-\alpha t}$, $\sin \omega t$, etc., and the antiderivative as being something unique: $-(1/\alpha)\varepsilon^{-\alpha t}$, $-(1/\omega)\cos \omega t$, etc., which is certainly not true in general. Also, in many cases the voltage or current may not be expressible in such a simple fashion for all t; the analytic expression for $v(t)$ or $i(t)$ may depend on the particular interval of the axis on which the point t falls. Some such wave shapes are shown in Fig. 4.

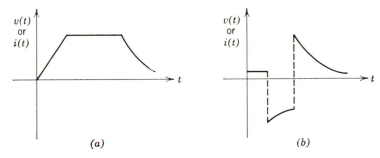

(a)   (b)

Fig. 4.   Signal waveshapes.

The origin of time t is arbitrary; it is usually chosen to coincide with some particular event, such as the opening or closing of a switch. In addition to the definite integral from 0 to t, the expression for the capacitor voltage, $v(t) = (1/C)\int_0^t i(x)\,dx + v(0)$, contains the initial value $v(0)$. This can be considered as a d-c voltage source (sources are discussed below) in series with an initially relaxed (no initial voltage) capacitor, as shown in Fig. 5. Similarly, for the inductor $i(t) = (1/L)\int_0^t v(x)\,dx + i(0)$,

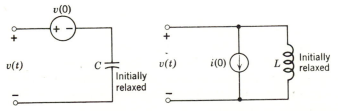

Fig. 5.   Initial values as sources.

where $i(0)$ is the initial value of the current. This can be considered as a d-c current source in parallel with an initially relaxed inductor, as shown in Fig. 5. If these sources are shown explicitly, they will account for all initial values, and all capacitors and inductors can be considered to be initially relaxed. Such initial-value sources can be useful for some methods of analysis but not for others, such as the state-equation formulation.

THE TRANSFORMER

The R, L, and C elements all have two terminals; other components have more than two terminals. The next element we shall introduce is the *ideal transformer* shown in Fig. 6. It has two pairs of terminals and is

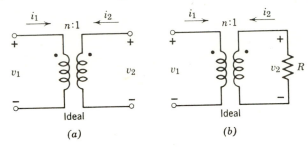

Fig. 6.   An ideal transformer.

defined in terms of the following *v-i* relationships:

$$v_1 = nv_2 \tag{70a}$$

$$i_2 = -ni_1 \tag{70b}$$

or

$$\begin{bmatrix} v_1 \\ i_2 \end{bmatrix} = \begin{bmatrix} 0 & n \\ -n & 0 \end{bmatrix} \begin{bmatrix} i_1 \\ v_2 \end{bmatrix}. \tag{70c}$$

The ideal transformer is characterized by a single parameter n called the *turns ratio*. The ideal transformer is an abstraction arising from coupled coils of wire. The *v-i* relationships are idealized relations expressing Faraday's law and Ampere's law, respectively. The signs in these equations apply for the references shown. If any one reference is changed, the corresponding sign will change.

An ideal transformer has the property that a resistance R connected to one pair of terminals appears as R times the turns ratio squared at the other pair of terminals. Thus in Fig. 6b, $v_2 = -Ri_2$. When this is used in the *v-i* relationships, the result becomes

$$v_1 = nv_2 = -nRi_2 = (n^2R)i_1. \tag{71}$$

At the input terminals, then, the equivalent resistance is $n^2 R$.

Observe that the total energy delivered to the ideal transformer from connections made at its terminals will be

$$E(t) = \int_{-\infty}^{t} [v_1(x)\, i_1(x) + v_2(x)\, i_2(x)]\, dx = 0 \tag{72}$$

The right-hand side results when the v-i relations of the ideal transformer are inserted in the middle. Thus the device is passive; it transmits—but neither stores nor dissipates—energy.

A less abstract model of a physical transformer is shown in Fig. 7.

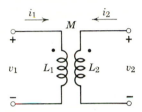

Fig. 7.   A transformer.

The diagram is almost the same except that the diagram of the ideal transformer shows the turns ratio directly on it. The transformer is characterized by the following v-i relationships for the references shown in Fig. 7:

$$v_1 = L_1 \frac{di_1}{dt} + M \frac{di_2}{dt} \tag{73a}$$

and

$$v_2 = M \frac{di_1}{dt} + L_2 \frac{di_2}{dt}. \tag{73b}$$

Thus it is characterized by three parameters: the two self-inductances L_1 and L_2, and the mutual inductance M.

The total energy delivered to the transformer from external sources is

$$E(t) = \int_{-\infty}^{t} [v_1(x)i_1(x) + v_2(x)i_2(x)]\, dx$$

$$= \int_{0}^{i_1} L_1 i_1' di_1' + \int_{0}^{i_1 i_2} M d(i_1' i_2') + \int_{0}^{i_2} L_2 i_2' di_2' \tag{74}$$

$$= \tfrac{1}{2}(L_1 i_1{}^2 + 2M i_1 i_2 + L_2 i_2{}^2)$$

It is easy to show* that the last line will be non-negative if

$$\frac{M^2}{L_1 L_2} = k^2 \leq 1. \tag{75}$$

Since physical considerations require the transformer to be passive, this condition must apply. The quantity k is called the *coefficient of coupling*. Its maximum value is unity.

A transformer for which the coupling coefficient takes on its maximum value $k = 1$ is called a *perfect*, or *perfectly coupled*, transformer. A perfect transformer is not the same thing as an ideal transformer. To find the difference, turn to the transformer equations (73) and insert the perfect-transformer condition $M = \sqrt{L_1 L_2}$; then take the ratio v_1/v_2. The result will be

$$\frac{v_1}{v_2} = \frac{L_1 \dfrac{di_1}{dt} + \sqrt{L_1 L_2} \dfrac{di_2}{dt}}{\sqrt{L_1 L_2} \dfrac{di_1}{dt} + L_2 \dfrac{di_2}{dt}} = \sqrt{L_1/L_2}. \tag{76}$$

This expression is identical with $v_1 = nv_2$ for the ideal transformer† if

$$n = \sqrt{L_1/L_2}. \tag{77}$$

Next let us consider the current ratio. Since (73) involve the derivatives of the currents, it will be necessary to integrate. The result of inserting the perfect-transformer condition $M = \sqrt{L_1 L_2}$ and the value $n = \sqrt{L_1/L_2}$, and integrating (73a) from 0 to t will yield, after rearranging,

$$i_1(t) = -\frac{1}{n} i_2(t) + \left\{ \frac{1}{L_1} \int_0^t v_1(x)\, dx + \left[i_1(0) + \frac{1}{n} i_2(0) \right] \right\}. \tag{78}$$

* A simple approach is to observe (with L_1, L_2, and M all non-negative) that the only way $L_1 i_1{}^2 + 2M i_1 i_2 + L_2 i_2{}^2$ can become negative is for i_1 and i_2 to be of opposite sign. So set $i_2 = -x i_1$, with x any real positive number, and the quantity of interest becomes $L_1 - 2Mx + L_2 x^2$. If the minimum value of this quadratic in x is non-negative, then the quantity will be non-negative for any value of x. Differentiate the quadratic with respect to x and find the minimum value; it will be $L_1 - M^2/L_2$, from which the result follows.

† Since, for actual coils of wire, the inductance is approximately proportional to the square of the number of turns in the coil, the expression $\sqrt{L_1/L_2}$ equals the ratio of the turns in the primary and secondary of a physical transformer. This is the origin of the name "turns ratio" for n.

This is to be compared with $i_1 = -i_2/n$ for the ideal transformer. The form of the expression in brackets suggests the $v\text{-}i$ equation for an inductor. The diagram shown in Fig. 8 satisfies both (78) and (76). It shows how a perfect transformer is related to an ideal transformer. If, in a perfect transformer, L_1 and L_2 are permitted to approach infinity, but in such a way that their ratio remains constant, the result will be an ideal transformer.

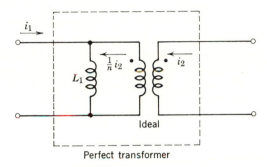

Fig. 8.   Relationship between a perfect and an ideal transformer.

THE GYRATOR

Another component having two pairs of terminals is the *gyrator*, whose diagrammatic symbol is shown in Fig. 9. It is defined in terms of

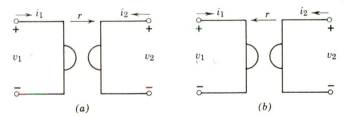

(a)   (b)

Fig. 9.   A gyrator.

the following $v\text{-}i$ relations:

For Fig. 9a

$$\begin{aligned} v_1 &= -ri_2 \\ v_2 &= ri_1 \end{aligned} \quad \text{or} \quad \begin{bmatrix} v_1 \\ v_2 \end{bmatrix} = \begin{bmatrix} 0 & -r \\ r & 0 \end{bmatrix} \begin{bmatrix} i_1 \\ i_2 \end{bmatrix}. \tag{79a}$$

For Fig. 9b

$$\begin{aligned} v_1 &= ri_2 \\ v_2 &= -ri_1 \end{aligned} \quad \text{or} \quad \begin{bmatrix} v_1 \\ v_2 \end{bmatrix} = \begin{bmatrix} 0 & r \\ -r & 0 \end{bmatrix} \begin{bmatrix} i_1 \\ i_2 \end{bmatrix} \tag{79b}$$

The gyrator, like the ideal transformer, is characterized by a single parameter r, called the *gyration resistance*. The arrow to the right or the left in Fig. 9 shows the *direction of gyration*.

The gyrator is a hypothetical device that is introduced to account for physical situations in which the reciprocity condition does not hold. Indeed, if first the right-hand side is short-circuited and a voltage $v_1 = v$ is applied to the left side, and if next the left side is shorted and the same voltage $(v_2 = v)$ is applied to the right side, then it will be found that $i_2 = -i_1$. Thus the gyrator is not a reciprocal device. In fact, it is antireciprocal.

On the other hand, the total energy input to the gyrator is

$$E(t) = \int_{-\infty}^{t} (v_1 i_1 + v_2 i_2)\, dx = \int_{-\infty}^{t} [(-ri_2)i_1 + (ri_1)i_2]\, dx = 0. \quad (80)$$

Hence it is a passive device that neither stores nor dissipates energy. In this respect it is similar to an ideal transformer.

In the case of the ideal transformer, it was found that the resistance at one pair of terminals, when the second pair is terminated in a resistance R, is n^2R. The ideal transformer thus changes a resistance by a factor n^2. What does the gyrator do in the corresponding situation? If a gyrator is terminated in a resistance R (Fig. 10), the output voltage and current

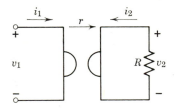

Fig. 10.   Gyrator terminated in a resistance R.

will be related by $v_2 = -Ri_2$. When this is inserted into the *v-i* relations, the result becomes

$$v_1 = -ri_2 = -r\left(-\frac{v_2}{R}\right) = r\left(\frac{ri_1}{R}\right) = (r^2G)i_1. \quad (81)$$

Thus the equivalent resistance at the input terminals equals r^2 times the conductance terminating the output terminals. The gyrator thus has the property of *inverting*.

The inverting property brings about more unusual results when the

gyrator is terminated in a capacitor or an inductor; for example, suppose
a gyrator is terminated in a capacitor, as shown in Fig. 11. We know that

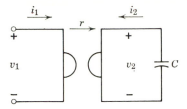

Fig. 11.   Gyrator terminated in a capacitance C.

$i_2 = -C \, dv_2/dt$. Therefore, upon inserting the v-i relations associated with
the gyrator, we observe that

$$v_1 = -ri_2 = -r\left(-C\frac{dv_2}{dt}\right) = rC\frac{d(ri_1)}{dt} = r^2C\frac{di_1}{dt}. \qquad (82)$$

Thus at the input terminals the v-i relationship is that of an inductor,
with inductance r^2C. In a similar manner it can be shown that the v-i
relationship at the input terminals of an inductor-terminated gyrator is
that of a capacitor.

INDEPENDENT SOURCES

All the devices introduced so far have been passive. Other network
components are needed to account for the ability to generate voltage,
current, or power.

Two types of sources are defined as follows:

1. A *voltage source* is a two-terminal device whose voltage at any instant
of time is independent of the current through its terminals. No matter
what network may be connected at the terminals of a voltage source, its
voltage will maintain its magnitude and waveform. (It makes no sense to
short-circuit the terminals of a voltage source, because this imposes two
idealized conflicting requirements at the terminals.) The current in the
source, on the other hand, will be determined by this network. The
diagram is shown in Fig. 12a.

2. A *current source* is a two-terminal device whose current at any instant
of time is independent of the voltage across its terminals. No matter what
network may be connected at the terminals of a current source, the current will maintain its magnitude and waveform. (It makes no sense to

open-circuit the terminals of a current source because this, again, imposes two conflicting requirements at the terminals.) The voltage across the source, on the other hand, will be determined by this network. The diagram is shown in Fig. 12*b*.

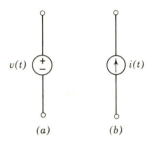

$v(t)$ $i(t)$

(a)   (b)

Fig. 12.   Voltage and current sources.

Everyone is familiar enough with the dimming of the house lights when a large electrical appliance is switched on the line to know that the voltage of a physical source varies under load. Also, in an actual physical source the current or voltage generated may depend on some nonelectrical quantity, such as the speed of a rotating machine, or the concentration of acid in a battery, or the intensity of light incident on a photoelectric cell. These relationships are of no interest to us in network analysis, since we are not concerned with the internal operation of sources, but only with their terminal behavior. Thus our idealized sources take no cognizance of the dependence of voltage or current on nonelectrical quantities; they are called *independent sources*.

CONTROLLED OR DEPENDENT SOURCES

The independent sources just introduced cannot account for our ability to amplify signals. Another class of devices is now introduced; these are called *controlled*, or *dependent*, sources. A *controlled voltage source* is a source whose terminal voltage is a function of some other voltage or current. A *controlled current source* is defined analogously. The four possibilities are shown in Table 2. These devices have two pairs of terminals—one pair designating the controlling quantity; the other, the controlled quantity. In each of the components in Table 2, the controlled voltage or current is directly proportional to the controlling quantity, voltage or current. This is the simplest type of dependence; it would be possible to introduce a dependent source whose voltage or current is proportional to the derivative of some other voltage or current, for example. However, detailed consideration will not be given to any other type of dependence.

For certain ranges of voltage and current the behavior of certain
vacuum tubes and transistors can be approximated by a model consisting

Table 2

Device	Symbol	Voltage-Current relationship
Voltage-controlled voltage source	v_1 μv_1 v_2	$\begin{bmatrix} i_1 \\ v_2 \end{bmatrix} = \begin{bmatrix} 0 & 0 \\ \mu & 0 \end{bmatrix} \begin{bmatrix} v_1 \\ i_2 \end{bmatrix}$ (hybrid g)
Current-controlled voltage source	i_1 $r_m i_1$ v_2	$\begin{bmatrix} v_1 \\ v_2 \end{bmatrix} = \begin{bmatrix} 0 & 0 \\ r_m & 0 \end{bmatrix} \begin{bmatrix} i_1 \\ i_2 \end{bmatrix}$ (impedance)
Voltage-controlled current source	v_1 $g_m v_1$ i_2	$\begin{bmatrix} i_1 \\ i_2 \end{bmatrix} = \begin{bmatrix} 0 & 0 \\ g_m & 0 \end{bmatrix} \begin{bmatrix} v_1 \\ v_2 \end{bmatrix}$ (admittance)
Current-controlled current source	i_1 αi_1 i_2	$\begin{bmatrix} v_1 \\ i_2 \end{bmatrix} = \begin{bmatrix} 0 & 0 \\ \alpha & 0 \end{bmatrix} \begin{bmatrix} i_1 \\ i_2 \end{bmatrix}$ (hybrid h)

of interconnected dependent sources and other network elements. Figure
13 shows two such models. These models are not valid representations
of the physical devices under all conditions of operation; for example, at
high enough frequency, the interelectrode capacitances of the tube would
have to be included in the model.

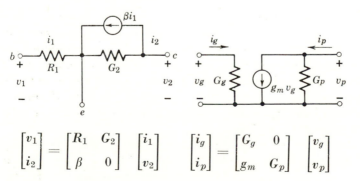

$$\begin{bmatrix} v_1 \\ i_2 \end{bmatrix} = \begin{bmatrix} R_1 & G_2 \\ \beta & 0 \end{bmatrix} \begin{bmatrix} i_1 \\ v_2 \end{bmatrix} \qquad \begin{bmatrix} i_g \\ i_p \end{bmatrix} = \begin{bmatrix} G_g & 0 \\ g_m & G_p \end{bmatrix} \begin{bmatrix} v_g \\ v_p \end{bmatrix}$$

Fig. 13. Models of transistor and triode.

The last point brings up a question. When an engineer is presented with a physical problem concerned with calculating certain voltages and currents in an interconnection of various physical electrical devices, his first task must be one of representing each device by a model. This model will consist of interconnections of the various components that have been defined in this chapter. The extent and complexity of the model will depend on the type of physical devices involved and the conditions under which they are to operate. Considerations involved in choosing an appropriate model to use, under various given conditions, do not form a proper part of network analysis. This is not to say that such considerations and the ability to choose an appropriate model are not important; they are. However, many other things are important in the total education of an engineer, and they certainly cannot all be treated in one book. In this book we will make no attempt to construct a model of a given physical situation before proceeding with the analysis. Our starting point will be a model.

THE NEGATIVE CONVERTER

The last component we shall introduce is the *negative converter* (NC for short). It is a device with two pairs of terminals and is defined by the following *v-i* equations:

$$\begin{bmatrix} v_1 \\ i_2 \end{bmatrix} = \begin{bmatrix} 0 & k \\ k & 0 \end{bmatrix} \begin{bmatrix} i_1 \\ v_2 \end{bmatrix} \tag{83a}$$

or

$$\begin{bmatrix} v_1 \\ i_2 \end{bmatrix} = \begin{bmatrix} 0 & -k \\ -k & 0 \end{bmatrix} \begin{bmatrix} i_1 \\ v_2 \end{bmatrix}. \tag{83b}$$

There is no special diagram for the NC, so it is shown by the general symbol in Fig. 14. The NC is characterized by a single parameter k, called the *conversion ratio*. If the left-hand terminals are considered the input and the right-hand ones the output, it is seen from the first set of equations that when i_1 is in its reference direction, i_2 will also be in its reference direction; hence the current will be "inverted" in going through the NC. On the other hand, the voltage will not be inverted. This type is therefore called a *current NC*, or INC.

For the second set of relations the opposite is true: the voltage is inverted, but the current is not. This type is called a *voltage NC*, or VNC.

When either of these devices is terminated in a passive component at one pair of terminals, something very interesting happens at the other pair.

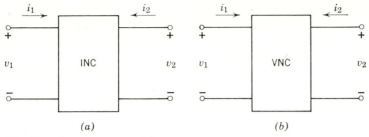

Fig. 14. Negative converters: (*a*) current-inverting variety: $v_1 = kv_2$ and $i_2 = ki_1$; (*b*) voltage-inverting variety: $v_1 = -kv_2$ and $i_2 = -ki_1$.

Thus, let an inductance L terminate the output; then $v_2 = -L\, di_2/dt$. When this is inserted into the *v-i* relations, there results

$$v_1 = \pm kv_2 = \pm k\left(-L\frac{di_2}{dt}\right) = \pm k(-L)\left(\pm k\frac{di_1}{dt}\right) = -k^2 L\frac{di_1}{dt}.$$

(84)

Thus at the input terminals the equivalent inductance is proportional to the *negative* of the terminating inductance. (Hence its name.) Similar conclusions would follow if the terminating element were a resistor or capacitor.

The introduction of the NC expands the number of network building blocks considerably, because it is now possible to include negative R, L, and C elements in the network.

PROBLEMS

1. Is $(A + B)^2 = A^2 + 2AB + B^2$ in matrix algebra? if not, give the correct formula.

2. Let

$$A = \begin{bmatrix} 1 & 0 & -1 \\ 0 & 1 & 2 \end{bmatrix}, \quad B = \begin{bmatrix} 4 & 2 \\ 1 & 2 \\ 2 & 1 \end{bmatrix}, \quad C = \begin{bmatrix} 2 & 5 \\ 5 & -4 \\ 0 & 4 \end{bmatrix}.$$

Compute **AB** and **AC** and compare them. Deduce, thereby, which law of ordinary algebra fails to hold for matrices?

3. Under what conditions can we conclude $B = C$ from $AB = AC$?

4. Let

$$\mathbf{A} = \begin{bmatrix} 1 & -1 \\ -1 & 1 \end{bmatrix} \quad \text{and} \quad \mathbf{B} = \begin{bmatrix} 1 & -1 \\ 1 & -1 \end{bmatrix}.$$

Compute **AB**. What theorem of ordinary algebra is not true for matrices?

5. Let **A** and **B** be conformable and let the submatrices $\mathbf{A}_{ij}$ and $\mathbf{B}_{jk}$ be conformable for all i and k. Verify the statement that the (i, k) submatrix of the product **AB** is $\sum_j \mathbf{A}_{ij} \mathbf{B}_{jk}$.

6. Show that

$$\int^x \mathbf{A}(y) \frac{d\mathbf{B}(y)}{dy} \, dy = \mathbf{A}(x) \mathbf{B}(x) - \int^x \frac{d\mathbf{A}(y)}{dy} \mathbf{B}(y) \, dy$$

7. Prove that $\overline{(\mathbf{AB})} = \overline{\mathbf{A}}\,\overline{\mathbf{B}}$ and $(\mathbf{AB})^* = \mathbf{B}^*\mathbf{A}^*$.

8. Verify the statement that any square matrix **A** can be expressed as the sum of a Hermitian matrix $\mathbf{A}_H$ and a skew-Hermitian matrix $\mathbf{A}_{SH}$. Find $\mathbf{A}_H$ and $\mathbf{A}_{SH}$.

9. Prove that if **A** is skew-Hermitian, $\text{Re}\,(a_{ii}) = 0$ for all i.

10. Prove $\sum_{k=1}^{n} a_{ik} \Delta_{jk} = 0$ if $i \neq j$.

11. Define the inverse of **A** as the matrix **B** such that $\mathbf{BA} = \mathbf{AB} = \mathbf{U}$. Show that, if the inverse exists, it is unique. (Assume two inverses and show they are equal.)

12. Check whether any of the following matrices are nonsingular. Find the inverses of the nonsingular matrices

$$\mathbf{A} = \begin{bmatrix} 2 & 3 \end{bmatrix},$$

$$\mathbf{B} = \begin{bmatrix} 1 & 0 & 5 & 2 \\ 0 & 1 & 4 & 6 \\ 0 & 0 & 1 & 3 \\ 0 & 0 & 0 & 1 \end{bmatrix}, \quad \mathbf{D} = \begin{bmatrix} 1 & 0 & 3 \\ 0 & 2 & 1 \\ 3 & 1 & 2 \end{bmatrix},$$

$$\mathbf{C} = \begin{bmatrix} 1 & 0 & 0 & 0 \\ 2 & 2 & 0 & 0 \\ 3 & 0 & 4 & 0 \\ 2 & 1 & 3 & 1 \end{bmatrix}, \quad \mathbf{E} = \begin{bmatrix} 1 & 0 & 1 \\ -1 & 1 & 0 \\ 0 & -1 & -1 \end{bmatrix}.$$

13. Prove that the inverse of a symmetric matrix is symmetric.

14. Prove that $(\mathbf{A}^{-1})' = (\mathbf{A}')^{-1}$.

15. Prove that if **Z** is symmetric, so is $(\mathbf{BZB}')$.

16. Prove that adj $(\mathbf{AB}) = (\text{adj } \mathbf{B})(\text{adj } \mathbf{A})$ when $\mathbf{A}$ and $\mathbf{B}$ are nonsingular square matrices.

17. Prove that $\dfrac{d\mathbf{A}^{-1}}{dx} = -\mathbf{A}^{-1}\dfrac{d\mathbf{A}}{dx}\mathbf{A}^{-1}$.

18. Prove that $\dfrac{d}{dx}\mathbf{AB} = \dfrac{d\mathbf{A}}{dx}\mathbf{B} + \mathbf{A}\dfrac{d\mathbf{B}}{dx}$.

19. Prove that $\dfrac{d|\mathbf{A}|}{dx} = \text{tr}\left\{(\text{adj } \mathbf{A})\dfrac{d\mathbf{A}}{dx}\right\} = \text{tr}\left\{\dfrac{d\mathbf{A}}{dx}(\text{adj } \mathbf{A})\right\}$.

20. Show that

$$
\frac{d|\mathbf{A}|}{dx} = \sum_{j=1}^{n}
\begin{vmatrix}
a_{11} \cdots a_{1j-1} & \dfrac{da_{1j}}{dx} & a_{1j+1} \cdots a_{1n} \\
\vdots & \vdots & \vdots \\
a_{n1} \cdots a_{nj-1} & \dfrac{da_{nj}}{dx} & a_{nj+1} \cdots a_{nn}
\end{vmatrix}.
$$

21. If $\mathbf{A}$ and $\mathbf{B}$ are nonsquare matrices, is it true that $\mathbf{AB}$ can never equal $\mathbf{BA}$? Explain.

22. $\mathbf{A}$ is of order n and rank $n - 1$. Prove that adj $\mathbf{A}$ is of rank 1.

23. Let $\mathbf{D}$ be a diagonal matrix with diagonal elements d_{ii}, and let $\mathbf{A} = [a_{ij}]$ be a square matrix of the same order. Show that (a) When $\mathbf{D}$ premultiplies $\mathbf{A}$, the elements of the ith row of $\mathbf{A}$ are multiplied by d_{ii}; and (b) When $\mathbf{D}$ postmultiplies $\mathbf{A}$, the elements of the ith column of $\mathbf{A}$ are multiplied by d_{ii}.

24. Prove that (a) $(\mathbf{AB})' = \mathbf{B}'\mathbf{A}'$ and (b) $(\mathbf{A} + \mathbf{B})' = \mathbf{A}' + \mathbf{B}'$.

25. Let $\mathbf{A}$ and $\mathbf{B}$ be symmetric and of order n. Prove that (a) the product $\mathbf{AB}$ is symmetric if $\mathbf{A}$ and $\mathbf{B}$ commute, and (b) $\mathbf{A}$ and $\mathbf{B}$ commute if the product $\mathbf{AB}$ is symmetric.

26. In the matrix product $\mathbf{A} = \mathbf{BC}$, $\mathbf{A}$ and $\mathbf{C}$ are square nonsingular matrices. Prove that $\mathbf{B}$ is nonsingular.

27. Use pivotal condensation to evaluate the determinant of the following matrices:

$$
\mathbf{A} = \begin{bmatrix}
1 & 6 & -2 & 0 \\
0 & 3 & 3 & 2 \\
-1 & -2 & 4 & 5 \\
4 & 5 & 0 & -3
\end{bmatrix}
\quad \text{and} \quad
\mathbf{B} = \begin{bmatrix}
0 & 6 & 0 & 1 & 5 \\
-1 & 4 & 5 & 0 & 2 \\
0 & 0 & -3 & 2 & 3 \\
2 & 1 & 2 & -1 & -1 \\
3 & 3 & -1 & -2 & 0
\end{bmatrix}.
$$

28. For a set of homogeneous equations the solution as given in (45) will become $x_1 = -A_{11}^{-1}A_{12}x_2$. If $m = n$ and the matrix A is of rank $r = n - 1$, determine an expression for each of the x_i variables in x_1 in terms of cofactors of A.

29. Prove the statement: A determinant is zero if and only if the rows and columns are linearly dependent.

30. For the following set of equations verify (42).

$$x_1 + 2x_2 + 3x_3 = 2$$
$$-3x_1 - 2x_2 - x_3 = -2$$
$$x_2 + 2x_3 = 1$$

31. Solve the following systems of equations:

(a) $x_1 + x_2 - x_3 = 2$ (b) $2x_1 + x_2 + 4x_3 = 3$
$\quad\ \ 2x_1 - x_2 + x_3 = 7$ $\quad\ -x_1 + 2x_2 - 7x_3 = 1$
$\quad\ \ x_1 + 4x_2 - 4x_3 = -1$ $\quad\ 3x_1 + 4x_2 + x_3 = 7$

(c) $x_1 + x_2 = -1$ (d) $x_1 + 2x_2 - 3x_3 + 4x_4 = 1$
$\quad\ \ x_1 + 3x_2 = 3$ $\quad\ x_1 + x_2 + x_3 - x_4 = 2$
$\quad\ \ x_1 - 3x_2 = -9$ $\quad\ x_1 + 3x_2 - 7x_3 + 9x_4 = 0.$

32. Evaluate det A by applying the definition of a determinant when

$$A = \begin{bmatrix} a_{11} & a_{12} & a_{13} \\ a_{21} & a_{22} & a_{23} \\ a_{31} & a_{32} & a_{33} \end{bmatrix}.$$

33. Show the maximum number of linearly independent n-vectors from the set of all n-vectors x that satisfy $0 = Ax$ is equal to the nullity of A.

34. If q_P, q_Q, and q_{PQ} denote the nullity of P, Q, and PQ, respectively, show that

$$q_Q \leq q_{PQ} \leq q_P + q_Q.$$

35. Show that the determinant of a triangular matrix equals the product of the main diagonal elements.

36. Let

$$A = \begin{bmatrix} A_{11} & A_{12} \\ 0 & A_{22} \end{bmatrix}$$

where A_{11} and A_{22} are square submatrices. Show that

$$\det A = (\det A_{11})(\det A_{22}).$$

37. Find the eigenvalues and eigenvectors of the following matrices:

(a) $\begin{bmatrix} -4 & 2 \\ -3 & 1 \end{bmatrix}$, (b) $\begin{bmatrix} 1 & 2 & 0 \\ 0 & 3 & 0 \\ 3 & 1 & 2 \end{bmatrix}$, (c) $\begin{bmatrix} 2 & 1 & 1 \\ 0 & 3 & 0 \\ 3 & 0 & 4 \end{bmatrix}$, (d) $\begin{bmatrix} -3 & 0 & 0 \\ -1 & 1 & 3 \\ 2 & -2 & -4 \end{bmatrix}$.

38. Each of the following matrices A is similar to a diagonal matrix $\Lambda = S^{-1}AS$. Find S in each case.

(a) $A = \begin{bmatrix} 1 & 1 & -3 \\ 1 & 1 & 3 \\ -3 & 3 & 3 \end{bmatrix}$, (b) $A = \begin{bmatrix} -1 & 2 & 4 \\ 0 & -3 & 0 \\ 0 & 3 & -2 \end{bmatrix}$,

(c) $A = \begin{bmatrix} 1 & j2 \\ -j2 & 4 \end{bmatrix}$.

39. Evaluate the matrix norms $\|A\|_1$, $\|A\|_2$, and $\|A\|_\infty$ when

(a) $A = \begin{bmatrix} 1 & 3 & -2 \\ 4 & -2 & 0 \end{bmatrix}$, (b) $A = \begin{bmatrix} 1 & 4 \\ 0 & -3 \\ 2 & -1 \end{bmatrix}$,

(c) $A = \begin{bmatrix} -1 & -1 & 2 \\ 0 & -2 & 0 \\ 4 & 0 & -3 \end{bmatrix}$.

40. A network has the excitation and response pair shown in Fig. P–40. A second excitation is also shown. If the network is linear and time-invariant, sketch the response for this second excitation.

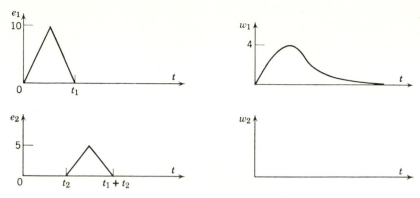

Fig. P–40

41. Suppose the output current of a linear, time-invariant, and reciprocal network, subject to excitation only at the input, is as illustrated in Fig. P–41a. Find the input current i_1 when the network is excited as shown in Fig. P–41b.

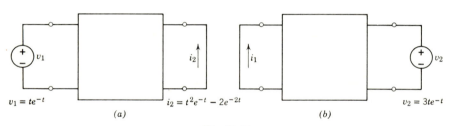

$v_1 = te^{-t}$ $i_2 = t^2 e^{-t} - 2e^{-2t}$ $v_2 = 3te^{-t}$

(a)   (b)

Fig. P–41

42. Show that the controlled-voltage source shown in the circuit of Fig. P–42 is not a passive device. Comment on the activity or passivity of both independent and dependent sources.

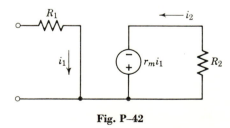

Fig. P–42

43. Show that a negative converter is not passive.

44. Establish the terminal equations for the networks shown in Fig. P–44.

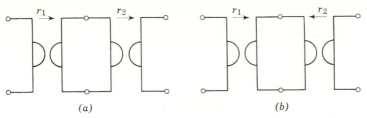

(a)   (b)

Fig. P–44

45. Draw a diagram containing only controlled sources to represent: (a) an ideal transformer; (b) a gyrator; (c) a negative converter.

46. Find the v-i relations at the input terminals of a gyrator when the output terminals are terminated in an inductor L. In a particular application a customer requires a 1000 μf capacitor. Can you suggest what to ship the customer?

.2.

GRAPH THEORY
AND NETWORK EQUATIONS

2.1 INTRODUCTORY CONCEPTS

When two or more of the components defined in the previous chapter are interconnected, the result is an *electric network*. (A more abstract definition is given in Section 2.3.) Such networks store energy, dissipate energy, and transmit signals from one point to another. A component part of a network lying between two terminals to which connections can be made is called a *branch*. The position where two or more branches are connected together is called a *node*, or *junction*. A simple closed path in a network is called a *loop*.

In the first section of this chapter, we shall briefly discuss a number of ideas with which you are no doubt familiar to a greater or lesser degree. Most of these will be amplified subsequently, but an early introduction in rather a simple form will serve to focus the discussion of these concepts before a fuller treatment is given.

KIRCHHOFF'S LAWS

In network theory, the fundamental laws, or postulates, are Kirchhoff's two laws. These can be stated as follows.

Kirchhoff's current law (KCL) states that in any electric network *the sum of all currents leaving any node equals zero at any instant of time.* When this law is applied at a node in a network, an equation relating the branch currents will result. Proper attention must, of course, be given to the current references. Thus, in Fig. 1, KCL applied at node A leads to the following equation:

$$-i_1 + i_2 - i_3 + i_4 = 0. \tag{1}$$

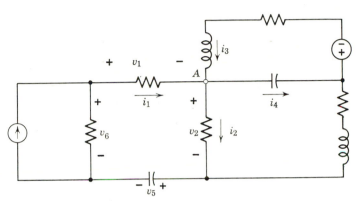

Fig. 1.   Illustrating KCL and KVL.

Since the reference of i_1 is oriented toward the node, the current "leaving" the node through branch 1 is $-i_1$; similarly, the current leaving the node through branch 3 is $-i_3$.

Kirchhoff's voltage law (KVL) states that in any electric network, *the sum of voltages of all branches forming any loop equals zero at any instant of time.* Application of this law to a loop in an electric network leads to an equation relating the branch voltages on the loop. In stating KCL, it was arbitrarily chosen that currents "leaving the node" were to be summed. It could, alternatively, have been decided to sum the currents "entering the node." Similarly, in applying KVL, the summing of voltages can be performed in either of the two ways one can traverse a loop. Thus, going clockwise in the loop formed by branches 1, 2, 5, and 6 in Fig. 1 leads to the equation

$$v_1 + v_2 + v_5 - v_6 = 0. \tag{2}$$

Since the reference of v_6 is oriented opposite to the orientation of the loop, the contribution of this voltage to the summation will be $-v_6$.

It should be noted that KCL and KVL lead to algebraic equations that form constraints on the branch currents and voltages. There will be as many KCL equations as there are nodes in the network and as many KVL equations as there are loops. Later we shall prove that these equations are not all independent; if the number of nodes is $n + 1$ and the number of branches is b, then we shall prove that the number of independent KCL equations is n and the number of independent KVL equations is $b - n$. We shall not dwell on these matters now, but simply observe them and note that together there are $n + (b - n) = b$ independent KCL and KVL equations.

Now each branch in the network also contributes a relationship between its voltage and current. This may be an algebraic relationship like $v = Ri$ or a dynamic relationship like $v = Ldi/dt$. In any case there will be as many such equations as branches, or b equations. Altogether there will be $b + b = 2b$ equations relating b currents and b voltages, or $2b$ variables. (As discussed in a later section, independent sources will not be counted as branches in this context.) Hence the three sets of relations together—namely, KCL, KVL, and the branch v-i relationships—provide an adequate set of equations to permit a solution for all voltages and currents.

Now $2b$ is a relatively large number, and a prudent person will try to avoid having to solve that many equations simultaneously. There are a number of systematic ways of combining the three basic sets of equations, each leading to a different formulation requiring the solution of less than $2b$ simultaneous equations. In this introductory section, we shall briefly discuss three such procedures and illustrate each one. The bridge in Fig. 2

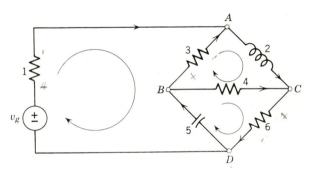

Fig. 2.   Example for network equations.

will be used as an example. The voltage source in series with a resistor is counted as a single branch. Standard branch references are assumed and are shown by the arrows on the branches. The time following which we are interested in this network is taken to be $t = 0$, and we assume there is an initial voltage $v_5(0) = V_0$ on the capacitor and an initial current $i_2(0) = I_0$ through the inductor.

LOOP EQUATIONS

In this network there are six branches ($b = 6$) and four nodes ($n = 3$). Hence there will be $b - n = 3$ independent KVL equations and $n = 3$ independent KCL equations. In Fig. 2, the circular arrows indicate the orientations of the loops we have chosen for writing KVL equations. They do not carry any implication of current (as yet). But suppose we conceive of fictitious circulating *loop currents*, with references given by the loop orientations. Examination of the figure shows that these loop currents are identical with the branch currents i_1, i_2, and i_6.

The following equations result if KVL is applied to the loops shown:

$$v_1 - v_g - v_3 - v_5 = 0$$
$$v_2 + v_3 - v_4 = 0 \tag{3}$$
$$v_4 + v_5 + v_6 = 0.$$

These are three equations in six unknowns, and they are observed to be independent. Into these equations are next substituted the *v-i* relationships of the branches, leading to

$$R_1 i_1 - v_g - R_3 i_3 - \frac{1}{C_5} \int_0 i_5(x)\, dx - V_0 = 0$$

$$L_2 \frac{di_2}{dt} + R_3 i_3 - R_4 i_4 = 0 \tag{4}$$

$$R_4 i_4 + \frac{1}{C_5} \int_0 i_5(x)\, dx + V_0 + R_6 i_6 = 0.$$

These are still three equations in six unknowns, this time the branch current unknowns.

There remains KCL. Application of KCL at the nodes labeled A, C, and D results in

$$i_3 = -i_1 + i_2$$
$$i_4 = -i_2 + i_6 \tag{5}$$
$$i_5 = -i_1 + i_6.$$

These equations are seen to be independent. In writing KCL equations any one node of a network can be omitted, and the resulting equations will be independent. It is observed that all branch currents are expressed in terms of the three currents i_1, i_2, and i_6, which are identical with the loop currents. When these equations are substituted into (4) and terms are collected, the result becomes

$$(R_1 + R_3)i_1 + \frac{1}{C_5} \int_0^t i_1 \, dx - R_3 i_2 - \frac{1}{C_5} \int_0^t i_6 \, dx = v_g + V_0$$

$$-R_3 i_1 + L_2 \frac{di_2}{dt} + (R_3 + R_4)i_2 - R_4 i_6 = 0$$

$$-\frac{1}{C_5} \int_0^t i_1 \, dx - R_4 i_2 + (R_4 + R_6)i_6 + \frac{1}{C_5} \int_0^t i_6 \, dx = -V_0.$$

This is a set of three simultaneous equations in the three loop-current unknowns. When the terms are collected in this manner, the equations are said to be in standard form. They are called the *loop equations*. Since both integrals and derivatives of variables appear, they are integrodifferential equations. Once the equations are solved for the loop currents, the remaining currents can be found from the KCL relationships in (5).

Let us review the procedure for writing loop equations. The first step consists of writing a set of (independent) KVL equations in the branch-voltage variables. Into these equations are next inserted the *v-i* relationships, changing the variables to branch currents. Finally, the branch currents are expressed in terms of loop currents, resulting in a set of integrodifferential equations in the loop-current unknowns.

NODE EQUATIONS

Suppose now that the order in which the steps were taken is modified. Suppose we first write KCL equations as in (5); then we insert the *i-v* relationships. The result (when the terms are written on the same side) becomes

$$G_1 v_1 - \frac{1}{L_2} \int_0^t v_2 \, dx - I_0 + G_3 v_3 = 0$$

$$\frac{1}{L_2} \int_0^t v_2 \, dx + I_0 + G_4 v_4 - G_6 v_6 = 0 \qquad (6)$$

$$G_1 v_1 + C_5 \frac{dv_5}{dt} - G_6 v_6 = 0.$$

These are three equations in the six branch-voltage variables. When writing the KCL equations, node B was omitted from consideration.

Suppose node B is selected as a datum with respect to which the voltages of all other nodes in the network are to be referred. Let these voltages be called the *node voltages*. In Fig. 2, the node voltages are v_{AB}, v_{CB}, and v_{DB}. All the branch voltages can be expressed in terms of the node voltages by applying KVL. Thus

$$v_1 = v_{DB} - v_{AB} + v_g \qquad v_4 = -v_{CB}$$

$$v_2 = v_{AB} - v_{CB} \qquad v_5 = v_{DB} \qquad\qquad (7)$$

$$v_3 = -v_{AB} \qquad v_6 = v_{CB} - v_{DB}.$$

When these expressions are inserted into (5), and the terms are collected, the result becomes

$$(G_1 + G_3)v_{AB} + \frac{1}{L_2}\int_0^t v_{AB}\,dx - \frac{1}{L_2}\int_0^t v_{CB}\,dx - G_1 v_{DB} = G_1 v_g - I_0$$

$$-\frac{1}{L_2}\int_0^t v_{AB}\,dx + (G_4 + G_6)v_{CB} + \frac{1}{L_2}\int_0^t v_{CB}\,dx - G_6 v_{DB} = I_0 \qquad (8)$$

$$-G_1 v_{AB} - G_6 v_{CB} + C_5\frac{dv_{DB}}{dt} + (G_1 + G_6)v_{DB} = -G_1 v_g.$$

These equations are called the *node equations*. Like the loop equations, they are integrodifferential equations. Once these equations are solved for the node voltages v_{AB}, v_{CB}, and v_{DB}, all the branch voltages will be known from (7).

To review, the first step in writing node equations is to write KCL equations at all nodes of a network but one. This particular node is chosen as a datum, and node voltages are defined as the voltages of the other nodes relative to this datum. The *i-v* relationships are inserted into the KCL equations, changing the variables to branch voltages. The branch voltages are then expressed in terms of the node voltages. Thus the order in which KCL, KVL, and the *v-i* relationships are used for writing node equations is the reverse of the order for writing loop equations.

STATE EQUATIONS—A MIXED SET

The presence of the integral of an unknown in the loop and node equations presents some difficulties of solution. Such integrals can, of course, be eliminated by differentiating the equations in which they appear. But this will increase the order of the equations. It would be better to avoid the appearance of the integrals in the first place.

In the present example it may be seen that an integral appears in the loop equations when the voltage of a capacitor is eliminated in a KVL equation by substituting its v-i relationship. Similarly, an integral appears in the node equations when the current of an inductor is eliminated in a KCL equation by substituting its i-v relationship. These integrals will not arise if we leave the capacitor voltages and inductor currents as variables in the equations.

With this objective in mind, return to the KVL equations in (3). Eliminate all branch voltages except the capacitor voltage v_5 by using the branch v-i relations. Since the branch relationship of the capacitor is not used in this process, add it as another equation to the set. The result will be

$$R_1 i_1 - v_g - R_3 i_3 - v_5 = 0$$

$$L_2 \frac{di_2}{dt} + R_3 i_3 - R_4 i_4 = 0$$

$$R_4 i_4 + v_5 + R_6 i_6 = 0 \tag{9}$$

$$C_5 \frac{dv_5}{dt} - i_5 = 0.$$

These are four equations in six unknowns, but now one of the unknowns is a voltage. As before, the KCL equations can be used to eliminate some of the currents. Substituting them from (5) into (9) and rearranging terms yields

$$L_2 \frac{di_2}{dt} = -(R_3 + R_4)i_2 + R_3 i_1 + R_4 i_6$$

$$C_5 \frac{dv_5}{dt} = -i_1 + i_6$$

$$0 = v_5 + R_3 i_2 - (R_1 + R_3)i_1 + v_g$$

$$0 = v_5 - R_4 i_2 + (R_4 + R_6)i_6.$$

Here we have four equations in four unknowns, and they can now be solved. But we seem to have complicated matters by increasing the number of equations that must be solved simultaneously. However, note that the last two equations in this set are algebraic; they contain no derivatives or integrals. The first of them can be solved for i_1, the second for i_6, and the results inserted into the previous two equations. The result of this manipulation will be

$$L_2 \frac{di_2}{dt} = -(R_3 + R_4)i_2 + \frac{R_3}{R_1 + R_3}(v_5 + R_3 i_2 + v_g)$$

$$+ \frac{R_4}{R_4 + R_6}(R_4 i_2 - v_5)$$

$$C_5 \frac{dv_5}{dt} = -\frac{1}{R_1 + R_3}(v_5 + R_3 i_2 + v_g) + \frac{1}{R_4 + R_6}(R_4 i_2 - v_5)$$

or, collecting terms,

$$\frac{di_2}{dt} = ai_2 + bv_5 + cv_g$$

$$\frac{dv_5}{dt} = di_2 + ev_5 + fv_g$$

or, in matrix form,

$$\frac{d}{dt}\begin{bmatrix} i_2 \\ v_5 \end{bmatrix} = \begin{bmatrix} a & b \\ d & e \end{bmatrix}\begin{bmatrix} i_2 \\ v_5 \end{bmatrix} + \begin{bmatrix} c \\ f \end{bmatrix}v_g, \tag{10}$$

where

$$a = \frac{1}{L_2}\left(-R_3 - R_4 + \frac{R_3{}^2}{R_1 + R_3} + \frac{R_4{}^2}{R_4 + R_6}\right)$$

$$= -\frac{1}{L_2}\left(\frac{R_1 R_3}{R_1 + R_3} + \frac{R_4 R_6}{R_4 + R_6}\right)$$

$$b = \frac{1}{L_2}\left(\frac{R_3}{R_1 + R_3} - \frac{R_4}{R_4 + R_6}\right)$$

$$c = \frac{R_3}{L_2(R_1 + R_3)}$$

$$d = \frac{1}{C_5} \left(\frac{R_4}{R_4 + R_6} - \frac{R_3}{R_1 + R_3} \right)$$

$$e = -\frac{1}{C_5} \left(\frac{1}{R_1 + R_3} + \frac{1}{R_4 + R_6} \right)$$

$$f = \frac{-1}{C_5(R_1 + R_3)}. \tag{11}$$

The resulting matrix equation (10) represents two first-order differential equations in two unknowns. It is called the *state equation*, for reasons which will be discussed in a later chapter. The variables i_2 and v_5 are called the *state variables*.

In looking back at the procedure used to write the state equations, note that the basic ingredients are the same as those for writing loop or node equations. This time, however, we do not insist that the eventual variables be either all voltages or all currents; we settle for a mixed set of variables. Integrals are avoided if we choose capacitor voltages and inductor currents as variables. The starting point is a set of KVL equations into which all branch relationships except for capacitor branches are inserted. Then KCL is used to eliminate some of the branch currents. In the resulting equations, some resistor currents also appear as variables, but it is possible to eliminate them because a sufficient number of equations are algebraic in nature and not differential.

SOLUTIONS OF EQUATIONS

The loop, node, and state equations of linear, time-invariant, lumped networks are ordinary differential equations with constant coefficients. (Integrals that may appear initially can be removed by differentiation.) There remains the task of solving such equations. A number of different methods exist for this purpose. A time-domain method for solving the state equations is discussed in a later chapter, and the Laplace-transform method is discussed in Appendix 3.

In the transform method, the Laplace transform of the simultaneous differential equations is taken, converting them to simultaneous algebraic equations in the complex variable s. These algebraic equations are solved for the transform of the desired variables, whether loop currents, node voltages, or state variables. Finally, the inverse transform is taken. This yields the solution as a function of time following the initial instant, $t_0 = 0$.

The solution has contributions from two places: the exciting signal sources and the initial conditions. The initial conditions are the values of capacitor voltages and inductor currents immediately after t_0. The principles of continuity of charge and of flux linkage impose constraints on the capacitor voltages and inductor currents—constraints that serve to determine their values just after t_0 from their values just before t_0.* The network is said to be *initially relaxed* if the values of capacitor voltages and inductor currents are initially zero.

In the interest of concreteness, we shall carry out the remainder of the discussion with the state equations (10) previously derived for Fig. 2 as an illustration. Taking Laplace transforms of these equations and rearranging leads to

$$
\begin{aligned}
(s - a)I_2(s) - bV_5(s) &= cV_g(s) + I_0 \\
-d\,I_2(s) + (s - e)V_5(s) &= fV_g(s) + V_0,
\end{aligned}
\tag{12}
$$

where I_0 and V_0 are the initial values. These can now be solved for $I_2(s)$ or $V_5(s)$. For $I_2(s)$ we get

$$
I_2(s) = \frac{(c\,(s - e) + bf)}{\Delta}\,V_g(s) + \frac{(s - e)I_0 + bV_0}{\Delta}
\tag{13}
$$

where $\Delta = s^2 - (a + e)s + ae - bd$ is the determinant of the equations. The contributions of the source and of the initial conditions are clearly evident.

Again for concreteness, suppose that

$$
v_g(t) = \sin t \quad \text{or} \quad V_g(s) = \frac{1}{s^2 + 1}
$$

and

$$
\Delta = (s + 2)(s + 3)
\tag{14}
$$

---

* For a detailed discussion of initial conditions, see S. Seshu and N. Balabanian, *Linear Network Analysis*, John Wiley & Sons, Inc., New York, 1959, pp. 101–112.

and suppose the initial values are such that

$$I_2(s) = \frac{s^2 + s + 1}{(s^2 + 1)(s + 2)(s + 3)} = \frac{1}{10}\left(\frac{6}{s + 2} - \frac{7}{s + 3}\right) + \frac{1}{10}\left(\frac{s + 1}{s^2 + 1}\right).$$

(15)

Then

$$i_2(t) = \mathcal{L}^{-1}[I_2(s)] = \frac{\sqrt{2}}{10}\sin\left(t + \frac{\pi}{4}\right) + \frac{6}{10}\epsilon^{-2t} - \frac{7}{10}\epsilon^{-3t}.$$

(16)

The partial fraction expansion of $I_2(s)$ puts into evidence all of its poles. Some of these poles [the second term in (15)] are contributed by the exciting (source) function, whereas the remainder are contributed by the network. In the inverse transform we find terms that resemble the driving function and also other terms that are exponentials. There is an abundance of terminology relating to these terms that has been accumulated from the study of differential equations in mathematics, from the study of vibrations in mechanics, and from the study of a-c circuit theory, so that today we have a number of names to choose from. These are the following:

1. Forced response—natural, or free, response;
2. Particular integral—complementary function; and
3. Steady state—transient.

Perhaps you are most familiar with the terms "steady state" and "transient." When the driving function is a sinusoid, as in our example, there will be a sinusoidal term in the response that continues indefinitely. In our example the other terms present die out with time; they are transient. Thus the sinusoidal term will eventually dominate. This leads to the concept of the steady state. If the driving function is not periodic the concept of the steady state loses its significance. Nevertheless, the poles of the transform of the driving function contribute terms to the partial-fraction expansion of the response transform, and so the response will contain terms due to these poles. These terms constitute the *forced response*. In form they resemble the driving function. The remaining terms represent the *natural* response. They will be present in the solution (with different coefficients) no matter what the driving function is, even if there is no driving function except initial capacitance voltages or initial inductance currents. This leads to the name "natural," or "free," response. The exponents in the natural response are called the *natural frequencies*.

In the illustrative example the exponents in the natural response, the natural frequencies, are negative real numbers. If there were positive exponents, or complex ones with positive real parts, then the natural response would increase indefinitely with time instead of dying out. A network with such a behavior is said to be *unstable*. We define a *stable* network as *one whose natural frequencies lie in the closed left half s-plane*; that is, in the left half-plane or on the *j*-axis.* Actually, some people prefer to exclude networks with *j*-axis natural frequencies from the class of stable networks.

Let us now clearly define the various classes of response. The *complete response* of a network consists of two parts; the *forced* response and the *natural*, or *free*, response. The forced response consists of all those terms that are contributed by poles of the driving functions, whereas the free response consists of all the terms that are contributed by the natural frequencies [the zeros of $\Delta(s)$]. If the driving function is periodic, the forced response is also called the steady state. If there are no *j*-axis natural frequencies, then the free response is also called the transient.

2.2 LINEAR GRAPHS

As discussed briefly in the previous introductory section, the fundamental "laws," or postulates, of network theory are the two laws of Kirchhoff and the relationships between the voltages and the currents of the components whose interconnection constitutes the network. Kirchhoff's two laws express constraints imposed on the currents and voltages of network elements by the very arrangement of these elements into a structure. Network *topology* is a generic name that refers to all properties arising from the structure or geometry of a network.

The topological properties of a network are independent of the types of components that constitute the branches. So it is convenient to replace each network element by a simple line segment, thereby not committing oneself to a specific component. The resulting structure consists of nodes interconnected by line segments. There is a branch of mathematics, called the theory of linear graphs, that is concerned with the study of just such structures.

We shall begin a thorough study of network analysis by focusing first on linear graphs and those of their properties that are important in this study. The discussion of linear graphs will not be exhaustive, and it will

---

* This definition is applicable to all lumped, linear, time-invariant systems. More general and precise definitions of stability will be discussed in later chapters.

be necessary to consider rapidly the definitions of many terms without, perhaps, adequate motivation for their introduction.

INTRODUCTORY DEFINITIONS

A *linear graph* is defined as a collection of points, called *nodes*, and line segments called *branches*, the nodes being joined together by the branches. Sometimes it is convenient to view the nodes, which coincide with the end points of the branches, as parts of the branches themselves. At other times it is convenient to view the nodes as being detached from the branches.

A correspondence between a network and a linear graph can immediately be made. Thus the graph associated with the network in Fig. 3a is shown in Fig. 3b. The nodes and branches are each numbered. In

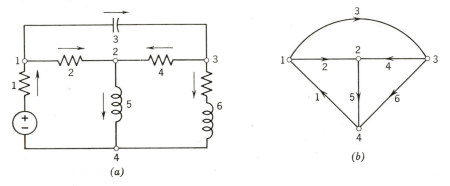

Fig. 3. A network and its associated linear graph.

what follows, we will use this graph to make some observations from which generalizations can be drawn. Also, properties to be defined will be illustrated with this graph as an example.

Branches whose ends fall on a node are said to be *incident* at the node. In the example, branches 2, 4, and 5 are incident at node 2.

Each branch of the graph in the example carries an arrow to indicate its orientation. A graph whose branches are oriented is called an *oriented graph*. The elements of a network with which a graph is associated have both a voltage and a current variable, each with its own reference. In order to relate the orientation of the branches of a graph to these references, we shall make the convention that the voltage and current of an element have the standard reference—voltage-reference " plus " at the tail of the current-reference arrow. The branch orientation of a graph will be assumed to coincide with the associated current reference. Of course, the properties of the graph have nothing to do with those conventions related to a network.

A *subgraph* is a subset of the branches and nodes of a graph. The subgraph is said to be *proper* if it consists of strictly less than all the branches and nodes of the graph.

A *path* is a particular subgraph consisting of an ordered sequence of branches having the following properties:

1. At all but two of its nodes, called *internal nodes*, there are incident exactly two branches of the subgraph.

2. At each of the remaining two nodes, called the *terminal nodes*, there is incident exactly one branch of the subgraph.

3. No proper subgraph of this subgraph, having the same two terminal nodes, has properties 1 and 2.

In the example, branches 2, 5, and 6, together with all the nodes, constitute a path. Nodes 1 and 3 are the terminal nodes. Although three branches of the graph are incident at node 2, only two of these, 2 and 5, are members of the subgraph.

A graph is *connected* if there exists at least one path between any two nodes. The example is a connected graph. The graph associated with a network containing a transformer could be *unconnected*.

A *loop* is a particular connected subgraph of a graph at each node of which are incident exactly two branches of the subgraph. Thus, if the two terminal nodes of a path are made to coincide, the result (which can be called a *closed path*) will be a loop. In the example, branches 4, 5, and 6 together with nodes 2, 3, and 4 constitute a loop. When specifying a loop, either the set of all the branches in the loop or (when no two branches are in parallel) the set of all nodes can be listed. Each of these will uniquely specify the loop. Thus, in the example, to specify the loop just described it is sufficient to list either the set of branches $\{4, 5, 6\}$ or the set of nodes $\{2, 3, 4\}$.

A *tree* is a connected subgraph of a connected graph containing all the nodes of the graph but containing no loops. When specifying a tree, it is enough to list its branches. In the graph of the example, branches 2, 4, and 5 constitute a tree. The concept of a tree is a key concept in the theory of graphs. The branches of a tree are called *twigs*; those branches that are not on a tree are called *links*. Together they constitute the complement of the tree, or the *cotree*. This is not a unique decomposition of the branches of a graph. Figure 4 shows two trees for the graph of Fig. 3. In the first one, branches 2, 4, and 5 are twigs, so branches 1, 3, and 6 are links. In the second tree, branch 2 is still a twig, but branches 3 and 6, which were previously links, are now twigs. Whether a particular branch is a twig or a link cannot be uniquely stated for a graph; it makes sense to give such a designation only after a tree has been specified.

Each of the trees in Fig. 4 has a special structure. In the first one, all the twigs are incident at a common node. Such a tree is called a *starlike tree*, or a *star-tree* for short. In the second one, the nodes can be so ordered that the tree consists of a single path extending from the first node to the last. Such a tree is called a *linear tree*. In a linear tree there are exactly two terminal nodes, whereas in a starlike tree all nodes but one are terminal nodes.

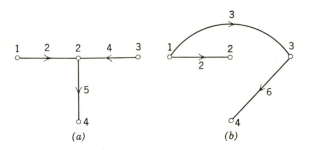

Fig. 4.   Two trees of a given graph.

The number of branches on a tree of a graph is one less than the number of nodes of the graph. This result can be proved by induction. Thus for a graph with two nodes the number of twigs is 1. Suppose the result is true for a graph with k nodes; that is, suppose the number of twigs equals $k - 1$. Now consider a connected graph having $k + 1$ nodes and focus on a tree of this graph. There will be at least one node on this tree at which is incident exactly one twig. (If not, two or more twigs will be incident at each node, which is impossible, because this would require that the tree contain at least one loop.) Let the node and the single twig incident at this node be removed, leaving a tree having k nodes. By hypothesis, the number of twigs on this tree is $k - 1$. Replacing the removed node and twig leads to the result. For future convenience the number of nodes in a graph will be denoted by $n + 1$. Then, the number of twigs on a tree will be n.

If a graph is unconnected, the concept corresponding to a tree for a connected graph is called a *forest*, which is defined as a set of trees, one for each of the separate parts. If $p + 1$ is the number of separate parts of an unconnected graph and $n + 1$ is the number of nodes, then a forest will contain $n - p$ twigs. This can be proved as before for a connected graph. The complement of the forest is the *coforest*.

THE INCIDENCE MATRIX

When a graph is given, say the example in Fig. 3, it is possible to tell completely which branches are incident at which nodes and what are the orientations, relative to the nodes. Conversely, the graph would be completely defined if this information (namely, which branches are incident at which nodes and with what orientation) is given. The most convenient form in which this incidence information can be given is in matrix form.

For a graph having $n + 1$ nodes and b branches, the *complete incidence matrix* (or more completely, the *complete node-branch incidence matrix*) $\mathbf{A}_a = [a_{ij}]$ is an $(n + 1) \times b$ rectangular matrix whose elements have the following values:

$a_{ij} = 1$ if branch j is incident at node i and oriented away from it;
$a_{ij} = -1$ if branch j is incident at node i and oriented toward it;
$a_{ij} = 0$ if branch j is not incident at node i.

The subscript a on $\mathbf{A}_a$ stands for *all* nodes.

For the example in Fig. 3, the complete incidence matrix is

$$
\mathbf{A}_a = \begin{array}{c} \\ \\ 1 \\ 2 \\ 3 \\ 4 \end{array}
\begin{bmatrix}
-1 & 1 & 1 & 0 & 0 & 0 \\
0 & -1 & 0 & -1 & 1 & 0 \\
0 & 0 & -1 & 1 & 0 & 1 \\
1 & 0 & 0 & 0 & -1 & -1
\end{bmatrix}. \tag{17}
$$

nodes↓   branches→ 1 2 3 4 5 6

In this example it is observed that each column contains a single $+1$ and a single -1. This is a general property for any linear graph because each branch is incident at exactly two nodes, and it must perforce be oriented away from one of them and toward the other. Thus, if all other rows are added to the last row, the result will be a row of zeros, indicating that the rows are not all independent. At least one of them can be eliminated, since it can be obtained as the negative sum of all the others. Thus the rank of $\mathbf{A}_a$ can be no more than $(n + 1) - 1 = n$.

The matrix obtained from $\mathbf{A}_a$ by eliminating one of the rows is called the incidence matrix and is denoted by $\mathbf{A}$. (For emphasis, it is sometimes called the *reduced* incidence matrix.) It is of order $n \times b$. We shall now discuss the rank of matrix $\mathbf{A}$ and how to determine its nonsingular submatrices.

For a given graph, select a tree. In the incidence matrix arrange the columns so that the first n columns correspond to the twigs for the selected tree and the last $b - n$ correspond to the links.

In terms of the example, let $\mathbf{A}$ be obtained from (17) by eliminating the last row. Select the first tree in Fig. 4. Then the $\mathbf{A}$ matrix will become

$$
\begin{array}{cc}
\text{twigs} & \text{links} \\
\overbrace{2 \quad 4 \quad 5} & \overbrace{1 \quad 3 \quad 6}
\end{array}
$$

$$
\mathbf{A} = \begin{bmatrix} 1 & 0 & 0 & -1 & 1 & 0 \\ -1 & -1 & 1 & 0 & 0 & 0 \\ 0 & 1 & 0 & 0 & -1 & 1 \end{bmatrix}. \tag{18}
$$

In general terms, the matrix $\mathbf{A}$ can be partitioned in the form

$$
\mathbf{A} = [\mathbf{A}_t \qquad \mathbf{A}_l]. \tag{19}
$$

where $\mathbf{A}_t$ is a square matrix of order n whose columns correspond to the twigs, and $\mathbf{A}_l$ is a matrix of order $n \times (b - n)$ whose columns correspond to the links.

For the example,

$$
\mathbf{A}_t = \begin{bmatrix} 1 & 0 & 0 \\ -1 & -1 & 1 \\ 0 & 1 & 0 \end{bmatrix}.
$$

The determinant of this matrix is found to equal -1, and so the matrix is nonsingular. Hence, for this example, the $\mathbf{A}$ matrix is seen to be of rank n.

We shall now show that this is a general result. Specifically, if a graph has $n + 1$ nodes, the rank of its incidence matrix equals n. This will be established by showing that *an nth order submatrix of* $\mathbf{A}$ *whose columns correspond to the twigs for any tree is nonsingular.*

Proof. Given a connected graph and its complete incidence matrix $\mathbf{A}_a$, eliminate one row to get $\mathbf{A}$ and partition it in the form $[\mathbf{A}_t \quad \mathbf{A}_l]$, where $\mathbf{A}_t$ is square, of order n, and its columns correspond to twigs. Since the tree is connected, there is at least one twig incident at the node corresponding to the eliminated row. The column in $\mathbf{A}_t$ corresponding to this twig contains only one nonzero element (which equals ± 1). Hence det $\mathbf{A}_t$ equals plus or minus the cofactor of this element. The matrix associated with this cofactor corresponds to a connected subgraph with $n-1$ branches; this matrix contains neither of the rows in which the eliminated column had nonzero elements. Since this subgraph is connected, it must contain at least one twig incident at one of the two eliminated nodes. The column in the matrix corresponding to this twig contains only one non-zero element. Hence its determinant equals plus or minus the correspond-ing cofactor, whose associated matrix is of order $n-2$. Continue this process until a cofactor of order 1 remains. This corresponds to the last node, and, since the graph is connected, the cofactor is nonzero. As a conclusion, not only has det $\mathbf{A}_t$ been found to be nonzero—and so $\mathbf{A}_t$ is nonsingular—but its value has been found to be ± 1. And since an $n \times n$ submatrix of $\mathbf{A}_a$ is nonsingular, $\mathbf{A}_a$ is of rank n.

The preceding is a very useful result. The converse is also true. That is, given an $n \times n$ nonsingular submatrix of the incidence matrix $\mathbf{A}$, its columns correspond to the twigs for some choice of tree. The proof will be left as a problem.

The preceding has also shown that the determinant of every $n \times n$ nonsingular submatrix of the incidence matrix equals $+1$ or -1.

With the preceding result, it is now possible to find a measure of the number of trees in a graph. Since each nonsingular $n \times n$ submatrix of $\mathbf{A}$ corresponds to a tree, all we have to do is to count all such nonsingular submatrices. This implies evaluating the determinants of all $n \times n$ submatrices of $\mathbf{A}$, which is quite tedious. The problem can be simplified by using the Binet-Cauchy theorem, which was discussed in Chapter 1. According to this theorem,

$$\det (\mathbf{AA}') = \sum_{\substack{\text{all} \\ \text{majors}}} \ \text{(products of corresponding majors of } \mathbf{A} \text{ and } \mathbf{A}')$$

$$= \sum \ \text{(all nonzero majors of } \mathbf{A})^2 \qquad\qquad (20)$$

$$= \text{number of trees.}$$

The second line follows from the fact that a nonsingular submatrix of $\mathbf{A}'$ has the same determinant as the corresponding submatrix of $\mathbf{A}$. Since each nonzero major equals ± 1, and there are as many nonzero majors as trees, the last line follows.

Thus, to find the number of trees of a graph, it is required only to evaluate det $(\mathbf{AA}')$. For the example of Fig. 3, the incidence matrix was given in (18). Hence the number of trees will be

$$\det(\mathbf{AA'}) = \det \begin{bmatrix} 1 & 0 & 0 & -1 & 1 & 0 \\ -1 & -1 & 1 & 0 & 0 & 0 \\ 0 & 1 & 0 & 0 & -1 & 1 \end{bmatrix} \begin{bmatrix} 1 & -1 & 0 \\ 0 & -1 & 1 \\ 0 & 1 & 0 \\ -1 & 0 & 0 \\ 1 & 0 & -1 \\ 0 & 0 & 1 \end{bmatrix}$$

$$= \det \begin{bmatrix} 3 & -1 & -1 \\ -1 & 3 & -1 \\ -1 & -1 & 3 \end{bmatrix}$$

$$= 3 \begin{bmatrix} 3 & -1 \\ -1 & 3 \end{bmatrix} + \begin{bmatrix} -1 & -1 \\ -1 & 3 \end{bmatrix} - \begin{bmatrix} -1 & 3 \\ -1 & -1 \end{bmatrix} = 16.$$

*   *   *

Given a graph, it is a simple matter to write the incidence matrix. The problem might often be the converse: given an incidence matrix (or complete incidence matrix), draw the graph. In an abstract sense, the incidence matrix *defines* the graph. It is one representation of the graph, whereas the drawing of lines joining nodes is another representation. It is desired to obtain this second representation from the first.

The procedure is quite straightforward. Given the matrix **A**, place on the paper one more node than there are rows in **A** and number them according to the rows. Then consider the columns one at a time. There are at most two nonzero elements in each column; place a branch between the two nodes corresponding to the two rows having nonzero elements in that column. If there is only one nonzero element, the branch goes between the node corresponding to this row and the extra node. The orientations will be determined by the signs of the elements.

To illustrate, let the given **A** matrix be

$$\mathbf{A} = \begin{bmatrix} 1 & 1 & 0 & 0 & 0 & 0 & 0 & -1 \\ 0 & -1 & 1 & 1 & 0 & 0 & 0 & 0 \\ 0 & 0 & 0 & -1 & 1 & 1 & 0 & 0 \\ 0 & 0 & 0 & 0 & 0 & -1 & 1 & 1 \end{bmatrix}.$$

Two different people studying network analysis were given this job, and each came up with a different-looking graph, as shown in Fig. 5. The apparent problem was that right at the start they placed the nodes on the paper in a different pattern. However, both graphs have as incidence matrix the given matrix.

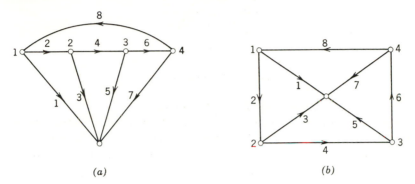

Fig. 5.   Isomorphic graphs.

We say *two graphs are isomorphic if they have the same incidence matrix*. This requires that they have the same number of nodes and branches, and that there be a one-to-one correspondence between the nodes and a one-to-one correspondence between the branches in a particular way—namely, in a way that leads to the same incidence matrix.

THE LOOP MATRIX

The incidence matrix gives information about the incidences of branches at nodes, but it does not explicitly tell anything about the way in which the branches constitute loops. This latter information can also be given conveniently in matrix form. For this purpose we first endow each loop of a graph with an orientation, which is done by giving its nodes a cyclic order. This order is most easily shown by a curved arrow, as in Fig. 6 where two possible loops are shown. To avoid cluttering the diagram, it is sometimes necessary simply to list the set of nodes in the chosen order. For the loops shown in Fig. 6, this listing would be $\{1, 3, 2\}$ and $\{1, 2, 3, 4\}$.

For a graph having $n + 1$ nodes and b branches, the *complete loop matrix* (also sometimes called the *complete circuit matrix*) $\mathbf{B}_a = [b_{ij}]$ is a rectangular matrix having b columns and as many rows as there are loops; its elements have the following values:

$b_{ij} = 1$ if branch j is in loop i, and their orientations coincide;
$b_{ij} = -1$ if branch j is in loop i, and their orientations do not coincide;
$b_{ij} = 0$ if branch j is not in loop i.

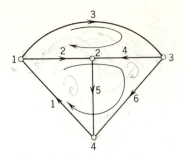

Fig. 6. Loop orientations.

The subscript a again stands for *all* the loops.

Unlike the number of rows in the complete incidence matrix (equal to the number of nodes of the graph), the number of rows in $\mathbf{B}_a$ is not simply expressible in terms of n and b. For the example in Fig. 6 there are seven loops specified by the nodes as:

loop 1: $\{1, 3, 2\}$ loop 4: $\{1, 3, 4\}$

loop 2: $\{1, 2, 4\}$ loop 5: $\{1, 2, 3, 4\}$

loop 3: $\{2, 3, 4\}$ loop 6: $\{1, 2, 4, 3\}$

loop 7: $\{3, 2, 4, 1\}$.

The loop matrix will therefore be

$$
\mathbf{B}_a = \begin{array}{c} \\ 1 \\ 2 \\ 3 \\ 4 \\ 5 \\ 6 \\ 7 \end{array}
\begin{array}{c} \text{loops branches} \rightarrow \\ \begin{array}{cccccc} 1 & 2 & 3 & 4 & 5 & 6 \\ \end{array} \\
\left[\begin{array}{cccccc}
0 & -1 & 1 & 1 & 0 & 0 \\
1 & 1 & 0 & 0 & 1 & 0 \\
0 & 0 & 0 & -1 & -1 & 1 \\
1 & 0 & 1 & 0 & 0 & 1 \\
1 & 1 & 0 & -1 & 0 & 1 \\
0 & 1 & -1 & 0 & 1 & -1 \\
1 & 0 & 1 & 1 & 1 & 0 \\
\end{array} \right]
\end{array}. \qquad (21)
$$

The set of all loops in a graph is quite a large set, as illustrated by this

example. There is a smaller subset of this set of all loops; it has some
interesting properties and will be discussed next.

Given a graph, first select a tree and remove all the links. Then replace
each link in the graph, one at a time. As each link is replaced, it will form
a loop. (If it does not, it must have been a twig.) This loop will be characterized by the fact that all but one of its branches are twigs of the chosen
tree. Loops formed in this way will be called *fundamental loops*, or *f*-loops
for short. The orientation of an *f*-loop will be chosen to coincide with that
of its defining link. There are as many *f*-loops as there are links; in a graph
having b branches and $n + 1$ nodes, this number will be $b - n$.

For the example, let the chosen tree be the second one in Fig. 4. The
f-loops formed when the links are replaced one at a time are shown in
Fig. 7. (Note the orientation.) In writing the loop matrix for the *f*-loops,

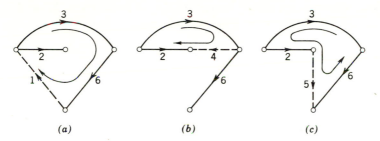

(a)   (b)   (c)

Fig. 7.   Fundamental loops.

let the columns be arranged in the same order as for the reduced incidence
matrix for the same tree; that is, with the twigs first, then the links. Also,
let the order of the loops be the same as the order of the columns of the
corresponding links.
The matrix of *f*-loops will then be

$$
\mathbf{B}_f = \begin{array}{c@{}c}
 & \begin{array}{ccc@{\;\;}ccc}
\overbrace{\hspace{3.5em}}^{\text{twigs}} & & & \overbrace{\hspace{3.5em}}^{\text{links}} & & \\
2 & 3 & 6 & 1 & 4 & 5
\end{array} \\
\left[\begin{array}{rrrrrr}
0 & 1 & 1 & 1 & 0 & 0 \\
-1 & 1 & 0 & 0 & 1 & 0 \\
1 & -1 & -1 & 0 & 0 & 1
\end{array}\right] &
\end{array}. \tag{22}
$$

The subscript *f* stands for fundamental. The square matrix formed by the
last three columns corresponding to the links is seen to be a unit matrix;
hence it is nonsingular, and the rank of $\mathbf{B}_f$ for this example equals the
number of links, $b - n$.

In general terms, the matrix of f-loops for an arbitrary connected graph can be partitioned in the form

$$\mathbf{B}_f = [\mathbf{B}_t \quad \mathbf{B}_l] = [\mathbf{B}_t \quad \mathbf{U}]. \tag{23}$$

The $(b-n) \times (b-n)$ square submatrix whose columns correspond to the links for a particular tree will be a unit matrix from the very way in which it is formed. Hence the rank of $\mathbf{B}_f$ will be $b-n$.

Now the matrix of fundamental loops is a submatrix of the matrix of all loops. Hence the rank of $\mathbf{B}_a$ is no less than that of $\mathbf{B}_f$; namely, $b-n$. We shall next show that the rank of $\mathbf{B}_a$ is no more than $b-n$, and so it is exactly $b-n$. To do this, we shall use a result that is of great importance in its own right.

Given a graph, let the columns of the two matrices $\mathbf{A}_a$ and $\mathbf{B}_a$ be arranged in the same order. Then it will be true that

$$\mathbf{A}_a \mathbf{B}_a' = 0 \tag{24}$$

and

$$\mathbf{B}_a \mathbf{A}_a' = 0. \tag{25}$$

Of course, the second one will be true if the first one is, since $\mathbf{B}_a \mathbf{A}_a' = (\mathbf{A}_a \mathbf{B}_a')'$. The relationships in (24) and (25) are called the *orthogonality relations* and can be proved as follows.

The matrices $\mathbf{A}_a$ and $\mathbf{B}_a'$ will have the following forms:

$$
\mathbf{A}_a =
\begin{matrix}
 & \text{branches} \rightarrow \\
\text{nodes} & 1 \quad 2 \quad \ldots \quad b \\
 & \downarrow \\
1 & \\
2 & \\
\cdot & \\
\cdot & \\
n+1 &
\end{matrix}
\left[\right]
\qquad
\mathbf{B}_a' =
\begin{matrix}
 & \text{loops} \rightarrow \\
\text{branches} & 1 \quad 2 \quad \ldots \\
 & \downarrow \\
1 & \\
2 & \\
\cdot & \\
\cdot & \\
b &
\end{matrix}
\left[\right].
$$

Focus attention on any one of the columns of $\mathbf{B}_a'$ and on any one of the rows of $\mathbf{A}_a$; that is, focus attention on a loop and a node. Either the node is on the loop or not. If not, then none of the branches on the loop can be incident at the node. This means that corresponding to any nonzero element in a column of $\mathbf{B}'_a$ there will be a zero element in the row of $\mathbf{A}_a$; so the product will yield zero. If the node is on the loop, then exactly two of the branches incident at the node will lie on the loop. If these two branches are similarly oriented relative to the node (either both oriented away or both toward), they will be oppositely oriented relative to the loop, and vice versa. In terms of the matrices, if the elements in a row of $\mathbf{A}_a$ corresponding to two branches are both $+1$ or both -1, the corresponding two elements in a column of $\mathbf{B}'_a$ will be of opposite sign, and vice versa. When the product is formed, the result will be zero. The theorem is thus proved.

With the preceding result it is now possible to determine the rank of B_a by invoking Sylvester's law of nullity, which was discussed in Chapter 1. According to this law, if the product of two matrices equals zero, the sum of the ranks of the two matrices is not greater than the number of columns of the first matrix in the product. In the present case the number of columns equals the number of branches b of the graph. So, since the rank of a matrix is the same as the rank of its transpose,

$$(\text{rank of } \mathbf{A}_a) + (\text{rank of } \mathbf{B}_a) \leq b. \qquad (26)$$

The rank of $\mathbf{A}_a$ has already been determined to be n. Hence

$$(\text{rank of } \mathbf{B}_a) \leq (b - n). \qquad (27)$$

Since it was previously established that the rank of $\mathbf{B}_a$ is no less than $b - n$ and it is now found that it can be no greater than $b - n$, then *the rank of* $\mathbf{B}_a$ *is exactly* $b - n$.

Observe that the removal of any number of rows from $\mathbf{A}_a$ or any number of rows from $\mathbf{B}_a$ will not invalidate the result of (26) and (27). Let $\mathbf{B}$ be any submatrix of $\mathbf{B}_a$ having $b - n$ rows and of rank $b - n$. (One possibility is the matrix of f-loops, $\mathbf{B}_f$.) Then the orthogonality relations can be written as

$$\mathbf{AB}' = \mathbf{0}, \qquad \mathbf{BA}' = \mathbf{0}. \qquad (28)$$

RELATIONSHIPS BETWEEN SUBMATRICES OF $\mathbf{A}$ AND $\mathbf{B}$

Let the columns of $\mathbf{B}$ be arranged, as were the columns of $\mathbf{A}$ earlier, with the twigs for a given tree first and then the links. The matrix can then be partitioned in the form

$$\mathbf{B} = [\mathbf{B}_t \quad \mathbf{B}_l] \qquad (29)$$

where $\mathbf{B}_l$ is square of order $b - n$. [If $\mathbf{B}$ is the matrix of f-loops, then $\mathbf{B}_l$ is a unit matrix, as in (23).] We shall now show that, with $\mathbf{B}$ partitioned as shown, the submatrix $\mathbf{B}_l$, whose columns are links for a tree, will be nonsingular.

To prove this, let $\mathbf{A}$ be partitioned as in (19) and use (28) to write

$$\mathbf{AB}' = [\mathbf{A}_t \quad \mathbf{A}_l]\begin{bmatrix}\mathbf{B}_t' \\ \mathbf{B}_l'\end{bmatrix} = \mathbf{A}_t\mathbf{B}_t' + \mathbf{A}_l\mathbf{B}_l' = \mathbf{0}. \tag{30}$$

Since $\mathbf{A}_t$ is nonsingular,

$$\mathbf{B}_t' = -\mathbf{A}_t^{-1}\mathbf{A}_l\mathbf{B}_l' \quad \text{or} \quad \mathbf{B}_t = -\mathbf{B}_l(\mathbf{A}_t^{-1}\mathbf{A}_l)'. \tag{31}$$

Finally matrix $\mathbf{B}$ becomes

$$\mathbf{B} = [-\mathbf{B}_l(\mathbf{A}_t^{-1}\mathbf{A}_l)' \quad \mathbf{B}_l] = \mathbf{B}_l[-(\mathbf{A}_t^{-1}\mathbf{A}_l)' \quad \mathbf{U}]. \tag{32}$$

Now let the same procedure be carried out starting at (30), but this time with the matrix $\mathbf{B}_f$ of the f-loops for some tree, with $\mathbf{B}_f$ partitioned in the form

$$\mathbf{B}_f = [\mathbf{B}_{ft} \quad \mathbf{U}], \tag{33}$$

the subscript f on $\mathbf{B}_{ft}$ being used to avoid confusion. The details will be left for you to carry out; the result will be

$$\mathbf{B}_f = [-(\mathbf{A}_t^{-1}\mathbf{A}_l)' \quad \mathbf{U}]. \tag{34}$$

By comparing this with (32), it follows that

$$\mathbf{B} = \mathbf{B}_l\mathbf{B}_f \quad \text{or} \quad \mathbf{B}_f = \mathbf{B}_l^{-1}\mathbf{B}. \tag{35}$$

Since $\mathbf{B}$ and $\mathbf{B}_f$ are both of rank $b - n$, then $\mathbf{B}_l$ must be nonsingular. This follows from (52) in Chapter 1. The result is thus proved.

The converse is also true; that is, if the loop-matrix $\mathbf{B}$ is partitioned into two matrices as in (29), one of them being square of order $b - n$ and nonsingular, the columns of this matrix will correspond to the links for some tree. The proof is left for you to carry out. (See Problem 6.)

Since $\mathbf{B}_l$ in (35) is nonsingular, the matrices $\mathbf{B}$ and $\mathbf{B}_f$ are *row-equivalent matrices*. (For a discussion of equivalent matrices, see Chapter 7.) Hence, the rows of $\mathbf{B}$ are linear combinations of the rows of $\mathbf{B}_f$, and vice versa.

Additional useful results are obtained by solving (30) for $\mathbf{A}_l$. Since $\mathbf{B}_l$ is nonsingular, we get

$$\mathbf{A}_l = -\mathbf{A}_t \mathbf{B}_t'(\mathbf{B}_l')^{-1} = -\mathbf{A}_t \mathbf{B}_t'(\mathbf{B}_l^{-1})' = -\mathbf{A}_t(\mathbf{B}_l^{-1}\mathbf{B}_t)'. \qquad (36)$$

The last step follows because the transpose of a product equals the product of transposes in the reverse order. The preceding step follows because the operations of transpose and inverse are commutative for a nonsingular matrix.

When this is inserted into the partitioned form of the $\mathbf{A}$ matrix, the result will be

$$\mathbf{A} = \mathbf{A}_t[\mathbf{U} \quad -(\mathbf{B}_l^{-1}\mathbf{B}_t)']. \qquad (37)$$

This should be compared with (32), which gives the loop matrix in a similar form.

CUT-SETS AND THE CUT-SET MATRIX

In the example of Fig. 3, suppose branches 1 and 5 are removed. The result is shown in Fig. 8a. (By "removing" a branch we mean deleting it

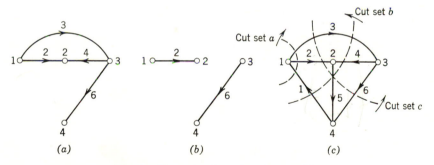

Fig. 8.   Removing branches for a graph.

or "open-circuiting" it, but leaving intact the nodes at which it is incident.) The graph is still a connected graph. Now if branches 3 and 4 are also removed, the result becomes Fig. 8b. The graph is now unconnected: it has been "cut" into two parts. This development leads to the notion of a cut set, which is defined as follows: *A cut-set is a set of branches of a connected graph whose removal causes the graph to become unconnected into exactly two connected subgraphs, with the further stipulation that the removal of any proper subset of this set leaves the graph connected.*

In the example, $\{1, 3, 4, 5\}$ was seen to be a cut-set. Set $\{1, 2, 3\}$ is also a cut-set. (A single isolated node, which is one of the two parts in this case, is considered to be a bona fide " part.") But $\{1, 2, 3, 6\}$ is not a cut-set even though the graph is cut into two parts, because the removal of the proper subset $\{1, 2, 3\}$ does not leave the graph connected.

The cut-set classifies the nodes of a graph into two groups, each group being in one of the two parts. Each branch of the cut-set has one of its terminals incident at a node in one group and its other end incident at a node in the other group. A cut-set is oriented by selecting an orientation from one of the two parts to the other. The orientation can be shown on the graph as in Fig. 8c. The orientations of the branches in a cut-set will either coincide with the cut-set orientation or they will not.

Just as the incidence matrix describes the incidences and the orientations of branches at nodes, so a cut-set matrix can be defined to describe the presence of branches in a cut-set and their orientation relative to that of the cut-set. We define a cut-set matrix $\mathbf{Q}_a = [q_{ij}]$ whose rows correspond to cut-sets and whose columns are the branches of a graph. The elements have the following values:

$q_{ij} = 1$ if branch j is in cut-set i, and the orientations coincide;
$q_{ij} = -1$ if branch j is in cut-set i, and the orientations do not coincide;
$q_{ij} = 0$ if branch j is not in cut-set i.

The subscript a stands for all cut-sets.

Since cutting all branches incident at a node separates this node from the rest of the graph, this set of branches will be a cut-set, *provided the rest of the graph is not itself cut into more than one part.* In the graph shown in Fig. 9, cutting the set of branches incident at node 1 will separate

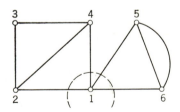

Fig. 9. A hinged, or separable, graph.

the graph into three parts, of which the isolated node 1 will be one part. So this set of branches is not a cut-set.

However, the graph shown in Fig. 9 is a peculiar kind of graph, and node 1 is a peculiar node. We define a *hinged graph as a graph in which there is at least one subgraph which has only one node in common with its* complement subgraph in the graph. A node having this property is called a *hinged node*. In a hinged graph the nodes can be grouped in two sets such that every path from a member of one set to a member of the other must pass through a hinged node. In Fig. 9, nodes 2, 3, and 4 form one set; and nodes 5 and 6, the other. If the branches incident at a hinged node are cut, there will be no path from a node in one set to a node in the other. Hence the remaining subgraph, not counting the hinged node, will not be connected; so the set of branches incident at a hinged node will not be a cut-set. For all other nodes, the set of branches incident there will be a cut-set.

For nonhinged graphs the orientation of the cut-set that consists of all branches incident at a node is chosen to be away from the node. Thus the cut-set matrix will include the incidence matrix for nonhinged graphs.

For the example of Fig. 3, in addition to the cut-sets consisting of the sets of branches incident at each node, there are three other cut-sets: $\{1, 3, 4, 5\}$, $\{2, 3, 5, 6\}$, and $\{1, 2, 4, 6\}$. The cut-set matrix $\mathbf{Q}_a$ is then

$$
\begin{array}{c}
\text{cut sets} \\
\downarrow
\end{array}
\quad
\begin{array}{c}
\text{branches} \rightarrow
\end{array}
$$

$$
\mathbf{Q}_a =
\begin{array}{c}
1 \\ 2 \\ 3 \\ 4 \\ 5 \\ 6 \\ 7
\end{array}
\begin{bmatrix}
-1 & 1 & 1 & 0 & 0 & 0 \\
0 & -1 & 0 & -1 & 1 & 0 \\
0 & 0 & -1 & 1 & 0 & 1 \\
1 & 0 & 0 & 0 & -1 & -1 \\
1 & 0 & -1 & 1 & -1 & 0 \\
0 & 1 & 1 & 0 & -1 & -1 \\
-1 & 1 & 0 & 1 & 0 & 1
\end{bmatrix}
$$

(columns: 1 2 3 4 5 6)

where the first four rows are identical with $\mathbf{A}_a$ and the last three rows correspond to the cut-sets $\{1, 3, 4, 5\}$, $\{2, 3, 5, 6\}$, and $\{1, 2, 4, 6\}$, respectively.

The cut-set matrix $\mathbf{Q}_a$ of a graph is seen to have more rows than its incidence matrix. The question of the rank of the cut-set matrix then arises. To answer this question, consider a special kind of cut-set formed as follows. Given a connected graph, first select a tree and focus on a branch b_k of the tree. Removing this branch from the tree unconnects the tree into two pieces. All the links which go from one part of this unconnected tree to the other part, together with b_k, will constitute a cut-set. We call this a *fundamental cut-set*, or *f-cut-set* for short. For each twig, there will be an f-cut-set, so for a graph having $n + 1$ nodes (hence, n twigs) there will be n fundamental cut-sets. The orientation of an f-cut-set is chosen to coincide with that of its defining twig.

As an illustration, take the graph of Fig. 10. The tree is shown in

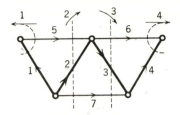

Fig. 10.   Example of fundamental cut-set.

heavy lines. Each f-cut-set is uniquely determined. Let us write the cut-set matrix for the f-cut-sets, arranging the columns so that the first n of them correspond to the twigs and are arranged in the same order as the associated cut-sets.

$$
\mathbf{Q}_f =
\begin{array}{c}
\overbrace{}^{\text{twigs}} \quad \overbrace{}^{\text{links}} \\
\begin{array}{ccccccc}
1 & 2 & 3 & 4 & 5 & 6 & 7
\end{array} \\
\begin{bmatrix}
1 & 0 & 0 & 0 & -1 & 0 & 0 \\
0 & 1 & 0 & 0 & 1 & 0 & 1 \\
0 & 0 & 1 & 0 & 0 & 1 & 1 \\
0 & 0 & 0 & 1 & 0 & 1 & 0
\end{bmatrix}
\end{array}.
$$

The subscript f stands for fundamental. The square submatrix corresponding to the first four columns is seen to be a unit matrix; hence it is nonsingular, and the rank of this cut-set matrix equals the number of its rows, or the number of twigs in a tree.

This is, in fact, a specific illustration of a general result. In the general case let the columns of the f-cut-set matrix be arranged for a given tree with the twigs first, then the links, the twigs being in the same order as the cut-sets they define. The matrix can be partitioned in the form

$$\mathbf{Q}_f = [\mathbf{Q}_t \quad \mathbf{Q}_l] = [\mathbf{U} \quad \mathbf{Q}_l]. \tag{38}$$

From the very way in which it is constructed, the $n \times n$ submatrix $\mathbf{Q}_t$ whose columns correspond to the twigs will be a unit matrix. Hence the

Q_f matrix will be of rank n. This still does not tell us everything about the rank of Q_a. But since the matrix of f-cut-sets is just a submatrix of Q_a, the rank of Q_a can be no less than that of Q_f, or rank of $Q_a \geq n$.

When seeking to find the rank of B_a, it became necessary to use the orthogonality relation $A_a B_a' = 0$. But Q_a is a matrix that contains A_a as a submatrix, and it might be suspected that this same relationship is satisfied with A_a replaced by Q_a.* This is true and can be proved in the same way as before. It is only necessary to establish that if a cut-set has any branches in common with a loop, it must have an even number. This fact can be readily appreciated by reference to Fig. 11, which shows a

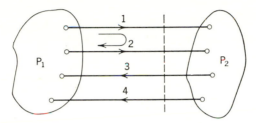

Fig. 11.   A cut-set has an even number of branches in common with a loop.

cut-set separating a graph into two parts. Suppose branch 1 of the cut-set is in a loop. If we start at its P_1 end and traverse the branch to P_2, it will be necessary to return to P_1 via another branch of the cut-set in order to form a closed path. The path need not be closed by only one excursion between P_1 and P_2, but any single excursion will use up two branches of the cut-set. If these two branches have the same orientation relative to the cut-set, they will have the opposite orientation relative to the loop, and vice versa. Hence, by the same reasoning used in obtaining (24), it follows that

$$Q_a B' = 0 \quad \text{and} \quad BQ_a' = 0. \tag{39}$$

The rank of Q_a can now be determined. Using Sylvester's law of nullity and the known rank of B, it will follow that the rank of Q_a is no greater than n. (You may carry out the details.) And since Q_f is a submatrix of Q_a having a rank of n, the rank of Q_a is no less than n. Hence the rank of Q_a is exactly n.

Removal of any number of rows from Q_a will not invalidate (39). Let Q be any n-rowed submatrix of Q_a of rank n. (One possibility is the matrix of f-cut-sets, Q_f.) Then

$$QB' = 0 \quad \text{and} \quad BQ' = 0. \tag{40}$$

* This statement applies only for a nonhinged graph.

In particular, let $\mathbf{Q}$ be $\mathbf{Q}_f$ and let it be partitioned as in (38). Then

$$[\mathbf{U} \quad \mathbf{Q}_l]\begin{bmatrix}\mathbf{B}_t' \\ \mathbf{B}_l'\end{bmatrix} = \mathbf{B}_t' + \mathbf{Q}_l\mathbf{B}_l' = \mathbf{0}$$

or

$$\mathbf{Q}_l = -\mathbf{B}_t'(\mathbf{B}_l')^{-1} = -(\mathbf{B}_l^{-1}\mathbf{B}_t)' \tag{41}$$

and, finally,

$$\mathbf{Q}_f = [\mathbf{U} \quad -(\mathbf{B}_l^{-1}\mathbf{B}_t)']. \tag{42}$$

Something very interesting follows from this expression. Comparing it with (37) shows that

$$\mathbf{A} = \mathbf{A}_t\mathbf{Q}_f \quad \text{or} \quad \mathbf{Q}_f = \mathbf{A}_t^{-1}\mathbf{A} = [\mathbf{U} \quad \mathbf{A}_t^{-1}\mathbf{A}_l]. \tag{43}$$

Since $\mathbf{A}_t$ is a nonsingular matrix, the incidence matrix of a graph is row equivalent to the fundamental cut-set matrix for some tree. Thus the rows of $\mathbf{A}$ are linear combinations of the rows of $\mathbf{Q}_f$, and vice versa.

PLANAR GRAPHS

All of the properties of graphs that have been discussed up to this point do not depend on the specifically geometrical or topological character of the graph, only on its abstract characteristics. We shall now discuss some properties that depend on the topological structure.

Topological graphs can be drawn, or mapped, on a plane. Either they can be drawn so that no branches cross each other or they cannot. We define a *planar graph* as a graph that can be mapped on a plane in such a way that no two branches cross each other (i.e., at any point that is not a node). Figure 12 shows two graphs having the same number of nodes and branches; the first is planar; the second, nonplanar.

The branches of a planar graph separate a plane into small regions; each of these is called a *mesh*. Specifically, a mesh is a sequence of branches of a planar graph that enclose no other branch of the graph within the boundary formed by these branches. In Fig. 12a, the set {1, 2, 3} is a mesh, whereas {1, 2, 4, 5} is not. The outermost set of branches separates the plane into two regions: the finite region in which the remaining branches lie and the infinite region. The infinite region can be looked upon as the "interior" of this set of branches. It is the complement of the finite region. Hence this set of branches can also be considered a mesh and is called the *outside mesh*. In Fig. 12a the outside mesh is {1, 2, 6, 7, 8, 5}. However, when the meshes of a graph are enumerated, the outside mesh is not counted.

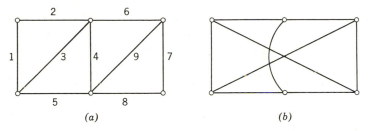

Fig. 12.   Planar (*a*) and nonplanar (*b*) graphs.

The set of meshes of a planar graph is a special set of loops. A question arises as to whether the meshes can be the *f*-loops for some tree; or, stated differently, is it possible to find a tree, the *f*-loops corresponding to which are meshes? To answer this question, note that each *f*-loop contains a branch (a link) that is in no other *f*-loop. Hence any branches that are common between two meshes cannot be a link and must be a twig. A tree can surely be found for which the meshes are *f*-loops if the branches common between meshes form no closed paths. For some planar graphs it will be possible; and for others, not. To illustrate, Fig. 13 shows two

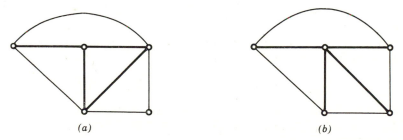

Fig. 13.   Meshes may or may not be *f*-loops for some tree.

very similar planar graphs having the same number of nodes, branches, and meshes. The branches common between meshes are shown darkened. These must be twigs for a tree if the meshes are to be *f*-loops. However, in the first graph, these branches form a loop, and so the desired result is not possible; whereas it is possible for the second graph.

This completes the discussion of linear graphs as such.

2.3   BASIC LAWS OF ELECTRIC NETWORKS

Broadly, an electric network consists of an interconnection of two or more elements, or branches. These branches may consist of the components discussed in Chapter 1 or other, more general (nonlinear, time-varying, etc.) components. Each branch has a voltage and a current variable, and these variables are related to each other by specific relationships.

To tie in the analysis of networks with linear graphs, we shall make the following definition:

An *electric network* is an oriented linear graph with each branch of which there are associated two functions of time t: the current $i(t)$ and the voltage $v(t)$. These functions are constrained by Kirchhoff's two laws and the branch relationships to be described.

KIRCHHOFF'S CURRENT LAW

Kirchhoff's current law (abbreviated as KCL) states that in any electric network the sum of all currents leaving a node equals zero, at each instant of time and for each node of the network. For a connected network (graph) having $n+1$ nodes and b branches, the KCL equations can be written as

$$\sum_{k=1}^{b} a_{jk} i_k(t) = 0, \quad j = 1, 2, \cdots, (n+1), \tag{44}$$

where a_{jk} has the same definition as the elements of the incidence matrix. Hence, in matrix form, KCL becomes

$$\mathbf{A}i(t) = \mathbf{0} \quad \text{or} \quad \mathbf{A}I(s) = \mathbf{0}. \tag{45}$$

where $\mathbf{A}$ is the incidence matrix, $\mathbf{i}(t)$ is a column matrix of branch currents, and $\mathbf{I}(s)$ is the column matrix of Laplace transforms of the branch currents.

$$\mathbf{i}(t) = \begin{bmatrix} i_1(t) \\ i_2(t) \\ \vdots \\ i_b(t) \end{bmatrix} \quad \text{and} \quad \mathbf{I}(s) = \begin{bmatrix} I_1(s) \\ I_2(s) \\ \vdots \\ I_b(s) \end{bmatrix}. \tag{46}$$

(Of course, it is also true that $\mathbf{A}_a \mathbf{i}(t) = \mathbf{0}$ if all the nodes are included.) Since the rank of $\mathbf{A}$ is n, all of the equations in this set are linearly independent.

Let the incidence matrix be partitioned in the form $\mathbf{A} = [\mathbf{A}_t \quad \mathbf{A}_l]$ for some choice of tree, and let the matrix $\mathbf{i}$ be similarly partitioned as

$$\mathbf{i} = \begin{bmatrix} \mathbf{i}_t \\ \mathbf{i}_l \end{bmatrix}.$$

Then KCL yields

$$[\mathbf{A}_t \quad \mathbf{A}_l] \begin{bmatrix} \mathbf{i}_t \\ \mathbf{i}_l \end{bmatrix} = \mathbf{A}_t \mathbf{i}_t + \mathbf{A}_l \mathbf{i}_l = \mathbf{0} \tag{47}$$

or

$$\mathbf{i}_t(t) = -\mathbf{A}_t^{-1} \mathbf{A}_l \mathbf{i}_l(t) \tag{48}$$

since $\mathbf{A}_t$ is a nonsingular matrix.

The message carried by this expression is that, given a tree, there is a linear relationship by which the twig currents are determined from the link currents. This means that, if the link currents can be determined by some other means, the twig currents become known from (48). Of all the b branch currents, only $b - n$ of them need be determined independently.

Using (48), the matrix of all currents can now be written as

$$\mathbf{i} = \begin{bmatrix} \mathbf{i}_t \\ \mathbf{i}_l \end{bmatrix} = \begin{bmatrix} -\mathbf{A}_t^{-1} \mathbf{A}_l \\ \mathbf{U} \end{bmatrix} \mathbf{i}_l. \tag{49}$$

$B_f = [-(A_t^{-1} A_l)' \; U]$ ③④

B_f

Comparing the matrix on the right with the one in (34) and also using (35), there follows that ③⑤

$B = B_l \, B_f$

$$\mathbf{i} = \mathbf{B}_f' \mathbf{i}_l \quad \text{or} \quad \mathbf{I}(s) = \mathbf{B}_f' \mathbf{I}_l(s) \qquad B_f = B_l^{-1} B \tag{50a}$$

$$\mathbf{i} = \mathbf{B}'(\mathbf{B}_l^{-1})' \mathbf{i}_l. \qquad B_f' = B'(B_l')^{-1} \tag{50b}$$

Each of these equations expresses all the branch currents of a network in terms of the link currents for some tree by means of a transformation that is called a *loop transformation*. The link currents for a tree are seen to be a *basis* for the set of all currents. We shall shortly discuss sets of basis currents other than the link currents for a tree.

Other sets of equations *equivalent* to KCL in (45) can be obtained. (Recall that two sets of equations are equivalent if they have the same solution.) Consider a particular cut-set of a network. It will separate the network into two parts, P_1 and P_2. Write the KCL equations at all the nodes in one of the parts, say P_1, and consider the columns. If both ends of a branch are incident at nodes in P_1, the corresponding column will have two nonzero elements, a $+1$ and a -1. But if one end of a branch is incident at a node in P_1 and the other end at a node in P_2 (i.e., if this branch is in the cut-set), this column will have but a single nonzero element. Now suppose these KCL equations are added; only the currents of the cut-set will have nonzero coefficients in the sum. The result will be called a *cut-set equation*. A cut-set equation is, then, a linear combination of KCL equations. The set of all such cut-set equations will be precisely $\mathbf{Q}_a \mathbf{i}(t) = \mathbf{0}$, where $\mathbf{Q}_a$ is the previously defined cut-set matrix for all cut-sets. But the rank of $\mathbf{Q}_a$ is n, which is less than the number of equations. These equations are thus not independent. Let $\mathbf{Q}$ be a cut-set matrix of n cut-sets and of rank n. (One possibility is the matrix of f-cut-sets, $\mathbf{Q}_f$.) Then

$$\mathbf{Q}\mathbf{i}(t) = \mathbf{0} \quad \text{or} \quad \mathbf{Q}\mathbf{I}(s) = \mathbf{0} \tag{51}$$

will be equivalent to the KCL equations.

In particular, if the matrix of f-cut-sets is partitioned in the form $\mathbf{Q}_f = [\mathbf{U} \quad \mathbf{Q}_l]$, then

$$\mathbf{Q}_f \mathbf{i} = [\mathbf{U} \quad \mathbf{Q}_l] \begin{bmatrix} \mathbf{i}_t \\ \mathbf{i}_l \end{bmatrix} = \mathbf{i}_t + \mathbf{Q}_l \mathbf{i}_l = \mathbf{0}$$

or

$$\mathbf{i}_t = -\mathbf{Q}_l \mathbf{i}_l, \tag{52}$$

which is the same as (48) in view of (43). This expression can be inserted into the partitioned form of the matrix of all currents as in the case of (48) to yield

$$\mathbf{i}(t) = \begin{bmatrix} \mathbf{i}_t \\ \mathbf{i}_l \end{bmatrix} = \begin{bmatrix} -\mathbf{Q}_l \\ \mathbf{U} \end{bmatrix} \mathbf{i}_l(t) = \mathbf{B}_f' \mathbf{i}_l(t) \quad \text{or} \quad \mathbf{I}(s) = \mathbf{B}_f' \mathbf{I}_l(s). \tag{53}$$

(The final step comes from the results of Problem 17.) This is again a loop transformation identical with (50a). Note that the matrix of the transformation is the transpose of the matrix of fundamental loops.

We have seen that the link currents for a given tree are basis currents in terms of which all currents in a network can be expressed. Another set of basis currents are the *loop currents*, which are fictitious circulating currents on the contours of closed loops. This can best be illustrated by means of an example. Figure 14 is a redrawing of Fig. 13a, with the branches and nodes appropriately numbered. This graph is planar, but it is not possible to find a tree for which the meshes are f-loops, as discussed earlier. Let a loop matrix be written for the loops specified in the figure. (This set of loops is neither a set of f-loops nor a set of meshes.) The orientation of the loops is given by the ordering of the nodes; it is also shown by means of the arrows on the diagram. The **B** matrix will be

$$
\begin{array}{c}
\text{branches}\rightarrow \\[2pt]
\begin{array}{c}
\text{loops}\\ \downarrow
\end{array}
\quad
\begin{array}{cccccccc}
1 & 2 & 3 & 4 & 5 & 6 & 7 & 8
\end{array}
\end{array}
$$

$$
\mathbf{B} =
\begin{array}{c}
1\\2\\3\\4
\end{array}
\left[
\begin{array}{cccccccc}
1 & 0 & 0 & 0 & 1 & 0 & -1 & 0\\
0 & 1 & 1 & 1 & -1 & 0 & 0 & 0\\
0 & 0 & 1 & 1 & 0 & -1 & 0 & 0\\
-1 & -1 & 0 & 0 & 0 & 0 & 0 & 1
\end{array}
\right]
$$

$$
\mathbf{B'} =
\begin{bmatrix}
1 & 0 & 0 & -1\\
0 & 1 & 0 & -1\\
0 & 1 & 1 & 0\\
0 & 1 & 1 & 0\\
1 & -1 & 0 & 0\\
0 & 0 & -1 & 0\\
-1 & 0 & 0 & 0\\
0 & 0 & 0 & 1
\end{bmatrix}.
\qquad (54)
$$

(This matrix is of rank 4 since the submatrix consisting of the last four columns is nonsingular.) Now suppose a set of circulating currents, i_{m1}, i_{m2}, etc., is defined on the contours of the same loops for which the **B** matrix is written and having the same orientation. By inspection of the

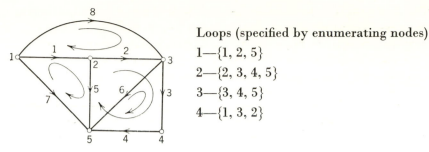

Loops (specified by enumerating nodes)

1—$\{1, 2, 5\}$

2—$\{2, 3, 4, 5\}$

3—$\{3, 4, 5\}$

4—$\{1, 3, 2\}$

Fig. 14.   Illustrative example.

graph, it is possible to express the branch currents in terms of these circulating currents as follows:

$$
\begin{bmatrix} i_1 \\ i_2 \\ i_3 \\ i_4 \\ i_5 \\ i_6 \\ i_7 \\ i_8 \end{bmatrix}
=
\begin{bmatrix}
1 & 0 & 0 & -1 \\
0 & 1 & 0 & -1 \\
0 & 1 & 1 & 0 \\
0 & 1 & 1 & 0 \\
1 & -1 & 0 & 0 \\
0 & 0 & -1 & 0 \\
-1 & 0 & 0 & 0 \\
0 & 0 & 0 & 1
\end{bmatrix}
\begin{bmatrix} i_{m1} \\ i_{m2} \\ i_{m3} \\ i_{m4} \end{bmatrix} .
\tag{55}
$$

By comparing the matrix of this transformation with the transpose of **B** in (54), it is seen that they are the same.

This is a general result that follows by observing that each row of a **B** matrix tells the incidence of branches of the graph on the corresponding loop. Similarly, each column of **B** focuses on a branch; the entries in the column specify those loops on which that branch is incident and with what orientation. If circulating loop currents are defined on the same contours as the loops, and with the same orientation, then each column of **B** will specify the corresponding branch current in terms of the loop currents.

In a graph having b branches and $n + 1$ nodes, let $\mathbf{i}_m$ be a vector of loop currents defined by $b - n$ loops for which the **B** matrix is of rank $b - n$. Then the branch-current matrix **i** is given in terms of $\mathbf{i}_m$ by the transformation

$$
\mathbf{i} = \mathbf{B}'\mathbf{i}_m .
\tag{56}
$$

For a planar network the currents defined by the meshes will be an adequate basis. (The proof of this is left as a problem.) For this case the transformation in (56) is called the *mesh transformation*.

KIRCHHOFF'S VOLTAGE LAW

The second of Kirchhoff's laws is *Kirchhoff's voltage law* (*KVL*), which states that *in any electric network the sum, relative to the loop orientation, of the voltages of all branches on the loop equals zero, at each instant of time and for each loop in the network.* For a connected network having b branches the KVL equations can be written as

$$\sum_{k=1}^{b} b_{jk} v_k(t) = 0, \qquad j = 1, 2, \cdots, \text{all loops}, \qquad (57)$$

where b_{jk} has the same definition as the elements of the loop matrix. In matrix form, KVL becomes

$$\mathbf{B}\mathbf{v}(t) = \mathbf{0} \quad \text{or} \quad \mathbf{B}\mathbf{V}(s) = \mathbf{0}, \qquad (58)$$

where $\mathbf{B}$ is a loop matrix, $\mathbf{v}(t)$ is a column matrix of branch voltages, and $\mathbf{V}(s)$ is a column matrix of their Laplace transforms.

$$\mathbf{v}(t) = \begin{bmatrix} v_1(t) \\ v_2(t) \\ \vdots \\ v_b(t) \end{bmatrix} \quad \text{and} \quad \mathbf{V}(s) = \begin{bmatrix} V_1(s) \\ V_2(s) \\ \vdots \\ V_b(s) \end{bmatrix} \qquad (59)$$

If all the loops in the network are included, the coefficient matrix will be $\mathbf{B}_a$. However, the rank of $\mathbf{B}_a$ is $b - n$, and the equations in this set will not be independent.

Let $\mathbf{B}$ have $b - n$ rows and be of rank $b - n$. (One possibility is the matrix of f-loops.) It can be partitioned in the form $\mathbf{B} = [\mathbf{B}_t \quad \mathbf{B}_l]$ for some choice of tree. Let $\mathbf{v}$ also be partitioned conformally as

$$\mathbf{v} = \begin{bmatrix} \mathbf{v}_t \\ \mathbf{v}_l \end{bmatrix}.$$

Then KVL can be written as

$$[\mathbf{B}_t \quad \mathbf{B}_l] \begin{bmatrix} \mathbf{v}_t \\ \mathbf{v}_l \end{bmatrix} = \mathbf{B}_t \mathbf{v}_t + \mathbf{B}_l \mathbf{v}_l = \mathbf{0}$$

from which

$$\mathbf{v}_l = -\mathbf{B}_l^{-1}\mathbf{B}_t\,\mathbf{v}_t \quad \text{or} \quad \mathbf{v}_l = -\mathbf{B}_{ft}\,\mathbf{v}_t \tag{60}$$

since $\mathbf{B}_l$ is a nonsingular matrix.

The message carried by this expression is that, given a tree of a graph, there is a linear relationship by which the link voltages are determined from the twig voltages. If the twig voltages can be determined by some other means, the link voltages become known by (60). Of all the b branch voltages, only n of them, those of the twigs, need be determined independently.

Now let (60) be inserted into the partitioned form of the branch voltage matrix $\mathbf{v}(t)$. Then

$$\mathbf{v}(t) = \begin{bmatrix} \mathbf{v}_t \\ \mathbf{v}_l \end{bmatrix} = \begin{bmatrix} \mathbf{U} \\ -\mathbf{B}_l^{-1}\mathbf{B}_t \end{bmatrix}\mathbf{v}_t(t) = \mathbf{Q}_f'\mathbf{v}_t(t) \quad \text{or} \quad \mathbf{V}(s) = \mathbf{Q}_f'\mathbf{V}_t(s). \tag{61}$$

The last step follows from (42).

Another form is obtained by inserting (43) here. There results

$$\mathbf{v} = \mathbf{A}'(\mathbf{A}_t^{-1})'\mathbf{v}_t. \tag{62}$$

If $\mathbf{A}_t$ is a unit matrix, then

$$\mathbf{v}(t) = \mathbf{A}'\mathbf{v}_t(t). \tag{63}$$

Thus the branch-voltage matrix is expressed in terms of the twig-voltage matrix for some tree by means of a transformation. The matrix of the transformation can be the transpose of the f-cut-set matrix $\mathbf{Q}_f$ or the transpose of the $\mathbf{A}$ matrix, provided $\mathbf{A}_t$ is a unit matrix. (See Problem 5 concerning the condition for which $\mathbf{A}_t$ will be a unit matrix.) The twig voltages for a tree are seen to be a basis for the set of all voltages. Since a twig voltage is the difference in voltage between a pair of nodes, twig voltages are *node-pair voltages*. Not all node-pair voltages are twig voltages; Nevertheless, a suitable set of node-pair, but not necessarily twig, voltages may constitute a basis set of voltages. Let us consider this point further.

If one of the two nodes of each node pair is a common node, then each node-pair voltage will simply equal the voltage of a node relative to that of this common, or datum, node. These voltages are referred to as the *node voltages*. Since each branch of a graph is incident at two nodes, its voltage will necessarily be the difference between two node voltages (with the datum-node voltage being zero). Thus all voltages of a graph are expressible in terms of only the node voltages, of which there are n.

When writing the **A** matrix of a graph, one of the nodes is omitted. If this node is also taken as the datum node for the definition of node voltages, then the branch-voltage matrix can be expressed in terms of the node-voltage matrix $\mathbf{v}_n$ as

$$\mathbf{v}(t) = \mathbf{A}'\mathbf{v}_n(t) \quad \text{or} \quad \mathbf{V}(s) = \mathbf{A}'\mathbf{V}_n(s). \tag{64}$$

This follows from the fact that each column of the **A** matrix pertains to a specific branch. The nonzero elements in a column specify the nodes on which that branch is incident, the sign indicating its orientation. Hence each column of **A** will specify the corresponding branch voltage in terms of the node voltages.

The following example will illustrate this result. Figure 15 is the same

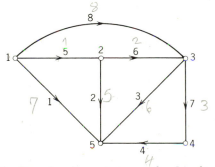

Fig. 15. Branch voltages in terms of node voltages.

graph as in Fig. 14, with a different branch numbering. The **A** matrix with node 5 omitted is

$$\mathbf{A} = \begin{matrix} & 1 & 2 & 3 & 4 & 5 & 6 & 7 & 8 \\ & \begin{bmatrix} 1 & 0 & 0 & 0 & 1 & 0 & 0 & 1 \\ 0 & 1 & 0 & 0 & -1 & 1 & 0 & 0 \\ 0 & 0 & 1 & 0 & 0 & -1 & 1 & -1 \\ 0 & 0 & 0 & 1 & 0 & 0 & -1 & 0 \end{bmatrix} \end{matrix}$$

and

$$\mathbf{A'} = \begin{bmatrix} 1 & 0 & 0 & 0 \\ 0 & 1 & 0 & 0 \\ 0 & 0 & 1 & 0 \\ 0 & 0 & 0 & 1 \\ 1 & -1 & 0 & 0 \\ 0 & 1 & -1 & 0 \\ 0 & 0 & 1 & -1 \\ 1 & 0 & -1 & 0 \end{bmatrix}. \tag{65}$$

Now take node 5 as datum node and let v_{n1}, v_{n2}, v_{n3}, and v_{n4} be the node voltages, the voltages of the other nodes relative to that of node 5. By inspection of the graph, it is possible to express the branch voltages relative to the node voltages. Thus

$$\begin{bmatrix} v_1 \\ v_2 \\ v_3 \\ v_4 \\ v_5 \\ v_6 \\ v_7 \\ v_8 \end{bmatrix} = \begin{bmatrix} 1 & 0 & 0 & 0 \\ 0 & 1 & 0 & 0 \\ 0 & 0 & 1 & 0 \\ 0 & 0 & 0 & 1 \\ 1 & -1 & 0 & 0 \\ 0 & 1 & -1 & 0 \\ 0 & 0 & 1 & -1 \\ 1 & 0 & -1 & 0 \end{bmatrix} \begin{bmatrix} v_{n1} \\ v_{n2} \\ v_{n3} \\ v_{n4} \end{bmatrix}. \tag{66}$$

The matrix of this transformation is seen to be the transpose of the $\mathbf{A}$ matrix.

Note that the first 4 columns of $\mathbf{A}$ in (65) form a unit matrix. This agrees with Problem 5, since the tree consisting of branches 1 through 4 is a star tree. Hence, it happens that the node-pair voltages defined by a tree in this case are the same as the node voltages (relative to node 5 as datum). To allay any worries on this score, choose node 4 as datum. There is no star tree with node 4 as the common node. The $\mathbf{A}$ matrix will be the same as that in (65) but with the last row replaced by $[-1 \quad -1$ $-1 \quad -1 \quad 0 \quad 0 \quad 0 \quad 0]$. Whatever tree is now chosen, A_t will not be a unit matrix, and (63) will not apply. However (64), in terms of the new node voltages, will still apply. You are urged to verify this.

THE BRANCH RELATIONS

Kirchhoff's two laws expressing constraints on the voltages and currents of the branches of a network are independent of the specific nature of the branches, whether capacitor, resistor, source, etc. They apply to nonlinear as well as to linear elements, and to time-varying as well as to time-invariant elements. They express constraints imposed by the topology of a network.

However, the manner in which the voltage of a particular branch is related to its current does depend on the constituents of the branch. There is a considerable amount of flexibility in selecting the makeup of a network branch. One possibility is to let each element itself (resistor, capacitor, etc.) constitute a branch. This would require us to count as a node the junction between elements that are connected in series. It may sometimes be convenient to consider series-connected elements as a single branch or parallel-connected elements as a single branch. In the network of Fig. 16, the series connection of R_a and L_a can be taken as a single branch or R_a and L_a can be considered as two separate branches.

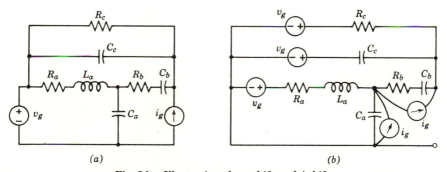

(a)   (b)

Fig. 16. Illustrating the v-shift and i-shift.

There is similar flexibility in the manner in which sources are treated. Let a voltage source be said to be *accompanied* if there is a passive branch in series with it. Likewise, let a current source be said to be accompanied if there is a passive branch in parallel with it. In Fig. 16a, neither source is accompanied. For a passive branch, both the current and voltage are unknowns whose variations with time are to be determined. However.

either the voltage or the current of a source is known. Thus general statements about the number of unknowns in terms of the number of branches cannot be made if unaccompanied sources are treated as branches. For this purpose it is convenient to use some equivalences, which will now be discussed in order to remove unaccompanied sources from consideration.

Consider the network in Fig. 16b. The source v_g has been shifted through one of its terminals into each branch incident there, maintaining its proper reference and leaving its original position short-circuited. Applications of KVL to any loop show that these equations have not been changed. Now, however, each voltage source is accompanied; it is in series with a passive branch. In the case of the current source, it has been shifted and placed across each branch of a loop that contained the original source, maintaining the proper reference and leaving its original position open-circuited. Application of KCL to all nodes will show that these equations have remained invariant. Thus the solutions for the other branch variables should be expected to be the same in this new network as in the old one. We shall refer to these two equivalences as the *voltage shift* (or *v*-shift) and the *current shift* (or *i*-shift), respectively. As a result it is always possible to make all sources accompanied sources. Sometimes it is convenient to treat all independent sources as accompanied; and at other times, it is not. For the development of the loop and node equations, the former is the case. Hence in this chapter we will assume that all independent sources are accompanied. Later, when convenient, this requirement will be relaxed.

As we start discussing the *v-i* relationships of branches, aside from independent sources, we shall first consider passive, reciprocal networks only. After the basic procedures have been established, active and nonreciprocal components will also be introduced.

The most general branch will be assumed to have the form shown in Fig. 17, containing both a voltage source in series with a passive branch

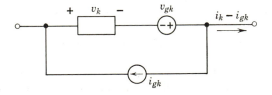

Fig. 17.   A general branch.

and a current source in parallel with this combination. (In Fig. 16b, for example, the left-hand current source can be considered to be in parallel with the series connection of v_g, R_a, and L_a.) Thus the current of this kth branch that is to be used in KCL will be $i_k - i_{gk}$. Likewise the voltage that is to be used in KVL will be $v_k - v_{gk}$. Hence (45) and (58) will be replaced by

$$\mathbf{A}\mathbf{i}(t) = \mathbf{A}\mathbf{i}_g(t), \quad \mathbf{A}\mathbf{I}(s) = \mathbf{A}\mathbf{I}_g(s) \tag{67}$$

$$\mathbf{B}\mathbf{v}(t) = \mathbf{B}\mathbf{v}_g(t), \quad \mathbf{B}\mathbf{V}(s) = \mathbf{B}\mathbf{V}_g(s) \tag{68}$$

$$\mathbf{Q}\mathbf{i}(t) = \mathbf{Q}\mathbf{i}_g(t), \quad \mathbf{Q}\mathbf{I}(s) = \mathbf{Q}\mathbf{I}_g(s). \tag{69}$$

where $\mathbf{i}_g$ and $\mathbf{v}_g$ are column matrices of source currents and voltages.

Similarly, the transformations from branch variables to loop currents or node voltages must be replaced by the following:

$$\mathbf{i}(t) - \mathbf{i}_g(t) = \mathbf{B}'\mathbf{i}_m(t), \quad \mathbf{I}(s) - \mathbf{I}_g(s) = \mathbf{B}'\mathbf{I}_m(s) \tag{70}$$

$$\mathbf{v}(t) - \mathbf{v}_g(t) = \mathbf{Q}'_f\mathbf{v}_t(t), \quad \mathbf{V}(s) - \mathbf{V}_g(s) = \mathbf{Q}'_f\mathbf{V}_t(s) \tag{71}$$

$$\mathbf{v}(t) - \mathbf{v}_g(t) = \mathbf{A}'\mathbf{v}_n(t), \quad \mathbf{V}(s) - \mathbf{V}_g(s) = \mathbf{A}'\mathbf{V}_n(s). \tag{72}$$

Since sources can be handled in this way independently of the passive parts of a branch, we shall henceforth concentrate on the passive components. When doing so, the sources will be made to vanish, which is done by replacing the voltage sources by short circuits and the current sources by open circuits.

Now we can turn to a consideration of the relationships between voltage and current of the branches of a graph. At the outset, we shall make no special conventions regarding the manner in which branches are selected and numbered. We shall deal with the Laplace-transformed variables and assume that initial conditions have been represented as equivalent sources. The impedance and admittance of branch k will be represented by lower case z_k and y_k, respectively, whereas the corresponding matrices will be $\mathbf{Z}$ and $\mathbf{Y}$.

The branch relationships can be written as follows:

Branch k	*Network*	
$V_k(s) = z_k\,I_k(s)$	$\mathbf{V}(s) = \mathbf{Z}(s)\mathbf{I}(s)$	(73)
$I_k(s) = y_k\,V_k(s)$	$\mathbf{I}(s) = \mathbf{Y}(s)\mathbf{V}(s).$	

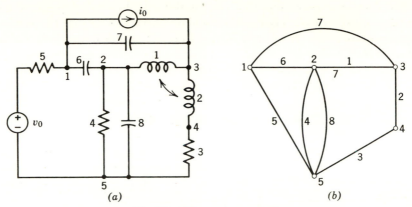

Fig. 18.   Illustrative example.

The nature of these branch matrices can be illustrated with the example shown in Fig. 18. The two inductive branches 1 and 2 are mutually coupled. Now (73) with the branch-impedance matrix $\mathbf{Z}$ shown in detail can be written as follows:

$$\mathbf{V}(s) = \begin{bmatrix} sL_{11} & sL_{12} & & & & & & & \\ & & & \mathbf{0} & & & & \mathbf{0} & \\ sL_{21} & sL_{22} & & & & & & & \\ & & & R_3 & \bigcirc & & & & \\ \mathbf{0} & & & R_4 & & & \mathbf{0} & \\ & & \bigcirc & R_5 & & & & \\ & & & & & 1/sC_6 & & \bigcirc \\ \mathbf{0} & & \mathbf{0} & & & 1/sC_7 & \\ & & & & \bigcirc & & 1/sC_8 \end{bmatrix} \mathbf{I}(s)$$

$$\mathbf{V}(s) = \begin{bmatrix} s\mathbf{L}_p & & \bigcirc \\ & \mathbf{R}_p & \\ \bigcirc & & \dfrac{1}{s}\mathbf{D}_p \end{bmatrix} \mathbf{I}(s). \tag{74}$$

Note that, because of the way the branches were numbered—with each element a separate branch and with inductance first, then resistance, then capacitance—the matrix can be partitioned as shown, with obvious meanings for the submatrices. The resistance and inverse-capacitance matrices are diagonal because there is no coupling from one branch to the other. This is not the case for the $\mathbf{L}_p$ matrix because of the inductive coupling.

It is clear that these properties of the submatrices of the impedance matrix are quite general if the above numbering scheme for branches is used. Such a numbering scheme is useful, so we shall adopt it when convenient for the purpose of determining properties of the corresponding matrices. There will be times, however, when we will want greater flexibility and will number the branches differently.

In the general case, then, if each element is counted as a separate branch and if the inductive branches are numbered first, then the resistive branches, and then the capacitive branches, the branch impedance and admittance matrices can be written as follows:

$$\mathbf{Z} = \begin{bmatrix} s\mathbf{L}_p & & \bigcirc \\ & \mathbf{R}_p & \\ \bigcirc & & \frac{1}{s}\mathbf{D}_p \end{bmatrix} \quad \text{and} \quad \mathbf{Y} = \begin{bmatrix} \frac{1}{s}\boldsymbol{\Gamma}_p & & \bigcirc \\ & \mathbf{G}_p & \\ \bigcirc & & s\mathbf{C}_p \end{bmatrix} \tag{75}$$

where $\mathbf{R}_p$, $\mathbf{G}_p$, $\mathbf{C}_p$, and $\mathbf{D}_p$ are diagonal matrices with $\mathbf{G}_p = \mathbf{R}_p^{-1}$, $\mathbf{D}_p = \mathbf{C}_p^{-1}$, and $\boldsymbol{\Gamma}_p = \mathbf{L}^{-1}$. (The subscript p in these matrices stands for "partial.") They are called the *partial branch-parameter matrices*. In case there are perfectly coupled transformers in the network, $\mathbf{L}_p$ will be a singular matrix and $\boldsymbol{\Gamma}_p$ will not exist.

Sometimes it is convenient in referring to the branch-parameter matrices to assume that each one extends over the total dimensions of the $\mathbf{Z}$ matrix; that is, we could write

$$\mathbf{R} = \begin{bmatrix} \mathbf{0} & \mathbf{0} & \mathbf{0} \\ \mathbf{0} & \mathbf{R}_p & \mathbf{0} \\ \mathbf{0} & \mathbf{0} & \mathbf{0} \end{bmatrix} \tag{76}$$

for the branch resistance matrix and similarly for the others. If this is done, then the branch impedance and admittance matrices can be simply expressed as follows:

$$\mathbf{Z} = s\mathbf{L} + \mathbf{R} + \frac{1}{s}\mathbf{D} \tag{77a}$$

and

$$\mathbf{Y} = s\mathbf{C} + \mathbf{G} + \frac{1}{s}\mathbf{\Gamma}. \tag{77b}$$

It should be observed that the order of each of the branch-parameter matrices equals the number of branches of the graph. For convenience in writing (77), we have increased the dimensions of these matrices. For computational purposes, the increased dimensions of the matrices will be a disadvantage. In this case it would be better to use the partial matrices.

Although a special branch-numbering scheme was used in arriving at the branch-parameter matrices given by (76) and others like it, these matrices can be defined without using that numbering scheme. The only difference will be that the nonzero elements will not all be concentrated in a single submatrix as in (76). In a later chapter we shall investigate the properties of these parameter matrices and shall discuss their realizability conditions.

2.4   LOOP, NODE, AND NODE-PAIR EQUATIONS

The basic relationships presented in the last section are Kirchhoff's current law (KCL), Kirchhoff's voltage law (KVL), and the branch voltage-current relationships. For a network containing b branches and $n + 1$ nodes, there are n independent KCL equations and $b - n$ independent KVL equations for a total of b. There are also b branch v-i relationships, which combined with the other b independent equations are sufficient to solve for the $2b$ branch variables, b currents and b voltages. However, solving $2b$ simultaneous equations is a substantial task, and anything that can be done to reduce the work will be of advantage.

In the last section it was observed that the branch currents could all be determined in terms of a smaller subset—for example, the link currents for a tree or the loop currents. Similarly, the branch voltages could all be determined in terms of a smaller subset of voltages. We shall now consider a number of procedures for utilizing these results to carry out an analysis of a network problem. The outcome depends on the order in which the three basic relationships are used.

LOOP EQUATIONS

Given a network, let us first apply KVL, arriving at (68) which is repeated here (in Laplace-transformed form).

$$\mathbf{B}\mathbf{V}(s) = \mathbf{B}\mathbf{V}_g(s). \tag{78}$$

Here $\mathbf{B}$ is of order $(b-n) \times b$ and of rank $b-n$. Into this expression we next insert the branch relations of (73), obtaining

$$\mathbf{B}\mathbf{Z}(s)\mathbf{I}(s) = \mathbf{B}\mathbf{V}_g(s). \tag{79}$$

Finally, we express the branch currents in terms of a set of $b-n$ other currents, which may be loop currents or link currents for a tree—if $\mathbf{B}$ is the matrix of fundamental loops for that tree. Let us say the former; that is, we substitute for $\mathbf{I}(s)$ the loop transformation in (70). The result is

$$\{\mathbf{B}\mathbf{Z}(s)\mathbf{B}'\}\mathbf{I}_m(s) = \mathbf{B}\{\mathbf{V}_g - \mathbf{Z}(s)\mathbf{I}_g\} \tag{80a}$$

or

$$\mathbf{Z}_m(s)\mathbf{I}_m(s) = \mathbf{E}(s) \tag{80b}$$

where $\mathbf{E}$ is shorthand for $\mathbf{B}\{\mathbf{V}_g - \mathbf{Z}\mathbf{I}_g\}$ and

$$\mathbf{Z}_m(s) = \mathbf{B}\mathbf{Z}(s)\mathbf{B}'. \tag{81}$$

This matrix equation represents a set of $b-n$ equations, called the *loop equations*, in the $b-n$ loop-current variables. The coefficient matrix $\mathbf{Z}_m(s)$ is called the *loop-impedance* matrix, not to be confused with the *branch-impedance* matrix $\mathbf{Z}$. For a passive reciprocal network, $\mathbf{Z}$ is a symmetric matrix. Hence (see Problem 1.15) $\mathbf{Z}_m$ is also symmetric.

The loop-impedance matrix can be written explicitly in terms of the branch-parameter matrices by inserting (77) into (81). Thus

$$\mathbf{Z}_m = \mathbf{BZB}' = s\mathbf{L}_m + \mathbf{R}_m + \frac{1}{s}\mathbf{D}_m \tag{82}$$

where

$$\mathbf{L}_m = \mathbf{BLB}' \tag{83a}$$

$$\mathbf{R}_m = \mathbf{BRB}' \tag{83b}$$

$$\mathbf{D}_m = \mathbf{BDB}' \tag{83c}$$

are the loop-parameter matrices.

To illustrate (80), consider the network of Fig. 18 for which the branch-impedance matrix was given in (74). Its graph is redrawn in Fig. 19 to

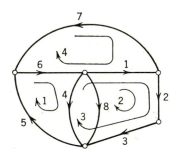

Fig. 19.   Illustrative example for loop equations.

show the loops chosen. This is a planar graph, and perhaps the simplest loops would be the meshes. However, for illustrative purposes another set of loops is chosen. The **B** matrix for this choice is the following:

$$
\mathbf{B} = \begin{array}{c} \\ 1 \\ 2 \\ 3 \\ 4 \end{array}
\begin{array}{cccccccc}
1 & 2 & 3 & 4 & 5 & 6 & 7 & 8 \\
\left[\begin{array}{cccccccc}
0 & 0 & 0 & 1 & 1 & 1 & 0 & 0 \\
1 & 1 & 1 & 0 & 0 & 0 & 0 & -1 \\
1 & 1 & 1 & -1 & 0 & 0 & 0 & 0 \\
-1 & 0 & 0 & 0 & 0 & -1 & -1 & 0
\end{array}\right]
\end{array}.
$$

Then the loop-impedance matrix becomes

$$
\mathbf{Z}_m =
\begin{bmatrix}
0 & 0 & 0 & 1 & 1 & 0 & 0 \\
1 & 1 & 1 & 0 & 0 & -1 & 0 \\
1 & 1 & 1 & 0 & 0 & 0 & 0 \\
-1 & 0 & 0 & 0 & -1 & -1 & 0
\end{bmatrix}
\begin{bmatrix}
sL_{11} & sL_{12} & & & & & & \\
sL_{21} & sL_{22} & & & & & & \\
& & R_3 & & & & & \\
& & & R_4 & & & & \\
& & & & R_5 & & & \\
& & & & & \dfrac{1}{sC_6} & & \\
& & & & & & \dfrac{1}{sC_7} & \\
& & & & & & & \dfrac{1}{sC_8}
\end{bmatrix}
\begin{bmatrix}
0 & 0 & 1 & 1 & -1 \\
0 & 0 & 1 & 1 & 0 \\
0 & 0 & 1 & 1 & 0 \\
1 & 1 & 0 & 1 & 0 \\
1 & -1 & 0 & 0 & 0 \\
1 & 0 & 0 & 0 & -1 \\
0 & 0 & 0 & 0 & -1 \\
0 & -1 & 0 & 0 & 0
\end{bmatrix}
$$

$$
\mathbf{Z}_m =
\begin{bmatrix}
R_4+R_5+\dfrac{1}{sC_6} & 0 & -R_4 & -\dfrac{1}{sC_6} \\[2ex]
0 & s(L_{11}+L_{12}+L_{21}+L_{22})+R_3+\dfrac{1}{sC_8} & s(L_{11}+L_{12}+L_{21}+L_{22})+R_3 & -s(L_{11}+L_{21}) \\[2ex]
-R_4 & s(L_{11}+L_{12}+L_{21}+L_{22})+R_3 & s(L_{11}+L_{12}+L_{21}+L_{22})+R_3+R_4 & -s(L_{11}+L_{21}) \\[2ex]
-\dfrac{1}{sC_6} & -s(L_{11}+L_{12}) & -s(L_{11}+L_{12}) & sL_{11}+\dfrac{1}{sC_6}+\dfrac{1}{sC_7}
\end{bmatrix}
$$

107

By relating this matrix to the graph, it is observed that the elements of the loop-impedance matrix can be interpreted in the following straightforward way. Each term on the main diagonal is the sum of the impedances of the branches on the corresponding loop, with due regard to the impedance coupled in from other loops by mutual coupling. Each off-diagonal term is plus or minus the impedance of branches common between two loops; the sign is positive if the loop currents traverse the common branch with the same orientation, and negative if they traverse the common branch with opposite orientations. Verify the loop-impedance matrix for the example by using this interpretation.

A similar interpretation applies to the loop-parameter matrices $\mathbf{L}_m$, $\mathbf{R}_m$, and $\mathbf{D}_m$. Thus from the $\mathbf{Z}_m$ matrix we can write the loop-resistance matrix as follows:

$$\mathbf{R}_m = \begin{bmatrix} R_4 + R_5 & 0 & -R_4 & 0 \\ 0 & R_3 & R_3 & 0 \\ -R_4 & R_3 & R_3 + R_4 & 0 \\ 0 & 0 & 0 & 0 \end{bmatrix}.$$

From the network we observe that the main diagonal elements in this matrix are simply the total resistance on the contour of the corresponding loop; the off-diagonal elements are plus or minus the resistance common to the corresponding loops: plus if the orientations of the two loops are the same through the common resistance, minus if they are opposite.

The source vectors will be

$$\mathbf{V}_g = \begin{bmatrix} 0 \\ 0 \\ 0 \\ 0 \\ V_0 \\ 0 \\ 0 \\ 0 \end{bmatrix} \quad \text{and} \quad \mathbf{I}_g = \begin{bmatrix} 0 \\ 0 \\ 0 \\ 0 \\ 0 \\ 0 \\ I_0 \\ 0 \end{bmatrix}.$$

Then $\mathbf{ZI}_g$ has a nonzero entry only in the seventh row, and its value is $-I_0/sC_7$. Hence the right-hand side of (80) becomes

$$
\mathbf{E} = \mathbf{B}(\mathbf{V}_g - \mathbf{ZI}_g) =
\begin{bmatrix}
0 & 0 & 0 & 1 & 1 & 1 & 0 & 0 \\
1 & 1 & 1 & 0 & 0 & 0 & 0 & -1 \\
1 & 1 & 1 & -1 & 0 & 0 & 0 & 0 \\
-1 & 0 & 0 & 0 & 0 & -1 & -1 & 0
\end{bmatrix}
\begin{bmatrix}
0 \\ 0 \\ 0 \\ 0 \\ V_0 \\ 0 \\ -I_0/sC_7 \\ 0
\end{bmatrix}
$$

$$
=
\begin{bmatrix}
V_0 \\ 0 \\ 0 \\ I_0/sC_7
\end{bmatrix}.
$$

The quantity I_0/sC_7 is the Thévenin equivalent voltage of the current source in parallel with C_7. Thus the quantity $\mathbf{E}$ is the equivalent *loop voltage-source vector* whose entries are the algebraic sums of voltage sources (including Thévenin equivalents of current sources) on the contour of the corresponding loop, with the references chosen so that they are opposite to the corresponding loop reference.

Once the loop equations have been obtained in the form

$$\mathbf{Z}_m \mathbf{I}_m = \mathbf{E} \tag{84}$$

the solution is readily obtained as

$$\mathbf{I}_m = \mathbf{Z}_m^{-1}\mathbf{E} \tag{85}$$

This, of course, is essentially a symbolic solution in matrix form. The actual solution for the elements of $\mathbf{I}_m$ requires a considerable amount of

further work. We shall postpone until the next chapter further consideration of this subject.

In reviewing the preceding discussion of loop equations it should be
noted that, at least for the example considered, less effort will be required
to write the final loop equations if a straightforward scalar approach is
used. In fact, it is possible to write the loop-impedance matrix $\mathbf{Z}_m$ and
the equivalent source matrix $\mathbf{E}$ by no more than inspection of the network,
once a set of loops is chosen. We seem to have introduced a matrix procedure that is more complicated than necessary. Three comments are
appropriate to this point. In the first place, the general approach discussed
here should not be used for writing loop equations for networks with a
$\mathbf{B}$ matrix of relatively low rank. The general procedure becomes preferable
when dealing with networks having large $\mathbf{B}$ matrices—with tens of rows.
Secondly, the general approach using topological relationships is amenable
to digital computation, which makes it quite valuable. Finally, the
general form constitutes an "existence theorem"; it is a verification
that loop equations can always be written for the networks under consideration.

NODE EQUATIONS

In writing loop equations, the branch v-i relationships were inserted
into the KVL equations, after which a loop transformation was used to
transform to loop-current variables. Now, given a network, let us first
apply KCL, arriving at (67), which is repeated here:

$$\mathbf{AI}(s) = \mathbf{AI}_g(s). \tag{86}$$

Into this expression we next insert the branch relations in (73), getting

$$\mathbf{AY}(s)\,\mathbf{V}(s) = \mathbf{AI}_g(s). \tag{87}$$

Finally, we express the branch voltages in terms of the node voltages
through the node transformation in (72). The result is

$$\mathbf{AY}(s)\mathbf{A}'\mathbf{V}_n(s) = \mathbf{A}\{\mathbf{I}_g(s) - \mathbf{YV}_g(s)\} \tag{88a}$$

or

$$\mathbf{Y}_n\mathbf{V}_n(s) = \mathbf{J}(s) \tag{88b}$$

where $\mathbf{J}$ is shorthand for $\mathbf{A}(\mathbf{I}_g - \mathbf{Y}\mathbf{V}_g)$ and where

$$\mathbf{Y}_n(s) = \mathbf{A}\mathbf{Y}(s)\mathbf{A}'. \tag{89}$$

This matrix equation represents a set of n equations, called the *node equations*, in the n node-voltage variables. The coefficient matrix $\mathbf{Y}_n(s)$ is called the *node-admittance matrix*.

This time the right-hand side, $\mathbf{J}$, is the equivalent *node current-source vector* whose entries are algebraic sums of current sources (including the Norton equivalents of voltage sources) incident at the corresponding nodes, with the references chosen so that they enter the node.

The node-admittance matrix can be written explicitly in terms of the branch-parameter matrices by inserting (77) into (89). The result will be

$$\mathbf{Y}_n(s) = \mathbf{A}\mathbf{Y}(s)\mathbf{A}' = s\mathbf{C}_n + \mathbf{G}_n + \frac{1}{s}\mathbf{\Gamma}_n \tag{90}$$

where

$$\mathbf{C}_n = \mathbf{A}\mathbf{C}\mathbf{A}' \tag{91a}$$

$$\mathbf{G}_n = \mathbf{A}\mathbf{G}\mathbf{A}' \tag{91b}$$

$$\mathbf{\Gamma}_n = \mathbf{A}\mathbf{\Gamma}\mathbf{A}' \tag{91c}$$

are the *node-parameter matrices*.

Once the node equations are available in the form

$$\mathbf{Y}_n(s)\mathbf{V}_n(s) = \mathbf{J}(s) \tag{92}$$

the solution is readily obtained by inverting:

$$\mathbf{V}_n(s) = \mathbf{Y}_n^{-1}\mathbf{J}(s). \tag{93}$$

Again, this is essentially a symbolic solution. In the next chapter we shall consider the details of the solution.

Let us now illustrate the use of node equations with the example of

Fig. 18, which is redrawn as Fig. 20. Let node 5 be chosen as the datum node and as the node that is omitted in writing the **A** matrix.

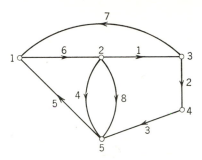

Fig. 20.   Illustrative example for node equations.

The **A** matrix and the branch-admittance matrix will be

$$
\mathbf{A} =
\begin{bmatrix}
0 & 0 & 0 & 0 & -1 & 1 & -1 & 0 \\
1 & 0 & 0 & 1 & 0 & -1 & 0 & 1 \\
-1 & 1 & 0 & 0 & 0 & 0 & 1 & 0 \\
0 & -1 & 1 & 0 & 0 & 0 & 0 & 0
\end{bmatrix}
$$

$$
\mathbf{Y} =
\begin{bmatrix}
\dfrac{L_{22}}{s\Delta} & -\dfrac{L_{12}}{s\Delta} & & & & & & \\[2ex]
-\dfrac{L_{21}}{s\Delta} & \dfrac{L_{11}}{s\Delta} & & & \bigcirc & & & \\[2ex]
 & & G_3 & & & & & \\
 & & & G_4 & & & & \\
 & & & & G_5 & & & \\
 & \bigcirc & & & & sC_6 & & \\
 & & & & & & sC_7 & \\
 & & & & & & & sC_8
\end{bmatrix}
$$

where $\Delta = L_{11}L_{22} - L_{12}L_{21}$. Then

$$\mathbf{Y}_n = \mathbf{AYA'} = \begin{bmatrix} G_5 + s(C_6 + C_7) & -sC_6 & -sC_7 & 0 \\[2mm] -sC_6 & \dfrac{L_{22}}{s\Delta} + G_4 + s(C_6 + C_8) & -\dfrac{(L_{22}+L_{12})}{s\Delta} & \dfrac{L_{12}}{s\Delta} \\[3mm] -sC_7 & -\dfrac{(L_{22}+L_{21})}{s\Delta} & \dfrac{L_{11}+2L_{12}+L_{21}+L_{22}}{s\Delta} + sC_7 & -\dfrac{(L_{11}+L_{12})}{s\Delta} \\[3mm] 0 & \dfrac{L_{21}}{s\Delta} & -\dfrac{(L_{11}+L_{21})}{s\Delta} & \dfrac{L_{11}}{s\Delta} + G_3 \end{bmatrix} .$$

The source matrices $\mathbf{V}_g$ and $\mathbf{I}_g$ are the same as before. Hence we get

$$\mathbf{J} = \mathbf{A}(\mathbf{I}_g - \mathbf{Y}\mathbf{V}_g) = \begin{bmatrix} G_5 V_0 - I_0 \\ 0 \\ I_0 \\ 0 \end{bmatrix}.$$

The quantity $G_5 V_0$ is the Norton equivalent of the voltage source in series with G_5. Thus, $\mathbf{J}$ is the equivalent current-source vector whose elements are the algebraic sum of current sources (including Norton equivalents of accompanied voltage sources) incident at the corresponding node, with the references chosen to be directed toward the node.

As in the case of loop equations, the node equations can be written directly from the network by inspection, for networks without mutual coupling. The elements of the node-admittance matrix can be found as follows. Each element on the main diagonal is the sum of the admittances of branches incident at the corresponding node. Each off-diagonal element is the negative of the admittance common between two nodes. In this case all signs of off-diagonal terms are negative, unlike the loop-impedance case, because the voltage of a branch is always the difference between two node voltages, since the node voltage references are uniformly positive relative to the datum node.

A similar interpretation applies to the node-parameter matrices $\mathbf{C}_n$, $\mathbf{G}_n$, and $\mathbf{\Gamma}_n$. Let us, for example, construct the node-capacitance matrix. From the diagram, there are two capacitors, C_4 and C_7, incident at node 1, C_6 being common between nodes 1 and 2, and C_7 being common between nodes 1 and 3. Hence the diagonal term in the first row of $\mathbf{C}_n$ will be $C_6 + C_7$, and the three off-diagonal terms will be $-C_6$, $-C_7$, and 0. Continuing in this fashion, $\mathbf{C}_n$ is found to be

$$\mathbf{C}_n = \begin{bmatrix} C_6 + C_7 & -C_6 & -C_7 & 0 \\ -C_6 & C_6 + C_8 & 0 & 0 \\ -C_7 & 0 & C_7 & 0 \\ 0 & 0 & 0 & 0 \end{bmatrix}.$$

This agrees with the node-capacitance matrix obtained from the previously found $\mathbf{Y}_n$ matrix.

NODE-PAIR EQUATIONS

The variables in terms of which the node equations are written are the voltages of the nodes all related to a datum node. This set of variables is a basis for all branch variables. It was observed earlier that the twig voltages for a given tree also constitute a basis for all branch voltages. Hence we should expect the possibility of another set of equations similar to the node equations, but with twig voltages for a tree as the variables; this expectation is fulfilled.

Given a network, the first task is to select a tree and to apply KCL to the fundamental cut sets, arriving at (69), which is repeated here:

$$\mathbf{QI}(s) = \mathbf{QI}_g(s). \tag{94}$$

Here $\mathbf{Q}$ is of order $n \times b$ and of rank n. (The subscript f is omitted for simplicity.) Into this expression we next insert the branch relations of (73) getting

$$\mathbf{QY}(s)\mathbf{V}(s) = \mathbf{QI}_g(s). \tag{95}$$

Finally, we express the branch voltages in terms of twig voltages through the transformation in (71). The result is

$$\mathbf{QY}(s)\mathbf{Q}'\mathbf{V}_t(s) = \mathbf{Q}\{\mathbf{I}_g - \mathbf{Y}(s)\mathbf{V}_g\} \tag{96a}$$

or

$$\mathbf{Y}_t(s)\mathbf{V}_t(s) = \mathbf{J}_t \tag{96b}$$

where $\mathbf{J}_t$ is simply shorthand for $\mathbf{Q}\{\mathbf{I}_g - \mathbf{Y}(s)\mathbf{V}_g\}$ and

$$\mathbf{Y}_t(s) = \mathbf{QY}(s)\mathbf{Q}'. \tag{97}$$

Note that this expression is quite similar to the node equations given in (88), the difference being that the f-cut-set matrix $\mathbf{Q}$ replaces the incidence matrix $\mathbf{A}$, and the variables here are not node voltages but *node-pair* voltages. We shall call these equations the *node-pair equations*.

The coefficient matrix of the node-pair equations $\mathbf{Y}_t(s)$ is called the *node-pair admittance* matrix; it can be written explicitly in terms of the branch-parameter matrices by inserting (77). The result will be

$$\mathbf{Y}_t(s) = \mathbf{QY}(s)\mathbf{Q}' = s\mathbf{C}_t + \mathbf{G}_t + \frac{1}{s}\mathbf{\Gamma}_t \tag{98}$$

where

$$\mathbf{C}_t = \mathbf{QCQ'} \tag{99a}$$

$$\mathbf{G}_t = \mathbf{QGQ'} \tag{99b}$$

$$\mathbf{\Gamma}_t = \mathbf{Q\Gamma Q'} \tag{99c}$$

are the node-pair parameter matrices.

The same example will be used (Fig. 18) to illustrate the node-pair equations as used earlier for the loop and node equations, except that it will be assumed there is no mutual coupling between branches 1 and 2. The diagram is repeated here as Fig. 21. The tree consisting of branches

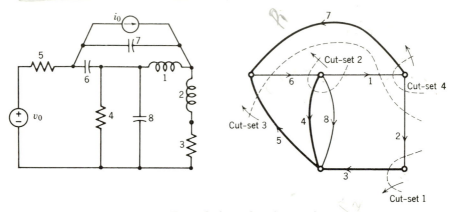

Fig. 21.   Example for node-pair equations.

3, 4, 5, and 7 shown in heavy lines is selected. The branches in each cut set and the f-cut-set matrix for this tree are as follows:

cut-set 1: $\{3, 2\}$

2: $\{4, 1, 6, 8\}$

3: $\{5, 1, 2, 6\}$

4: $\{7, 1, 2\}$

$$\mathbf{Q} = \begin{array}{c} 3 \\ 4 \\ 5 \\ 7 \end{array} \begin{bmatrix} 0 & -1 & 1 & 0 & 0 & 0 & 0 & 0 \\ 1 & 0 & 0 & 1 & 0 & -1 & 0 & 1 \\ 1 & -1 & 0 & 0 & 1 & -1 & 0 & 0 \\ -1 & 1 & 0 & 0 & 0 & 0 & 1 & 0 \end{bmatrix}.$$

The order of the columns is the same as the original numbering of the branches, not the order for which $\mathbf{Q}$ can be partitioned into $[\mathbf{U} \quad \mathbf{Q}_l]$. The reason for this is that the branch-admittance matrix was already written for that order when the node equations were written. Refer back to the branch-admittance matrix and note that L_{22}/Δ is now replaced by $1/L_1$, and L_{11}/Δ by $1/L_2$; also the off-diagonal elements are zero, since there is no mutual coupling. The node-pair equations are found to be

$$\mathbf{QYQ'V}_t = \begin{bmatrix} G_3 + \dfrac{1}{L_2 s} & 0 & \dfrac{1}{L_2 s} & -\dfrac{1}{L_2 s} \\[2ex] 0 & G_4 + \dfrac{1}{L_1 s} + s(C_6 + C_8) & \dfrac{1}{L_1 s} + sC_6 & -\dfrac{1}{L_1 s} \\[2ex] \dfrac{1}{L_2 s} & \dfrac{1}{L_1 s} + sC_6 & G_5 + \dfrac{1}{L_1 s} + \dfrac{1}{L_2 s} + sC_6 & -\left(\dfrac{1}{L_1 s} + \dfrac{1}{L_2 s}\right) \\[2ex] -\dfrac{1}{L_2 s} & -\dfrac{1}{L_1 s} & -\left(\dfrac{1}{L_1 s} + \dfrac{1}{L_2 s}\right) & sC_7 + \dfrac{1}{L_1 s} + \dfrac{1}{L_2 s} \end{bmatrix} \begin{bmatrix} V_3 \\[2ex] V_4 \\[2ex] V_5 \\[2ex] V_7 \end{bmatrix} = \begin{bmatrix} 0 \\[2ex] 0 \\[2ex] -G_5 V_0 \\[2ex] I_0 \end{bmatrix}$$

Again we find that a simple interpretation can be given to the cut-set admittance matrix. By observing the network in Fig. 21, we see, for example, that $sC_7 + 1/L_1s + 1/L_2s$, which is the (4, 4) term in $\mathbf{Y}_t$, is the sum of the admittances of the branches in cut-set 4. Similar interpretations apply to the other diagonal elements. We also observe that some of the off-diagonal elements have positive signs, others, negative signs; for example, the (1, 3) element in $\mathbf{Y}_t$ is $1/L_2s$. This is seen to be the admittance of a branch common to cut-sets 1 and 3. The orientation of this common branch is the same relative to both cut-sets, so the sign of the term is positive.

As a general rule, the elements of the cut-set admittance matrix $\mathbf{Y}_t$ have the following interpretations. Each diagonal element is the sum of the admittances of branches which are in the corresponding cut-set. Each off-diagonal term is plus or minus the admittance of a branch common to two cut-sets. The sign is plus if the branch orientation is the same relative to both cut-sets, minus if it is not the same. You are urged to verify the $\mathbf{Y}_t$ matrix of the example by using this interpretation.

As for the source term, $\mathbf{Q}(\mathbf{I}_g - \mathbf{YV}_g)$, this is the equivalent *cut-set current-source vector*, each of whose entries is the algebraic sum of source currents (including Norton equivalent of voltage sources) that lie in the corresponding cut-set.

2.5   DUALITY

There is a striking parallelism between the loop and node systems of equations. This observation raises the following interesting question. Is it possible to find two networks such that the loop equations for one network are the same as the node equations of the other, except for the symbols? In other words, can the loop equations for one network become the node equations for the other if we interchange the symbols v and i throughout? To answer this question, note that the loop equations result when the branch relations are substituted into KVL, and then KCL is used (in the form of the loop transformation). On the other hand, the node equations result when this order is reversed; that is, the branch relations are inserted into KCL, and then KVL is used (in the form of the node transformation). On this basis we see that the question can be answered affirmatively if two networks N_1 and N_2 exist that satisfy the following two conditions:

1. The KCL equations of N_1 are a suitable set of KVL equations for N_2 on replacing i_j by v_j for all j.

2. The expression for branch voltage v_j of N_2 in terms of branch current i_j becomes the expression for branch current i_j of N_1 in terms of branch voltage v_j on interchanging i_j and v_j.

If these conditions are satisfied, N_2 is said to be the *dual* of N_1. In fact, it is easy to see that if N_1 and N_2 are interchanged, the above conditions will still be satisfied (if they were originally satisfied). Hence N_1 is also the dual of N_2. The property of duality is a mutual property; N_1 and N_2 are *dual networks*.

In matrix form, condition 1 can be stated as follows: Let $\mathbf{A}_1 = [a_{ij}]$ be the incident matrix of N_1. Then

$$\mathbf{A}_1 = \mathbf{B}_2 \tag{100}$$

where $\mathbf{B}_2$ is a loop matrix of N_2. Clearly, the number of branches of the two networks must be equal, and the rank of $\mathbf{A}_1$ must equal the rank of $\mathbf{B}_2$. Thus

$$b_1 = b_2 \tag{101a}$$

$$n_1 = b_2 - n_2 \tag{101b}$$

where b_1 and b_2 refer to the number of branches; $n_1 + 1$ and $n_2 + 1$ refer to the number of nodes of the two networks, respectively.

Evidently these relationships constitute conditions on the structure of the two networks. First, there must be a correspondence between the branches of the two networks, as defined by the ordering of the columns in the matrices $\mathbf{A}_1$ and $\mathbf{B}_2$ to satisfy (100). Secondly, there must be a correspondence between the nodes of N_1 (rows of $\mathbf{A}_1$) and the loops of N_2 (rows of $\mathbf{B}_2$).

Two structures that are related by (100) are called *dual graphs*. We shall not discuss the abstract properties of dual graphs here but shall state some of the simpler results.*

The basic result is that a network will have a geometrical (structural) dual if and only if it is planar. If two planar networks can be superimposed such that each junction but one of N_1 lies inside a mesh of N_2 and the references of corresponding branches are suitably oriented, then a row of $\mathbf{A}_1$ and the corresponding row of $\mathbf{B}_2$ will be identical. Such a loop and node are shown in Fig. 22. The branches and loops of N_2 are primed, and the branches and nodes of N_1 are unprimed. Node 1 in N_1 corresponds to loop $1'$ in N_2. You should verify that the KCL equation at node 1 has the same coefficients as those of the KVL equation for loop $1'$. The coefficient of i_2 in the KCL equation at node 2 is $+1$. In order to make the

---

* For a more detailed account, see H. Whitney, "Nonseparable and Planar Graphs," *Trans. Amer. Math. Soc.*, vol. 34, No. 2, pp. 339–362, 1932, and C. Kuratowski, "Sur le probleme des courbes gauches en topologie," *Fundamenta Mathematicae*, vol. 15, pp. 271–283, 1930.

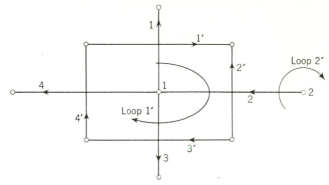

Fig. 22. Dual loop and node.

coefficient of v_2 in the KVL equation for loop $2'$ the same, namely $+1$, loop $2'$ must be oriented as shown. By following through the entire graph in the same manner, it can be seen that all the loops must be oriented in the same sense (all clockwise if the branch references are as shown in the figure, or all counterclockwise if the branch references are reversed).

If the meshes of a planar graph are chosen for writing KVL equations, and all loops are oriented in the same sense, then the off-diagonal terms in the loop equations will all carry negative signs, just as they do in node equations.

The second condition of duality has to do with the branch relationships. Figure 23 shows dual pairs of v-i relationships. For mutual inductance there is no dual relationship. From the definition, then, only planar networks without mutual inductance have duals.

Given such a network, say N_1, construction of the dual network

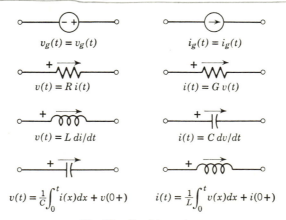

$$v_g(t) = v_g(t) \qquad\qquad i_g(t) = i_g(t)$$

$$v(t) = R\,i(t) \qquad\qquad i(t) = G\,v(t)$$

$$v(t) = L\,di/dt \qquad\qquad i(t) = C\,dv/dt$$

$$v(t) = \frac{1}{C}\int_0^t i(x)dx + v(0+) \qquad i(t) = \frac{1}{L}\int_0^t v(x)dx + i(0+)$$

Fig. 23. Dual branches.

proceeds as follows. Within each mesh of N_1 we place a node of what will
be the dual network N_2. An additional node, which will be the datum
node, is placed outside N_1. Across each branch of N_1 is placed the dual
branch joining the two nodes located inside the two meshes to which that
particular branch of N_1 is common.* Finally, the branch references of N_2
are chosen so that the matrix of KVL equations for N_1 (with all loops
similarly oriented) is the same as the matrix of KCL equations for N_2.
Because the two networks are mutually dual, the node-admittance
matrix of one network equals the loop-impedance matrix of the other,
and vice versa; that is,

$$\mathbf{Y}_{n1} = \mathbf{Z}_{m2} \qquad (102a)$$

and

$$\mathbf{Z}_{m1} = \mathbf{Y}_{n2}. \qquad (102b)$$

Since these matrices are equal, their determinants and cofactors will be
equal.

As an illustration, consider the diagram in Fig. 24a. A node is placed

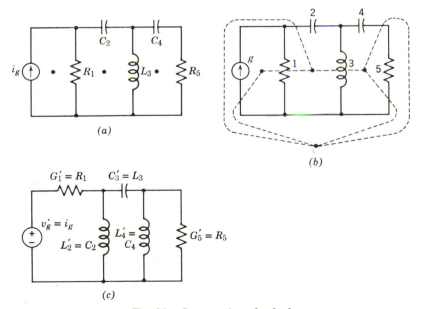

Fig. 24. Construction of a dual.

* In this process it is convenient to consider the sources as separate branches for the
purpose of constructing the dual.

within each mesh, and an additional one is placed outside. In part (*b*) of the figure, dashed lines are used to represent dual branches crossing each of the branches of N_1. Finally, the dual is shown in part (*c*). You should verify (100) and (102) for this example.

2.6   NONRECIPROCAL AND ACTIVE NETWORKS

The coefficient matrices of the loop equations and the node equations are, respectively, **BZB′** and **AYA′**. The success of carrying out a loop analysis or node analysis in this general form, then, depends on the existence of a branch-impedance matrix **Z** or a branch-admittance matrix **Y**. For the passive, reciprocal networks dealt with up to this point, both of these matrices exist. (This statement must be qualified for a perfectly coupled transformer, for which **Y** does not exist.)

Now we shall consider networks containing active and/or nonreciprocal devices, in addition to passive ones having more than two terminals. Table 1 shows such components, together with their representations.

There are two points to consider when dealing with these components. One has to do with how the graphs of such multiterminal components are to be represented. This will influence the number of KCL and KVL equations and, hence, matrices **A** and **B**. The other point concerns the existence of a branch representation that can be used in a loop or node analysis. We shall initially take up the former point and consider the graphs of the components.

Each of the components shown in Table 1 has four terminals. However, the terminals are always taken in pairs, so that it is more appropriate to consider them as having two pairs of terminals. It is possible, of course, to identify (connect together) one terminal from each pair without influencing the *v-i* relationships of the components. Thus each component can be looked upon as a three-terminal component. The behavior of each component in the table is specified by two relationships among two pairs of variables—two currents and two voltages. This is a special case of a general condition; namely, that the behavior of an *n*-terminal component can be completely specified in terms of $n-1$ relationships among $n-1$ pairs of voltage and current variables. (This condition is, in effect, a postulate and, as such, not susceptible to proof.)

For a component with one pair of terminals, described by one voltage and one current, the graph is represented by a single branch. The components in Table 1 have two pairs of terminals, and two voltages and currents; their graph will be represented by two branches across the

Table 1

Device	Symbol	Equations	Type of Representation[a]	Graph
Voltage-controlled current-source		$\begin{bmatrix} i_1 \\ i_2 \end{bmatrix} = \begin{bmatrix} 0 & 0 \\ g_{21} & 0 \end{bmatrix} \begin{bmatrix} v_1 \\ v_2 \end{bmatrix}$	Admittance	
Current-controlled voltage-source		$\begin{bmatrix} v_1 \\ v_2 \end{bmatrix} = \begin{bmatrix} 0 & 0 \\ r_{21} & 0 \end{bmatrix} \begin{bmatrix} i_1 \\ i_2 \end{bmatrix}$	Impedance	
Voltage-controlled voltage-source		$\begin{bmatrix} i_1 \\ v_2 \end{bmatrix} = \begin{bmatrix} 0 & 0 \\ \mu_{21} & 0 \end{bmatrix} \begin{bmatrix} v_1 \\ i_2 \end{bmatrix}$	Hybrid g	
Current-controlled current-source		$\begin{bmatrix} v_1 \\ i_2 \end{bmatrix} = \begin{bmatrix} 0 & 0 \\ \alpha_{21} & 0 \end{bmatrix} \begin{bmatrix} i_1 \\ v_2 \end{bmatrix}$	Hybrid h	
Gyrator		$\begin{bmatrix} v_1 \\ v_2 \end{bmatrix} = \begin{bmatrix} 0 & \mp r \\ \pm r & 0 \end{bmatrix} \begin{bmatrix} i_1 \\ i_2 \end{bmatrix}$ or $\begin{bmatrix} i_1 \\ i_2 \end{bmatrix} = \begin{bmatrix} 0 & \pm g \\ \mp g & 0 \end{bmatrix} \begin{bmatrix} v_1 \\ v_2 \end{bmatrix}$	Impedance or admittance	
Negative converter		$\begin{bmatrix} v_1 \\ i_2 \end{bmatrix} = \begin{bmatrix} 0 & \pm k \\ \pm 1/k & 0 \end{bmatrix} \begin{bmatrix} i_1 \\ v_2 \end{bmatrix}$ or $\begin{bmatrix} i_1 \\ v_2 \end{bmatrix} = \begin{bmatrix} 0 & \pm 1/k \\ \pm 1/k & 0 \end{bmatrix} \begin{bmatrix} v_1 \\ i_2 \end{bmatrix}$	Hybrid h or Hybrid g	
Ideal transformer		$\begin{bmatrix} v_1 \\ i_2 \end{bmatrix} = \begin{bmatrix} 0 & n \\ -n & 0 \end{bmatrix} \begin{bmatrix} i_1 \\ v_2 \end{bmatrix}$ $\begin{bmatrix} i_1 \\ v_2 \end{bmatrix} = \begin{bmatrix} 0 & -1/n \\ 1/n & 0 \end{bmatrix} \begin{bmatrix} v_1 \\ i_2 \end{bmatrix}$	Hybrid h or Hybrid g	

[a] Each of these components also has a chain matrix representation.

pairs of terminals, as shown in Fig. 25. If one terminal from each pair is

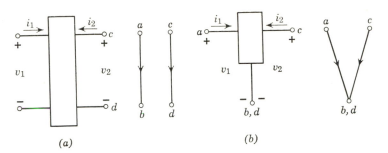

Fig. 25. Two-port and its terminal graph.

a common terminal, the corresponding branches of the graph will also have a common terminal.

The above discussion has related to general three- and four-terminal components. But the components in Table 1 have certain degeneracies, as shown by the number of zeros in their characterizing equations. For the controlled sources the input pair of terminals is either open or shorted. It will support either an arbitrary voltage with zero current or an arbitrary current with zero voltage. There is a row of zeros in the v-i relationship. This means that only one representation exists for each type of component. To write loop (node) equations, however, requires an impedance (admittance) representation. If loop (node) equations are to be written for networks containing any of these components, it will, therefore, be necessary to make some preliminary adjustments.

This can be done in the manner illustrated by the following example. The subnetwork of Fig. 26 contains a controlled source. The graph of the subnetwork is also shown, the heavy lines representing the two branches of the controlled source. Branch 3 of the graph does not even physically appear in the network; there, it is simply an open circuit. However, the voltage of this branch, v_3, is nonzero; and, since it appears explicitly in the branch equations of the component, it cannot simply be disregarded. But observe that v_3 can be expressed in terms of other branch voltages; in this example, $v_3 = v_1 - v_2$. Hence, if this relationship is inserted for v_3 in the branch relationship, neither v_3 nor i_3 will appear any longer. Consequently, branch 3 can simply be removed from the graph. Thus the controlled source is now represented by a single branch (branch 4) in the graph, and its branch equation becomes

$$i_4 = g_m v_3 = g_m v_1 - g_m v_2. \tag{103}$$

$$
\begin{bmatrix} i_3 \\ i_4 \end{bmatrix} = \begin{bmatrix} 0 & 0 \\ g_m & 0 \end{bmatrix} \begin{bmatrix} v_3 \\ v_4 \end{bmatrix}
$$

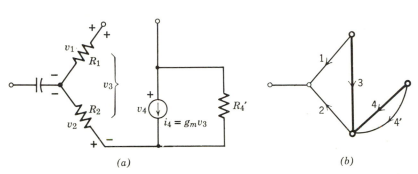

(a)   (b)

Fig. 26. Subnetwork with v-controlled i-source.

The preceding example dealt with a single component of the seven listed in Table 1. Since each of them has a different type of representation, somewhat different approaches are appropriate for each. Let us now examine them systematically to see how an impedance and/or admittance representation can be obtained for each.

First consider the gyrator; it has both an admittance and an impedance representation. Hence there need be no special techniques for carrying out a loop or a node analysis. As an illustration, consider the network containing a gyrator shown in Fig. 27. The graph of the gyrator will

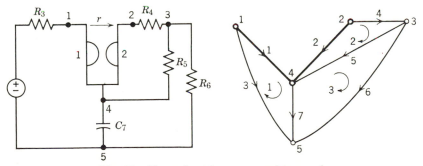

Fig. 27. Network with gyrator and its graph.

have the two branches shown in heavy lines. The branch-impedance matrix of the entire network, including the gyrator, is easily written.

With the branch and loop numbering schemes shown, the **B** matrix and the branch-impedance matrix become the following:

$$
\mathbf{Z} =
\begin{bmatrix}
0 & -r & 0 & 0 & 0 & 0 & 0 \\
r & 0 & 0 & 0 & 0 & 0 & 0 \\
0 & 0 & R_3 & 0 & 0 & 0 & 0 \\
0 & 0 & 0 & R_4 & 0 & 0 & 0 \\
0 & 0 & 0 & 0 & R_5 & 0 & 0 \\
0 & 0 & 0 & 0 & 0 & R_6 & 0 \\
0 & 0 & 0 & 0 & 0 & 0 & 1/sC_7
\end{bmatrix}
$$

and

$$
\mathbf{B} =
\begin{bmatrix}
1 & 0 & -1 & 0 & 0 & 0 & 1 \\
0 & -1 & 0 & 1 & 1 & 0 & 0 \\
0 & 0 & 0 & 0 & -1 & 1 & -1
\end{bmatrix}.
$$

From these the loop-impedance matrix follows:

$$
\mathbf{Z}_m = \mathbf{BZB'} =
\begin{bmatrix}
R_3 + 1/sC_7 & r & -1/sC_7 \\
-r & R_4 + R_5 & -R_5 \\
-1/sC_7 & -R_5 & R_5 + R_6 + R_7
\end{bmatrix}.
$$

(Verify these relationships.) Notice how the gyrator parameter enters to make both the branch- and loop-impedance matrices nonsymmetric.

Since the gyrator also has an admittance representation, node equations can be readily written. With node 5 as the datum node, the incidence matrix, the branch-admittance matrix, and, from these, the node-admittance matrix are

$$
\mathbf{A} =
\begin{bmatrix}
1 & 0 & 1 & 0 & 0 & 0 & 0 \\
0 & 1 & 0 & 1 & 0 & 0 & 0 \\
0 & 0 & 0 & -1 & 1 & 1 & 0 \\
-1 & -1 & 0 & 0 & -1 & 0 & 1
\end{bmatrix}
$$

$$\mathbf{Y} = \begin{bmatrix} 0 & g & & & & & \\ -g & 0 & & & & & \\ & & G_3 & & & \bigcirc & \\ & & & G_4 & & & \\ & & \bigcirc & & G_5 & & \\ & & & & & G_6 & \\ & & & & & & sC_7 \end{bmatrix}$$

$$\mathbf{Y}_n = \mathbf{AYA}' = \begin{bmatrix} G_3 & g & 0 & -g \\ -g & G_4 & -G_4 & g \\ 0 & -G_4 & G_4 + G_5 + G_6 & -G_5 \\ g & -g & -G_5 & G_5 + sC_7 \end{bmatrix}.$$

You are urged to verify these. The presence of a gyrator, then, requires nothing special for writing loop and node equations.

Next let us consider the v-controlled i-source. This was already done in the illustration in Fig. 26. We saw that the input branch of the graph of this device can be eliminated when the controlling voltage is expressed in terms of other branch voltages. The device is then represented by a single branch, and a branch-admittance matrix is easily written, leading to a successful node analysis.

If a loop analysis is required, however, a problem will arise. Loop equations require an impedance representation of the components. In Fig. 26, what is needed is to write an expression for v_4 explicitly in terms of currents. Since v_4 does not appear explicitly in (103), which is the pertinent branch equation, we appear to be at an impasse. However, a remedy can be found if branch $4'$, which is in parallel with 4 in Fig. 26, is lumped together with 4 into a single branch. If i_4 is the current of the combined branch, then the following equation will replace (103):

$$i_4 = g_m v_1 - g_m v_2 + G_4' v_4. \tag{104}$$

Now it is possible to solve for v_4, after each of the other voltages has been eliminated by inserting its branch relationship. Thus

$$v_4 = -g_m R_1 R_4' i_1 + g_m R_2 R_4' i_2 + R_4' i_4. \tag{105}$$

The branch-impedance matrix can now be completed.

What has happened to the graph in carrying out the preceding? The one branch that was left representing the controlled source has been eliminated by combining it with a parallel branch. Clearly, this required that the controlled source be accompanied, which can always be arranged by using the i-shift.

The i-controlled v-source can be handled in a completely dual way. Here, the input branch of the graph of the device can be eliminated when the controlling current is expressed in terms of other branch currents. The branch-impedance matrix can then be written, and a loop analysis can be carried out. This time, however, if a node analysis is required, it will become necessary to combine into a single branch the controlled source and an accompanying branch. An accompanying branch can always be provided by the v-shift. After this step, the branch relation can be inverted, and an admittance representation can be written. The details of this development will be left to you.

This leaves us in Table 1 with four components that have only a hybrid representation. Before an impedance or admittance representation can be found, the branches of these components must be combined with accompanying branches, and the controlling voltages or currents must be expressed in terms of other branch currents.

Let us illustrate these comments by considering the network of Fig. 28,

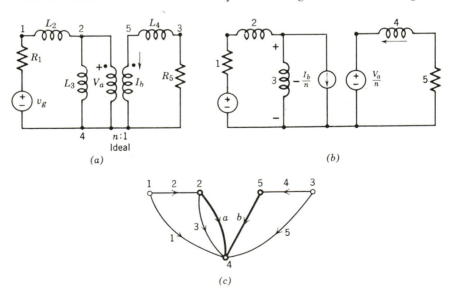

Fig. 28.   Network with ideal transformer, and its graph.

which contains an ideal transformer. The ideal transformer can be replaced by an equivalent consisting of two controlled sources, as shown. This step is not essential; all further steps can be carried out with reference to the graph, without considering this equivalent network. In the graph of the network, the branches representing the transformer are the heavy ones. Note that each of the controlled sources is accompanied. The voltage across the left-hand branch, V_a, is the same as the voltage across its accompanying branch, V_3. Similarly, the current in the right-hand branch, I_b, is the same as the current in its accompanying branch, I_4. The transformer equations are

$$V_b = \frac{V_a}{n} \quad \text{and} \quad I_a = -\frac{I_b}{n} \tag{106}$$

Now let us write the branch relations for the transformer, this time combining them with their accompanying branches. Let the sum of I_a and the current in branch 3 be written I_3, and the sum of V_b and the voltage across branch 4 be V_4. Then

$$V_4 = sL_4 I_4 + \frac{V_a}{n} = sL_4 I_4 + \frac{V_3}{n} \tag{107a}$$

$$I_3 = \frac{1}{sL_3} V_3 - \frac{I_b}{n} = \frac{1}{sL_3} V_3 - \frac{I_4}{n} \tag{107b}$$

These equations can be rewritten to provide either an impedance or an admittance representation. The results are

Impedance Representation	*Admittance Representation*
$V_3 = sL_3 I_3 + \dfrac{sL_3}{n} I_4$	$I_3 = \left(\dfrac{1}{sL_3} + \dfrac{1}{n^2 sL_4} \right) V_3 - \dfrac{V_4}{nsL_4}$
$V_4 = \dfrac{sL_3}{n} I_3 + s\left(\dfrac{L_3}{n^2} + L_4 \right) I_4$	$I_4 = -\dfrac{V_3}{nsL_4} + \dfrac{1}{sL_4} V_4.$

As a consequence of the preceding steps, branches a and b in the graph merge with their accompanying branches and disappear from the graph.

(Node 5 also disappears.) The branch impedance will now be

$$
\mathbf{Z} =
\begin{bmatrix}
R_1 & & & & \\
 & sL_2 & & & \bigcirc \\
 & & sL_3 & \dfrac{sL_3}{n} & \\
 & \bigcirc & \dfrac{sL_3}{n} & s\left(L_4 + \dfrac{L_3}{n^2}\right) & \\
 & & & & R_5
\end{bmatrix}.
\tag{180}
$$

You can complete the problem by determining the loop-impedance matrix $\mathbf{Z}_m$.

The preceding discussion makes it clear that the presence of certain four-terminal devices presents some problems in the writing of loop or node equations, but the problems are not insurmountable. If we insist on writing such equations, it can be done, with some preliminary manipulations. These manipulations involve the lumping of certain branches, with a corresponding effect on the graph. Although the graphs of these devices are represented in general by two branches, we have sometimes found ourselves discarding these branches as soon as we have put them in. The controlling branches of the controlled sources were removed (after appropriately expressing the controlling voltage or current). The controlled branches were also removed from the graph whenever it was desired to carry out an analysis that did not fit the natural representation

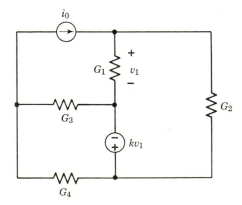

Fig. 29.   Potentially singular impedance matrix.

of the device. In later work, we shall find other ways of dealing with the branches of the graphs of these devices.

The comment should be made that for low-order networks the general systematic formulation we have carried out requires more effort than needed to write loop or node equations using a straightforward scalar approach. Furthermore, the work is increased by the requirement that the components be represented in a fashion that is not natural to them. In the next section we shall consider an alternative approach, which utilizes the hybrid representation directly.

One final point: the presence of nonreciprocal and/or active devices in a network may mean that a unique solution cannot be found for certain values of network parameters. This is illustrated by the network in Fig. 29, which contains a v-controlled v-source. For this network the node-admittance matrix will be singular if $G_1 = (k-1)G_2$. Hence a solution will not exist in this case. Verify this.

2.7 MIXED-VARIABLE EQUATIONS

We observed in the last section that the desire to write loop or node equations for networks containing multiterminal elements meets with some difficulty for certain kinds of elements. The success of writing loop or node equations depends on the existence of an impedance or admittance representation for the branch relationships. When a multiterminal device has no such representation, certain preliminary manipulations and combinations of branches can be performed in order to write the desired equations.

We shall now discuss a scheme that avoids these preliminary manipulations and permits all branch relationships to appear in their natural form. This desirable result must be purchased at a price, however, as will be evident as we proceed.

Of all the elements in our arsenal, some (including R, L, C, and gyrators) have both impedance and admittance representations; one (the v-controlled i-source) has only an admittance; one (the i-controlled v-source) has only an impedance; and some (transformer, NC, v-controlled v-source and i-controlled i-source) have only a mixed, or hybrid, representation. Whatever scheme we come up with must accommodate this fact. We should also recall that twig voltages form a basis set of voltages in terms of which all branch voltages can be expressed. Similarly, link currents form a basis set of currents in terms of which all branch currents can be expressed.

With these thoughts in mind, given a network, we first choose a tree. Instead of expressing the branch relationships as $\mathbf{V} = \mathbf{ZI}$ or $\mathbf{I} = \mathbf{YV}$, we write mixed branch relationships in the following form:

$$\begin{array}{cc} & b-n \quad n \\ \begin{array}{c} b-n \\ n \end{array} & \begin{bmatrix} \mathbf{V}_l \\ \mathbf{I}_t \end{bmatrix} = \begin{bmatrix} \mathbf{Z}_l & \mathbf{H}_{12} \\ \mathbf{H}_{21} & \mathbf{Y}_t \end{bmatrix} \begin{bmatrix} \mathbf{I}_l \\ \mathbf{V}_t \end{bmatrix} \end{array} \quad (109)$$

or

$$\mathbf{V}_l = \mathbf{Z}_l \mathbf{I}_l + \mathbf{H}_{12} \mathbf{V}_t \quad (110a)$$

$$\mathbf{I}_t = \mathbf{H}_{21} \mathbf{I}_l + \mathbf{Y}_t \mathbf{V}_t . \quad (110b)$$

(The orders of the submatrices are as indicated.) The notation for the coefficient matrices is chosen as an aid to the memory. For uniformity the first submatrix in the first row should be written $\mathbf{H}_{11}$, and the second one in the second row should be $\mathbf{H}_{22}$. But since they have the dimensions of admittance and impedance, respectively, the simpler and more suggestive notation was chosen.

In these equations we express twig currents and link voltages in terms of a mixed set of variables consisting of twig voltages and link currents. Here we see the possibility of accommodating hybrid branch representations if the branches having such representation are chosen as links or twigs appropriately. This point will be amplified shortly.

There remain KVL and KCL to apply. Let us assume that KVL is applied to the fundamental loops for the given tree, and KCL is applied at the f-cut-sets. Using (68) and (69) for KVL and KCL; partitioning the $\mathbf{B}$ and $\mathbf{Q}$ matrices in the usual manner, with $\mathbf{B} = [\mathbf{B}_t \quad \mathbf{U}]$ and $\mathbf{Q} = [\mathbf{U} \quad \mathbf{Q}_l]$; and remembering that $\mathbf{B}_t = -\mathbf{Q}'_l$ from (41), we get

$$[\mathbf{B}_t \quad \mathbf{U}]\begin{bmatrix} \mathbf{V}_t \\ \mathbf{V}_l \end{bmatrix} = -\mathbf{Q}'_l \mathbf{V}_t + \mathbf{V}_l = \mathbf{B}\mathbf{V}_g \quad (111a)$$

and

$$[\mathbf{U} \quad \mathbf{Q}_l]\begin{bmatrix} \mathbf{I}_t \\ \mathbf{I}_l \end{bmatrix} = \mathbf{I}_t + \mathbf{Q}_l \mathbf{I}_l = \mathbf{Q}\mathbf{I}_g . \quad (111b)$$

(The f subscript, denoting "fundamental," has been omitted for simplicity.) The first of these can be solved for $\mathbf{V}_l$, the second can be solved for $\mathbf{I}_t$, and the result can be inserted into (110). After rearranging, the result will be

$$\begin{bmatrix} \mathbf{Y}_t & (\mathbf{H}_{21} + \mathbf{Q}_l) \\ (\mathbf{H}_{12} - \mathbf{Q}_l') & \mathbf{Z}_l \end{bmatrix} \begin{bmatrix} \mathbf{V}_t \\ \mathbf{I}_l \end{bmatrix} = \begin{bmatrix} \mathbf{QI}_g \\ \mathbf{BV}_g \end{bmatrix}. \tag{112}$$

This is a set of equations in mixed voltage and current variables; namely, twig voltages and link currents. There are as many variables as there are branches in the graph, which is the price we have to pay. On the right-hand side, the $\mathbf{Q}$ matrix can also be partitioned, and $\mathbf{B}_t$ can be replaced by $-\mathbf{Q}_l'$. Thus, for a given problem, it is enough to form $\mathbf{Q}$ and to write the V-I relationships as in (109). From these, $\mathbf{Q}_l$ and submatrices of the branch-parameter matrix $\mathbf{H}$ are identified; then the final equations in (112) follow.

Note that when multiterminal elements are not present in the network, $\mathbf{H}_{12} = \mathbf{0}$ and $\mathbf{H}_{21} = \mathbf{0}$. Then, after the matrix equation (112) is expanded into two separate equations, substituting the second equation into the first will lead to the node-pair equations; and substituting the first into the second will lead to the loop equations. The truth of this is left for you to demonstrate.

The forms of the equations in (110) are guides to the proper selection of a tree. Any R, L, or C branch has both an impedance and an admittance representation and can be chosen as either a twig or a link. (Refer to Table 1.) The gyrator also has either representation, so its two branches can be either twigs or links. However, since the voltage of one branch is related to the current of the other, both branches must be twigs or links together.

Since the i-controlled v-source has only an impedance representation, both of its branches must be links. Just the opposite is true for the v-controlled i-source; it has only an admittance representation, so both of its branches must be twigs. In the case of the ideal transformer and negative converter, only hybrid representations exist, but there are two possibilities: hybrid h or hybrid g; so that input voltage and output current can be expressed in terms of output voltage and input current, or vice versa. Hence one of the two branches must be a twig, the other a link, and it does not matter which is which. Finally, the v-controlled v-source and the i-controlled i-source each have only one hybrid representation. For the first of these, the controlling branch current is explicitly specified (to be zero), and so the branch must be a twig; the controlled branch voltage is specified, so it must be a link. The situation is just the opposite for the i-controlled i-source. These results are summarized in Table 2.

Table 2

| Device | Type of Representation | Twig or Link | |
		Controlling Branch	Controlled Branch
Gyrator	Admittance or impedance	Twig or link	Twig or link
Voltage-controlled current source	Admittance	Twig	Twig
Current-controlled voltage source	Impedance	Link	Link
Ideal transformer or NC	Hybrid h or g	Twig or link	Link or twig
Voltage-controlled voltage source	Hybrid g	Twig	Link
Current-controlled current source	Hybrid h	Link	Twig

Since there are limitations on the nature of what is permitted (twig or link) for the branches of some of the multiterminal components for some networks, it might be impossible to find a tree that permits each branch to become what Table 2 decrees. In this case the present approach will fail.

However, observe that there was no a priori reason for choosing to write the branch relationships in the form of (109). We can instead write them in the inverse form, as follows:

$$\begin{bmatrix} \mathbf{I}_l \\ \mathbf{V}_t \end{bmatrix} = \begin{bmatrix} G_{11} & G_{12} \\ G_{21} & G_{22} \end{bmatrix} \begin{bmatrix} \mathbf{V}_l \\ \mathbf{I}_t \end{bmatrix}. \tag{113}$$

In this case, the twig and link designations in the last two columns of

Table 2 will have to be reversed. It turns out that the equation corresponding to (112) using the G system will be somewhat more complicated. The details of obtaining this equation will be left as a problem. In any case, if one of these hybrid representations fails, the other one might succeed.*

To illustrate the mixed-variable equations, consider again the network of Fig. 28, which is redrawn here as Fig. 30. A possible tree is shown in the graph by heavy lines. For purposes of comparison, the same branch numbering is made as before, but this is not the natural one for the tree selected; care must be exercised about the ordering of branches when writing the equations. The matrix of f-cut sets, taken in the order 1, a, 4, 5, is found to be

$$
\mathbf{Q} = \begin{array}{c}
 \\ 1 \\ a \\ 4 \\ 5
\end{array}
\begin{array}{cccccccc}
1 & a & 4 & 5 & b & 2 & 3 \\
\end{array}
\left[\begin{array}{cccc:ccc}
1 & 0 & 0 & 0 & 0 & 1 & 0 \\
0 & 1 & 0 & 0 & 0 & -1 & 1 \\
0 & 0 & 1 & 0 & -1 & 0 & 0 \\
0 & 0 & 0 & 1 & 1 & 0 & 0
\end{array}\right], \quad
\mathbf{Q}_l = \left[\begin{array}{ccc}
0 & 1 & 0 \\
0 & -1 & 1 \\
-1 & 0 & 0 \\
1 & 0 & 0
\end{array}\right].
$$

To write the $V\text{-}I$ relations in the form of (109), the twig and link variables are listed in column matrices in the same order as for $\mathbf{Q}$. Thus

$$
\begin{bmatrix}
V_b \\ V_2 \\ V_3 \\ \cdots \\ I_1 \\ I_a \\ I_4 \\ I_5
\end{bmatrix}
=
\left[\begin{array}{c:c}
\mathbf{Z}_l & \mathbf{H}_{12} \\
\hdotsfor{2} \\
\mathbf{H}_{21} & \mathbf{Y}_t
\end{array}\right]
\begin{bmatrix}
I_b \\ I_2 \\ I_3 \\ \cdots \\ V_1 \\ V_a \\ V_4 \\ V_5
\end{bmatrix}.
$$

* Matters can be salvaged, even if both fail, by a modification that uses two trees. It should be clear, however, that failure of both representations can occur only in networks having a proliferation of controlled sources in peculiar connections. Hence the need for such a modification is so rare that further discussion is not warranted.

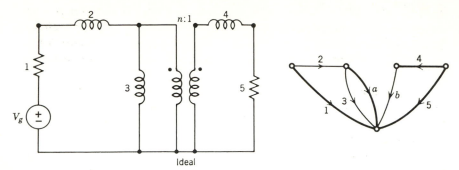

Fig. 30. Example for mixed-variable equations.

Now the **H** matrix is filled in. The only branches that will yield off-diagonal terms are those of the transformer. The appropriate equations are $V_b = V_a/n$, and $I_a = -I_b/n$. The result will be

$$
\begin{bmatrix} V_b \\ V_2 \\ V_3 \\ \cdots \\ I_1 \\ I_a \\ I_4 \\ I_5 \end{bmatrix}
=
\begin{bmatrix}
0 & 0 & 0 & \vdots & 0 & 1/n & 0 & 0 \\
0 & sL_2 & 0 & \vdots & 0 & 0 & 0 & 0 \\
0 & 0 & sL_3 & \vdots & 0 & 0 & 0 & 0 \\
\cdots & & & & & & & \\
0 & 0 & 0 & \vdots & G_1 & 0 & 0 & 0 \\
-1/n & 0 & 0 & \vdots & 0 & 0 & 0 & 0 \\
0 & 0 & 0 & \vdots & 0 & 0 & 1/sL_4 & 0 \\
0 & 0 & 0 & \vdots & 0 & 0 & 0 & G_5
\end{bmatrix}
\begin{bmatrix} I_b \\ I_2 \\ I_3 \\ \cdots \\ V_1 \\ V_a \\ V_4 \\ V_5 \end{bmatrix}.
$$

From this expression the **H** submatrices are easily identified.

As for the sources, assume there are no initial currents in the inductors; so $\mathbf{I}_g = \mathbf{0}$, and $\mathbf{V}_g$ is nonzero only in the first element, which is a twig. Hence $\mathbf{B V}_g$ reduces to $\mathbf{B}_t \mathbf{V}_{gt} = -\mathbf{Q}'_l \mathbf{V}_{gt}$, which, with the $\mathbf{Q}_l$ previously found, becomes

$$
\mathbf{B V}_g = -\mathbf{Q}'_l \mathbf{V}_{gt} =
\begin{bmatrix}
0 & 0 & 1 & -1 \\
-1 & 1 & 0 & 0 \\
0 & -1 & 0 & 0
\end{bmatrix}
\begin{bmatrix} V_g \\ 0 \\ 0 \\ 0 \end{bmatrix}
=
\begin{bmatrix} 0 \\ -V_g \\ 0 \end{bmatrix}.
$$

Now all the submatrices that make up (112) are determined. Putting everything together leads to

$$
\begin{bmatrix}
G_1 & 0 & 0 & 0 & \vdots & 0 & 1 & 0 \\
0 & 0 & 0 & 0 & \vdots & -1/n & -1 & 1 \\
0 & 0 & 1/sL_4 & 0 & \vdots & -1 & 0 & 0 \\
0 & 0 & 0 & G_5 & \vdots & 1 & 0 & 0 \\
\cdots & \cdots & \cdots & \cdots & \vdots & \cdots & \cdots & \cdots \\
0 & 1/n & 1 & -1 & \vdots & 0 & 0 & 0 \\
-1 & 1 & 0 & 0 & \vdots & 0 & sL_2 & 0 \\
0 & -1 & 0 & 0 & \vdots & 0 & 0 & sL_3
\end{bmatrix}
\begin{bmatrix}
V_1 \\ V_a \\ V_4 \\ V_5 \\ \cdots \\ I_b \\ I_2 \\ I_3
\end{bmatrix}
=
\begin{bmatrix}
0 \\ 0 \\ 0 \\ 0 \\ \cdots \\ 0 \\ V_g \\ 0
\end{bmatrix}.
$$

(You should verify this result.) This is a seventh-order matrix equation. However, it is not as bad as it looks, since the coefficient matrix is sparse; that is, many entries are zero.

Let us consider one more example. In the network in Fig. 31, let the transistor have the simple model of a current-controlled current source shown in Fig. 31b. The heavy lines in the graph designate the tree, as usual.

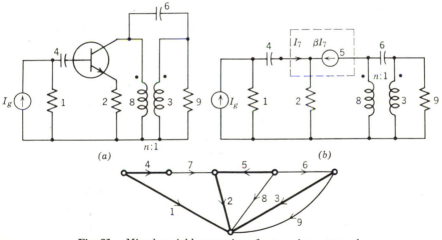

Fig. 31. Mixed variable equations for transistor network.

Branch 7 is the controlling branch of the source. According to Table 2, this branch must be a link, and the controlled current branch (branch 5) must be a twig. For the ideal transformer one branch must be a twig; and the other, a link. Once the branches of the multiterminal components have been assigned their appropriate character, the other branches of the tree are selected. The branch relationship and the **Q** matrix can now be written.

$$
\begin{bmatrix} V_6 \\ V_7 \\ V_8 \\ V_9 \\ \cdots \\ I_1 \\ I_2 \\ I_3 \\ I_4 \\ I_5 \end{bmatrix}
=
\begin{bmatrix}
1/sC_6 & 0 & 0 & 0 & \vdots & 0 & 0 & 0 & 0 & 0 \\
0 & 0 & 0 & 0 & \vdots & 0 & 0 & 0 & 0 & 0 \\
0 & 0 & 0 & 0 & \vdots & 0 & 0 & n & 0 & 0 \\
0 & 0 & 0 & R_9 & \vdots & 0 & 0 & 0 & 0 & 0 \\
\cdots & & & & & & & & & \\
0 & 0 & 0 & 0 & \vdots & G_1 & 0 & 0 & 0 & 0 \\
0 & 0 & 0 & 0 & \vdots & 0 & G_2 & 0 & 0 & 0 \\
0 & 0 & -n & 0 & \vdots & 0 & 0 & 0 & 0 & 0 \\
0 & 0 & 0 & 0 & \vdots & 0 & 0 & 0 & sC_4 & 0 \\
0 & \beta & 0 & 0 & \vdots & 0 & 0 & 0 & 0 & 0
\end{bmatrix}
\begin{bmatrix} I_6 \\ I_7 \\ I_8 \\ I_9 \\ \cdots \\ V_1 \\ V_2 \\ V_3 \\ V_4 \\ V_5 \end{bmatrix}
$$

$$
\mathbf{Q} =
\begin{bmatrix}
1 & 0 & 0 & 0 & 0 & 0 & 1 & 0 & 0 \\
0 & 1 & 0 & 0 & 0 & 1 & -1 & 1 & 0 \\
0 & 0 & 1 & 0 & 0 & -1 & 0 & 0 & 1 \\
0 & 0 & 0 & 1 & 0 & 0 & -1 & 0 & 0 \\
0 & 0 & 0 & 0 & 1 & 1 & 0 & 1 & 0
\end{bmatrix},
\quad
\mathbf{Q}_l =
\begin{bmatrix}
0 & 1 & 0 & 0 \\
1 & -1 & 1 & 0 \\
-1 & 0 & 0 & 1 \\
0 & -1 & 0 & 0 \\
1 & 0 & 1 & 0
\end{bmatrix}.
$$

As for the sources, assume that the capacitors have initial voltages v_{40} and v_{60}, respectively. These can be represented as voltage sources v_{40}/s and v_{60}/s. The source matrices are, therefore,

$$
\mathbf{I}_g = \begin{bmatrix} \mathbf{I}_{gt} \\ \mathbf{I}_{gl} \end{bmatrix} =
\begin{bmatrix} I_g \\ 0 \\ 0 \\ 0 \\ 0 \\ \cdots \\ 0 \\ 0 \\ 0 \\ 0 \end{bmatrix},
\quad
\mathbf{V}_g = \begin{bmatrix} \mathbf{V}_{gt} \\ \mathbf{V}_{gl} \end{bmatrix} =
\begin{bmatrix} 0 \\ 0 \\ 0 \\ v_{40}/s \\ 0 \\ \cdots \\ v_{60}/s \\ 0 \\ 0 \\ 0 \end{bmatrix}.
$$

Hence $\mathbf{QI}_g$ becomes $\mathbf{I}_{gt} = [I_g \quad 0 \quad 0 \quad 0 \quad 0]'$, and $\mathbf{BV}_g$ becomes $-\mathbf{Q}_l'\mathbf{V}_{gt} + \mathbf{V}_{gl} = [v_{60}/s \quad v_{40}/s \quad 0 \quad 0]'$.

When all of the above are inserted into (112), the final equations become

$$
\begin{bmatrix}
G_1 & 0 & 0 & 0 & 0 & 0 & 1 & 0 & 0 \\
0 & G_2 & 0 & 0 & 0 & 1 & -1 & 1 & 0 \\
0 & 0 & 0 & 0 & 0 & -1 & 0 & -n & 0 \\
0 & 0 & 0 & sC_4 & 0 & 0 & -1 & 0 & 0 \\
0 & 0 & 0 & 0 & 0 & 1 & \beta & 1 & 0 \\
0 & -1 & 1 & 0 & -1 & 1/sC_6 & 0 & 0 & 0 \\
-1 & 1 & 0 & 1 & 0 & 0 & 0 & 0 & 0 \\
0 & -1 & n & 0 & -1 & 0 & 0 & 0 & 0 \\
0 & 0 & -1 & 0 & 0 & 0 & 0 & 0 & R_9
\end{bmatrix}
\begin{bmatrix}
V_1 \\ V_2 \\ V_3 \\ V_4 \\ V_5 \\ I_6 \\ I_7 \\ I_8 \\ I_9
\end{bmatrix}
=
\begin{bmatrix}
I_g \\ 0 \\ 0 \\ 0 \\ 0 \\ v_{60}/s \\ v_{40}/s \\ 0 \\ 0
\end{bmatrix}.
$$

Again the size of the matrix may appear to be excessively large but again it is quite sparse.

To summarize the features of the mixed-variable equations, after a tree has been selected, the *V-I* relationships of the branches are written as a single matrix equation in which twig currents and link voltages are expressed in terms of twig voltages and link currents. Next, KCL applied to the *f*-cut-sets for the chosen tree and KVL applied to the *f*-loops are used to eliminate the twig currents and link voltages from this expression. The result is rearranged to give a set of equations in the twig-voltage and link-current variables. The number of equations equals the number of branches, which is the same as the sum of the number of loop equations and node equations. This is the major drawback of the approach. The virtue is the fact that multiterminal elements can be accommodated quite naturally.

One other observation should be made here. In selecting a tree the only restrictions involve the branches of multiterminal components. No special pains are taken in assigning two-terminal elements to the tree or the cotree, because the equations require no special distinctions among capacitors, inductors, or resistors. As we shall see in a later chapter, this is not the case when dealing with the state equations, and there are reasons for assigning capacitors and inductors uniquely to the tree or cotree.

PROBLEMS

1. When writing the state equations for the bridge network in Fig. 2, the branch relations were inserted into KVL to eliminate all branch voltages except that of the capacitor. Then KCL was used to eliminate some of the branch currents.

Instead, start with KCL and use the branch relations to eliminate appropriate currents. Then use KVL to eliminate some of the branch voltages. The final result should be the same as (10).

2. Let A_t be an nth order nonsingular submatrix of the incidence matrix of a connected linear graph, where $n + 1$ is the number of nodes. Prove that the columns of A_t correspond to twigs for some tree. (This is the converse of the theorem on p. 74).

3. Suppose a set of branches of a linear graph contains a loop. Show that the corresponding columns of the A matrix are linearly dependent.

4. Prove that any set of n branches in a connected graph that contains no loops is a tree.

5. The incidence matrix of a graph can be partitioned as $A = [A_t \quad A_l]$, where the columns of A_t correspond to twigs and the columns of A_l correspond to links for a given tree. The submatrix A_t is nonsingular; in some cases it will be a unit matrix. What must be the structure of the tree for which $A_t = U$? Prove and illustrate with an example.

6. Let B_l be a $(b - n)$th order nonsingular submatrix of a loop matrix B that is of order $(b - n) \times b$. Prove that the columns of B_l correspond to the links for some cotree of the graph which is assumed to be connected.

7. A linear graph has five nodes and seven branches. The reduced incidence matrix for this graph is given as

$$A = \begin{bmatrix} 1 & 1 & 0 & 0 & 0 & 0 & 1 \\ -1 & -1 & 1 & 0 & 0 & 0 & 0 \\ 0 & 0 & -1 & 1 & 0 & 0 & 0 \\ 0 & 0 & 0 & -1 & -1 & -1 & 0 \end{bmatrix}.$$

(a)   It is claimed that branches $\{1, 3, 4, 5\}$ constitute a tree. Without drawing the graph, verify the truth of this claim.

(b)   For this tree write the matrix of f-loops, $\mathbf{B}_f$ (again, without drawing a graph).

(c)   For the same tree determine the matrix of f-cut-sets, $\mathbf{Q}_f$. (No graphs, please.)

(d)   Determine the number of trees in the graph.

(e)   Draw the graph and verify the preceding results.

8. Repeat Problem 7 for the following incidence matrices and the specified branches:

(a) $$\mathbf{A} = \begin{bmatrix} 0 & 0 & 1 & 1 & 1 & 0 & -1 \\ 0 & 1 & 0 & 0 & -1 & 1 & 1 \\ -1 & 0 & -1 & 0 & 0 & -1 & 0 \end{bmatrix} \quad \text{branches: } \{2, 3, 4\}$$

(b) $$\mathbf{A} = \begin{bmatrix} 0 & 0 & -1 & 1 & 1 & 0 & 0 & -1 \\ 0 & 1 & 0 & 0 & 0 & 1 & 0 & 1 \\ 0 & -1 & 0 & 0 & 0 & 0 & -1 & 0 \\ -1 & 0 & 0 & -1 & 0 & 0 & 1 & 0 \end{bmatrix} \quad \text{branches: } \{1, 3, 5, 6\}$$

9. Prove this statement: In a linear graph every cut-set has an even number of branches in common with every loop.

10. With $\mathbf{A} = [\mathbf{A}_t \quad \mathbf{A}_l]$ and $\mathbf{B}_f = [\mathbf{B}_t \quad \mathbf{U}]$, it is known that $\mathbf{B}_t = -(\mathbf{A}_t^{-1}\mathbf{A}_l)'$. Thus, to find $\mathbf{B}_f$, it is necessary to know $\mathbf{A}_t^{-1}$. The following matrices are specified as candidates for $\mathbf{A}_t^{-1}$. State whether or not they are suitable.

(a) $$\mathbf{A}_t^{-1} = \begin{bmatrix} 0 & -1 & 0 \\ 1 & 1 & 0 \\ 0 & 0 & 1 \end{bmatrix}, \qquad (b) \quad \mathbf{A}_t^{-1} = \begin{bmatrix} 1 & 0 & 1 \\ 0 & 1 & -1 \\ 1 & -1 & 0 \end{bmatrix},$$

(c) $$\mathbf{A}_t^{-1} = \begin{bmatrix} 1 & -1 & 0 \\ 0 & 1 & -1 \\ 0 & 0 & 1 \end{bmatrix}.$$

11. Let $\mathbf{A} = [\mathbf{A}_t \quad \mathbf{A}_l]$, where $\mathbf{A}_t$ is nonsingular. Prove that the nonzero elements in each row of $\mathbf{A}_t^{-1}$ must have the same sign.

12. Define a *path matrix* $\mathbf{P} = [p_{ij}]$ of a tree as follows:

$p_{ij} = +1[-1]$ if branch j is in the (unique) directed path in the tree from node i to the datum node and its orientation agrees [disagrees] with that of the path, and $p_{ij} = 0$ otherwise.

The path matrix **P** is a square matrix whose rows correspond to paths from the corresponding nodes to the datum node and whose columns correspond to branches; for example, the nonzero entries in, say the third column, specify the paths on which branch 3 lies; and the nonzero entries in, say the fourth row, specify the branches that lie on the path from node 4 to datum. Write the path matrix for each of the indicated trees in the graph of Fig. P–12.

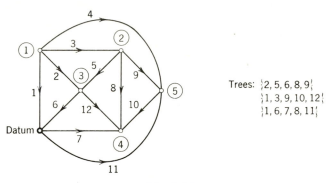

Trees: $\{2, 5, 6, 8, 9\}$
$\{1, 3, 9, 10, 12\}$
$\{1, 6, 7, 8, 11\}$

Fig. P–12

13. When forming $\mathbf{B}_f$ or $\mathbf{Q}_f$ from the incidence matrix, it is necessary to find the inverse of $\mathbf{A}_t$. A simple method of doing this is required. Prove that $\mathbf{A}_t^{-1} = \mathbf{P}'$, where **P** is the path matrix of the tree whose branches are the columns of $\mathbf{A}_t$.

14. Use the result of Problem 13 to find $\mathbf{A}_t^{-1}$ for the trees of Fig. P–12. Verify by evaluation of $\mathbf{A}_t\mathbf{P}'$. For each case verify the result of Problem 11.

15. (a) Two branches are in parallel in a graph. Determine the relationship of the columns corresponding to these two branches in the $\mathbf{Q}_f$ matrix.
(b) Repeat if the two branches are in series.

16. Let $\mathbf{Q}_f = [\mathbf{U} \quad \mathbf{Q}_l]$. Suppose a column of $\mathbf{Q}_l$, say the jth, has a single nonzero entry, and this is in the kth row. What can you say about the structure of the graph as it relates to branches j and k?

17. Let $\mathbf{Q}_f = [\mathbf{U} \quad \mathbf{Q}_l]$ and $\mathbf{B}_f = [\mathbf{B}_t \quad \mathbf{U}]$. Suppose the $\mathbf{Q}_f$ matrix of a graph is given. (a) Determine the $\mathbf{B}_f$ matrix for this graph; (b) discuss the possibility of determining the **A** matrix of the graph and the uniqueness of this graph from a given $\mathbf{Q}_f$ matrix or a given $\mathbf{B}_f$ matrix.

18. In the graphs of Fig. P–18 determine the number of trees

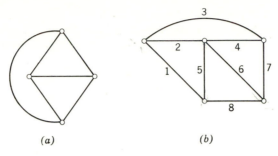

(a)   (b)

Fig. P–18

19. Prove that a graph contains at least one loop if two or more branches are incident at each node.

20. For a connected planar graph of b branches and $n + 1$ nodes show that the **B** matrix of the meshes is of rank $b - n$. This will mean that the mesh currents will be an adequate basis for expressing all branch currents.

21. Using the concept of duality, show that KVL equations written for the meshes of a planar graph are linearly independent.

22. Let a branch t of a graph be a twig for some tree. The f-cut-set determined by t contains a set of links $l_1, l_2, \cdots$ for that tree. Each of these links defines an f-loop. Show that every one of the f-loops formed by links l_1, l_2, etc. contains twig t.

23. For the graph in Fig. P–23, let **B** be the loop matrix for the meshes.

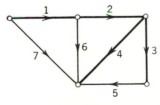

Fig. P–23

(a)   Form $\mathbf{B}'(\mathbf{B}_l^{-1})'$ and verify that the branch currents are correctly given in terms of the link currents for the tree shown.

(b)   For the same tree determine $\mathbf{B}_f$ directly and verify that $\mathbf{B}_f = \mathbf{B}'(\mathbf{B}_l^{-1})'$.

(c)   Verify the mesh transformation.

24. In the graph of Fig. P–18b, branches $\{2, 3, 7, 6\}$ form a loop. Verify the result of Problem 3 for this set of branches.

25. In the graph of Fig. P–18b, branches $\{2, 4, 5, 6\}$ form a tree. Partition the $\mathbf{A}$ matrix into $[\mathbf{A}_t \quad \mathbf{A}_l]$ by using this tree. From this, letting $\mathbf{B}_f = [\mathbf{B}_t \quad \mathbf{U}]$ and $\mathbf{Q}_f = [\mathbf{U} \quad \mathbf{Q}_l]$, determine $\mathbf{B}_t$ and $\mathbf{Q}_l$ and verify that $\mathbf{B}_t = -\mathbf{Q}_l'$.

26. It is possible for two f-loops of a graph for a given tree to have twigs and nodes in common. Prove that it is possible for two f-loops to have two nodes in common only if the path in the tree between the nodes is common to the two loops.

27. In Fig. P–27 the following loops exist:

$$\text{loop 1: } a, e, g, b$$
$$\text{loop 2: } d, c, g, f$$
$$\text{loop 3: } a, d, f, j, h, b$$
$$\text{loop 4: } e, c, h, j$$

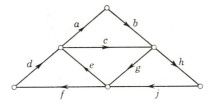

Fig. P–27

(a) Do the KVL equations for these loops constitute an independent set?

(b) Either (1) find a set of links that form f-loops that are meshes of the graph or (2) prove that there is no such set.

28. In a linear network let the power in branch j be defined as $p_j(t) = v_j(t)\, i_j(t)$, with standard references for the variables.

(a) For a given network assume that KCL and KVL are satisfied. Show that $\Sigma\, p_j(t) = \mathbf{v}(t)'\, \mathbf{i}(t) = 0$, where the summation is over all branches of the network. The result $\mathbf{v}'\mathbf{i} = 0$ is called *Tellegen's theorem*.

(b) Next assume that Tellegen's theorem and KVL are both true. Show that KCL follows as a consequence.

(c) Finally, assume that Tellegen's theorem and KCL are both true. Show that KVL follows as a consequence.

This problem demonstrates that, of the three laws, KCL, KVL and Tellegen's theorem, any two can be taken as fundamental; the third will follow as a consequence.

29. Construct the duals of the networks in Fig. P–29. The numbers shown are values of R, L, or C.

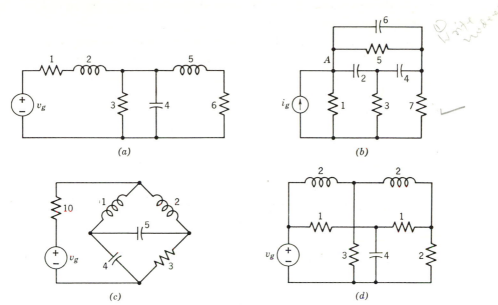

(a)   (b)

(c)   (d)

Fig. P–29

30. (a)   If two branches are parallel or in series in a graph, how are the corresponding branches in the dual graph related? Verify in terms of Fig. P–29.
(b)   Figure P–30 represents, the network in Fig. P–29b. Within the box

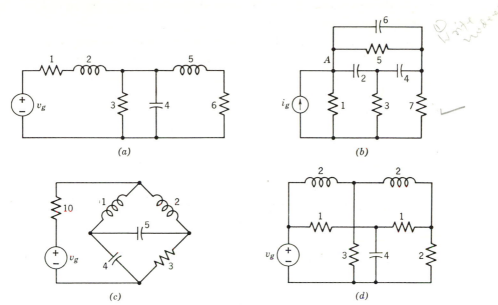

Fig. P–30

is a bridged-tee. In the dual network how is this bridged-tee modified?
(c)   What kind of structure is the dual of the bridge network in Fig. P–29c?

31. For the networks in Fig. P–29 (a) write the loop equations by using the meshes for loops and the mesh currents as a set of basis currents. For which of these networks do the meshes constitute f-loops for some tree?

(b) Choose some tree [different from part (a)] and write loop equations for the f-loops.

32. For the networks of Fig. P–29 choose the lowest node as datum and write a set of node equations.

33. For the networks of Fig. P–29 choose a tree and write a set of node-pair equations with the twig voltages for this tree as basis variables.

34. For the network in Fig. P–34 (a) write a set of node-pair equations for the tree shown. (b) Write a set of node equations using a convenient node for datum.

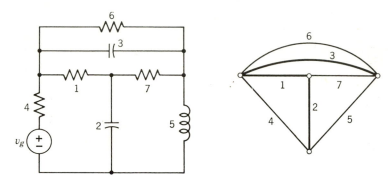

Fig. P–34

35. For the network in Fig. 29 in the text write a set of node equations and verify that the node-admittance matrix will be singular when $G_1 = (k - 1)G_2$.

36. Fig. P–36 shows a network having a potentially singular node-admittance or loop-impedance matrix. Determine conditions among the parameters that will make it singular

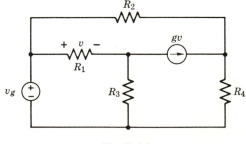

Fig. P–36

37. For the network in Fig. P–37 set up in matrix form **(a)** the node equations; **(b)** the mixed-variable equations, and **(c)** for the latter, show all possible trees of the graph.

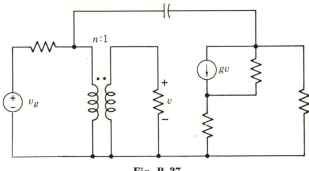

Fig. P–37

38. For the network of Fig. P–38 set up **(a)** the loop equations, **(b)** the mixed-variable equations (use the usual small-signal linear equivalent for the triode), and **(c)** for the latter, show all possible trees.

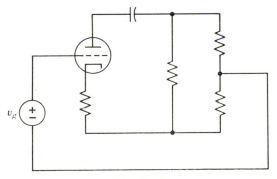

Fig. P–38

39. Find all possible trees for which mixed-variable equations can be written for the network in Fig. 31 in the text.

40. Show that the loop-impedance matrix $\mathbf{Z}_m = \mathbf{B}_f \mathbf{Z} \mathbf{B}_f'$ can be written as $\mathbf{Z}_l + \mathbf{Q}_l' \mathbf{Z}_t \mathbf{Q}_l$, where $\mathbf{Q}_f = [\mathbf{U} \quad \mathbf{Q}_l]$ and $\mathbf{Z}$ is conformally partitioned.

41. Show by doing it that, for *RLC* networks, the mixed-variable equations in (112) can be converted into loop equations or node-pair equations.

The next six problems involve the preparation of a computer program to help in implementing the solution of some problems. In each case prepare a program flow chart and a set of program instructions, in some user language like FORTRAN IV, for a digital computer program to carry out the job specified in the problem. Include a set of user instructions for the program.

*42. Prepare a program to identify a tree of a connected network when each branch and its orientation in the graph are specified by a sequence of triplets of numbers: The first number identifies the branch, the second number identifies the node at the branch tail, and the third number identifies the node at the branch head. The program must also renumber the branches such that twigs are numbered from 1 to n and links are numbered from $n + 1$ to b. An example of a typical set of data is given in Fig. P–42 for the network graph shown there.

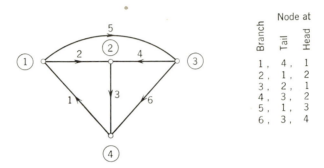

Fig. P–42

*43. Prepare a program to determine a reduced-incidence matrix of a connected network using, as input data, the output data from Problem 42.

*44. Prepare a program to determine the following:

(a) An f-loop matrix by using (34).
(b) An f-cut-set matrix by using (43).

The output data from Problem 43 should be taken as the input data for this problem.

*45. Prepare a program to determine the number of trees in a network by (20) with pivotal condensation used to evaluate the determinant. Specify the input data format.

*46. Prepare a program to determine the (a) node-admittance matrix, (b) loop-impedance matrix, and (c) node-pair-admittance matrix of an RLC network by evaluating the (1) node-parameter matrices of (91), (2) loop-parameter matrices of (83), and (3) node-pair-parameter matrices of (99), respectively. Assume the input data is presented as a sequence of quintuplets of numbers: The first identifies the branch, the second identifies the node at the branch tail, and the third identifies the node at the branch head, the fourth identifies the branch type according to the following schedule, and the fifth is the value of the branch parameter. Assume also that (1) a reduced-incidence matrix, (2) an f-loop matrix, and (3) an f-cut-set matrix have already been evaluated by the programs of Problems 43 and 44.

Fourth Number	Branch Type
1	Capacitor
2	Resistor
3	Inductor

*47. Combine the programs of Problems 43, 44, and 46 to create a single program that will determine at the program user's option the node-admittance matrix, loop-impedance matrix, and/or node-pair-admittance matrix of an RLC network.

. 3 .

NETWORK FUNCTIONS

In the last chapter we described a number of systematic methods for applying the fundamental laws of network theory to obtain sets of simultaneous equations: loop, node, node-pair, and mixed-variable equations. Of course, these formal procedures are not necessarily the simplest to use for all problems. In many problems involving networks of only moderate structural complexity, inspection, Thévenin's theorem, and other shortcuts may doubtless provide answers more easily than setting up and solving, say, the loop equations. The value of these systematic procedures lies in their generality and in our ability to utilize computers in setting them up and solving them.

The equations to which these systematic methods lead are differential or integrodifferential. Classical methods for solving such equations can be employed, but we have so far assumed that solutions will be obtained by the Laplace transform. With this in mind, the formulation was often carried out in Laplace-transformed form.

Assuming that a network and the Laplace-transformed equations describing its behavior are available, we now turn to the solution of these equations and the network functions in terms of which the network behavior is described.

3.1   DRIVING-POINT AND TRANSFER FUNCTIONS

Given a linear, time-invariant network, excited by any number of independent voltage-and-current sources, and with arbitrary initial capacitor voltages and inductor currents, (which can also be represented

150

as independent sources) a set of loop, node, or node-pair equations can be written. The network may be nonpassive and nonreciprocal. In matrix form these equations will all be similar. Thus

$$\mathbf{Z}_m(s)\mathbf{I}_m(s) = \mathbf{E}(s) \qquad \text{(loop)}, \tag{1a}$$

$$\mathbf{Y}_n(s)\mathbf{V}_n(s) = \mathbf{J}(s) \qquad \text{(node)}, \tag{1b}$$

$$\mathbf{Y}_t(s)\mathbf{V}_t(s) = \mathbf{J}_t(s) \qquad \text{(node-pair)}. \tag{1c}$$

The right-hand sides are the contributions of the sources, including the initial-condition equivalent sources; for example, $\mathbf{J} = [J_i]$, where J_i is the sum of current sources (including Norton equivalents of accompanied voltage sources) connected at node i, with due regard for the orientations.

The symbolic solution of these equations can be written easily and is obtained by multiplying each equation by the inverse of the corresponding coefficient matrix. Thus

$$\mathbf{I}_m(s) = \mathbf{Z}_m^{-1}\mathbf{E}(s), \tag{2a}$$

$$\mathbf{V}_n(s) = \mathbf{Y}_n^{-1}\mathbf{J}(s), \tag{2b}$$

$$\mathbf{V}_t(s) = \mathbf{Y}_t^{-1}\mathbf{J}_t(s). \tag{2c}$$

Each of these has the same form. For purposes of illustration the second one will be shown in expanded form. Thus

$$
\begin{bmatrix} V_1 \\ V_2 \\ \vdots \\ V_m \end{bmatrix}
=
\begin{bmatrix}
\dfrac{\Delta_{11}}{\Delta} & \dfrac{\Delta_{21}}{\Delta} & \cdots & \dfrac{\Delta_{m1}}{\Delta} \\[2ex]
\dfrac{\Delta_{12}}{\Delta} & \dfrac{\Delta_{22}}{\Delta} & \cdots & \dfrac{\Delta_{m2}}{\Delta} \\[2ex]
\vdots & \vdots & & \vdots \\[2ex]
\dfrac{\Delta_{1m}}{\Delta} & \dfrac{\Delta_{2m}}{\Delta} & \cdots & \dfrac{\Delta_{mm}}{\Delta}
\end{bmatrix}
\begin{bmatrix} J_1 \\ J_2 \\ \vdots \\ J_m \end{bmatrix}
\tag{3}
$$

where Δ is the determinant of the node admittance matrix and the Δ_{jk} are its cofactors. The expression for just one of the node voltages, say V_k, is

$$V_k(s) = \frac{\Delta_{1k}}{\Delta}J_1 + \frac{\Delta_{2k}}{\Delta}J_2 + \cdots + \frac{\Delta_{mk}}{\Delta}J_m. \tag{4}$$

This gives the transform of a node voltage as a linear combination of the equivalent sources. The J's are not the actual sources; for example, in Fig. 3.1 suppose there are no initial conditions. The J-matrix will be

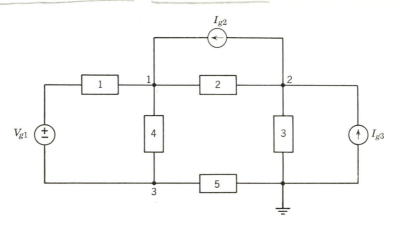

Fig. 1.   Illustration for equivalent sources.

$$\mathbf{J} = \begin{bmatrix} J_1 \\ J_2 \\ J_3 \end{bmatrix} = \begin{bmatrix} I_{g2} + Y_1 V_{g1} \\ -I_{g2} + I_{g3} \\ -Y_1 V_{g1} \end{bmatrix}.$$

When the appropriate expressions in terms of actual sources are substituted for the J's, it is clear that (4) can be arranged to give $V_k(s)$ as a linear combination of the actual sources. Thus for Fig. 1 the expression for V_k would become

$$V_k(s) = \left(\frac{Y_1 \Delta_{1k} - Y_1 \Delta_{3k}}{\Delta} \right) V_{g1} + \left(\frac{\Delta_{1k} - \Delta_{2k}}{\Delta} \right) I_{g2} + \left(\frac{\Delta_{2k}}{\Delta} \right) I_{g3}. \quad (5)$$

As a general statement, then, it can be said that any response transform can be written as a linear combination of excitation transforms. The coefficients of this linear combination are themselves linear combinations of some functions of s. These functions are ratios of two determinants, the denominator one being the determinant of $\mathbf{Z}_m$, $\mathbf{Y}_n$, or $\mathbf{Y}_t$, and the numerator one being some cofactor of these matrices; for example, in (5) the denominator determinant is det $\mathbf{Y}_n$, and the coefficient of I_{g2} is the difference between two such ratios of determinants.

Once the functions relating any response transform (whether voltage or current) to any excitation transform (V or I) are known, then the response to any given excitation can be determined. Thus in (5) knowledge of the quantities in parentheses is enough to determine the response $V_k(s)$ for any given values of V_{g1}, I_{g2}, and I_{g3}.

We shall define the general term *network function* as the ratio of a response transform to an excitation transform. Both the response and the excitation may be either voltage or current. If the response and excitation refer to the same terminals (in which case one must be a voltage, the other a current), then the function is called a *driving-point function*, either impedance or admittance. If the excitation and response refer to different terminals, then the function is a *transfer function*.

DRIVING-POINT FUNCTIONS

To be more specific, consider Fig. 2a, in which attention is focused

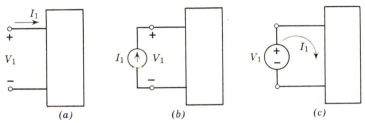

(a)   (b)   (c)

Fig. 2. Driving-point functions.

on one pair of terminals to which external connections can be made. We assume that the network (1) contains no independent sources, and (2) is initially relaxed.

By the " driving-point impedance " and the " driving-point admittance " of a network at a pair of terminals we mean

$$Z(s) = \frac{V_1(s)}{I_1(s)}, \quad \text{and} \quad Y(s) = \frac{I_1(s)}{V_1(s)} = \frac{1}{Z(s)}, \tag{6}$$

where V_1 and I_1 are the transforms of the terminal voltage and current with the references as shown in Fig. 2. In making this definition nothing is said about how the terminals are excited or what is connected to them. The implication is that it makes no difference. (Is this completely intuitive or does it require proof?) The conditions of no independent sources and zero initial conditions are essential to the definition. Clearly, if the network contains independent sources or initial conditions, then Z or Y can take on different values, depending on what is connected at the terminals; thus it will not be an invariant characteristic of the network itself.

Another factor to be noted is that no assumption has been made about the nature of the time functions $v_1(t)$ and $i_1(t)$. Whatever they may be, the definition of Z or Y involves the ratio of their Laplace transforms.

Now let us turn to an evaluation of Z and Y for a network and let us initially assume that the network contains no *nonpassive, nonreciprocal devices*. Since it makes no difference, suppose the network is excited with a current source, as in Fig. 2b. Let us write a set of node equations. We choose one of the two external terminals as the datum node in order that the source appear in only one of the node equations. Under these circumstances the solution for V_1 can be obtained from (4), in which only J_1 is nonzero and its value is I_1. Hence for the impedance we find

$$Z(s) = \frac{V_1(s)}{I_1(s)} = \frac{\Delta_{11}(s)}{\Delta(s)}\bigg|_y \qquad (7)$$

where Z is the *driving-point impedance* at the terminals of the network. The notation y is used on the determinants to indicate that Δ is the determinant of the node equations.

A dual formula for the driving-point admittance Y can be obtained, quite evidently, by considering that the network is excited by a voltage source, as in Fig.2c; a set of loop equations are then written. The loops are chosen such that the voltage source appears only in the first loop; thus V_1 will appear in only one of the loop equations. The solution for the loop equations will be just like (3), except that the sources will be the equivalent voltage sources E_k (of which all but the first will be zero in this case and this one will equal V_1) and the variables will be the loop currents. Solving for I_1 then gives

$$Y(s) = \frac{I_1(s)}{V_1(s)} = \frac{\Delta_{11}(s)}{\Delta(s)}\bigg|_z \qquad (8)$$

where Y is the *driving-point admittance* and the notation z means that Δ is the determinant of the loop impedance matrix.

Expressions 7 and 8 are useful for calculating Z and Y, but it should be remembered that they apply when there are no controlled sources or other such devices. They may also apply in some cases when controlled sources are present, but not always.

As an illustration of a simple case in which (7) and (8) do not apply when a controlled source is present, consider the grounded-grid amplifier shown in Fig. 3a. The linear model is shown in Fig. 3b, in which a voltage-controlled voltage source appears. Since this does not have either an

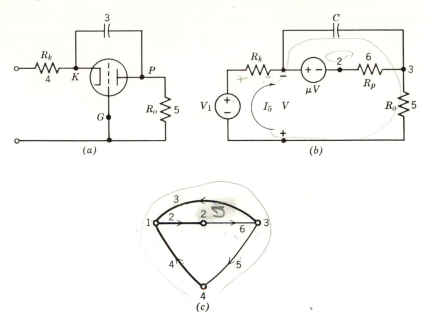

Fig. 3.   Grounded grid amplifier.

impedance or an admittance representation, let us express the control-ling voltage V in terms of a branch current; from Fig. 3b the proper expression is $V = R_k I_5 - V_1$. With this expression replacing the control-ling voltage, the controlling branch is not even shown in the graph of Fig. 3c. The V-I relation of the *controlled* branch (branch 2 in the graph) is $V_2 = \mu V = \mu R_k I_5 - \mu V_1$. This gives an impedance representation of the branch. (μV_1 is simply an independent accompanying source and will appear in the source-voltage matrix.) The tree in solid lines is chosen so that the exciting source appears in only one loop. The loop matrix $\mathbf{B}$, the branch impedance matrix $\mathbf{Z}$, and the source-voltage matrix $\mathbf{V}_g$ can be written as follows:

$$\mathbf{Z} = \begin{bmatrix} 0 & 0 & \mu R_k & 0 & 0 \\ 0 & \dfrac{1}{sC} & 0 & 0 & 0 \\ 0 & 0 & R_k & 0 & 0 \\ 0 & 0 & 0 & R_0 & 0 \\ 0 & 0 & 0 & 0 & R_p \end{bmatrix}, \quad \mathbf{V}_g = \begin{bmatrix} \mu V_1 \\ 0 \\ V_1 \\ 0 \\ 0 \end{bmatrix}, \quad \mathbf{B} = \begin{bmatrix} 0 & -1 & 1 & 1 & 0 \\ 1 & 1 & 0 & 0 & 1 \end{bmatrix}.$$

The loop equations ($\mathbf{BZB'I}_m = \mathbf{BV}_g$) now become

$$\begin{bmatrix} R_k + R_0 + \dfrac{1}{sC} & -\dfrac{1}{sC} \\[2mm] -\dfrac{1}{sC} + \mu R_k & R_p + \dfrac{1}{sC} \end{bmatrix} \begin{bmatrix} I_5 \\ I_6 \end{bmatrix} = \begin{bmatrix} V_1 \\ \mu V_1 \end{bmatrix}.$$

The driving-point admittance is I_5/V_1. This can be found by solving the loop equations for I_5. The result is

$$Y = \frac{I_5}{V_1} = \frac{\Delta_{11} + \mu \Delta_{21}}{\Delta}.$$

Thus, even though we were careful to choose only one loop through the exciting source, the source voltage V_1 appears in both loop equations, and the final result differs from (8). It may be concluded that one should not rely on special formulas such as (7) and (8) but should go to definitions such as (6).

TRANSFER FUNCTIONS

In the definition of a network function, when the excitation and response are at different terminals, the function is a transfer function. Let the response be the voltage or current of some branch. We can focus attention on the branch by drawing it separately, as in Fig. 4. For the notion of transfer function to be meaningful, we continue to assume that there are no internal independent sources and the network is initially relaxed. In Fig. 4a, four different transfer functions can be defined with

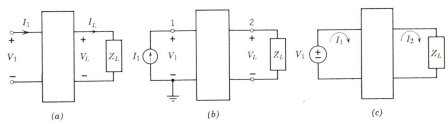

$$(a) \qquad\qquad (b) \qquad\qquad (c)$$

Fig. 4.   Transfer functions.

either $V_L(s)$ or $I_L(s)$ as the response and either $V_1(s)$ or $I_1(s)$ as the excitation. Again, in making these definitions there is no stipulation as to the manner in which the network is excited. Since it makes no difference let us apply a current source, as shown in Fig. 4b, and use node equations. The solutions for the node voltages are given by (4). The only things left to settle are the sources. Even though there is only one actual source, this may appear in more than one node equation if the network contains controlled sources. Hence let us temporarily assume there are no controlled sources. Then all the J's in (4) are zero except J_1, which equals I_1. The result will be

$$V_1(s) = \frac{\Delta_{11}}{\Delta}\bigg|_y I_1(s),$$

$$V_L(s) = V_2 - V_3 = \frac{\Delta_{12} - \Delta_{13}}{\Delta}\bigg|_y I_1(s).$$

From these, and from the fact that $I_L = Y_L V_L$, each of the transfer functions V_L/V_1, V_L/I_1, I_L/V_1, and I_L/I_1 can be obtained. In a similar manner, by assuming a voltage source is applied, as shown in Fig. 4c, the loop equations may be written. The resulting expressions from both the node and loop equations will be

Transfer impedance:

$$\frac{V_L(s)}{I_1(s)} = \frac{\Delta_{12} - \Delta_{13}}{\Delta}\bigg|_y = Z_L \frac{\Delta_{12}}{\Delta_{11}}\bigg|_z . \tag{9a}$$

Transfer admittance:

$$\frac{I_L(s)}{V_1(s)} = Y_L \frac{\Delta_{12} - \Delta_{13}}{\Delta_{11}}\bigg|_y = \frac{\Delta_{12}}{\Delta}\bigg|_z . \tag{9b}$$

Voltage gain, or transfer voltage ratio:

$$\frac{V_L(s)}{V_1(s)} = \frac{\Delta_{12} - \Delta_{13}}{\Delta_{11}}\bigg|_y = Z_L \frac{\Delta_{12}}{\Delta}\bigg|_z . \tag{9c}$$

Current gain, or transfer current ratio:

$$\frac{I_L(s)}{I_1(s)} = Y_L \frac{\Delta_{12} - \Delta_{13}}{\Delta}\bigg|_y = \frac{\Delta_{12}}{\Delta_{11}}\bigg|_z . \tag{9d}$$

Let us emphasize that—

1. These formulas are valid in the absence of controlled sources and other nonpassive, nonreciprocal devices.

2. The transfer impedance *is not* the reciprocal of the transfer admittance.

3. The references of voltage and current must be as shown in Fig. 4. When controlled sources are present, these particular formulas may not apply. Nevertheless, the transfer functions will still be linear combinations of similar ratios of determinants and cofactors.

3.2 MULTITERMINAL NETWORKS

In our study of electric networks up to this point we have assumed that the internal structure of the network is available and that an analysis is to be carried out for the purpose of determining the currents and voltages anywhere in the network. However, very often there is no interest in all the branch voltages and currents. Interest is limited to only a number of these; namely, those corresponding to terminals to which external connections to the network are to be made. As far as the outside world is concerned, the details of the internal structure of a network are not important; it is only important to know the relationships among the voltages and currents at the external terminals. The external behavior of the network is completely determined once these relationships are known.

Consider the network, shown in Fig. 5a, having six terminals to which

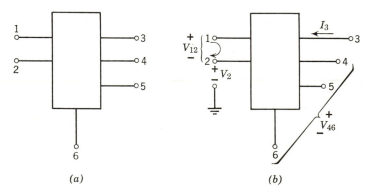

Fig. 5. Six-terminal network.

external connections can be made. Exactly how should the voltage-and-current variables be defined? Consider Fig. 5b. Should the voltages be defined so that each terminal voltage is referred to some arbitrary datum or ground, such as V_2? Should they be defined as the voltages between pairs of terminals, such as V_{12}, or V_{46}? Should the currents be the terminal currents, such as I_3, or should they be like the loop current J_1 shown flowing into one terminal and flowing out another? In fact, each of these may be useful for different purposes, as we shall observe.

In many applications external connections are made to the terminals of the network only in pairs. Each pair of terminals, or *terminal-pair*, represents an entrance to, and exit from, a network and is quite descriptively called a *port*. The six-terminal network of Fig. 5 is shown as a *three-port* in Fig. 6a and as a *five-port* in Fig. 6b. Note that no other

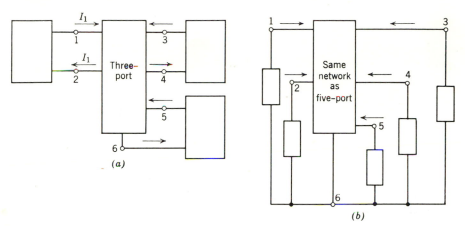

Fig. 6. Six-terminal network connected (*a*) as a three-port and (*b*) as a five-port.

external connections are to be made except at the ports shown; for example, in Fig. 6a no connection is to be made between terminals 1 and 3 or 1 and 5. Connections must be so made that the same current enters one terminal of the port as leaves the network through the second terminal of the port. The port voltages are the voltages between the pairs of terminals that constitute the port.

There is a basic difference (besides the number of ports) between the two types of multiport network shown in Fig. 6. In Fig. 6b, one of the terminals of each port is common to all the ports. The port voltages are therefore the same as the terminal voltages of all but one terminal, relative to this last one. Such a network is called a *common-terminal*, or *grounded*, multiport. In the first network in Fig. 6 there is no such identified common ground.

It is possible that other kinds of external connections may be required to a multiterminal network besides the terminal-pair kind. In such a case a port description is not possible. An alternative means of description, which will be necessary in this case, will be discussed in Section 3.6.

3.3 TWO-PORT NETWORKS

At this point it is possible to proceed by treating the general multiport network and discussing sets of equations relating the port variables. After this is done, the results can be applied to the special case of a two-port. An alternative approach is to treat the simplest multiport (namely, the two-port) first. This might be done because of the importance of the two-port in its own right, and because treating the simplest case first can lead to insights into the general case that will not be obvious without experience with the simplest case. We shall take this second approach.

A two-port network is illustrated in Fig. 7. Because of the application

Fig. 7.   Two-port network.

of two-ports as transmission networks, one of the ports—normally the port labeled 1—is called the *input*; the other, port 2, is called the *output*. The port variables are two port currents and two port voltages, with the standard references shown in Fig. 7. (In some of the literature the reference for I_2 is taken opposite to our reference. When comparing any formulas in other publications, verify the references of the port parameters.) External networks that may be connected at the input and output are called the *terminations*. We shall deal throughout with the transformed variables and shall assume the two-port to be initially relaxed and to contain no independent sources.

The discussion that follows may appear somewhat unmotivated, since in restricting ourselves to analysis we have lost much of the motivation for finding various ways of describing the behavior of two-port networks. The need for these various schemes arises from the demands made by the many applications of two-ports. The usefulness of the different methods of description comes clearly into evidence when the problem is one of synthesizing or designing networks—filters, matching networks, wave-shaping networks, and a host of others. A method of description that is convenient for a power system, say, may be less so for a filter network, and may be completely unsuited for a transistor amplifier. For this reason we shall describe many alternative, but equivalent, ways of describing two-port behavior.

In the problem of synthesizing a network for a specific application, it is often very convenient to break down a complicated problem into several parts. The pieces of the overall network are designed separately and then put together in a manner consistent with the original decomposition. In order to carry out this procedure it is necessary to know how the description of the behavior of the overall network is related to the behavior of the components. For this reason we shall spend some time on the problem of interconnecting two-ports.

Many of the results obtained in this section require a considerable amount of algebraic manipulation that is quite straightforward. We shall not attempt to carry through all the steps, but shall merely outline the desired procedure, leaving to you the task of filling in the omitted steps.

OPEN- AND SHORT-CIRCUIT PARAMETERS

To describe the relationships among the port voltages and currents of a linear multiport requires as many linear equations as there are ports. Thus for a two-port two linear equations are required among the four variables. However, which two variables are considered "independent" and which "dependent" is a matter of choice and convenience in a given application. To return briefly to the general case, in an n-port, there will be $2n$ voltage-and-current variables. The number of ways in which these $2n$ variables can be arranged in two groups of n each equals the number of ways in which $2n$ things can be taken n at a time; namely, $(2n)!/(n!)^2$. For a two-port this number is 6.

One set of equations results when the two-port currents are expressed in terms of the two-port voltages:

$$\begin{bmatrix} I_1(s) \\ I_2(s) \end{bmatrix} = \begin{bmatrix} y_{11} & y_{12} \\ y_{21} & y_{22} \end{bmatrix} \begin{bmatrix} V_1(s) \\ V_2(s) \end{bmatrix}. \tag{10}$$

It is a simple matter to obtain interpretations for these parameters by letting each of the voltages be zero in turn. It follows from the equation that

$$y_{11}(s) = \frac{I_1(s)}{V_1(s)} \bigg|_{V_2=0}, \qquad y_{12}(s) = \frac{I_1(s)}{V_2(s)} \bigg|_{V_1=0},$$

$$y_{21}(s) = \frac{I_2(s)}{V_1(s)} \bigg|_{V_2=0}, \qquad y_{22}(s) = \frac{I_2(s)}{V_2(s)} \bigg|_{V_1=0}. \tag{11}$$

Dimensionally, each parameter is an admittance. Setting a port voltage to zero means short-circuiting the port. Hence the y's (for which the lower case letter y will be reserved) are called the *short-circuit admittance parameters* (the y-parameters for short). The matrix of y's is designated $\mathbf{Y}_{sc}$ and is called *the short-circuit admittance matrix*. The terms y_{11} and y_{22} are the short-circuit driving-point admittances at the two ports, and y_{21} and y_{12} are short-circuit transfer admittances. In particular, y_{21} is the *forward* transfer admittance—that is, the ratio of a current response in port 2 to a voltage excitation in port 1—and y_{12} is the *reverse* transfer admittance.

A second set of relationships can be written by expressing the port voltages in terms of the port currents:

$$\begin{bmatrix} V_1(s) \\ V_2(s) \end{bmatrix} = \begin{bmatrix} z_{11} & z_{12} \\ z_{21} & z_{22} \end{bmatrix} \begin{bmatrix} I_1(s) \\ I_2(s) \end{bmatrix}. \tag{12}$$

This time interpretations are obtained by letting each current be zero in turn. Then

$$z_{11}(s) = \frac{V_1(s)}{I_1(s)}\bigg|_{I_2=0}, \qquad z_{12}(s) = \frac{V_1(s)}{I_2(s)}\bigg|_{I_1=0},$$

$$z_{21}(s) = \frac{V_2(s)}{I_1(s)}\bigg|_{I_2=0}, \qquad z_{22}(s) = \frac{V_2(s)}{I_2(s)}\bigg|_{I_1=0}. \tag{13}$$

Dimensionally, each parameter is an impedance. Setting a port current equal to zero means open-circuiting the port. Hence the z's (for which the lower case letter z will be reserved) are called the *open-circuit impedance parameters* (the z-parameters for short). The matrix of z's is designated $\mathbf{Z}_{oc}$ and is called the *open-circuit impedance matrix*. The elements z_{11} and z_{22} are the driving-point impedances at the two ports, and z_{21} and z_{12} are the transfer impedances; z_{21} is the *forward* transfer impedance, and z_{12} is the *reverse* transfer impedance.

It should be clear from (10) and (12) that the $\mathbf{Y}_{sc}$ and $\mathbf{Z}_{oc}$ matrices are inverses of each other; for example,

$$\mathbf{Y}_{sc} = \begin{bmatrix} y_{11} & y_{12} \\ y_{21} & y_{22} \end{bmatrix} = \mathbf{Z}_{oc}^{-1} = \frac{1}{\det \mathbf{Z}_{oc}} \begin{bmatrix} z_{22} & -z_{12} \\ -z_{21} & z_{11} \end{bmatrix}. \tag{14}$$

From this it follows that

$$\det \mathbf{Y}_{sc} = \frac{1}{\det \mathbf{Z}_{oc}}. \tag{15}$$

Demonstration of this is left as an exercise.

The results developed so far apply whether the network is passive or active, reciprocal or nonreciprocal. Now consider the two transfer functions y_{21} and y_{12}. If the network is reciprocal, according to the definition in Section 1.4, they will be equal. So also will z_{12} and z_{21}; that is, for a reciprocal network

$$y_{12} = y_{21}, \qquad z_{12} = z_{21}, \tag{16}$$

which means that both $\mathbf{Y}_{sc}$ and $\mathbf{Z}_{oc}$ are symmetrical for reciprocal networks.

HYBRID PARAMETERS

The z and y representations are two of the ways in which the relationships among the port variables can be expressed. They express the two voltages in terms of the two currents, and vice versa. Two other sets of equations can be obtained by expressing a current and voltage from opposite ports in terms of the other voltage and current. Thus

$$\begin{bmatrix} V_1 \\ I_2 \end{bmatrix} = \begin{bmatrix} h_{11} & h_{12} \\ h_{21} & h_{22} \end{bmatrix} \begin{bmatrix} I_1 \\ V_2 \end{bmatrix} \tag{17}$$

and

$$\begin{bmatrix} I_1 \\ V_2 \end{bmatrix} = \begin{bmatrix} g_{11} & g_{12} \\ g_{21} & g_{22} \end{bmatrix} \begin{bmatrix} V_1 \\ I_2 \end{bmatrix}. \tag{18}$$

The interpretations of these parameters can be easily determined from the preceding equations to be the following:

$$
h_{11} = \frac{V_1(s)}{I_1(s)} \bigg|_{V_2=0}, \qquad h_{12} = \frac{V_1(s)}{V_2(s)} \bigg|_{I_1=0},
$$

$$
h_{21} = \frac{I_2(s)}{I_1(s)} \bigg|_{V_2=0}, \qquad h_{22} = \frac{I_2(s)}{V_2(s)} \bigg|_{I_1=0},
$$

$$
g_{11} = \frac{I_1(s)}{V_1(s)} \bigg|_{I_2=0}, \qquad g_{12} = \frac{I_1(s)}{I_2(s)} \bigg|_{V_1=0}, \tag{19}
$$

$$
g_{21} = \frac{V_2(s)}{V_1(s)} \bigg|_{I_2=0}, \qquad g_{22} = \frac{V_2(s)}{I_2(s)} \bigg|_{V_1=0}.
$$

Thus we see that the h- and g-parameters are interpreted under a mixed set of terminal conditions, some of them under open-circuit and some under short-circuit conditions. They are called the hybrid h- and hybrid g-parameters. From these interpretations we see that h_{11} and g_{22} are impedances, whereas h_{22} and g_{11} are admittances. They are related to the z's and y's by

$$h_{11} = \frac{1}{y_{11}}, \qquad g_{11} = \frac{1}{z_{11}},$$

$$h_{22} = \frac{1}{z_{22}}, \qquad g_{22} = \frac{1}{y_{22}}. \tag{20}$$

The transfer g's and h's are dimensionless. The quantity h_{21} is the *forward short-circuit current gain*, and g_{12} is the *reverse short-circuit current gain*. The other two are voltage ratios: g_{21} is the *forward open-circuit voltage gain*, whereas h_{12} is the *reverse open-circuit voltage gain*. We shall use $\mathbf{H}$ and $\mathbf{G}$ to represent the corresponding matrices.

By direct computation we find the following relations among the transfer parameters:

$$h_{12} = \frac{-z_{12}}{z_{21}} h_{21}, \tag{21a}$$

$$g_{12} = \frac{-y_{12}}{y_{21}} g_{21}. \tag{21b}$$

In the special case of reciprocal networks these expressions simplify to $h_{12} = -h_{21}$ and $g_{12} = -g_{21}$. In words this means that for reciprocal networks the open-circuit voltage gain for transmission in one direction through the two-port equals the negative of the short-circuit current gain for transmission in the opposite direction.

Just as $\mathbf{Z}_{oc}$ and $\mathbf{Y}_{sc}$ are each the other's inverse, so also $\mathbf{H}$ and $\mathbf{G}$ are each the other's inverse. Thus

$$\mathbf{G}(s) = \mathbf{H}^{-1}(s), \qquad \det \mathbf{G} = \frac{1}{\det \mathbf{H}}. \tag{22}$$

You should verify this.

CHAIN PARAMETERS

The remaining two sets of equations relating the port variables express the voltage and current at one port in terms of the voltage and current at the other. These were, in fact, historically the first set used—in the analysis of transmission lines. One of these equations is

$$\begin{bmatrix} V_1(s) \\ I_1(s) \end{bmatrix} = \begin{bmatrix} A & B \\ C & D \end{bmatrix} \begin{bmatrix} V_2(s) \\ -I_2(s) \end{bmatrix}. \tag{23}$$

They are called the *chain*, or *ABCD*, parameters. The first name comes from the fact that they are the natural ones to use in a *cascade*, or *tandem*, or *chain* connection typical of a transmission system. Note the negative sign in $-I_2$, which is a consequence of the choice of reference for I_2.

Note that we are using the historical symbols for these parameters rather than using, say, a_{ij} for i and j equal 1 and 2, to make the system of notation uniform for all the parameters. We are also not introducing further notation to define the inverse parameters obtained by inverting (23), simply to avoid further proliferation of symbols.

The determinant of the chain matrix can be computed in terms of z's and y's. It is found to be

$$\det \begin{bmatrix} A & B \\ C & D \end{bmatrix} = AD - BC = \frac{z_{12}}{z_{21}} = \frac{y_{12}}{y_{21}}, \tag{24}$$

which is equal to 1 for reciprocal two-ports.

The preceding discussion is rather detailed and can become tedious if one loses sight of the objective of developing methods of representing the external behavior of two-ports by giving various relationships among the port voltages and currents. Each of these sets of relationships finds useful applications. For future reference we shall tabulate the inter-relationships among the various parameters. The result is given in Table 1. Note that these relationships are valid for a general nonreciprocal two-port.

TRANSMISSION ZEROS

There is an important observation that can be made concerning the locations of the zeros of the various transfer functions. This can be seen most readily, perhaps, by looking at one of the columns in Table 1; for

Table 1

	Open-Circuit Impedance Parameters	Short-Circuit Admittance Parameters	Chain Parameters	Hybrid h-Parameters	Hybrid g-Parameters
z	$\begin{matrix} z_{11} & z_{12} \\ z_{21} & z_{22} \end{matrix}$	$\begin{matrix} \dfrac{y_{22}}{\lvert y \rvert} & \dfrac{-y_{12}}{\lvert y \rvert} \\[2mm] \dfrac{-y_{21}}{\lvert y \rvert} & \dfrac{y_{11}}{\lvert y \rvert} \end{matrix}$	$\begin{matrix} \dfrac{A}{C} & \dfrac{AD-BC}{C} \\[2mm] \dfrac{1}{C} & \dfrac{D}{C} \end{matrix}$	$\begin{matrix} \dfrac{\lvert h \rvert}{h_{22}} & \dfrac{h_{12}}{h_{22}} \\[2mm] \dfrac{-h_{21}}{h_{22}} & \dfrac{1}{h_{22}} \end{matrix}$	$\begin{matrix} \dfrac{1}{g_{11}} & \dfrac{-g_{12}}{g_{11}} \\[2mm] \dfrac{g_{21}}{g_{11}} & \dfrac{\lvert g \rvert}{g_{11}} \end{matrix}$
y	$\begin{matrix} \dfrac{z_{22}}{\lvert z \rvert} & \dfrac{-z_{12}}{\lvert z \rvert} \\[2mm] \dfrac{-z_{21}}{\lvert z \rvert} & \dfrac{z_{11}}{\lvert z \rvert} \end{matrix}$	$\begin{matrix} y_{11} & y_{12} \\ y_{21} & y_{22} \end{matrix}$	$\begin{matrix} \dfrac{D}{B} & \dfrac{-(AD-BC)}{B} \\[2mm] \dfrac{-1}{B} & \dfrac{A}{B} \end{matrix}$	$\begin{matrix} \dfrac{1}{h_{11}} & \dfrac{-h_{12}}{h_{11}} \\[2mm] \dfrac{h_{21}}{h_{11}} & \dfrac{\lvert h \rvert}{h_{11}} \end{matrix}$	$\begin{matrix} \dfrac{\lvert g \rvert}{g_{22}} & \dfrac{g_{12}}{g_{22}} \\[2mm] \dfrac{-g_{21}}{g_{22}} & \dfrac{1}{g_{22}} \end{matrix}$
$ABCD$	$\begin{matrix} \dfrac{z_{11}}{z_{21}} & \dfrac{\lvert z \rvert}{z_{21}} \\[2mm] \dfrac{1}{z_{21}} & \dfrac{z_{22}}{z_{21}} \end{matrix}$	$\begin{matrix} \dfrac{-y_{22}}{y_{21}} & \dfrac{-1}{y_{21}} \\[2mm] \dfrac{-\lvert y \rvert}{y_{21}} & \dfrac{-y_{11}}{y_{21}} \end{matrix}$	$\begin{matrix} A & B \\ C & D \end{matrix}$	$\begin{matrix} \dfrac{-\lvert h \rvert}{h_{21}} & \dfrac{-h_{11}}{h_{21}} \\[2mm] \dfrac{-h_{22}}{h_{21}} & \dfrac{-1}{h_{21}} \end{matrix}$	$\begin{matrix} \dfrac{1}{g_{21}} & \dfrac{g_{11}}{g_{21}} \\[2mm] \dfrac{g_{22}}{g_{21}} & \dfrac{\lvert g \rvert}{g_{21}} \end{matrix}$
h	$\begin{matrix} \dfrac{\lvert z \rvert}{z_{22}} & \dfrac{z_{12}}{z_{22}} \\[2mm] \dfrac{-z_{21}}{z_{22}} & \dfrac{1}{z_{22}} \end{matrix}$	$\begin{matrix} \dfrac{1}{y_{11}} & \dfrac{-y_{12}}{y_{11}} \\[2mm] \dfrac{y_{21}}{y_{11}} & \dfrac{\lvert y \rvert}{y_{11}} \end{matrix}$	$\begin{matrix} \dfrac{B}{D} & \dfrac{AD-BC}{D} \\[2mm] \dfrac{-1}{D} & \dfrac{C}{D} \end{matrix}$	$\begin{matrix} h_{11} & h_{12} \\ h_{21} & h_{22} \end{matrix}$	$\begin{matrix} \dfrac{g_{22}}{\lvert g \rvert} & \dfrac{-g_{12}}{\lvert g \rvert} \\[2mm] \dfrac{-g_{21}}{\lvert g \rvert} & \dfrac{g_{11}}{\lvert g \rvert} \end{matrix}$
g	$\begin{matrix} \dfrac{1}{z_{11}} & \dfrac{-z_{12}}{z_{11}} \\[2mm] \dfrac{z_{21}}{z_{11}} & \dfrac{\lvert z \rvert}{z_{11}} \end{matrix}$	$\begin{matrix} \dfrac{\lvert y \rvert}{y_{22}} & \dfrac{y_{12}}{y_{22}} \\[2mm] \dfrac{-y_{21}}{y_{22}} & \dfrac{1}{y_{22}} \end{matrix}$	$\begin{matrix} \dfrac{C}{A} & \dfrac{-(AD-BC)}{A} \\[2mm] \dfrac{1}{A} & \dfrac{B}{A} \end{matrix}$	$\begin{matrix} \dfrac{h_{22}}{\lvert h \rvert} & \dfrac{-h_{12}}{\lvert h \rvert} \\[2mm] \dfrac{-h_{21}}{\lvert h \rvert} & \dfrac{h_{11}}{\lvert h \rvert} \end{matrix}$	$\begin{matrix} g_{11} & g_{12} \\ g_{21} & g_{22} \end{matrix}$

example, the column in which all parameters are expressed in terms of the y-parameters. We see that

$$z_{21}(s) = \frac{-y_{21}(s)}{y_{11}y_{22} - y_{12}y_{21}}, \tag{25a}$$

$$h_{21}(s) = \frac{y_{21}(s)}{y_{11}(s)}, \tag{25b}$$

$$g_{21}(s) = \frac{-y_{21}(s)}{y_{22}(s)}. \tag{25c}$$

Except for possible cancellations, all of these transfer functions will have the same zeros. We use the generic term *transmission zero* to refer to a value of s for which there is a transfer-function zero, without having to specify which transfer function—whether current gain, transfer admittance, or any other.

Example

As an illustrative example of the computation of two-port parameters, consider the network shown in Fig. 8, which can be considered as a

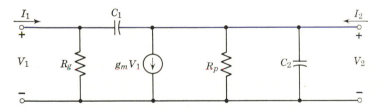

Fig. 8.   Example for calculating two-port parameters.

model for a vacuum triode under certain conditions. (The capacitances are the grid-to-plate and plate-to-cathode capacitances.) Let us compute the y-parameters for this network. The simplest procedure is to use the interpretations in (11). If the output terminals are short-circuited, the resulting network will take the form shown in Fig. 9. As far as the input terminals are concerned, the controlled source has no effect. Hence y_{11} is the admittance of the parallel combination of R_g and C_1:

$$y_{11}(s) = \frac{1}{R_g} + sC_1.$$

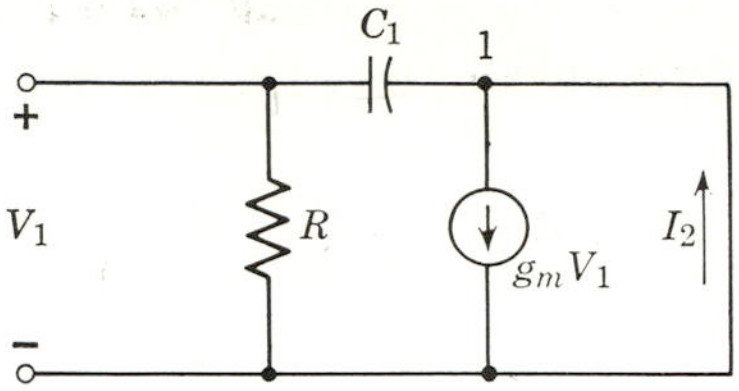

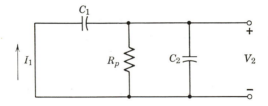

Fig. 9.   Network with output terminals shorted.

To find y_{21}, assume that a voltage source with transform $V_1(s)$ is applied at the input terminals. By applying Kirchhoff's current law at the node labeled 1 in Fig. 9, we find that $I_2 = g_m V_1 - sC_1 V_1$. Hence y_{21} becomes

$$y_{21} = \frac{I_2(s)}{V_1(s)}\bigg|_{V_2=0} = g_m - sC_1.$$

Now short-circuit the input terminals of the original network. The result will take the form in Fig. 10. Since V_1 is zero, the dependent source

Fig. 10.   Network with input terminals shorted.

current is also zero. It is now a simple matter to compute y_{22} and y_{12}:

$$y_{22} = \frac{I_2}{V_2}\bigg|_{V_1=0} = s(C_1 + C_2) + \frac{1}{R_p},$$

$$y_{12} = \frac{I_1}{V_2}\bigg|_{V_1=0} = -sC_1.$$

We see that y_{12} is different from y_{21}, as it should be, because of the presence of the controlled source.

If the y-parameters are known, any of the other sets of parameters can be computed by using Table 1. Note that even under the conditions that C_1 and C_2 are zero and R_g infinite, the y-parameters exist, but the z-parameters do not (z_{11}, z_{22}, and z_{21} become infinite).

3.4   INTERCONNECTION OF TWO-PORT NETWORKS

A given two-port network having some degree of complexity can be viewed as being constructed from simpler two-port networks whose ports are interconnected in certain ways. Conversely, a two-port network that is to be built can be designed by combining simple two-port structures as building blocks. From the designer's standpoint it is much easier to design simple blocks and to interconnect them than to design a complex network in one piece. A further practical reason for this approach is that it is much easier to shield smaller units and thus reduce parasitic capacitances to ground.

CASCADE CONNECTION

There are a number of ways in which two-ports can be interconnected. In the simplest interconnection of 2 two-ports, called the *cascade*, or tandem-connection, one port of each network is involved. Two two-ports are said to be connected *in cascade* if the output port of one is the input port of the second, as shown in Fig. 11.

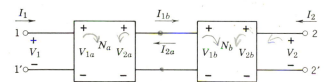

Fig. 11.   Cascade connection of two-ports.

Our interest in the problem of "interconnection" is, from the analysis point of view, to study how the parameters of the overall network are related to the parameters of the individual building blocks. The tandem combination is most conveniently studied by means of the $ABCD$-parameters. From the references in the figure we see that

$$\begin{bmatrix} V_1 \\ I_1 \end{bmatrix} = \begin{bmatrix} V_{1a} \\ I_{1a} \end{bmatrix}, \qquad \begin{bmatrix} V_{2a} \\ -I_{2a} \end{bmatrix} = \begin{bmatrix} V_{1b} \\ I_{1b} \end{bmatrix}, \qquad \begin{bmatrix} V_{2b} \\ -I_{2b} \end{bmatrix} = \begin{bmatrix} V_2 \\ -I_2 \end{bmatrix}.$$

Hence for the $ABCD$ system of equations of the network N_b we can write

$$\begin{bmatrix} V_{2a} \\ -I_{2a} \end{bmatrix} = \begin{bmatrix} V_{1b} \\ I_{1b} \end{bmatrix} = \begin{bmatrix} A_b & B_b \\ C_b & D_b \end{bmatrix}\begin{bmatrix} V_2 \\ -I_2 \end{bmatrix}.$$

Furthermore, if we write the $ABCD$ system of equations for the network N_a and substitute in the last equation, we get

$$\begin{bmatrix} V_1 \\ I_1 \end{bmatrix} = \begin{bmatrix} A_a & B_a \\ C_a & D_a \end{bmatrix} \begin{bmatrix} V_{2a} \\ -I_{2a} \end{bmatrix} = \begin{bmatrix} A_a & B_a \\ C_a & D_a \end{bmatrix} \begin{bmatrix} A_b & B_b \\ C_b & D_b \end{bmatrix} \begin{bmatrix} V_2 \\ -I_2 \end{bmatrix}.$$

Thus the ABCD-matrix of two-ports in cascade is equal to the product of the ABCD-matrices of the individual networks; that is,

$$\begin{bmatrix} A & B \\ C & D \end{bmatrix} = \begin{bmatrix} A_a & B_a \\ C_a & D_a \end{bmatrix} \begin{bmatrix} A_b & B_b \\ C_b & D_b \end{bmatrix}. \tag{26}$$

Once the relationships between the parameters of the overall two-port and those of the components are known for any one set of parameters, it is merely algebraic computation to get the relationships for any other set; for example, the open-circuit parameters of the overall two-port can be found in terms of those for each of the two cascaded ones by expressing the z-parameters in terms of the $ABCD$-parameters for the overall network, using (26) and then expressing the $ABCD$-parameters for each network in the cascade in terms of their corresponding z-parameters. The result will be

$$\begin{bmatrix} z_{11} & z_{12} \\ z_{21} & z_{22} \end{bmatrix} = \begin{bmatrix} z_{11a} - \dfrac{z_{12a} z_{21a}}{z_{22a} + z_{11b}} & \dfrac{z_{12a} z_{12b}}{z_{22a} + z_{11b}} \\ \dfrac{z_{21a} z_{21b}}{z_{22a} + z_{11b}} & z_{22b} - \dfrac{z_{12b} z_{21b}}{z_{22a} + z_{11b}} \end{bmatrix}. \tag{27}$$

The details of this computation are left to you.

A word of caution is necessary. When it is desired to determine some specific parameter of an overall two-port in terms of parameters of the components in the interconnection, it may be simpler to use a direct analysis than to rely on relationships such as those in Table 1. As an example, suppose it is desired to find the expression for z_{21} in Fig. 11. The term z_{21} is the ratio of open-circuit output voltage to input current: $z_{21} = V_2/I_1$. Suppose a current source I_1 is applied; looking into the output terminals of N_a, let the network be replaced by its Thévenin equivalent. The result is shown in Fig. 12. By definition, $z_{21b} = V_2/I_{1b}$

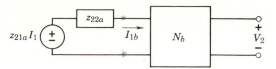

Fig. 12.   Replacement of network N_a by its Thévenin equivalent.

with the output terminals open. Now I_{1b} can easily be found from the network in Fig. 12 to be

$$I_{1b} = \frac{z_{21a} I_1}{z_{22a} + z_{11b}}.$$

Hence

$$z_{21b} = \frac{V_2}{I_{1b}} = \frac{V_2}{\dfrac{z_{21a} I_1}{z_{22a} + z_{11b}}} = \left(\frac{z_{22a} + z_{11b}}{z_{21a}}\right)\frac{V_2}{I_1}.$$

Finally,

$$z_{21} = \frac{V_2}{I_1} = \frac{z_{21a} z_{21b}}{z_{22a} + z_{11b}}, \tag{28}$$

which agrees with (27).

An important feature of cascaded two-ports is observed from the expressions for the transfer impedances in (27). The zeros of z_{21} are the zeros of z_{21a} and z_{21b}. (A similar relationship holds for z_{12}.) Thus the transmission zeros of the overall cascade consist of the transmission zeros of each of the component two-ports. This is the basis of some important methods of network synthesis. It permits individual two-ports to be designed to achieve certain transmission zeros before they are connected together. It also permits independent adjustment and tuning of elements within each two-port to achieve the desired null without influencing the adjustment of the cascaded two-ports.

PARALLEL AND SERIES CONNECTIONS

Now let us turn to other interconnections of two-ports, which, unlike the cascade connection, involve both ports. Two possibilities that immediately come to mind are parallel and series connections. Two *two-ports*

*are said to be connected in parallel if corresponding (input and output)
ports are connected in parallel as in Fig. 13a. In the parallel connection*

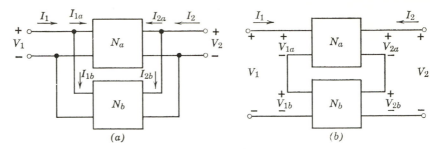

Fig. 13.   Parallel and series connections of two-ports.

the input and output voltages of the component two-ports are forced to
be the same, whereas the overall port currents equal the sums of the
corresponding component port currents. This statement assumes that
the port relationships of the individual two-ports are not altered when
the connection is made. In this case the overall port relationship can be
written as

$$\begin{bmatrix} I_1 \\ I_2 \end{bmatrix} = \begin{bmatrix} I_{1a} \\ I_{2a} \end{bmatrix} + \begin{bmatrix} I_{1b} \\ I_{2b} \end{bmatrix} = \begin{bmatrix} y_{11a} & y_{12a} \\ y_{21a} & y_{22a} \end{bmatrix} \begin{bmatrix} V_{1a} \\ V_{2a} \end{bmatrix} + \begin{bmatrix} y_{11b} & y_{12b} \\ y_{21b} & y_{22b} \end{bmatrix} \begin{bmatrix} V_{1b} \\ V_{2b} \end{bmatrix}$$

$$= \begin{bmatrix} y_{11a} + y_{11b} & y_{12a} + y_{12b} \\ y_{21a} + y_{21b} & y_{22a} + y_{22b} \end{bmatrix} \begin{bmatrix} V_1 \\ V_2 \end{bmatrix}.$$

That is, *the short-circuit admittance matrix of two-ports connected in parallel
equals the sum of the short-circuit admittance matrices of the component two-
ports*:

$$\mathbf{Y}_{sc} = \mathbf{Y}_{sca} + \mathbf{Y}_{scb}. \tag{30}$$

The dual of the parallel connection is the series connection. Two
*two-ports are connected in series if corresponding ports (input and output)
are connected in series*, as shown in Fig. 13b. In this connection the input
and output port currents are forced to be the same, whereas the overall
port voltages equal the sums of the corresponding port voltages of the
individual two-ports. Again, it is assumed that the port relationships of
the individual two-ports are not altered when the connection is made. In
this case the overall port relationship can be written as

$$\begin{bmatrix} V_1 \\ V_2 \end{bmatrix} = \begin{bmatrix} V_{1a} \\ V_{2a} \end{bmatrix} + \begin{bmatrix} V_{1b} \\ V_{2b} \end{bmatrix} = \begin{bmatrix} z_{11a} & z_{12a} \\ z_{21a} & z_{22a} \end{bmatrix} \begin{bmatrix} I_{1a} \\ I_{2a} \end{bmatrix} + \begin{bmatrix} z_{11b} & z_{12b} \\ z_{21b} & z_{22b} \end{bmatrix} \begin{bmatrix} I_{1b} \\ I_{2b} \end{bmatrix}$$

$$= \begin{bmatrix} z_{11a} + z_{11b} & z_{12a} + z_{12b} \\ z_{21a} + z_{21b} & z_{22a} + z_{22b} \end{bmatrix} \begin{bmatrix} I_1 \\ I_2 \end{bmatrix}. \tag{31}$$

That is, *the open-circuit impedance matrix of two-ports connected in series equals the sum of the open-circuit impedance matrices of the component two-ports*:

$$\mathbf{Z}_{oc} = \mathbf{Z}_{oca} + \mathbf{Z}_{ocb}. \tag{32}$$

Of these two—parallel and series connections—the parallel connection is more useful and finds wider application in synthesis. One reason for this is the practical one that permits two common-terminal (grounded) two-ports to be connected in parallel, the result being a common-terminal two-port. An example of this is the *parallel-ladders network* (of which the twin-tee null network is a special case) shown in Fig. 14.

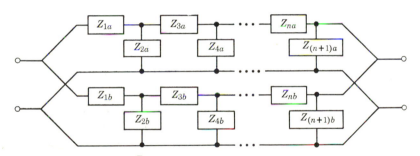

Fig. 14.   Parallel-ladders network.

On the other hand, the series connection of two common-terminal two-ports is not a common-terminal two-port unless one of them is a tee network. Consider two grounded two-ports connected in series, as in Fig. 15a. It is clear that this is inadmissible, since the ground terminal of N_a will short out parts of N_b, thus violating the condition that the individual two-ports be unaltered by the interconnection. The situation is remedied by making the common terminals of both two-ports common to each other, as in Fig. 15b. In this case the resulting two-port is not a common-terminal one. If one of the component two-ports is a tee, the series connection takes the form shown in Fig. 15c. This can be redrawn, as in Fig. 15d, as a common-terminal two-port. That the last two networks have the same z-parameters is left for you to demonstrate.

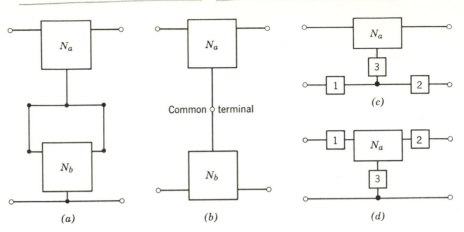

Fig. 15.   Series connection of common-terminal two-ports.

Variations of the series and parallel types of interconnections are possible by connecting the ports in series at one end and in parallel at the other. These are referred to as the *series-parallel* and *parallel-series* connections. As one might surmise, it is the *h*- and *g*-parameters of the individual two-ports that are added to give the overall *h*- and *g*-parameters, respectively. This also is left as an exercise.

PERMISSIBILITY OF INTERCONNECTION

It remains for us to inquire into the conditions under which two-ports can be interconnected without causing the port relationships of the individual two-ports to be disturbed by the connection. For the parallel connection, consider Fig. 16. A pair of ports, one from each two-port, is

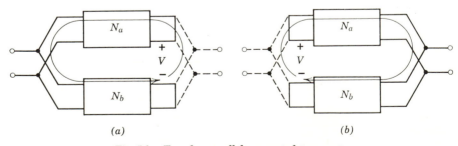

Fig. 16.   Test for parallel-connected two-ports.

connected in parallel, whereas the other ports are individually short-circuited. The short circuits are employed because the parameters characterizing the individual two-ports and the overall two-port are the short-circuit admittance parameters. If the voltage V shown in Fig. 16 is nonzero, then when the second ports are connected there will be a circulating current, as suggested in the diagram. Hence the condition that the current leaving one terminal of a port be equal to the current entering the other terminal of each individual two-port is violated, and the port relationships of the individual two-ports are altered.

For the case of the series connection, consider Fig. 17. A pair of ports,

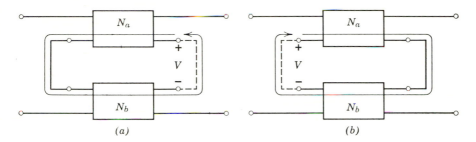

Fig. 17.   Test for series-connected two-ports.

one from each two-port, is connected in series, whereas the other ports are left open. The open circuits are employed because the parameters characterizing the individual two-ports and the overall two-port are the open-circuit impedance parameters. If the voltage V is nonzero, then when the second ports are connected in series there will be a circulating current, as suggested in the diagram. Again, the port relationships of the individual two-ports will be modified by the connection, and hence the addition of impedance parameters will not be valid for the overall network.

Obvious modifications of these tests apply to the series-parallel and parallel-series connections. The preceding discussion of the conditions under which the overall parameters for interconnected two-ports can be obtained by adding the component two-port parameters has been in rather skeletal form. We leave to you the task of supplying details.

When it is discovered that a particular interconnection cannot be made because circulating currents will be introduced, there is a way of stopping such currents and thus permitting the connection to be made. The approach is simply to put an isolating ideal transformer of $1:1$ turns ratio at one of the ports, as illustrated in Fig. 18 for the case of the parallel connection.

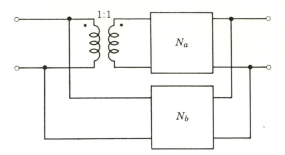

Fig. 18.   Isolating transformer to permit interconnection.

3.5   MULTIPORT NETWORKS

The preceding section has dealt with two-port networks in considerable detail. Let us now turn our attention to networks having more than two ports. The ideas discussed in the last section apply also to multiports with obvious extensions.

Consider the n-port network shown in Fig. 19. The external behavior

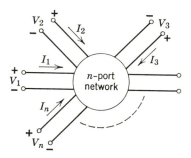

Fig. 19.   Multiport network.

of this network is completely described by giving the relationships among the port voltages and currents. One such relationship expresses all the port voltages in terms of the port currents:

$$
\begin{bmatrix} V_1 \\ V_2 \\ \vdots \\ V_n \end{bmatrix} = \begin{bmatrix} z_{11} & z_{12} & \cdots & z_{1n} \\ z_{21} & z_{22} & \cdots & z_{2n} \\ \vdots & \vdots & & \vdots \\ z_{n1} & z_{n2} & \cdots & z_{nn} \end{bmatrix} \begin{bmatrix} I_1 \\ I_2 \\ \vdots \\ I_n \end{bmatrix}
\tag{33a}
$$

or

$$\mathbf{V} = \mathbf{Z}_{\mathrm{oc}}\,\mathbf{I}. \qquad (33b)$$

By direct observation, it is seen that the parameters can be interpreted as

$$z_{jk} = \frac{V_j}{I_k}\bigg|_{\substack{\text{all other} \\ \text{currents}\,=\,0}} \qquad (34)$$

which is simply the extension of the open-circuit impedance representation of a two-port. The matrix $\mathbf{Z}_{\mathrm{oc}}$ is the same as that in (12) except that it is of order n.

The short-circuit admittance matrix for a two-port can also be directly extended to an n-port. Thus

$$\mathbf{I} = \mathbf{Y}_{\mathrm{sc}}\,\mathbf{V}, \qquad \mathbf{Y}_{\mathrm{sc}} = [y_{jk}] \qquad (35a)$$

where

$$y_{jk} = \frac{I_j}{V_k}\bigg|_{\substack{\text{all other} \\ \text{voltages}\,=\,0}} \qquad (35b)$$

If we now think of extending the hybrid representations of a two-port, we encounter some problems. In a hybrid representation the variables are mixed voltage and current. For a network of more than two ports, how are the "independent" and "dependent" variables to be chosen? In a three-port network, for example, the following three choices can be made:

$$\begin{bmatrix} V_1 \\ V_2 \\ I_3 \end{bmatrix} = \mathbf{M}_1 \begin{bmatrix} I_1 \\ I_2 \\ V_3 \end{bmatrix}, \qquad \begin{bmatrix} V_1 \\ I_2 \\ V_3 \end{bmatrix} = \mathbf{M}_2 \begin{bmatrix} I_1 \\ V_2 \\ I_3 \end{bmatrix}, \qquad \begin{bmatrix} I_1 \\ V_2 \\ V_3 \end{bmatrix} = \mathbf{M}_3 \begin{bmatrix} V_1 \\ I_2 \\ I_3 \end{bmatrix}$$

as well as their inverses. In these choices each vector contains exactly one variable from each port. It would also be possible to make such selections as

$$\begin{bmatrix} V_1 \\ V_2 \\ I_2 \end{bmatrix} = \mathbf{M}_4 \begin{bmatrix} I_1 \\ I_3 \\ V_3 \end{bmatrix}$$

where each vector contains both the current and voltage of one particular port. The former category are like the hybrid h and hybrid g representations of a two-port. The latter has some of the features of the chain-matrix representation. It is clearly not very productive to pursue this topic of possible representations in the general case.

Just as in the case of two-ports, it is possible to interconnect multiports. Two multiports are said to be connected in parallel if their ports are connected in parallel in pairs. It is not, in fact, necessary for the two multiports to have the same number of ports. The ports are connected in parallel in pairs until we run out of ports. It does not matter whether we run out for both networks at the same time or earlier for one network. Similarly, two multiports are said to be connected in series if their ports are connected in series in pairs. Again, the two multiports need not have the same number of ports.

As in the case of two-ports, the overall y-matrix for two n-ports connected in parallel equals the sum of the y-matrices of the individual n-ports. Similarly, the overall z-matrix of two n-ports connected in series equals the sum of the z-matrices of the individual n-ports. This assumes, of course, that the interconnection does not alter the parameters of the individual n-ports.

3.6   THE INDEFINITE ADMITTANCE MATRIX

The port description of networks is possible only when external connections are to be made to the terminals of the network taken in pairs. More generally, the terminals need not be paired into ports. In such a case it would be useful to have a description of the external behavior as a multiterminal network rather than a multiport network. In this section we shall introduce such a description.

Let us return to Fig. 6. The six-terminal network shown there is viewed as a common-terminal five-port by defining the voltages of five of the terminals with reference to the voltage of the sixth one as a datum. For any such common-terminal multiport, suppose the datum for voltage is taken as an arbitrary point external to the network, as shown for an n-terminal network in Fig. 20. We assume that the network is connected, implying that none of the terminals is isolated from the rest of the network.

The currents are not port currents but terminal currents. Clearly, they satisfy Kirchhoff's current law, so that

$$\sum_{k=1}^{n} I_k(s) = 0.$$

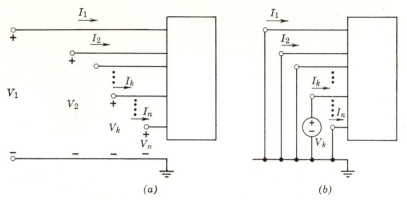

Fig. 20.   Definition of terminal variables.

Since the network is linear, currents can be expressed as a linear combination of the terminal voltages to yield

$$
\begin{bmatrix} I_1 \\ I_2 \\ \vdots \\ I_n \end{bmatrix} = \begin{bmatrix} y_{11} & y_{12} & \cdots & y_{1n} \\ y_{21} & y_{22} & \cdots & y_{2n} \\ \vdots & \vdots & & \vdots \\ y_{n1} & y_{n2} & \cdots & y_{nn} \end{bmatrix} \begin{bmatrix} V_1 \\ V_2 \\ \vdots \\ V_n \end{bmatrix}. \tag{36}
$$

The elements of the coefficient matrix of this equation are, dimensionally, admittance. They are, in fact, short-circuit admittances. Figure 20b shows all terminals but one grounded to the arbitrary datum; to the kth terminal is connected a voltage source. Each of the terminal currents can now be found. The parameters of the matrix will be

$$
y_{jk} = \frac{I_j}{V_k} \Big|_{\substack{\text{all other terminals} \\ \text{grounded to datum.}}} \tag{37}
$$

They are almost like the y-parameters of a multiport. We shall examine the relationships below.

The coefficient matrix in (36) is called the *indefinite admittance matrix* and is designated $\mathbf{Y}_i$. A number of the properties of this matrix will now be established.

First, suppose the scalar equations represented by the matrix equation 36 are all added. By Kirchhoff's current law, the sum of the currents is zero. Hence

$$
(y_{11} + y_{21} + \cdots + y_{n1})V_1 + (y_{12} + y_{22} + \cdots + y_{n2})V_2 + \cdots
$$
$$
+ (y_{1n} + y_{2n} + \cdots + y_{nn})V_n = 0.
$$

The quantity within each pair of parentheses is the sum of the elements in a column of $\mathbf{Y}_i$. The terminal voltages are all independent. Suppose all terminals but one, say the kth one, are short-circuited. This expression then reduces to

$$(y_{1k} + y_{2k} + \cdots + y_{nk})V_k = 0. \tag{38}$$

Since $V_k \neq 0$, this means the sum of elements in each column of the indefinite admittance matrix equals zero. Thus the columns are not all independent and $\mathbf{Y}_i$ is a singular matrix.

What is true of the columns is also true of the rows. This can be shown as follows. Suppose all but the kth terminal are left open, and to the kth is applied a voltage source V_k. Assuming, as we did, that none of the terminals is isolated, the voltages of all other terminals will also equal V_k. All terminal currents will be zero—obviously, all but I_k because the terminals are open, and I_k because of Kirchhoff's current law. With all the voltages equal in (36), the jth current can be written as

$$I_j = (y_{j1} + y_{j2} + \cdots + y_{jn})V_k = 0.$$

Since $V_k \neq 0$, the sum of elements in each row of $\mathbf{Y}_i$ equals zero.

To make a common-terminal n-port out of a network with $n+1$ terminals is simple once the indefinite admittance matrix is known. If the terminal that is to be common to the n-ports, say terminal $n+1$, is taken as the arbitrary datum, its voltage will be zero. Hence the last column of $\mathbf{Y}_i$ in (36) can be removed, since its elements are the coefficients of this voltage, which is zero. Also, the current at this terminal is redundant, by Kirchhoff's current law, and hence the last row of $\mathbf{Y}_i$ can also be removed. Thus, to make one of the terminals of a network the common terminal of a grounded n-port, simply delete the row and column of the indefinite admittance matrix corresponding to that terminal.

The converse operation permits the formation of a $\mathbf{Y}_i$ matrix from the short-circuit admittance matrix of a grounded n-port; that is, given the short-circuit admittance matrix of a common-terminal n-port, add to the matrix another row each of whose elements is the negative sum of all elements in the corresponding column. Then add another column, each of whose elements is the negative of the sum of all elements in the corresponding row.

Let us illustrate this process with the common-terminal two-port shown in Fig. 21a. The port voltages that would normally be labeled V_1 and V_2 are labeled V_{ac} and V_{bc} to emphasize the two terminals of each port. In Fig. 21b the voltage datum is taken as a point other than one of the

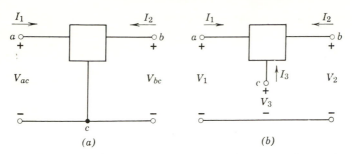

Fig. 21. Grounded two-port represented as three-terminal network.

terminals. Let us first write the two-port y-equations; then replace V_{ac} by $V_1 - V_3$ and V_{bc} by $V_2 - V_3$:

$$I_1 = y_{11}V_{ac} + y_{12}V_{bc} = y_{11}(V_1 - V_3) + y_{12}(V_2 - V_3),$$
$$I_2 = y_{21}V_{ac} + y_{22}V_{bc} = y_{21}(V_1 - V_3) + y_{22}(V_2 - V_3).$$

By Kirchhoff's current law, I_3 in Fig. 21b equals $-(I_1 + I_2)$. When this equation is added to the previous two the result becomes

$$I_1 = y_{11}V_1 + y_{12}V_2 - (y_{11} + y_{12})V_3,$$
$$I_2 = y_{21}V_1 + y_{22}V_2 - (y_{21} + y_{22})V_3, \qquad (39)$$
$$I_3 = -(y_{11} + y_{21})V_1 - (y_{12} + y_{22})V_2 + (y_{11} + y_{12} + y_{21} + y_{22})V_3.$$

The coefficient matrix of these equations is $\mathbf{Y}_i$. Notice how it could have been formed immediately from the original $\mathbf{Y}_{sc}$-matrix by the process of adding a row and column, using the zero-sum property of rows and columns.

The preceding discussion provides a method for taking the $\mathbf{Y}_{sc}$-matrix of a common-terminal multiport with one terminal as the common terminal and from it easily writing the $\mathbf{Y}_{sc}$-matrix of the common-terminal multiport with any other terminal taken as the common ground. This is especially useful in obtaining, say, the grounded-grid representation of a triode amplifier from the grounded-cathode representation or the common-base representation of a transistor amplifier from the common-emitter representation. To illustrate the approach, consider the common-terminal two-port shown in Fig. 22a. The short-circuit admittance matrix of this two-port is the following:

$$\mathbf{Y}_{sc1} = \begin{matrix} & G & P \\ & \begin{bmatrix} G_g + sC & -sC \\ g_m - sC & G_p + sC \end{bmatrix} & \begin{matrix} G \\ P \end{matrix} \end{matrix}$$

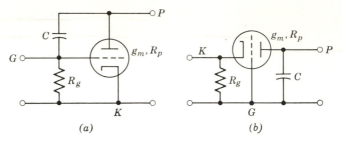

Fig. 22.   Indefinite admittance matrix for electronic amplifier.

(The letters at the top and the end identify the columns and the rows with specific terminals.) From this the indefinite admittance matrix is immediately written as

$$
\mathbf{Y}_i =
\begin{array}{ccc}
\;\;G & \;\;P & \;\;K
\end{array}
\left[
\begin{array}{ccc}
G_g + sC & -sC & -G_g \\
g_m - sC & G_p + sC & -(g_m + G_p) \\
-(g_m + G_g) & -G_p & g_m + G_g + G_p
\end{array}
\right]
\begin{array}{c}
G \\
P \\
K
\end{array}
\qquad (40)
$$

To find the short-circuit admittance matrix of the grounded-grid configuration shown in Fig. 22b, all that is necessary is to delete the row and column corresponding to the grid terminal, which is the first one in (40). Of course, it should be ensured that the order of the rows and columns that remain correspond to the desired order of input and output—in this case, cathode-plate. Hence

$$
\mathbf{Y}_{\text{sc2}} =
\begin{array}{cc}
\;\;K & \;\;P
\end{array}
\left[
\begin{array}{cc}
g_m + G_g + G_p & -G_p \\
-(G_p + g_m) & G_p + sC
\end{array}
\right]
\begin{array}{c}
K \\
P
\end{array}
$$

Once the indefinite admittance matrix of a multiterminal network is at hand, the results of further manipulations on the network easily show up as changes in the $\mathbf{Y}_i$-matrix. Some of these will now be discussed.

CONNECTING TWO TERMINALS TOGETHER

Suppose two terminals of an n-terminal network are connected together into a single external terminal. The two external currents are replaced by one, equal to the sum of the original two. The two voltages are now identical. Hence the $\mathbf{Y}_i$-matrix of the resulting $(n-1)$ terminal network is obtained by adding the two corresponding rows and columns of the original $\mathbf{Y}_i$-matrix. (This sum replaces the original two rows and columns.) The extension to more than two terminals is obvious.

SUPPRESSING TERMINALS

Suppose one of the terminals is to be made an internal terminal to which external connections are not to be made. This procedure is called *suppressing* a terminal. The current at that terminal, say the nth one, will be zero. The equation for $I_n = 0$ can be solved for V_n (assuming $y_{nn} \neq 0$) and the result substituted into the remaining equations. This will eliminate V_n and will leave $n-1$ equations in $n-1$ voltages. This can be extended in matrix form to more than one terminal as follows. With $\mathbf{I} = \mathbf{Y}_i \mathbf{V}$, partition the matrices as follows:

$$\begin{bmatrix} \mathbf{I}_a \\ \mathbf{I}_b \end{bmatrix} = \begin{bmatrix} \mathbf{Y}_{11} & \mathbf{Y}_{12} \\ \mathbf{Y}_{21} & \mathbf{Y}_{22} \end{bmatrix} \begin{bmatrix} \mathbf{V}_a \\ \mathbf{V}_b \end{bmatrix}$$

or

$$\begin{aligned} \mathbf{I}_a &= \mathbf{Y}_{11}\mathbf{V}_a + \mathbf{Y}_{12}\mathbf{V}_b, \\ \mathbf{I}_b &= \mathbf{Y}_{21}\mathbf{V}_a + \mathbf{Y}_{22}\mathbf{V}_b, \end{aligned} \tag{41}$$

where $\mathbf{I}_b$ and $\mathbf{V}_b$ correspond to the terminals that are to be suppressed; that is, vector $\mathbf{I}_b = 0$. From the second equation solve for $\mathbf{V}_b$ and substitute into the first equations. The result will be

$$\begin{aligned} \mathbf{V}_b &= -\mathbf{Y}_{22}^{-1}\mathbf{Y}_{21}\mathbf{V}_a, \\ \mathbf{I}_a &= (\mathbf{Y}_{11} - \mathbf{Y}_{12}\mathbf{Y}_{22}^{-1}\mathbf{Y}_{21})\mathbf{V}_a. \end{aligned} \tag{42}$$

The new indefinite admittance matrix is

$$\mathbf{Y}_i = \mathbf{Y}_{11} - \mathbf{Y}_{12}\mathbf{Y}_{22}^{-1}\mathbf{Y}_{21}. \tag{43}$$

NETWORKS IN PARALLEL

The indefinite admittance matrix of two networks connected in parallel equals the sum of the $\mathbf{Y}_i$ matrices of each. By "connected in parallel" we mean that each terminal of one network is connected to a terminal of the other and both have a common datum for voltage. It is not necessary that the two networks have the same number of terminals. If they do not, then rows and columns of zeros are appended to the $\mathbf{Y}_i$-matrix of the network having the fewer terminals. In particular, a simple two-terminal branch connected across two terminals of a multi-terminal network can be considered to be connected in parallel with it. Note that the indefinite admittance matrix of a branch having admittance Y is

$$\mathbf{Y}_i \text{ of a single branch} = \begin{bmatrix} Y & -Y \\ -Y & Y \end{bmatrix}.$$

THE COFACTORS OF THE DETERMINANT OF $\mathbf{Y}_i$

A very interesting property of the determinant of the indefinite admittance matrix results from the fact that the sum of elements of each row or column equals zero. Suppose $\det \mathbf{Y}_i$ is expanded along the jth row. The result is

$$\det \mathbf{Y}_i = y_{j1}\Delta_{j1} + y_{j2}\Delta_{j2} + \cdots + y_{jn}\Delta_{jn},$$

where Δ_{jk} is the (j, k)th cofactor of $\det \mathbf{Y}_i$. Because the elements of the jth row sum to zero, we can write one of the elements as the negative sum of all the others. Thus

$$y_{j1} = -(y_{j2} + y_{j3} + \cdots + y_{jn}).$$

When this is inserted into the preceding equation and terms are collected, the result becomes

$$\det \mathbf{Y}_i = y_{j2}(\Delta_{j2} - \Delta_{j1}) + y_{j3}(\Delta_{j3} - \Delta_{j1}) + \cdots + y_{jn}(\Delta_{jn} - \Delta_{j1}) = 0.$$

The determinant equals zero because $\mathbf{Y}_i$ is singular. Furthermore, it is zero no matter what the values of the elements y_{jk} may be. Hence the last equation can be satisfied only if each of the parenthetical terms are zero; that is,

$$\Delta_{jk} = \Delta_{j1}. \tag{44}$$

This means all cofactors of elements of any row are equal.

The same procedure, starting with an expansion of $\det \mathbf{Y}_i$ along a column, will yield a similar result concerning the equality of cofactors of elements of any column. Since each row and column has a common element, the cofactor of this element equals all cofactors of that row and column. The conclusion is that *all (first) cofactors of the indefinite admittance matrix are equal.* This property has led to the name *equi-cofactor matrix* for $\mathbf{Y}_i$.

Example

Let us now illustrate with an example how the indefinite admittance matrix can be used for certain network calculations. Consider the network shown in Fig. 23a. It is desired to find the short-circuit admittance matrix of the common-terminal two-port shown. We shall do this by (1) finding the indefinite admittance matrix of the four-terminal network in Fig. 23b, (2) adding to it that of the single branch, (3) suppressing terminal 3, and finally (4) making terminal 4 the datum.

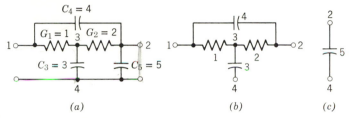

(a)   (b)   (c)

Fig. 23. Finding short-circuit admittances by using $\mathbf{Y}_i$.

To find $\mathbf{Y}_i$ in Fig. 23b, we shall first treat that network as a common-terminal three-port with terminal 4 as datum. The y-parameters of this three-port can be found from the definitions; for example, apply a voltage to the left-hand port and short-circuit the other two, as in Fig. 24. Three

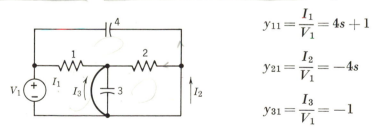

$$y_{11} = \frac{I_1}{V_1} = 4s + 1$$

$$y_{21} = \frac{I_2}{V_1} = -4s$$

$$y_{31} = \frac{I_3}{V_1} = -1$$

Fig. 24. Calculating $\mathbf{Y}_{sc}$ for the three-port.

of the y-parameters are easily found from the diagram shown there. The remaining y's are found in a similar way, with the result

$$\mathbf{Y}_{sc} = \begin{bmatrix} 4s+1 & -4s & -1 \\ -4s & 4s+2 & -2 \\ -1 & -2 & 3s+3 \end{bmatrix}.$$

The indefinite admittance matrix is easily found by adding a row and column whose elements are determined by the zero-sum property of rows and columns. To this is added the $\mathbf{Y}_i$-matrix of branch 5. Since it is connected between terminals 2 and 4, its nonzero elements appear in those two rows and columns. The overall $\mathbf{Y}_i$-matrix is

$$\mathbf{Y}_i = \begin{bmatrix} 4s+1 & -4s & -1 & 0 \\ -4s & 4s+2 & -2 & 0 \\ -1 & -2 & 3s+3 & -3s \\ 0 & 0 & -3s & 3s \end{bmatrix} + \begin{bmatrix} 0 & 0 & 0 & 0 \\ 0 & 5s & 0 & -5s \\ 0 & 0 & 0 & 0 \\ 0 & -5s & 0 & 5s \end{bmatrix}$$

$$= \begin{bmatrix} 4s+1 & -4s & -1 & 0 \\ -4s & 9s+2 & -2 & -5s \\ -1 & -2 & 3s+3 & -3s \\ 0 & -5s & -3s & 8s \end{bmatrix}.$$

The next step is to suppress terminal 3. For this purpose we interchange rows and columns 3 and 4 in order to make 3 the last one. Then we partition as follows, in order to identify the submatrices in (41):

$$\mathbf{Y}_i = \left[\begin{array}{ccc:c} 4s+1 & -4s & 0 & -1 \\ -4s & 9s+2 & -5s & -2 \\ 0 & -5s & 8s & -3s \\ \hdashline -1 & -2 & -3s & 3s+3 \end{array} \right] = \begin{bmatrix} \mathbf{Y}_{11} & \mathbf{Y}_{12} \\ \mathbf{Y}_{21} & \mathbf{Y}_{22} \end{bmatrix}.$$

Then

$$\mathbf{Y}_{12}\mathbf{Y}_{22}^{-1}\mathbf{Y}_{21} = \begin{bmatrix} -1 \\ -2 \\ -3s \end{bmatrix}\left[\frac{1}{3s+3}\right][-1 \quad -2 \quad -3s] = \frac{1}{3s+3}\begin{bmatrix} 1 & 2 & 3s \\ 2 & 4 & 6s \\ 3s & 6s & 9s^2 \end{bmatrix}.$$

The new indefinite admittance matrix with terminal 3 suppressed, using (43), is

$$
\mathbf{Y}_{i\ new} =
\begin{bmatrix}
4s+1 & -4s & 0 \\
-4s & 9s+2 & -5s \\
0 & -5s & 8s
\end{bmatrix}
- \frac{1}{3s+3}
\begin{bmatrix}
1 & 2 & 3s \\
2 & 4 & 6s \\
3s & 6s & 9s^2
\end{bmatrix}
$$

$$
=
\begin{bmatrix}
4s+1-\dfrac{1}{3s+3} & -\left(4s+\dfrac{2}{3s+3}\right) & \dfrac{-3s}{3s+3} \\[3mm]
-\left(4s+\dfrac{2}{3s+3}\right) & 9s+2-\dfrac{4}{3s+3} & -\left(5s+\dfrac{6s}{3s+3}\right) \\[3mm]
\dfrac{-3s}{3s+3} & -\left(5s+\dfrac{6s}{3s+3}\right) & 8s-\dfrac{9s^2}{3s+3}
\end{bmatrix}
$$

Finally, terminal 4 is made the common terminal by deleting the last row and column. The desired y-matrix of the two-port is

$$
\mathbf{Y}_{sc} =
\begin{bmatrix}
4s+1-\dfrac{1}{3s+3} & -\left(4s+\dfrac{2}{3s+3}\right) \\[3mm]
-\left(4s+\dfrac{2}{3s+3}\right) & 9s+2-\dfrac{4}{3s+3}
\end{bmatrix}.
$$

It may appear that a conventional approach would have required less work. It is true that more steps are involved here, but each step is almost trivial; many of them are simply written by inspection. Also, the last row and column of $\mathbf{Y}_{i\ new}$ need not be calculated; it is done here only for completeness.

3.7   THE INDEFINITE IMPEDANCE MATRIX

After learning about the indefinite admittance matrix, natural curiosity probably impels you to think of a dual, which might be called an *indefinite impedance matrix*. All our notions of duality dispose us favorably to such an idea.

Look back at Fig. 20, in terms of which the notion of $\mathbf{Y}_i$ was developed. The important condition that led to the zero-sum property of rows and columns is the fact that the sum of all terminal currents equals 0, by Kirchhoff's current law. Clearly, for the dual situation we would need to find that the sum of some set of voltages equals zero, by Kirchhoff's voltage law. Obviously, the terminal voltages are not such a set of voltages, since Kirchhoff's voltage law is satisfied only on a closed path, and the terminal voltages are not a set of voltages encountered on a closed path. However, if we choose as variables the voltages between adjacent pairs of terminals (as in Fig. 25a), we see that they do satisfy Kirchhoff's voltage law. There will be as many of these voltage variables as there are terminals.

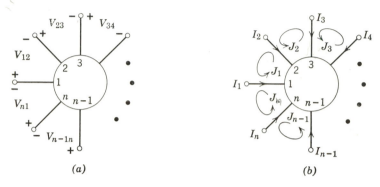

Fig. 25.   Variables for the indefinite impedance matrix.

For the linear networks with which we are concerned, linear relations expressing these voltage variables in terms of the terminal currents can be written. However, it is more convenient to define a set of currents other than the terminal currents, as shown in Fig. 25b. The currents J_k are external loop currents between pairs of terminals. There is a simple linear relationship between the terminal currents and these loop currents; namely, $I_k = J_k - J_{k+1}$. Hence a linear relationship expressing the voltages in terms of these loop currents can be written just as readily as one expressing them in terms of the terminal currents. Such a relationship will be

$$
\begin{bmatrix} V_{12} \\ V_{23} \\ \cdot \\ \cdot \\ V_{n1} \end{bmatrix} = \begin{bmatrix} z_{11} & z_{12} & \cdots & z_{1n} \\ z_{21} & z_{22} & \cdots & z_{2n} \\ \cdot & \cdot & & \cdot \\ \cdot & \cdot & & \cdot \\ z_{n1} & z_{n2} & \cdots & z_{nn} \end{bmatrix} \begin{bmatrix} J_1 \\ J_2 \\ \cdot \\ \cdot \\ J_n \end{bmatrix} \quad \text{or} \quad \begin{bmatrix} V_{12} \\ V_{23} \\ \cdot \\ \cdot \\ V_{n1} \end{bmatrix} = \mathbf{Z}_i \begin{bmatrix} J_1 \\ J_2 \\ \cdot \\ \cdot \\ J_n \end{bmatrix}. \quad (45)
$$

The matrix of these equations is designated $\mathbf{Z}_i$ and is called the *indefinite impedance matrix.*

Exactly as for the $\mathbf{Y}_i$-matrix, it is possible to show that the *sum of the elements of each row and column of* $\mathbf{Z}_i$ *equals zero.* This is done for the columns by adding the equations and using Kirchhoff's voltage law. It is done for the rows by setting all voltages but one equal to zero and applying a current source between the terminals associated with that voltage.

This implies shorting all terminals together, which makes (1) the last voltage also zero, by Kirchhoff's voltage law, and (2) all the external loop currents equal. The details of these steps are left to you. Then, as for Y_i, the indefinite impedance matrix is singular.

Recall that the indefinite admittance matrix was simply related to the short-circuit admittance matrix of the common terminal $(n-1)$ port derived by short-circuiting one terminal of the network in Fig. 20 to the datum. The corresponding situation here is: The indefinite impedance matrix Z_i is simply related to the open-circuit impedance matrix Z_{oc} of the $(n-1)$ port derived by opening one pair of terminals of the network of Fig. 25. (The open-circuit impedance matrix thus established will be called the *common-loop* open-circuit impedance matrix.) In fact, Z_i is obtained by first adding a row to Z_{oc}, each element of that row being the negative sum of all elements in the corresponding column, and then adding a column to the resulting matrix, each element of that column being the negative sum of all the elements in the corresponding row.

For the Y_i-matrix the zero-sum property of rows and columns led to the equicofactor property. In the same way *the (first) cofactors of the indefinite impedance matrix are all equal*.

Because common-terminal port relations are often encountered, let us examine the relation of a common-terminal open-circuit impedance matrix to a common-loop open-circuit impedance matrix. In Fig. 25, suppose terminal n is to be the common terminal and the voltage variables for the port representation are to be the terminal voltages taken with respect to terminal n as datum. Thus $V_n = 0$. Since the sum of the terminal currents equals zero, by Kirchhoff's current law, one of them is redundant. Since each terminal current is the difference between two external loop currents, the terminal currents will not change if all the external currents are increased or reduced by the same amount. Suppose J_n is taken to be zero. This is equivalent to subtracting J_n from each external current, so the terminal currents remain unchanged. From Fig. 25, we can write the following:

$$I_1 = J_1, \qquad\qquad \text{or} \qquad J_1 = I_1.$$
$$I_2 = J_2 - J_1 = J_2 - I_1, \qquad \text{or} \qquad J_2 = I_1 + I_2.$$
$$I_3 = J_3 - J_2 = J_3 - I_1 - I_2, \qquad \text{or} \qquad J_3 = I_1 + I_2 + I_3.$$
$$\vdots \qquad\qquad\qquad\qquad\qquad\qquad \vdots$$
$$I_{n-1} = J_{n-1} - J_{n-2}$$
$$= J_{n-1} - (I_1 + I_2 + \cdots + I_{n-2}) \qquad \text{or} \qquad J_{n-1} = I_1 + I_2 + \cdots + I_{n-1}.$$

The equations for the J's can be written

$$\mathbf{J} = \mathbf{MI}, \tag{46}$$

where $\mathbf{M}$ is the lower triangular matrix of order $n - 1$.

$$\mathbf{M} = \begin{bmatrix} 1 & 0 & 0 & \cdots & 0 \\ 1 & 1 & 0 & \cdots & 0 \\ 1 & 1 & 1 & \cdots & 0 \\ \vdots & \vdots & \vdots & & \vdots \\ 1 & 1 & 1 & \cdots & 1 \end{bmatrix}. \tag{47}$$

A similar development applies for the voltages. The port voltage V_k is the voltage of terminal k relative to that of terminal n. These port voltages are expressible as

$$\begin{bmatrix} V_1 \\ V_2 \\ \vdots \\ V_{n-1} \end{bmatrix} = \mathbf{M}' \begin{bmatrix} V_{12} \\ V_{23} \\ \vdots \\ V_{n-1,\,n} \end{bmatrix}. \tag{48}$$

The details are left for you.

Since $J_n = 0$, the last column in $\mathbf{Z}_i$ in (45) can be deleted. Similarly, the last row can be deleted because V_{n1} does not appear in (48). In this way we obtain the common-loop open-circuit impedance matrix $\mathbf{Z}_{oc(l)}$. Hence, using (48), (46), and (45) with the last row and column deleted, we get

$$\begin{bmatrix} V_1 \\ V_2 \\ \vdots \\ V_{n-1} \end{bmatrix} = \mathbf{M}' \begin{bmatrix} V_{12} \\ V_{13} \\ \vdots \\ V_{n-1,\,n} \end{bmatrix} = \mathbf{M}' \mathbf{Z}_{oc(l)} \begin{bmatrix} J_1 \\ J_2 \\ \vdots \\ J_{n-1} \end{bmatrix} = \mathbf{M}' \mathbf{Z}_{oc(l)} \mathbf{M} \begin{bmatrix} I_1 \\ I_2 \\ \vdots \\ I_{n-1} \end{bmatrix}. \tag{49}$$

(Keep in mind that $\mathbf{Z}_{oc(l)}$ here is the $\mathbf{Z}_i$ of (45) with its last row and column removed.) This equation relates the port voltages to the port currents. Hence the common-terminal open-circuit impedance matrix $\mathbf{Z}_{oc}$ is

$$\mathbf{Z}_{oc} = \mathbf{M}' \mathbf{Z}_{oc(l)} \mathbf{M}, \tag{50a}$$

and

$$\mathbf{Z}_{oc(l)} = (\mathbf{M}')^{-1} \mathbf{Z}_{oc} \mathbf{M}^{-1}. \tag{50b}$$

The last follows from the fact that a triangular matrix with nonzero diagonal terms is nonsingular. The relationship here is seen to be much more complicated than the corresponding one relating the common-terminal short-circuit admittance matrix $\mathbf{Y}_{sc}$ to $\mathbf{Y}_i$. Given $\mathbf{Z}_i$, $(50a)$ permits finding $\mathbf{Z}_{oc}$ after first writing $\mathbf{Z}_{oc(l)}$. Conversely, given $\mathbf{Z}_{oc}$ for a common-terminal $(n-1)$ port, $(50b)$ gives $\mathbf{Z}_{oc(l)}$. From this, $\mathbf{Z}_i$ is obtained by adding a row and column, using the zero-sum property of rows and columns.

3.8 TOPOLOGICAL FORMULAS FOR NETWORK FUNCTIONS

Let us pause briefly and review what has been done in this chapter. In the first section we defined network functions as the ratios of Laplace transforms of a response to an excitation. These may be driving-point functions or transfer functions; but for linear, lumped networks they will all be rational functions of the complex frequency variable s. Expressions for any of these functions can be found by solving the loop or node equations. In all cases the network functions can be expressed in terms of ratios of the determinant of the node admittance matrix and its cofactors or the loop impedance matrix and its cofactors. The subsequent sections of the chapter were devoted to a discussion of different ways of describing the external behavior of networks. These descriptions all entail sets of network functions defined under various conditions imposed on the terminals (open-circuit impedance, short-circuit admittance, etc.). Any of these functions can be evaluated according to the discussion in the first section. They all boil down to calculating a determinant and its cofactor.

We shall now turn our attention to the task of finding relatively simple means for evaluating determinants. The usual methods of evaluating determinants (e.g., cofactor expansion or pivotal condensation) require that many terms be multiplied together and then added. In this process many terms eventually cancel, but only after extensive calculations have been made. It would be of tremendous value to know which terms will cancel in the end. The method we shall now discuss achieves exactly this result.

DETERMINANT OF THE NODE ADMITTANCE MATRIX

We start by considering the node equations $\mathbf{Y}_n(s)\mathbf{V}_n(s) = \mathbf{J}(s)$, where

$$\mathbf{Y}_n = \mathbf{AYA'}, \tag{51}$$

in which $\mathbf{A}$ is the incidence matrix and $\mathbf{Y}$ is the branch admittance matrix. In this section we shall restrict ourselves to passive reciprocal networks without mutual coupling; that is, to *RLC* networks with no transformers. Our interest is in evaluating det (**AYA'**). To this task we shall apply the Binet-Cauchy theorem, which we have used with profit before. According to this theorem

$$\det \mathbf{AYA'} = \sum_{\text{all majors}} \text{products of corresponding majors of } (\mathbf{AY}) \text{ and } \mathbf{A'}.$$
(52)

We have here taken the product **AY** as one of the two matrices of the theorem.

You should here recall some of the properties of **A** and **Y**. Remember that the branch admittance matrix **Y** for the networks to which we are here limited is diagonal; denote the diagonal elements by y_j for $j = 1$, $2, \ldots, b$. Remember also that the nonsingular submatrices of **A** correspond to trees and that their determinants equal ± 1.

The matrix **AY** has the same structure as **A** except that the jth column is multiplied by y_j. Hence the nonsingular submatrices of **AY** still correspond to trees of the network, but the value of each nonzero major is no longer ± 1 but plus or minus the product of branch admittances on the corresponding tree. As discussed in the previous chapter, a submatrix of **A'** is simply the transpose of the corresponding submatrix of **A**. A nonsingular submatrix of **A'** will have the same determinant (± 1) as the corresponding submatrix of **A**. Consequently, each term in the summation in (52) will simply be the product of admittances of all twigs of a tree, which we shall call a *tree admittance product* and designate $T(y)$. Hence

$$\Delta|_y = \det \mathbf{AYA'} = \sum_{\substack{\text{all} \\ \text{trees}}} T(y)$$

$$= \sum_{\substack{\text{all} \\ \text{trees}}} \text{tree admittance products.}$$
(53)

This is a very interesting result, first developed by Maxwell. It says that to calculate the determinant of the node admittance matrix of a network we must first locate all the trees of the network, multiply together the branch admittances of each tree, then add the resulting products for all trees. (For simplicity we shall often say " tree products " instead of " tree admittance products.")

To illustrate this result, consider the example shown in Fig. 26. We

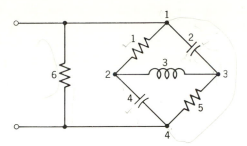

Fig. 26. Example for evaluation of det $\mathbf{AYA}'$.

shall assume the network is to be excited across R_6 with a current source. There are four nodes in the network, so each tree will have three twigs. The trees are the following:

$$
\begin{array}{cccccc}
124 & 134 & 145 & 234 & 245 & 346 & 456 \\
125 & 135 & 156 & 235 & 246 & 356 \\
126 & 136 &   & 236 &
\end{array}
$$

Note that, in this example, locating all the trees may not be a difficult job. Nevertheless, it is worthwhile to tackle the problem systematically. Since each tree contains n twigs, in a network having $n+1$ nodes, one procedure is to list all combinations of b branches taken n at a time. From this set are then eliminated all those combinations that form a loop and hence cannot be a tree.

To return to the example, once the trees are listed, the tree admittance products are formed. In fact, this can be done in such a way that terms with like powers of s are written together. The result will be

$$
\det \mathbf{AYA}' = s^2 C_2 C_4 (G_1 + G_5 + G_6) + s \left[C_2 G_1 (G_5 + G_6) + C_4 G_5 (G_1 + G_6) \right.
$$

$$
\left. + \frac{C_2 C_4}{L_3} \right] + G_1 G_5 G_6 + \frac{C_2}{L_3}(G_5 + G_6) + \frac{C_4}{L_3}(G_1 + G_6) + \frac{1}{sL_3}(G_1 G_5 + G_1 G_6
$$

$$
+ G_5 G_6).
$$

SYMMETRICAL COFACTORS OF THE NODE ADMITTANCE MATRIX

Next we turn to the cofactors of det $\mathbf{AYA}'$. We shall treat these in two groups: the symmetrical cofactors, like Δ_{jj}, and the unsymmetrical ones like Δ_{jk}. We consider first the symmetrical ones.

The cofactor Δ_{jj} is obtained from the node admittance matrix by deleting row j and column j. The same result is obtained if, in $\mathbf{AYA'}$, we delete row j of the first matrix and column j of the last one. But column j of $\mathbf{A'}$ is row j of $\mathbf{A}$. Let us denote by $\mathbf{A}_{-j}$ the matrix $\mathbf{A}$ with row j deleted. Hence

$$\Delta_{jj} = \det(\mathbf{A}_{-j}\mathbf{Y}\mathbf{A'}_{-j}). \tag{54}$$

How is $\mathbf{A}_{-j}$ related to the network? Since each row of $\mathbf{A}$ corresponds to a nondatum node, deleting a row means short-circuiting the corresponding node to the datum node. If the original network is labeled N, let the network that results by short-circuiting node j to the datum node d be labeled N_{-j}. Then $\mathbf{A}_{-j}$ is simply the incidence matrix of network N_{-j}. Consequently,

$$\Delta_{jj} = \det(\mathbf{A}_{-j}\mathbf{Y}\mathbf{A'}_{-j}) = \sum_{\substack{\text{all} \\ \text{trees} \\ \text{of } N_{-j}}} \text{tree admittance products of } N_{-j}. \tag{55}$$

This expression can be used to find Δ_{jj}; but it would be even more useful to relate Δ_{jj} to the original network N.

Now N_{-j} has one less node than N; hence, one less twig in a tree. A tree of N_{-j} cannot be a tree of N, and it cannot contain a loop of N. This statement is an obvious consequence of the fact that a tree of N_{-j} contains no closed path in N_{-j} and, thus, cannot contain one in N. Then, since a tree of N_{-j} has one less twig than a tree of N, it must be contained in a tree of N. Let T_{-j} denote a tree of N_{-j} and let it be contained in T, a tree of N. Clearly, T_{-j} is a two-part subgraph of T. (This does not contradict the fact that it is a one-part subgraph of N_{-j}. Why?) Since node j and the datum node d are short-circuited in T_{-j} as a subgraph of N_{-j}, there is no path between them in T_{-j} as a subgraph of T. Hence nodes j and d are each in a different part of T_{-j} as a subgraph of T. Such a structure is called a two-tree. Specifically, in a network of $n + 1$ nodes, *a two-tree is a set of $n - 1$ branches that forms no loops and separates some tree of the network into two connected parts.* The product of branch admittances constituting a two-tree is called a *two-tree admittance product* and is labeled $^2T(y)$. Subscripts are used to indicate the nodes that are required to be in different parts. Thus $^2T_{j,d}(y)$ means a two-tree admittance product in which nodes j and d are in separate parts.

For the example in Fig. 26, the network N_{-1} is formed by short-circuiting nodes 1 and 4 together, as shown in Fig. 27. Branch sets 13, 34, 45, and 24 are four of the trees of this network. In the original network N these branch sets have the configurations shown in Fig. 28. Each of these is a two-tree with nodes 1 and 4 in different parts. In some of them

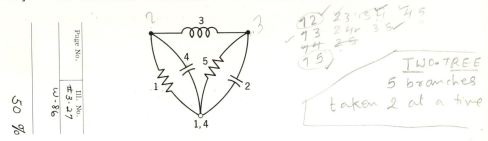

Fig. 27. Network N_{-j} corresponding to N of Fig. 26.

node 1 or 4 is isolated; in another, not. Besides these, there are four other two-trees (12, 15, 23, and 35) in which nodes 1 and 4 are in different parts. (Verify them.) All of these contribute to Δ_{11}.

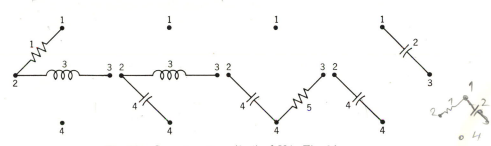

Fig. 28. Some two-trees (1, 4) of N in Fig. 26.

With the introduction of the concept of a two-tree, the expression for the cofactor Δ_{jj} in (55) can be rewritten as

$$\Delta_{jj} = \sum_{\substack{\text{all}\\ \text{two-trees}}} T_{j,\,d}(y) = \sum_{\substack{\text{all}\\ \text{two-trees}}} \text{two-tree } (j,\,d) \text{ admittance products,} \quad (56)$$

where d is the datum node.

With formulas for both Δ and a symmetrical cofactor available, we are now in position to evaluate driving-point functions. Let the network N in Fig. 29 be a passive, reciprocal network without transformers and

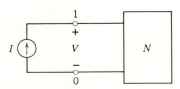

Fig. 29. Calculation of driving-point function.

assume that it is excited by a current source. Let the lower terminal of the source (labeled 0) be chosen as a datum for node voltages. The driving-point impedance from (7) is

$$Z(s) = \frac{\Delta_{11}(s)}{\Delta(s)}\bigg|_y .$$

(57)

Substituting for Δ and Δ_{11} from (53) and (56) leads to

$$Z(s) = \frac{\sum {}^2T_{1,0}(y)}{\sum T(y)} = \frac{\sum \text{two-tree } (1, 0) \text{ products}}{\sum \text{tree products}}$$

(58)

This is truly a powerful result. It permits the evaluation of driving-point functions of simple networks and even those of moderate complexity essentially by inspection, without extensive analysis. Furthermore, it provides an approach by which the digital computer can be applied in more extensive networks, first for searching out all the trees and two-trees and, second, in forming the required products.

Let us apply the formula to the high-pass filter network in Fig. 30

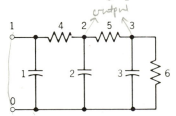

Fig. 30.   High-pass filter network.

for which it is desired to find the driving-point function. There are four nodes, hence three twigs in a tree and two branches in a two-tree. The trees and two-trees $(1, 0)$ are the following:

Trees				Two-trees $(1, 0)$		
123	125	145	456	23	25	45
124	126	146		24	26	46
	134	156			34	56
	135	245			35	
	234	246			36	
	235	345				

(Branch 1 connects the two input nodes, so it can immediately be ruled out of a two-tree.) The admittance products are now formed. They can be written in any order, of course, but we can note the powers of s to which each tree or two-tree will lead and can group those with like powers of s, as we have actually done in the listing. The result will be

$$Z(s) = \frac{C_2 C_3 s^2 + sC_2(G_5 + G_6) + sC_3(G_4 + G_5) + G_4 G_5 + G_4 G_6 + G_5 G_6}{\left\{ \begin{array}{l} C_1 C_2 C_3 s^3 + s^2[C_1 C_2(G_5 + G_6) + C_1 C_3(G_4 + G_5) + C_2 C_3 G_4] \\ + s[G_4(G_5 + G_6)(C_1 + C_2) + G_5(C_1 G_6 + C_3 G_4)] + G_4 G_5 G_6 . \end{array} \right.}$$ (59)

From the preceding development, and as illustrated in the examples, the topological formulas that we have developed can be called *minimum-effort* formulas. There is no subtraction of terms; every term that is evaluated appears in the final result.

UNSYMMETRICAL COFACTORS OF THE NODE ADMITTANCE MATRIX

As far as the node admittance matrix is concerned, there remains only for us to discuss the unsymmetrical cofactors of the form Δ_{ij}. Now $\Delta_{ij} = (-1)^{i+j} M_{ij}$, where M_{ij} is the corresponding minor. To form M_{ij} we delete row i and column j from the node admittance matrix. Hence we need to examine $\det (\mathbf{A}_{-i} \mathbf{Y} \mathbf{A}'_{-j})$. By the Binet-Cauchy theorem,

$$(-1)^{i+j} \Delta_{ij} = \det \mathbf{A}_{-i} \mathbf{Y} \mathbf{A}'_{-j}$$

$$= \sum_{\substack{\text{all} \\ \text{majors}}} \text{products of corresponding majors of } \mathbf{A}_{-i}\mathbf{Y} \text{ and } \mathbf{A}'_{-j} .$$ (60)

As before, a nonzero major of $\mathbf{A}_{-i}\mathbf{Y}$ corresponds to a two-tree in which nodes i and datum are in separate parts. Similarly, a nonzero major of $\mathbf{A}_{-j}$ (which equals ± 1) corresponds to a two-tree in which nodes j and datum are in separate parts. Since, to be of interest, each factor of the product in (60) must be nonzero, the subnetworks that contribute to Δ_{ij} must be two-trees with nodes i and d, as well as j and d, in separate parts. Since there are only two parts to a two-tree and d is in one of them, we conclude that both i and j must be in the same part. Thus two-trees with nodes i and j in one part and the datum node in the other are the only ones that contribute to Δ_{ij}. Such two-tree products are designated $^2T_{ij,\,d}(y)$. The only reservation lies in the sign. Since $\mathbf{A}_{-i}$ and $\mathbf{A}_{-j}$ are different, there is no assurance concerning the signs of corresponding majors of $\mathbf{A}_{-i}\mathbf{Y}$ and $\mathbf{A}'_{-j}$. However, it turns out* that the sign of the

---

* For a proof see S. Seshu and M. B. Reed, *Linear Graphs and Electric Networks* Addison-Wesley, 1961.

product is $(-1)^{i+j}$. Hence the final result becomes

$$\Delta_{ij} = \sum {}^2T_{ij,\,d}(y) = \sum \text{ two-tree } (ij,\,d) \text{ admittance products.}\quad (61)$$

With this result it is now possible to evaluate some transfer functions. Consider the situation in Fig. 31. It is desired to find the transfer voltage

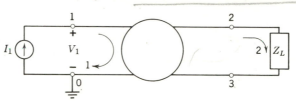

Fig. 31.   Calculation of the transfer function.

ratio $V_{23}(s)/V_1(s)$. Suppose a current source is applied as shown and node 0 is taken as datum. The node equations can be solved for node voltages V_1, V_2, and V_3. Since $V_{23} = V_2 - V_3$, we get

$$\frac{V_{23}}{V_1} = \frac{(V_2 - V_3)/I_1}{V_1/I_1} = \left.\frac{(\Delta_{21} - \Delta_{31})}{\Delta_{11}}\right|_y$$

$$= \frac{\sum [{}^2T_{12,\,0}(y) - {}^2T_{13,\,0}(y)]}{\sum {}^2T_{1,\,0}(y)}.\qquad (62)$$

As a specific example, look back at the network in Fig. 30. Let the output voltage be the voltage across branch 5. The ${}^2T_{1,\,0}(y)$ products were already determined and given as the numerator in (59), so let us concentrate on the numerator two-tree products in (62). For the two-tree containing both nodes 1 and 2, branch 4 must be present. Branch 1 connects nodes 0 and 1; branch 2 connects nodes 0 and 2. Neither of these branches can be present in ${}^2T_{12,\,0}(y)$. Similarly, in order for nodes 1 and 3 to be in one connected part, branches 4 and 5 must both be present. Since the two-tree has only two branches, no others are possible. Hence

$$^2T_{12,\,0}(y) = sC_3\,G_4 + G_4\,G_5 + G_4\,G_6,\qquad (63a)$$

$$^2T_{13,\,0}(y) = G_4\,G_5.\qquad (63b)$$

The transfer function, therefore, is

$$\frac{V_{23}}{V_1} = \frac{sC_3\,G_4 + G_4\,G_6}{C_2\,C_3\,s^2 + s[C_2(G_5 + G_6) + C_3(G_4 + G_5)] + G_4\,G_5 + G_4\,G_6 + G_5\,G_6}.$$

$$(64)$$

The occurrence of something very interesting can be observed here. There is a common term in the two-trees in (63) that cancels when they are substituted into (62). Hence that is not a minimum-effort formula. Another interesting observation can be made by comparing the two-tree products in (63) with $^2T_{1,\,0}(y)$, which is the denominator of (64); $^2T_{1,\,0}(y)$ contains both $^2T_{12,\,0}(y)$ and $^2T_{13,\,0}(y)$ entirely. These observations motivate the following discussion.

Consider a two-tree (j, d) and a node i. This node must be in either the part containing j or the part containing d. Thus the sum of $^2T_{ij,\,d}(y)$ and $^2T_{j,\,di}(y)$ must contain all the terms contained in $^2T_{j,\,d}(y)$; that is,

$$\sum {}^2T_{j,\,d}(y) = \sum {}^2T_{ij,\,d}(y) + \sum {}^2T_{j,\,di}(y). \tag{65}$$

This relationship can be used to write the following identities:

$$\sum {}^2T_{12,\,0}(y) \equiv \sum {}^2T_{12,\,03}(y) + \sum {}^2T_{123,\,0}(y), \tag{66a}$$

$$\sum {}^2T_{13,\,0}(y) \equiv \sum {}^2T_{13,\,02}(y) + \sum {}^2T_{123,\,0}(y). \tag{66b}$$

Note that they have some common terms. When these are inserted into (62), the result becomes

$$\frac{V_{23}}{V_1} = \frac{\sum [{}^2T_{12,\,03}(y) - {}^2T_{13,\,02}(y)]}{\sum {}^2T_{1,\,0}(y)}. \tag{67}$$

In contrast with (63), this is a minimum-effort formula, since nodes 2 and 3 are in separate parts in both of the two-tree products in the numerator. This also makes it evident that *the load admittance cannot appear in the numerator.*

As a final illustration, let us use this formula to compute the transfer voltage ratio V_{23}/V_{14} for the network in Fig. 26. In this example node 4 plays the role of node 0. The two-trees $(1, 4)$ were discussed in connection with Fig. 28. As for the other two-trees, there are very few of them. In fact,

$$\sum {}^2T_{12,\,34}(y) = G_1 G_5,$$

$$\sum {}^2T_{13,\,24}(y) = C_4 C_2 s^2,$$

and so

$$\frac{V_{23}}{V_{14}} = \frac{G_1 G_5 - C_4 C_2 s^2}{s^2 C_1 C_2 + s(G_1 C_2 + C_4 G_5) + \left(G_1 G_5 + \dfrac{C_2 + C_4}{L_3}\right) + \dfrac{G_1 G_5}{s L_3}}.$$

THE LOOP IMPEDANCE MATRIX AND ITS COFACTORS

From notions of duality one would expect that what has been done for the node admittance matrix can also be done for the loop impedance matrix. This is true, with some characteristic differences, as we shall now discuss. One of these differences is that determinants of nonsingular submatrices of the loop matrix **B** do not necessarily equal ± 1. They do, however, if **B** is the matrix of fundamental loops. Hence we shall make the assumption that fundamental loops have been chosen in writing the **B** matrix.

The beginning point here is the loop impedance matrix **BZB'**. Again the Binet-Cauchy theorem is applied. This time the nonsingular submatrices (of **B**) correspond to *cotrees* (complements of trees) instead of trees. We define a *cotree impedance product* as the product of link impedances for some tree. We use the symbol $C[T(z)]$ to indicate this product. By following the same kind of proof as for the node admittance matrix (whose details you should supply) we find that

$$\Delta\big|_z = \det \mathbf{BZB'} = \sum_{\substack{\text{all} \\ \text{trees}}} C[T(z)]$$

$$= \sum_{\substack{\text{all} \\ \text{trees}}} \text{cotree impedance products;} \qquad (68)$$

that is, to find the determinant Δ, we must locate all the trees, from which we find all the cotrees, multiply together the link impedances of each cotree, and then add the resulting products.

The question no doubt has occurred to you as to whether there is any relationship between the determinants of the loop impedance matrix and the node admittance matrix. We shall now examine this question. Suppose a tree admittance product is multiplied by the product of all the branch impedances of the network. The twig impedances will cancel with the twig admittances, leaving a cotree impedance product. If we do this for all the tree admittance products and add, the result will be a sum of all cotree impedance products; that is to say

$$\Delta\big|_z = (z_1 z_2 z_3 \cdots z_b)\Delta\big|_y. \qquad (69)$$

Since the branch impedance matrix **Z** is diagonal, the product of impedances in this expression is simply the determinant of **Z**. Hence it can be rewritten as

$$\Delta\big|_z = (\det \mathbf{Z})\Delta\big|_y. \qquad (70)$$

This is a very significant result. It states that the loop and node determinants, although arising from different matrices (which are, in general, of different orders), are related in a very simple way. In particular, if we take each R, L, and C to be a branch of the network, the two determinants can differ at most by a multiplicative factor ks^p. This means the loop and the node determinants always have the same zeros, except possibly at $s = 0$; that is, the nonzero natural frequencies of a network are independent of whether the loop or the node basis is chosen for analysis. (In this form, the relationship applies also when mutual inductance and transformers are present.)

It must be emphasized that this relationship between the determinants applies when they refer to the same network. There is the possibility of going astray on this point when different sources are applied to a network. Consider the situation in Fig. 32a. Suppose a voltage source is applied

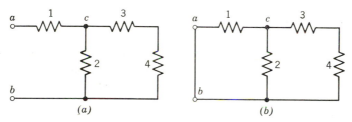

Fig. 32. Importance of terminal conditions.

at the terminals a and b. As far as the loop impedance matrix is concerned, the voltage source is a short circuit, and the appropriate network for evaluating $\Delta|_z$ is the one in part (b). This even has one less node and hence one less twig in a tree than the original. It would be a mistake to imagine that $\Delta|_y$ for Fig. 32a is related to $\Delta|_z$ for Fig. 32b by (70). If the first network is labeled N, then the second one is obtained by short-circuiting node a to node b. This is what we called network N_{-a}.

Now let us turn to the cofactors of $\Delta|_z$ and consider first the symmetrical cofactor $\Delta_{jj} = \det \mathbf{B}_{-j}\mathbf{Z}\mathbf{B}'_{-j}$, where $\mathbf{B}_{-j}$ is the matrix obtained from $\mathbf{B}$ when row j is deleted. Deleting row j from the matrix means destroying loop j in the network. To destroy a loop, we must simply open it, without at the same time opening any other loop. This is possible if loop j contains a branch that no other loop contains. Since we are assuming f-loops, this is the case, and so opening loop j alone is possible. The resulting network when loop j is opened has one less loop than the original. The determinant Δ_{jj} is simply the determinant of the loop impedance matrix of this new network. Hence (68) applies for its evaluation, except that the network is the new one with one loop less.

The ideas in the preceding development should become clearer as we use them in finding the driving-point impedance of the network in Fig. 33.

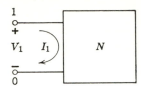

Fig. 33.   Driving-point function.

Let N be the network when the terminals are open; that is, when loop 1 is opened. Then, with a voltage source applied, the resulting network is N_{-1}, since the source is just a short circuit for the evaluation of determinants. This means $\Delta|_z$ is evaluated from (68) for network N_{-1}. Recall that tree admittance products for N_{-1} are two-tree admittance products for N. Hence cotree impedance products for N_{-1} are *co-two-tree impedance products* for N.

The driving-point admittance at the terminals of N is given by (8) as $Y(s) = \Delta_{11}/\Delta|_z$, where Δ_{11} is the determinant of the network N that results when loop 1 in network N_{-1} is opened. Then, since $Z(s) = 1/Y(s)$, we find from (68) that

$$Z(s) = \frac{\Delta}{\Delta_{11}}\bigg|_z = \frac{\sum \text{co-two-tree }(1,0)\text{ impedance products}}{\sum \text{cotree-impedance products}}$$

$$= \frac{\sum C[^2T_{1,0}(z)]}{\sum C[T(z)]}. \tag{71}$$

The notation appears cumbersome, but the ideas are simple. Thus $C[^2T_{1,0}(z)]$ is a blueprint for certain operations. It says: Find a two-tree in which nodes 1 and 0 are in separate parts; take the branches that are *not* in the two-tree (they are in the complement of the two-tree, the co-two-tree) and multiply together their impedances. The numerator of (71) is simply the sum of such terms for all two-trees $(1, 0)$.

Note that this result could have been anticipated from the expression for the impedance in (58) arrived at from the node admittance matrix. Suppose numerator and denominator of that expression are multiplied by det $\mathbf{Z}$. As discussed above, each term, which consists of a product of admittances of certain branches, is converted to a product of impedances of the complements of those branches. Hence (71) follows.

Finally we turn to the unsymmetrical cofactors of the loop impedance matrix. Specifically, we look back at Fig. 31 with the change that a voltage source is applied instead of a current source. We assume that the source and load Z_L do not lie on any other loops except loops 1 and 2, respectively. The transfer functions V_{23}/V_1, V_{23}/I_1, I_2/V_1, and I_2/I_1 *all* contain Δ_{12} in the numerator. One would intuitively expect that the topological formula for $\Delta_{12}|_z$ will be the dual of that obtained for the corresponding cofactors of the node admittance matrix. This turns out to be the case, the result being

$$\Delta_{12} = \sum C[^2 T_{12,\,03}(z)] - \sum C[^2 T_{13,\,02}(z)], \tag{72}$$

in which the complements are computed without the load Z_L.*

Consider, for example, the transfer voltage ratio for Fig. 31. In terms of the loop equations, this is given by

$$\frac{V_{23}}{V_1} = Z_L \frac{I_2}{V_1} = Z_L \frac{\Delta_{12}}{\Delta}\bigg|_z. \tag{73}$$

An expression for this voltage ratio was previously given in (67). Suppose the numerator and denominator of that expression are multiplied by det **Z** for the network with loop 1 open. (Why is det **Z** for this particular network used?) Now Z_L is a factor of det **Z**. Since Y_L did not appear in the numerator, Z_L will not cancel in any term of the numerator. Hence it can be factored out. The result will be

$$
\begin{aligned}
\frac{V_{23}}{V_1} &= \frac{\det \mathbf{Z} \sum [^2 T_{12,\,03}(y) - {}^2 T_{13,\,02}(y)]}{\det \mathbf{Z} \sum {}^2 T_{1,\,0}(y)} \\
&= \frac{Z_L \sum \{C[^2 T_{12,\,03}(z)] - C[^2 T_{13,\,02}(z)]\}}{\sum C[^2 T_{1,\,0}(z)]},
\end{aligned}
\tag{74}
$$

in which the complements in the numerator are computed without Z_L. Comparison of this result with (73) leads to the formula for Δ_{12} given in (72). The above, in fact, constitutes a proof of the formula.

TWO-PORT PARAMETERS

Since short-circuit admittance and open-circuit impedance parameters of a two-port are frequently used, it would be of value to have available topological formulas for their evaluation. We shall rapidly develop such formulas here, not pausing for detailed examination of all the steps.

* For a lengthy proof, see Seshu and Reed, *op. cit.*

The prototype two-port is shown in Fig. 34. Terminal 0 is chosen as

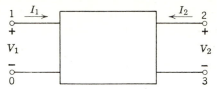

Fig. 34.   References and datum node used in calculation of two-port parameters.

a datum for defining node voltages. Note that the port voltage V_2 is the difference between node voltages V_{20} and V_{30}. Assume that current sources are applied at the input and output ports. The node equations will be

$$\mathbf{AYA'} \begin{bmatrix} V_1 \\ V_{20} \\ V_{30} \\ \vdots \end{bmatrix} = \begin{bmatrix} I_1 \\ I_2 \\ -I_2 \\ 0 \\ \vdots \end{bmatrix}. \tag{75}$$

Next we solve for node voltages V_1, V_{20}, and V_{30}:

$$V_1 = \frac{1}{\Delta} (\Delta_{11} I_1 + \Delta_{21} I_2 - \Delta_{31} I_2), \tag{76a}$$

$$V_{20} = \frac{1}{\Delta} (\Delta_{12} I_1 + \Delta_{22} I_2 - \Delta_{32} I_2), \tag{76b}$$

$$V_{30} = \frac{1}{\Delta} (\Delta_{13} I_1 + \Delta_{23} I_2 - \Delta_{33} I_2). \tag{76c}$$

Observing that $V_2 = V_{20} - V_{30}$, the open-circuit equations become

$$\begin{bmatrix} V_1 \\ V_2 \end{bmatrix} = \frac{1}{\Delta} \begin{bmatrix} \Delta_{11} & \Delta_{12} - \Delta_{13} \\ \Delta_{12} - \Delta_{13} & \Delta_{22} + \Delta_{33} - 2\Delta_{23} \end{bmatrix} \begin{bmatrix} I_1 \\ I_2 \end{bmatrix} = \mathbf{Z}_{oc} \begin{bmatrix} I_1 \\ I_2 \end{bmatrix}. \tag{77}$$

To find the short-circuit admittance matrix we must invert $\mathbf{Z}_{oc}$. Let Δ_{oc} be the determinant of $\mathbf{Z}_{oc}$. From (77), this determinant is

$$\Delta_{oc} = \frac{1}{\Delta^2} (\Delta_{11}\Delta_{22} + \Delta_{11}\Delta_{33} - 2\Delta_{11}\Delta_{23} - \Delta_{12}{}^2 - \Delta_{13}{}^2 + 2\Delta_{12}\Delta_{13}). \tag{78}$$

This expression can be simplified by using Jacobi's theorem,* which says that

$$\Delta_{ii}\Delta_{jj} - \Delta_{ij}^2 = \Delta\Delta_{iijj},$$
$$\Delta_{ii}\Delta_{jk} - \Delta_{ij}\Delta_{ik} = \Delta\Delta_{iijk}, \qquad (79)$$

where Δ_{iijk} is the cofactor formed by deleting rows i and j, and columns i and k from Δ. When these identities are used (78) becomes

$$\Delta_{oc} = \frac{1}{\Delta}(\Delta_{1122} + \Delta_{1133} - 2\Delta_{1123}). \qquad (80)$$

With this the inverse of (77) can now be written as

$$\begin{bmatrix} I_1 \\ I_2 \end{bmatrix} = \frac{1}{\Delta_{1122} + \Delta_{1133} - 2\Delta_{1123}} \begin{bmatrix} \Delta_{22} + \Delta_{33} - 2\Delta_{23} & \Delta_{13} - \Delta_{12} \\ \Delta_{13} - \Delta_{12} & \Delta_{11} \end{bmatrix} \begin{bmatrix} V_1 \\ V_2 \end{bmatrix}. \qquad (81)$$

We have topological formulas for all the cofactors except those with four subscripts. Let us define a *three-tree of a graph having $n+1$ nodes as a set of three unconnected subgraphs having a total of $n-2$ branches and containing no loops*. We denote a three-tree with the symbol ^3T together with subscripts indicating which nodes are in the separate parts. Now just as Δ was shown to equal the sum of the admittance products over all trees of the graph and Δ_{ij} was shown to equal two-tree (ij, d) admittance products over all two-trees of the type in which i and j are in one part and d in the other, in the same way it can be shown that

$$\Delta_{1122} = \sum {}^3T_{1,\,2,\,0}(y),$$
$$\Delta_{1133} = \sum {}^3T_{1,\,3,\,0}(y), \qquad (82)$$
$$\Delta_{1123} = \sum {}^3T_{1,\,23,\,0}(y),$$

where $^3T_{1,\,2,\,0}(y)$ is a three-tree admittance product with nodes 1, 2, and 0 in separate parts, and the other three-trees have similar interpretations. But

$$^3T_{1,\,2,\,0} = {}^3T_{1,\,23,\,0} + {}^3T_{13,\,2,\,0} + {}^3T_{1,\,2,\,03},$$
$$^3T_{1,\,3,\,0} = {}^3T_{1,\,23,\,0} + {}^3T_{12,\,3,\,0} + {}^3T_{1,\,3,\,02}. \qquad (83)$$

* See, for example, A. C. Aitken, *Determinants and Matrices*, 9th ed., Interscience Publishers, New York, 1956.

Using the last two sets of equations, we find that

$$\Delta_{1122} + \Delta_{1133} - 2\Delta_{1123} = \sum {}^3T_{1,\,2,\,30}(y) + \sum {}^3T_{13,\,2,\,0}(y) + \sum {}^3T_{12,\,3,\,0}(y)$$
$$+ \sum {}^3T_{1,\,3,\,20}(y)$$
$$\equiv \sum {}^3T(y) \qquad \text{(for short).} \tag{84}$$

Before inserting the topological formulas for the cofactors into (81), note also that it is possible to simplify the expression for the first element in the matrix there. Thus

$$\Delta_{22} = \sum {}^2T_{2,\,0} = \sum {}^2T_{23,\,0} + \sum {}^2T_{2,\,03},$$
$$\Delta_{33} = \sum {}^2T_{3,\,0} = \sum {}^2T_{23,\,0} + \sum {}^2T_{3,\,02},$$
$$\Delta_{23} = \sum {}^2T_{23,\,0}.$$

Hence

$$\Delta_{22} + \Delta_{33} - 2\Delta_{23} = \sum {}^2T_{2,\,03} + \sum {}^2T_{02,\,3} = \sum {}^2T_{2,\,3}. \tag{85}$$

When all the above is collected and inserted into (77) and (81), the results are

$$\mathbf{Z}_{\text{oc}} = \frac{1}{\sum T(y)} \begin{bmatrix} \sum {}^2T_{1,\,0} & \sum ({}^2T_{12,\,30} - {}^2T_{13,\,20}) \\ \sum ({}^2T_{12,\,30} - {}^2T_{13,\,20}) & \sum {}^2T_{2,\,3} \end{bmatrix} \tag{86}$$

$$\mathbf{Y}_{\text{sc}} = \frac{1}{\sum {}^3T(y)} \begin{bmatrix} \sum {}^2T_{2,\,3} & -\sum ({}^2T_{12,\,30} - {}^2T_{13,\,20}) \\ -\sum ({}^2T_{12,\,30} - {}^2T_{13,\,20}) & \sum {}^2T_{1,\,0} \end{bmatrix} \tag{87}$$

where $\sum {}^3T(y)$ is defined in (84). Note that the numerators of the transfer impedances and of the transfer admittances differ in sign only. This verifies our earlier observation that these functions have the same zeros—unless there is a cancellation with the denominators.

As an illustration, the network of Fig. 30 is redrawn as a two-port in Fig. 35. Let us find $\mathbf{Y}_{\text{sc}}$. For this network the two-trees $(1, 0)$ were already listed under (58). We shall repeat them here together with the other required two-trees:

two-trees $(1, 0)$:   23, 25, 26, 34, 35, 45, 46, and 56
two-trees $(12, 30)$:   34 and 46
two-trees $(13, 20)$:   None
two-trees $(2, 3)$:   12, 13, 14, 16, 24, 34, and 46

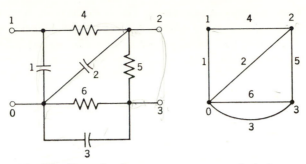

Fig. 35. Example of two-port parameter calculation.

In this example there are four nodes. Hence a three-tree has only one branch. The three-trees are, therefore, easily determined. They are

$$\text{three-trees } (1, 2, 30): \quad 3 \text{ and } 6$$
$$\text{three-trees } (13, 2, 0): \quad \text{None}$$
$$\text{three-trees } (12, 3, 0): \quad 4$$
$$\text{three-trees } (1, 3, 20): \quad 2$$

The denominator of the short-circuit admittance parameters, therefore, is

$$\sum {}^3T(y) = s(C_2 + C_3) + G_4 + G_6 .$$

The parameters themselves are

$$y_{11} = \frac{s^2 C_1 (C_2 + C_3) + s[G_4 (C_1 + C_2 + C_3) + C_1 G_6] + G_4 G_6}{s(C_2 + C_3) + G_4 + G_6} ,$$

$$y_{12} = y_{21} = \frac{-G_4 (sC_3 + G_6)}{s(C_2 + C_3) + G_4 + G_6} ,$$

$$y_{22} = \frac{s^2 C_2 C_3 + s[C_2 (G_5 + G_6) + C_3 (G_4 + G_5)] + G_4 G_5 + G_4 G_6 + G_5 G_6}{s(C_2 + C_3) + G_4 + G_6} .$$

As a final observation we should note that, although the development of the topological formulas was couched in terms of the node equations and the loop equations, we do not write these equations when applying the formulas. Given a network, what we do is to enumerate trees, two-trees, and three-trees and then form products of branch impedances or admittances. Thus what looked like unnecessary complications when we were

setting up loop and node equations in the last chapter, requiring extensive matrix multiplications to obtain **AYA′** and **BZB′** turns out to be of
great value, because the Cauchy-Binet theorem permits some further
mathematical derivations culminating in some simple formulas that
require no matrix operations.

PROBLEMS

1. In the network of Fig. P1 solve for the voltage-gain function $V_2(s)/V_1(s)$.
   Do this by (a) using mixed-variable equations; (b) using node equations
   after expressing I_1 in terms of appropriate voltages.

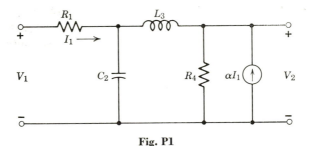

Fig. P1

2. Figure P2 shows an amplifier together with appropriate linear equivalents. It is desired to find the output impedance Z_o for both cases shown,
   when the output is taken from the plate and when it is taken from the
   cathode.

   (a)   Do this by using node equations and an admittance representation
   for the controlled source.
   (b)   Repeat by using loop equations. (How many are there?)
   (c)   Repeat by using mixed-variable equations.
3. Repeat Problem 2 for the amplifier of Fig. P3.
4. The diagram in Fig. P4 shows a difference amplifier. Assume that each
   transistor can be represented by the linear equivalent circuit shown. It
   is desired to find values for R_L, R_f, and R_e in order that the output
   voltage V_o, will equal approximately $K(I_2 - I_1)$. Use any convenient
   set of equations.
5. The diagram in Fig. P5 is an approximate hybrid π model of a transistor.
   Find the h-parameters.

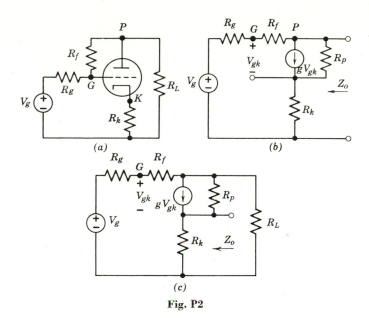

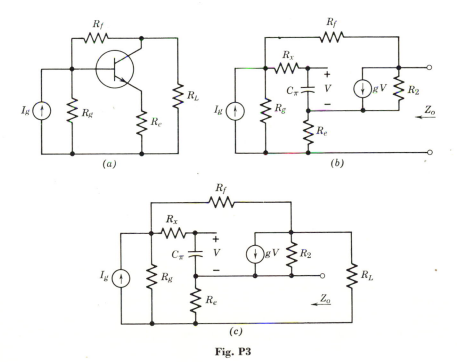

Fig. P2

Fig. P3

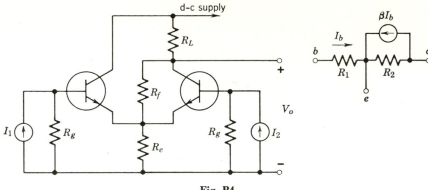

Fig. P4

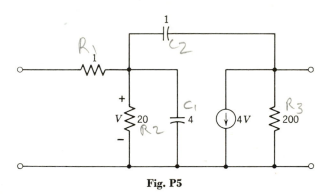

Fig. P5

6. The two-port shown in Fig. P6 is a potential negative converter.

(a) Find the hybrid *h*-parameters.
(b) Specify the ratio R_2/R_1 in terms of β to make $h_{12} h_{21} = 1$.
(c) Comment on the relative values of R_1 and R_2 for $\beta = 50$. ($\beta = 50$ is an easily realizable current gain.)
(d) Is this a voltage or a current negative converter?

7. The two-port of Fig. P7 is a potential negative converter.

(a) Find the hybrid *h*-parameters.
(b) Find the ratio of R_2/R_1 in terms of β to make $h_{12} h_{21} = 1$.
(c) Draw an equivalent network based on the hybrid *g*-parameters for this two-port. Show all component values using the condition found in (b).
(d) Let $\beta = 50$. Design possible compensating networks to be placed in series or shunt at the ports in order to convert this two-port to an ideal negative converter.

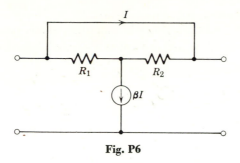

Fig. P6

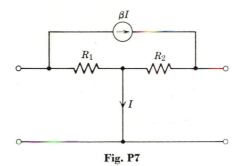

Fig. P7

8. (a)   Find the h-parameters of the two-port in Fig. P8.

(b)   Let $\beta_1 = 1$. Can you find values of β, R_2 and R_1 to make the two-port an ideal negative converter?

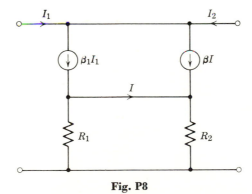

Fig. P8

9. In Fig. P9a, a common-terminal gyrator has terminal 3 as the common terminal.

(a)   Determine the short-circuit admittance matrix of the two-port obtained by making terminal 1 common instead.

(b)   Repeat if terminal 2 is made common.

(c)   The symbol for the gyrator is sometimes drawn as in Fig. P9b. Comment on the appropriateness of this symbol, in view of **(a)** and **(b)**.

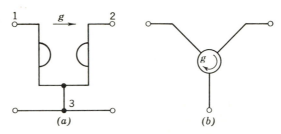

Fig. P9

10. A certain nonreciprocal network can be represented by the network shown in Fig. P10a. It is desired to connect a resistor R_1 as shown in Fig. P10b in order to stop reverse transmission (from right to left). Determine the required value of R_1.

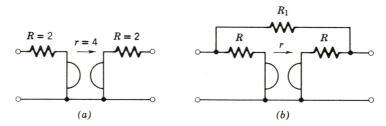

Fig. P10

11. Figure P11 shows a two-port network terminated in an impedance Z_L. Show that

$$Z_{21}(s) = \frac{V_2(s)}{I_1(s)} = \frac{z_{21} Z_L}{z_{22} + Z_L},$$

$$Y_{21}(s) = \frac{I_2(s)}{V_1(s)} = \frac{y_{21} Y_L}{y_{22} + Y_L}.$$

12. Verify that $\det \mathbf{Y}_{sc} = 1/\det \mathbf{Z}_{oc}$ for a two-port.

13. Show that the two-ports in Fig. 15c and d have the same open-circuit impedance parameters.

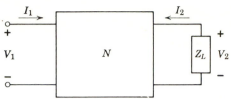

Fig. P11

14. Two two-ports N_a and N_b are connected in cascade, as in Fig. P14. Using a direct analysis, determine the overall short-circuit transfer admittance $y_{21}(s)$ in terms of the short-circuit parameters of N_a and N_b.

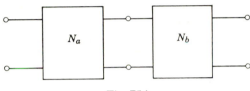

Fig. P14

15. Repeat Problem 14 with a voltage negative converter cascaded between N_a and N_b, as in Fig. P15.

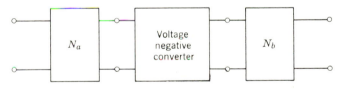

Fig. P15

16. Figure P16 shows an interconnection of two-ports, one of which is a current negative converter. Obtain an expression for the voltage transfer ratio $V_2(s)/V_1(s)$ in terms of the y parameters of N_a and N_b and the conversion ratio k of the negative converter. Compare, if the negative converter is not present.

17. An ideal transformer is cascaded with a two-port in the two possible ways shown in Fig. P17. Write the open-circuit impedance parameters of the combination in terms of n and the z-parameters of the two-port.

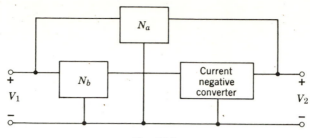

Fig. P16

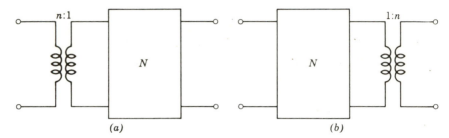

(a)   (b)

Fig. P17

18. Show that the $\mathbf{Y}_{sc}$ matrix of the two-port shown in Fig. P18a, considering each transistor as an ideal current-dependent current source as shown in Fig. P18b, is

$$\mathbf{Y}_{sc} = \begin{bmatrix} g - G & -(g - G) \\ G & -G \end{bmatrix}$$

where $G = R/R_{e1}R_{e2}$ and $g = G + 1/R_{e1}$, under the assumption that $R_{e2}/\beta_1 \ll R \ll \beta_2\,R_{e2}$. Verify that the two-port in Fig. P18c is equivalent to this two-port.

19. The hybrid h-matrix of a two-port device has one of the following forms:

$$\mathbf{H}_1 = \begin{bmatrix} 0 & 1 \\ \dfrac{Z_1(s)}{Z_2(s)} & 0 \end{bmatrix} \quad \text{or} \quad \mathbf{H}_2 = \begin{bmatrix} 0 & -\dfrac{Z_1(s)}{Z_2(s)} \\ -1 & 0 \end{bmatrix}.$$

This two-port is terminated in an impedance Z_L. Find the impedance at the other port. (The device is called a *general converter.*)

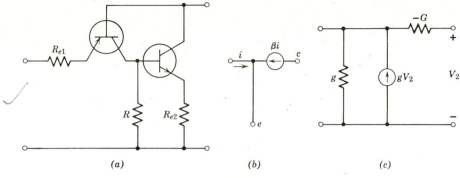

Fig. P18

20. Find the hybrid h-matrices for each of the networks in Fig. P20.
a   (Replace each transistor by the simplest possible small-signal equivalent.)

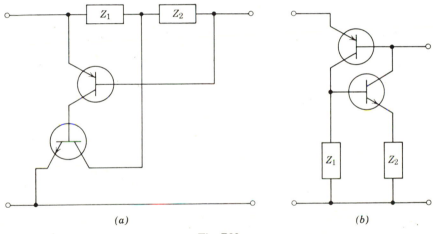

Fig. P20

21. In Fig. P21, the two-port is a general converter having the hybrid h-matrix shown. Find the impedance Z.

22. For the diagram of Fig. P22, show that the voltage transfer ratio is

$$\frac{V_2(s)}{V_1(s)} = \frac{-y_{21}}{y_{22} + G_1 - (g-1)Y}.$$

(Observe that the conductances G_1, G_2, and G_3 can account for the input, output, and feedback impedances of an actual amplifier of which the controlled source is the idealization.)

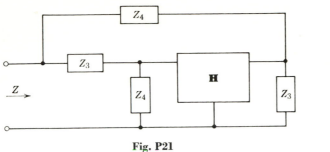

$$H = \begin{bmatrix} 0 & 1 \\ \dfrac{Z_1}{Z_2} & 0 \end{bmatrix}$$

Fig. P21

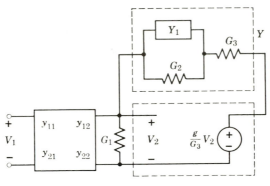

Fig. P22

23. For each of the networks in Fig. P23 find the short-circuit current-gain function h_{21}. Use the y-parameters of the two-port N.

24. Find the voltage transfer ratio for the two networks in Fig. P24 in terms of the y-parameters of the two-ports N_a and N_b, and the amplifier gain μ. Verify for Fig. P24b that the limiting value as $\mu \to \infty$ is $V_2/V_1 = -y_{21b}/y_{21a}$.

25. Find the open-circuit impedance parameters of the feedback amplifier shown in Fig. P25 in terms of g and of the z-parameters of two-port N_a. The limiting values as $g \to \infty$ should be

$$z_{21} \to 1/y_{21a}, \qquad z_{11} \to 0, \qquad z_{22} \to 0, \qquad z_{12} \to 0.$$

26. A two-port N_c with a resistance R across both its input and output ports is shown in Fig. P26a. The resulting two-part is denoted N_b, and the z-parameters of N_b are z_{11b}, z_{12b}, z_{21b}, and z_{22b}. The network N_c, after introducing either a series or shunt resistance R at its ports, is to be cascaded with the feedback amplifier of Problem 25. The two cascade configurations to be considered are shown in Fig. P26b and c. Show that

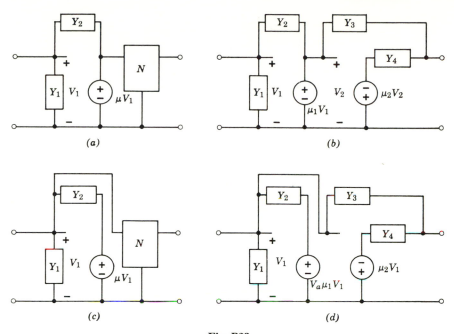

Fig. P23

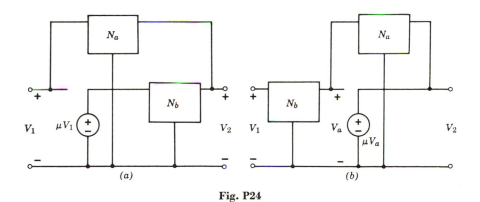

Fig. P24

the open-circuit transfer impedance in both cascade cases is given by

$$z_{21} = \frac{z_{21b}}{Ry_{21a}}$$

as $g \to \infty$. (N_f denotes the entire feedback structure of Fig. P25.)

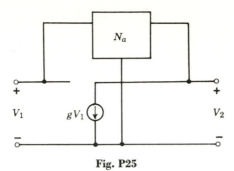

Fig. P25

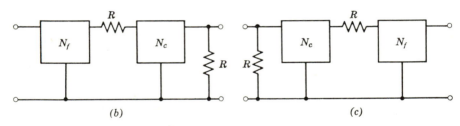

Fig. P26

27. Let a transmission zero of a two-port be defined as a zero of the short-circuit transfer admittance $y_{21}(s)$. Show that the output current or voltage of the terminated two-port in Fig. P27 will be zero with either a voltage or current excitation, even if y_{11} or y_{22} also have a zero at this frequency, leading to a cancellation in (25) and causing z_{21}, h_{21}, or g_{21}, or all three to be nonzero. Comment on the appropriateness of the term "transmission zero".

28. (a) For the series-parallel and parallel-series connections of two-ports in Fig. P28, show that the h- and g-parameters of the components are added to give the overall h- and g-parameters, respectively.
    (b) State and prove conditions under which the series-parallel and

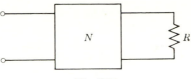

Fig. P27

parallel-series connections can be made without violating the condition that the same current leave one terminal of a port as enters the other terminal.

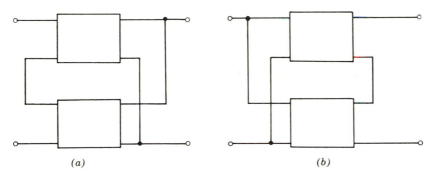

Fig. P28(*a*) Series-parallel connection; (*b*) parallel-series connection.

29. Find the chain matrix of the two-ports in Fig. P29. The transformers are perfect.

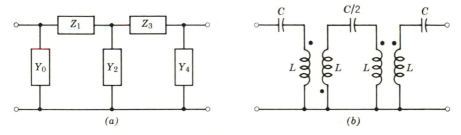

Fig. P29

30. Treat the bridged-tee network of Fig. P30 first as the parallel connection of 2 two-ports and then as the series connection of 2 two-ports to find the overall *y*-parameters. (The answers should be the same.)

31. Find the *y*-parameters of the two-ports in Fig. P31 by decomposing them into suitable parallel-connected two-ports.

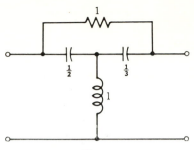

Fig. P30

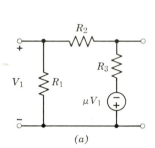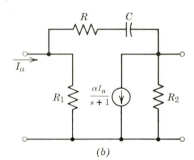

Fig. P31

32. The short-circuit admittance matrix of the π network in Fig. P32 with terminal 3 as the common terminal is

$$\mathbf{Y}_{sc3} = \begin{bmatrix} s+2 & -2 \\ -2 & 4s+2 \end{bmatrix}.$$

Find the short-circuit admittance matrices when each of the other terminals is made the common terminal of a two-port.

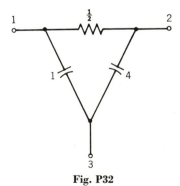

Fig. P32

33. Figure P33a shows a common-terminal three-port with terminal 4 as the common terminal. The short-circuit admittance matrix of this configuration is the one given. (Take the elements of the matrix to be values of conductance.) It is desired to reconnect this network as a two-port, as shown in Fig. P33b, the input port being 3, 2; the output port 1, 2. Find the corresponding short-circuit admittance matrix.

$$\begin{bmatrix} I_1 \\ I_2 \\ I_3 \end{bmatrix} = \begin{bmatrix} 1 & 2 & 3 \\ 6 & 5 & 4 \\ 7 & 8 & 9 \end{bmatrix} \begin{bmatrix} V_1 \\ V_2 \\ V_3 \end{bmatrix}$$

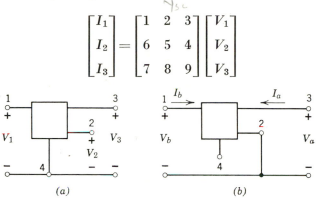

(a)   (b)

Fig. P33

34. Figure P34a shows a four-terminal network connected as a common-terminal three-port. The short-circuit equations of this three-port are as shown. It is desired to connect a unit capacitor between terminals 1 and 2, as shown in Fig. P34b. Find the short-circuit admittance matrix of the network when it is considered as a two-port with the ports shown in Fig. P34b.

$$\begin{bmatrix} I_1 \\ I_2 \\ I_3 \end{bmatrix} = \begin{bmatrix} 5 & -1 & -2 \\ -3 & 6 & -1 \\ -2 & -1 & 4 \end{bmatrix} \begin{bmatrix} V_1 \\ V_2 \\ V_3 \end{bmatrix}$$

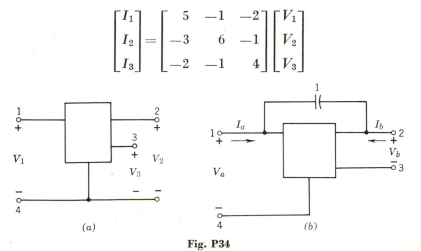

(a)   (b)

Fig. P34

35. The n-terminal network shown in Fig. P35 is linear, lumped, and time invariant. It is represented by $\mathbf{I} = \mathbf{Y}_i \mathbf{V}$, where $\mathbf{Y}_i$ is the indefinite admittance matrix, and the currents and voltages are defined on the diagram. It is proposed to retain the first k terminals as terminals and to connect the remaining ones to ground through impedances, as shown. Let $\mathbf{Z}$ be the diagonal matrix whose diagonal elements are the impedances Z_j. Find an expression relating the new terminal currents to the voltages in terms of $\mathbf{Z}$, $\mathbf{Y}_i$, and submatrices thereof.

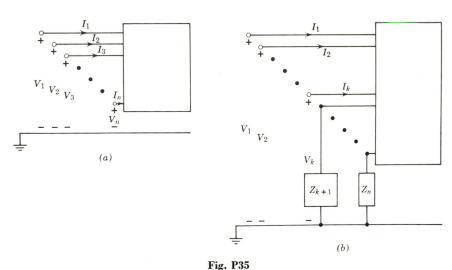

Fig. P35

36. The diagram in Fig. P36 is a linear RLC network with no transformers. Determine a relationship for the voltage transform $V(s)$ using topological expressions for the determinant of the node admittance matrix and its cofactors, taking node 5 as the datum. Simplify as much as possible.

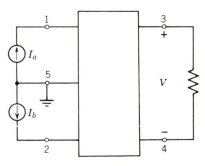

Fig. P36

37. Figure P37 is an *RLC* network without transformers. Find an expression for the voltage transform $V(s)$ in terms of the node admittance determinant and appropriate cofactors, carefully specifying the precise network structure to which this matrix is pertinent. Write the result in terms of topological formulas and simplify as much as possible.

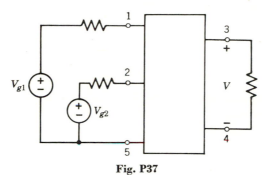

Fig. P37

38. Discuss the modification in the indefinite impedance matrix $\mathbf{Z}_i$ when (a) two terminals are connected together and (b) a terminal is suppressed. Compare with $\mathbf{Y}_i$.

39. Given the indefinite impedance matrix for an *n*-terminal network, with $n = 3$ and $n = 4$, find the open-circuit impedance matrix of the common terminal multiport resulting when terminal *n* is made the common terminal.

40. Find the driving-point admittance of each of the networks in Fig. P40 by using topological formulas. Do it twice, once with the node admittance matrix and once with the loop impedance matrix.

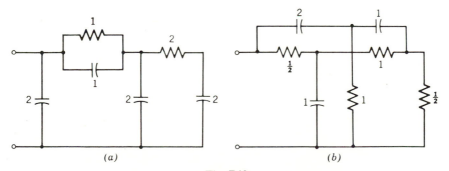

Fig. P40

41. In the networks of Fig. P41 find the driving-point admittance at the left-hand port by using topological formulas. Do it twice, once with the node admittance matrix and once with the loop impedance matrix.

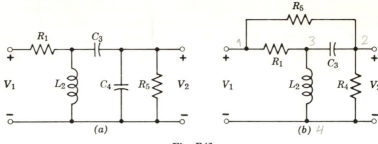

Fig. P41

23-11-70
3-42b →**42.** In the networks of Fig. P41, find the transfer voltage ratio V_2/V_1 by
using topological formulas.

43. For a common-terminal two-port the topological formulas for $\mathbf{Y}_{sc}$ and
$\mathbf{Z}_{oc}$ will simplify to some extent. Find these simplified formulas.

23-11-70
3-44-e **44.** For the networks in Fig. P44, find the open-circuit impedance matrix
$\mathbf{Z}_{oc}$ using topological formulas.

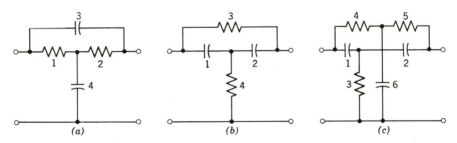

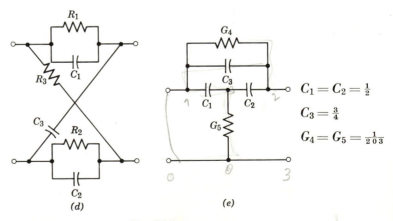

$C_1 = C_2 = \frac{1}{2}$

$C_3 = \frac{3}{4}$

$G_4 = G_5 = \frac{1}{203}$

Fig. P44

45. (a)   Prove that the *impedance* of an *RLC* network without mutual inductance will have a pole at $s = 0$ if and only if there is an all-capacitor cut-set that separates the two terminals.

(b)   Prove that the impedance will have a pole at infinity if and only if there is an all-inductor cut-set separating the terminals as in (a).

46. (a)   Prove that the *admittance* of an RLC network without mutual inductance will have a pole at $s = 0$ if and only if there is an all-inductor path between the terminals.

(b)   Prove that the admittance will have a pole at infinity if and only if there is an all-capacitor path between the terminals.

47. Let

$$z_{11} = \frac{a_0 + a_1 s + \cdots + a_n s^n}{D(s)},$$

$$z_{22} = \frac{b_0 + b_1 s + \cdots + b_m s^m}{D(s)},$$

$$z_{21} = \frac{c_0 + c_1 s + \cdots + c_r s^r}{D(s)}$$

be the open-circuit parameters of an RLC two-port without mutual inductance.

(a)   Using topological formulas show that

$$a_k \geq |c_k| \quad \text{and} \quad b_k \geq |c_k|.$$

This means that if a power of s is present in the numerator of z_{21}, it must also be present in the numerators of both z_{11} and z_{22}. Furthermore, the coefficients in z_{11} and z_{22} will be positive and greater than the magnitude of the corresponding coefficients in z_{21}. which can be negative.

(b)   What further conclusions can you draw if the two-port is a common-terminal one?

(c)   Suppose the three functions given refer to y_{11}, y_{22}, and $-y_{21}$. What is the corresponding result?

These conditions on the coefficients are called the *Fialkow condition*.

48. (a)   For the network of Fig. P48 find the short circuit admittance parameters by direct application of the definition. Fialkow's condition is apparently not satisfied.

(b)   Find the parameters again, using topological formulas and compare the two answers. State a condition that must be assured in order for Fialkow's condition to be valid.

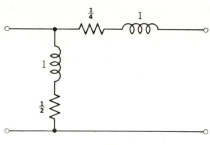

Fig. P48

49. Let the hybrid h-matrix for a transistor in the common-emitter connection be

$$\mathbf{h} = \begin{bmatrix} h_{11} & h_{12} \\ h_{21} & h_{22} \end{bmatrix}.$$

Find the h-matrix of the transistor in the common-base and common-
collector configurations through the agency of the indefinite admittance
matrix.

50. The diagram in Fig. P50 is a passive, reciprocal network in which the
resistor R_k is shown explicitly. The driving-point impedance of the
network is $Z(s)$. Suppose the branch containing R_k is opened and the
terminals so formed constitute the input port of a two-port whose other
port is the original pair of terminals of the network. Let $g_{21k}(s)$ be the
forward voltage-gain function of this two-port. Show that, if the network contains n resistors,

$$\text{Re } Z(j\omega) = \sum_{k=1}^{n} R_k |g_{21k}(j\omega)|^2.$$

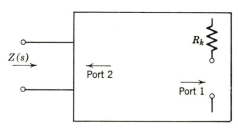

Fig. P50

51. Consider a reciprocal two-port N that is both structurally and electri-
cally symmetrical. Its z- and y-parameters are denoted by $z_{11} = z_{22}$,

z_{12} and $y_{11} = y_{22}$, y_{12}, respectively. If we consider bisecting the two-port at its structural line of symmetry, a number of terminals (two or more) will be created at the junction between the two halves. Assume that none of the leads from which these terminals are formed are crossed. Now consider the two cases shown in Fig. P51 in which these terminals are left open and short-circuited, respectively. The input impedance and input admittance are designated z_{11h} and y_{11h} in the two cases, respectively, where the subscript h stands for "half." Show that

$$z_{11h} = z_{11} + z_{12} \quad \text{and} \quad \frac{1}{y_{11h}} = z_{11} - z_{12}$$

or

$$\frac{1}{z_{11h}} = y_{11} + y_{12} \quad \text{and} \quad y_{11h} = y_{11} - y_{12}.$$

(*Hint:* Apply voltages $V_1 = V_2 = V$ at the terminals of the original network and show that no current will flow across the structural line of symmetry. Then apply voltages $V_1 = -V_2 = V$ and show that the voltage at each point on the structural line of symmetry will be the same.) This result is known as *Bartlett's bisection theorem.*

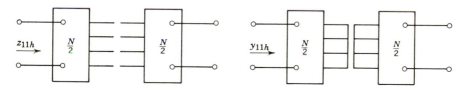

Fig. P51

52. The current and voltage variables in the loop, node and node-pair equations in (1) are Laplace transforms. The solution, say, for one of the node voltages is as given in (4).

$$V_k(s) = \sum_i J_i(s) \frac{\Delta_{ik}(s)}{\Delta(s)}$$

Now suppose the excitations are all exponentials, so that the ith equivalent current source is $I_i e^{j\omega_o t}$. I_i is a complex number, called a *phasor.* Assume that $s_k = j\omega_0$ is not a natural frequency of the network and assume the network is initially relaxed. The forced response to the exponential excitation will also be exponential and the forced component of the node voltage $v_k(t)$ will be

$$v_k(t)|_{\text{forced}} = U_k \epsilon^{j\omega_o t},$$

where U_k is also a phasor.

Find an expression for U_k. Compare it with the expression for $V_k(s)$ above, in which the excitation is arbitrary.

53. A two-port has the following hybrid V–I relationship.

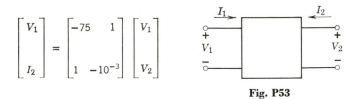

$$\begin{bmatrix} V_1 \\ I_2 \end{bmatrix} = \begin{bmatrix} -75 & 1 \\ 1 & -10^{-3} \end{bmatrix} \begin{bmatrix} V_1 \\ V_2 \end{bmatrix}$$

Fig. P53

Design compensating networks to be placed in series or shunt at the ports in Fig. P53 in order to convert the two-port to an ideal negative converter.

54. Repeat Problem 53 for the two-ports having the following V–I relationships.

(a) $$\begin{bmatrix} V_1 \\ I_2 \end{bmatrix} = \begin{bmatrix} 100 & -2 \\ -2 & 2 \times 10^{-3} \end{bmatrix} \begin{bmatrix} I_1 \\ V_2 \end{bmatrix}$$

(c) $$\begin{bmatrix} V_1 \\ I_2 \end{bmatrix} = \begin{bmatrix} -100 & 1 \\ 1 & 10^{-4} \end{bmatrix} \begin{bmatrix} I_1 \\ V_2 \end{bmatrix}$$

(b) $$\begin{bmatrix} V_1 \\ I_2 \end{bmatrix} = \begin{bmatrix} 100 & 1 \\ 1 & -10^{-4} \end{bmatrix} \begin{bmatrix} I_1 \\ V_2 \end{bmatrix}$$

· 4 ·

STATE EQUATIONS

In Chapter 2 we developed loop-current, node-voltage, and mixed variable representations of electrical networks. In the general case each scalar loop or node equation is an integrodifferential equation of the second order. Each scalar mixed-variable equation, on the other hand, is of the first order. However, unless care is exercised in the selection of a tree, some of the mixed-variable equations may contain integrals, rather than derivatives, of the variables.

There are some distinct advantages in describing the network in such a way that first-order differential equations, without integrals, result. When expressed in matrix form the result is a first-order vector differential equation that governs the dynamical behavior of the network. Some of the reasons for seeking such a network description are the following:

1. There is a wealth of mathematical knowledge on solving such equations and on the properties of their solutions that can be directly applied to the case at hand.

2. The representation is easily and naturally extended to time-varying and nonlinear networks and is, in fact, the approach most often used in characterizing such networks.

3. The first-order differential equation is easily programmed for computer solution.

In this chapter we shall formulate and solve the first-order vector differential equations that are known as *state equations*. We shall be

229

limited here to linear, time-invariant networks that may be passive or
nonpassive, reciprocal or nonreciprocal. In preceding chapters we considered only the Laplace-transformed equations. In this chapter we shall
revert to the basic equations with the variables expressed as functions of
time. This may require a reorientation in your patterns of thought; for
example, if in the present context we say an equation is algebraic, we
mean that no derivatives of the variables appear in the equations. In
terms of the Laplace-transformed equations this would mean that the
coefficients are independent of the complex frequency variable.

4.1   ORDER OF COMPLEXITY OF A NETWORK

Related to the network description (state equations) we shall develop
in this chapter is a question we shall discuss first. The number of independent Kirchhoff's current law (KCL) and Kirchhoff's voltage law (KVL)
equations in a network, n and $b - n$, respectively, is determined only by
the graph of the network, and not by the types of the branches. The same
is true of the number of independent node-voltage variables (n) and loop-
current variables ($b - n$). These numbers would not be influenced if the
branches were all resistors, or if some were capacitors or inductors. However, in an all-resistor network the loop or node equations would be
algebraic, with no variation in time; that is to say, *static*. On the other
hand, when capacitors or inductors are present, the equations will be
dynamic. The question arises as to how many *dynamically independent*
variables there are; that is, how many variables are there such that, when
these variables are determined (as a function of time), the remaining
variables can be found purely algebraically?

We know that each capacitor and each inductor introduces a dynamic
variable, since the *v-i* relationship of each contains a derivative. We also
know that all initial voltages and currents in a network become known if
the initial capacitor voltages and inductor currents are specified. The
maximum number of initial conditions that can be specified independently,
therefore, equals the number of independent energy-storing branches
(capacitors plus inductors). This motivates us to introduce the notion of
order of complexity by the following definition:

The *order of complexity* of a network is equal to the number of independent initial conditions that can be specified in a network.

This is also the number of arbitrary constants appearing in the general solution of the network equations. Hence it is equal to the number of natural frequencies, if we count each one according to its multiplicity; for example, suppose the free response of a network consists of the following:

$$w^f(t) = A_1 \epsilon^{s_1 t} + (A_2 + A_3 t)\epsilon^{s_2 t} + A_4 \epsilon^{s_3 t} + A_5 \epsilon^{s_4 t} \tag{1}$$

The natural frequency s_2 is of multiplicity two; hence the total number of natural frequencies is stated to be five. This is also the order of complexity.

Clearly the order of complexity cannot exceed the number of energy-storing elements. Suppose, however, that there is an algebraic constraint relationship among capacitor voltages or inductor currents. Such constraints can be caused by loops containing only capacitors or only capacitors and independent voltage sources, and cut-sets containing only inductors or only inductors and independent current sources.*

In the first case, KVL applied around the loop will give a linear relationship among the capacitor voltages, and in the second case, the KCL equation for the cut-set will give a linear relationship among the inductor currents. In Fig. 1 there are five energy-storing elements. However, in

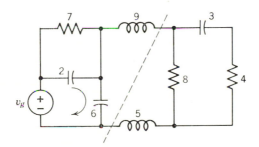

Fig. 1.   Network with an all-capacitor loop and an all-inductor cut-set.

this network there is an all-capacitor loop consisting of two capacitors and a voltage source. There is also an all-inductor cut-set consisting of two inductors. Thus the capacitor voltages and inductor currents will be restricted by the following constraints:

$$v_2 + v_6 = v_g, \tag{2a}$$

$$i_5 + i_9 = 0 \tag{2b}$$

* To avoid repetition, we shall use the term "all-capacitor loop" to mean a loop containing only capacitors or only capacitors and independent voltage sources. Likewise, we shall use the term "all-inductor cut-set" to mean a cut-set containing only inductors or only inductors and independent current-sources.

(with appropriate orientation of the variables). This means that initial values of both v_2 and v_6 cannot be prescribed independently, nor can initial values of both i_5 and i_9.

Each such constraint relationship reduces the number of independent initial conditions by one. In a network having only two-terminal components, the component equations cannot introduce additional algebraic relationships between capacitor voltages or inductor currents. We conclude, therefore, that:

The order of complexity of an *RLC* network equals the total number of reactive elements, less the number of independent all capacitor loops and the number of independent all-inductor cut-sets.

In the network of Fig. 1 the order of complexity is $5 - 1 - 1 = 3$.

The question might arise as to the influence of loops containing inductors only or cut-sets of capacitors only. Consider, for example, the network in Fig. 2, which contains an all-inductor loop; KVL around the loop leads to

$$L_3 \frac{di_3}{dt} + L_5 \frac{di_5}{dt} + L_6 \frac{di_6}{dt} = \frac{d}{dt} (L_3 i_3 + L_5 i_5 + L_6 i_6) = 0. \qquad (3)$$

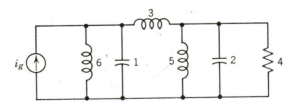

Fig. 2.   Network with an all-inductor loop.

Integration of this expression from 0 to t leads to

$$L_3 i_3(t) + L_5 i_5(t) + L_6 i_6(t) = L_3 i_3(0) + L_5 i_5(0) + L_6 i_6(0) = K, \qquad (4)$$

where we understand $t = 0$ to mean $0+$. It might appear that this also represents a constraint on the inductor currents. However, the constant K is not specified. In fact, its determination requires an independent relationship. This is provided by the *principle of conservation of flux linkages*, which states that $\sum L_j i_j$ over any closed loop is continuous. (This principle *cannot* be derived from Kirchhoff's laws.) The continuity condition

requires that the value of flux linkage just before $t = 0$ (i.e., at $0-$)
equal its value just after $t = 0$. Thus

$$K = L_3\, i_3(0-) + L_5\, i_5(0-) + L_6\, i_6(0-). \tag{5}$$

The $0-$ values of all three inductor currents can certainly be independently specified, without violating Kirchhoff's laws; such specification will
fix the $0+$ value of the flux linkage. Hence we conclude that an all-
inductor loop does not reduce the number of initial conditions that can be
independently specified and so has no influence on the order of complexity.

A similar conclusion follows concerning an all-capacitor cut-set;
namely, that it will have no influence on the order of complexity. An
equation like (5) will be obtained for an all capacitor cut-set except that
the terms will be $C_j\, v_j = q_j$ (charge) rather than flux linkage. In this case,
the role analogous to the principle of conservation of flux linkage is the
principle of conservation of charge, which, applied to a network, states that
$\sum c_j v_j = \sum q_j$ summed over any cut-set is continuous.

Although all-capacitor cut-sets and all-inductor loops do not influence
the *number* of natural frequencies, they do influence the *values* of natural
frequencies. In Fig. 2, for example, suppose $i_3(t)$ is the desired response.
It is clear that a constant current can circulate around the all-inductor
loop. Hence one of the terms in $i_3(t)$ will be a constant, which corresponds
to a natural frequency $s = 0$. Thus an all-inductor loop leads to a zero
natural frequency. A similar conclusion follows for an all-capacitor cut-
set. However, natural frequencies at $s = 0$ are somewhat peculiar in that
whether or not the corresponding term appears in the response depends on
(1) what specific variable constitutes the response and (2) the location
of the excitation. In Fig. 2, if the response is $v_3(t)$ rather than $i_3(t)$, a
constant term will not appear, since $v_3 = di_3/dt$, and differentiation will
remove the constant. All other natural frequencies will contribute to v_3,
since the derivative of an exponential is proportional to that exponential.

The preceding discussion indicates that what may be of interest in some
cases is not the total number of natural frequencies but the number of
nonzero natural frequencies. This can be obtained from the total number
by subtracting the number of all-capacitor cut-sets and all-inductor loops.
Hence

The number of nonzero natural frequencies equals the order of complexity minus the number of independent all-inductor loops and the
number of independent all-capacitor cut-sets.

The word "independent," both here and in the definition of the order of complexity given previously, is important. We can confirm this by reference to Fig. 3. Of the three all-inductor cut-sets only two are independent; the KCL equation for one of the cut-sets can be obtained from the

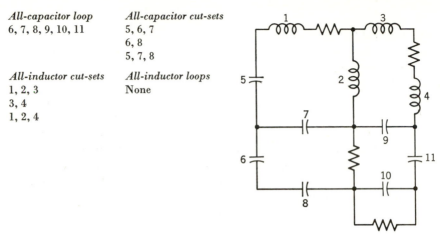

All-capacitor loop	All-capacitor cut-sets
6, 7, 8, 9, 10, 11	5, 6, 7
	6, 8
	5, 7, 8

All-inductor cut-sets	All-inductor loops
1, 2, 3	None
3, 4	
1, 2, 4	

Fig. 3.   Network with many degeneracies.

other two. The same is true of the three all-capacitor cut-sets: only two are independent. Since there are a total of 11 inductors and capacitors, and 3 linear constraints (one all-capacitor loop and two all-inductor cut-sets), the order of complexity and the number of natural frequencies is $11 - 3 = 8$. Of these natural frequencies, two are zero, corresponding to the two independent all-capacitor cut-sets. Thus there are $8 - 2 = 6$ nonzero natural frequencies.

4.2 BASIC CONSIDERATIONS IN WRITING STATE EQUATIONS

We are now ready to begin the development of the state equations. The basic equations at our disposal are still KVL, KCL, and the v-i relationships. It is the particular combination and the particular order in which these are invoked that we must choose. The decision is made on the basis of a number of considerations:

① We want the final equations to contain no integrals. Integrals arise from the substitution of $i = \int_0^t v\,dx/L + i(0)$ for an inductor current in KCL and the substitution of $v = \int_0^t i\,dx/C + v(0)$ for a capacitor voltage in KVL. So, we shall simply not make these eliminations of inductor currents and capacitor voltages but keep them as variables.

② We want the final equations to be first-order differential equations.

Derivatives arise from the substitution of $v = L \, di/dt$ for an inductor voltage in KVL and $i = C \, dv/dt$ for a capacitor current in KCL. We shall make these substitutions, thus removing capacitor currents and inductor voltages from the final set of variables.

3. Of the two capacitor variables, voltage and current, it is the voltages whose initial values can be independently specified in a network—*except when there is an all-capacitor loop*, as discussed in the last section. Likewise, for inductors, the initial currents can be independently specified—*except when there is an all-inductor cut-set*. This gives added impetus to the retention of capacitor voltages and inductor currents among the variables.

4. All the above considerations are nontopological; they have no bearing on how a tree is selected and what type of branches are twigs or links. Topologically, we know that twig voltages are a basis for all voltages; that is, knowledge of twig voltages determines all other voltages. Since we wish to have capacitor voltages among the final variables, we should place capacitors in a tree, as much as possible. Likewise, link currents are a basis for all currents. Since we wish to have inductor currents among the final variables, we should place inductors in the cotree, as much as possible.

5. Up to this point we have not counted independent sources as separate branches but have assumed they are always accompanied; we have lumped them with their accompanying branches. For reasons of convenience we shall here reverse this procedure and shall count independent sources as separate branches. The strain of the reorientation should not be too great. Since the voltage of a voltage source is a "known," it cannot be determined from other voltages. So, a voltage source cannot be made a link, because then its voltage would be fixed in terms of twig voltages. Similarly, a current source cannot be made a twig, since its current would then be fixed in terms of link currents. One might conceive of a network having a loop containing only independent voltage sources, in which case one of them would have to be a link. These sources could not truly be independent, since their voltages would have to satisfy KVL around the loop. If they did, then one of the sources could be a link, and its voltage would be determined from the other sources. Similar considerations apply to a cut-set containing only independent current sources. We assume, therefore, that our networks have no all-independent-voltage-source loops and no all-independent-current-source cut-sets.

The convergence of the above considerations leads to the following approach: Define a *normal tree* as a tree having as twigs all of the independent voltage sources, the maximum possible number of capacitors, the minimum possible number of inductors, and none of the independent current sources. (If the network is not connected, the corresponding term is "normal forest." For simplicity, we shall later refer to a normal tree, sometimes reverting to the use of "normal forest" for emphasis.) If there are no all-capacitor loops, all the capacitors will be twigs of the normal tree. Also, if there are no all-inductor cut-sets, none of the inductors will be twigs of the normal tree; they will all be links. This can be proved by contradiction. Suppose there is a capacitor link in the absence of an all-capacitor loop. Both end nodes of the capacitor link lie on the corresponding normal tree. If this capacitor link is added to the tree, it will form a loop that, by hypothesis, is not an all-capacitor loop. If from this loop a noncapacitor branch is removed, the result will be a new tree that will have one more capacitor than the preceding one. This is not possible, since the preceding normal tree has the maximum number of capacitors, by definition. A similar proof applies for the inductors.

Given a network, we first select a normal tree (or a normal forest if the network is not connected). We then write KVL equations for the f-loops and KCL equations for the f-cut-sets of this tree. We use the branch v-i relationships to eliminate capacitor currents and inductor voltages but we have not yet discussed how to handle the variables of the resistive elements, including controlled sources, gyrators, as well as resistors. Before considering this problem in a general way, let us spend some time in discussing some examples through which the general approach can evolve.

As a first illustration consider the network shown in Fig. 4a. It is desired to find the output voltage $v_0(t)$ when voltages $v_{g1}(t)$ and $v_{g2}(t)$ are the inputs. There are three reactive elements and no degeneracies. Hence the order of complexity is 3. There are six nodes and nine branches. (Remember that the two voltage sources are counted as separate branches.) A normal tree must contain both voltage sources and both

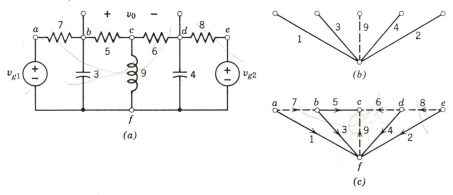

Fig. 4.   Illustrative example for writing state equations.

capacitors, but not the inductor. This information is shown in Fig. 4b, where the solid lines show a partial normal tree and the dashed line represents the inductor link. We need one more twig, and this must connect node c to the rest. There are clearly two possibilities among the resistive branches—branch 5 or 6. We have chosen 5 to complete the tree, as shown in Fig. 4c. Notice that the branches have been numbered with the twigs first, then the links.

Now let us write the KCL equations for the f-cut-sets and the KVL equations for the f-loops. With the usual partitioning, these can be written as

$$\mathbf{Qi} = [\mathbf{U} \quad \mathbf{Q}_l]\begin{bmatrix} \mathbf{i}_t \\ \mathbf{i}_l \end{bmatrix} = \mathbf{i}_t + \mathbf{Q}_l\mathbf{i}_l = \mathbf{0}, \tag{6a}$$

$$\mathbf{Bv} = [\mathbf{B}_t \quad \mathbf{U}]\begin{bmatrix} \mathbf{v}_t \\ \mathbf{v}_l \end{bmatrix} = \mathbf{B}_t\mathbf{v}_t + \mathbf{v}_l = \mathbf{0} \tag{6b}$$

or

$$\mathbf{i}_t = -\mathbf{Q}_l\mathbf{i}_l, \tag{7a}$$

$$\mathbf{v}_l = -\mathbf{B}_t\mathbf{v}_t. \tag{7b}$$

Note that, because the sources are counted as separate branches, the right-hand sides in (6) are $\mathbf{0}$ and not $\mathbf{Qi}_g$ and $\mathbf{Bv}_g$, as we have been used to writing them. In scalar form (7) leads to the following:

$$i_1 = \quad -i_7 \tag{8a}$$

$$i_2 = \quad\quad -i_8 \tag{8b}$$

$$i_3 = i_6 + i_7 \quad +i_9 \tag{8c}$$

$$i_4 = -i_6 \quad\quad +i_8 \tag{8d}$$

$$i_5 = -i_6 \quad\quad -i_9 \tag{8e}$$

$$v_6 = \quad -v_3 + v_4 \quad +v_5 \tag{8f}$$

$$v_7 = v_1 \quad -v_3 \tag{8g}$$

$$v_8 = \quad v_2 \quad -v_4 \tag{8h}$$

$$v_9 = \quad -v_3 \quad +v_5 \tag{8i}$$

There remain the *v-i* relationships of the components. We assume that $t = t_0$ is the point in time at which the network is established. The source voltages v_{g1} and v_{g2} are specified for $t \geq t_0$. The initial inductor current and capacitor voltages are also specified; that is, $v_3(t_0)$, $v_4(t_0)$, and $i_9(t_0)$ are given. The *v-i* relationships of the sources and reactive elements are

$$v_1 = v_{g1}(t), \tag{9a}$$

$$v_2 = v_{g2}(t), \tag{9b}$$

$$i_3 = C_3 \frac{dv_3}{dt}, \tag{9c}$$

$$i_4 = C_4 \frac{dv_4}{dt}, \tag{9d}$$

$$v_9 = L_9 \frac{di_9}{dt}. \tag{9e}$$

The resistor equations are either of the form $v = Ri$ or $i = Gv$, but we have no guidelines yet as to how they should be written. According to our earlier discussion, we want to eliminate capacitor currents and inductor voltages. Hence we substitute i_3, i_4, and v_9 from the last three equations into the appropriate KCL and KVL equations in (8). This step leads to

$$C_3 \frac{dv_3}{dt} = i_6 + i_7 + i_9, \tag{10a}$$

$$C_4 \frac{dv_4}{dt} = -i_6 + i_8, \tag{10b}$$

$$L_9 \frac{di_9}{dt} = -v_3 + v_5. \tag{10c}$$

There are two classes of variables on the right-hand sides: (1) capacitor voltages and inductor currents (v_3 and i_9), which we want to keep; and (2) resistor currents and voltages. There are four of the latter kind of variable; namely, v_5, i_6, i_7, and i_8. Note that none of these appears explicitly solved for in the Kirchhoff equations in (8), but their complementary variables do. To change these complementary variables to the desired ones, we can combine the appropriate *v-i* relationships with (8).

In fact, now we have a guide as to a suitable form of these relationships. They are

$$i_5 = G_5 v_5, \tag{11a}$$

$$v_6 = R_6 i_6, \tag{11b}$$

$$v_7 = R_7 i_7, \tag{11c}$$

$$v_8 = R_8 i_8. \tag{11d}$$

Thus for the twigs we use the form $v = Ri$; and for the links $i = Gv$. When these are inserted into the appropriate four equations in (8), the result can be rewritten in the following form:

$$G_5 v_5 + i_6 = -i_9 \tag{12a}$$

$$-v_5 + R_6 i_6 = -v_3 + v_4 \tag{12b}$$

$$R_7 i_7 = v_1 - v_3 \tag{12c}$$

$$R_8 i_8 = v_2 - v_4 \tag{12d}$$

These are purely algebraic equations giving resistor voltages or currents in terms of (1) source voltages, (2) capacitor voltages, and (3) inductor currents. These algebraic equations can be easily solved (the last two trivially) to yield

$$v_5 = \frac{1}{1 + G_5 R_6} v_3 - \frac{1}{1 + G_5 R_6} v_4 - \frac{R_6}{1 + G_5 R_6} i_9 \tag{13a}$$

$$i_6 = -\frac{G_5}{1 + G_5 R_6} v_3 + \frac{G_5}{1 + G_5 R_6} v_4 - \frac{1}{1 + G_5 R_6} i_9 \tag{13b}$$

$$i_7 = G_7 v_1 - G_7 v_3 \tag{13c}$$

$$i_8 = G_8 v_2 - G_8 v_4 \tag{13d}$$

Finally, these equations can be substituted into (10) to yield, after rearrangement,

$$C_3 \frac{dv_3}{dt} = -\left(G_7 + \frac{G_5}{1 + G_5 R_6}\right) v_3 + \frac{G_5}{1 + G_5 R_6} v_4 + \frac{G_5 R_6}{1 + G_5 R_6} i_9 - G_7 v_{g1},$$

$$\tag{14a}$$

$$C_4 \frac{dv_4}{dt} = \left(\frac{G_5}{1 + G_5 R_6}\right) v_3 - \left(G_8 + \frac{G_5}{1 + G_5 R_6}\right) v_4 + \frac{1}{1 + G_5 R_6} i_9 + G_8 v_{g2},$$

(14b)

$$L_9 \frac{di_9}{dt} = \frac{-G_5 R_6}{1 + G_5 R_6} v_3 - \frac{1}{1 + G_5 R_6} v_4 - \frac{R_6}{1 + G_5 R_6} i_9.$$

(14c)

This is a set of first-order differential equations. Let us write them in matrix form. After dividing by the coefficients on the left, the result will be

$$\frac{d}{dt}\begin{bmatrix} v_3 \\ v_4 \\ i_9 \end{bmatrix} = \begin{bmatrix} -\left(\frac{G_7}{C_3} + \frac{G_5/C_3}{1 + G_5 R_6}\right) & \frac{G_5/C_3}{1 + G_5 R_6} & \frac{G_5 R_6/C_3}{1 + G_5 R_6} \\ \frac{G_5/C_4}{1 + G_5 R_6} & -\left(\frac{G_8}{C_4} + \frac{G_5/C_4}{1 + G_5 R_6}\right) & \frac{1/C_4}{1 + G_5 R_6} \\ -\frac{G_5 R_6/L_9}{1 + G_5 R_6} & -\frac{1/L_9}{1 + G_5 R_6} & -\frac{R_6/L_9}{1 + G_5 R_6} \end{bmatrix} \begin{bmatrix} v_3 \\ v_4 \\ i_9 \end{bmatrix}$$

$$+ \begin{bmatrix} -\frac{G_7}{C_3} & 0 \\ 0 & \frac{G_8}{C_4} \\ 0 & 0 \end{bmatrix} \begin{bmatrix} v_{g1} \\ v_{g2} \end{bmatrix}. \quad (15)$$

This is the equation toward which we have been striving. It is a matrix differential equation of the first order. The following terminology is used:

State vector   *Input vector*

$$\mathbf{x}(t) = \begin{bmatrix} v_3(t) \\ v_4(t) \\ i_9(t) \end{bmatrix} \qquad \mathbf{e}(t) = \begin{bmatrix} v_{g1}(t) \\ v_{g2}(t) \end{bmatrix}$$

The elements of the state vector are *state variables*. We refer to the matrix equation as a *state equation*. Equation (15) can be written in compact matrix notation as

$$\frac{d\mathbf{x}}{dt} = \mathcal{A}\mathbf{x} + \mathcal{B}\mathbf{e}, \tag{16}$$

where the meanings of the matrices $\mathcal{A}$ and $\mathcal{B}$ are obvious. This is called the *normal form* of the state equation. The derivative of the state vector is given as a linear combination of the state vector itself and the input, or excitation, vector. (The letter " $\mathbf{e}$ " stands for excitation.)

The desired output quantities may be state variables or any other variables in the network. In the present case we had wanted the output to be $v_0(t)$. From the network it is found that $v_0 = v_3 - v_4$, which can be written in matrix form as

$$v_0 = \begin{bmatrix} 1 & -1 & 0 \end{bmatrix} \begin{bmatrix} v_3 \\ v_4 \\ i_9 \end{bmatrix}$$

or more compactly as

$$\mathbf{w}(t) = \mathbb{C}\mathbf{x}, \tag{17}$$

where $\mathbf{w}$ is the *output vector*.

The next step would be to solve the state equations, a project that will occupy a major part of the remainder of this chapter. However, before undertaking this major effort, we shall consider another example, which will introduce some features not present in this past example.

The network for this example is shown in Fig. 5. It has an all-capacitor loop (capacitors and voltage sources). A normal tree must contain both voltage sources but clearly cannot include all three capacitors. The ones numbered 3 and 4 have been placed on the tree. The normal tree can be completed by any one of the three resistors; the one numbered 5 is chosen here.

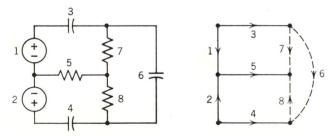

Fig. 5.   Illustrative example of circuit with all-C loop for writing state equations.

The next step is to write KCL equations for the f-cut-sets and KVL equations for the f-loops in the form of (7). The result is

$$i_1 = -i_6 - i_7 , \tag{18a}$$

$$i_2 = i_6 - i_8 , \tag{18b}$$

$$i_3 = i_6 + i_7 , \tag{18c}$$

$$i_4 = -i_6 + i_8 , \tag{18d}$$

$$i_5 = -i_7 - i_8 , \tag{18e}$$

$$v_6 = v_1 - v_2 - v_3 + v_4 , \tag{18f}$$

$$v_7 = v_1 - v_3 + v_5 , \tag{18g}$$

$$v_8 = v_2 - v_4 + v_5 . \tag{18h}$$

If we followed the approach of the preceding example, we would next write the v-i equations of all the capacitors (the only reactive elements in this case) in order to eliminate the capacitor currents from (18) and retain only the capacitor voltages. However, because of the all-capacitor loop, not all these voltages are dynamically independent. Therefore we write the v-i equations of only those capacitors that are in the normal tree. Thus

$$i_3 = C_3 \frac{dv_3}{dt} , \tag{19a}$$

$$i_4 = C_4 \frac{dv_4}{dt} . \tag{19b}$$

These are now inserted into the appropriate KCL equations in (18) to yield

$$C_3 \frac{dv_3}{dt} = i_6 + i_7 , \tag{20a}$$

$$C_4 \frac{dv_4}{dt} = -i_6 + i_8 . \tag{20b}$$

Of the three variables on the right sides, the capacitor link current i_6 and the resistor link currents i_7 and i_8 are treated separately. The KVL equation for v_6 in (18f) is inserted into the v-i equation for i_4, yielding

$$i_6 = C_6 \frac{dv_6}{dt} = C_6 \frac{dv_1}{dt} - C_6 \frac{dv_2}{dt} - C_6 \frac{dv_3}{dt} + C_6 \frac{dv_4}{dt}, \qquad (21)$$

which gives i_6 in terms of " desirable " variables only.

This leaves us with i_7 and i_8 to eliminate from (20). These do not appear explicitly solved for in (18). Again we write the v-i relations for the resistor links as $v = Ri$ and for the twig (branch 5) as $i = Gv$. The appropriate three equations in (18) can then be rewritten as follows:

$$G_5 v_5 + \quad i_7 + \quad i_8 = 0 \qquad (22a)$$

$$-v_5 + R_7 i_7 \qquad = v_1 - v_3 \qquad (22b)$$

$$-v_5 \qquad + R_8 i_8 = v_2 - v_4 \qquad (22c)$$

This is simply a set of algebraic equations that can be solved to yield

$$v_5 = -\frac{1}{K} v_1 - \frac{R_7 G_8}{K} v_2 + \frac{1}{K} v_3 + \frac{R_7 G_8}{K} v_4 , \qquad (23a)$$

$$i_7 = \frac{(G_5 + G_8)}{K} v_1 - \frac{G_8}{K} v_2 - \frac{(G_5 + G_8)}{K} v_3 + \frac{G_8}{K} v_4 , \qquad (23b)$$

$$i_8 = -\frac{G_8}{K} v_1 + G_8 \frac{(1 + G_5 R_7)}{K} v_2 + \frac{G_8}{K} v_3 - G_8 \frac{(1 + G_5 R_7)}{K} v_4 . \qquad (23c)$$

where $K = 1 + R_7(G_5 + G_8)$. The last two of these equations together with (21), when inserted into (20), will eliminate the unwanted variables. After rearrangement, the result in matrix form becomes

$$\begin{bmatrix} C_3 + C_6 & -C_6 \\ -C_6 & C_4 + C_6 \end{bmatrix} \frac{d}{dt} \begin{bmatrix} v_3 \\ v_4 \end{bmatrix} = \begin{bmatrix} \dfrac{-(G_5 + G_8)}{K} & \dfrac{-G_8}{K} \\[2mm] \dfrac{G_8}{K} & -G_8 \dfrac{(1 + G_5 R_7)}{K} \end{bmatrix} \begin{bmatrix} v_3 \\ v_4 \end{bmatrix}$$

$$(24)$$

$$+ \begin{bmatrix} \dfrac{(G_5 + G_8)}{K} & -\dfrac{G_8}{K} \\[2mm] -\dfrac{G_8}{K} & G_8 \dfrac{(1 + G_5 R_7)}{K} \end{bmatrix} \begin{bmatrix} v_{g1} \\ v_{g2} \end{bmatrix} + \begin{bmatrix} C_6 & C_6 \\ -C_6 & -C_6 \end{bmatrix} \frac{d}{dt} \begin{bmatrix} v_{g1} \\ v_{g2} \end{bmatrix} .$$

The final result is obtained upon premultiplying by the inverse of the coefficient matrix on the left. Again we have a first-order matrix differential equation, but this time the right-hand side has a term containing the derivative of the input vector, in addition to the input vector itself.

As a final calculation, suppose the outputs are taken as $-i_1$ and i_2. These can be expressed in terms of the link currents i_6, i_7, and i_8 by means of (18a) and (18b). These currents are, in turn, eliminated by using (21) and (23). The result of these operations for the output equation in matrix form will be

$$
\begin{bmatrix} -i_1 \\ i_2 \end{bmatrix} = \begin{bmatrix} -\dfrac{(G_5 + G_8)}{K} & \dfrac{G_8}{K} \\[2ex] -\dfrac{G_8}{K} & \dfrac{G_8(1 + G_5 R_7)}{K} \end{bmatrix} \begin{bmatrix} v_3 \\ v_4 \end{bmatrix} + \begin{bmatrix} -C_6 & C_6 \\ -C_6 & C_6 \end{bmatrix} \frac{d}{dt}\begin{bmatrix} v_3 \\ v_4 \end{bmatrix}
$$

$$
\tag{25}
$$

$$
+ \begin{bmatrix} \dfrac{(G_5 + G_8)}{K} & -\dfrac{G_8}{K} \\[2ex] \dfrac{G_8}{K} & -\dfrac{G_8(1 + G_5 R_7)}{K} \end{bmatrix} \begin{bmatrix} v_{g1} \\ v_{g2} \end{bmatrix} + \begin{bmatrix} C_6 & -C_6 \\ C_6 & -C_6 \end{bmatrix} \frac{d}{dt}\begin{bmatrix} v_{g1} \\ v_{g2} \end{bmatrix}.
$$

The derivative of

$$
\begin{bmatrix} v_3 \\ v_4 \end{bmatrix}
$$

can be eliminated by using (24). The detailed result is algebraically complicated and is not given here. Nevertheless, it is clear that the kinds of terms present on the right side of the output equation will be the same as those present on the right side of (24).

Equations (24) and (25) can be written in compact form as follows:

$$
\frac{d\mathbf{x}}{dt} = \mathscr{A}\mathbf{x} + \mathscr{B}_1\mathbf{e} + \mathscr{B}_2\frac{d\mathbf{e}}{dt}, \tag{26a}
$$

$$
\mathbf{w} = \mathcal{C}\mathbf{x} + \mathscr{D}_1\mathbf{e} + \mathscr{D}_2\frac{d\mathbf{e}}{dt}. \tag{26b}
$$

The first of these is a differential equation. Once it is solved, the output variables are determined algebraically from the second one. As a matter of terminology, the two equations together are called the *state equations*. The second one is called the *output equation*. Our next task is to carry out a solution of these equations.

4.3   TIME-DOMAIN SOLUTIONS OF THE STATE EQUATIONS

In the examples of the last section we found that the input and output variables are related through equations such as (26). We shall find in later sections that such equations result for all networks of the class we are considering. Observe that a somewhat simpler form is obtained by setting

$$\mathbf{x} = \bar{\mathbf{x}} + \mathscr{B}_2 \mathbf{e} \tag{27}$$

in (26), which then becomes

$$\frac{d\bar{\mathbf{x}}}{dt} = \mathscr{A}\bar{\mathbf{x}} + (\mathscr{B}_1 + \mathscr{A}\mathscr{B}_2)\mathbf{e},$$

$$\mathbf{w} = \mathrm{C}\bar{\mathbf{x}} + (\mathscr{D}_1 + \mathrm{C}\mathscr{B}_2)\mathbf{e} + \mathscr{D}_2 \frac{d\mathbf{e}}{dt}.$$

The derivative of the excitation has been eliminated in the first equation but not in the second. For simplicity we shall simply remove the bar and write $\mathbf{x}$ instead of $\bar{\mathbf{x}}$. Furthermore, we shall replace $\mathscr{B}_1 + \mathscr{A}\mathscr{B}_2$ by $\mathscr{B}$, $\mathscr{D}_1 + \mathrm{C}\mathscr{B}_2$ by $\mathscr{D}$, and $\mathscr{D}_2$ by $\widehat{\mathscr{D}}$. The equations we shall treat will, therefore, have the forms

$$\frac{d\mathbf{x}}{dt} = \mathscr{A}\mathbf{x} + \mathscr{B}\mathbf{e}, \tag{28a}$$

$$\mathbf{w} = \mathrm{C}\mathbf{x} + \mathscr{D}\mathbf{e} + \widehat{\mathscr{D}} \frac{d\mathbf{e}}{dt}. \tag{28b}$$

If the first of these equations resulting from a network is not initially in this form (because it contains a term involving the derivative of $\mathbf{e}$), the transformation of (27) will put it in this form. Even with this transformation, however, we see that the derivative of $\mathbf{e}$ will be present in the output equation unless $\widehat{\mathscr{D}} \equiv 0$. Whether or not this condition is true will depend on the specific network and on the variables that are the outputs.

The vector $\mathbf{x}$ is assumed to be an n-vector (n components). The number of components in e will generally be different from n. So, $\mathscr{A}$ is a square matrix, but $\mathscr{B}$ is generally not square.

We now turn our attention to the solution of (28) for $\mathbf{x}(t)$, assuming the initial values are expressed by the vector $\mathbf{x}(t_0)$. For this purpose we shall use the method of *variation of parameter*. Let

$$\mathbf{x}(t) = \mathbf{Y}(t)\, \mathbf{x}_1(t), \tag{29}$$

in which $\mathbf{Y}(t)$ is a square matrix of order n that is assumed to be non-singular for all finite $t \geq t_0$.* Insert this transformation into (28). The result after suitable arrangement of terms will be

$$\left(\frac{d\mathbf{Y}}{dt} - \mathscr{A}\mathbf{Y}\right)\mathbf{x}_1 = -\,\mathbf{Y}\frac{d\mathbf{x}_1}{dt} + \mathscr{B}\mathbf{e}. \tag{30}$$

It is clear that the solution is simplified if the quantity in parentheses is assumed to be zero. This will then lead to a homogeneous matrix differential equation for $\mathbf{Y}$. After that equation is solved, $\mathbf{Y}$ can be inserted into the right side of (30). The result can then be directly integrated to find $\mathbf{x}_1$. After both $\mathbf{Y}$ and $\mathbf{x}_1$ are found, $\mathbf{x}$ is determined from (29).

Proceding in this way, the two equations that result from (30) by setting the quantity in parenthesis equal to zero are

$$\frac{d\mathbf{Y}(t)}{dt} - \mathscr{A}\mathbf{Y}(t) = \mathbf{0}, \tag{31}$$

$$\frac{d\mathbf{x}_1(t)}{dt} = \mathbf{Y}(t)^{-1}\,\mathscr{B}\mathbf{e}(t). \tag{32}$$

The second equation comes from premultiplying by the inverse of $\mathbf{Y}$, which exists because $\mathbf{Y}$ was assumed to be nonsingular. We shall temporarily postpone solving the first equation and assume a solution has been obtained. The second equation can be solved for $\mathbf{x}_1$ by direct integration from t_0 to t. The result will be

$$\mathbf{x}_1(t) = \mathbf{x}_1(t_0) + \int_{t_0}^{t} \mathbf{Y}(\tau)^{-1}\,\mathscr{B}\mathbf{e}(\tau)\,d\tau. \tag{33}$$

---

* The development of this chapter is extravagant in the use of symbols. The need far outstrips the availability. This forces us to use a symbol in one context when it already has a well-defined meaning elsewhere. Thus in earlier work $\mathbf{Y}$ is an admittance matrix. Its use here with a different meaning will hopefully cause no confusion.

The initial-value matrices are related by

$$\mathbf{x}(t_0) = \mathbf{Y}(t_0)\, \mathbf{x}_1(t_0). \tag{34}$$

Therefore $\mathbf{x}_1(t_0) = \mathbf{Y}(t_0)^{-1}\mathbf{x}(t_0)$ is the specified initial condition for (33).

Now premultiply both sides of (33) by $\mathbf{Y}(t)$. Since the integration is with respect to τ, $\mathbf{Y}(t)$ can be taken under the integral sign. Furthermore, since $\mathbf{x}_1(t_0) = \mathbf{Y}(t_0)^{-1}\, \mathbf{x}(t_0)$, the result will be

$$\mathbf{x}(t) = \mathbf{Y}(t)\, \mathbf{Y}(t_0)^{-1}\, \mathbf{x}(t_0) + \int_{t_0}^{t} \mathbf{Y}(t)\, \mathbf{Y}(\tau)^{-1}\, \mathscr{B}\mathbf{e}(\tau)\, d\tau. \tag{35}$$

This is a very significant result. It tells us that in order to solve (28) we first solve (31) with some nonsingular initial condition, such as $\mathbf{Y}(t_0) = \mathbf{U}$. We then carry out the indicated integration in (35). However, the integrand requires us first to find the inverse of $\mathbf{Y}$, which is a considerable chore.

It turns out that this can be avoided, because $\mathbf{Y}(t)\,\mathbf{Y}(\tau)^{-1}$ is a matrix function of $t - \tau$, which, as will be seen shortly, is easily determined. We express this relationship symbolically as

$$\mathbf{Y}(t)\ \mathbf{Y}(\tau)^{-1} = \Phi(t - \tau). \tag{36}$$

When this is inserted into (35), the result becomes

$$\mathbf{x}(t) = \Phi(t - t_0)\, \mathbf{x}(t_0) + \int_{t_0}^{t} \Phi(t - \tau)\mathscr{B}\mathbf{e}(\tau)\, d\tau. \tag{37}$$

The matrix Φ is called the *state-transition matrix*. The name derives from the idea that when $\mathbf{e} \equiv \mathbf{0}$ the transition from the " state " of the network at time t_0 to the " state " at time t is governed by Φ, as (37) illustrates.

Equation (37) constitutes the time-domain solution of the original nonhomogeneous differential equation in (28). Its importance cannot be overemphasized. However, it is really a symbolic solution, because we are still required to solve the homogeneous equation (31) before the job is complete. This will be our task now.

SOLUTION OF HOMOGENEOUS EQUATION

Consider a first-order homogeneous differential equation

$$\frac{dy}{dt} - ay = 0,$$

where a is a constant and the initial condition is $y(t_0) = 1$. The solution that satisfies the initial condition is

$$y(t) = \epsilon^{a(t-t_0)},$$

which may be verified by direct substitution into the equation. Since the form of the matrix equation (31) is identical with that of the scalar equation, it is tempting to seek an exponential solution:

$$\mathbf{Y}(t) = \epsilon^{\mathscr{A}(t-t_0)}. \tag{38}$$

The only trouble is, we do not know the meaning of an exponential with a matrix in the exponent. You have no doubt encountered a similar difficulty in defining an exponential with a complex-number exponent. This is handled by defining a complex exponential in terms of the series expansion of the real exponential. (See Appendix 2.) We shall do the same thing here and define

$$\epsilon^{\mathscr{A}t} = \sum_{k=0}^{\infty} \frac{\mathscr{A}^k t^k}{k!} = \mathbf{U} + \mathscr{A}t + \frac{\mathscr{A}^2 t^2}{2!} + \cdots + \frac{\mathscr{A}^k t^k}{k!} + \cdots. \tag{39}$$

Since $\mathscr{A}$ is a square matrix of order n, $\epsilon^{\mathscr{A}t}$ is also a square matrix of order n.

As an example, suppose

$$\mathscr{A} = \begin{bmatrix} -1 & 0 \\ 1 & -2 \end{bmatrix}, \quad \mathscr{A}^2 = \mathscr{A}\mathscr{A} = \begin{bmatrix} 1 & 0 \\ -3 & 4 \end{bmatrix}, \quad \mathscr{A}^3 = \begin{bmatrix} -1 & 0 \\ 7 & -8 \end{bmatrix}.$$

Then

$$e^{\mathscr{A}t} = \begin{bmatrix} 1 & 0 \\ 0 & 1 \end{bmatrix} + \begin{bmatrix} -1 & 0 \\ 1 & -2 \end{bmatrix} t + \begin{bmatrix} 1 & 0 \\ -3 & 4 \end{bmatrix} \frac{t^2}{2} + \begin{bmatrix} -1 & 0 \\ 7 & -8 \end{bmatrix} \frac{t^3}{6} + \cdots$$

$$= \begin{bmatrix} 1 - t + \dfrac{t^2}{2} - \dfrac{t^3}{6} + \cdots & 0 \\ t - \dfrac{3t^2}{2} + \dfrac{7t^3}{6} + \cdots & 1 - 2t + 2t^2 - \dfrac{4t^3}{3} + \cdots \end{bmatrix}. \tag{40}$$

It can be shown that each of the elements of the matrix $\epsilon^{\mathscr{A}t}$ converges to a continuous function of t, absolutely for any finite t and uniformly over any finite time interval. Hence term-by-term differentiation of the series is permitted. Thus

$$\frac{d}{dt}\left(\epsilon^{\mathscr{A}t}\right) = \mathscr{A} + \mathscr{A}^2 t + \frac{\mathscr{A}^3 t^2}{2!} + \frac{\mathscr{A}^4 t^3}{3!} + \cdots$$

$$= \mathscr{A}\left(\mathbf{U} + \mathscr{A}t + \frac{\mathscr{A}^2 t^2}{2!} + \frac{\mathscr{A}^3 t^3}{3!} + \cdots\right) = \mathscr{A}\,\epsilon^{\mathscr{A}t}\,;$$

(41)

that is, the formula for the derivative of a matrix exponential is the same as it is for a scalar exponential. When this result is used it is found that $\mathbf{Y}(t) = \epsilon^{\mathscr{A}(t-t_0)}$ given in (38) is the (unique) solution satisfying (31) and the initial condition $\mathbf{Y}(t_0) = \mathbf{U}$.

Recall that in obtaining (32) it was assumed that $\mathbf{Y}(t)$ is nonsingular for all finite time following t_0. We must now show that it is, in fact, nonsingular. This is not difficult. From the series definition of a matrix exponential, we can write

$$\epsilon^{-\mathscr{A}t} = \mathbf{U} - \mathscr{A}t + \mathscr{A}^2 \frac{t^2}{2!} - \cdots + (-1)^k \mathscr{A}^k \frac{t^k}{k!} + \cdots.$$

(42)

Now let this series be multiplied by the series for the positive exponential in (39). The result will be

$$\epsilon^{\mathscr{A}t}\epsilon^{-\mathscr{A}t} = \mathbf{U}.$$

(43)

All other terms cancel. This term-by-term multiplication is permissible because of the absolute convergence of the two series for all finite t. The result tells us that we have found a matrix $(\epsilon^{-\mathscr{A}t})$, which, when multiplied by $\epsilon^{\mathscr{A}t}$, gives a unit matrix. By definition, it is the inverse of $\epsilon^{\mathscr{A}t}$. Hence $\mathbf{Y}(t)$ is nonsingular for $t \geq t_0$.

There is only one thing left to do. We must give an explicit expression for the state-transition matrix $\Phi(t - \tau) = \mathbf{Y}(t)\,\mathbf{Y}(\tau)^{-1}$. This is an easy task. We know that $\mathbf{Y}(t) = \epsilon^{\mathscr{A}(t-t_0)}$ and $\mathbf{Y}(\tau)^{-1} = \epsilon^{-\mathscr{A}(\tau-t_0)}$; therefore

$$\Phi(t - \tau) = \epsilon^{\mathscr{A}(t-\tau)}.$$

(44)

and is, like $\mathbf{Y}(t)$, a matrix exponential—only the scalar time variable is different. This relation can now be inserted into (37) to yield

$$\mathbf{x}(t) = \epsilon^{\mathscr{A}(t-t_0)}\,\mathbf{x}(t_0) + \int_{t_0}^{t} \epsilon^{\mathscr{A}(t-\tau)}\,\mathscr{B}\mathbf{e}(\tau)\,d\tau.$$

(45)

The solution is now complete. Starting with a vector differential equation of the form of (28), we first solve the homogeneous equation (31) subject to the initial condition $\mathbf{Y}(t_0) = \mathbf{U}$. The solution is $\epsilon^{\mathscr{A}(t-t_0)}$. We then insert this into (45), changing t_0 to τ under the integral, carry out the integration, and the job is done.

ALTERNATIVE METHOD OF SOLUTION

We have just treated the solution of the state equation in the general case. In a particular situation, suppose there is no excitation ($\mathbf{e} \equiv \mathbf{0}$) or for a given network $\mathscr{B} \equiv \mathbf{0}$. Then the state equation reduces to the homogeneous equation

$$\frac{d}{dt}\mathbf{x} = \mathscr{A}\mathbf{x}. \tag{46}$$

Comparing this with (31) shows that they are of the same form. There is, however, one difference: whereas $\mathbf{Y}$ (or, equivalently, the state-transition matrix Φ) is a square matrix, in the present equation $\mathbf{x}$ is a column vector. Among other things, this means the initial value in the present case cannot be a unit matrix but must be represented by the initial-value vector $\mathbf{x}(t_0)$.

From the general solution in (45) the solution of (46) can be written as

$$\mathbf{x}(t) = \epsilon^{\mathscr{A}(t-t_0)}\mathbf{x}(t_0). \tag{47}$$

This is, of course, much simpler than the general solution when $\mathscr{B}\mathbf{e} \neq 0$. It would, therefore, be of considerable value if, by some modification of variables, it would be possible to convert the general nonhomogeneous state equation into a homogeneous one. This is what we shall pursue in this section.

Consider the state equation (28a), repeated here for convenience:

$$\frac{d}{dt}\mathbf{x} = \mathscr{A}\mathbf{x} + \mathscr{B}\mathbf{e}. \tag{48}$$

Suppose there exists a vector $\mathbf{f}$ that satisfies the differential equation

$$\frac{d}{dt}\mathbf{f} = \mathscr{F}\mathbf{f} \tag{49}$$

with the initial value $\mathbf{f}(t_0)$ and that is related to $\mathbf{e}$ by

$$\mathbf{e} = \mathcal{K}\mathbf{f}. \tag{50}$$

In these expressions $\mathcal{F}$ and $\mathcal{K}$ are matrices to be determined. Now substitute (50) into (48) and combine the resulting equation with (49). The result can be put in the following form:

$$\frac{d}{dt}\begin{bmatrix} \mathbf{x} \\ \mathbf{f} \end{bmatrix} = \begin{bmatrix} \mathcal{A} & \mathcal{B}\mathcal{K} \\ \mathbf{0} & \mathcal{F} \end{bmatrix}\begin{bmatrix} \mathbf{x} \\ \mathbf{f} \end{bmatrix}, \tag{51}$$

which is homogeneous like (46). Consequently the solution will be

$$\begin{bmatrix} \mathbf{x}(t) \\ \mathbf{f}(t) \end{bmatrix} = \exp\left\{\begin{bmatrix} \mathcal{A} & \mathcal{B}\mathcal{K} \\ \mathbf{0} & \mathcal{F} \end{bmatrix}(t - t_0)\right\}\begin{bmatrix} \mathbf{x}(t_0) \\ \mathbf{f}(t_0) \end{bmatrix} \tag{52}$$

just as (47) was the solution of (46). (The notation $\exp(u)$ stands for ϵ^u.) The solution for $\mathbf{x}(t)$ is, of course, just the first n-elements of this solution for $[\mathbf{x} \ \ \mathbf{f}]'$.

There is one major drawback to this method of obtaining a homogeneous differential equation equivalent to the nonhomogeneous state equation. Suppose $\mathbf{f}$ is an m-vector. Then the matrix exponential in (52) is of order $n + m$. In the solution (45) of the original state equation the order of the matrix exponential is just n. Since m may very easily be large, the increase in the order of the matrix exponential by m can result in a substantial computing effort merely to eliminate computing the integral appearing in (45).

It is possible to use an alternative procedure that will lead to a homogeneous differential equation without increasing the order of the matrix exponential to be evaluated. As might be expected, this is achieved at a price. Let

$$\mathbf{y} = \dot{\mathbf{x}} + \mathcal{S}\mathbf{f} \tag{53}$$

or, equivalently,

$$\mathbf{x} = \mathbf{y} - \mathcal{S}\mathbf{f}, \tag{54}$$

where $\mathcal{S}$ is a matrix to be determined. Substitute (54) into the state equation (48) to get

$$\frac{d}{dt}\mathbf{y} - \mathcal{S}\frac{d}{dt}\mathbf{f} = \mathcal{A}\mathbf{y} - \mathcal{A}\mathcal{S}\mathbf{f} + \mathcal{B}\mathbf{e}. \tag{55}$$

Then substitute (49) and (50) into this equation and rearrange terms to obtain

$$\frac{d}{dt}\mathbf{y} = \mathscr{A}\mathbf{y} + [\mathscr{S}\mathscr{F} - \mathscr{A}\mathscr{S} + \mathscr{B}\mathscr{K}]\mathbf{f}. \tag{56}$$

If an $\mathscr{S}$ can be found that satisfies the following two-sided, linear, algebraic matrix equation

$$\mathscr{A}\mathscr{S} - \mathscr{S}\mathscr{F} = \mathscr{B}\mathscr{K}, \tag{57}$$

then (56) becomes

$$\frac{d}{dt}\mathbf{y} = \mathscr{A}\mathbf{y}, \tag{58}$$

which is the same homogeneous differential equation as in (46). Its solution, therefore, is

$$\mathbf{y}(t) = \epsilon^{\mathscr{A}(t-t_0)}\mathbf{y}(t_0), \tag{59}$$

in which

$$\mathbf{y}(t_0) = \mathbf{x}(t_0) + \mathscr{S}\mathbf{f}(t_0) \tag{60}$$

from the definition of $\mathbf{y}$ in (53). The solution for $\mathbf{x}(t)$ is obtained by substituting $\mathbf{y}(t)$ from (59) and $\mathbf{f}(t)$ from the solution of (49) into (54).

The solution of the two-sided matrix equation (57) is not a trivial matter.* Since $\mathscr{S}$ will be an $n \times m$ matrix, (57) is equivalent to nm linear algebraic equations for the nm unknown elements of $\mathscr{S}$.

To illustrate these two methods of obtaining an equivalent homogeneous differential equation, let us start with the state equation

$$\frac{d}{dt}\mathbf{x} = \begin{bmatrix} 1 & 4 \\ -2 & -5 \end{bmatrix}\mathbf{x} + \begin{bmatrix} 1 \\ -1 \end{bmatrix}[2\sin 2t - 3\cos 2t].$$

---

* It turns out that, if the eigenvalues of $\mathscr{F}$ are different from those of $\mathscr{A}$, the solution for $\mathscr{S}$ can be expressed in closed form using some of the results of the next section. This closed-form solution will be given in Problem 17. You will find a proof for that solution in: J. S. Frame, "Matrix Functions and Applications—Part IV," *IEEE Spectrum*, Vol. 1, No. 6, June 1964, pp. 123–131. A second closed form solution will be given in Problem 35. You will find a proof for that solution in: A. Jameson, "Solution of the Equation $\mathbf{AX} + \mathbf{XB} = \mathbf{C}$ by Inversion of an $M \times M$ or $N \times N$ Matrix," *SIAM Jour. of Applied Mathematics*, Vol. 16, No. 5, Sept. 1968, pp. 1020–1023.

SKIP except one method upto 280

It may easily be verified that

$$\mathbf{f} = \begin{bmatrix} \sin 2t \\ \cos 2t \end{bmatrix}$$

is the solution of the differential equation

$$\frac{d}{dt}\mathbf{f} = \begin{bmatrix} 0 & 2 \\ -2 & 0 \end{bmatrix}\mathbf{f}, \qquad \mathbf{f}(0) = \begin{bmatrix} 0 \\ 1 \end{bmatrix}.$$

Observe that

$$e = (2\sin 2t - 3\cos 2t) = \begin{bmatrix} 2 & -3 \end{bmatrix}\begin{bmatrix} \sin 2t \\ \cos 2t \end{bmatrix} = \mathscr{K}\mathbf{f}.$$

Therefore

$$\mathscr{K} = \begin{bmatrix} 2 & -3 \end{bmatrix}.$$

The matrices $\mathscr{A}$, $\mathscr{B}$, and $\mathscr{F}$ are obvious from the state equation and the differential equation for **f**. The vector differential equation corresponding to (51) is, therefore,

$$\frac{d}{dt}\begin{bmatrix} \mathbf{x} \\ \mathbf{f} \end{bmatrix} = \left[\begin{array}{cc:cc} 1 & 4 & 2 & -3 \\ -2 & -5 & -2 & 3 \\ \hdashline 0 & 0 & 0 & 2 \\ 0 & 0 & -2 & 0 \end{array}\right]\begin{bmatrix} \mathbf{x} \\ \mathbf{f} \end{bmatrix}.$$

The solution of this equation is now easily written.

The alternative method requires solution of the two-sided matrix equation (57). Since the order of $\mathscr{A}$ is $n = 2$ and the order of $\mathscr{F}$ is $m = 2$, $\mathscr{S}$ will be a 2×2 matrix. For this example (57) will be

$$\begin{bmatrix} 1 & 4 \\ -2 & -5 \end{bmatrix}\begin{bmatrix} S_{11} & S_{12} \\ S_{21} & S_{22} \end{bmatrix} - \begin{bmatrix} S_{11} & S_{12} \\ S_{21} & S_{22} \end{bmatrix}\begin{bmatrix} 0 & 2 \\ -2 & 0 \end{bmatrix} = \begin{bmatrix} 2 & -3 \\ -2 & 3 \end{bmatrix}.$$

As you may easily verify, this is equivalent to the following algebraic equation:

$$
\begin{bmatrix}
1 & 4 & 2 & 0 \\
-2 & -5 & 0 & 2 \\
-2 & 0 & 1 & 4 \\
0 & -2 & -2 & -5
\end{bmatrix}
\begin{bmatrix}
S_{11} \\
S_{21} \\
S_{12} \\
S_{22}
\end{bmatrix}
=
\begin{bmatrix}
2 \\
-2 \\
-3 \\
3
\end{bmatrix}
$$

whose solution is

$$
\begin{bmatrix}
S_{11} \\
S_{21} \\
S_{12} \\
S_{22}
\end{bmatrix}
=
\begin{bmatrix}
0 \\
0 \\
1 \\
-1
\end{bmatrix}.
$$

Using these values for the S_{ij}, the matrix $\mathscr{S}$ is

$$
\mathscr{S} = \begin{bmatrix} 0 & 1 \\ 0 & -1 \end{bmatrix}.
$$

Thus $\mathscr{S}$ exists, and the solution for $\mathbf{y}(t)$, and then $\mathbf{x}(t)$, can be obtained by the use of this method.

In this example we converted the two-sided matrix equation for $\mathscr{S}$ into an equivalent vector equation for a vector with the same elements as $\mathscr{S}$. Let us indicate how this is accomplished in general. Let $\mathbf{s}_i$ and $\mathbf{k}_i$ denote the ith column vectors of $\mathscr{S}$ and $\mathscr{K}$, respectively. Then the vector equation

$$
\begin{bmatrix}
\mathbf{R}_{11} & \mathbf{R}_{12} & \cdots & \mathbf{R}_{1m} \\
\mathbf{R}_{21} & \mathbf{R}_{22} & \cdots & \mathbf{R}_{2m} \\
\vdots & \vdots & & \vdots \\
\mathbf{R}_{m1} & \mathbf{R}_{m2} & \cdots & \mathbf{R}_{mm}
\end{bmatrix}
\begin{bmatrix}
\mathbf{s}_1 \\
\mathbf{s}_2 \\
\vdots \\
\mathbf{s}_m
\end{bmatrix}
=
\begin{bmatrix}
\mathscr{B}\mathbf{k}_1 \\
\mathscr{B}\mathbf{k}_2 \\
\vdots \\
\mathscr{B}\mathbf{k}_m
\end{bmatrix}
\tag{61}
$$

where

$$
\mathbf{R}_{ij} = \delta_{ij}\,\mathscr{A} - f_{ji}\,\mathbf{U}
\tag{62}
$$

is equivalent to the two-sided matrix equation (57) in the sense that the solution of (61) above yields values for all the elements of $\mathscr{S}$.

Many excitation functions are encountered over and over again in analysis. There are, in a sense, a standard set of excitations used in analyzing the performance of a network. The sine and cosine of the past example are only two functions from that set. Some of the other often-used excitations are step functions, ramp functions, exponential functions, and exponentially damped sinusoids. To eliminate the necessity of constructing the $\mathscr{F}$ matrix and the corresponding initial vector $\mathbf{f}(0)$ each time these standard excitation functions are encountered, we have constructed a table (Table 1) of the most often encountered $\mathbf{f}(t)$'s and the associated $\mathscr{F}$'s and $\mathbf{f}(0)$'s. Observe that $t_0 = 0$ in this table; there is no loss of generality in doing this, and it is decidedly convenient to do so. Note also that the constant α appearing in Table 1 can be zero. If this is done, the elements of $\mathbf{f}(t)$ become simply powers of t in the one case, and ordinary sine and cosine functions in the other.

Table 1

$\mathbf{f}(t)$	$\mathscr{F}$	$\mathbf{f}(0)$
$\begin{bmatrix} \epsilon^{-\alpha t} \\ t\epsilon^{-\alpha t} \\ t^2\epsilon^{-\alpha t} \\ \vdots \\ t^{k-1}\epsilon^{-\alpha t} \\ t^k\epsilon^{-\alpha t} \end{bmatrix}$	$\begin{bmatrix} -\alpha & 0 & 0 \cdots & 0 & 0 \\ 1 & -\alpha & 0 \cdots & 0 & 0 \\ 0 & 2 & -\alpha \cdots & 0 & 0 \\ \vdots & \vdots & \vdots & \vdots & \vdots \\ 0 & 0 & 0 \cdots -\alpha & 0 \\ 0 & 0 & 0 \cdots & k & -\alpha \end{bmatrix}$	$\begin{bmatrix} 1 \\ 0 \\ 0 \\ \vdots \\ 0 \\ 0 \end{bmatrix}$
$\begin{bmatrix} \epsilon^{-\alpha t} \sin \omega t \\ \epsilon^{-\alpha t} \cos \omega t \end{bmatrix}$	$\begin{bmatrix} -\alpha & \omega \\ -\omega & -\alpha \end{bmatrix}$	$\begin{bmatrix} 0 \\ 1 \end{bmatrix}$

Since a network excitation vector is very apt to have elements that are combinations of the elements of standard excitation vectors with different values of α, ω, and k, it may be necessary to combine the several differential equations for the different standard excitation vectors into a single differential equation. To make this point clear we shall consider a simple example. Let

$$\mathbf{e}(t) = \begin{bmatrix} 4 - 3\epsilon^{-2t} \\ 2 - 2t\epsilon^{-3t} \\ \epsilon^{-t} \cos t + 4 \sin 1.5t \end{bmatrix}.$$

The elements of **e** can be related to the standard excitation functions, appearing as elements of **f**, as follows:

$$\mathbf{e}(t) = \begin{bmatrix} 4 - 3\epsilon^{-2t} \\ 2 - 2t\epsilon^{-3t} \\ \epsilon^{-t}\cos t + 4\sin 1.5t \end{bmatrix}$$

$$= \begin{bmatrix} 4 & \vdots & -3 & \vdots & 0 & 0 & \vdots & 0 & 0 & \vdots & 0 & 0 \\ 2 & \vdots & 0 & \vdots & 0 & -2 & \vdots & 0 & 0 & \vdots & 0 & 0 \\ 0 & \vdots & 0 & \vdots & 0 & 0 & \vdots & 0 & 1 & \vdots & 4 & 0 \end{bmatrix} \begin{bmatrix} 1 \\ \cdots \\ \epsilon^{-2t} \\ \cdots \\ \epsilon^{-3t} \\ t\epsilon^{-3t} \\ \cdots \\ \epsilon^{-t}\sin t \\ \epsilon^{-t}\cos t \\ \cdots \\ \sin 1.5t \\ \cos 1.5t \end{bmatrix} = \mathscr{K}\mathbf{f}(t).$$

Based on Table 1, there are five differential equations, each for a part of the **f**-vector on the right; they are

$$\frac{d}{dt}[f_1] = [0][f_1] \qquad\qquad [f_1(0)] = [1]$$

$$\frac{d}{dt}[f_2] = [-2][f_2], \qquad\qquad [f_2(0)] = [1],$$

$$\frac{d}{dt}\begin{bmatrix} f_3 \\ f_4 \end{bmatrix} = \begin{bmatrix} -3 & 0 \\ 1 & -3 \end{bmatrix}\begin{bmatrix} f_3 \\ f_4 \end{bmatrix}, \qquad \begin{bmatrix} f_3(0) \\ f_4(0) \end{bmatrix} = \begin{bmatrix} 1 \\ 0 \end{bmatrix},$$

$$\frac{d}{dt}\begin{bmatrix} f_5 \\ f_6 \end{bmatrix} = \begin{bmatrix} -1 & 1 \\ -1 & -1 \end{bmatrix}\begin{bmatrix} f_5 \\ f_6 \end{bmatrix}, \qquad \begin{bmatrix} f_5(0) \\ f_6(0) \end{bmatrix} = \begin{bmatrix} 0 \\ 1 \end{bmatrix},$$

$$\frac{d}{dt}\begin{bmatrix} f_7 \\ f_8 \end{bmatrix} = \begin{bmatrix} 0 & 1.5 \\ -1.5 & 0 \end{bmatrix}\begin{bmatrix} f_7 \\ f_8 \end{bmatrix}, \qquad \begin{bmatrix} f_7(0) \\ f_8(0) \end{bmatrix} = \begin{bmatrix} 0 \\ 1 \end{bmatrix}.$$

They combine into one vector differential equation for $\mathbf{f}(t)$:

$$\frac{d}{dt}\begin{bmatrix} f_1 \\ f_2 \\ f_3 \\ f_4 \\ f_5 \\ f_6 \\ f_7 \\ f_8 \end{bmatrix} = \begin{bmatrix} 0 & 0 & 0 & 0 & 0 & 0 & 0 & 0 \\ 0 & -2 & 0 & 0 & 0 & 0 & 0 & 0 \\ 0 & 0 & -3 & 0 & 0 & 0 & 0 & 0 \\ 0 & 0 & 1 & -3 & 0 & 0 & 0 & 0 \\ 0 & 0 & 0 & 0 & -1 & 1 & 0 & 0 \\ 0 & 0 & 0 & 0 & -1 & -1 & 0 & 0 \\ 0 & 0 & 0 & 0 & 0 & 0 & 0 & 1.5 \\ 0 & 0 & 0 & 0 & 0 & 0 & -1.5 & 0 \end{bmatrix}\begin{bmatrix} f_1 \\ f_2 \\ f_3 \\ f_4 \\ f_5 \\ f_6 \\ f_7 \\ f_8 \end{bmatrix}, \qquad \begin{bmatrix} f_1(0) \\ f_2(0) \\ f_3(0) \\ f_4(0) \\ f_5(0) \\ f_6(0) \\ f_7(0) \\ f_8(0) \end{bmatrix} = \begin{bmatrix} 1 \\ 1 \\ 1 \\ 0 \\ 0 \\ 1 \\ 0 \\ 1 \end{bmatrix}.$$

MATRIX EXPONENTIAL

These formal solutions have buried a major difficulty. The matrix exponential is just a symbolic solution—it does not tell us much. Although the series form of the exponential may permit some approximate numerical answers, it does not lead to a closed form. Thus in the simple example shown in (40), each element of the matrix is an infinite series, and we do not know what function it represents. Clearly, we need some means for finding closed-form equivalents for the exponential $\epsilon^{\mathscr{A}t}$.

One equivalent of the exponential can be found by Laplace transforms. To simplify matters, suppose the initial time is $t_0 = 0$. If we take the Laplace transform of the homogeneous equation in (31), we shall get

$$s\bar{\mathbf{Y}}(s) - \mathscr{A}\bar{\mathbf{Y}}(s) = \mathbf{Y}(0) = \mathbf{U},$$

where $\bar{\mathbf{Y}}$ is the Laplace transform of $\mathbf{Y}(t)$. This can be written as follows:

$$(s\mathbf{U} - \mathscr{A})\,\bar{\mathbf{Y}}(s) = \mathbf{U}$$

or

$$\bar{\mathbf{Y}}(s) = (s\mathbf{U} - \mathscr{A})^{-1}.$$

Finally, we take the inverse transform to get $\mathbf{Y}(t)$. Because we have taken $t_0 = 0$, $\mathbf{Y}(t)$ will also equal $\epsilon^{\mathscr{A}t}$. Hence

$$\epsilon^{\mathscr{A}t} = \mathcal{L}^{-1}\{(s\mathbf{U} - \mathscr{A})^{-1}\}. \tag{63}$$

This is very interesting. Let us apply it to the simple matrix considered earlier in (40). The matrix $(s\mathbf{U} - \mathscr{A})$, its determinant, and its inverse are easily obtained as

$$s\mathbf{U} - \mathscr{A} = \begin{bmatrix} s+1 & 0 \\ -1 & s+2 \end{bmatrix}, \qquad \det(s\mathbf{U} - \mathscr{A}) = (s+1)(s+2),$$

$$(s\mathbf{U} - \mathscr{A})^{-1} = \begin{bmatrix} \dfrac{1}{s+1} & 0 \\[2ex] \dfrac{1}{(s+1)(s+2)} & \dfrac{1}{s+2} \end{bmatrix} = \begin{bmatrix} \dfrac{1}{s+1} & 0 \\[2ex] \dfrac{1}{s+1} - \dfrac{1}{s+2} & \dfrac{1}{s+2} \end{bmatrix}. \tag{64}$$

A partial-fraction expansion was made in the last step. The inverse transform of this expression is

$$\mathcal{L}^{-1}(s\mathbf{U} - \mathscr{A})^{-1} = \begin{bmatrix} \epsilon^{-t} & 0 \\ \epsilon^{-t} - \epsilon^{-2t} & \epsilon^{-2t} \end{bmatrix}.$$

It is left as an exercise for you to expand the exponentials here and to verify that the series are the same as in (40).

The Laplace transform is one way of evaluating the matrix exponential $\epsilon^{\mathscr{A}t}$. However, if we are going to use Laplace transforms, we can do so on the original nonhomogeneous equations and avoid going through all the intervening steps. This, of course, can be done; but we will not be using matrix mathematics to advantage in so doing. We need some additional means for finding the matrix exponential.

4.4 FUNCTIONS OF A MATRIX

The matrix exponential $\epsilon^{\mathscr{A}t}$ is a particular function of a matrix; it is a member of a general class that can be called *functions of a matrix*. It is possible to learn much about the particular function $\epsilon^{\mathscr{A}t}$ by studying the theory of the general class. This is what we shall do in this section.

The simplest functions of an ordinary scalar variable are powers of the variable and polynomials. These are also the simplest functions of a matrix. Consider a polynomial $f(s)$ of the complex variable s.

$$f(s) = s^k + a_{k-1}s^{k-1} + \cdots + a_1 s + a_0 .$$

Suppose the variable s is replaced by a square matrix $\mathscr{A}$ of order n. The corresponding function will be a matrix polynomial:

$$f(\mathscr{A}) = \mathscr{A}^k + a_{k-1}\mathscr{A}^{k-1} + \cdots + a_1\mathscr{A} + a_0 \mathbf{U}.$$

The generalization of a polynomial is an infinite series:

$$f(s) = a_0 + a_1 s + \cdots + a_k s^k + \cdots = \sum_{k=0}^{\infty} a_k s^k. \qquad (65)$$

Such a series, in fact, can represent any analytic function of a complex variable, within its domain of convergence. With s replaced by $\mathscr{A}$, the series becomes

$$f(\mathscr{A}) = a_0 \mathbf{U} + a_1 \mathscr{A} + a_2 \mathscr{A}^2 + \cdots + a_k \mathscr{A}^k + \cdots = \sum_{k=0}^{\infty} a_k \mathscr{A}^k. \quad (66)$$

The function $f(\mathscr{A})$ is itself a matrix, each of whose elements is an infinite series. This matrix series is said to converge if each of the element series converges. We shall not show it, but it turns out that the matrix series will, in fact, converge if the eigenvalues of $\mathscr{A}$—that is, the zeros of the characteristic polynomial, $\det(s\mathbf{U} - \mathscr{A})$—lie within the circle of convergence of the scalar series in (65).*

Transcendental functions of a matrix can be defined by means of infinite series. One such function is the exponential, for which the series definition has already been given in (39). A series definition of a function of a matrix is not of much value in evaluating the function, except for an approximate numerical value. Furthermore, a series definition will not always be suitable, as when the zeros of the characteristic polynomial do not lie within the circle of convergence. Fortunately, it turns out that if $f(s)$ is an analytic function that is regular at the zeros of the characteristic polynomial of $\mathscr{A}$, then $f(\mathscr{A})$ can be expressed as a polynomial function; that is, a finite series. Let us see how this comes about.

* For a proof, see L. Minsky, *An Introduction to Linear Algebra*, Oxford University Press, London, 1955, pp. 332–334.

THE CAYLEY-HAMILTON THEOREM AND ITS CONSEQUENCES

To start, let us define an *annihilating polynomial* of a matrix $\mathscr{A}$ as a polynomial $a(s)$, which reduces to zero when s is replaced by $\mathscr{A}$; that is, $a(\mathscr{A}) = 0$. The characteristic polynomial, $d(s) = \det(s\mathbf{U} - \mathscr{A})$, of a square matrix $\mathscr{A}$ is an annihilating polynomial of $\mathscr{A}$. This follows by observing that the inverse of $(s\mathbf{U} - \mathscr{A})$ is given by

$$(s\mathbf{U} - \mathscr{A})^{-1} = \frac{1}{d(s)} \operatorname{adj}(s\mathbf{U} - \mathscr{A}) \tag{67}$$

or

$$d(s)\mathbf{U} = (s\mathbf{U} - \mathscr{A})\operatorname{adj}(s\mathbf{U} - \mathscr{A}). \tag{68}$$

Now suppose s is replaced by $\mathscr{A}$. On the right side a factor $\mathscr{A} - \mathscr{A}$ appears; hence $d(\mathscr{A}) = 0$, and $d(s)$ is an annihilating polynomial. This result is known as the *Cayley-Hamilton theorem*:

Theorem. *Any square matrix satisfies it own characteristic equation.*

The Cayley-Hamilton theorem permits us to reduce the order of a matrix polynomial of any (high) order to one of order no greater than $n - 1$, where n is the order of the matrix. Suppose $\mathscr{A}$ is a square matrix of order 3. Its characteristic equation will have the form $d(s) = s^3 + d_1 s^2 + d_2 s + d_3$. Hence, by the Cayley-Hamilton theorem,

$$d(\mathscr{A}) = \mathscr{A}^3 + d_1 \mathscr{A}^2 + d_2 \mathscr{A} + d_3 \mathbf{U} = \mathbf{0}$$

and

$$\mathscr{A}^3 = -(d_1 \mathscr{A}^2 + d_2 \mathscr{A} + d_3 \mathbf{U}).$$

Given a polynomial of order greater than 3, all powers of 3 or more can be replaced by quadratics in $\mathscr{A}$ by using this expression for $\mathscr{A}^3$. Hence the entire polynomial will reduce to a polynomial of order 2.

As a collateral result, the Cayley-Hamilton theorem permits the evaluation of the inverse of a matrix as a matrix polynomial. Thus, if the characteristic equation of a matrix $\mathscr{A}$ is

$$d(s) = s^n + d_1 s^{n-1} + \cdots + d_{n-1} s + d_n,$$

then

$$d(\mathscr{A}) = \mathscr{A}^n + d_1 \mathscr{A}^{n-1} + \cdots + d_{n-1} \mathscr{A} + d_n \mathbf{U} = \mathbf{0}.$$

If the equation is multiplied through by $\mathscr{A}^{-1}$, the last term becomes $d_n \mathscr{A}^{-1}$. Hence

$$\mathscr{A}^{-1} = -\frac{1}{d_n} \left(\mathscr{A}^{n-1} + d_1 \mathscr{A}^{n-2} + \cdots + d_{n-2} \mathscr{A} + d_{n-1} \mathbf{U} \right). \qquad (69)$$

This explicit relationship is valid only when zero is not an eigenvalue of $\mathscr{A}$, so that $d(s)$ does not have a factor s and $d_n \neq 0$.

We are mainly interested in functions $f(\mathscr{A})$ other than polynomials; in particular, exponentials. How do we deal with such functions? A clue is obtained by considering polynomials again. Suppose two polynomials $p_1(s)$ and $p_2(s)$ are given, the order of $p_1(s)$ being less than that of $p_2(s)$. The latter can be divided by the former, yielding a quotient $q(s)$ and a remainder $r(s)$ whose order is one less than that of the divisor polynomial, $p_1(s)$. The result, after multiplying through by $p_1(s)$, can be written as

$$p_2(s) = q(s)\, p_1(s) + r(s).$$

Instead of polynomial $p_2(s)$, suppose we have an analytic function $f(s)$ and we replace $p_1(s)$ by $a(s)$. Then, in analogy with the preceding equation, we hope that

$$f(s) = q(s)\, a(s) + g(s), \qquad (70)$$

where $q(s)$ is an analytic "quotient" function, which is regular at the zeros of the polynomial $a(s)$, and where $g(s)$ is a "remainder" polynomial whose order is less than the order of $a(s)$.

Suppose the polynomial $a(s)$ is an annihilating polynomial of matrix $\mathscr{A}$; that is, $a(\mathscr{A}) = 0$. This means that, with s replaced by $\mathscr{A}$ in (70), we get

$$f(\mathscr{A}) = g(\mathscr{A}), \qquad (71)$$

where $f(\mathscr{A})$ is a function and $g(\mathscr{A})$ is a polynomial.

This is a very interesting result. Remember that $f(s)$ is an arbitrary function; thus this result states that *any analytic function of a matrix $\mathscr{A}$ can be expressed as a polynomial in $\mathscr{A}$ of order no greater than one less than the order of $\mathscr{A}$*.

We still have the job of determining the "remainder" polynomial $g(s)$. Before doing this, let us look at annihilating polynomials a little further. The Cayley-Hamilton theorem assures as that a square matrix has at least one annihilating polynomial. (You should show that this implies there are an infinite number of annihilating polynomials of the matrix.) Let the one having the lowest degree and with unity leading coefficient be labeled $m(s)$ and be called the *minimal polynomial*.

One interesting fact about a minimal polynomial is given in the following theorem:

Theorem. *The minimal polynomial of any square matrix $\mathscr{A}$ is a factor of every annihilating polynomial of $\mathscr{A}$.*

This is easy to prove. Given $a(s)$ and $m(s)$, where $m(s)$ is of no higher degree than $a(s)$, we can divide $a(s)$ by $m(s)$ to obtain a quotient $q_1(s)$ and a remainder $r_1(s)$ of degree lower than that of $m(s)$. After multiplying through by $m(s)$ the result will be

$$a(s) = q_1(s)\, m(s) + r_1(s).$$

Now replace s by $\mathscr{A}$ and observe that $a(\mathscr{A}) = \mathbf{0}$ and $m(\mathscr{A}) = \mathbf{0}$. Hence $r_1(\mathscr{A}) = \mathbf{0}$. But this is a contradiction, unless r_1 is identically zero, because $r_1(\mathscr{A}) = \mathbf{0}$ means $r_1(s)$ is an annihilating polynomial of lower degree than $m(s)$. Hence $r_1(s) = 0$, and $m(s)$ is a factor of $a(s)$.

It is unnecessary to spend effort seeking to find the minimal polynomial in anything we shall do. In any of the calculations to be carried out it is possible to use the easily determined characteristic polynomial $d(s)$, which is an annihilating polynomial and which may sometimes be the corresponding minimal polynomial.

We are now ready to take up the job of determining the polynomial $g(s)$, which can be written as

$$g(s) = g_0 + g_1 s + g_2 s^2 + \cdots + g_{n-1} s^{n-1}, \tag{72}$$

in which the coefficients are unknown. The starting point is (70).

Let us deal with the characteristic polynomial $d(s)$ of matrix $\mathscr{A}$ which we know to be an annihilating polynomial, and rewrite (70) as

$$f(s) = q(s)\, d(s) + g(s). \tag{73}$$

DISTINCT EIGENVALUES

We shall first assume that the eigenvalues of $\mathscr{A}$ are distinct and write $d(s)$ in factored form as

$$d(s) = (s - s_1)(s - s_2) \cdots (s - s_n). \tag{74}$$

Now let us evaluate (73) at each of the eigenvalues s_i. Since $d(s_i) = 0$, we find

$$f(s_1) = g(s_i). \tag{75}$$

There are n such relationships. When (72) is used for $g(s)$ these n relationships become

$$
\begin{aligned}
g_0 + s_1 g_1 + s_1{}^2 g_2 + \cdots + s_1{}^{n-1} g_{n-1} &= f(s_1), \\
g_0 + s_2 g_1 + s_2{}^2 g_2 + \cdots + s_2{}^{n-1} g_{n-1} &= f(s_2), \\
&\cdots \cdots \cdots \cdots \cdots \cdots \cdots \\
g_0 + s_n g_1 + s_n{}^2 g_2 + \cdots + s_n{}^{n-1} g_{n-1} &= f(s_n).
\end{aligned}
\tag{76}
$$

The right-hand sides are known quantities, since $f(s)$ is the originally given function. This is a set of n equations in n unknown g_i coefficients. Inversion of this set of equations gives the solution.

Let us illustrate the process with the same simple example considered before. For the $\mathscr{A}$ in (40), the characteristic polynomial was given in (64). They are repeated here:

$$
\mathscr{A} = \begin{bmatrix} -1 & 0 \\ 1 & -2 \end{bmatrix}
\qquad
\begin{aligned}
d(s) &= (s+1)(s+2), \\
s_1 &= -1, \\
s_2 &= -2.
\end{aligned}
$$

The desired matrix function is $\epsilon^{\mathscr{A}t}$, so $f(s) = \epsilon^{st}$. By substituting into (76) we obtain

$$
\begin{aligned}
g_0 - g_1 &= \epsilon^{s_1 t} = \epsilon^{-t}, \\
g_0 - 2g_1 &= \epsilon^{s_2 t} = \epsilon^{-2t},
\end{aligned}
$$

from which

$$
\begin{aligned}
g_1 &= \epsilon^{-t} - \epsilon^{-2t}, \\
g_0 &= 2\epsilon^{-t} - \epsilon^{-2t}.
\end{aligned}
$$

Hence,

$$g(s) = (2\epsilon^{-t} - \epsilon^{-2t}) + (\epsilon^{-t} - \epsilon^{-2t})s.$$

The next step is to replace s by $\mathscr{A}$ to get $g(\mathscr{A})$, which, by (71), equals $f(\mathscr{A})$:

$$f(\mathscr{A}) = g(\mathscr{A}) = (2\epsilon^{-t} - \epsilon^{-2t})\mathbf{U} + (\epsilon^{-t} - \epsilon^{-2t})\mathscr{A}$$

$$= (2\epsilon^{-t} - \epsilon^{-2t})\begin{bmatrix} 1 & 0 \\ 0 & 1 \end{bmatrix} + (\epsilon^{-t} - \epsilon^{-2t})\begin{bmatrix} -1 & 0 \\ 1 & -2 \end{bmatrix}.$$

By an obvious rearrangement, this becomes

$$\epsilon^{\mathscr{A}t} = \begin{bmatrix} \epsilon^{-t} & 0 \\ \epsilon^{-t} - \epsilon^{-2t} & \epsilon^{-2t} \end{bmatrix},$$

which agrees with the previously determined result.

A glance back at the set of equations in (76) reveals a certain uniformity in the matrix of coefficients. It should be possible to solve the equations in literal form and take advantage of the uniformity in the matrix to arrive at an easily interpreted result.

If we let Δ be the determinant and Δ_{ij} the (i, j)th cofactor of the co-efficient matrix in (76), the solution for the g_i's can be written

$$g_{i-1} = \sum_{j=1}^{n} \frac{\Delta_{ji}}{\Delta} f(s_j).$$

With these coefficients inserted, the polynomial $g(s)$ now becomes

$$g(s) = \sum_{i=1}^{n} g_{i-1} s^{i-1} = \sum_{i=1}^{n} s^{i-1} \left\{ \sum_{j=1}^{n} \frac{\Delta_{ji}}{\Delta} f(s_j) \right\}$$

$$= \sum_{j=1}^{n} \frac{\Delta_{j1}}{\Delta} f(s_j) + s \left\{ \sum_{j=1}^{n} \frac{\Delta_{j2}}{\Delta} f(s_j) \right\} + \cdots + s^{n-1} \left\{ \sum_{j=1}^{n} \frac{\Delta_{jn}}{\Delta} f(s_j) \right\},$$

or

$$g(s) = \sum_{j=1}^{n} \left\{ \frac{\Delta_{j1}}{\Delta} + \frac{s\Delta_{j2}}{\Delta} + \frac{s^2\Delta_{j3}}{\Delta} + \cdots + \frac{s^{n-1}\Delta_{jn}}{\Delta} \right\} f(s_j).$$

What we have done in the last step is to rearrange the terms so that it is not the powers of s on which we focus, but on the values $f(s_j)$. Since the equations of (76) are all so similar, it must be possible to write this result in a simpler form, and indeed it is. The result was first given by Lagrange in the context of passing a polynomial of degree $n - 1$ through n points. It is called the *Lagrange interpolation formula* and converts the summation within the braces above to a product, as follows:

$$g(s) = \sum_{j=1}^{n} \left(\prod_{\substack{k=1 \\ k \neq j}}^{n} \frac{s - s_k}{s_j - s_k} \right) f(s_j). \tag{77}$$

(It is suggested that you verify this.) From this, $g(\mathscr{A})$ is easily obtained. Finally, since $f(\mathscr{A}) = g(\mathscr{A})$, we get

$$f(\mathscr{A}) = \sum_{j=1}^{n} \left(\prod_{\substack{k=1 \\ k \neq j}}^{n} \frac{\mathscr{A} - s_k \mathbf{U}}{s_j - s_k} \right) f(s_j). \tag{78}$$

From this result a very interesting observation can be made. Given a matrix $\mathscr{A}$ the eigenvalues s_i are uniquely determined from $\mathscr{A}$. Hence everything within parentheses in (78) is a function only of $\mathscr{A}$ and is independent of the specific function $f(s)$ under consideration. Once the quantity within the parentheses is determined, any function of a matrix can be determined merely by evaluating the function at the eigenvalues of $\mathscr{A}$. We shall make note of this again in the more general case to be treated next.

MULTIPLE EIGENVALUES

If the eigenvalues of $\mathscr{A}$ are not distinct and there are repeated values, a modification of this procedure is necessary. Let $d(s)$ be written as

$$d(s) = (s - s_1)^{r_1} (s - s_2)^{r_2} \cdots (s - s_l)^{r_l}, \tag{79}$$

where the multiplicities are obviously r_i for the ith eigenvalue.

Let us now consider differentiating (73), after which we shall evaluate the result for $s = s_k$. Except for the derivatives of the product $q(s)d(s)$, the derivatives of f and g will be equal. Thus,

$$\frac{d^j f(s)}{ds^j} = \frac{d^j g(s)}{ds^j} + \sum_{i=0}^{j} \binom{j}{i} \frac{d^{j-1} q(s)}{ds^{j-1}} \frac{d^i d(s)}{ds^i}. \tag{80}$$

What happens to the summation when $s = s_k$? If the order of the derivative is less than the multiplicity of the eigenvalue s_k (i.e., if $j < r_k$), then from (79) it is clear that $d^i d(s)/ds^i = 0$ for $s = s_k$ and for $i < j$. This means all the terms under the summation sign will vanish, and so

$$\frac{d^j g(s)}{ds^j} = \frac{d^j f(s)}{ds^j} \quad \text{for} \quad s = s_k \tag{81}$$

and for derivative orders $j = 0, 1, 2, \cdots, (r_k - 1)$; that is, for derivatives of any order up to one less than the multiplicity of s_k. This equation is a generalization of, and includes, (75). It gives as many relationships as there are eigenvalues when each eigenvalue is counted according to its multiplicity. Since $g(s)$ in (72) also has that many coefficients, they can be determined by applying (81). Thus the first r_1 relationships evaluated for $s = s_1$ will be

$$g_0 + s_1 g_1 + s_1^2 g_2 + \cdots + s_1^{n-1} g_{n-1} = f(s_1)$$

$$g_1 + 2s_1 g_2 + \cdots + (n-1)s_1^{n-2} g_{n-1} = f^{(1)}(s_1)$$

$$\vdots$$

$$(r_1 - 1)! \, g_{r_1 - 1} + \cdots + \frac{(n-1)!}{(n-r_1)!} \, s_1^{n-r_1} g_{n-1} = f^{(r_1 - 1)}(s_1).$$

Similar sets of equations will result for each distinct eigenvalue. The entire result in matrix form will be

$$
\begin{bmatrix}
1 & s_1 & s_1^2 & \cdots & s_1^{r_1} & \cdots & s_1^{n-1} \\
0 & 1 & 2s_1 & \cdots & r_1 s_1^{r_1-1} & \cdots & (n-1)s_1^{n-2} \\
\vdots & \vdots & \vdots & & \vdots & & \vdots \\
0 & 0 & 0 & \cdots & (r_1-1)! & \cdots & \dfrac{(n-1)!}{(n-r_1)!} s_1^{n-r_1} \\
\hdotsfor{7} \\
1 & s_2 & s_2^2 & \cdots & s_2^{r_2} & \cdots & s_2^{n-1} \\
0 & 1 & 2s_2 & \cdots & r_2 s_2^{r_2-1} & \cdots & (n-1)s_2^{n-2} \\
\vdots & \vdots & \vdots & & \vdots & & \vdots \\
0 & 0 & 0 & \cdots & (r_2-1)! & \cdots & \dfrac{(n-1)!}{(n-r_2)!} s_2^{n-r_2} \\
\hdotsfor{7} \\
& & & \vdots \\
\hdotsfor{7} \\
1 & s_k & s_k^2 & \cdots & s_k^{r_k} & \cdots & s_k^{n-1} \\
0 & 1 & 2s_k & \cdots & r_k s_k^{r_k-1} & \cdots & (n-1)s_k^{n-2} \\
\vdots & \vdots & \vdots & & \vdots & & \vdots \\
0 & 0 & 0 & \cdots & (r_k-1)! & \cdots & \dfrac{(n-1)!}{(n-r_k)!} s_k^{n-r_k}
\end{bmatrix}
\begin{bmatrix}
g_0 \\ g_1 \\ g_2 \\ \vdots \\ g_{n-1}
\end{bmatrix}
=
\begin{bmatrix}
f(s_1) \\
f^{(1)}(s_1) \\
f^{(2)}(s_1) \\
\vdots \\
f^{(r_1-1)}(s_1) \\
f(s_2) \\
f^{(1)}(s_2) \\
f^{(2)}(s_2) \\
\vdots \\
f^{(r_2-1)}(s_2) \\
\vdots \\
f(s_k) \\
f^{(1)}(s_k) \\
f^{(2)}(s_k) \\
\vdots \\
f^{(r_k-1)}(s_k)
\end{bmatrix}
\qquad (82)
$$

This is a horrible-looking expression; it is inflicted on you by our desire
to be general. Actual cases will rarely have such generality, and the actual
equations will look considerably simpler than this. In any case, the g_i
coefficients are obtained by inverting this matrix equation. This is the
generalization for multiple eigenvalues of (76) which applies for simple
eigenvalues.

As an example of the determination of a function $f(\mathscr{A})$ when $\mathscr{A}$ has
multiple eigenvalues, let $f(s) = \epsilon^{st}$ and

$$\mathscr{A} = \begin{bmatrix} -2 & 1 & 3 \\ 0 & -3 & 0 \\ 0 & 2 & -2 \end{bmatrix}, \qquad d(s) = \begin{vmatrix} s+2 & -1 & -3 \\ 0 & s+3 & 0 \\ 0 & -2 & s+2 \end{vmatrix}$$

$$= (s+2)^2(s+3).$$

Take $s_1 = -2$ and $s_2 = -3$; then the multiplicities are $r_1 = 2$ and $r_2 = 1$.
Let us use $d(s)$ as the annihilating polynomial in determining $g(s)$. For
this example (82) becomes

$$\begin{bmatrix} \epsilon^{-2t} \\ t\epsilon^{-2t} \\ \epsilon^{-3t} \end{bmatrix} = \begin{bmatrix} 1 & -2 & 4 \\ 0 & 1 & -4 \\ 1 & -3 & 9 \end{bmatrix} \begin{bmatrix} g_0 \\ g_1 \\ g_2 \end{bmatrix}$$

since $df(s_1)/ds = t\epsilon^{s_1 t} = t\epsilon^{-2t}$. The solution for the g's is easily found by
inversion, as

$$\begin{bmatrix} g_0 \\ g_1 \\ g_2 \end{bmatrix} = \begin{bmatrix} 1 & -2 & 4 \\ 0 & 1 & -4 \\ 1 & -3 & 9 \end{bmatrix}^{-1} \begin{bmatrix} \epsilon^{-2t} \\ t\epsilon^{-2t} \\ \epsilon^{-3t} \end{bmatrix} = \begin{bmatrix} -3 & 6 & 4 \\ -4 & 5 & 4 \\ -1 & 1 & 1 \end{bmatrix} \begin{bmatrix} \epsilon^{-2t} \\ t\epsilon^{-2t} \\ \epsilon^{-3t} \end{bmatrix}$$

$$= \begin{bmatrix} -3\epsilon^{-2t} + 6t\epsilon^{-2t} + 4\epsilon^{-3t} \\ -4\epsilon^{-2t} + 5t\epsilon^{-2t} + 4\epsilon^{-3t} \\ -\ \epsilon^{-2t} + \ t\epsilon^{-2t} + \ \epsilon^{-3t} \end{bmatrix}.$$

With g_0, g_1, and g_2 now known, $g(s) = g_0 + g_1 s + g_2 s^2$ will be

$$g(s) = (-3\epsilon^{-2t} + 6t\epsilon^{-2t} + 4\epsilon^{-3t}) + (-4\epsilon^{-2t} + 5t\epsilon^{-2t} + 4\epsilon^{-3t})s$$

$$+ (-\epsilon^{-2t} + t\epsilon^{-2t} + \epsilon^{-3t})s^2,$$

or, after an obvious rearrangement of terms,

$$g(s) = (-3 - 4s - s^2)\epsilon^{-2t} + (6 + 5s + s^2)t\epsilon^{-2t} + (4 + 4s + s^2)\epsilon^{-3t}.$$

The next step is to form $g(\mathscr{A})$ by replacing s by $\mathscr{A}$:

$$g(\mathscr{A}) = (-3\mathbf{U} - 4\mathscr{A} - \mathscr{A}^2)\epsilon^{-2t} + (6\mathbf{U} + 5\mathscr{A} + \mathscr{A}^2)t\epsilon^{-2t}$$
$$+ (4\mathbf{U} + 4\mathscr{A} + \mathscr{A}^2)\epsilon^{-3t}.$$

The remaining work is just arithmetic, after $\mathscr{A}$ is inserted here. The final result is obtained from (71): $f(\mathscr{A}) = g(\mathscr{A})$ leads to

$$\epsilon^{\mathscr{A}t} = \begin{bmatrix} 1 & -5 & 0 \\ 0 & 0 & 0 \\ 0 & 2 & 1 \end{bmatrix}\epsilon^{-2t} + \begin{bmatrix} 0 & 6 & 3 \\ 0 & 0 & 0 \\ 0 & 0 & 0 \end{bmatrix}t\epsilon^{-2t} + \begin{bmatrix} 0 & 5 & 0 \\ 0 & 1 & 0 \\ 0 & -2 & 0 \end{bmatrix}\epsilon^{-3t}.$$

This completes the example. (You should verify that this equation follows from the immediately preceding one upon the insertion of $\mathscr{A}$.)

CONSTITUENT MATRICES

Let us look back at (82). This equation can be solved for the g_i coefficients which are then inserted into the $g(s)$ polynomial, just as in the case of (76). Again we rearrange the terms so that the focus is on the elements of the right-hand vector in (82), rather than on the powers of s. The rearranged expression can be written as follows:

$$g(s) = K_{11}(s)\, f(s_1) + K_{12}(s)\frac{f^{(1)}(s_1)}{1!} + K_{13}(s)\frac{f^{(2)}(s_1)}{2!}$$
$$+ \cdots + K_{1r_1}(s)\frac{f^{(r_1-1)}(s_1)}{(r_1-1)!}$$
$$+ K_{21}(s)\, f(s_2) + K_{22}(s)\frac{f^{(1)}(s_2)}{1!} + K_{23}(s)\frac{f^{(2)}(s_2)}{2!} \qquad (83)$$
$$+ \cdots + K_{2r_2}(s)\frac{f^{(r_2-1)}(s_2)}{(r_2-1)!}$$
$$+ \cdots$$
$$+ K_{k1}(s)\, f(s_k) + K_{k2}(s)\frac{f^{(1)}(s_k)}{1!} + K_{k3}(s)\frac{f^{(2)}(s_k)}{2!}$$
$$+ \cdots + K_{kr_k}(s)\frac{f^{(r_k-1)}(s_k)}{(r_k-1)!}$$

It is only for later convenience that the K_{ij}'s have been chosen as the coefficients of the derivatives divided by the factorials rather than just the derivatives themselves. When the eigenvalues are of single multiplicity this complicated expression reduces to just the first column, which is simply (77).

The next step is to replace s by $\mathscr{A}$. Recalling that $g(\mathscr{A}) = f(\mathscr{A})$, we now get

$$
\begin{aligned}
f(\mathscr{A}) = \; &K_{11}(\mathscr{A})f(s_1) + K_{12}(\mathscr{A})\frac{f^{(1)}(s_1)}{1!} + \cdots + K_{1r_k}(\mathscr{A})\frac{f^{(r_1-1)}(s_1)}{(r_1-1)!} \\
&+ \cdots \\
&+ K_{k1}(\mathscr{A})f(s_k) + K_{k2}(\mathscr{A})\frac{f^{(1)}(s_k)}{1!} + \cdots + K_{kr_k}(\mathscr{A})\frac{f^{(r_1-1)}(s_k)}{(r_k-1)!}
\end{aligned}
\tag{84}
$$

assuming the functions $f^i(s)$ not singular for $s = s_i$. The coefficients $K_{ij}(\mathscr{A})$ in (84) are matrices, which are often expressed as $\mathbf{K}_{ij}$. They are called the *constituent matrices of* $(\mathscr{A})$ and depend only on $\mathscr{A}$, not on the function $f(s)$. This can be observed by looking at (82). The nonzero elements of the coefficient matrix there are proportional to various powers of the eigenvalues of $\mathscr{A}$; $K_{ij}(s)$ is simply a linear combination of cofactors of that coefficient matrix. Since the eigenvalues and, hence, the entries of that coefficient matrix depend only on $\mathscr{A}$, the result in verified. This is a very powerful point. It means that the constituent matrices $K_{ij}(\mathscr{A})$ of a square matrix $\mathscr{A}$ can be determined once and for all, independent of any specific function. For any given function f, the expression in (84) can then be formed simply by evaluating the various derivatives of f at the eigenvalues of $\mathscr{A}$.

So far the only way we know to find the constituent matrices when $\mathscr{A}$ has multiple eigenvalues is to set up (82), solve for the g_i coefficients, insert into $g(s)$, and then rearrange to put it in the form of (83). It would be in order to look for simpler methods, and fortunately the search pays off. When the eigenvalues of $\mathscr{A}$ are simple, of course, we have the Lagrange interpolation formula. We need something like that for multiple eigenvalues.

THE RESOLVENT MATRIX

Since the constituent matrices $\mathbf{K}_{ij} = K_{ij}(\mathscr{A})$ do not depend on the specific function f, if we can find a simple function for which (84) can be written, then the $\mathbf{K}_{ij}$'s thus determined will be the same for any function. The success of this approach depends on finding a convenient function. Consider the function $f(s') = 1/(s - s') = (s - s')^{-1}$, where s' is a complex variable that is to play the role formerly played by s; for example, it is s' that is to be replaced by $\mathscr{A}$. At the risk of confusion, we have used the symbol s to stand for another complex variable. Another symbol, say z, should have been used here instead of s, but our choice results from the desire to end with an equation containing s. You might avoid confusion by thinking "z" when you see "s" in this development. If we take derivatives with respect to s', we obtain

$$\frac{f^{j-1}(s_i)}{(j-1)!} = \frac{1}{(s-s_i)^j}, \tag{85}$$

where s_i are specific values of s'. Now substitute (85) in (84) and replace s' by $\mathscr{A}$ in $f(s') = (s-s')^{-1}$. The result will be $(s\mathbf{U} - \mathscr{A})^{-1}$ in terms of the constituent matrices. Next, from (67) this can be written

$$(s\mathbf{U} - \mathscr{A})^{-1} = \frac{\text{adj } (s\mathbf{U} - \mathscr{A})}{d(s)}. \tag{86}$$

The numerator of the right side is a matrix whose elements are polynomials in s since they are cofactors of the matrix $(s\mathbf{U} - \mathscr{A})$. Since each element of the numerator is divided by $d(s)$, the whole thing is a matrix of rational functions. A partial-fraction expansion of the right-hand side can be carried out and leads to

$$\begin{aligned} (s\mathbf{U} - \mathscr{A})^{-1} &= \frac{\mathbf{K}_{11}}{(s-s_1)} + \frac{\mathbf{K}_{12}}{(s-s_1)^2} + \cdots + \frac{\mathbf{K}_{1r_1}}{(s-s_1)^{r_1}} \\ &+ \cdots \\ &+ \frac{\mathbf{K}_{k1}}{(s-s_k)} + \frac{\mathbf{K}_{k2}}{(s-s_k)^2} + \cdots + \frac{\mathbf{K}_{kr_k}}{(s-s_k)^{r_k}}. \end{aligned} \tag{87}$$

In view of (85), this expression is exactly of the form of (84), and our anticipation in using the same symbols $\mathbf{K}_{ij} = K_{ij}(\mathscr{A})$ for the coefficients of this partial-fraction expansion as for the constituent matrices is justified. That is, the constituent matrices are the coefficient matrices in the partial-fraction expansion of $(s\mathbf{U} - \mathscr{A})^{-1}$. The matrix $(s\mathbf{U} - \mathscr{A})^{-1}$ is called the *resolvent matrix*.

Given a matrix $\mathscr{A}$ and a function $f(s)$, the determination of $f(\mathscr{A})$ in the form of (84) is carried out by expanding the resolvent matrix $(s\mathbf{U} - \mathscr{A})^{-1}$ into partial fractions. The coefficients of the expansion (which are the residues if the eigenvalues are simple) are the constituent matrices.

Let us illustrate by means of the example considered before. Let $f(s) = \epsilon^{st}$ and

$$\mathscr{A} = \begin{bmatrix} -2 & 1 & 3 \\ 0 & -3 & 0 \\ 0 & 2 & -2 \end{bmatrix}, \quad (s\mathbf{U} - \mathscr{A}) = \begin{bmatrix} s+2 & -1 & -3 \\ 0 & s+3 & 0 \\ 0 & -2 & s+2 \end{bmatrix}.$$

To find the inverse of $s\mathbf{U} - \mathscr{A}$, we need its determinant and cofactors. When these are determined, we get

$$(s\mathbf{U} - \mathscr{A})^{-1} = \frac{\begin{bmatrix} (s+2)(s+3) & (s+8) & 3(s+3) \\ 0 & (s+2)^2 & 0 \\ 0 & 2(s+2) & (s+2)(s+3) \end{bmatrix}}{(s+2)^2(s+3)}.$$

Let $s_1 = -2$ and $s_2 = -3$. The partial-fraction expansion is carried out next. The result is

$$(s\mathbf{U} - \mathscr{A})^{-1} = \frac{\overset{\mathbf{K}_{11}}{\begin{bmatrix} 1 & -5 & 0 \\ 0 & 0 & 0 \\ 0 & 2 & 1 \end{bmatrix}}}{s+2} + \frac{\overset{\mathbf{K}_{12}}{\begin{bmatrix} 0 & 6 & 3 \\ 0 & 0 & 0 \\ 0 & 0 & 0 \end{bmatrix}}}{(s+2)^2} + \frac{\overset{\mathbf{K}_{21}}{\begin{bmatrix} 0 & 5 & 0 \\ 0 & 1 & 0 \\ 0 & -2 & 0 \end{bmatrix}}}{s+3}.$$

Finally, these constituent matrices are inserted into (84) to yield for the matrix exponential

$$\epsilon^{\mathscr{A}t} = \mathbf{K}_{11}\epsilon^{-2t} + \mathbf{K}_{12}t\epsilon^{-2t} + \mathbf{K}_{21}\epsilon^{-3t}.$$

This agrees with the earlier answer.

THE RESOLVENT MATRIX ALGORITHM

Let us review the procedure for determining the constituent matrices. It requires first the determination of the eigenvalues, which is true for any other method also. It requires next the inversion of the matrix $(s\mathbf{U} - \mathscr{A})$, which we have done by determining the cofactors of this matrix. But this will be an onerous task for large n. It requires, finally, the partial-fraction expansion of the resolvent matrix $(s\mathbf{U} - \mathscr{A})^{-1}$. Any assistance in reducing the tedium of the computation would be quite valuable. There is available an algorithm that provides just this required assistance. We shall call it the *resolvent matrix algorithm.**

Observe from (86) that the resolvent matrix is expressed in terms of the characteristic polynomial in the denominator and the adjoint matrix of $(s\mathbf{U} - \mathscr{A})$ in the numerator. The elements of this matrix are polynomials in s. We can focus attention on the powers of s by rewriting this matrix as a sum of matrices, one for each power of s. Let.

$$\text{adj}(s\mathbf{U} - \mathscr{A}) = \mathbf{P}(s) = \mathbf{P}_0 s^{n-1} + \mathbf{P}_1 s^{n-2} + \cdots + \mathbf{P}_{n-2} s + \mathbf{P}_{n-1}, \quad (88)$$

$$d(s) = s^n + d_1 s^{n-1} + \cdots + d_{n-1} s + d_n. \quad (89)$$

Multiplication of (86) by $d(s)(s\mathbf{U} - \mathscr{A})$ yields

$$(s\mathbf{U} - \mathscr{A})\mathbf{P}(s) = d(s)\mathbf{U}, \quad (90)$$

which, upon inserting (88) and (89), becomes

$$(\mathbf{P}_0 s^n + \mathbf{P}_1 s^{n-1} + \cdots + \mathbf{P}_{n-1} s)$$
$$- (\mathscr{A}\mathbf{P}_0 s^{n-1} + \mathscr{A}\mathbf{P}_1 s^{n-2} + \cdots + \mathscr{A}\mathbf{P}_{n-1})$$
$$= \mathbf{U}s^n + d_1 \mathbf{U}s^{n-1} + \cdots + d_n \mathbf{U}.$$

Equating coefficients of like powers of s on the two sides leads to

$$\mathbf{P}_0 = \mathbf{U},$$
$$\mathbf{P}_1 = \mathscr{A}\mathbf{P}_0 + d_1\mathbf{U},$$
$$\mathbf{P}_2 = \mathscr{A}\mathbf{P}_1 + d_2\mathbf{U},$$
$$\vdots$$
$$\mathbf{P}_k = \mathscr{A}\mathbf{P}_{k-1} + d_k\mathbf{U}, \quad (91)$$
$$\vdots$$
$$\mathbf{P}_{n-1} = \mathscr{A}\mathbf{P}_{n-2} + d_{n-1}\mathbf{U},$$
$$\mathbf{0} = \mathscr{A}\mathbf{P}_{n-1} + d_n\mathbf{U}.$$

* Early reports of this algorithm may be found in: J. M. Souriau, "Une méthode pour la Décomposition spectrale à l'inversion des matrices," *Compt. Rend.*, Vol. 227, pp. 1010–1011, 1948; D. K. Fadeev and I. S. Sominskii, "Collection of Problems on Higher Algebra," 2nd ed. (in Russian), Gostekhizdat, Moscow, 1949; J. S. Frame, "A Simple Recursion Formula for Inverting a Matrix," *Bull. Am. Math. Soc.*, Vol. 55, p. 1045, 1949. H. E. Fettis, "A Method for Obtaining the Characteristic Equation of a Matrix and Computing the Associated Model Columns," *Quart. Appl. Math.*, Vol. 8, pp. 206–212, 1950.

It is clear that, if we knew the d_i coefficients of the characteristic polynomial, these equations would permit us to determine the $\mathbf{P}_i$ matrices one at a time. Of course, the d_i coefficients can be determined by evaluating the determinant of $(s\mathbf{U} - \mathscr{A})$. We shall now show that even this is not necessary.

By taking the trace of the matrices on both sides of (90) we find that

$$s \operatorname{tr} \mathbf{P}(s) - \operatorname{tr} [\mathscr{A}\mathbf{P}(s)] = nd(s). \tag{92}$$

We shall now show that $\operatorname{tr} [\mathbf{P}(s)]$ equals the derivative of $d(s)$. Write $d(s)$ as

$$d(s) = \sum_{j=1}^{n} (s\delta_{ij} - a_{ij})\, \Delta_{ij}, \tag{93}$$

where δ_{ij} is the Kronecker delta, $(s\delta_{ij} - a_{ij})$ is an element of $(s\mathbf{U} - \mathscr{A})$, and Δ_{ij} is the cofactor of $(s\delta_{ij} - a_{ij})$. Refer back to (25) in Chapter 1, where the derivative of a determinant was discussed. For an arbitrary matrix $\mathbf{B}(s) = [b_{kj}(s)]$ it was shown that

$$\frac{d}{ds} \det \mathbf{B}(s) = \sum_{i=1}^{n} \sum_{j=1}^{n} \frac{db_{ij}}{ds}\, \Delta_{ij}. \tag{94}$$

In the present case the determinant is $d(s) = \det (s\mathbf{U} - \mathscr{A})$. Hence, using (94), we get

$$\frac{d}{ds}[d(s)] = \sum_{i=1}^{n} \left(\sum_{j=1}^{n} \frac{d(s\delta_{ij} - a_{ij})}{ds}\, \Delta_{ij} \right)$$

$$= \sum_{i=1}^{n} \left(\sum_{j=1}^{n} \delta_{ij}\, \Delta_{ij} \right) = \sum_{i=1}^{n} \Delta_{ii} \tag{95}$$

$$= \operatorname{tr}[\operatorname{adj}(s\mathbf{U} - \mathscr{A})] = \operatorname{tr} \mathbf{P}(s).$$

Using this relationship, we can substitute for $\operatorname{tr} \mathbf{P}(s)$ in (92) to get, after rearrangement,

$$s\frac{d}{ds}[d(s)] - nd(s) = \operatorname{tr} [\mathscr{A}\mathbf{P}(s)]. \tag{96}$$

Finally, we substitute the expressions for $\mathbf{P}(s)$ and $d(s)$ from (88) and (89) into this equation to get

$$(ns^n - (n-1)\,d_1\,s^{n-1} + (n-2)\,d_2\,s^{n-2} + \cdots + d_{n-1}s)$$
$$- (ns^n + nd_1s^{n-1} + \cdots + nd_n)$$
$$= s^{n-1}\,\mathrm{tr}\,(\mathscr{A}\mathbf{P}_0) + s^{n-2}\,\mathrm{tr}\,(\mathscr{A}\mathbf{P}_1) + \cdots + s\,\mathrm{tr}\,(\mathscr{A}\mathbf{P}_{n-2}) + \mathrm{tr}\,(\mathscr{A}\mathbf{P}_{n-1}).$$

Again we equate coefficients of like powers of s on both sides to find solutions for the d_i coefficients:

$$
\begin{aligned}
d_1 &= -\ \mathrm{tr}\,(\mathscr{A}\mathbf{P}_0), \\
d_2 &= -\tfrac{1}{2}\,\mathrm{tr}\,(\mathscr{A}\mathbf{P}_1), \\
d_3 &= -\tfrac{1}{3}\,\mathrm{tr}\,(\mathscr{A}\mathbf{P}_2), \\
&\ \ \vdots \\
d_k &= -\frac{1}{k}\,\mathrm{tr}\,(\mathscr{A}\mathbf{P}_{k-1}), \\
&\ \ \vdots \\
d_n &= -\frac{1}{n}\,\mathrm{tr}\,(\mathscr{A}\mathbf{P}_{n-1}).
\end{aligned}
\tag{97}
$$

This set of expressions for the d_k coefficients, together with (91) for the $\mathbf{P}_k$ matrices constitute an algorithm, with a finite number of steps, to compute the resolvent matrix, $(s\mathbf{U} - \mathscr{A})^{-1}$. We shall write them again, side by side, showing the sequence of steps:

$$
\begin{aligned}
\mathbf{P}_0 &= \mathbf{U} & &\rightarrow & d_1 &= -\,\mathrm{tr}\,(\mathscr{A}), \\
\mathbf{P}_1 &= \mathscr{A} + d_1\mathbf{U} & &\rightarrow & d_2 &= -\tfrac{1}{2}\,\mathrm{tr}\,(\mathscr{A}\mathbf{P}_1), \\
\mathbf{P}_2 &= \mathscr{A}\mathbf{P}_1 + d_2\mathbf{U} & &\rightarrow & d_3 &= -\tfrac{1}{3}\,\mathrm{tr}\,(\mathscr{A}\mathbf{P}_2), \\
&\ \ \vdots & & & &\ \ \vdots \\
\mathbf{P}_k &= \mathscr{A}\mathbf{P}_{k-1} + d_k\mathbf{U} & &\rightarrow d_{k+1} &= -\frac{1}{k+1}\,\mathrm{tr}\,(\mathscr{A}\mathbf{P}_k), & &(98) \\
&\ \ \vdots & & & &\ \ \vdots \\
\mathbf{P}_{n-1} &= \mathscr{A}\mathbf{P}_{n-2} + d_{n-1}\mathbf{U} & &\rightarrow d_n &= -\frac{1}{n}\,\mathrm{tr}\,(\mathscr{A}\mathbf{P}_{n-1}), \\
\end{aligned}
$$

$$\mathbf{0} = \mathscr{A}\mathbf{P}_{n-1} + d_n\mathbf{U} \quad \text{(check)}.$$

The last equation in this set can be used as a check since all its components have already been determined in the preceding steps. If the equation is not satisfied, then an error (or more than one) has been made.

The important point concerning the resolving matrix algorithm is the fact that all of the steps involve purely numerical operations; the variable s does not appear. Consequently, although it might appear that there is a prodigious amount of matrix arithmetic, the algorithm can be easily programmed for a computer.

A side result of the algorithm is an evaluation of the inverse of $\mathscr{A}$ when zero is not an eigenvalue of $\mathscr{A}$. In that case $d_n = d(0) \neq 0$. From (86), $(s\mathbf{U} - \mathscr{A})^{-1} = \mathbf{P}(s)/d(s)$. Setting $s = 0$ gives $-\mathscr{A}^{-1} = \mathbf{P}(0)/d(0)$, or

$$\mathscr{A}^{-1} = -\frac{\mathbf{P}_{n-1}}{d_n}. \tag{99}$$

To illustrate the resolvent matrix algorithm, consider again the example treated earlier. The flow of the algorithm is as follows:

$$\mathscr{A} = \begin{bmatrix} -2 & 1 & 3 \\ 0 & -3 & 0 \\ 0 & 2 & -2 \end{bmatrix},$$

$$d_1 = -\operatorname{tr}\mathscr{A} = (2 + 3 + 2) = 7,$$

$$\mathbf{P}_1 = \mathscr{A} + 7\mathbf{U} = \begin{bmatrix} 5 & 1 & 3 \\ 0 & 4 & 0 \\ 0 & 2 & 5 \end{bmatrix}, \qquad \mathscr{A}\mathbf{P}_1 = \begin{bmatrix} -10 & 8 & 9 \\ 0 & -12 & 0 \\ 0 & 4 & -10 \end{bmatrix},$$

$$d_2 = -\tfrac{1}{2}\operatorname{tr}(\mathscr{A}\mathbf{P}_1) = 16,$$

$$\mathbf{P}_2 = \mathscr{A}\mathbf{P}_1 + 16\mathbf{U} = \begin{bmatrix} 6 & 8 & 9 \\ 0 & 4 & 0 \\ 0 & 4 & 6 \end{bmatrix}, \qquad \mathscr{A}\mathbf{P}_2 = \begin{bmatrix} -12 & 0 & 0 \\ 0 & -12 & 0 \\ 0 & 0 & -12 \end{bmatrix},$$

$$d_3 = -\tfrac{1}{3}\operatorname{tr}(\mathscr{A}\mathbf{P}_2) = 12.$$

As a check, we find that $\mathscr{A}\mathbf{P}_2 + 12\mathbf{U} = 0$. Collecting the information from the preceding, we can write the resolvent matrix:

$$(s\mathbf{U} - \mathscr{A})^{-1} = \frac{\mathbf{U}s^3 + \mathbf{P}_1 s + \mathbf{P}_2}{s^3 + d_1 s^2 + d_2 s + d_3}$$

$$= \frac{\begin{bmatrix} s^2 + 5s + 6 & s + 8 & 3s + 9 \\ 0 & s^2 + 4s + 4 & 0 \\ 0 & 2s + 4 & s^2 + 5s + 6 \end{bmatrix}}{s^2 + 7s^2 + 16s + 12}.$$

In the last step the result was again rearranged into a single matrix. Compare this with the earlier solution. Incidentally, since zero is not an eigenvalue of $\mathscr{A}$, (99) gives the collateral result that

$$\mathscr{A}^{-1} = -\frac{1}{d_3}\mathbf{P}_2 = -\frac{1}{12}\begin{bmatrix} 6 & 8 & 9 \\ 0 & 4 & 0 \\ 0 & 4 & 6 \end{bmatrix}.$$

Note that the algorithm gives the characteristic polynomial $d(s)$ in expanded form. To find the constituent matrices it is still necessary (1) to factor $d(s)$ in order to find the eigenvalues and (2) to obtain a partial-fraction expansion. Computational algorithms for the first of these are readily available.*

RESOLVING POLYNOMIALS

Looking back again at (84), we remember that the constituent matrices $\mathbf{K}_{ij}$ depend only on $\mathscr{A}$, and not on any specific function. As we mentioned earlier, if these matrices can be evaluated for some specific functions, the results so obtained will be good for any other function. We found one function, leading to the resolvent matrix $(s\mathbf{U} - \mathscr{A})^{-1}$, from which the constituent matrices could be evaluated. We shall now discuss a set of functions that can also do the job.

Consider a set of functions $f_1(s)$, $f_2(s)$, $\cdots$, $f_n(s)$, each of which is a polynomial. Each of these polynomials can be inserted, in turn, into (84) and will lead to an equation in which the unknowns are the constituent

* The quotient-difference algorithm is one of the most widely known of the methods for determining the zeros of a polynomial. The algorithm is described in: P. Henrici, *Elements of Numerical Analysis*, John Wiley, New York, 1964, Chap. 8.

matrices. There will be as many equations as there are unknowns. In matrix form these equations will be as follows:

$$
\begin{bmatrix}
f_1(s_1) & \dfrac{f_1^{(1)}(s_1)}{1!} & \cdots & \dfrac{f^{r_{(1}-1)}(s_1)}{(r_1-1)!} & \vdots & f_1(s_2) & \cdots & \dfrac{f_1^{(r_k-1)}(s_k)}{(r_k-1)!} \\
f_2(s_1) & \dfrac{f_2^{(1)}(s_1)}{1!} & \cdots & \dfrac{f_2^{(r_1-1)}(s_1)}{(r_1-1)!} & \vdots & f_2(s_2) & \cdots & \dfrac{f_2^{(r_k-1)}(s_k)}{(r_k-1)} \\
\vdots & \vdots & & \vdots & \vdots & \vdots & & \vdots \\
f_n(s_1) & \dfrac{f_n^{(1)}(s_1)}{1!} & \cdots & \dfrac{f_n^{(r_1-1)}(s_1)}{(r_1-1)!} & \vdots & f_n(s_2) & \cdots & \dfrac{f_n^{(r_k-1)}(s_k)}{(r_k-1)!}
\end{bmatrix}
\begin{bmatrix}
\mathbf{K}_{11} \\
\mathbf{K}_{12} \\
\vdots \\
\mathbf{K}_{1r_1} \\
\cdots \\
\mathbf{K}_{21} \\
\vdots \\
\mathbf{K}_{kr_k}
\end{bmatrix}
=
\begin{bmatrix}
f_1(\mathscr{A}) \\
f_2(\mathscr{A}) \\
\vdots \\
f_n(\mathscr{A})
\end{bmatrix}.
$$

(100)

We shall call this the *resolving equation*. Although the vector elements are square matrices of order n, these matrices are treated as single quantities when the indicated matrix multiplication is interpreted. Thus the matrix multiplication, when performed, gives terms in which a matrix is multiplied by a scalar—a perfectly legitimate operation; for example, in the first row of the product.

$$
f_1(s_1)\mathbf{K}_{11} + \frac{f_1^{(1)}(s_1)}{1!}\mathbf{K}_{12} + \cdots + \frac{f_1^{(r_1-1)}(s_1)}{(r_1-1)!}\mathbf{K}_{1r_1}
$$

$$
+ f_1(s_2)\mathbf{K}_{21} + \cdots + \frac{f_1^{(r_k-1)}(s_k)}{(r_k-1)!}\mathbf{K}_{kr_k} \quad (101)
$$

the matrix $\mathbf{K}_{21}$ is multiplied by the scalar $f_1(s_2)$.

If this approach is to work, the coefficient matrix of (100) must be nonsingular and easily inverted. It becomes a problem of selecting an appropriate set of polynomials, which will be called *resolving polynomials*.

The simplest polynomial is a power of s. Thus one possible set of resolving polynomials is

$$
f_i(s) = s^{i-1} \qquad i = 1, 2, \cdots, n. \quad (102)
$$

Rather than writing the general expression for this case, suppose, for example, the characteristic polynomial is $d(s) = (s - s_1)^2(s - s_2)^2$. Then $n = 4$ and $f_1 = 1, f_2 = s, f_3 = s^2$, and $f_4 = s^3$. Hence (100) becomes

$$
\begin{bmatrix}
1 & 0 & 1 & 0 \\
s_1 & 1 & s_2 & 1 \\
s_1^2 & 2s_1 & s_2^2 & 2s_2 \\
s_1^3 & 3s_1^2 & s_2^3 & 3s_2^2
\end{bmatrix}
\begin{bmatrix}
\mathbf{K}_{11} \\
\mathbf{K}_{12} \\
\mathbf{K}_{21} \\
\mathbf{K}_{22}
\end{bmatrix}
=
\begin{bmatrix}
\mathbf{U} \\
\mathscr{A} \\
\mathscr{A}^2 \\
\mathscr{A}^3
\end{bmatrix}.
\tag{103}
$$

It is clear that the elements of the right-hand vector are easy to determine in this case but that inverting the matrix will require considerable effort, especially if n is much larger.

As a more explicit illustration let us consider the example given earlier in which

$$
\mathscr{A} =
\begin{bmatrix}
-2 & 1 & 3 \\
0 & -3 & 0 \\
0 & 2 & -2
\end{bmatrix}
\qquad
\begin{aligned}
& d(s) = (s + 2)^2(s + 3), \\
& s_1 = -2, \\
& s_2 = -3.
\end{aligned}
\tag{104}
$$

With the resolving polynomials chosen according to (102), we get for (100)

$$
\begin{bmatrix}
1 & 0 & 1 \\
-2 & 1 & -3 \\
4 & -4 & 9
\end{bmatrix}
\begin{bmatrix}
\mathbf{K}_{11} \\
\mathbf{K}_{12} \\
\mathbf{K}_{21}
\end{bmatrix}
=
\begin{bmatrix}
\mathbf{U} \\
\mathscr{A} \\
\mathscr{A}^2
\end{bmatrix}.
$$

We now invert this equation to get

$$
\begin{bmatrix}
\mathbf{K}_{11} \\
\mathbf{K}_{12} \\
\mathbf{K}_{21}
\end{bmatrix}
=
\begin{bmatrix}
-3 & -4 & -1 \\
6 & 5 & 1 \\
4 & 4 & 1
\end{bmatrix}
\begin{bmatrix}
\mathbf{U} \\
\mathscr{A} \\
\mathscr{A}^2
\end{bmatrix}
$$

which in expanded form gives

$$
\begin{aligned}
\mathbf{K}_{11} &= -3\mathbf{U} - 4\mathscr{A} - \mathscr{A}^2, \\
\mathbf{K}_{12} &= 6\mathbf{U} + 5\mathscr{A} + \mathscr{A}^2, \\
\mathbf{K}_{21} &= 4\mathbf{U} + 4\mathscr{A} + \mathscr{A}^2.
\end{aligned}
\tag{105}
$$

Now you can complete the problem by inserting $\mathscr{A}$ and $\mathscr{A}^2$ in these equations and verifying that the same constituent matrices are obtained as before.

Another choice of the set of polynomials is the following:

$$f_1(s) = 1,$$
$$f_2(s) = (s - s_1),$$
$$f_3(s) = (s - s_1)^2,$$
$$\vdots$$
$$f_{r_1+1}(s) = (s - s_1)^{r_1},$$
$$f_{r_1+2}(s) = (s - s_1)^{r_1}(s - s_2),$$
$$\vdots$$
$$f_{r_1+r_2+1}(s) = (s - s_1)^{r_1}(s - s_2)^{r_2},$$
$$f_{r_1+r_2+2}(s) = (s - s_1)^{r_1}(s - s_2)^{r_2}(s - s_3),$$
$$\vdots$$
$$f_n(s) = (s - s_1)^{r_1}(s - s_2)^{r_2} \cdots (s - s_k)^{r_k-1},$$

where the s_i's are the eigenvalues. In this case the evaluation of $f_i(\mathscr{A})$ will require a large effort, but the matrix in (100) will be easy to invert. Again we shall consider the particular case in which $d(s) = (s - s_1)^2$ $(s - s_2)^2$. Then

$$f_1(s) = 1, \qquad\qquad\qquad f_1(\mathscr{A}) = \mathbf{U},$$
$$f_2(s) = (s - s_1), \qquad\qquad f_2(\mathscr{A}) = (\mathscr{A} - s_1\mathbf{U}),$$
$$f_3(s) = (s - s_1)^2, \qquad\qquad f_2(\mathscr{A}) = (\mathscr{A} - s_1\mathbf{U})^2,$$
$$f_4(s) = (s - s_1)^2(s - s_2), \quad f_3(\mathscr{A}) = (\mathscr{A} - s_1\mathbf{U})^2(\mathscr{A} - s_2\mathbf{U}),$$

and (100) becomes

$$\begin{bmatrix} 1 & 0 & 1 & 0 \\ 0 & 1 & (s_2 - s_1) & 1 \\ 0 & 0 & (s_2 - s_1)^2 & 2(s_2 - s_1) \\ 0 & 0 & 0 & (s_2 - s_1)^2 \end{bmatrix} \begin{bmatrix} \mathbf{K}_{11} \\ \mathbf{K}_{12} \\ \mathbf{K}_{21} \\ \mathbf{K}_{22} \end{bmatrix} = \begin{bmatrix} \mathbf{U} \\ \mathscr{A} - s_1\mathbf{U} \\ (\mathscr{A} - s_1\mathbf{U})^2 \\ (\mathscr{A} - s_1\mathbf{U})^2(\mathscr{A} - s_2\mathbf{U}) \end{bmatrix}.$$

$$(107)$$

The coefficient matrix is seen to be upper triangular in this case and can be easily inverted. This is true in general for this selection of resolving polynomials.

For the specific example treated before, given in (104), the resolving equation becomes

$$
\begin{bmatrix} 1 & 0 & 1 \\ 0 & 1 & -1 \\ 0 & 0 & 1 \end{bmatrix}
\begin{bmatrix} \mathbf{K}_{11} \\ \mathbf{K}_{12} \\ \mathbf{K}_{21} \end{bmatrix}
=
\begin{bmatrix} \mathbf{U} \\ \mathscr{A} + 2\mathbf{U} \\ (\mathscr{A} + 2\mathbf{U})^2 \end{bmatrix}
$$

which can be readily inverted to yield

$$\mathbf{K}_{11} = \mathbf{U} - (\mathscr{A} + 2\mathbf{U})^2 = -3\mathbf{U} - 4\mathscr{A} - \mathscr{A}^2,$$

$$\mathbf{K}_{12} = (\mathscr{A} + 2\mathbf{U}) + (\mathscr{A} + 2\mathbf{U})^2 = 6\mathbf{U} + 5\mathscr{A} + \mathscr{A}^2,$$

$$\mathbf{K}_{21} = (\mathscr{A} + 2\mathbf{U})^2 = 4\mathbf{U} + 4\mathscr{A} + \mathscr{A}^2.$$

On comparing the last step with (105) we find they are identical, as they should be.

In this section we have treated a number of methods for evaluating a function of a matrix. Each method has certain advantages and disadvantages. Some are more readily applied to low-order matrices; others lend themselves to numerical evaluation by computer. Our basic interest is in determining equivalent closed-form expressions for the function $\epsilon^{\mathscr{A}t}$, which constitutes the solution to a homogeneous state equation.

4.5 SYSTEMATIC FORMULATION OF THE STATE EQUATIONS

Let us briefly review what has been done in this chapter. We started by considering the order of complexity of a network. We discovered that the number of dynamically independent variables for RLC networks equals the number of reactive elements, minus the number of all-capacitor loops and the number of all-inductor cut-sets. For a network containing multiterminal components (controlled sources, etc.), additional algebraic constraints among capacitor voltages and inductor currents may be introduced, thus further reducing the order of complexity. We shall here assume that in all cases the order of complexity is the same as would be computed for an RLC network. If it turns out that the assumption is false in a particular network, then it will be impossible to obtain the equations in the desired form by the process we shall describe. An illustration will be provided shortly.

Next, we defined a normal tree of a graph as one containing the maximum number of capacitors and the minimum number of inductors—as well as all the independent voltage sources, but none of the independent current sources. We showed by means of examples that a set of network equations could be written having as variables the capacitor twig voltages and inductor link currents for a normal tree. The equations had the general form

$$\frac{d\mathbf{x}}{dt} = \mathscr{A}\mathbf{x} + \mathscr{B}_1\mathbf{e} + \mathscr{B}_2\frac{d\mathbf{e}}{dt}, \tag{108a}$$

$$\mathbf{w} = C\mathbf{x} + \mathscr{D}_1\mathbf{e} + \mathscr{D}_2\frac{d\mathbf{e}}{dt}. \tag{108b}$$

However, by the transformation $\mathbf{x} \rightarrow \mathbf{x} + \mathscr{B}_2\mathbf{e}$, these could be reduced to

$$\frac{d\mathbf{x}}{dt} = \mathscr{A}\mathbf{x} + \mathscr{B}\mathbf{e}, \tag{109a}$$

$$\mathbf{w} = C\mathbf{x} + \mathscr{D}\mathbf{e} + \widehat{\mathscr{D}}\frac{d\mathbf{e}}{dt}, \tag{109b}$$

where

$$\mathscr{B} = \mathscr{B}_1 + \mathscr{A}\mathscr{B}_2, \mathscr{D} = \mathscr{D}_1 + C\mathscr{B}_2, \text{ and } \widehat{\mathscr{D}} = \mathscr{D}_2.$$

Equation (109a) is the normal form for the state equation; $\mathbf{x}$ is the state vector, and its elements are the state variables. Actually the state vector in the last pair of equations is a linear combination of the original " state vector " (with capacitor twig voltages and inductor link currents as variables) and the source vector $\mathbf{e}$. Even with this transformation the second equation of the pair—the output equation—may still contain the $d\mathbf{e}/dt$ term. We shall shortly clarify the conditions under which this will occur. For purposes of accurate reference we shall refer to the first equation of either pair as the *state equation* and the second equation of either pair as the *output equation*. The two equations together will be referred to as the *state equations.*

Our next task was to solve the state equation, and this was done by finding a solution first for the homogeneous equation (with $\mathbf{e} = \mathbf{0}$). Symbolically, this solution involves the matrix exponential $e^{\mathscr{A}t}$, so we devoted some effort to determine methods for evaluating such functions of a matrix. Once the matrix exponential is evaluated, the state vector $\mathbf{x}$ is found from (45). We shall defer further consideration of the evaluation of this integral and its ramifications to the next chapter.

We must now formalize the writing of the state equations (109) for a given network and show that this is the general form. Let us observe at the outset that it is possible to choose some variables other than the capacitor voltages and inductor currents as state variables. In Fig. 6,

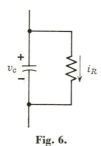

Fig. 6.

for example, the resistor current i_R might be chosen as a state variable—rather than the capacitor voltage—since v_C is directly proportional to i_R. Nevertheless, while recognizing this flexibility, we shall proceed in the manner to be described now.

TOPOLOGICAL CONSIDERATIONS

The first step is the selection of a normal tree (or normal forest). Generally this is not unique. If there are no degeneracies (no all-capacitor loops or all-inductor cut-sets), at least the reactive elements will be uniquely assignable to the normal tree and cotree—but not the resistive elements. However, when there are degeneracies, there will be a choice even among the reactive elements.

According to our usual convention, in writing a loop or cut-set matrix we first number the twigs and then the links. We shall here make a more detailed convention of branch numbering and adopt the following order within the twig and link categories:

Twigs	*Links*
1. Voltage-source twigs	5. Capacitor links
2. Capacitor twigs	6. Resistor links
3. Resistor twigs	7. Inductor links
4. Inductor twigs	8. Current-source links

Note that the terms "resistor twig" and "resistor link" include branches of multiterminal devices, such as gyrators and controlled sources, whose *v-i* relationships are algebraic like that of a resistor. We are imposing no specific order in the numbering of such branches but are including them among the branches representing resistors. This numbering of branches leads to a partitioning of the current-and-voltage vectors as follows:

$$\mathbf{v} = \begin{bmatrix} \mathbf{v}_t \\ \mathbf{v}_l \end{bmatrix} \Rightarrow \mathbf{v}_t = \begin{bmatrix} \mathbf{v}_E \\ \mathbf{v}_{Ct} \\ \mathbf{v}_{Rt} \\ \mathbf{v}_{Lt} \end{bmatrix} \quad \text{and} \quad \mathbf{v}_l = \begin{bmatrix} \mathbf{v}_{Cl} \\ \mathbf{v}_{Rl} \\ \mathbf{v}_{Ll} \\ \mathbf{v}_J \end{bmatrix}, \tag{110a}$$

$$\mathbf{i} = \begin{bmatrix} \mathbf{i}_t \\ \mathbf{i}_l \end{bmatrix} \Rightarrow \mathbf{i}_t = \begin{bmatrix} \mathbf{i}_E \\ \mathbf{i}_{Ct} \\ \mathbf{i}_{Rt} \\ \mathbf{i}_{Lt} \end{bmatrix} \quad \text{and} \quad \mathbf{i}_l = \begin{bmatrix} \mathbf{i}_{Cl} \\ \mathbf{i}_{Rl} \\ \mathbf{i}_{Ll} \\ \mathbf{i}_J \end{bmatrix}, \tag{110b}$$

where, for example, $\mathbf{v}_{Ct}$ is the vector of capacitor twig voltages and $\mathbf{i}_{Rl}$ is the vector of resistor link currents. We have placed no twig subscript on $\mathbf{v}_E$ and $\mathbf{i}_E$, and no link subscript on $\mathbf{i}_J$ and $\mathbf{v}_J$, because the voltage sources are always twigs and the current sources are always links.

Our next step is to write KVL equations for the f-loops and KCL equations for the f-cut-sets, as in (6) and (7). They are repeated here as

$$\mathbf{Q}\mathbf{i} = 0, \quad \text{or} \quad \mathbf{i}_t = -\mathbf{Q}_l \mathbf{i}_l, \tag{111a}$$

$$\mathbf{B}\mathbf{v} = 0, \quad \text{or} \quad \mathbf{v}_l = -\mathbf{B}_t \mathbf{v}_t = \mathbf{Q}_l' \mathbf{v}_t, \tag{111b}$$

where the usual partitioning is $\mathbf{Q} = [\mathbf{U} \quad \mathbf{Q}_l]$, $\mathbf{B} = [\mathbf{B}_t \quad \mathbf{U}]$. The last step follows from $\mathbf{B}_t = -\mathbf{Q}_l'$. If the current-and-voltage vectors partitioned according to (110) are to be inserted here, we must also partition the $\mathbf{Q}_l$ matrix conformally; that is, into four rows and columns. Now each row of $\mathbf{Q}$ corresponds to an f-cut-set defined by a twig for the normal tree. The columns correspond to links. If we arrange the columns and rows in the conventional order decided upon, $\mathbf{Q}_l$ must take the form

$$\mathbf{Q}_l = \begin{array}{c} \text{links} \rightarrow \\ \text{twigs} \quad C \quad R \quad L \quad J \\ \begin{array}{c} \downarrow \ E \\ C \\ R \\ L \end{array} \begin{bmatrix} - & - & - & - \\ - & - & - & - \\ - & - & - & - \\ - & - & - & - \end{bmatrix} \end{array}.$$

If there is a capacitor link, it will be by virtue of an all-capacitor loop. Since there will be no resistors or inductors in such a loop, the column corresponding to a capacitor link cannot have a nonzero entry in the rows corresponding to R and L twigs; that is, the entries in the first column, third and fourth rows must be zero. Similarly, if there is an inductor twig, it is by virtue of an all-inductor cut-set. Since there can be no resistors or capacitors in such a cut-set, the row corresponding to inductor twigs cannot have nonzero entries in the columns corresponding to C and R links. Hence $\mathbf{Q}_l$ can be written as

$$
\mathbf{Q}_l =
\begin{matrix} \text{links} \rightarrow \\ \\ C \text{ twigs} \\ \\ C \\ \\ R \\ \\ L \end{matrix}
\overset{\begin{matrix} C & C & R & L \end{matrix}}{
\begin{bmatrix}
\mathbf{Q}_{EC} & \mathbf{Q}_{ER} & \mathbf{Q}_{EL} & \mathbf{Q}_{EJ} \\
\mathbf{Q}_{CC} & \mathbf{Q}_{CR} & \mathbf{Q}_{CL} & \mathbf{Q}_{CJ} \\
\mathbf{0} & \mathbf{Q}_{RR} & \mathbf{Q}_{RL} & \mathbf{Q}_{RJ} \\
\mathbf{0} & \mathbf{0} & \mathbf{Q}_{LL} & \mathbf{Q}_{LJ}
\end{bmatrix}}.
\tag{112}
$$

When this is inserted into the Kirchhoff equations in (111) the result can be expanded into

$$
\mathbf{i}_E = -\mathbf{Q}_{EC}\mathbf{i}_{Cl} - \mathbf{Q}_{ER}\mathbf{i}_{Rl} - \mathbf{Q}_{EL}\mathbf{i}_{Ll} - \mathbf{Q}_{EJ}\mathbf{i}_J
\tag{113a}
$$

$$
\mathbf{i}_{Ct} = -\mathbf{Q}_{CC}\mathbf{i}_{Cl} - \mathbf{Q}_{CR}\mathbf{i}_{Rl} - \mathbf{Q}_{CL}\mathbf{i}_{Ll} - \mathbf{Q}_{CJ}\mathbf{i}_J
\tag{113b}
$$

$$
\mathbf{i}_{Rt} = \qquad -\mathbf{Q}_{RR}\mathbf{i}_{Rl} - \mathbf{Q}_{RL}\mathbf{i}_{Ll} - \mathbf{Q}_{RJ}\mathbf{i}_J
\tag{113c}
$$

$$
\mathbf{i}_{Lt} = \qquad\qquad -\mathbf{Q}_{LL}\mathbf{i}_{Ll} - \mathbf{Q}_{LJ}\mathbf{i}_J
\tag{113d}
$$

$$
\mathbf{v}_{Cl} = \mathbf{Q}'_{EC}\mathbf{v}_E + \mathbf{Q}'_{CC}\mathbf{v}_{Ct}
\tag{113e}
$$

$$
\mathbf{v}_{Rl} = \mathbf{Q}'_{ER}\mathbf{v}_E + \mathbf{Q}'_{CR}\mathbf{v}_{Ct} + \mathbf{Q}'_{RR}\mathbf{v}_{Rt}
\tag{113f}
$$

$$
\mathbf{v}_{Ll} = \mathbf{Q}'_{EL}\mathbf{v}_E + \mathbf{Q}'_{CL}\mathbf{v}_{Ct} + \mathbf{Q}'_{RL}\mathbf{v}_{Rt} + \mathbf{Q}'_{LL}\mathbf{v}_{Lt}
\tag{113g}
$$

$$
\mathbf{v}_J = \mathbf{Q}'_{EJ}\mathbf{v}_E + \mathbf{Q}'_{CJ}\mathbf{v}_{Ct} + \mathbf{Q}'_{RJ}\mathbf{v}_{Rt} + \mathbf{Q}'_{LJ}\mathbf{v}_{Lt}
\tag{113h}
$$

To illustrate this partitioning return to the examples considered earlier in the chapter. For Fig. 4 the $\mathbf{Q}_l$ matrix is

$$
\mathbf{Q}_l = \begin{array}{c} E \\ \\ C \\ \\ R \end{array}
\overset{\begin{array}{cccc} R & & & L \\ 6 & 7 & 8 & 9 \end{array}}{
\begin{bmatrix}
0 & 1 & 0 & \vdots & 0 \\
0 & 0 & 1 & \vdots & 0 \\
\hdashline
-1 & -1 & 0 & \vdots & -1 \\
1 & 0 & -1 & \vdots & 0 \\
\hdashline
1 & 0 & 0 & \vdots & 1
\end{bmatrix}}
,
$$

$$
\mathbf{Q}_{ER} = \begin{bmatrix} 0 & 1 & 0 \\ 0 & 0 & 1 \end{bmatrix}, \qquad \mathbf{Q}_{EL} = \begin{bmatrix} 0 \\ 0 \end{bmatrix}
$$

$$
\mathbf{Q}_{CR} = \begin{bmatrix} -1 & -1 & 0 \\ 1 & 0 & -1 \end{bmatrix}, \; \mathbf{Q}_{CL} = \begin{bmatrix} -1 \\ 0 \end{bmatrix}
$$

$$
\mathbf{Q}_{RR} = \begin{bmatrix} 1 & 0 & 0 \end{bmatrix}, \qquad \mathbf{Q}_{RL} = [1].
$$

Since there are no inductor twigs, capacitor links, or current sources in the network, the $\mathbf{Q}_l$ matrix is less than full.

ELIMINATING UNWANTED VARIABLES

Up to this point the discussion has been topological. We must now bring in the *v-i* relationships. First, let us write them for the reactive elements; thus

$$
\begin{bmatrix} \mathbf{i}_{Ct} \\ \mathbf{i}_{Cl} \end{bmatrix} = \frac{d}{dt} \begin{bmatrix} \mathbf{C}_t & \mathbf{0} \\ \mathbf{0} & \mathbf{C}_l \end{bmatrix} \begin{bmatrix} \mathbf{v}_{Ct} \\ \mathbf{v}_{Cl} \end{bmatrix} \tag{114a}
$$

and

$$
\begin{bmatrix} \mathbf{v}_{Ll} \\ \mathbf{v}_{Lt} \end{bmatrix} = \frac{d}{dt} \begin{bmatrix} \mathbf{L}_{ll} & \mathbf{L}_{lt} \\ \mathbf{L}_{tl} & \mathbf{L}_{tt} \end{bmatrix} \begin{bmatrix} \mathbf{i}_{Ll} \\ \mathbf{i}_{Lt} \end{bmatrix}. \tag{114b}
$$

In these expressions $\mathbf{C}_t$ and $\mathbf{C}_l$ are the matrices of capacitor twigs and links; they are both diagonal. Because of the possibility of mutual inductance, there may be coupling between inductor twigs and links, as well as between inductor twigs themselves and inductor links themselves. Hence the inductor matrices need not be diagonal ($\mathbf{L}_{tl}$ and $\mathbf{L}_{lt}$ are not even square), but $\mathbf{L}_{tt}$ and $\mathbf{L}_{ll}$ are symmetric and $\mathbf{L}_{lt} = \mathbf{L}'_{tl}$. Note that by keeping the capacitance and inductance matrices under the derivative sign these expressions apply equally well to time-varying networks.

The eventual variables of interest are $\mathbf{v}_{Ct}$ and $\mathbf{i}_{Ll}$; all others must be eliminated. For the capacitors this means eliminating $\mathbf{i}_{Ct}$, $\mathbf{v}_{Cl}$, and $\mathbf{i}_{Cl}$. Let us start this process by rewriting (113b) as follows:

$$\mathbf{i}_{Ct} + \mathbf{Q}_{CC}\mathbf{i}_{Cl} = [\mathbf{U} \quad \mathbf{Q}_{CC}]\begin{bmatrix} \mathbf{i}_{Ct} \\ \mathbf{i}_{Cl} \end{bmatrix}$$

$$= -\mathbf{Q}_{CR}\mathbf{i}_{Rl} - \mathbf{Q}_{CL}\mathbf{i}_{Ll} - \mathbf{Q}_{CJ}\mathbf{i}_{J} . \tag{115}$$

Into the left-hand side we insert the capacitor v-i relationship from (114). This side then becomes

$$[\mathbf{U} \quad \mathbf{Q}_{CC}]\begin{bmatrix} \mathbf{i}_{Ct} \\ \mathbf{i}_{Cl} \end{bmatrix} = [\mathbf{U} \quad \mathbf{Q}_{CC}]\frac{d}{dt}\begin{bmatrix} \mathbf{C}_t & \mathbf{0} \\ \mathbf{0} & \mathbf{C}_l \end{bmatrix}\begin{bmatrix} \mathbf{v}_{Ct} \\ \mathbf{v}_{Cl} \end{bmatrix}$$

$$= \frac{d}{dt}[\mathbf{U} \quad \mathbf{Q}_{CC}]\begin{bmatrix} \mathbf{C}_t & \mathbf{0} \\ \mathbf{0} & \mathbf{C}_l \end{bmatrix}\begin{bmatrix} \mathbf{v}_{Ct} \\ \mathbf{Q}'_{CC}\mathbf{v}_{Ct} + \mathbf{Q}'_{EC}\mathbf{v}_E \end{bmatrix}$$

$$= \frac{d}{dt}[\mathbf{U} \quad \mathbf{Q}_{CC}]\begin{bmatrix} \mathbf{C}_t & \mathbf{0} \\ \mathbf{0} & \mathbf{C}_l \end{bmatrix}\begin{bmatrix} \mathbf{U} \\ \mathbf{Q}'_{CC} \end{bmatrix}\mathbf{v}_{Ct} \tag{116}$$

$$+ \frac{d}{dt}[\mathbf{U} \quad \mathbf{Q}_{CC}]\begin{bmatrix} \mathbf{C}_t & \mathbf{0} \\ \mathbf{0} & \mathbf{C}_l \end{bmatrix}\begin{bmatrix} \mathbf{0} \\ \mathbf{Q}'_{EC} \end{bmatrix}\mathbf{v}_E .$$

The next-to-last step follows by substituting for $\mathbf{v}_{Cl}$ from (113e). To simplify, define

$$\mathscr{C} = [\mathbf{U} \quad \mathbf{Q}_{CC}]\begin{bmatrix} \mathbf{C}_t & \mathbf{0} \\ \mathbf{0} & \mathbf{C}_l \end{bmatrix}\begin{bmatrix} \mathbf{U} \\ \mathbf{Q}'_{CC} \end{bmatrix} = \mathbf{C}_t + \mathbf{Q}_{CC}\mathbf{C}_l\mathbf{Q}'_{CC}, \tag{117}$$

which equals $\mathbf{C}_t$ when there are no all-capacitor loops, and

$$\hat{\mathscr{C}} = -[\mathbf{U} \quad \mathbf{Q}_{CC}]\begin{bmatrix} \mathbf{C}_t & \mathbf{0} \\ \mathbf{0} & \mathbf{C}_l \end{bmatrix}\begin{bmatrix} \mathbf{0} \\ \mathbf{Q}'_{EC} \end{bmatrix} = -\mathbf{Q}_{CC}\mathbf{C}_l\mathbf{Q}'_{EC}, \tag{118}$$

which is the zero matrix when there are no loops containing just capacitors *and* independent voltage sources. Then, with the last two equations inserted into (115), we get

$$\frac{d}{dt}(\mathscr{C}\mathbf{v}_{Ct}) = -\mathbf{Q}_{CR}\mathbf{i}_{Rl} - \mathbf{Q}_{CL}\mathbf{i}_{Ll} - \mathbf{Q}_{CJ}\mathbf{i}_J + \frac{d}{dt}(\hat{\mathscr{C}}\mathbf{v}_E). \tag{119}$$

There is still an unwanted variable here, $\mathbf{i}_{Rl}$, but before we discuss its elimination, let us arrive at a similar result for inductors.

We start the process by rewriting (113g) as

$$\mathbf{v}_{Ll} - \mathbf{Q}'_{LL}\mathbf{v}_{Lt} = [\mathbf{U} \quad -\mathbf{Q}'_{LL}] \begin{bmatrix} \mathbf{v}_{Ll} \\ \mathbf{v}_{Lt} \end{bmatrix} = \mathbf{Q}'_{EL}\mathbf{v}_E + \mathbf{Q}'_{CL}\mathbf{v}_{Ct} + \mathbf{Q}'_{RL}\mathbf{v}_{Rt}. \quad (120)$$

Into this we next insert the inductor $v\text{-}i$ relationship from (114). The left-hand side becomes

$$[\mathbf{U} \quad -\mathbf{Q}'_{LL}] \begin{bmatrix} \mathbf{v}_{Ll} \\ \mathbf{v}_{Lt} \end{bmatrix} = [\mathbf{U} \quad -\mathbf{Q}'_{LL}] \frac{d}{dt} \begin{bmatrix} \mathbf{L}_{ll} & \mathbf{L}_{lt} \\ \mathbf{L}_{tl} & \mathbf{L}_{tt} \end{bmatrix} \begin{bmatrix} \mathbf{i}_{Ll} \\ \mathbf{i}_{Lt} \end{bmatrix}$$

$$= \frac{d}{dt}[\mathbf{U} \quad -\mathbf{Q}'_{LL}] \begin{bmatrix} \mathbf{L}_{ll} & \mathbf{L}_{lt} \\ \mathbf{L}_{tl} & \mathbf{L}_{tt} \end{bmatrix} \begin{bmatrix} \mathbf{i}_{Ll} \\ -\mathbf{Q}_{LL}\mathbf{i}_{Ll} - \mathbf{Q}_{LJ}\mathbf{i}_J \end{bmatrix}$$

$$= \frac{d}{dt}[\mathbf{U} \quad -\mathbf{Q}'_{LL}] \begin{bmatrix} \mathbf{L}_{ll} & \mathbf{L}_{lt} \\ \mathbf{L}_{tl} & \mathbf{L}_{tt} \end{bmatrix} \begin{bmatrix} \mathbf{U} \\ -\mathbf{Q}_{LL} \end{bmatrix} \mathbf{i}_{Ll} \qquad (121)$$

$$+ \frac{d}{dt}[\mathbf{U} \quad -\mathbf{Q}'_{LL}] \begin{bmatrix} \mathbf{L}_{ll} & \mathbf{L}_{lt} \\ \mathbf{L}_{tl} & \mathbf{L}_{tt} \end{bmatrix} \begin{bmatrix} \mathbf{0} \\ -\mathbf{Q}_{LJ} \end{bmatrix} \mathbf{i}_J.$$

The next-to-last step follows by substituting for $\mathbf{i}_{Lt}$ from (113). To simplify, define

$$\mathscr{L} = [\mathbf{U} \quad -\mathbf{Q}'_{LL}] \begin{bmatrix} \mathbf{L}_{ll} & \mathbf{L}_{lt} \\ \mathbf{L}_{tl} & \mathbf{L}_{tt} \end{bmatrix} \begin{bmatrix} \mathbf{U} \\ -\mathbf{Q}_{LL} \end{bmatrix}$$

$$= \mathbf{L}_{ll} - \mathbf{L}_{lt}\mathbf{Q}_{LL} - \mathbf{Q}'_{LL}\mathbf{L}_{tl} + \mathbf{Q}'_{LL}\mathbf{L}_{tt}\mathbf{Q}_{LL}, \qquad (122)$$

which equals $\mathbf{L}_{ll}$ when there are no all-inductor cut-sets, and

$$\hat{\mathscr{L}} = -[\mathbf{U} \quad -\mathbf{Q}'_{LL}] \begin{bmatrix} \mathbf{L}_{ll} & \mathbf{L}_{lt} \\ \mathbf{L}_{tl} & \mathbf{L}_{tt} \end{bmatrix} \begin{bmatrix} \mathbf{0} \\ -\mathbf{Q}_{LJ} \end{bmatrix} = -\mathbf{Q}'_{LL}\mathbf{L}_{tt}\mathbf{Q}_{LJ} + \mathbf{L}_{lt}\mathbf{Q}_{LJ},$$

$$= (\mathbf{L}_{lt} - \mathbf{Q}'_{LL}\mathbf{L}_{tt})\mathbf{Q}_{LJ} \qquad (123)$$

which is the zero matrix when there are no cut-sets containing just inductors *and* independent current sources. With the last two equations inserted into (120) there results

$$\frac{d}{dt}(\mathscr{L}\mathbf{i}_{Ll}) = \mathbf{Q}'_{CL}\mathbf{v}_{Ct} + \mathbf{Q}'_{RL}\mathbf{v}_{Rt} + \mathbf{Q}'_{EL}\mathbf{v}_E + \frac{d}{dt}(\hat{\mathscr{L}}\mathbf{i}_J). \qquad (124)$$

This is the counterpart of (119). It also contains the unwanted variable $\mathbf{v}_{Rt}$, just as (119) contains $\mathbf{i}_{Rl}$.

To continue with the elimination process it will be necessary to express the *v-i* relationships of the resistor branches in terms of these two variables; namely, $\mathbf{i}_{Rl}$ and $\mathbf{v}_{Rt}$ (resistor link currents and resistor twig voltages). We shall assume that the resistor branch *v-i* relationships can be written as

$$\mathbf{i}_{Rl} = \mathbf{G}_{ll}\mathbf{v}_{Rl} + \mathbf{G}_{lt}\mathbf{i}_{Rt}, \qquad (125a)$$

$$\mathbf{v}_{Rt} = \mathbf{G}_{tl}\mathbf{v}_{Rl} + \mathbf{G}_{tt}\mathbf{i}_{Rt} \qquad (125b)$$

This is one of the hybrid-parameter forms.* It is the same form used in Chapter 2 for the mixed-variable equations. For simple RLC networks the $\mathbf{G}_{ll}$ and $\mathbf{G}_{tt}$ matrices will be diagonal, and $\mathbf{G}_{lt}$ and $\mathbf{G}_{tl}$ will be zero matrices. More generally, none of the matrices need be diagonal, and they may all be nonzero. There is no assurance at the outset that equations of the form of (125) exist for a given network; but unless they exist, the method we are developing will not work. This does not mean that state equations do not exist, but only that our method will fail.

As a simple example, consider the network of Fig. 7. The normal tree includes only the two capacitors. The resistor branches of the graph are all links, but their *v-i* relationships have the form

$$i_3 = 0,$$

$$v_4 = kv_3 = kv_2,$$

$$i_5 = G_5 v_5.$$

The second one prevents us from writing an equation of the form of (125). This comes about because the controlled voltage source, controlled by a capacitor voltage, introduces an additional algebraic constraint

---

* We could have started with the resistor branch *v-i* relations in the other hybrid-parameter form. (Or, indeed, in the *y*-parameter, the *z*-parameter, or the $ABCD$-parameter forms.) We shall leave these alternate representations of the resistor branch *v-i* relationships to you as an exercise.

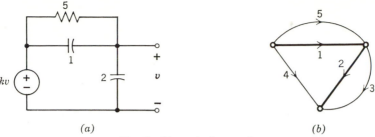

Fig. 7.   Numerical example.

among capacitor voltages, thus reducing the order of complexity. This means that the normal tree must contain only a single capacitor. With very little effort we can write the following state equation for this network, demonstrating that one exists:

$$\frac{dv_2}{dt} = -\frac{G_5}{C_1 + C_2/(1-k)}\, v_2.$$

Returning now to the task of eliminating $\mathbf{i}_{Rl}$ and $\mathbf{v}_{Rt}$ from (119) and (124), we have available 'for this purpose the $v\text{-}i$ relations in (125) and the topological relations in (113c) and (113f). When the latter two are substituted into the former two and the terms are rearranged, the result will be

$$(\mathbf{U} + \mathbf{G}_{lt}\mathbf{Q}_{RR})\mathbf{i}_{Rl} - \mathbf{G}_{ll}\mathbf{Q}'_{RR}\,\mathbf{v}_{Rt}$$
$$= \mathbf{G}_{ll}\mathbf{Q}'_{CR}\mathbf{v}_{Ct} - \mathbf{G}_{lt}\mathbf{Q}_{RL}\mathbf{i}_{Ll} + \mathbf{G}_{ll}\mathbf{Q}'_{ER}\mathbf{v}_E - \mathbf{G}_{lt}\mathbf{Q}_{RJ}\mathbf{i}_J, \quad (126a)$$

$$\mathbf{G}_{tt}\mathbf{Q}_{RR}\mathbf{i}_{Rl} + (\mathbf{U} - \mathbf{G}_{tl}\mathbf{Q}'_{RR})\mathbf{v}_{Rt}$$
$$= \mathbf{G}_{tl}\mathbf{Q}'_{CR}\mathbf{v}_{Ct} - \mathbf{G}_{tt}\mathbf{Q}_{RL}\mathbf{i}_{Ll} + \mathbf{G}_{tl}\mathbf{Q}'_{ER}\mathbf{v}_E - \mathbf{G}_{tt}\mathbf{Q}_{RJ}\mathbf{i}_J. \quad (126b)$$

These are a pair of vector algebraic equations in the two variables $\mathbf{i}_{Rl}$ and $\mathbf{v}_{Rt}$. They will have a solution if the following matrices have inverses:

$$(\mathbf{U} + \mathbf{G}_{lt}\mathbf{Q}_{RR}) \qquad\qquad (127a)$$

and

$$\mathscr{K}_1 = (\mathbf{U} - \mathbf{G}_{tl}\mathbf{Q}'_{RR}) + \mathbf{G}_{tt}\mathbf{Q}_{RR}(\mathbf{U} + \mathbf{G}_{lt}\mathbf{Q}_{RR})^{-1}\mathbf{G}_{ll}\mathbf{Q}'_{RR} \qquad (127b)$$

or

$$(\mathbf{U} - \mathbf{G}_{tl}\mathbf{Q}'_{RR}) \qquad\qquad (128a)$$

and

$$\mathscr{K}_2 = (\mathbf{U} + \mathbf{G}_{ll}\mathbf{Q}_{RR}) + \mathbf{G}_{ll}\mathbf{Q}'_{RR}(\mathbf{U} - \mathbf{G}_{tl}\mathbf{Q}'_{RR})^{-1}\mathbf{G}_{tt}\mathbf{Q}_{RR}. \quad (128b)$$

It is left as an exercise for you to verify this statement.

If these inverses do not exist, we cannot proceed, and our procedure fails. So we assume that they exist and that (126) can be solved for i_{Rl} and v_{Rt}. When the solutions are substituted back into the equations for the derivatives of $\mathscr{C}v_{Ct}$ and $\mathscr{L}i_{Ll}$ in (119) and (124), the resulting expressions are extremely complicated and do not give any particular insight. It is clear, however, that there are terms involving the state variables v_{Ct} and i_{Ll}, the sources v_E and i_J, and their derivatives. We shall not write out the details here but simply indicate the final form:

$$\frac{d}{dt}(\mathscr{C}v_{Ct}) = [-\mathscr{Y} \quad \mathscr{H}]\begin{bmatrix} v_{Ct} \\ i_{Ll} \end{bmatrix} + [-\hat{\mathscr{Y}} \quad \hat{\mathscr{H}}]\begin{bmatrix} v_E \\ i_J \end{bmatrix} + \frac{d}{dt}[\hat{\mathscr{C}} \quad \mathbf{0}]\begin{bmatrix} v_E \\ i_J \end{bmatrix},$$

$$(129a)$$

$$\frac{d}{dt}(\mathscr{L}i_{Ll}) = [\mathscr{G} \quad -\mathscr{Z}]\begin{bmatrix} v_{Ct} \\ i_{Ll} \end{bmatrix} + [\hat{\mathscr{G}} \quad -\hat{\mathscr{Z}}]\begin{bmatrix} v_E \\ i_J \end{bmatrix} + \frac{d}{dt}[\mathbf{0} \quad \hat{\mathscr{L}}]\begin{bmatrix} v_E \\ i_J \end{bmatrix}$$

$$(129b)$$

The symbols we have used for the matrices take into account the dimensions. Thus $\mathscr{Y}$ and $\hat{\mathscr{Y}}$ relate a current vector to a voltage vector and so have the dimensions of admittance. The $\mathscr{H}$ and $\mathscr{G}$ are dimensionless; they correspond to the hybrid h- and hybrid g-matrices. In this form the equations apply to time-varying as well as time-invariant networks. Also, they can more readily be generalized to nonlinear networks. Look over the preceding development and note that, in arriving at these equations, we have used all the v-i relationships and all the topological (Kirchhoff) relationships in (113) except the first and last, relating to the voltage-source currents and the current-source voltages. These two will be used in the determination of output variables, assuming that elements of i_E and v_J are output variables.

In fact, we should establish that, once the state equation and its solution is available, all other variables can be expressed in terms of the state variables v_{Ct} and i_{Ll}, the source quantities v_E and i_J, and the derivatives of the latter, as in (109). It is a matter only of looking over the previously developed equations to verify that this is the case. It will be left as an exercise for you to do at this point.

One point should be clear after contemplating (129) and the way in which any output variable is expressed in terms of the state variables. This is that source-voltage derivatives will appear only when there is an all-capacitor loop—and even then only when this loop includes a voltage source, making $\widehat{\mathscr{C}} = -\mathbf{Q}_{CC}\,\mathbf{C}_l\,\mathbf{Q}'_{EC}$ nonzero. Similarly, source-current derivatives will appear only when there is an all-inductor cut-set—and only when this cut-set includes a current source, making $\widehat{\mathscr{L}} = -\mathbf{Q}'_{LL}\,\mathbf{L}_{tt}\,\mathbf{Q}_{LJ} + \mathbf{L}_{lt}\,\mathbf{Q}_{LJ}$ nonzero. It is only in these cases that derivatives of source quantities can appear in the state equations.

TIME-INVARIANT NETWORKS

Let us now limit ourselves to time-invariant networks. In this case (129) can be rewritten as follows:

$$
\begin{bmatrix} \mathscr{C} & 0 \\ 0 & \mathscr{L} \end{bmatrix} \frac{d}{dt} \begin{bmatrix} \mathbf{v}_{Ct} \\ \mathbf{i}_{Ll} \end{bmatrix} = \begin{bmatrix} -\mathscr{Y} & \mathscr{H} \\ \mathscr{G} & -\mathscr{Z} \end{bmatrix} \begin{bmatrix} \mathbf{v}_{Ct} \\ \mathbf{i}_{Ll} \end{bmatrix}
$$

$$
+ \begin{bmatrix} -\widehat{\mathscr{Y}} & \widehat{\mathscr{H}} \\ \widehat{\mathscr{G}} & -\widehat{\mathscr{Z}} \end{bmatrix} \begin{bmatrix} \mathbf{v}_E \\ \mathbf{i}_J \end{bmatrix} + \begin{bmatrix} \widehat{\mathscr{C}} & 0 \\ 0 & \widehat{\mathscr{L}} \end{bmatrix} \frac{d}{dt} \begin{bmatrix} \mathbf{v}_E \\ \mathbf{i}_J \end{bmatrix}. \qquad (130)
$$

Finally, assuming that $\mathscr{C}$ and $\mathscr{L}$ are nonsingular matrices, we get

$$
\frac{d\mathbf{x}}{dt} = \mathscr{A}\mathbf{x} + \mathscr{B}_1\mathbf{e} + \mathscr{B}_2\frac{d\mathbf{e}}{dt}, \qquad (131)
$$

where

$$
\mathbf{x} = \begin{bmatrix} \mathbf{v}_{Ct} \\ \mathbf{i}_{Ll} \end{bmatrix}, \qquad \mathbf{e} = \begin{bmatrix} \mathbf{v}_E \\ \mathbf{i}_J \end{bmatrix},
$$

$$
\mathscr{A} = \begin{bmatrix} \mathscr{C}^{-1} & 0 \\ 0 & \mathscr{L}^{-1} \end{bmatrix} \begin{bmatrix} -\mathscr{Y} & \mathscr{H} \\ \mathscr{G} & -\mathscr{Z} \end{bmatrix}, \qquad (132)
$$

$$
\mathscr{B}_1 = \begin{bmatrix} \mathscr{C}^{-1} & 0 \\ 0 & \mathscr{L}^{-1} \end{bmatrix} \begin{bmatrix} -\widehat{\mathscr{Y}} & \widehat{\mathscr{H}} \\ \widehat{\mathscr{G}} & -\widehat{\mathscr{Z}} \end{bmatrix}, \qquad (133)
$$

$$
\mathscr{B}_2 = \begin{bmatrix} \mathscr{C}^{-1} & 0 \\ 0 & \mathscr{L}^{-1} \end{bmatrix} \begin{bmatrix} \widehat{\mathscr{C}} & 0 \\ 0 & \widehat{\mathscr{L}} \end{bmatrix}. \qquad (134)
$$

This is the desired result. What has been done here is to present a procedure for arriving at a first-order vector differential equation for a given network in the form of (131). However, we have not derived formulas for the $\mathscr{A}$, $\mathscr{B}_1$, and $\mathscr{B}_2$ matrices directly in terms of branch-parameter matrices and submatrices of $\mathbf{Q}_l$, because such formulas would be extremely complicated and impossible to use. The result depends crucially on the existence of the inverse of the matrices in (127) or (128) and of the matrices $\mathscr{C}$ and $\mathscr{L}$. Unfortunately, there are no simple necessary and sufficient conditions to tell us when these inverses exist and when this procedure will work.

RLC NETWORKS

There is, however, one class of networks for which the above-mentioned procedure will always work. This is the class of time-invariant *RLC* networks. It is of interest to carry through the development for this class, because the results can be written out explicitly and provide insight into the more general case.

The first simplification comes in the *v-i* relations of the resistor branches in (125). There will be no coupling terms in the parameter matrices. Hence $\mathbf{G}_{tl}$ and $\mathbf{G}_{lt}$ are both zero matrices, and the matrices $\mathbf{G}_{ll}$ and $\mathbf{G}_{tt}$ are diagonal and, hence, nonsingular. Let us rename these matrices according to the dimensions of their elements; $\mathbf{G}_{ll}$ is dimensionally conductance, and $\mathbf{G}_{tt}$ is resistance. Set

$$\mathbf{G}_{ll} = \mathbf{G}_l = \mathbf{R}_l^{-1}$$

$$\mathbf{G}_{tt} = \mathbf{R}_t = \mathbf{G}_t^{-1}$$

from which

$$\mathbf{i}_{Rl} = \mathbf{G}_l \mathbf{v}_{Rl}, \tag{135a}$$

$$\mathbf{v}_{Rt} = \mathbf{R}_t \mathbf{i}_{Rt}. \tag{135b}$$

Equations (126) reduces to

$$\mathbf{i}_{Rl} - \mathbf{G}_l \mathbf{Q}'_{RR} \mathbf{v}_{Rt} = \mathbf{G}_l \mathbf{Q}'_{CR} \mathbf{v}_{Ct} + \mathbf{G}_l \mathbf{Q}'_{ER} \mathbf{v}_E, \tag{136a}$$

$$\mathbf{R}_t \mathbf{Q}_{RR} \mathbf{i}_{Rl} + \mathbf{v}_{Rt} = -\mathbf{R}_t \mathbf{Q}_{RL} \mathbf{i}_{Ll} - \mathbf{R}_t \mathbf{Q}_{RJ} \mathbf{i}_J. \tag{136b}$$

The conditions for the existence of a solution reduce to the existence of the inverse of $\mathscr{K}_1$ or $\mathscr{K}_2$ where these have become

$$\mathscr{K}_1 = \mathbf{U} + \mathbf{R}_t\,\mathbf{Q}_{RR}\,\mathbf{G}_l\,\mathbf{Q}'_{RR}, \tag{137a}$$

$$\mathscr{K}_2 = \mathbf{U} + \mathbf{G}_l\,\mathbf{Q}'_{RR}\,\mathbf{R}_t\,\mathbf{Q}_{RR}. \tag{137b}$$

Define

$$\mathbf{G} = \mathbf{G}_t + \mathbf{Q}_{RR}\,\mathbf{G}_l\,\mathbf{Q}'_{RR}, \tag{138a}$$

$$\mathbf{R} = \mathbf{R}_l + \mathbf{Q}'_{RR}\,\mathbf{R}_t\,\mathbf{Q}_{RR}, \tag{138b}$$

so that $\mathscr{K}_1 = \mathbf{R}_t\,\mathbf{G}$ and $\mathscr{K}_2 = \mathbf{G}_l\,\mathbf{R}$. Thus $\mathscr{K}_1$ and $\mathscr{K}_2$ will be nonsingular if $\mathbf{G}$ and $\mathbf{R}$ are nonsingular. We shall shortly show that $\mathbf{R}$ and $\mathbf{G}$ can be interpreted as loop-and-node parameter matrices and are consequently nonsingular. Accepting this fact here we conclude that a solution for (136) always exists. We now solve this equation for $\mathbf{i}_{Rl}$ and $\mathbf{v}_{Rt}$ and substitute into (119) and (124). The details of the process are tedious and will not be given here. The result will be

$$\frac{d}{dt}\begin{bmatrix} \mathscr{C} & \mathbf{0} \\ \mathbf{0} & \mathscr{L} \end{bmatrix}\begin{bmatrix} \mathbf{v}_{Ct} \\ \mathbf{i}_{Ll} \end{bmatrix} = \begin{bmatrix} -\mathscr{Y} & \mathscr{H} \\ \mathscr{G} & -\mathscr{L} \end{bmatrix}\begin{bmatrix} \mathbf{v}_{Ct} \\ \mathbf{i}_{Ll} \end{bmatrix}$$
$$+ \begin{bmatrix} -\widehat{\mathscr{Y}} & \widehat{\mathscr{H}} \\ \widehat{\mathscr{G}} & -\widehat{\mathscr{L}} \end{bmatrix}\begin{bmatrix} \mathbf{v}_E \\ \mathbf{i}_J \end{bmatrix} + \frac{d}{dt}\begin{bmatrix} \widehat{\mathscr{C}} & \mathbf{0} \\ \mathbf{0} & \widehat{\mathscr{L}} \end{bmatrix}\begin{bmatrix} \mathbf{v}_E \\ \mathbf{i}_J \end{bmatrix}, \tag{139}$$

where

$$\mathscr{Y} = \mathbf{Q}_{CR}\,\mathbf{R}^{-1}\mathbf{Q}'_{CR}, \quad \mathscr{H} = -\mathbf{Q}_{CL} + \mathbf{Q}_{CR}\,\mathbf{R}^{-1}\mathbf{Q}'_{RR}\,\mathbf{R}_t\,\mathbf{Q}_{RL},$$
$$\mathscr{L} = \mathbf{Q}'_{RL}\,\mathbf{G}^{-1}\mathbf{Q}_{RL}, \quad \mathscr{G} = \mathbf{Q}'_{CL} - \mathbf{Q}'_{RL}\,\mathbf{G}^{-1}\mathbf{Q}_{RR}\,\mathbf{G}_l\,\mathbf{Q}'_{CR} = -\mathscr{H}' \tag{140}$$

and

$$\widehat{\mathscr{Y}} = \mathbf{Q}_{CR}\,\mathbf{R}^{-1}\mathbf{Q}_{ER}, \quad \widehat{\mathscr{H}} = -\mathbf{Q}_{CJ} + \dot{\mathbf{Q}}_{CR}\,\mathbf{R}^{-1}\mathbf{Q}'_{RR}\,\mathbf{R}_t\,\mathbf{Q}_{RJ},$$
$$\widehat{\mathscr{L}} = \mathbf{Q}'_{RL}\,\mathbf{G}^{-1}\mathbf{Q}_{RJ}, \quad \widehat{\mathscr{G}} = \mathbf{Q}'_{EL} - \mathbf{Q}'_{RL}\,\mathbf{G}^{-1}\,\mathbf{Q}_{RR}\,\mathbf{G}_l\,\mathbf{Q}'_{ER}. \tag{141}$$

Note that the matrix $\mathscr{G}$ in the case of the reciprocal networks under consideration is the negative transpose of $\mathscr{H}$, which is something we would expect. The form of (139) is the same as that of (129) for the general network. The difference in the present case is that we have explicit expressions for the coefficient matrices of the state and source vectors.

Now, in a time-invariant RLC network, $\mathscr{C}$ and $\mathscr{L}$ will be time-invariant, diagonal matrices. Therefore, their inverses exist and (139) can be rewritten in the desired form

$$\frac{d}{dt}\begin{bmatrix} \mathbf{v}_{Ct} \\ \mathbf{i}_{Ll} \end{bmatrix} = \mathscr{A}\begin{bmatrix} \mathbf{v}_{Ct} \\ \mathbf{i}_{Ll} \end{bmatrix} + \mathscr{B}_1\begin{bmatrix} \mathbf{v}_E \\ \mathbf{i}_J \end{bmatrix} + \mathscr{B}_2\frac{d}{dt}\begin{bmatrix} \mathbf{v}_E \\ \mathbf{i}_J \end{bmatrix}. \qquad (142)$$

where $\mathscr{A}$, $\mathscr{B}_1$, and $\mathscr{B}_2$ are as indicated in (132), (133), and (134).

The innocent-looking simplicity of this final equation masks the extensive matrix operations that go to make up the $\mathscr{A}$, $\mathscr{B}_1$, and $\mathscr{B}_2$ matrices. For ease of reference and as an aid to the memory, the essential results are summarized in Table 2.

PARAMETER MATRICES FOR RCL NETWORKS

In arriving at the final equation, a number of matrices such as $\mathscr{C}$, $\mathbf{R}$, and $\mathscr{Y}$ were introduced for notational simplicity. It is possible to give rather simple interpretations for these matrices—which we shall now outline.

First consider the parameter matrices $\mathbf{R}$, $\mathbf{G}$, $\mathscr{C}$, $\mathscr{L}$. Although we are here dealing with the state equations, let us temporarily switch our attention to the loop-impedance matrix $\mathbf{Z}_m = \mathbf{BZB}'$, where $\mathbf{Z}$ is the branch-impedance matrix formed after removing the sources—replacing v-sources by short circuits and replacing i-sources by open circuits. Let us arrange the rows and columns of $\mathbf{Z}$ in the following order: C, R, and L twigs; then L, R, and C links. The branch-impedance matrix can then be written as follows:

$$\mathbf{Z} = \begin{bmatrix} \frac{1}{s}\mathbf{C}_t^{-1} & & & & & \\ & \mathbf{R}_t & & & \bigcirc & \\ & & s\mathbf{L}_{tt} & & & \\ & & & s\mathbf{L}_{ll} & & \\ & \bigcirc & & & \mathbf{R}_l & \\ & & & & & \frac{1}{s}\mathbf{C}_l^{-1} \end{bmatrix}.$$

Next we partition $\mathbf{B}$ in the usual form $[\mathbf{B}_t \quad \mathbf{B}_l]$. Then we further partition $\mathbf{B}_t(=-\mathbf{Q}_l')$ in accordance with the partitioning of $\mathbf{Q}_l$ in (112), keeping

Table 2

RLC Networks	**Topological Relationships**	$\mathbf{Q} = [\mathbf{U} \quad \mathbf{Q}_l] \qquad \mathbf{Q}_l = $ $\begin{array}{c} \text{links} \to \\ \text{twigs} \\ \downarrow \end{array}\quad \begin{array}{c} E \\ C \\ R \\ L \end{array}\begin{array}{cccc} C & R & L & J \\ \begin{bmatrix} \mathbf{Q}_{EC} & \mathbf{Q}_{ER} & \mathbf{Q}_{EL} & \mathbf{Q}_{EJ} \\ \mathbf{Q}_{CC} & \mathbf{Q}_{CR} & \mathbf{Q}_{CL} & \mathbf{Q}_{CJ} \\ \mathbf{0} & \mathbf{Q}_{RR} & \mathbf{Q}_{RL} & \mathbf{Q}_{RJ} \\ \mathbf{0} & \mathbf{0} & \mathbf{Q}_{LL} & \mathbf{Q}_{LJ} \end{bmatrix} \end{array}$
	Voltage-Current Relationships	$\begin{bmatrix} \mathbf{i}_{Ct} \\ \mathbf{i}_{Cl} \end{bmatrix} = \dfrac{d}{dt}\begin{bmatrix} \mathbf{C}_t & \mathbf{0} \\ \mathbf{0} & \mathbf{C}_l \end{bmatrix}\begin{bmatrix} \mathbf{v}_{Ct} \\ \mathbf{v}_{Cl} \end{bmatrix}$ $\begin{bmatrix} \mathbf{v}_{Ll} \\ \mathbf{v}_{Lt} \end{bmatrix} = \dfrac{d}{dt}\begin{bmatrix} \mathbf{L}_{ll} & \mathbf{0} \\ \mathbf{0} & \mathbf{L}_{tt} \end{bmatrix}\begin{bmatrix} \mathbf{i}_{Ll} \\ \mathbf{i}_{Lt} \end{bmatrix}$ $\begin{bmatrix} \mathbf{i}_{Rl} \\ \mathbf{v}_{Rt} \end{bmatrix} = \begin{bmatrix} \mathbf{G}_l & \mathbf{0} \\ \mathbf{0} & \mathbf{R}_t \end{bmatrix}\begin{bmatrix} \mathbf{v}_{Rl} \\ i_{Rt} \end{bmatrix}$
	Parameter Matrices	$\mathscr{C} = \mathbf{C}_t + \mathbf{Q}_{CC}\,\mathbf{C}_l\,\mathbf{Q}'_{CC} \qquad\qquad \widehat{\mathscr{C}} = -\mathbf{Q}_{CC}\,\mathbf{C}_l\,\mathbf{Q}'_{EC}$ $\mathscr{L} = \mathbf{L}_{ll} + \mathbf{Q}'_{LL}\,\mathbf{L}_{tt}\,\mathbf{Q}_{LL} \qquad\quad \mathbf{R} = \mathbf{R}_l + \mathbf{Q}'_{RR}\,\mathbf{R}_t\,\mathbf{Q}_{RR}$ $\widehat{\mathscr{L}} = -\mathbf{Q}'_{LL}\,\mathbf{L}_{tt}\,\mathbf{Q}_{LJ} \qquad\qquad \mathbf{G} = \mathbf{G}_t + \mathbf{Q}_{RR}\,\mathbf{G}_l\,\mathbf{Q}'_{RR}$
	Resistor Multiport Matrices	$\mathscr{Y} = \mathbf{Q}_{CR}\,\mathbf{R}^{-1}\,\mathbf{Q}'_{CR}, \quad \mathscr{H} = -\mathbf{Q}_{CL} + \mathbf{Q}_{CR}\,\mathbf{R}^{-1}\,\mathbf{Q}'_{RR}\,\mathbf{R}_t\,\mathbf{Q}_{RL}$ $\mathscr{Z} = \mathbf{Q}'_{RL}\,\mathbf{G}^{-1}\,\mathbf{Q}_{RL}, \quad \mathscr{G} = \mathbf{Q}'_{CL} - \mathbf{Q}'_{RL}\,\mathbf{G}^{-1}\,\mathbf{Q}_{RR}\,\mathbf{G}_l\,\mathbf{Q}'_{CR} = -\mathscr{H}'$ $\widehat{\mathscr{Y}} = \mathbf{Q}_{CR}\,\mathbf{R}^{-1}\,\mathbf{Q}'_{ER}, \quad \widehat{\mathscr{H}} = -\mathbf{Q}_{CJ} + \mathbf{Q}_{CR}\,\mathbf{R}^{-1}\,\mathbf{Q}'_{RR}\,\mathbf{R}_t\,\mathbf{Q}_{RJ}$ $\widehat{\mathscr{Z}} = \mathbf{Q}'_{RL}\,\mathbf{G}^{-1}\,\mathbf{Q}_{RJ}, \quad \widehat{\mathscr{G}} = \mathbf{Q}'_{EL} - \mathbf{Q}'_{RL}\,\mathbf{G}^{-1}\,\mathbf{Q}_{RR}\,\mathbf{G}_l\,\mathbf{Q}'_{ER}$
	State Equation $\dfrac{dx}{dt} = \mathscr{A}\mathbf{x} + \mathscr{B}_1\mathbf{e} + \mathscr{B}_2\dfrac{de}{dt}$	$\mathbf{x} = \begin{bmatrix} \mathbf{v}_{Ct} \\ \mathbf{i}_{Ll} \end{bmatrix}, \quad \mathbf{e} = \begin{bmatrix} \mathbf{v}_E \\ \mathbf{i}_J \end{bmatrix}, \quad \mathscr{B}_2 = \begin{bmatrix} \mathscr{C}^{-1} & \mathbf{0} \\ \mathbf{0} & \mathscr{L}^{-1} \end{bmatrix}\begin{bmatrix} \widehat{\mathscr{C}} & \mathbf{0} \\ \mathbf{0} & \widehat{\mathscr{L}} \end{bmatrix}$ $\mathscr{A} = \begin{bmatrix} \mathscr{C}^{-1} & \mathbf{0} \\ \mathbf{0} & \mathscr{L}^{-1} \end{bmatrix}\begin{bmatrix} -\mathscr{Y} & \mathscr{H} \\ \mathscr{G} & -\mathscr{Z} \end{bmatrix} \quad \mathscr{B}_1 = \begin{bmatrix} \mathscr{C}^{-1} & \mathbf{0} \\ \mathbf{0} & \mathscr{L}^{-1} \end{bmatrix}\begin{bmatrix} -\widehat{\mathscr{Y}} & \widehat{\mathscr{H}} \\ \widehat{\mathscr{G}} & -\widehat{\mathscr{Z}} \end{bmatrix}$

in mind that the row and column corresponding to sources are now absent. Now, because of the order in which the elements are arranged in $\mathbf{Z}$, $\mathbf{B}_l \neq \mathbf{U}$ but is a rearrangement of the columns of $\mathbf{U}$. Thus the partitioning of $\mathbf{B}$ becomes

$$
\begin{array}{cccccccc}
 & & \text{twigs} & & & \text{links} & & \\
\text{links} & C & R & L & L & R & C & \\
\mathbf{B} = \begin{array}{c} C \\ R \\ L \end{array} & \left[\begin{array}{cccccc}
-\mathbf{Q}'_{CC} & 0 & 0 & 0 & 0 & \mathbf{U} \\
-\mathbf{Q}'_{CR} & -\mathbf{Q}'_{RR} & 0 & 0 & \mathbf{U} & 0 \\
-\mathbf{Q}'_{CL} & -\mathbf{Q}'_{RL} & -\mathbf{Q}'_{LL} & \mathbf{U} & 0 & 0
\end{array}\right].
\end{array}
$$

The loop-impedance matrix is now formed. This will be quite a complicated expression. The details will be left for you to work out. From the loop-impedance matrix, the loop-parameter matrices (resistance and inductance) can be written. When this is done, it is found that—

1. $\mathbf{R} = \mathbf{R}_l + \mathbf{Q}'_{RR}\mathbf{R}_t\mathbf{Q}_{RR}$ is a submatrix of the loop-resistance matrix for the f-loops defined by resistor links for a normal tree, with all sources removed.

2. $\mathscr{L} = \mathbf{L}_{ll} + \mathbf{Q}'_{LL}\mathbf{L}_{tt}\mathbf{Q}_{LL}$ is a submatrix of the loop-inductance matrix for the f-loops defined by inductor links for a normal tree, with all sources removed.

In a completely analogous way, by forming the node-pair admittance matrix $\mathbf{QYQ}'$ and partitioning in the same manner as above, it will be found that—

1. $\mathscr{C} = \mathbf{C}_t + \mathbf{Q}_{CC}\mathbf{C}_l\mathbf{Q}'_{CC}$ is a submatrix of the cut-set capacitance matrix for the f-cut-sets defined by capacitor twigs for a normal tree, with all sources removed.

2. $\mathbf{G} = \mathbf{G}_t + \mathbf{Q}_{RR}\mathbf{G}_l\mathbf{Q}'_{RR}$ is a submatrix of the cut-set conductance matrix for the f-cut-sets defined by conductance twigs for a normal tree, with all sources removed.

With these interpretations it is possible to evaluate these parameter matrices—without going through the extensive matrix multiplications involved—merely by inspection of the network after a normal tree has been selected.

To illustrate, look back at the example in Fig. 4. For this example, $\mathscr{L}$ is trivially given by $[L_9]$. To find $\mathbf{R}$, we note that the resistor links are branches 6, 7, 8. The f-loop defined by link 6, for example, contains R_5

and R_6. There are no resistors common to the f-loops formed by branches 6, 7, and 8. Hence $\mathbf{R}$ is

$$\mathbf{R} = \begin{bmatrix} R_5 + R_6 & 0 & 0 \\ 0 & R_7 & 0 \\ 0 & 0 & R_8 \end{bmatrix}.$$

As for $\mathbf{G}$, there is only one f-cut-set defined by a conductance twig; namely, branch 5. This cut-set contains G_5 and G_6. Hence $\mathbf{G} = (G_5 + G_6)$. Finally, for $\mathscr{C}$ there are two f-cut-sets defined by capacitor twigs (branches 3 and 4) and they contain no other capacitors. Hence

$$\mathscr{C} = \begin{bmatrix} C_3 & 0 \\ 0 & C_4 \end{bmatrix}.$$

This example is too simple to permit a meaningful illustration. Let us find the $\mathscr{C}$ matrix for the example of Fig. 5. Here also there are two f-cut-sets defined by capacitor twigs (branches 3 and 4 again). This time each cut-set contains two capacitors, and C_6 is common to both cut-sets. However, branch 6 is oriented in one way relative to one cut-set and the opposite way relative to the other. Hence the off-diagonal term in the matrix will carry a negative sign. Thus

$$\mathscr{C} = \begin{bmatrix} C_3 + C_6 & -C_6 \\ -C_6 & C_4 + C_6 \end{bmatrix}.$$

This checks with the result back in (24).

To illustrate the preceding development for the writing of a vector state equation, consider the network in Fig. 8. The first task is to find a

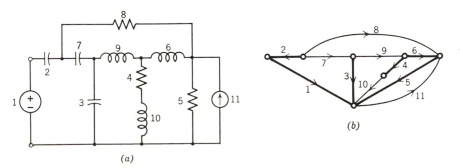

(a)

(b)

Fig. 8.   Illustrative example for state equations.

normal tree. There are a total of six reactive elements, but there is an all-C loop (including a voltage source) and an all-L cut-set. Hence the order of complexity will be 4, so there will be four state variables. One of the capacitors must be a link; and one of the inductors, a twig. A possible normal tree is shown in heavy lines in Fig. 8b. The branches are numbered according to the scheme: twig v-source, C, R, and L; then link C, R, L, and i-source. For purposes of simplification we will assume that the branch numbers are also the numerical values of the elements: resistance in ohms, inductance in henries, and capacitance in farads. With this choice of normal tree, the branch parameters are

$$\mathbf{C}_t = \begin{bmatrix} 2 & 0 \\ 0 & 3 \end{bmatrix}, \quad \mathbf{G}_t = \begin{bmatrix} \frac{1}{4} & 0 \\ 0 & \frac{1}{5} \end{bmatrix},$$

$$\mathbf{C}_l = [7], \quad\quad \mathbf{R}_l = [8],$$

$$\mathbf{L}_{ll} = \begin{bmatrix} 9 & 0 \\ 0 & 10 \end{bmatrix}, \quad \mathbf{L}_{tt} = [6],$$

The next step is to write the $\mathbf{Q}$ matrix and to partition it appropriately:

links→	E	C		R		L	C	R	L		J
twigs	1	2	3	4	5	6	7	8	9	10	11

$$\mathbf{Q} =$$

		E	C		R		L	C	R	L		J
E	1	1	0	0	0	0	0	1	1	0	0	0
C	2	0	1	0	0	0	0	1	1	0	0	0
	3	0	0	1	0	0	0	−1	0	1	0	0
R	4	0	0	0	1	0	0	0	0	0	−1	0
	5	0	0	0	0	1	0	0	−1	−1	1	−1
L	6	0	0	0	0	0	1	0	0	−1	1	0

The various submatrices are evident from the partitioning. The parameter matrices are now computed, as follows:

$$\mathscr{C} = \mathbf{C}_t + \mathbf{Q}_{CC}\mathbf{C}_l\mathbf{Q}'_{CC} = \begin{bmatrix} 2 & 0 \\ 0 & 3 \end{bmatrix} + \begin{bmatrix} 1 \\ -1 \end{bmatrix}[7][1 \quad -1] = \begin{bmatrix} 9 & -7 \\ -7 & 10 \end{bmatrix}$$

$$\mathscr{L} = \mathbf{L}_{ll} + \mathbf{Q}'_{LL}\mathbf{L}_{tt}\mathbf{Q}_{LL} = \begin{bmatrix} 9 & 0 \\ 0 & 10 \end{bmatrix} + \begin{bmatrix} -1 \\ 1 \end{bmatrix}[6][-1 \quad 1] = \begin{bmatrix} 15 & -6 \\ -6 & 16 \end{bmatrix},$$

$$\mathbf{R} = \mathbf{R}_l + \mathbf{Q}'_{RR}\mathbf{R}_t\mathbf{Q}_{RR} = [8] + [0 \quad -1]\begin{bmatrix} 4 & 0 \\ 0 & 5 \end{bmatrix}\begin{bmatrix} 0 \\ -1 \end{bmatrix} = [13],$$

$$\mathbf{G} = \mathbf{G}_t + \mathbf{Q}_{RR}\mathbf{G}_l\mathbf{Q}'_{RR} = \begin{bmatrix} \frac{1}{4} & 0 \\ 0 & \frac{1}{5} \end{bmatrix} + \begin{bmatrix} 0 \\ -1 \end{bmatrix}[\frac{1}{8}][0 \quad -1] = \begin{bmatrix} \frac{1}{4} & 0 \\ 0 & \frac{1}{5}+\frac{1}{8} \end{bmatrix}.$$

The next step is to compute the $\mathscr{Y}$, $\mathscr{L}$, and $\mathscr{H}$ matrices:

$$\mathscr{Y} = \mathbf{Q}_{CR}\mathbf{R}^{-1}\mathbf{Q}'_{CR} = \begin{bmatrix} 1 \\ 0 \end{bmatrix}[\frac{1}{13}][1 \quad 0] = \begin{bmatrix} \frac{1}{13} & 0 \\ 0 & 0 \end{bmatrix},$$

$$\mathscr{L} = \mathbf{Q}'_{RL}\mathbf{G}^{-1}\mathbf{Q}_{RL} = \begin{bmatrix} 0 & -1 \\ -1 & 1 \end{bmatrix}\begin{bmatrix} 4 & 0 \\ 0 & \frac{40}{13} \end{bmatrix}\begin{bmatrix} 0 & -1 \\ -1 & 1 \end{bmatrix} = \begin{bmatrix} \frac{40}{13} & -\frac{40}{13} \\ -\frac{40}{13} & \frac{92}{13} \end{bmatrix}$$

$$\mathscr{H} = -\mathbf{Q}_{CL} + \mathbf{Q}_{CR}\mathbf{R}^{-1}\mathbf{Q}'_{RR}\mathbf{R}_t\mathbf{Q}_{RL}$$

$$= -\begin{bmatrix} 0 & 0 \\ 1 & 0 \end{bmatrix} + \begin{bmatrix} 1 \\ 0 \end{bmatrix}[\frac{1}{13}][0 \quad -1]\begin{bmatrix} 4 & 0 \\ 0 & 5 \end{bmatrix}\begin{bmatrix} 0 & -1 \\ -1 & 1 \end{bmatrix} = \begin{bmatrix} \frac{5}{13} & -\frac{5}{13} \\ -1 & 0 \end{bmatrix}.$$

Next we compute the $\hat{\mathscr{Y}}$, $\hat{\mathscr{L}}$, $\hat{\mathscr{H}}$, and $\hat{\mathscr{G}}$ matrices:

$$\hat{\mathscr{Y}} = \mathbf{Q}_{CR}\mathbf{R}^{-1}\mathbf{Q}'_{ER} = \begin{bmatrix} 1 \\ 0 \end{bmatrix}[\frac{1}{13}][1] = \begin{bmatrix} \frac{1}{13} \\ 0 \end{bmatrix},$$

$$\hat{\mathscr{L}} = \mathbf{Q}'_{RL}\mathbf{G}^{-1}\mathbf{Q}_{RJ} = \begin{bmatrix} 0 & -1 \\ -1 & 1 \end{bmatrix}\begin{bmatrix} 4 & 0 \\ 0 & \frac{40}{13} \end{bmatrix}\begin{bmatrix} 0 \\ -1 \end{bmatrix} = \begin{bmatrix} \frac{40}{13} \\ -\frac{40}{13} \end{bmatrix},$$

$$\hat{\mathscr{H}} = \mathbf{Q}_{CR}\mathbf{R}^{-1}\mathbf{Q}'_{RR}\mathbf{R}_t\mathbf{Q}_{RL} = \begin{bmatrix} 1 \\ 0 \end{bmatrix}[\frac{1}{13}][0 \quad -1]\begin{bmatrix} 4 & 0 \\ 0 & 5 \end{bmatrix}\begin{bmatrix} 0 \\ -1 \end{bmatrix} = \begin{bmatrix} \frac{5}{13} \\ 0 \end{bmatrix},$$

$$\hat{\mathscr{G}} = -\mathbf{Q}'_{RL}\mathbf{G}^{-1}\mathbf{Q}_{RR}\mathbf{G}_l\mathbf{Q}'_{ER}$$

$$= -\begin{bmatrix} 0 & -1 \\ -1 & 1 \end{bmatrix}\begin{bmatrix} 4 & 0 \\ 0 & \frac{40}{13} \end{bmatrix}\begin{bmatrix} 0 \\ -1 \end{bmatrix}[\frac{1}{8}][1] = \begin{bmatrix} -\frac{5}{13} \\ \frac{5}{13} \end{bmatrix}.$$

Finally, the terms involving the source derivatives:

$$\hat{\mathscr{C}} = -\mathbf{Q}_{CC}\mathbf{C}_l\mathbf{Q}'_{EC} = -\begin{bmatrix} 1 \\ -1 \end{bmatrix}[7][1] = \begin{bmatrix} -7 \\ 7 \end{bmatrix},$$

$$\hat{\mathscr{L}} = -\mathbf{Q}'_{LL}\mathbf{L}_{tt}\mathbf{Q}_{LJ} = -\begin{bmatrix} -1 \\ 1 \end{bmatrix}[6][0] = \mathbf{0}.$$

When all of this is inserted into (139) and the resulting equation is pre-multiplied by

$$\begin{bmatrix} \mathscr{C} & \mathbf{0} \\ \mathbf{0} & \mathscr{L} \end{bmatrix}^{-1} = \begin{bmatrix} \frac{10}{41} & \frac{7}{41} & 0 & 0 \\ \frac{7}{41} & \frac{9}{41} & 0 & 0 \\ 0 & 0 & \frac{16}{204} & \frac{6}{204} \\ 0 & 0 & \frac{6}{204} & \frac{15}{204} \end{bmatrix}$$

we obtain

$$\frac{d}{dt}\begin{bmatrix} v_2 \\ v_3 \\ i_9 \\ i_{10} \end{bmatrix} = \frac{1}{533}\begin{bmatrix} -10 & 0 & -41 & -50 \\ -7 & 0 & -82 & -35 \\ -10 & -41.6 & 80 & -1.6 \\ 9 & 15.6 & -72 & 228 \end{bmatrix}\begin{bmatrix} v_2 \\ v_3 \\ i_9 \\ i_{10} \end{bmatrix}$$

$$+ \frac{1}{533}\begin{bmatrix} -10 & 50 \\ -7 & 35 \\ -10 & -80 \\ 9 & 72 \end{bmatrix}\begin{bmatrix} v_g \\ i_g \end{bmatrix} + \frac{1}{41}\begin{bmatrix} -21 & 0 \\ 14 & 0 \\ 0 & 0 \\ 0 & 0 \end{bmatrix}\frac{d}{dt}\begin{bmatrix} v_g \\ i_g \end{bmatrix}. \quad (143)$$

This is the state equation. Observe that it is not in normal form. If we set

$$\begin{bmatrix} x_1 \\ x_2 \\ x_3 \\ x_4 \end{bmatrix} = \begin{bmatrix} v_2 \\ v_3 \\ i_9 \\ i_{10} \end{bmatrix} - \frac{1}{41}\begin{bmatrix} -21 & 0 \\ 14 & 0 \\ 0 & 0 \\ 0 & 0 \end{bmatrix}\begin{bmatrix} v_g \\ i_g \end{bmatrix}$$

and substitute it into the above state equation, we obtain the following normal-form state equation:

$$
\frac{d}{dt}\begin{bmatrix} x_1 \\ x_2 \\ x_3 \\ x_4 \end{bmatrix} = \frac{1}{533}\begin{bmatrix} -10 & 0 & -41 & -50 \\ -7 & 0 & -82 & -35 \\ -10 & -41.6 & 80 & -1.6 \\ 9 & 15.6 & -72 & 228 \end{bmatrix}\begin{bmatrix} x_1 \\ x_2 \\ x_3 \\ x_4 \end{bmatrix}
$$

$$
+ \frac{1}{533}\begin{bmatrix} -4.9 & 50 \\ -3.4 & 35 \\ -19.1 & -80 \\ 9.7 & 72 \end{bmatrix}\begin{bmatrix} v_g \\ i_g \end{bmatrix}. \qquad (144)
$$

You should verify this result. The new state variables x_1 and x_2 are linear combinations of a capacitor voltage and the voltage of the source. They cannot be identified on the network diagram as measurable voltages.

In looking over the effort just completed, you may despair at the large amount of work involved. But observe the kinds of mathematical operations that occur. They are largely matrix multiplications and additions. Such operations are easily programmable for a computer, and so the work reduces to writing a convenient program.

Note, in this case, that the parameter matrices could have been written by inspection; for example, $\mathscr{L}$ is the inductance submatrix of f-loops defined by inductor links 9 and 10. Each loop also contains inductor 6 whose orientation coincides with that of the first loop but is opposite to that of the second. Hence

$$
\mathscr{L} = \begin{bmatrix} 6+9 & -6 \\ -6 & 6+10 \end{bmatrix},
$$

which is what we obtained before. You should verify the remaining parameter matrices in this manner.

The equation in (143) contains the derivative of the voltage source. This is why we had to make the change $\mathbf{x} \rightarrow (\mathbf{x} - \mathscr{B}_2\mathbf{e})$ in order to obtain a state equation in normal form. Observe that the presence of the source-voltage derivative is caused by the all-capacitor loop including the voltage source. The current source is not included in an all-L cut set, and so there is no derivative of the source current. Although we carried out the example by evaluating all the matrices previously defined and then inserting into the formulas, for any given problem we could also proceed by actually retracing the steps of the derivation. This may sometimes require less effort than inserting into formulas. You might carry out the solution for this example that way and compare the amount of work.

CONSIDERATIONS IN HANDLING CONTROLLED SOURCES

In writing the state equations for the preceding example we simply evaluated the matrices appearing in the final equations and substituted therein. When dealing with nonpassive, nonreciprocal networks this approach is not possible. Instead it will be necessary to return to a point midway in the development of the equations and to proceed step by step from there. The essential equations are (119), (124), (125), and (126). The steps to be carried out and the differences from the *RLC* case are as follows:

1. Write $\mathbf{Q}$ matrix and partition—same as for *RLC*.
2. Evaluate $\mathscr{C}$ and $\mathscr{L}$ matrices—same as for *RLC*.
3. Write resistor-branch v-i equations as in (125).
4. Form the pair of equations (126) and solve. This is the critical point. If there is no solution, stop.
5. Use solution of (126) to eliminate $\mathbf{v}_{Rt}$ from (119) and $\mathbf{i}_{Rl}$ from (124).

When setting up the normal tree care must be exercised in assigning the resistor branches corresponding to multiterminal devices, as discussed in Chapter 2. In the case of a gyrator both branches must be either twigs or links. For an ideal transformer and a negative converter one branch must be a twig and the other a link. In the case of controlled sources each branch is uniquely assignable either as a twig or a link. Whether a branch (controlling or controlled) is a twig or link depends on which variable is specified. Equation (125) expresses link *currents* and twig *voltages* explicitly. Hence, if the current of a branch is specified, it must be a link—and if its voltage is specified, it must be a twig; for example, take the current controlled voltage source. The controlling quantity is the current in a short circuit; but for a short circuit the voltage is specified (specified to be zero). Hence this branch must be a twig. For the controlled branch it is again the voltage that is specified. Hence this branch must also be a twig.

As an illustration consider the network in Fig. 9. The transformer is ideal. The graph of the network is in two parts because of the transformer. A normal forest must contain four branches, of which two are the voltage source and the capacitor. One of the two branches of the transformer must also be a twig. Both branches of the controlled source must be links, since it is the current that is specified in both cases. The normal forest is shown in heavy lines in Fig. 9b.

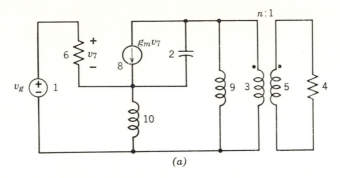

(a)

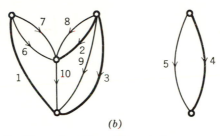

(b)

Fig. 9.   Example with multiterminal components.

The first task is to write the **Q** matrix and to partition it.

$$
\mathbf{Q} = \begin{array}{c}
\\
E \\
\\
C \\
\\
R \\
\\
\end{array}
\left[
\begin{array}{cc:c:cc:cccc:cc}
 & E & C & R & 5 & 6 & 7 & 8 & 9 & 10 \\
\end{array}
\right.
$$

		E	C	R	5	6	7	8	9	10
E		1	0	0 0	0	1	1	0	0	0
C		0	1	0 0	0	1	1	1	0	-1
R		0	0	1 0	0	-1	-1	0	1	1
		0	0	0 1	1	0	0	0	0	0

$$\mathbf{Q}_{ER} = [0 \ \ 1 \ \ 1 \ \ 0] \qquad \mathbf{Q}_{RR} = \begin{bmatrix} 0 & -1 & -1 & 0 \\ 1 & 0 & 0 & 0 \end{bmatrix}$$

$$\mathbf{Q}_{CR} = [0 \ \ 1 \ \ 1 \ \ 1] \qquad \mathbf{Q}_{RL} = \begin{bmatrix} 1 & 1 \\ 0 & 0 \end{bmatrix}$$

$$\mathbf{Q}_{CL} = [0 \ \ -1] \qquad \mathbf{Q}_{EL} = [0 \ \ 0]$$

The *v-i* relationships of the transformer and controlled source are

$$i_5 = -ni_3, \quad i_7 = 0,$$

$$v_3 = nv_5, \quad i_8 = g_m v_7.$$

Using these, the resistor branch *v-i* relations according to (125) become

$$
\begin{bmatrix} i_5 \\ i_6 \\ i_7 \\ i_8 \end{bmatrix} =
\begin{bmatrix} 0 & 0 & 0 & 0 \\ 0 & G_6 & 0 & 0 \\ 0 & 0 & 0 & 0 \\ 0 & 0 & g_m & 0 \end{bmatrix}
\begin{bmatrix} v_5 \\ v_6 \\ v_7 \\ v_8 \end{bmatrix} +
\begin{bmatrix} -n & 0 \\ 0 & 0 \\ 0 & 0 \\ 0 & 0 \end{bmatrix}
\begin{bmatrix} i_3 \\ i_4 \end{bmatrix}
$$

$$
\begin{bmatrix} v_3 \\ v_4 \end{bmatrix} =
\begin{bmatrix} n & 0 & 0 & 0 \\ 0 & 0 & 0 & 0 \end{bmatrix}
\begin{bmatrix} v_5 \\ v_6 \\ v_7 \\ v_8 \end{bmatrix} +
\begin{bmatrix} 0 & 0 \\ 0 & R_4 \end{bmatrix}
\begin{bmatrix} i_3 \\ i_4 \end{bmatrix}.
$$

We now have all the submatrices to insert into (126). These equations become

$$
\begin{bmatrix} 1 & n & n & 0 \\ 0 & 1 & 0 & 0 \\ 0 & 0 & 1 & 0 \\ 0 & 0 & 0 & 1 \end{bmatrix}
\mathbf{i}_{Rl} -
\begin{bmatrix} 0 & 0 \\ -G_6 & 0 \\ 0 & 0 \\ -g_m & 0 \end{bmatrix}
\mathbf{v}_{Rt} =
\begin{bmatrix} 0 \\ G_6 \\ 0 \\ g_m \end{bmatrix}
\mathbf{v}_{Ct} +
\begin{bmatrix} n & n \\ 0 & 0 \\ 0 & 0 \\ 0 & 0 \end{bmatrix}
\mathbf{i}_{Ll} +
\begin{bmatrix} 0 \\ G_6 \\ 0 \\ g_m \end{bmatrix}
\mathbf{v}_E
$$

$$
\begin{bmatrix} 0 & 0 & 0 & 0 \\ R_4 & 0 & 0 & 0 \end{bmatrix}
\mathbf{i}_{Rl} +
\begin{bmatrix} 1 & -n \\ 0 & 1 \end{bmatrix}
\mathbf{v}_{Rt} = \mathbf{0}.
$$

Since the coefficient of $\mathbf{v}_{Rt}$ in the second equation is nonsingular, we can solve for $\mathbf{v}_{Rt}$ in terms of $\mathbf{i}_{Rl}$, insert the solution into the first equation, and then solve for $\mathbf{i}_{Rl}$, from which $\mathbf{v}_{Rt}$ is then determined. The result of these steps is

$$\mathbf{v}_{Rt} = \begin{bmatrix} -nR_4 & 0 & 0 & 0 \\ -R_4 & 0 & 0 & 0 \end{bmatrix} \mathbf{i}_{Rl}$$

$$\begin{bmatrix} 1 & n & n & 0 \\ -nR_4G_6 & 1 & 0 & 0 \\ 0 & 0 & 1 & 0 \\ -nR_4g_m & 0 & 0 & 1 \end{bmatrix} \mathbf{i}_{Rl} = \begin{bmatrix} 0 \\ G_6 \\ 0 \\ g_m \end{bmatrix} \mathbf{v}_{Ct} + \begin{bmatrix} n & n \\ 0 & 0 \\ 0 & 0 \\ 0 & 0 \end{bmatrix} \mathbf{i}_{Ll} + \begin{bmatrix} 0 \\ G_6 \\ 0 \\ g_m \end{bmatrix} \mathbf{v}_E .$$

The matrix on the left is nonsingular, and its determinant is $\Delta = 1 + n^2 R_4 G_6$. Therefore, upon premultiplying by its inverse,

$$\mathbf{i}_{Rl} = \frac{1}{\Delta} \begin{bmatrix} -nG_6 \\ G_6 \\ 0 \\ g_m \end{bmatrix} \mathbf{v}_{Ct} + \frac{1}{\Delta} \begin{bmatrix} n & n \\ n^2R_4G_6 & n^2R_4G_6 \\ 0 & 0 \\ n^2R_4g_m & n^2R_4g_m \end{bmatrix} \mathbf{i}_{Ll} + \frac{1}{\Delta} \begin{bmatrix} -nG_6 \\ G_6 \\ 0 \\ g_m \end{bmatrix} \mathbf{v}_E$$

and hence

$$\mathbf{v}_{Rt} = \frac{1}{\Delta} \begin{bmatrix} n^2R_4G_6 \\ nR_4G_6 \end{bmatrix} \mathbf{v}_{Ct} + \frac{1}{\Delta} \begin{bmatrix} -n^2R_4 & -n^2R_4 \\ -nR_4 & -nR_4 \end{bmatrix} \mathbf{i}_{Ll} + \frac{1}{\Delta} \begin{bmatrix} n^2R_4G_6 \\ nR_4G_6 \end{bmatrix} \mathbf{v}_E .$$

For this example the reactive-element parameter matrices are simple. There is only one capacitor, so $\mathscr{C} = [\dot{C}_2]$ and $\mathscr{C}^{-1} = [1/C_2]$; the $\mathscr{L}$ matrix is the diagonal matrix

$$\mathscr{L} = \begin{bmatrix} L_9 & 0 \\ 0 & L_{10} \end{bmatrix} \quad \text{and} \quad \mathscr{L}^{-1} = \begin{bmatrix} 1/L_9 & 0 \\ 0 & 1/L_{10} \end{bmatrix} .$$

Finally, we substitute into (119) and (124), multiply by $\mathscr{C}^{-1}$ and $\mathscr{L}^{-1}$, respectively, and combine the two into a single equation. To simplify, let $R_4 = kR_6$, so that $R_4 G_6 = k$. The result will be

$$\frac{d}{dt}\begin{bmatrix} v_2 \\ i_9 \\ i_{10} \end{bmatrix} = \frac{1}{1+n^2k}\begin{bmatrix} \dfrac{-(g_m+G_6)}{C_2} & \dfrac{-n^2R_4(g_m+G_6)}{C_2} & \dfrac{1-n^2R_4g_m}{C_2} \\[2ex] \dfrac{n^2k}{L_9} & \dfrac{-n^2R_4}{L_9} & \dfrac{-n^2R_4}{L_9} \\[2ex] \dfrac{-1}{L_{10}} & \dfrac{-n^2R_4}{L_{10}} & \dfrac{-n^2R_4}{L_{10}} \end{bmatrix}\begin{bmatrix} v_2 \\ i_9 \\ i_{10} \end{bmatrix}$$

$$+ \begin{bmatrix} \dfrac{-(g_m+G_6)}{C_2(1+n^2k)} \\[2ex] \dfrac{n^2k}{L_9(1+n^2k)} \\[2ex] \dfrac{n^2k}{L_{10}(1+n^2k)} \end{bmatrix} v_g. \quad (145)$$

In looking over this example you will notice that many of the sub-matrices are *sparse* (meaning that many elements are zero). This results in the need for many operations whose result is zero. It is possible to carry out the same steps with the equations written in scalar form. This obviates the need for writing large numbers of zeros, but sacrifices compactness. You might parallel the steps of the solution with the equations in scalar form to observe the difference.

It should also be noted in this example that $\mathscr{G} \neq -\mathscr{H}'$ because of the presence of g_m; but if $g_m = 0$, then $\mathscr{G}$ will equal $-\mathscr{H}'$.

4.6 MULTIPORT FORMULATION OF STATE EQUATIONS

Let us now turn to an interpretation of the $\mathscr{Y}$, $\mathscr{Z}$, $\mathscr{H}$, and $\mathscr{G}$ matrices. We shall do this for the general case, not just the RLC case. For this purpose look back at (115) and (116), together with the definition of $\mathscr{C}$ and $\hat{\mathscr{C}}$ in (117) and (118), respectively. Combining these equations leads to

$$\mathbf{i}_{Ct} = \frac{d}{dt}(\mathscr{C}\mathbf{v}_{Ct}) - \mathbf{Q}_{CC}\mathbf{i}_{Cl} - \frac{d}{dt}(\hat{\mathscr{C}}\mathbf{v}_E).$$

Similarly, from (120) and (121), together with the definition of $\mathscr{L}$ and $\hat{\mathscr{L}}$ in (122) and (123), respectively, we get

$$\mathbf{v}_{Ll} = \frac{d}{dt}(\mathscr{L}\mathbf{i}_{Lt}) + \mathbf{Q}'_{LL}\mathbf{v}_{Lt} - \frac{d}{dt}(\hat{\mathscr{L}}\mathbf{i}_{J}).$$

These two can be combined into a single matrix equation. Then the derivatives can be eliminated by substituting the state equation of (129), which applies to the general network. The result will be

$$\begin{bmatrix} \mathbf{i}_{Ct} \\ \mathbf{v}_{Ll} \end{bmatrix} = \begin{bmatrix} -\mathscr{Y} & \mathscr{H} \\ \mathscr{G} & -\mathscr{L} \end{bmatrix} \begin{bmatrix} \mathbf{v}_{Ct} \\ \mathbf{i}_{Ll} \end{bmatrix} + \begin{bmatrix} -\hat{\mathscr{Y}} & \hat{\mathscr{H}} \\ \hat{\mathscr{G}} & -\hat{\mathscr{L}} \end{bmatrix} \begin{bmatrix} \mathbf{v}_{E} \\ \mathbf{i}_{J} \end{bmatrix}$$
$$+ \begin{bmatrix} -\mathbf{Q}_{CC} & \mathbf{0} \\ \mathbf{0} & \mathbf{Q}'_{LL} \end{bmatrix} \begin{bmatrix} \mathbf{i}_{Cl} \\ \mathbf{v}_{Lt} \end{bmatrix}. \quad (146)$$

This is a purely algebraic equation relating certain reactive-element voltages and currents, and the source quantities. Note that the derivatives of the sources have also disappeared. An interpretation of the various matrices can be obtained by first considering the network of Fig. 10.

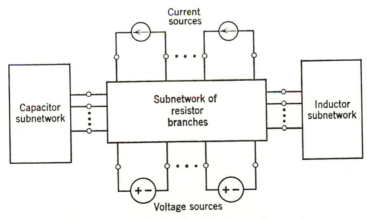

Fig. 10.   Network decomposition into subnetworks.

The overall network is shown as an interconnection of subnetworks. The central subnetwork consists of all the resistor branches (including controlled sources, etc.) to which are connected the subnetworks of capacitors, inductors, and independent sources. The resistor subnetwork can be considered a multiport with as many ports as there are reactive elements and independent sources. Now we must further distinguish between reactive elements that are twigs and those that are links.

This situation can be expressed symbolically as shown in Fig. 11, where

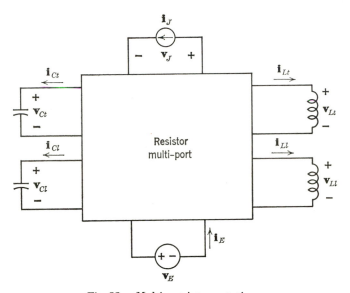

Fig. 11.   Multiport interpretation.

each element represents a subnetwork of its class. Each port variable shown stands for all the scalar variables of its class and so is a vector. The orientations of the variables are consistent with orientations for which the equations were developed. The current orientations are opposite to the standard references for port currents. Thus the power *into* the resistor multiport from any port is $-\mathbf{v}'\mathbf{i}$. The capacitor twigs and links, and the inductor twigs and links are shown as separate subnetworks. Let us reemphasize that, in order to arrive at (146) for the general case, we assume that no further algebraic constraints are introduced by the resistive branches to reduce the order of complexity below what it would be from reactive-element considerations alone.

Suppose we open-circuit all the reactive links and the independent current sources, and short-circuit the voltage sources. This will make $\mathbf{i}_{Cl}$, $\mathbf{i}_{Ll}$, $\mathbf{i}_J$, and $\mathbf{v}_E$ all zero vectors. But if all inductor *link* currents are zero, so also will all inductor *twig* currents be zero by KCL for the all-inductor cut-sets; that is, $\mathbf{i}_{Lt} = 0$. Under these circumstances from (146) we can write

$$-\mathbf{i}_{Ct} = \mathcal{Y}\mathbf{v}_{Ct}. \tag{147}$$

Remembering the opposite orientation of the port currents, we conclude that $\mathcal{Y}$ is the short-circuit admittance matrix of the resistive multiport whose ports are the terminals of the capacitor twigs, when all other reactive elements are open-circuited and all independent sources are removed (meaning that voltage sources are shorted and current sources are opened). This interpretation gives us a means for calculating the matrix $\mathcal{Y}$ for a purely resistive network without going through the formal development of the last section.

The other matrices can be evaluated in a similar manner. To find $\mathcal{Z}$, short-circuit all reactive twigs and the independent voltage sources, and open-circuit the current sources, so that $\mathbf{v}_{Ct}$, $\mathbf{v}_{Lt}$, $\mathbf{v}_{E}$, and $\mathbf{i}_{J}$ are all zero vectors. From (146) we can write

$$\mathbf{v}_{Ll} = \mathcal{Z}(-\mathbf{i}_{Ll}). \tag{148}$$

We conclude that $\mathcal{Z}$ is the open-circuit impedance matrix of the resistive multiport whose ports are the terminals of the inductor links, when all other reactive elements are shorted and all sources are removed.

Matrices $\mathcal{H}$ and $\mathcal{G}$ can be found by similar means. We shall state the results and ask you to supply the details.

$$\mathbf{i}_{Ct} = \mathcal{H}\mathbf{i}_{Ll}, \tag{149a}$$

$$\mathbf{v}_{Ll} = \mathcal{G}\mathbf{v}_{Ct}. \tag{149b}$$

Thus $\mathcal{H}$ is the current-transfer matrix of the resistive multiport, the input ports being the terminals of the inductor links that are replaced by current sources, and the output ports the *shorted* terminals of the capacitor twigs when the capacitor links are open-circuited and all independent sources are removed. The inductor twigs will be replaced by current sources as dictated by KCL at the all-inductor cut-sets.

Finally, $\mathcal{G}$ is the voltage-transfer matrix of the resistive multiport. the input ports being the terminals of the capacitor twigs that are replaced by voltage sources, and the output ports the *open-circuited* terminals of the inductor links, when the inductor twigs are shorted and all independent sources are removed.

To illustrate these interpretations return to the example considered in Fig. 9. We shall carry through the calculations of each of the matrices in turn. To find $\mathcal{Y}$, we are to open-circuit the two inductors, remove the independent voltage source by shorting it, replace the capacitor with a voltage source v_2 having the same polarity as the capacitor voltage, and then find the current i_2 in this voltage source. The appropriate diagram is shown in Fig. 12, where the ideal transformer terminated in R_4 is replaced by a resistor $n^2 R_4$, or $n^2 k R_6$ when, as previously, we set $R_4 = k R_6$. The result is a series connection of R_6 and $n^2 R_4$ across which there is a voltage v_2. The resulting current i is easily found, from which the voltage v_7 follows. The controlled source current is now known. Hence KCL applied at one of the terminals of v_2 yields i_2. The details of the computation are given in Fig. 12.

Calculation of $\mathcal{G}$ Calculation of $\mathcal{Y}$

$$i = \frac{v_2}{R_6 + n^2 R_4} = \frac{G_6 v_2}{1 + n^2 k}$$

$$v_9 = n^2 R_4 i \qquad\qquad v_7 = R_6 i$$

$$v_{10} = -R_6 i \qquad\qquad i_2 = -(g_m v_7 + i) = -(g_m R_6 + 1)i$$

$$\begin{bmatrix} v_9 \\ v_{10} \end{bmatrix} = \begin{bmatrix} \dfrac{n^2 k}{1 + n^2 k} \\ \dfrac{-1}{1 + n^2 k} \end{bmatrix} [v_2] \qquad\qquad -[i_2] = \begin{bmatrix} \dfrac{g_m + G_6}{1 + n^2 k} \end{bmatrix} [v_2]$$

$$\mathcal{G} = \begin{bmatrix} \dfrac{n^2 k}{1 + n^2 k} \\ \dfrac{-1}{1 + n^2 k} \end{bmatrix} \qquad\qquad \mathcal{Y} = \begin{bmatrix} \dfrac{g_m + G_6}{1 + n^2 k} \end{bmatrix}$$

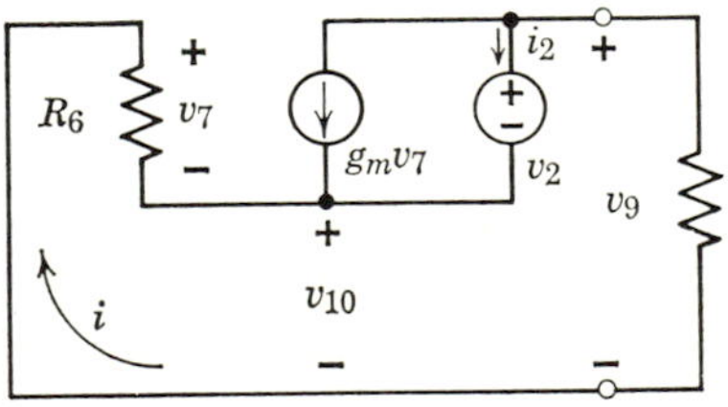

Fig. 12. Computation of $\mathcal{Y}$ and $\mathcal{G}$.

Since there are no inductor twigs, the diagram for the calculation of $\mathcal{G}$ is the same as the one in Fig. 12. However, now the desired outputs are the voltages across the open-circuited inductors. The computation is also shown in Fig. 12.

To find $\mathscr{L}$ we remove the voltage source, short the capacitor, and replace the two inductors with current sources i_9 and i_{10}, with appropriate references. The result, shown in Fig. 13, consists simply of R_6 in parallel with $n^2 R_4$ fed by the sum of i_9 and i_{10}. The voltages v_9 and v_{10} are equal and easily computed. The totally painless computation of $\mathscr{L}$ is shown in Fig. 13.

Finally, for $\mathscr{H}$ the diagram is the same as the one in Fig. 13, except that the desired output quantity is the current in the shorted capacitor. This can be found by applying KCL at the lower terminal of the controlled current source, leading to $i_2 = i_{10} - g_m v_7 - i_6$. But $i_6 = v_7/R_6$, $v_7 = -v_{10}$, and v_{10} was found in the diagram in terms of i_9 and i_{10}. The matrix $\mathscr{H}$ is obtained when all of these are inserted into the expression for i_2. The entire set of calculations is shown in Fig. 13.

Although the discussion of the computation of these matrices for this example appears to be somewhat lengthy, the actual effort involved is very small. (It takes longer to talk about it than to do it.) You should compare the results with the $\mathscr{A}$ matrix in (145) to verify that the same answers have been obtained.

Now let us turn to the submatrices $\hat{\mathfrak{Y}}$, $\hat{\mathscr{L}}$, $\hat{\mathscr{H}}$, and $\hat{\mathscr{G}}$ that make up $\mathscr{B}_1$, the coefficient matrix of the source quantities. Looking again at (146) and Fig. 11, suppose we open all reactive links and the independent current sources, and we short all *reactive* twigs. Under these circumstances (146) yields

$$\mathbf{i}_{Ct} = -\hat{\mathfrak{Y}}\mathbf{v}_E, \tag{150a}$$

$$\mathbf{v}_{Ll} = \hat{\mathscr{G}}\mathbf{v}_E. \tag{150b}$$

Thus $\hat{\mathfrak{Y}}$ is computed by finding the currents in the shorted capacitor twigs, and $\hat{\mathscr{G}}$ is computed by finding the voltages at the open inductor links, both resulting from the independent voltage sources.

Similarly, suppose we short all reactive twigs and the independent voltage sources, and open the *reactive* links. Then (146) yields

$$\mathbf{i}_{Ct} = \hat{\mathscr{H}}\mathbf{i}_J, \tag{151a}$$

$$\mathbf{v}_{Ll} = -\hat{\mathscr{L}}\mathbf{i}_J. \tag{151b}$$

Thus $\hat{\mathscr{H}}$ is computed by finding the currents in the shorted capacitor twigs, and $\hat{\mathscr{L}}$ is computed by finding the voltages across the open inductor links, both resulting from the independent current sources.

Calculation of $\mathscr{Z}$ **Calculation of $\mathscr{H}$**

$$v_9 = v_{10} = \frac{-n^2 R_4 R_6}{R_6 + n^2 R_4}(i_9 + i_{10}) = \frac{-n^2 R_4}{1 + n^2 k}(i_9 + i_{10})$$

$$\begin{bmatrix} v_9 \\ v_{10} \end{bmatrix} = \begin{bmatrix} \dfrac{n^2 R_4}{1+n^2 k} & \dfrac{n^2 R_4}{1+n^2 k} \\[2ex] \dfrac{n^2 R_4}{1+n^2 k} & \dfrac{n^2 R_4}{1+n^2 k} \end{bmatrix} \begin{bmatrix} -i_9 \\ -i_{10} \end{bmatrix}$$

$$\mathscr{Z} = \begin{bmatrix} \dfrac{n^2 R_4}{1+n^2 k} & \dfrac{n^2 R_4}{1+n^2 k} \\[2ex] \dfrac{n^2 R_4}{1+n^2 k} & \dfrac{n^2 R_4}{1+n^2 k} \end{bmatrix}$$

$$i_2 = i_{10} - g_m v_7 - i_6$$

$$i_6 = G_6 v_7$$

$$v_7 = -v_{10}$$

$$i_2 = i_{10} + (g_m + G_6)v_{10}$$

$$[i_2] = \begin{bmatrix} \dfrac{-n^2 R_4(g_m + G_6)}{1+n^2 k} & \dfrac{1 - n^2 R_4 g_m}{1+n^2 k} \end{bmatrix} \begin{bmatrix} i_9 \\ i_{10} \end{bmatrix}$$

$$\mathscr{H} = \begin{bmatrix} \dfrac{-n^2 R_4(g_m + G_6)}{1+n^2 k} & \dfrac{1 - n^2 R_4 g_m}{1+n^2 k} \end{bmatrix}$$

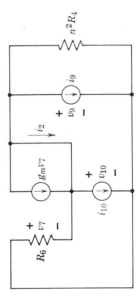

Fig. 13. Computation of $\mathscr{Z}$ and $\mathscr{H}$.

Again we shall illustrate by means of the example in Fig. 9. There are no current sources in this case, so $\widehat{\mathscr{L}}$ and $\widehat{\mathscr{H}}$ are both zero matrices. To find $\widehat{\mathscr{Y}}$ and $\widehat{\mathscr{G}}$, open both inductors and short the capacitor. The result is drawn in Fig. 14. The current i is trivially determined from the series connection of R_6 and n^2R_4. From this v_7 is obtained as $R_6 i$. This deter-

$$i = \frac{v_g}{R_6 + n^2R_4} = \frac{G_6 v_g}{1 + n^2k}$$

$$v_7 = R_6 i = \frac{v_g}{1 + n^2k}$$

$$i_2 = -g_m v_7 - i = -\frac{(g_m + G_6)}{1 + n^2k} v_g$$

$$v_9 = n^2R_4 i = \frac{n^2k}{1 + n^2k} v_g$$

$$v_{10} = -R_6 i + v_g = \frac{n^2k}{1 + n^2k} v_g$$

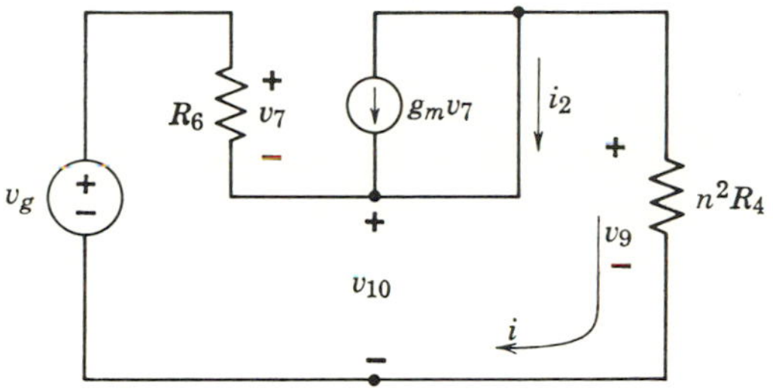

Fig. 14.   Computation of $\widehat{\mathscr{Y}}$ and $\widehat{\mathscr{G}}$.

mines the current of the controlled source, and KCL then gives i_2. The voltages v_9 and v_{10} are also trivially obtained. The details are shown in Fig. 14, with the result

$$\widehat{\mathscr{Y}} = \frac{g_m + G_6}{[1 + n^2k^2]}, \quad \widehat{\mathscr{G}} = \begin{bmatrix} \dfrac{n^2k}{1 + n^2k} \\[2mm] \dfrac{n^2k}{1 + n^2k} \end{bmatrix}.$$

Comparison with (145) shows agreement.

Let us see what we have to show for our efforts. Except for the terms containing source derivatives, we have been able to obtain the vector state equation completely from computations on a resistive multiport network, together with the simple evaluation of $\mathscr{C}$ and $\mathscr{L}$ as submatrices of the cut-set capacitance matrix and loop-inductance matrix. We have not obtained the terms contributed by the source derivatives when there are degeneracies. A glance back at (118) and (123) shows that these terms are rather simply given anyway, compared with the multiport matrices. To be accurate in stating that the state equations have been obtained by a series of calculations for port relations alone, however, we must assume that no source-derivative terms will appear. This means that we are assuming the network will have no voltage sources in an all-capacitor loop and no current sources in an all-inductor cut-set. In terms of the submatrices of $\mathbf{Q}_l$, we are assuming that $\mathbf{Q}_{EC} = \mathbf{0}$ and $\mathbf{Q}_{LJ} = \mathbf{0}$.

OUTPUT EQUATIONS

Having found an approach to determining the vector state equation by simple resistive multiport computations, let us turn our attention to similar calculations to find any variable that may be chosen as an output. Look again at Fig. 11. The port variables there encompass all possible output variables. For the reactive-element and source-element voltages and currents this statement is clear. The only other unknown branch variables are the resistive branch variables. Now current-source voltages and voltage-source currents can actually account for *any* resistor-branch variables for the following reasons. Any voltage variable of a resistor branch can be made the voltage across a current source simply by attaching a current source of zero value across those points. Similarly, any resistor-branch current can be considered the current in a voltage source simply by placing a voltage source of zero value in series with the branch. Of course, doing this increases the effort introduced, but it permits ease of interpretation.

Continue looking at Fig. 11. Now $\mathbf{v}_{Ct}$ and $\mathbf{i}_{Ll}$ are the state variables. The term $\mathbf{v}_{Cl}$ can be expressed in terms of $\mathbf{v}_{Ct}$ by KVL around the all-capacitor loops. Similarly, $\mathbf{i}_{Lt}$ can be expressed in terms of $\mathbf{i}_{Ll}$ by KCL at the all-inductor cut-sets. This leaves the following sets of variables:

$$\mathbf{i}_{Cl} \quad \mathbf{i}_{Ct} \quad \mathbf{i}_E$$

$$\mathbf{v}_{Lt} \quad \mathbf{v}_{Ll} \quad \mathbf{v}_J$$

Consider the first set. Into the *v-i* relation $\mathbf{i}_{Cl} = \mathbf{C}_l \, d\mathbf{v}_{Cl}/dt$ insert the KVL equation in (113e). Remembering the assumption that $\mathbf{Q}_{EC} = \mathbf{0}$, we obtain $\mathbf{i}_{Cl} = \mathbf{C}_l \mathbf{Q}'_{CC} \, d\mathbf{v}_{Ct}/dt$. But $d\mathbf{v}_{Ct}/dt$ is expressed in terms of state variables and sources by the state equation. When this is substituted from (129) there results

$$\mathbf{i}_{Cl} = \mathbf{C}_l \mathbf{Q}'_{CC} \mathscr{C}^{-1} \left\{ [-\mathscr{Y} \quad \mathscr{H}] \begin{bmatrix} \mathbf{v}_{Ct} \\ \mathbf{i}_{Ll} \end{bmatrix} + [-\hat{\mathscr{Y}} \quad \hat{\mathscr{H}}] \begin{bmatrix} \mathbf{v}_E \\ \mathbf{i}_J \end{bmatrix} \right\}. \quad (152a)$$

A similar development for $\mathbf{v}_{Lt}$ gives

$$\mathbf{v}_{Lt} = (\mathbf{L}_{tl} - \mathbf{L}_{tt}\mathbf{Q}_{LL})\mathscr{L}^{-1}\left\{ [\mathscr{G} \quad -\mathscr{L}]\begin{bmatrix} \mathbf{v}_{Ct} \\ \mathbf{i}_{Ll} \end{bmatrix} + [\hat{\mathscr{G}} \quad -\hat{\mathscr{L}}]\begin{bmatrix} \mathbf{v}_E \\ \mathbf{i}_J \end{bmatrix}\right\}.$$

(152b)

Thus both capacitor link voltages and inductor twig currents can be expressed as output variables in terms of state variables and source variables. Aside from the matrices already found when forming the state equations (i.e., $\mathscr{L}$, $\mathscr{C}$, $\mathscr{Y}$, and $\mathscr{H}$), we require a knowledge of the topological matrices $\mathbf{Q}_{CC}$ and $\mathbf{Q}_{LL}$, and the parameter matrices $\mathbf{C}_l$, $\mathbf{L}_{tt}$, and $\mathbf{L}_{tl}$.

As for $\mathbf{i}_{Ct}$ and $\mathbf{v}_{Ll}$, which are the variables complementary to the state variables, we already have (146). This, however, is not exactly in the desired form for output equations because of the term containing $\mathbf{i}_{Cl}$ and $\mathbf{v}_{Lt}$. However, (152) can be substituted for these, so that (146) can be rewritten as

$$\mathbf{i}_{Ct} = \mathbf{C}_t\mathscr{C}^{-1}\left\{[-\mathscr{Y} \quad \mathscr{H}]\begin{bmatrix} \mathbf{v}_{Ct} \\ \mathbf{i}_{Ll} \end{bmatrix} + [-\hat{\mathscr{Y}} \quad \hat{\mathscr{H}}]\begin{bmatrix} \mathbf{v}_E \\ \mathbf{i}_J \end{bmatrix}\right\},$$

(153a)

$$\mathbf{v}_{Ll} = (\mathbf{L}_{ll} - \mathbf{L}_{lt}\mathbf{Q}_{LL})\mathscr{L}^{-1}\left\{[\mathscr{G} \quad -\mathscr{L}]\begin{bmatrix} \mathbf{v}_{Ct} \\ \mathbf{i}_{Ll} \end{bmatrix} + [\hat{\mathscr{G}} \quad -\hat{\mathscr{L}}]\begin{bmatrix} \mathbf{v}_E \\ \mathbf{i}_J \end{bmatrix}\right\}.$$

(153b)

We observe that any reactive-component voltage or current variable can be written as an output in the standard form in terms of state variables and sources. Except for the matrices $\mathbf{C}_t$ and $\mathbf{L}_{ll} - \mathbf{L}_{lt}\mathbf{Q}_{LL}$, this is done in terms of matrices already found in writing the state equations.

This leaves the output variables $\mathbf{v}_J$ and $\mathbf{i}_E$. Recall that the topological equations expressing these variables in (113a) and (113h) had not been used in arriving at the state equations. When the solutions for $\mathbf{v}_{Rt}$ and $\mathbf{i}_{Rl}$ of (126) are inserted into these equations, the result will be in terms of state variables and sources. It will have the following form:

$$\begin{bmatrix} \mathbf{i}_E \\ \mathbf{v}_J \end{bmatrix} = \begin{bmatrix} -\overline{\mathscr{Y}} & \overline{\mathscr{H}} \\ \overline{\mathscr{G}} & -\overline{\mathscr{L}} \end{bmatrix}\begin{bmatrix} \mathbf{v}_{Ct} \\ \mathbf{i}_{Lt} \end{bmatrix} + \begin{bmatrix} -\overline{\overline{\mathscr{Y}}} & \overline{\overline{\mathscr{H}}} \\ \overline{\overline{\mathscr{G}}} & -\overline{\overline{\mathscr{L}}} \end{bmatrix}\begin{bmatrix} \mathbf{v}_E \\ \mathbf{i}_J \end{bmatrix}.$$

(154)

The interpretation of the matrices in these expressions can be obtained in terms of the multiport network in Fig. 11 in the same way as shown earlier; for example, $\overline{\mathcal{Y}}$ is obtained by opening the inductor and independent current-source links (and hence also the inductor twigs), replacing the capacitor twigs with voltage sources $\mathbf{v}_{Ct}$, shorting the independent voltage sources, and writing an expression relating the currents $(-\mathbf{i}_E)$ in these short circuits to $\mathbf{v}_{Ct}$.

The conclusion of this discussion is the following. By looking upon a network as made up of an interconnection of single-component types of subnetworks as in Fig. 11, we have a way of evaluating those matrices that are coefficients of the state variables and source variables in the state equation. We assume that the network contains no voltage sources in all-capacitor loops and no current sources in all-inductor cut-sets. By a similar approach we can compute the matrices that are coefficients in the output equation, whatever the output variables may be. When the output variables are reactive-element variables, no further calculations are needed on the resistive multiport. The pertinent equations are collected in Table 3.

Table 3

Multiport interpretation

$$\begin{bmatrix} \mathbf{i}_{Ct} \\ \mathbf{v}_{Ll} \end{bmatrix} = \begin{bmatrix} -\mathcal{Y} & \mathcal{H} \\ \mathcal{G} & -\mathcal{L} \end{bmatrix}\begin{bmatrix} \mathbf{v}_{Ct} \\ \mathbf{i}_{Ll} \end{bmatrix} + \begin{bmatrix} -\hat{\mathcal{Y}} & \hat{\mathcal{H}} \\ \mathcal{G} & -\hat{\mathcal{L}} \end{bmatrix}\begin{bmatrix} v_E \\ \mathbf{i}_J \end{bmatrix} + \begin{bmatrix} -\mathbf{Q}_{CC} & \mathbf{0} \\ \mathbf{0} & \mathbf{Q}'_{LL} \end{bmatrix}\begin{bmatrix} \mathbf{i}_{Cl} \\ \mathbf{v}_{Lt} \end{bmatrix}$$

Output equations

$$\begin{bmatrix} \mathbf{i}_{Ct} \\ \mathbf{i}_{Cl} \end{bmatrix} = \begin{bmatrix} \mathbf{C}_t\,\mathcal{C}^{-1} \\ \mathbf{C}_l\,\mathbf{Q}_{CC}\,\mathcal{C}^{-1} \end{bmatrix}\left\{[-\mathcal{Y} \quad \mathcal{H}]\begin{bmatrix} \mathbf{v}_{Ct} \\ \mathbf{i}_{Ll} \end{bmatrix} + [-\hat{\mathcal{Y}} \quad \hat{\mathcal{H}}]\begin{bmatrix} \mathbf{v}_E \\ \mathbf{i}_J \end{bmatrix}\right\}$$

$$\begin{bmatrix} \mathbf{v}_{Ll} \\ \mathbf{v}_{Lt} \end{bmatrix} = \begin{bmatrix} (\mathbf{L}_{ll} - \mathbf{L}_{lt}\,\mathbf{Q}_{LL})\mathcal{L}^{-1} \\ (\mathbf{L}_{tl} - \mathbf{L}_{tt}\,\mathbf{Q}_{LL})\mathcal{L}^{-1} \end{bmatrix}\left\{[\mathcal{G} \quad -\mathcal{L}]\begin{bmatrix} \mathbf{v}_{Ct} \\ \mathbf{i}_{Ll} \end{bmatrix} + [\hat{\mathcal{G}} \quad -\hat{\mathcal{L}}]\begin{bmatrix} \mathbf{v}_E \\ \mathbf{i}_J \end{bmatrix}\right\}$$

$$\begin{bmatrix} \mathbf{i}_E \\ \mathbf{v}_J \end{bmatrix} = \begin{bmatrix} -\overline{\mathcal{Y}} & \overline{\mathcal{H}} \\ \overline{\mathcal{G}} & -\overline{\mathcal{L}} \end{bmatrix}\begin{bmatrix} \mathbf{v}_{Ct} \\ \mathbf{i}_{Ll} \end{bmatrix} + \begin{bmatrix} -\overline{\overline{\mathcal{Y}}} & \overline{\overline{\mathcal{H}}} \\ \overline{\overline{\mathcal{G}}} & -\overline{\overline{\mathcal{L}}} \end{bmatrix}\begin{bmatrix} \mathbf{v}_E \\ \mathbf{i}_J \end{bmatrix}$$

To illustrate, consider the network shown in Fig. 15a. With the usual equivalent circuit for the triode, the result is redrawn in (15b). Let the desired output be the voltage v and the current i_3. Hence we insert a current source across the right-hand terminals, with $i_g = 0$. In the graph this current source is shown by branch 11. Branch 8 is the controlling

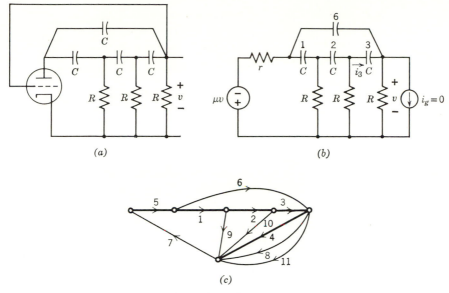

Fig. 15. RC oscillator network.

branch of the controlled source. There is an all-capacitor loop, so one capacitor becomes a link of the normal tree. The *f*-cut-sets defined by capacitor twigs 1, 2, and 3 each contain capacitor 6, and with the same orientation. Hence the $\mathscr{C}$ matrix is easily written as

$$\mathscr{C} = \begin{bmatrix} 2C & C & C \\ C & 2C & C \\ C & C & 2C \end{bmatrix}.$$

Since there are no inductors and voltage sources, (146) and (154) reduce to

$$\mathbf{i}_{Ct} = -\mathscr{Y}\mathbf{v}_{Ct} + \widehat{\mathscr{H}}\mathbf{i}_J - \mathbf{Q}'_{CC}\mathbf{i}_{Cl}, \tag{155a}$$

$$\mathbf{v}_J = \overline{\mathscr{G}}\mathbf{v}_{Ct} - \overline{\overline{\mathscr{Z}}}\mathbf{i}_J. \tag{155b}$$

Thus to find $\mathscr{Y}$ and $\overline{\mathscr{G}}$ we open-circuit the capacitor link 6 and the current source. (In this case the latter step is not needed since the only current source has zero value.) We then replace the capacitor twigs by voltage sources. The resulting network is shown in Fig. 16a. It is a matter of computing the currents i_1, i_2, and i_3 in this resistive network by any

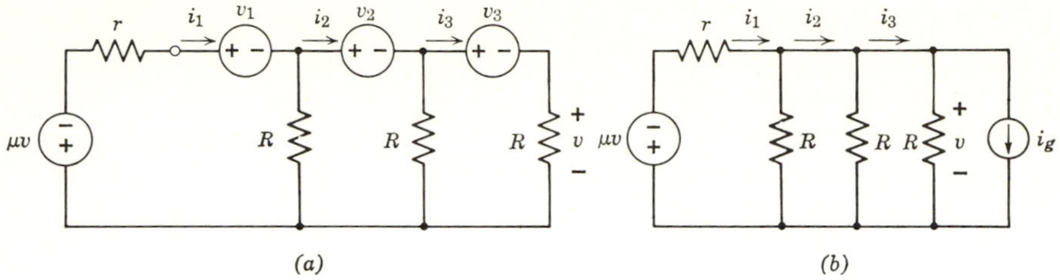

Fig. 16. Resistive network for computing $\mathcal{Y}$.

convenient method. The details of the computation will be left to you. The result is

$$-\mathbf{i}_{Ct} = \begin{bmatrix} -i_1 \\ -i_2 \\ -i_3 \end{bmatrix} = \frac{1}{3r + (\mu+1)R} \begin{bmatrix} 3 & 2-\mu & 1-2\mu \\ 2 & 2+2r/R & 1-\mu+r/R \\ 1 & 1+r/R & 1+2r/Rn \end{bmatrix} \begin{bmatrix} v_1 \\ v_2 \\ v_3 \end{bmatrix}$$
$$= \mathcal{Y}\mathbf{v}_{Ct}.$$

From the diagram it is clear that $\mathbf{v}_J = [v] = [Ri_3]$; so $\overline{\mathcal{G}}$ is also easily found as follows:

$$\mathbf{v}_J = v = \left[\frac{-R}{3r + (\mu+1)R} \quad \frac{-(R+r)}{3r + (\mu+1)R} \quad \frac{-(R+2r)}{3r + (\mu+1)R} \right] \begin{bmatrix} v_1 \\ v_2 \\ v_3 \end{bmatrix} = \overline{\mathcal{G}}\mathbf{v}_{Ct}.$$

For the remaining matrices we continue to open-circuit capacitor link 6, but this time we short-circuit the capacitor twigs. The details of the computation will be left to you. The result will be

$$\mathbf{i}_{Ct} = \begin{bmatrix} i_1 \\ i_2 \\ i_3 \end{bmatrix} = \frac{1}{3r + (\mu+1)R} \begin{bmatrix} (\mu+1)R \\ r + (\mu+1)R \\ 2r + (\mu+1)R \end{bmatrix} i_g = \widehat{\mathcal{H}}i_J,$$

$$\mathbf{v}_J = v = \frac{rR}{3r + (\mu+1)R}(-i_g) = -\overline{\overline{\mathcal{Z}}}i_J.$$

For this example, since the only source is a zero-value current source, there was no need of finding $\widehat{\mathscr{H}}$ and $\overline{\overline{\mathscr{Z}}}$; they will be multiplied by zero in the final equations anyway. We went to the trouble of finding them here only for the purpose of illustration.

We can now easily write the state equation and the output equation. For simplicity, let us use the following numerical values: $C = \frac{1}{4}$; $r = R = 10$; $\mu = 6$. Then

$$\mathscr{C} = \begin{bmatrix} \frac{1}{2} & \frac{1}{4} & \frac{1}{4} \\ \frac{1}{4} & \frac{1}{2} & \frac{1}{4} \\ \frac{1}{4} & \frac{1}{4} & \frac{1}{2} \end{bmatrix}, \quad \mathscr{C}^{-1} = \begin{bmatrix} 3 & -1 & -1 \\ -1 & 3 & -1 \\ -1 & -1 & 3 \end{bmatrix},$$

$$\mathscr{Y} = \tfrac{1}{100} \begin{bmatrix} 3 & -4 & -11 \\ 2 & 4 & -4 \\ 1 & 2 & 3 \end{bmatrix}, \quad \widehat{\mathscr{H}} = \tfrac{1}{100} \begin{bmatrix} 70 \\ 80 \\ 90 \end{bmatrix},$$

$$\overline{\mathscr{G}} = \tfrac{1}{100} [-10 \quad -20 \quad -30], \quad \overline{\overline{\mathscr{Z}}} = [1].$$

The output i_3 is a current in a capacitor twig. Hence, to get the corresponding output equation, we must use (153a), which reduces to $\mathbf{i}_{Ct} = -\mathbf{C}_t \mathscr{C}^{-1} \mathscr{Y} \mathbf{v}_{Ct}$, where $\mathbf{C}_t$ is a diagonal matrix with diagonal elements equal to $C = \frac{1}{4}$. We do not want all of $\mathbf{i}_{Ct}$ but only the third row. Thus the state and output equation will be

$$\frac{d}{dt} \begin{bmatrix} v_1 \\ v_2 \\ v_3 \end{bmatrix} = \mathscr{C}^{-1} \mathscr{Y} \begin{bmatrix} v_1 \\ v_2 \\ v_3 \end{bmatrix} = \tfrac{1}{100} \begin{bmatrix} 6 & -18 & -32 \\ 2 & 14 & -4 \\ -2 & 6 & 24 \end{bmatrix} \begin{bmatrix} v_1 \\ v_2 \\ v_3 \end{bmatrix},$$

$$\mathbf{w} = \begin{bmatrix} v \\ i_3 \end{bmatrix} = \tfrac{1}{100} \begin{bmatrix} -10 & -20 & -30 \\ -\frac{1}{2} & \frac{3}{2} & 6 \end{bmatrix} \begin{bmatrix} v_1 \\ v_2 \\ v_3 \end{bmatrix}.$$

This completes the example.

PROBLEMS

1. For the networks in Fig. P1 determine (a) the number of natural frequencies and (b) the number of nonzero natural frequencies. Verify your answers by considering topological formulas for the determinant of the node admittance matrix or the loop impedance matrix.

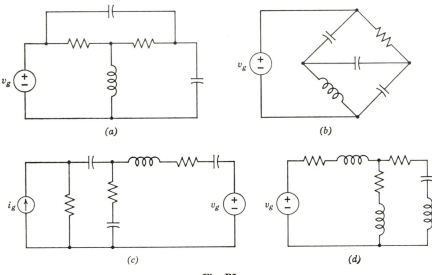

Fig. P1

2. For each of the networks in Fig. P1, draw at least one normal tree.
3. Show at least three normal trees for the network in Fig. 3 in the text.
4. How many normal trees are there in Fig. 5 in the text?
5. Equation 24 relating to Fig. 5 contains the derivative of the source voltages. Make a transformation of variables so that the corresponding equation in the new variables does not contain derivatives of source voltages. Can you interpret the new variables in terms of the network?
6. Assume **A** and **B** are square matrices of the same order. Under what condition is the following relation valid:

$$\epsilon^{A} \epsilon^{B} = \epsilon^{A+B}.$$

Use this result to show that (44) is a valid consequence of (36).

7. Show that each of the following matrices satisfies its characteristic equation:

(a) $\begin{bmatrix} 6 & 3 \\ -5 & 10 \end{bmatrix}$

(b) $\begin{bmatrix} -2 & 1 \\ -9 & 4 \end{bmatrix}$

(c) $\begin{bmatrix} 3 & -2 & -1 \\ 2 & -2 & 2 \\ 1 & 2 & -1 \end{bmatrix}$

(d) $\begin{bmatrix} 3 & -2 & 1 \\ 1 & -1 & 2 \\ -6 & 4 & -2 \end{bmatrix}$

(e) $\begin{bmatrix} 0 & 1 & 0 & 0 \\ 0 & 0 & 1 & 0 \\ 0 & 0 & 0 & 1 \\ -1 & -2 & -2 & -1 \end{bmatrix}$

(f) $\begin{bmatrix} 2 & 0 & 2 \\ 0 & 2 & 1 \\ 0 & 0 & 1 \end{bmatrix}$

8. Using the Cayley-Hamilton theorem, find the inverse of each of those matrices in Problem 7 that is nonsingular.

9. Let $f(s)$ be a polynomial and $d(s)$ the characteristic polynomial of a matrix $\mathscr{A}$. Let $g(s)$ be the remainder polynomial of least degree when $f(s)$ is divided by $d(s)$. In each of the following cases evaluate $f(\mathscr{A})$.

(a) $f(s) = s^6 + 3s^4 + 3s^2$ and $\mathscr{A} = \begin{bmatrix} 0 & 1 \\ 0 & 0 \end{bmatrix}$

(b) $f(s) = s^5 + s^4 + s^3 + s + 1$ and $\mathscr{A} = \begin{bmatrix} -2 & 2 \\ 1 & -3 \end{bmatrix}$

(c) $f(s) = s^4 + s^2 + 1$ and $\mathscr{A} = \begin{bmatrix} 1 & 3 & -4 \\ 0 & -23 & 36 \\ 0 & -18 & 28 \end{bmatrix}$

10. For a given matrix $\mathscr{A}$, observe that $d(s)$ equals the minimal polynomial $m(s)$ if the zeros of $d(s)$ are simple. For each of the following matrices

with simple eigenvalues evaluate $\epsilon^{\mathscr{A}t}$ using the Lagrange interpolation formula:

(a) $\mathscr{A} = \begin{bmatrix} 0 & 1 & 0 \\ 0 & 0 & 1 \\ -24 & -26 & -9 \end{bmatrix}$

(d) $\mathscr{A} = \begin{bmatrix} 0 & 1 & 0 & 1 \\ -1 & 0 & 1 & 0 \\ 0 & 0 & 0 & 1 \\ 0 & 0 & -1 & 0 \end{bmatrix}$

(b) $\mathscr{A} = \begin{bmatrix} 0 & 1 & 0 \\ 0 & 0 & 1 \\ -9 & -15 & -7 \end{bmatrix}$

(e) $\mathscr{A} = \begin{bmatrix} -5 & -5 & -3 \\ 1 & 0 & 1 \\ 2 & 3 & 0 \end{bmatrix}$

(c) $\mathscr{A} = \begin{bmatrix} 2 & 0 & 2 \\ 0 & 2 & 1 \\ 0 & 0 & 1 \end{bmatrix}$

(f) $\mathscr{A} = \begin{bmatrix} -2 & 7 & 6 \\ 0 & 8 & 0 \\ 18 & -3 & 10 \end{bmatrix}$

11. For each of the matrices of Problem 10, set up equations (76) or (82) if the eigenvalues are not simple, and solve them for the g_i coefficients using $f(s) = \epsilon^{st}$. From this determine $\epsilon^{\mathscr{A}t}$ and compare with the previous result.

12. For each of the matrices of Problem 10, use the resolving matrix algorithm to evaluate $[s\mathbf{U} - \mathscr{A}]^{-1}$. Then make a partial-fraction expansion of $[s\mathbf{U} - \mathscr{A}]^{-1}$ to determine the constituent matrices of $\mathscr{A}$.

13. For each of the matrices of Problem 10, determine the constituent matrices by the method of resolving polynomials, using $f_i(s) = s^{i-1}$.

14. For each of the matrices of Problem 10, determine the constituent matrices by the method of resolving polynomials using the set of polynomials in (106).

15. Evaluate the following matrix functions:

(a) $\ln \mathscr{A}, \quad \mathscr{A} = \begin{bmatrix} 1 & 2 & 0 \\ -2 & 1 & 0 \\ 1 & 0 & -2 \end{bmatrix}$

(b) $\sin \mathscr{A}t, \quad \mathscr{A} = \begin{bmatrix} 2 & 3 & 1 \\ -1 & -1 & -1 \\ 0 & -1 & 1 \end{bmatrix}$

(c) $\quad \cosh \mathscr{A}t, \quad \mathscr{A} = \begin{bmatrix} 1 & 2 \\ 4 & 3 \end{bmatrix}$

16. Solve the following sets of state equations with the state vector evaluated using first (45), then (52), and finally (54) with (49) and (59). In each case evaluate $\epsilon^{\mathscr{A}t}$ by first finding the constituent matrices of $\mathscr{A}$ and then applying (84).

(a) $\quad \dfrac{d}{dt} \begin{bmatrix} x_1 \\ x_2 \end{bmatrix} = \begin{bmatrix} 4 & 3 \\ -2 & -1 \end{bmatrix} \begin{bmatrix} x_1 \\ x_2 \end{bmatrix} + \begin{bmatrix} 1 & 0 & 1 \\ 0 & -1 & 1 \end{bmatrix} \begin{bmatrix} \epsilon^{-t} \\ 2 \\ t \end{bmatrix}$

$\begin{bmatrix} w_1 \\ w_2 \end{bmatrix} = \begin{bmatrix} 1 & 0 \\ -1 & 2 \end{bmatrix} \begin{bmatrix} x_1 \\ x_2 \end{bmatrix} + \begin{bmatrix} 1 & 0 & 1 \\ -1 & 2 & 0 \end{bmatrix} \begin{bmatrix} \epsilon^{-t} \\ 2 \\ t \end{bmatrix}$

$\begin{bmatrix} x_1(t_0) \\ x_2(t_0) \end{bmatrix} = \begin{bmatrix} 2 \\ -3 \end{bmatrix}$

(b) $\quad \dfrac{d}{dt} \begin{bmatrix} x_1 \\ x_2 \\ x_3 \end{bmatrix} = \begin{bmatrix} -1 & 0 & 1 \\ 2 & -3 & 2 \\ 0 & 0 & -4 \end{bmatrix} \begin{bmatrix} x_1 \\ x_2 \\ x_3 \end{bmatrix} + \begin{bmatrix} 1 & 2 \\ -2 & 3 \\ 1 & 0 \end{bmatrix} \begin{bmatrix} -1 \\ 3 \end{bmatrix}$

$\begin{bmatrix} w_1 \\ w_2 \end{bmatrix} = \begin{bmatrix} -1 & 2 & -1 \\ -2 & -3 & 0 \end{bmatrix} \begin{bmatrix} x_1 \\ x_2 \end{bmatrix} + \begin{bmatrix} 1 & -1 \\ -1 & 2 \end{bmatrix} \begin{bmatrix} -1 \\ 3 \end{bmatrix}$

$\begin{bmatrix} x_1(t_0) \\ x_2(t_0) \\ x_3(t_0) \end{bmatrix} = \begin{bmatrix} 0 \\ -3 \\ 1 \end{bmatrix}$

(c) $\quad \dfrac{d}{dt} \begin{bmatrix} x_1 \\ x_2 \end{bmatrix} = \begin{bmatrix} -2 & 2 \\ 0 & -3 \end{bmatrix} \begin{bmatrix} x_1 \\ x_2 \end{bmatrix} + \begin{bmatrix} 1 & 0 \\ 2 & -1 \end{bmatrix} \begin{bmatrix} t\epsilon^{-t} \\ \epsilon^{-2t} \end{bmatrix}$

$[w_1] = [1 \quad 0] \begin{bmatrix} x_1 \\ x_2 \end{bmatrix} + [2 \quad -2] \begin{bmatrix} t\epsilon^{-t} \\ \epsilon^{-2t} \end{bmatrix} + [0 \quad -2] \dfrac{d}{dt} \begin{bmatrix} t\epsilon^{-t} \\ \epsilon^{-2t} \end{bmatrix}$

$\begin{bmatrix} x_1(t_0) \\ x_2(t_0) \end{bmatrix} = \begin{bmatrix} 4 \\ 0 \end{bmatrix}$

(d)
$$\frac{d}{dt}\begin{bmatrix} x_1 \\ x_2 \\ x_3 \end{bmatrix} = \begin{bmatrix} 0 & 1 & 1 \\ -1 & 0 & 1 \\ 0 & 0 & -1 \end{bmatrix}\begin{bmatrix} x_1 \\ x_2 \\ x_3 \end{bmatrix} + \begin{bmatrix} 0 \\ 0 \\ 2 \end{bmatrix}[\sin t]$$

$$\begin{bmatrix} w_1 \\ w_2 \end{bmatrix} = \begin{bmatrix} 1 & 0 & 1 \\ 0 & 1 & -1 \end{bmatrix}\begin{bmatrix} x_1 \\ x_2 \\ x_3 \end{bmatrix}$$

$$\begin{bmatrix} x_1(0) \\ x_2(0) \\ x_3(0) \end{bmatrix} = \begin{bmatrix} 1 \\ 0 \\ -2 \end{bmatrix}$$

(e)
$$\frac{d}{dt}\begin{bmatrix} x_1 \\ x_2 \end{bmatrix} = \begin{bmatrix} 18 & -20 \\ 15 & -17 \end{bmatrix}\begin{bmatrix} x_1 \\ x_2 \end{bmatrix} + \begin{bmatrix} 1 & 0 & 2 \\ -2 & -1 & 0 \end{bmatrix}\begin{bmatrix} te^{-t} \\ 1-\epsilon^{-t} \\ \epsilon^{-2t} \end{bmatrix}$$

$$[w_1] = [-3 \quad 2]\begin{bmatrix} x_1 \\ x_2 \end{bmatrix} + \frac{d}{dt}[1 \quad 0 \quad -2]\begin{bmatrix} te^{-t} \\ 1-\epsilon^{-t} \\ \epsilon^{-2t} \end{bmatrix}$$

$$\begin{bmatrix} x_1(0) \\ x_2(0) \end{bmatrix} = \begin{bmatrix} -4 \\ -2 \end{bmatrix}$$

(f)
$$\frac{d}{dt}\begin{bmatrix} x_1 \\ x_2 \\ x_3 \end{bmatrix} = \begin{bmatrix} 0 & 5 & 4 \\ 0 & 8 & 0 \\ 12 & -3 & 8 \end{bmatrix}\begin{bmatrix} x_1 \\ x_2 \\ x_3 \end{bmatrix} + \begin{bmatrix} 1 & 0 \\ 0 & 2 \\ -1 & -2 \end{bmatrix}\begin{bmatrix} \sin t \\ \cos t \end{bmatrix}$$

$$\begin{bmatrix} w_1 \\ w_2 \\ w_3 \end{bmatrix} = \begin{bmatrix} 1 & -1 & 0 \\ 0 & 1 & -1 \\ 0 & 0 & 1 \end{bmatrix}\begin{bmatrix} x_1 \\ x_2 \\ x_3 \end{bmatrix} + \begin{bmatrix} 1 & 0 \\ 0 & 1 \\ -1 & -1 \end{bmatrix}\begin{bmatrix} \sin t \\ \cos t \end{bmatrix}$$

$$\begin{bmatrix} x_1(0) \\ x_2(0) \\ x_3(0) \end{bmatrix} = \begin{bmatrix} 0 \\ 1 \\ -1 \end{bmatrix}$$

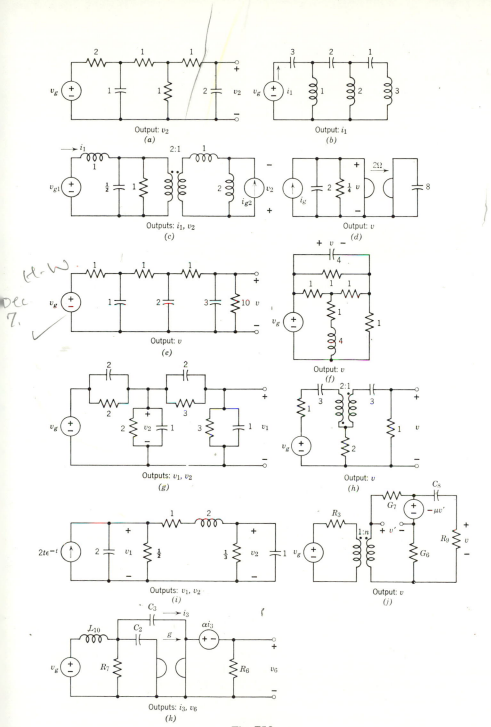

Output: v_2
(a)

Output: i_1
(b)

Outputs: i_1, v_2
(c)

Output: v
(d)

H-W
Dec
7.

Output: v
(e)

Output: v
(f)

Outputs: v_1, v_2
(g)

Output: v
(h)

Outputs: v_1, v_2
(i)

Output: v
(j)

Outputs: i_3, v_6
(k)

Fig. P18

17. If the eigenvalues of $\mathscr{A}$ are different from those of $\mathscr{F}$, then

$$\mathscr{S} = -\sum_{i=1}^{k}\sum_{j=1}^{r_i}\mathbf{K}_{ij}\mathscr{B}\mathscr{K}(\mathscr{F}-s_i\mathbf{U})^{-j}$$

is the unique solution of

$$\mathscr{A}\mathscr{S}-\mathscr{S}\mathscr{F}=\mathscr{B}\mathscr{K},$$

where the $\mathbf{K}_{ij}$ are the constituent matrices and the s_i are the eigenvalues of $\mathscr{A}$. We let r_i denote the multiplicity of s_i and k the number of distinct eigenvalues. In each of the parts of Problem 16 solve for $\mathscr{S}$ by using the above formula, when the eigenvalues of $\mathscr{A}$ and $\mathscr{F}$ are different.

18. For each of the networks in Fig. P18, derive the state equations by (a) the formal matrix approach and (b) the method of resistive multiport parameter evaluation.

19. Derive the state equations for the network in Fig. P19 by both the matrix method and the method of resistive multiport parameter evaluation. For the pentode use the equivalent circuit shown. The response variables are the volages across all the inductors.

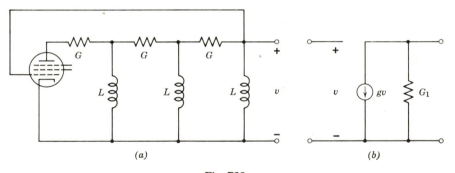

(a)   (b)

Fig. P19

20. Determine the order of complexity of the nonpassive, nonreciprocal networks in Fig. P20.

21. Using the method of Section 4.6, derive the state equations for the networks in Fig. P21. In (a) the ouput is $i_1(t)$; further, choose a normal tree so that all the C capacitors are included. In (b), the outputs are i_1 and v_2; further, choose the normal tree to include L_a.

22. Derive the state equations for the single-stage amplifier shown in Fig. P22a. Use the hybrid-pi-equivalent circuit for the transistor, as shown in Fig. P22b. The voltage v_2 is the network response.

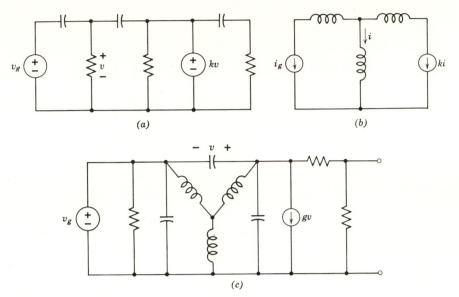

(a)

(b)

(c)

Fig. P20

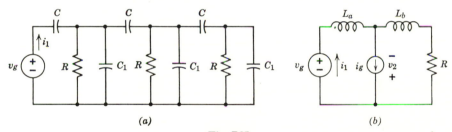

(a)

(b)

Fig. P21

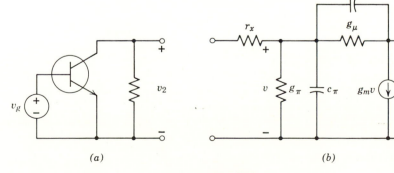

(a)

(b)

Fig. P22

23. Derive the state equations for the network of Fig. P23. Solve for the state-transition matrix. Set up the equation for the solution of the response vector $\mathbf{w} = [v_2 \quad v_3]'$. The capacitors are uncharged at time $t_0 = 0$. Carry out the solution.

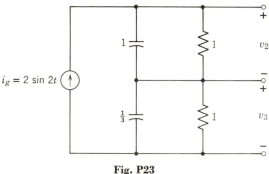

Fig. P23

24. In deriving the general formulation of the state equations from topological considerations, we assumed all branches could be classified as (1) independent sources, (2) capacitive, (3) inductive, or (4) resistive. The general formulation of the state equations can also be derived using the compound branch of Fig. P24a, instead of a capacitive branch; and

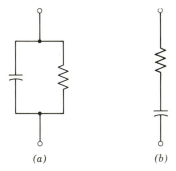

(a)   (b)

Fig. P24

the compound branch of Fig. P24b, instead of an inductive branch. Unaccompanied resistors would continue as branches. Carry through the details of this development. Discuss the advantages and disadvantages of deriving state equations in this manner.

25. Using the method developed in Problem 24, determine the state equations for each of the networks shown in Fig. P25. The output variables for each case are indicated.

26. Derive the state equations for the amplifier shown in Fig. P26a. Use the transistor model shown in Fig. P26b.

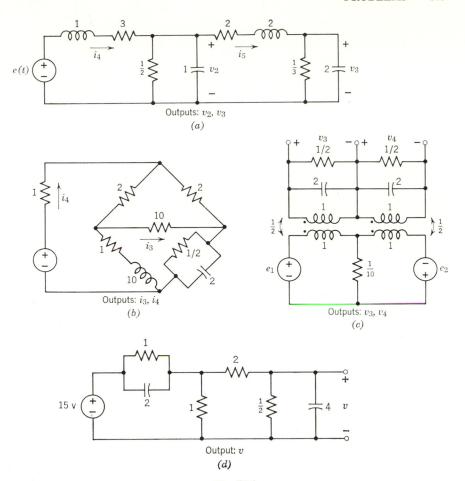

Outputs: v_2, v_3

(a)

Outputs: i_3, i_4

(b)

Outputs: v_3, v_4

(c)

Output: v

(d)

Fig. P25

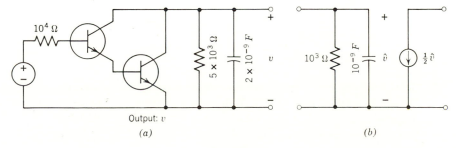

Output: v

(a)

(b)

Fig. P26

27. Derive the state equations for the three-stage amplifier network shown in Fig. P27a. Use the triode model shown in Fig. P27b.

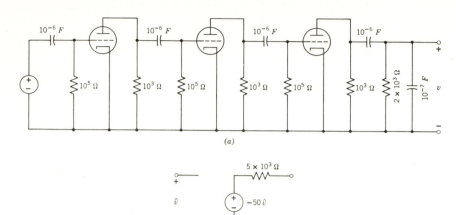

Fig. P27

28. In deriving the state equations in the text, a hybrid-g representation was used for the resistive v-i relationships. It is possible to use another representation instead. Replace the resistive equations (125) in the text with one of the following sets of equations and show how they, together with the Kirchhoff equations, yield the vectors i_{Rl} and v_{Rt} needed in (119) and (124). Explicitly state any conditions needed to guarantee a solution for i_{Rl} and v_{Rt}.

(a) $v_{Rl} = \mathbf{H}_{ll}\, i_{Rl} + \mathbf{H}_{lt}\, v_{Rt}$

   $i_{Rt} = \mathbf{H}_{tl}\, i_{Rl} + \mathbf{H}_{tt}\, v_{Rt}$

(b) $v_{Rl} = \mathbf{Z}_{ll}\, i_{Rl} + \mathbf{Z}_{lt}\, i_{Rt}$

   $v_{Rt} = \mathbf{Z}_{tl}\, i_{Rl} + \mathbf{Z}_{tt}\, i_{Rt}$

(c) $i_{Rl} = \mathbf{Y}_{ll}\, v_{Rl} + \mathbf{Y}_{lt}\, v_{Rt}$

   $i_{Rt} = \mathbf{Y}_{tl}\, v_{Rl} + \mathbf{Y}_{tt}\, v_{Rt}$

(d) $v_{Rl} = \mathbf{A} v_{Rt} - \mathbf{B} i_{Rt}$

   $i_{Rl} = \mathbf{C} v_{Rt} - \mathbf{D} i_{Rt}$

(e) $v_{Rt} = \widehat{\mathbf{A}} v_{Rl} - \widehat{\mathbf{B}} i_{Rl}$

   $i_{Rt} = \widehat{\mathbf{C}} v_{Rl} - \widehat{\mathbf{D}} i_{Rl}$

29. In the networks in Figs. 4 and 5 in the text, solve for i_{Rl} and v_{Rt} using the method based on (125) discussed in the text. Then solve for i_{Rt} and v_{Rt} using one of the other representations in Problem 28. Do some representations require less computational effort? Discuss.

30. Derive the state equations for the network shown in Fig. P30. Use the transistor model shown in Fig. P26*b*.

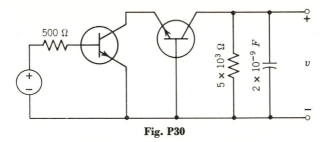

Fig. P30

31. Derive the state equations for the differential amplifier network shown in Fig. P31. Use the transistor model shown in Fig. P26*b*.

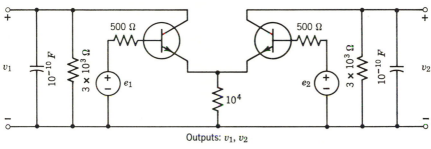

Outputs: v_1, v_2

Fig. P31

32. Derive the state equations for the network shown in Fig. P32 using the transistor model shown in Fig. P26*b*.

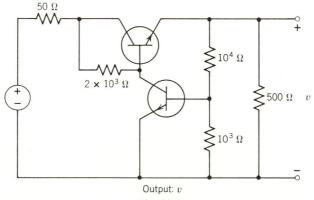

Output: v

Fig. P32

33. Determine the state equation for each of the oscillator networks shown
in Figs. P33*a* through *c* by using the transistor model shown in Fig. P33*d*.

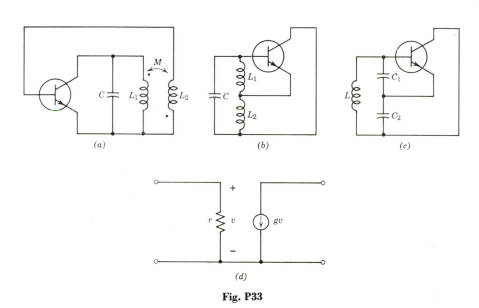

(a)   (b)   (c)

(d)

Fig. P33

34. The network shown in Fig. P34 is a simple *RC* oscillator circuit.
Determine the state equation and indicate for what value of α there will
be two imaginary eigenvalues of $\mathscr{A}$.

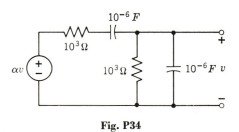

Fig. P34

35. Suppose $\mathscr{A}$ and $\mathscr{F}$ in (57) have no eigenvalues in common. Let $\sum\limits_{i=0}^{n} \alpha_i \lambda^{n-i}$
and $\sum\limits_{i=0}^{m} \beta_i \lambda^{m-i}$ be the characteristic polynomials of $\mathscr{A}$ and $\mathscr{F}$,
respectively.

Let

$$\mathbf{M}_0 = 0$$

$$\mathbf{M}_1 = \mathscr{B}\mathscr{K}$$

$$\mathbf{M}_k = \mathscr{A}\mathbf{M}_{k-1} + \mathbf{M}_{k-1}\mathscr{F} - \mathscr{A}\mathbf{M}_{k-2}\,\mathscr{F}, \qquad k = 2, 3, \cdots$$

Then the solution of (57) is given by either of the following equations:

$$\mathscr{S} = -\left\{\sum_{i=0}^{n-1}\alpha_i \mathbf{M}_{n-i}\right\}\left\{\sum_{i=0}^{n}\alpha_i \mathscr{F}^{n-i}\right\}^{-1}$$

$$\mathscr{S} = \left\{\sum_{i=0}^{m}\beta_i \mathscr{A}^{m-i}\right\}^{-1}\left\{\sum_{i=0}^{m-1}\beta_i \mathbf{M}_{m-i}\right\}$$

In each of the parts of Problem 16, solve for $\mathscr{S}$ using each of the above formulas, when the eigenvalues of $\mathscr{A}$ and $\mathscr{F}$ are different.

36. (a) For the network shown in Fig. P36a, specify the number of natural frequencies and the number of nonzero natural frequencies.

(b) Repeat for the network in Fig. P36b. State whether the values of the natural frequencies (not their number) are the same or different in the two cases. Explain.

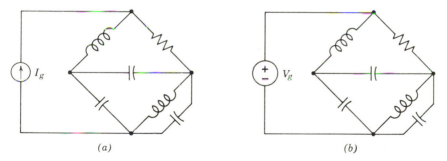

(a)   (b)

Fig. P36

The next six problems involve the preparation of a computer program to help in implementing the solution to some problems. In each case, prepare a program flow chart and a set of program instructions, in some user language such as FORTRAN IV, for a digital computer program to carry out the job specified in the problem. Include a set of user instructions for the program.

37*. Prepare a program to evaluate $(s\mathbf{U} + \mathscr{A})^{-1}$ by the resolvent-matrix algorithm of (98).

38*. Prepare a program to evaluate $\mathscr{A}^{-1}$ by (99).

39*. Prepare a program to identify a normal tree of a connected network when each branch is specified by a sequence of quadruplets of numbers: the first number identifies the branch; the second number identifies its type according to the schedule given below; and the third and fourth numbers identify the nodes to which the branch is attached. The program should also renumber the branches in accordance with the convention in Section 4.5 under "Topological considerations" and provide as output data the sequence of number quadruplets with the new branch number. An example of a typical set of data is given in Fig. P39 for the network shown there.

Number	Type of Branch
1	Independent voltage source
2	Independent current source
3	Capacitor
4	Resistor
5	Inductor

}Branch	}Type	}Nodes
1,	1,	1, 2
2,	3,	1, 5
3,	1,	3, 2
4,	4,	2, 4
5,	3,	3, 6
6,	5,	5, 6
7,	4,	5, 4
8,	3,	5, 6
9,	4,	4, 6

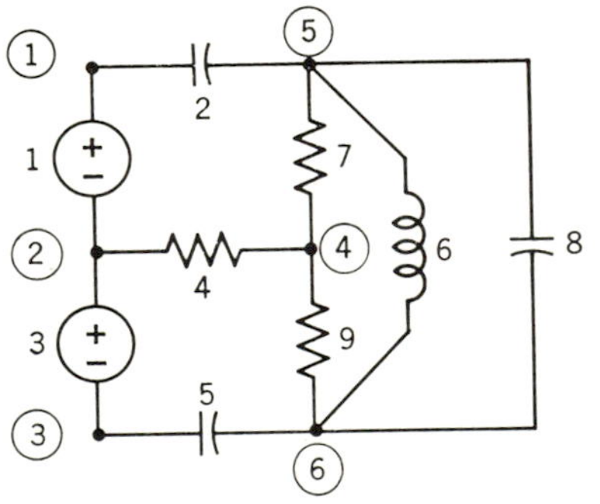

Fig. P39

40*. Prepare a program to determine, first, a reduced node-incidence matrix of a connected network and then an f-cut-set matrix associated with a normal tree. Take as input data the output data from Problem 39*. Assume the third number of each quadruplet of numbers denotes the node to which the tail of the branch is connected; then the node at the head of the branch will be the fourth number.

41*. Prepare a program to determine $\mathscr{A}$, $\mathscr{B}_1$, and $\mathscr{B}_2$ of (142) when the network is

(a) RC with no capacitor links and no resistor twigs

(b)   *RC* with no capacitor links

(c)   *RC* with no resistor twigs

(d)   *RC* (general)

(e)   *RL* with no inductor twigs and no resistor links

(f)   *RL* with no inductor twigs

(g)   *RL* with no resistor links

(h)   *RL* (general)

(i)   *LC* with no capacitor links and no inductor twigs

(j)   *LC* with no capacitor links

(k)   *LC* with no inductor twigs

(l)   *LC* (general)

(m)   *RLC* (general)

The input data are given as a sequence of quintuplets of numbers: the first number identifies the branch, the second number identifies its type according to the schedule of Problem 39*, the third number identifies the node at the branch tail, the fourth number identifies the node at the branch head, and the fifth number is the value of the parameter associated with the branch (zero will be entered for all independent sources). Assume the f-cut-set evaluated by the program of Problem 40* is available.

42*. Combine the programs of Problems 39* and 40* with each of those of problem 41* to create a single program which, starting with the input data of Problem 41*, determines the network state equation for each of the several types of networks listed in Problem 41*.

. 5 .

INTEGRAL SOLUTIONS

We have now reached a level of ability where given any linear, time-invariant network and an arbitrary excitation the complete response can be obtained. The complex-frequency-domain methods of Chapters 2 and 3 are very useful in determining an analytic expression for the response. In particular, when the network is initially relaxed, we have seen that it can be characterized by its transfer function. Hence it is not even necessary that the network be given, as long as its transfer function is known. The time-domain methods of Chapter 4 are also useful in establishing an analytic expression for the response, but they are particularly suited to numerical evaluation of the response in the time domain. As with the complex-frequency-domain methods, if the network is initially relaxed, it need not be given; it is sufficient to know the integral relation giving the response in terms of the network excitation.

In this chapter we shall be concerned first with the problem of determining the response of a network to an arbitrary excitation—not when the network is given, but when its response to some standard excitations is given. Step and impulse functions will be used in defining these standard excitations. We shall establish analytic results by using both time-domain and complex-frequency-domain methods. Furthermore, we shall treat the problem of obtaining numerical results in the time domain.

To start with, let us relate the network response in the complex-frequency domain to the response in the time domain. To achieve this goal we shall need a result from the theory of Laplace transforms. However, this result is probably less familiar than such standard things as partial-fraction expansions. Hence we shall spend some time discussing it.

5.1. CONVOLUTION THEOREM

Suppose an initially relaxed network is excited by voltage and/or current sources at its various inputs, and it is desired to determine the voltages and/or currents that are its outputs.

An illustration is given in Figure 1; in the amplifier network there is a voltage source and a current source. The desired responses are the two

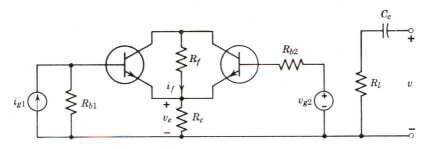

Fig. 1

voltages v_e and v_o, and the current i_f. The transforms of the excitation and response vectors for this network are as follows:

$$\mathbf{E}(s) = \mathcal{L}[\mathbf{e}(t)] = \begin{bmatrix} I_{g1}(s) \\ V_{g2}(s) \end{bmatrix}, \qquad \mathbf{W}(s) = \mathcal{L}\mathbf{w}(t) = \begin{bmatrix} V_o(s) \\ V_e(s) \\ I_f(s) \end{bmatrix},$$

assuming $\mathbf{e}(t)$ is transformable. Let $\mathbf{H}(s)$ be the matrix of transfer functions, called the *transfer matrix*, relating the excitation and response transforms. Then the response transform can be written

$$\mathbf{W}(s) = \mathbf{H}(s)\mathbf{E}(s). \tag{1}$$

In this example $\mathbf{H}$ is of order $(3, 2)$, but the relationship is quite general. In the general case $\mathbf{E}$ is a p-vector; and $\mathbf{W}$, an r-vector; hence $\mathbf{H}$ is of order (r, p).

Now, with $\mathbf{W}(s)$ known, $\mathbf{w}(t)$ is most often found by first making a partial-fraction expansion of $\mathbf{W}(s)$ and then inverting each term in the expansion. What we should like to do now is to express both $\mathbf{H}(s)$ and $\mathbf{E}(s)$ in terms of the time functions of which they are the transforms. We shall assume $\mathbf{H}(s)$ is the transform of a matrix of ordinary point functions. If we can express the result of subsequent manipulations on those time functions in the form

$$\mathbf{W}(s) = \int_0^\infty (\quad) \epsilon^{-st} \, dt, \tag{2}$$

then, from the definition of the Laplace transform, we can conclude that whatever is in the parentheses is the desired response vector. What we plan to do does not depend on the interpretations of $\mathbf{H}(s)$ as a transfer matrix and $\mathbf{E}(s)$ as an excitation vector. Hence we shall use a more general notation in developing this result.

Let $\mathbf{F}_1(s)$ and $\mathbf{F}_2(s)$ be the Laplace transforms of the matrix functions $\mathscr{F}_1(t) = [f_{1ij}(t)]$ and $\mathscr{F}_2(t) = [f_{2ij}(t)]$, respectively; that is,

$$\mathbf{F}_1(s) = \int_0^\infty \mathscr{F}_1(u) \epsilon^{-su} \, du, \tag{3a}$$

$$\mathbf{F}_2(s) = \int_0^\infty \mathscr{F}_2(v) \epsilon^{-sv} \, dv. \tag{3b}$$

We have used as dummy variables u and v rather than t to avoid confusion in the later development. Assume that the matrix product $\mathbf{F}_1(s)\,\mathbf{F}_2(s)$ is defined. Then we find

$$\mathbf{F}(s) = \mathbf{F}_1(s)\,\mathbf{F}_2(s)$$

$$= \left[\int_0^\infty \mathscr{F}_1(u)\epsilon^{-su}\,du\right]\left[\int_0^\infty \mathscr{F}_2(v)\epsilon^{-sv}\,dv\right] \tag{4}$$

$$= \int_0^\infty \int_0^\infty \mathscr{F}_1(u)\mathscr{F}_2(v)\epsilon^{-s(u+v)}\,du\,dv.$$

The last step is clearly justifiable, since each integral in the second line is a constant with respect to the other variable of integration. The product of integrals in the second line can be interpreted as a double integral over an area whose coordinate axes are u and v. The integration is to be performed over the entire first quadrant, as indicated in Fig. 2a.

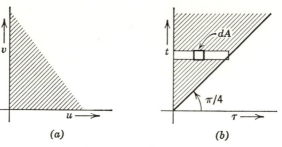

Fig. 2.   Region of integration.

Let us now make a transformation to a new set of variables, as follows:

$$t = u + v,$$
$$\tau = u.$$ (5)

Actually, the second one of these is an identity transformation and is included only for clarity. We now need to express the double integral in terms of the new variables. The element of area $du\, dv$ in the old variables is related to the element of area $d\tau\, dt$ in the new variables through the Jacobian of the transformation; thus,*

$$du\, dv = \left| \frac{\partial u}{\partial t} \frac{\partial v}{\partial \tau} - \frac{\partial u}{\partial \tau} \frac{\partial v}{\partial t} \right| d\tau\, dt.$$ (6)

Computing the partial derivatives from (5) and substituting here leads to the result that $d\tau\, dt = du\, dv$.

To complete the change of variables we must determine the new limits of integration. Note that, since $t = u + v = \tau + v$, and since v takes on only positive values, t can be no less than τ. The line $t = \tau$ in the τ-t plane bisects the first quadrant; thus the desired area of integration is the the area lying between this line and the t-axis, as shown in Fig. 2b. In order to cover this area we first integrate with respect to τ from $\tau = 0$ to $\tau = t$; then we integrate with respect to t from 0 to infinity.

With the change of variables given in (5) and with the limits changed as discussed, (4) yields

$$\mathbf{F}(s) = \int_0^\infty \left\{ \int_0^t \mathscr{F}_1(\tau) \mathscr{F}_2(t - \tau)\, d\tau \right\} \epsilon^{-st}\, dt.$$ (7)

---

* See Wilfred Kaplan, *Advanced Calculus*, Addison-Wesley, Cambridge, Mass., 1953, p. 200.

This is exactly the form of (2), so that we can identify the quantity in the parentheses as $\mathscr{F}(t) = \mathcal{L}^{-1}\{\mathbf{F}(s)\}$. It should be clear that, if in (3) we write $\mathbf{F}_1(s)$ in terms of the dummy variable v and $\mathbf{F}_2(s)$ in terms of u, then in the result given by (7) the arguments of $\mathscr{F}_1$ and $\mathscr{F}_2$ will be interchanged. The final result can therefore be written in the following two alternative forms:

$$\mathscr{F}(t) = \int_0^t \mathscr{F}_1(\tau)\mathscr{F}_2(t-\tau)\,d\tau, \tag{8}$$

$$\mathscr{F}(t) = \int_0^t \mathscr{F}_1(t-\tau)\mathscr{F}_2(\tau)\,d\tau. \tag{9}$$

The operation performed on the two matrices $\mathscr{F}_1(t)$ and $\mathscr{F}_2(t)$ represented by these expressions is called *convolution*. The two matrices are said to be convolved. The convolution of two matrices is often denoted by the short-hand notation $\mathscr{F}_1 * \mathscr{F}_2$. We can state the above result in the form of a theorem, as follows.

Convolution Theorem.   *Let the two matrices $\mathscr{F}_1(t)$ and $\mathscr{F}_2(t)$ be Laplace transformable and have the transforms $\mathbf{F}_1(s)$ and $\mathbf{F}_2(s)$, respectively. The product of $\mathbf{F}_1(s)$ with $\mathbf{F}_2(s)$, if they are conformable, is the Laplace transform of the convolution $\mathscr{F}_1(t)$ with $\mathscr{F}_2(t)$.*

$$\mathcal{L}\{\mathscr{F}(t)\} = \mathbf{F}(s) = \mathbf{F}_1(s)\,\mathbf{F}_2(s), \tag{10}$$

where

$$\mathscr{F}(t) = \mathscr{F}_1 * \mathscr{F}_2 = \int_0^t \mathscr{F}_1(\tau)\mathscr{F}_2(t-\tau)\,d\tau = \int_0^t \mathscr{F}_1(t-\tau)\,\mathscr{F}_2(\tau)\,d\tau. \tag{11}$$

While we are still in this general notation let us state another useful result concerning the derivative of the convolution of two matrices. If the matrices $\mathscr{F}_1(t)$ and $\mathscr{F}_2(t)$, in addition to being Laplace transformable, are also differentiable for $t > 0$ (they need be continuous only at $t = 0$), then their convolution will also be differentiable for $t > 0$. The derivative will be

$$\dot{\mathscr{F}}(t) = \int_0^t \mathscr{F}_1(\tau)\dot{\mathscr{F}}_2(t-\tau)\,d\tau + \mathscr{F}_1(t)\,\mathscr{F}_2(0) \tag{12}$$

or

$$\dot{\mathscr{F}}(t) = \int_0^t \dot{\mathscr{F}}_1(t-\tau)\mathscr{F}_2(\tau)\,d\tau + \mathscr{F}_1(0)\,\mathscr{F}_2(t), \qquad (13)$$

where the dot indicates differentiation with respect to t. These expressions can be found by applying the Leibniz formula for differentiation under an integral. In fact, we can observe that we do not really need the hypothesis that both $\mathscr{F}_1(t)$ and $\mathscr{F}_2(t)$ are differentiable. If *either function is differentiable and the other continuous, then the convolution $\mathscr{F}_1 * \mathscr{F}_2$ is differentiable.*

Although the preceding has been carried out in matrix form, the results are, of course, valid for the scalar case as well, a scalar being a one-dimensional vector. Thus, for scalars, (8) and (9) become

$$f(t) = \int_0^t f_1(\tau) f_2(t-\tau)\,d\tau,$$

$$f(t) = \int_0^t f_1(t-\tau) f_2(\tau)\,dt.$$

5.2 IMPULSE RESPONSE

Let us return to our original problem of finding the response $\mathbf{w}(t)$ of an initially relaxed network having the transfer matrix $\mathbf{H}(s)$ to the excitation $\mathbf{e}(t)$. Recall that $\mathbf{H}(s)$ must be the transform of a matrix of ordinary point functions in order to apply the convolution theorem. This implies that $\mathbf{H}(s)$ tends to $\mathbf{0}$ as s tends to infinity within the sector of convergence for $\mathbf{H}(s)$. Let $\mathbf{W}_\delta(t)$ denote the inverse transform of $\mathbf{H}(s)$; that is,

$$\mathcal{L}^{-1}\{\mathbf{H}(s)\} = \mathbf{W}_\delta(t). \qquad (14)$$

The reason for this choice of notation will be clear shortly. Now the convolution theorem applied to (1) provides us with the result

$$\mathbf{w}(t) = \int_0^t \mathbf{W}_\delta(\tau)\,\mathbf{e}(t-\tau)\,d\tau = \int_0^t \mathbf{W}_\delta(t-\tau)\,\mathbf{e}(\tau)\,d\tau. \qquad (15)$$

This is a very valuable result. By this expression we are able to express the time response of a network to an arbitrary excitation $e(t)$ in terms of the inverse transform of the transfer matrix of the network.

A further interpretation is possible if we are willing to admit the impulse function in our discussions.* Such an interpretation is not really needed, since (15) can stand on its own feet, so to speak. However, the interpretation may prove useful in some instances. For instance, by using this interpretation, one can find a close approximation to $\mathbf{W}_\delta(t)$ experimentally.

Now suppose that all excitations are zero except the jth one and that this one is an impulse. In this case the excitation vector, denoted by $\mathbf{e}_{\delta j}(t)$, has all elements equal to zero except the jth, this element being the impulse function $\delta(t)$. We may think of $\mathbf{e}_{\delta j}(t)$ as the jth column of the $p \times p$ excitation matrix $\mathbf{E}_\delta(t)$. Thus, for example,

$$
\mathbf{e}_{\delta 2}(t) = \begin{bmatrix} 0 \\ \delta(t) \\ 0 \\ \vdots \\ 0 \end{bmatrix}
$$

* The impulse function $\delta(t)$ is not an ordinary point function; rather, it is a generalized function. Symbolically, we may manipulate mathematical relations involving the impulse function and its derivatives as we would relations involving only ordinary point functions. On the other hand, mathematical precision requires that we view each function as a generalized function, and each operation as defined on the space of generalized functions. A short treatment of the theory of generalized functions is given in Appendix 1. Here it is enough to observe that the impulse function satisfies the following relations. With $a \leq \tau \leq b$,

$$
\int_a^b \delta(t - \tau) \, d\tau = 1,
$$

$$
\int_a^b \delta(t - \tau) f(\tau) \, d\tau = f(t),
$$

$$
\int_a^b \dot{\delta}(t - \tau) f(\tau) \, d\tau = \dot{f}(t).
$$

For other values of τ outside the range $[a, b]$, each of the above integrals yields zero.

is the second-column vector of the excitation matrix

$$\mathbf{E}_\delta(t) = \begin{bmatrix} \delta(t) & 0 & 0 & \ldots 0 \\ 0 & \delta(t) & 0 & \ldots 0 \\ 0 & 0 & \delta(t) & \ldots 0 \\ \vdots & \vdots & \vdots & \\ 0 & 0 & 0 & \ldots \delta(t) \end{bmatrix} = \delta(t)\mathbf{U}.$$

Similarly, let $\mathbf{w}_{\delta j}(t)$ be the response vector resulting from $\mathbf{e}_{\delta j}(t)$; that is, $\mathbf{w}_{\delta j}(t)$ is the collection of all the scalar responses when there is a single excitation and this excitation is an impulse. Let these $\mathbf{w}_{\delta j}$ vectors be arranged as the columns of an $r \times p$ matrix designated $\mathbf{W}_\delta(t)$ and called the *impulse response* of the network. Note carefully that $\mathbf{W}_\delta(t)$ is a set of response vectors, one column for each column of $\mathbf{E}_\delta(t)$. It is thus not an observable response in the same sense that each of its columns is an observable response. On the other hand, the sum of elements in each row of $\mathbf{W}_\delta(t)$ is an observable (scalar) response—the response to the sum of all excitations, these excitations all being impulses. Let us illustrate with the example of Fig. 1.

$$\mathbf{e}_{\delta 1}(t) = \begin{bmatrix} \delta(t) \\ 0 \end{bmatrix}, \qquad \mathbf{e}_{\delta 2}(t) = \begin{bmatrix} 0 \\ \delta(t) \end{bmatrix}, \qquad \mathbf{E}_\delta(t) = \begin{bmatrix} \delta(t) & 0 \\ 0 & \delta(t) \end{bmatrix},$$

$$\mathbf{w}_{\delta 1}(t) = \begin{bmatrix} v_{o\delta 1}(t) \\ v_{e\delta 1}(t) \\ i_{f\delta 1}(t) \end{bmatrix}, \qquad \mathbf{w}_{\delta 2}(t) = \begin{bmatrix} v_{o\delta 2}(t) \\ v_{e\delta 2}(t) \\ i_{f\delta 2}(t) \end{bmatrix}, \qquad \mathbf{W}_\delta(t) = \begin{bmatrix} v_{o\delta 1}(t) & v_{o\delta 2}(t) \\ v_{e\delta 1}(t) & v_{e\delta 2}(t) \\ i_{f\delta 1}(t) & i_{f\delta 2}(t) \end{bmatrix}.$$

Finally, the sum of the elements in the first row of $\mathbf{W}_\delta(t)$ [namely, $v_{o\delta 1}(t) + v_{o\delta 2}(t)$] is the voltage $v_o(t)$ when each of the two sources in the diagram is an impulse.

Now consider the Laplace transforms. Since $\mathbf{E}_\delta(t) = \delta(t)\mathbf{U}$, then

$$\mathcal{L}\{\mathbf{E}_\delta(t)\} = \mathcal{L}\{\delta(t)\mathbf{U}\} = \mathbf{U}. \tag{16}$$

Equation (1) relates corresponding columns of the transforms of $\mathbf{E}_\delta(t)$ and $\mathbf{W}_\delta(t)$. Hence, by using (16), we obtain

$$\mathcal{L}\{\mathbf{W}_\delta(t)\} = \mathbf{H}(s)\,\mathcal{L}\{\mathbf{E}_\delta(t)\} = \mathbf{H}(s) \tag{17}$$

or, equivalently,

$$\mathbf{W}_\delta(t) = \mathcal{L}^{-1}\{\mathbf{H}(s)\}. \tag{18}$$

In words, the last equation states that the inverse transform of the network transfer function is equal to the impulse response of the network. We anticipated this result in using the notation in (14).

Let us return now to (15). We see that this equation expresses the fact that, once the impulse response of an initially relaxed network is known, the response to any other excitation $\mathbf{e}(t)$ is determined. What we must do is premultiply the excitation at each point τ by the impulse response—not at the same point, but at a point $(t - \tau)$—and then integrate. Another viewpoint is that the input vector is "weighted" by the impulse response. This leads to the name of "weighting matrix" used by some authors for the impulse response.*

Let us elaborate a little on the concept of weighting. Perhaps a simple example would be more satisfying as a means of communicating this point. Therefore let us consider a single-input, single-output network with the transfer function

$$H(s) = \frac{\frac{3}{4}}{(s+1)^2},$$

which is the transfer function $V_2(s)/V_1(s)$ of the network of Fig. 3. Then the impulse response is given by

$$w_\delta(t) = \tfrac{3}{4}t\epsilon^{-t}.$$

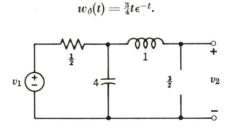

Fig. 3.   Example for the concept of the weighting function.

A plot of this function is given in Fig. 4.

Suppose we wish to compute the response of this network to some driving

---

* The term *weighting function* has been used as another name for the impulse response of a single-input, single-output network. *Weighting matrix* seemed to be the natural generalization for the multi-input, multi-output networks discussed here.

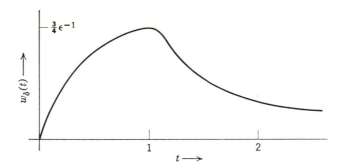

Fig. 4. Impulse response of the network of Fig. 3.

function $e(t)$. For convenience of interpretation let us take the convolution of $w_\delta(t)$ and $e(t)$ in the second form given in (15). We shall use τ as the running variable. To get the value of the response at any given time τ, we first reverse the impulse response and translate it along the τ-axis so as to obtain $w_\delta(t - \tau)$ as a function of τ. Compare $w_\delta(t)$ versus t in Fig. 4 to $w_\delta(t - \tau)$ versus τ in Fig. 5a. The excitation over the interval 0 to t is superimposed on $w_\delta(t - \tau)$ in Fig. 5a. Now according to (15) we must multiply the two curves $w_\delta(t - \tau)$ and $e(\tau)$ point by point on this interval. The resulting product is shown in Fig. 5b. Since $w_\delta(0) = 0$, the value of $e(\tau)$ at the point t contributes nothing to the response at t, in spite of the fact that $e(\tau)$ has a maximum value at this point. On the other hand, the *most important* neighborhood is around $(t - 1)$, because the values of $e(\tau)$ in this vicinity are multiplied by the largest values that w_δ assumes. Similarly, the values of $e(\tau)$ for τ less than $(t - 2)$ do virtually nothing to the response at t. Thus w_δ decides how much *weight* to attach to the values of e at various times. In this case the response, that by (15) is the integral of $w_\delta(t - \tau)\, e(\tau)$ from 0 to t, is decided almost entirely by the values of $e(t)$ for the previous 2 seconds; the most significant contribution coming from the values of $e(t)$ about the point 1 second prior to the time under consideration.

TRANSFER FUNCTION NONZERO AT INFINITY

The question now arises, What should we do if the network transfer function does not have a zero at infinity? In such a case $\mathbf{w}_\delta(t)$ will contain impulses and first derivatives of impulses. Since we are permitting impulses in the excitation, we might just as well relax the original condition on $\mathbf{H}(s)$ and permit it to be nonzero at infinity. Let us see what effect this will have.

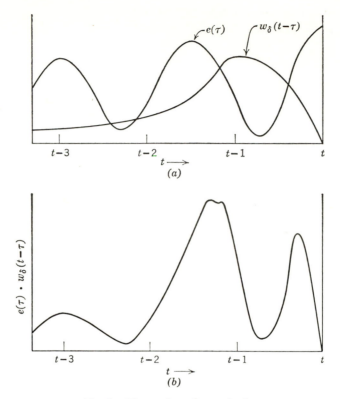

Fig. 5.   Illustration of convolution

The transfer function of a network can be more than just a nonzero constant at infinity, it can have a first-order pole there. Let $\mathbf{K}_\infty$ be the residue matrix of $\mathbf{H}(s)$ at the pole at infinity, and let $\mathbf{K}$ be the constant matrix that is the limit of $\mathbf{H}(s) - \mathbf{K}_\infty s$ as s approaches infinity along the real axis. Then we can write

$$\mathbf{H}(s) = \hat{\mathbf{H}}(s) + \mathbf{K} + \mathbf{K}_\infty s, \tag{19}$$

where $\hat{\mathbf{H}}(s)$ has a zero at infinity. The impulse response will then be

$$\mathbf{W}_\delta(t) = \hat{\mathbf{W}}_\delta(t) + \mathbf{K}\delta(t) + \mathbf{K}_\infty \dot{\delta}(t), \tag{20}$$

where $\hat{\mathbf{W}}_\delta(t)$ is a well-behaved matrix not containing impulses. Let us use this expression in the second form of (15) to find the response of the network to an excitation $e(t)$. The result will be

$$\mathbf{w}(t) = \int_0^t \widehat{\mathbf{W}}_\delta(t-\tau)\, \mathbf{e}(\tau)\, d\tau + \int_0^t \mathbf{K}\delta(t-\tau)\, \mathbf{e}(\tau)\, d\tau + \int_0^t \mathbf{K}_\infty \dot{\delta}(t-\tau)\, \mathbf{e}(\tau)\, d\tau$$

$$= \int_0^t \widehat{\mathbf{W}}_\delta(t-\tau)\, \mathbf{e}(\tau)\, d\tau + \mathbf{K}\mathbf{e}(t) + \mathbf{K}_\infty \dot{\mathbf{e}}(t). \tag{21}$$

The last step follows from the properties of impulse functions and their derivatives, as given in Appendix 1. This is the general form of the convolution integral.

ALTERNATIVE DERIVATION OF CONVOLUTION INTEGRAL

In deriving (21) we permitted $\mathbf{W}_\delta(t)$ in the second convolution integral to include impulses and derivatives of impulses; that is, all elements of $\mathbf{W}_\delta(t)$ were not necessarily ordinary point functions. Now it is a fact that the convolution theorem is valid when elements of $\mathbf{W}_\delta(t)$ are not ordinary point functions; however, the proof given in the last section does not hold in this more general case. It is necessary, therefore, to give an alternative derivation of (21). This is easily done in terms of the time-domain results of Chapter 4.

Recall that the state equations for a network may be written as

$$\frac{d}{dt}\mathbf{x} = \mathscr{A}\mathbf{x} + \mathscr{B}\mathbf{e}, \tag{22a}$$

$$\mathbf{w} = \mathbb{C}\mathbf{x} + \mathscr{D}\mathbf{e} + \widehat{\mathscr{D}}\frac{d}{dt}\mathbf{e}. \tag{22b}$$

Recall also that the solution of the differential equation for the state vector and, from (22b), the solution for the output vector, are

$$\mathbf{x}(t) = \epsilon^{\mathscr{A}(t-t_0)}\mathbf{x}(t_0) + \int_{t_0}^t \epsilon^{\mathscr{A}(t-\tau)}\mathscr{B}\,\mathbf{e}(\tau)\,d\tau, \tag{23a}$$

$$\mathbf{w}(t) = \mathbb{C}\epsilon^{\mathscr{A}(t-t_0)}\mathbf{x}(t_0) + \int_{t_0}^t \mathbb{C}\epsilon^{\mathscr{A}(t-\tau)}\mathscr{B}\,\mathbf{e}(\tau)\,d\tau + \mathscr{D}\,\mathbf{e} + \widehat{\mathscr{D}}\frac{d\mathbf{e}}{dt}. \tag{23b}$$

Observe that part of the response is a consequence of a nonzero initial state. We shall let

$$\mathbf{w}^f(t) = \mathbb{C}\epsilon^{\mathscr{A}(t-t_0)}\mathbf{x}(t_0) \tag{24}$$

denote this part. The superscript f was chosen because $\mathbf{w}^f(t)$ is known as the *free response* of the network; this is a sensible name, since $\mathbf{w}^f(t)$ is independent of the controlling influence of the network excitation. The remaining part of the response stems from nonzero excitation of the network. We shall let

$$\mathbf{w}^c(t) = \int_0^t \mathcal{C}e^{\mathcal{A}(t-\tau)}\,\mathcal{B}\,\mathbf{e}(\tau)\,d\tau + \mathcal{D}\,\mathbf{e}(t) + \widehat{\mathcal{D}}\,\frac{d}{dt}\,\mathbf{e}(t) \qquad (25)$$

denote this part. The superscript c was chosen because we may think of $\mathbf{w}^c(t)$ as the *controlled response*, since $\mathbf{w}^c(t)$ is controlled by the network excitation. As a consequence of these definitions we have

$$\mathbf{w}(t) = \mathbf{w}^f(t) + \mathbf{w}^c(t). \qquad (26)$$

When the network is initially relaxed, $\mathbf{x}(t_0) = \mathbf{0}$ and hence $\mathbf{w}^f(t) = \mathbf{0}$. Thus the total network response is simply the controlled response. Now, upon letting $t_0 = 0$, we get

$$\mathbf{w}(t) = \mathbf{w}^c(t) = \int_0^t \mathcal{C}\epsilon^{\mathcal{A}(t-\tau)}\,\mathcal{B}\,\mathbf{e}(\tau)\,d\tau + \mathcal{D}\,\mathbf{e}(t) + \widehat{\mathcal{D}}\,\dot{\mathbf{e}}(t). \qquad (27)$$

Observe that (21) and (27) are identical after the obvious identification of $\widehat{\mathbf{W}}_\delta(t)$ with $\mathcal{C}\epsilon^{\mathcal{A}t}\,\mathcal{B}$, $\mathbf{K}$ with $\mathcal{D}$, and $\mathbf{K}_\infty$ with $\widehat{\mathcal{D}}$; the theorem is thus proved.

One further comment on the need for generalized functions is in order. Suppose we restrict $\mathbf{e}(t)$ and $\mathbf{w}(t)$ to the set of ordinary vector-valued point functions. As (27) then makes evident, if $\widehat{\mathcal{D}} \neq 0$, we must also require that $\mathbf{e}(t)$ have a derivative for $t \geq 0$. A restriction such as this is not desirable, since it is often of value to examine $\mathbf{w}(t)$ when elements of $\mathbf{e}(t)$ are discontinuous functions or functions with slope discontinuities. Two very typical functions of this type are the unit step function

$$u(t) = 0 \qquad (t < 0)$$
$$\quad\;\; = 1 \qquad (0 \leq t)$$

which is discontinuous at $t = 0$; and the unit ramp function $t\,u(t)$, which has a slope discontinuity at $t = 0$. We can circumvent the differentiability restriction by removing the initial restriction on $\mathbf{e}(t)$ and $\mathbf{w}(t)$ to the set of ordinary vector-valued point functions, and by allowing $\mathbf{e}(t)$ and $\mathbf{w}(t)$

to be vector-valued generalized functions. This act removes the differ-entiability restriction because every generalized function has derivatives of all orders.

Example

The application of the convolution integral in the solution of a problem is quite straightforward. It is necessary first to find the impulse response $\mathbf{W}_\delta(t) = \mathcal{L}^{-1}\{\mathbf{H}(s)\}$ and then to substitute this, together with the excitation as specified by a functional expression—possibly different functional expressions over different intervals of time—in the convolution integral. We shall illustrate this process in the following example.

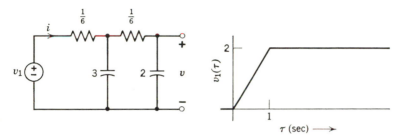

Fig. 6. Example.

In the initially relaxed network of Fig. 6, let the responses be v and i as shown. There is only a single source, having a voltage given in the diagram. Hence the excitation vector is 1×1, and the response vector is 2×1. The transfer-function matrix is

$$\mathbf{H}(s) = \mathcal{L}\begin{bmatrix} v_\delta(t) \\ i_\delta(t) \end{bmatrix} = \begin{bmatrix} \dfrac{6}{(s+1)(s+6)} \\ \dfrac{6s(s+5)}{(s+1)(s+6)} \end{bmatrix} = \begin{bmatrix} \dfrac{6/5}{s+1} - \dfrac{6/5}{s+6} \\ 6 - \dfrac{24/5}{s+1} - \dfrac{36/5}{s+6} \end{bmatrix}.$$

(You may go through the details of finding this; try it by finding the state equations.) We see that the first element of $\mathbf{H}(s)$ has a zero at infinity, whereas the second one is nonzero at infinity. The impulse response, therefore, is

$$\mathbf{W}_\delta(t) = \mathcal{L}^{-1}\mathbf{H}(s) = \tfrac{6}{5}\begin{bmatrix} \epsilon^{-t} - \epsilon^{-6t} \\ -4\epsilon^{-t} - 6\epsilon^{-6t} \end{bmatrix} + \begin{bmatrix} 0 \\ 6 \end{bmatrix}\delta(t).$$

Since the excitation has a different functional form for $0 \leq t \leq 1$ from the one it has for $t \geq 1$, the response for each of these ranges must be found separately. Thus

$$v_1(t) = 2t \qquad (0 \leq t \leq 1),$$
$$v_1(t) = 2 \qquad (t \geq 1),$$

Hence, from (21), for $0 \leq t \leq 1$

$$\mathbf{w}(t) = \int_0^t \frac{6}{5} \begin{bmatrix} \epsilon^{-(t-\tau)} - \epsilon^{-6(t-\tau)} \\ -4\epsilon^{-(t-\tau)} - 6\epsilon^{-6(t-\tau)} \end{bmatrix} 2\tau \, d\tau + \begin{bmatrix} 0 \\ 6 \end{bmatrix} 2t$$

$$= \tfrac{12}{5} \begin{bmatrix} \epsilon^{-t} \int_0^t \tau\epsilon^\tau \, d\tau - \epsilon^{-6t} \int_0^t \tau\epsilon^{6\tau} \, d\tau \\ -4\epsilon^{-t} \int_0^t \tau\epsilon^\tau \, d\tau - 6\epsilon^{-6t} \int_0^t \tau\epsilon^{6\tau} \, d\tau \end{bmatrix} + \begin{bmatrix} 0 \\ 12t \end{bmatrix}$$

$$= \begin{bmatrix} 2t - \tfrac{7}{3} + \tfrac{12}{5}\,\epsilon^{-t} - \tfrac{1}{15}\,\epsilon^{-6t} \\ -12t + 10 - \tfrac{48}{5}\,\epsilon^{-t} - \tfrac{2}{5}\,\epsilon^{-6t} \end{bmatrix} + \begin{bmatrix} 0 \\ 12t \end{bmatrix}.$$

In the range $t \geq 1$, the integral from 0 to t must be broken into two integrals, the first going from 0 to 1, the second from 1 to t. The excitation in the first interval is $v_1(t) = 2t$, the same as in the calculation just completed. Hence in this next calculation it is enough to replace the limit t by 1 in the preceding integral and then to evaluate the second integral. Note that t is not replaced by 1 everywhere in the evaluated integral above but only in the contribution coming from the upper limit. So, for $t \geq 1$,

$$\mathbf{w}(t) = \tfrac{12}{5} \begin{bmatrix} \epsilon^{-t} \int_0^1 \tau\epsilon^\tau \, d\tau - \epsilon^{-6t} \int_0^1 \tau\epsilon^{6\tau} \, d\tau \\ -4\epsilon^{-t} \int_0^1 \tau\epsilon^\tau \, d\tau - 6\epsilon^{-6t} \int_0^1 \tau\epsilon^{6\tau} \, d\tau \end{bmatrix}$$

$$+ \tfrac{12}{5} \int_1^t \begin{bmatrix} \epsilon^{-(t-\tau)} - \epsilon^{-6(t-\tau)} \\ -4\epsilon^{-(t-\tau)} - 6\epsilon^{-6(t-\tau)} \end{bmatrix} d\tau + \begin{bmatrix} 0 \\ 6 \end{bmatrix} 2$$

$$= \begin{bmatrix} -\tfrac{12}{5}(1 - \epsilon^{-1})\epsilon^{-(t-1)} + \tfrac{1}{15}(1 - \epsilon^{-6})\epsilon^{-6(t-1)} + 2 \\ \tfrac{48}{5}(1 - \epsilon^{-1})\epsilon^{-(t-1)} + \tfrac{2}{5}(1 - \epsilon^{-6})\epsilon^{-6(t-1)} - 12 \end{bmatrix} + \begin{bmatrix} 0 \\ 12 \end{bmatrix}.$$

You should carry out the details of these calculations and verify that both expressions give the same value of w at $t = 1$.

5.3 STEP RESPONSE

In the last section we established that the response of an initially relaxed network to any arbitrary excitation can be found simply from a knowledge of the impulse response of the same network. In this section we shall show that the same conclusion applies as well to a knowledge of what shall be called the step response of the network.

Suppose that all excitations are zero except the jth one, which is a unit step. Denote by $\mathbf{e}_{uj}(t)$ an excitation vector with all its elements equal to zero except the jth, this one being a unit step $u(t)$. We can think of $\mathbf{e}_{uj}$ as the jth column of a $p \times p$ matrix $\mathbf{E}_u(t)$. Thus, for example,

$$\mathbf{e}_{u3}(t) = \begin{bmatrix} 0 \\ 0 \\ u(t) \\ 0 \\ \vdots \\ 0 \end{bmatrix}$$

is the third column-vector of the excitation matrix

$$\mathbf{E}_u(t) = \begin{bmatrix} u(t) & 0 & 0 & \dots & 0 \\ 0 & u(t) & 0 & \dots & 0 \\ 0 & 0 & u(t) & \dots & 0 \\ \vdots & \vdots & \vdots & & \vdots \\ 0 & 0 & 0 & \dots & u(t) \end{bmatrix} = u(t)\mathbf{U}.$$

Similarly, let $\mathbf{w}_{uj}(t)$ denote the response vector resulting from $\mathbf{e}_{uj}(t)$; that is, $\mathbf{w}_{uj}(t)$ is the collection of all the scalar responses when there is a single excitation and this excitation is a unit step at the jth input. Suppose these $\mathbf{w}_{uj}$ vectors are arranged as the columns of an $r \times p$ matrix designated $\mathbf{W}_u(t)$ and called the *step response* of the network.

Now consider the Laplace transforms. From (1) it follows that

$$\mathcal{L}\{\mathbf{W}_u(t)\} = \mathbf{H}(s)\mathcal{L}\{\mathbf{E}_u(t)\}. \tag{28}$$

However, since $\mathcal{L}\{\mathbf{E}_u(t)\} = \mathcal{L}\{u(t)\mathbf{U}\} = \mathbf{U}/s$, it follows that

$$\mathcal{L}\{\mathbf{W}_u(t)\} = \frac{1}{s}\,\mathbf{H}(s). \tag{29}$$

This expression immediately tells us something about the relationship between the step response and the impulse response, since $\mathbf{H}(s) = \mathcal{L}\{\mathbf{W}_\delta(t)\}$. To get the relationship between the time responses, we take the inverse transform of (29) either as it stands or after multiplying through by s. The results will be

$$\mathbf{W}_u(t) = \int_0^t \mathbf{W}_\delta(\tau)\,d\tau, \tag{30}$$

$$\mathbf{W}_\delta(t) = \frac{d}{dt}\,\mathbf{W}_u(t) + \mathbf{W}_u(0)\,\delta(t). \tag{31}$$

The initial value of the step response is readily found from (29), by using the initial-value theorem. Provided $\mathbf{H}(\infty)$ is finite, it will be

$$\mathbf{W}_u(0) = \lim_{\sigma \to \infty}\left[s\mathcal{L}\{\mathbf{W}_u(t)\}_{s=\sigma}\right] = \lim_{\sigma \to \infty}\mathbf{H}(\sigma) = \mathbf{H}(\infty), \tag{32}$$

where σ is a real scalar. We conclude that *the initial value of the step response of a network will be zero if the transfer function has a zero at infinity.* Further, if $\mathbf{H}(\infty)$ is nonzero but finite, the initial value of the step response will be nonzero and, in addition, the impulse response will itself contain an impulse. If, on the other hand, $\mathbf{H}(\infty)$ is nonzero but infinite—$\mathbf{H}(s)$ has a simple pole at infinity—then the step response has an impulse at $t = 0$, and the impulse response contains the derivative of an impulse at $t = 0$. Note that this does violence to our ordinary concepts of calculus, as embodied in (30) and (31). If $\mathbf{W}_\delta(t)$ is an integrable matrix, then (30) tells us that $\mathbf{W}_u(0)$ should be zero (simply by putting zero as the upper limit); however, if we admit impulses, then our constitution must be strong to withstand the consequences. Note, however, that if $\mathbf{W}_\delta(t)$ contains a first-order impulse, $\mathbf{W}_u(t)$ is not impulsive; further, if $\mathbf{W}_\delta(t)$ contains the derivative of an impulse, $\mathbf{W}_u(t)$ has an impulse but not the derivative of an impulse. Hence $\mathbf{W}_u(t)$ is always better behaved than $\mathbf{W}_\delta(t)$.

Let us now return to our original task and assume that an arbitrary (but without impulses) Laplace-transformable excitation $e(t)$ is applied to the network. Equation 1 relates the transforms. This equation can be rewritten in one of several ways after multiplying numerator and denominator by s. Thus

$$\mathbf{W}(s) = s\left[\frac{\mathbf{H}(s)}{s}\,\mathbf{E}(s)\right] = s[\mathcal{L}\{\mathbf{W}_u(t)\}\mathbf{E}(s)], \qquad (33)$$

$$\mathbf{W}(s) = \left[s\,\frac{\mathbf{H}(s)}{s}\right]\mathbf{E}(s) = [s\mathcal{L}\{\mathbf{W}_u(t)\}]\mathbf{E}(s), \qquad (34)$$

$$\mathbf{W}(s) = \left[\frac{\mathbf{H}(s)}{s}\right][s\mathbf{E}(s)] = \mathcal{L}\{\mathbf{W}_u(t)\}[s\mathbf{E}(s)]. \qquad (35)$$

In each case we have used (29) to obtain the far-right side. To find $\mathbf{w}(t)$ we shall now use the convolution theorem. Focus attention on (33). This can be written

$$\mathbf{W}(s) = s\mathbf{F}(s), \qquad (36)$$

where

$$\mathbf{F}(s) = \mathcal{L}\{\mathbf{W}_u(t)\}\mathbf{E}(s). \qquad (37)$$

By using the convolution theorem, we can write

$$\mathbf{f}(t) = \int_0^t \mathbf{W}_u(\tau)\,\mathbf{e}(t-\tau)\,d\tau = \int_0^t \mathbf{W}_u(t-\tau)\,\mathbf{e}(\tau)\,d\tau. \qquad (38)$$

If we evaluate $\mathbf{f}(0)$, we shall find it to be zero, unless $\mathbf{W}_u(t)$ contains an impulse. We saw that this is not possible even if $\mathbf{H}(s)$ has a finite nonzero value at infinity. In fact, the step response will have an impulse only if $\mathbf{H}(s)$ has a pole at infinity. Hence, if we admit only those $\mathbf{H}(s)$ matrices that are regular at infinity, then $\mathbf{w}(t)$ will be the derivative of $\mathbf{f}(t)$, based on (36). Thus

$$\mathbf{w}(t) = \frac{d}{dt}\int_0^t \mathbf{W}_u(\tau)\,\mathbf{e}(t-\tau)\,d\tau$$

$$= \frac{d}{dt}\int_0^t \mathbf{W}_u(t-\tau)\,\mathbf{e}(\tau)\,d\tau. \qquad (39)$$

We now have an expression for the response of an initially relaxed network to an excitation $e(t)$ in terms of the step response. This result ranks in importance with (15). Using the results stated in (12) and (13), we can put the last equation in the following alternative forms:

$$\mathbf{w}(t) = \int_0^t \mathbf{W}_u(\tau)\, \dot{\mathbf{e}}(t - \tau)\, d\tau + \mathbf{W}_u(t)\, \mathbf{e}(0), \tag{40}$$

$$\mathbf{w}(t) = \int_0^t \dot{\mathbf{W}}_u(t - \tau)\, \mathbf{e}(\tau)\, d\tau + \mathbf{W}_u(0)\, \mathbf{e}(t). \tag{41}$$

This will require that $e(t)$ or $\mathbf{W}_u(t)$, as the case may be, be differentiable and that, correspondingly, $e(0)$ or $\mathbf{W}_u(0)$ be finite.

These same expressions can be obtained in an alternative manner, starting from (34) and (35). To use (34) let us first write

$$\mathcal{L}^{-1}\{s\mathcal{L}[\mathbf{W}_u(t)]\} = \mathcal{L}^{-1}\{s\mathcal{L}[\mathbf{W}_u(t)] - \mathbf{W}_u(0) + \mathbf{W}_u(0)\}$$

$$= \frac{d}{dt}\mathbf{W}_u(t) + \mathbf{W}_u(0)\delta(t). \tag{42}$$

We can now use the convolution theorem on (34). The result will be

$$\mathbf{w}(t) = \int_0^t \left[\frac{d}{d\tau}\mathbf{W}_u(\tau) + \mathbf{W}_u(0)\delta(\tau)\right]\mathbf{e}(t - \tau)\, d\tau$$

$$= \int_0^t \dot{\mathbf{W}}_u(\tau)\, \mathbf{e}(t - \tau)\, d\tau + \mathbf{W}_u(0)\, \mathbf{e}(t) \tag{43}$$

$$= \int_0^t \dot{\mathbf{W}}_u(t - \tau)\, \mathbf{e}(\tau)\, d\tau + \mathbf{W}_u(0)\, \mathbf{e}(t),$$

which is the same as (41). In a similar manner, (40) can be obtained starting from (35). The details are left to you.

For future reference we shall collect all of the forms of these expressions that have been derived. They are as follows:

$$\mathbf{w}(t) = \mathbf{W}_u(t)\, \mathbf{e}(0) + \int_0^t \mathbf{W}_u(t - \tau)\, \dot{\mathbf{e}}(\tau)\, d\tau$$

$$= \mathbf{W}_u(t)\, \mathbf{e}(0) + \int_0^t \mathbf{W}_u(\tau)\, \dot{\mathbf{e}}(t - \tau)\, d\tau, \tag{44}$$

$$\mathbf{w}(t) = \mathbf{W}_u(0)\,\mathbf{e}(t) + \int_0 \dot{\mathbf{W}}_u(\tau)\,\mathbf{e}(t-\tau)\,d\tau$$

$$(45)$$

$$= \mathbf{W}_u(0)\,\mathbf{e}(t) + \int_0^t \dot{\mathbf{W}}_u(t-\tau)\,\mathbf{e}(\tau)\,d\tau,$$

$$\mathbf{w}(t) = \frac{d}{dt}\int_0^t \mathbf{W}_u(\tau)\,\mathbf{e}(t-\tau)\,d\tau$$

$$(46)$$

$$= \frac{d}{dt}\int_0^t \mathbf{W}_u(t-\tau)\,\mathbf{e}(\tau)\,d\tau.$$

These expressions, as scalar rather than vector equations, were originally used by DuHamel in 1833 in dynamics. They are variously known as the DuHumel integrals, Carson integrals, and superposition integrals. Carson himself called (46) *the fundamental formula of circuit theory*.

Example

In applying the superposition integrals to the evaluation of a network response, the first step is to find the response $\mathbf{W}_u(t)$. A decision is then required as to which matrix, $\mathbf{e}$ or $\mathbf{W}_u$, should be reversed and translated to the argument $(t-\tau)$. This choice is guided by the simplicity of the resulting integrals. Then a decision is needed as to which one, $\mathbf{e}$ or $\mathbf{W}_u$, should be differentiated. Sometimes there may be no choice, since one of them may not be differentiable.

To illustrate, consider again the example of Fig. 6 which was earlier evaluated by using the impulse response. Since we already have $\mathbf{W}_\delta(t)$, (30) can be used to find $\mathbf{W}_u(t)$. The result is

$$\mathbf{W}_u(t) = \int_0^t \mathbf{W}_\delta(\tau)\,d\tau = \int_0^t \frac{6}{5}\begin{bmatrix} \epsilon^{-\tau} - \epsilon^{-6\tau} \\ -4\epsilon^{-\tau} - 6\epsilon^{-6\tau} \end{bmatrix} d\tau + \int_0^t \begin{bmatrix} 0 \\ 6 \end{bmatrix} \delta(\tau)\,d\tau$$

$$= \frac{1}{5}\begin{bmatrix} 5 - 6\epsilon^{-t} + \epsilon^{-6t} \\ 24\epsilon^{-t} + 6\epsilon^{-6t} \end{bmatrix}.$$

Because of the functional form of the excitation, it is much simpler to differentiate $\mathbf{e}(\tau) = [v_1(\tau)]$ then $\mathbf{W}_u(t)$. The derivative is shown in Fig. 7; its analytical form is also given.

Note that if we differentiate $\mathbf{W}_u(t)$ instead, then, by using (31) for $\dot{\mathbf{W}}_u$ and inserting it into (45), we get back the convolution integral.

Let us use the second form of (44), which should be simpler than the first form, since $\dot{v}_1(t-\tau) = 2[u(t-\tau) - u(t-\tau-1)]$, which is simply the

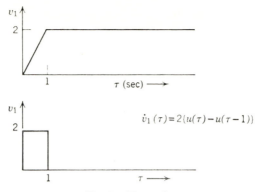

Fig. 7.   Example.

square pulse in Fig. 7 shifted by t units. Since $\mathbf{e}(0) = [v_1(0)] = 0$, we get for $0 \leq t < 1$

$$\mathbf{w}(t) = \int_0^t \frac{1}{5} \begin{bmatrix} 5 - 6\epsilon^{-\tau} + \epsilon^{-6\tau} \\ 24\epsilon^{-\tau} + 6\epsilon^{-6\tau} \end{bmatrix} 2d\tau$$

$$= \begin{bmatrix} 2t - \frac{7}{3} + \frac{12}{5}\,\epsilon^{-t} - \frac{1}{15}\,\epsilon^{-6t} \\ 10 - \frac{48}{5}\,\epsilon^{-t} - \frac{2}{5}\,\epsilon^{-6t} \end{bmatrix}.$$

This answer can be verified by comparing with the result previously found. Note that the integration was considerably simpler in this case.

This would be the response for all t if v_1' would stay constant at 2; but it does not; it takes a negative step downward at $t = 1$. Hence, to find the response for $t \geq 1$, we simply replace t by $t - 1$ in the above expression, to get the response to the negative step at $t = 1$, and subtract the result from the above expression. The result for $t \geq 1$ will be

$$\mathbf{w}(t) = \begin{bmatrix} 2t - \frac{7}{3} + \frac{12}{5}\,\epsilon^{-t} - \frac{1}{15}\,\epsilon^{-6t} \\ 10 - \frac{48}{5}\,\epsilon^{-t} - \frac{2}{5}\,\epsilon^{-6t} \end{bmatrix}$$

$$- \begin{bmatrix} 2(t-1) - \frac{7}{3} + \frac{12}{5}\,\epsilon^{-(t-1)} - \frac{1}{15}\,\epsilon^{-6(t-1)} \\ 10 - \frac{48}{5}\,\epsilon^{-(t-1)} - \frac{2}{5}\,\epsilon^{-6(t-1)} \end{bmatrix}$$

$$= \begin{bmatrix} -\frac{12}{5}\,(1 - \epsilon^{-1})\,\epsilon^{-(t-1)} + \frac{1}{15}\,(1 - \epsilon^{-6})\,\epsilon^{-6(t-1)} + 2 \\ \frac{48}{5}\,(1 - \epsilon^{-1})\,\epsilon^{-(t-1)} + \frac{2}{5}\,(1 - \epsilon^{-6})\,\epsilon^{-6(t-1)} \end{bmatrix}.$$

This again agrees with the previously found result.

5.4 SUPERPOSITION PRINCIPLE

In the preceding sections of this chapter we obtained, in a formal way, expressions that relate the response of an initially relaxed network to an excitation $e(t)$ and the impulse response or step response through a convolution integral. It is possible to interpret these integrals as statements of the superposition principle. This will be the subject of the present section.

Concepts relating to the superposition principle are best developed with illustrations. Unfortunately, it is difficult to illustrate vector and matrix functions. Therefore we shall develop our results for a single-input, single output network, since only scalar equations are then involved, and the scalar functions encountered can be illustrated. At the appropriate place we shall state the corresponding results for the vector equations characterizing multi-input, multi-output networks.

SUPERPOSITION IN TERMS OF IMPULSES

Consider the excitation function sketched in Fig. 8a. The positive time axis is divided into a sequence of equal-length intervals, the length being $\Delta\tau$. It is not necessary that the intervals be equal, but the task ahead is easier to formulate if they are.

Now consider the sequence of impulses labeled $f(t)$ shown in Fig. 8b. The impulse at the point $k\Delta\tau$ has a strength $e(k\Delta\tau)\Delta\tau$, which is the area of a rectangle formed by the base $\Delta\tau$ and the height of the curve of Fig. 8a at the point $k\Delta\tau$. The rectangle is shown shaded. The heights of the arrows in the figure have been drawn proportional to this strength. However, remember that the impulses are all of infinite height. Hence, for any finite $\Delta\tau$, no matter how small, the string of impulses is not a good pointwise representation of the excitation function, which is everywhere finite. Nevertheless, let us compute the response of the network to this sequence of impulses. Now $f(t)$ is not a point function; it is a generalized function that can be expressed as

$$f(t) = \sum_{k=1}^{\infty} \left[e(k\Delta\tau)\Delta\tau\right] \delta(t - k\Delta\tau). \tag{47}$$

Let us denote by Δw_k the response at time t to one of these impulses. Then

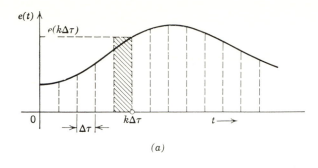

(a)

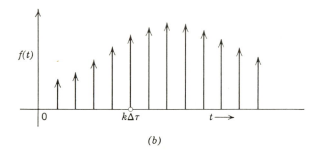

(b)

Fig. 8.   Decomposition of function into impulse train.

Δw_k will be equal to the strength of the impulse times the impulse response, suitably displaced. Thus

$$\Delta w_k = w_\delta(t - k\Delta\tau)[e(k\Delta\tau)\,\Delta\tau] \tag{48}$$

is the response at time t to the impulse at time $k\Delta\tau$.

Let us now concentrate on a particular point on the axis, which we can call τ. For a given value of $\Delta\tau$, this point will be $k\Delta\tau$. If we let $\Delta\tau$ get smaller, we shall have to increase k proportionately so that the value $\tau = k\Delta\tau$ will stay the same, since it refers to a fixed point on the axis. Hence, (47) and (48) can be rewritten as follows:

$$f(t) = \sum e(\tau)\,\delta(t - \tau)\,\Delta\tau, \tag{49}$$

$$\Delta w_k = w_\delta(t - \tau)\,e(\tau)\,\Delta\tau. \tag{50}$$

The response at any time t is obtained by adding the responses to each of the impulses up to time t. Let us denote by $w(t)$ the response to the sequence of impulses as we let $\Delta\tau$ approach zero.

$$w(t) = \lim_{\Delta\tau \to 0} \sum_{\tau=0}^{t} \Delta w_k = \lim_{\Delta\tau \to 0} \sum_{\tau=0}^{t} w_\delta(t-\tau)\, e(\tau)\, \Delta\tau$$

$$= \int_0^t w_\delta(t-\tau)\, e(\tau)\, d\tau. \tag{51}$$

(The summation has been indicated as extending from $\tau = 0$ to $\tau = t$. Actually, it should be $k = 1$ to n, where n is the largest integer such that $n\Delta\tau \le t$, with the limit taken as $\Delta\tau$ tends to zero. Since $\tau = k\Delta\tau$, the notation we have used is equivalent to this.) The indicated limit is, by definition, the integral written in the last line.

The question that remains to be answered is whether the sum of impulse functions $f(t)$ given in (49) can represent the original excitation $e(t)$ in the limit as $\Delta\tau$ approaches zero. In a formal way, the summation in (49) will become an integral which, by the sampling property of impulse functions, becomes $e(t)$. Thus, in the limit, the series of impulses represents the excitation.

In view of the preceding discussion, we can, for a single-input, single-output network, interpret the convolution integrals in (15) as expressing the response to an excitation $e(t)$ as the superposition of responses to a sequence of impulses that make up the function $e(t)$.

Now let us turn our attention to the multi-input, multi-output case. Precisely the same type of development leading to the convolution of the impulse response with the excitation can be carried out. Thus

$$\mathbf{w}(t) = \lim_{\Delta\tau \to 0} \sum_{\tau=0}^{t} \Delta\mathbf{w}_k = \lim_{\Delta\tau \to 0} \sum_{\tau=0}^{t} \mathbf{W}_\delta(t-\tau)\, \mathbf{e}(\tau)\, \Delta\tau$$

$$= \int_0^t \mathbf{W}_\delta(t-\tau)\, \mathbf{e}(\tau)\, d\tau, \tag{52}$$

where

$$\Delta\mathbf{w}_k = \mathbf{W}_\delta(t-\tau)\, \mathbf{e}(\tau)\, \Delta\tau \tag{53}$$

is the response to the vector of impulses occurring at time $\tau = k\Delta\tau$ in the representation of $\mathbf{e}(t)$ given by

$$\mathbf{f}(t) = \sum \mathbf{e}(\tau)\, \delta(t-\tau)\, \Delta\tau. \tag{54}$$

To give a suitable interpretation, we must reformulate the result. Let $\mathbf{w}_{\delta j}$ denote the jth column vector of $\mathbf{W}_\delta$. Then from (52) we get

$$\mathbf{w}(t) = \int_0^t \sum_{j=1}^p \mathbf{w}_{\delta j}(t-\tau)\, e_j(\tau)\, d\tau$$

$$= \sum_{j=1}^p \int_0^t \mathbf{w}_{\delta j}(t-\tau)\, e_j(\tau)\, d\tau. \qquad (55)$$

Now $\mathbf{w}_{\delta j}(t)$ is the response when the jth input is excited by a unit impulse at time zero and all other inputs have zero excitation. Thus, in the multi-input, multi-output case, an appropriate interpretation of (52) is the following: The response to the excitation $e(t)$ is the superposition of a set of responses, each of which is the response of the network to excitation at a single input, the excitation consisting of a sequence of impulses.

SUPERPOSITION IN TERMS OF STEPS

A development similar to that just completed can be carried out by representing an excitation function as a sum of step functions. Let us, as before, initiate the discussion in terms of a single-input, single-output network. The positive time axis is divided again into equal intervals of length $\Delta\tau$. The excitation function represented as a sum of step functions is illustrated in Fig. 9. The resulting "staircase" function is not a very

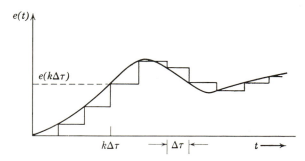

Fig. 9.   Decomposition of a function into step functions.

good approximation to $e(t)$, but it gets better as $\Delta\tau$ is made smaller. When $\Delta\tau$ is very small, the value of each step in the staircase can be approximated by the product of $\Delta\tau$ and the slope of the curve at the jump, since each of the little figures between the curve and the staircase function approaches a triangle.

The response of the network to the excitation $e(t)$ can be approximated by the response to the staircase function. However, this is nothing but a sum of step-function responses, suitably displaced and multiplied by the value of the discontinuity. Let Δw_k be the response at time t to the step occurring at $k\Delta\tau$. It will be given by

$$\Delta w_k = w_u(t - k\Delta\tau)[\dot{e}(k\Delta\tau)\,\Delta\tau], \tag{56}$$

where the dot indicates differentiation. The factor in brackets is the value of the step, whereas $w_u(t - k\Delta\tau)$ is the response to a displaced step function.

The total response will be the sum of the contributions for each step. Again, if we focus attention on the point $\tau = k\Delta\tau$ and take the limit as $\Delta\tau$ approaches zero, we shall get

$$w(t) = \lim_{\Delta\tau \to 0} \sum_{\tau=0}^{t} \Delta w_k = \lim_{\Delta\tau \to 0} \sum_{\tau=0}^{t} w_u(t - \tau)\,\dot{e}(\tau)\,\Delta\tau$$

$$= \int_0^t w_u(t - \tau)\,\dot{e}(\tau)\,d\tau. \tag{57}$$

In this development we have assumed that the excitation is a continuous function and that the initial value is zero. Now suppose it has discontinuities of value γ_i occurring at times t_i, respectively. We shall consider a nonzero initial value to be a discontinuity at $t = 0$. The total excitation will then be $\hat{e}(t) + \sum \gamma_i\,u(t - t_i)$, where $\hat{e}(t)$ is the continuous part of the excitation. We have already found the response to this part; to this we must now add the response due to the discontinuities. The complete response will be

$$w(t) = \sum_i w_u(t - t_i)\,\gamma_i + \int_0^t w_u(t - \tau)\,\dot{e}(\tau)\,d\tau. \tag{58}$$

In particular, if there are no discontinuities except at $t = 0$, then the total response will be [with $e(0)$ written for γ_0]

$$w(t) = w_u(t)\,e(0) + \int_0^t w_u(t - \tau)\,\dot{e}(\tau)\,d\tau. \tag{59}$$

This expression is identical with the first one in (44). We have now demonstrated that the response to an excitation $e(t)$ can be regarded as the superposition of the responses to a series of step functions that represent the excitation.

In the case of the multi-input, multi-output network, the corresponding result is achieved by simply changing $e(t)$ to $\mathbf{e}(t)$, γ_i to $\boldsymbol{\gamma}_i$, Δw_k to $\Delta \mathbf{w}_k$, $w(t)$ to $\mathbf{w}(t)$, and $w_u(t)$ to $\mathbf{W}_u(t)$ in (56) through (59). Corresponding to (58) we get

$$\mathbf{w}(t) = \sum_i \mathbf{W}_u(t - t_i)\,\boldsymbol{\gamma}_i + \int_0^t \mathbf{W}_u(t - \tau)\,\dot{\mathbf{e}}(\tau)\,d\tau, \tag{60}$$

which may be rewritten as

$$\mathbf{w}(t) = \sum_{j=1}^{p} \left[\sum_i \mathbf{w}_{uj}(t - t_i)(\gamma_i)_j + \int_0^t \mathbf{w}_{uj}(t - \tau)\,\dot{e}_j(\tau)\,d\tau \right], \tag{61}$$

where $(\gamma_i)_j$ denotes the jth element of $\boldsymbol{\gamma}_i$ and $\mathbf{w}_{uj}(t)$ denotes the jth column vector of $\mathbf{W}_u(t)$. Now $\mathbf{w}_{uj}(t)$ is the response when the jth input is excited by a unit step at time zero and all other inputs have zero excitation. Therefore the following interpretation of (61) can be given: The response to an excitation is the superposition of a set of responses, each of which is the response of the network to an excitation at a single input, the excitation consisting of a sequence of step functions.

5.5 NUMERICAL SOLUTIONS

These interpretations that we have given for the convolution-integral representations have an important application in the numerical computation of network response (e.g., by using a computing machine). It makes little difference to the final results whether we take the impulse or the step representation. Therefore, for the present discussion, let us take the former.

Suppose we wish to find the response of a single-input, single-output network to a time function that is not easily represented as a sum of elementary functions; for instance, the time function may be given simply as a curve, or its analytical formula may be very complicated. In such cases the Laplace transform $E(s)$ may be either difficult to find or be so involved as to be useless. If we approximate $E(s)$ by a rational function, we shall not know how good an approximation of the response function we shall get in the time domain. In such cases it is more meaningful to approximate $e(t)$ in the time domain by an impulse sequence, as in Fig. 8, or by a staircase function, as in Fig. 9.

Let us once again resort to an example. Suppose we have a network that has the impulse response shown in Fig. 10a. This impulse response

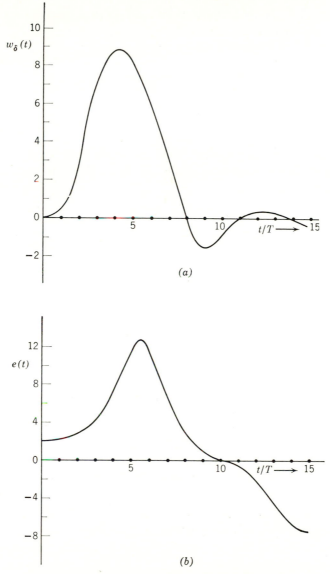

Fig. 10. Numerical computation of network response.

may have been found experimentally by using a short pulse as an "approximation" to the impulse. Suppose we wish to find the response of the network to the excitation in Fig. 10*b*, which again may be an experimental curve or the result of some other numerical computation.

We now select a suitable interval T such that the variations of $w_\delta(t)$ and $e(t)$ over an interval T are small enough to be negligible. Then we use the approximate representation

$$\hat{e}(t) = \sum_{k=1}^{\infty} e(kT)\,\delta(t - kT)\,T \tag{62}$$

for the excitation. This expression is usually interpreted as the result of multiplying $e(t)$ by the impulse train $\sum_{k=1}^{\infty} \delta(t - kT)\,T$. The set of values $e(kT)$ is referred to as a *time sequence*, and the function $e(t)$ is referred to as a *time series*. It can be shown that the same final results as we shall obtain here using these concepts can also be obtained by approximating

$$f(t) = \int_0^t e(x)\,dx \tag{63}$$

by a staircase function and using Laplace-Stieltjes transforms, without using the impulse function. Thus our final results can be justified in the realm of rigorous mathematics.

Now, by using the convolution theorem, the response of the network to the time series (62) can be written as follows:

$$w_{\hat{e}}(t) = \sum_{k=1}^{n} w_\delta(t - kT)\,e(kT)\,T, \tag{64}$$

where

$$nT \le t < (n+1)T.$$

In particular, the value of the response at our chosen points nT will be given by*

$$w_{\hat{e}}(nT) = \sum_{k=1}^{n} w_\delta[(n-k)T]\,e(kT)\,T. \tag{65}$$

Let us see the implication of this equation. We notice that the sum on the right is simply a sum of *real numbers*, *not functions*. Thus we can get an approximate idea of the response simply by adding these numbers, without integrating functions.

---

* Since n is the only variable in this equation, we can write this in more conventional form as

$$w_{\hat{e}n}/T = \sum_{k=1}^{n} e_k w_{n-k}$$

and observe that it is the Cauchy product of two time sequences, for e and w_δ.

To make this point a little clearer, let us find the approximate response at $t/T = 2, 4, 6, 8$, and 10 for the example of Fig. 10, by using the values given in Table 1. (The invervals chosen are too large for any accuracy, but the example suffices as an illustration.) Lines 1 and 2 of this table are found by reading the values of $e(kT)$ and $w_\delta(kT)$ from the graphs. The remaining odd-numbered lines, the lines associated with $w_\delta[(n-k)T]$, $n = 2, 4, 6, 8$, and 10, are obtained by copying line 1 backwards, starting at the column corresponding to $k = n - 1$. The elements in each of these lines, when multiplied by their corresponding entries in line 2, give the entries in the even-numbered lines, those associated with $w_\delta[(n-k)T]e(kT)$. The sum of the entries from $k = 1$ to $k = n$ in each of the even-numbered lines is $w_{\hat{e}}(nT)/T$. Thus

n	2	4	6	8	10
$w_{\hat{e}}(nT)/T$	1.0	24.1	99.6	247.2	304.54

This computation has given a numerical value for the response at a few selected points. The final tabulated values are a time *sequence*. With these values, the time *series* can be written as

$$\hat{w}(t) = \sum_n w_{\hat{e}}(nT)\,\delta(t - nT)\,T = \sum_{n=1}^{\infty} \sum_{k=1}^{n} \{w_\delta[(n-k)T]\,e(kT)\,T\}\delta(t - nT)\,T.$$

$$(66)$$

This method of representing everything by time series leads to the so-called z-transform method of analysis used in sampled-data systems.

The same concept of time series is also used in time-domain synthesis. Quite often the synthesis problem is specified by means of a curve for the excitation $e(t)$ and a curve for the response $w(t)$ that is desired. Then one of the procedures in use is to represent $e(t)$ and $w(t)$ as time series and to use simultaneous equations derived from (66) to find the time series for $w_\delta(t)$. The synthesis then proceeds by finding $H(s)$. The mathematical problems that arise are too numerous for us to consider this question any further, and so we leave this application to advanced courses on synthesis.

MULTI-INPUT, MULTI-OUTPUT NETWORKS

Let us now examine the changes needed to handle the multi-input, multi-output network. The time series for $e(t)$ will be

$$\hat{\mathbf{e}}(t) = \sum_{k=1}^{\infty} \mathbf{e}(kT)\,\delta(t - kT)\,T.$$

$$(67)$$

Table 1

Line No.	k	0	1	2	3	4	5	6	7	8	9	10
1	$w_\delta(kT)$	0.0	0.5	3.0	7.0	9.0	8.5	6.0	2.5	-0.3	-1.7	—
2	$e(kT)$		2.0	2.7	4.0	8.0	12.0	11.0	6.0	2.5	0.5	0.0
3	$w_\delta[(2-k)T]$		0.5	0.0	—	—						
4	$w_\delta[(2-k)T]\,e(kT)$		1.0	0.0	—	—						
5	$w_\delta[(4-k)T]$		7.0	3.0	0.5	0.0	—					
6	$w_\delta[(4-k)T]\,e(kT)$		14.0	8.1	2.0	0.0	—	—				
7	$w_\delta[(6-k)T]$		8.5	9.0	7.0	3.0	0.5	0.0				
8	$w_\delta[(6-k)T]\,e(kT)$		17.0	24.6	28.0	24.0	6.0	0.0				
9	$w_\delta[(8-k)T]$		2.5	6.0	8.5	9.0	7.0	3.0	0.5	0.0		
10	$w_\delta[(8-k)T]\,e(kT)$		5.0	16.2	34.0	72.0	84.0	33.0	3.0	0.0		
11	$w_\delta[(10-k)T]$		-1.7	-0.3	2.5	6.0	8.5	9.0	7.0	3.0	0.5	0.0
12	$w_\delta[(10-k)T]\,e(kT)$		-3.4	-0.81	10.0	48.0	102.0	99.0	42.0	7.5	0.25	0.0

The response to that time series for $nT \leq 1 < (n+1)T$, and specifically at the points nT, will be

$$\mathbf{w}_{\hat{e}}(t) = \sum_{k=1}^{n} \mathbf{W}_{\delta}(t - kT)\, \mathbf{e}(kT)\,T, \qquad (68a)$$

$$\mathbf{w}_{\hat{e}}(nT) = \sum_{k=1}^{n} \mathbf{W}_{\delta}[(n-k)T]\, \mathbf{e}(kT)\,T, \qquad (68b)$$

and hence the associated time series will be

$$\hat{\mathbf{w}}(t) = \sum_{n=1}^{\infty} \left\{ \sum_{k=1}^{n} \mathbf{W}_{\delta}[(n-k)T]\, \mathbf{e}(kT)\,T \right\} \delta(t - nT)\,T. \qquad (69)$$

Conceptually, the multi-input, multi-output network time-series response is quite easily computed. The time required to perform the calculations, however, is considerably greater than that needed to evaluate the response of a single-input, single-output network. It is significant, therefore, that (68) is easily programmed for solution by a digital computer. One drawback to such a program and its execution will be the large amount of storage space needed to retain $\mathbf{W}_{\delta}(nT)$ and $\mathbf{e}(nT)$, for $n = 0, 1, \ldots, N$, in memory, or, if not stored, the large amount of time required to recompute $\mathbf{W}_{\delta}(nT)$ and $\mathbf{e}(nT)$ each time they are needed in a calculation. The latter alternative is only possible if $\mathbf{W}_{\delta}(nT)$ and $\mathbf{e}(nT)$ are known analytically and not just experimentally. The technique to be developed now overcomes some of these problems.

STATE RESPONSE

Thus far we have sought to solve for the network response directly without consideration of the network state. Since the network response could be, without loss of generality, the network state, all that we have concluded till now concerning the response vector applies equally well to the state vector. Because, in the method about to be developed, no significant simplicity is achieved by assuming that the network is initially relaxed, we shall no longer make this assumption.

According to (23) the state vector is the sum of the *free-state response*

$$\mathbf{x}^{f}(t) = \epsilon^{\mathscr{A}(t - t_0)}\, \mathbf{x}(t_0) \qquad (70)$$

and the *controlled-state response*

$$\mathbf{x}^{c}(t) = \int_{t_0}^{t} \epsilon^{\mathscr{A}(t - \tau)}\, \mathscr{B}\, \mathbf{e}(\tau)\, d\tau; \qquad (71)$$

that is,

$$\mathbf{x}(t) = \mathbf{x}^c(t) + \mathbf{x}^f(t). \tag{72}$$

We shall take $t_0 = 0$ and suppose that $\mathbf{e}(t)$ is approximated by the time series (67). Remember that the network response in (68) is the controlled response. Then for the controlled state response for $nT \leq t < (n+1)T$ and at the point $t = nT$, we can write

$$\mathbf{x}_{\hat{e}}^c(t) = \sum_{k=1}^{n} \epsilon^{\mathscr{A}(t-kT)} \, \mathscr{B} \, \mathbf{e}(kT)T, \tag{73a}$$

$$\mathbf{x}_{\hat{e}}^c(nT) = \sum_{k=1}^{n} \epsilon^{\mathscr{A}(n-k)T} \, \mathscr{B} \, \mathbf{e}(kT)T. \tag{73b}$$

At $t = nT$, the free-state response is simply

$$\mathbf{x}^f(nT) = \epsilon^{\mathscr{A}nT}\mathbf{x}(0). \tag{74}$$

Substituting (73) and (74) into (72) we get

$$\mathbf{x}_{\hat{e}}(nT) = \epsilon^{\mathscr{A}nT}\mathbf{x}(0) + \sum_{k=1}^{n} \epsilon^{\mathscr{A}(n-k)T} \, \mathscr{B} \, \mathbf{e}(kT)T$$

$$= \epsilon^{\mathscr{A}T}\left\{ \epsilon^{\mathscr{A}(n-1)T} \, \mathbf{x}(0) + \sum_{k=1}^{n-1} \epsilon^{\mathscr{A}(n-k-1)T} \, \mathscr{B} \, \mathbf{e}(nT)T \right\} + \mathscr{B} \, \mathbf{e}(nT)T.$$

The last rearrangement is obtained by writing the term of the summation corresponding to $k = n$ separately, and then factoring $\epsilon^{\mathscr{A}T}$ from the other terms. By comparing the quantity within braces with the previous line, the only difference is seen to be a replacement of n by $n-1$. Hence

$$\mathbf{x}_{\hat{e}}(nT) = \epsilon^{\mathscr{A}T}\mathbf{x}_{\hat{e}}[(n-1)T] + \mathscr{B} \, \mathbf{e}(nT)T. \tag{75}$$

This is an extremely valuable recursion formula. Observe that this expression could have been deduced directly from (23) by setting $\mathbf{x}(t_0) = \mathbf{x}_{\hat{e}}[(n-1)T]$ and by assuming $\epsilon^{\mathscr{A}(nT-\tau)}\mathscr{B}\mathbf{e}(\tau)$ is essentially constant on the interval $(n-1)T \leq \tau \leq nT$ and equal to its value at $\tau = nT$. The present approach was adopted in order to be consistent with the previously formulated results.

Notice that the recursion relation established by (75) requires that we have knowledge of $\epsilon^{\mathscr{A}T}$ at only one point in time; namely, $t = T$. This fact largely alleviates the computer storage-space problem referred to earlier.

The solution for the network response now proceeds as follows: We first let $\mathbf{x}_{\hat{e}}(0)$ equal the initial state $\mathbf{x}(0)$; then (75) is used to solve for the state vector $\mathbf{x}_{\hat{e}}(nT)$ as n successively assumes the values $1, 2, 3, \ldots$. For each value of n, the network response at $t = nT$, according to (22b), is given by

$$\mathbf{w}_{\hat{e}}(nT) = \mathbb{C}\mathbf{x}_{\hat{e}}(nT) + \mathscr{D}\mathbf{e}(nT) + \widehat{\mathscr{D}}\dot{\mathbf{e}}(nT). \tag{76}$$

This gives the time sequence of the network at the sampling points nT. The time-series approximation to the network response then follows directly.

Example

Let us now illustrate these results with a simple example. Suppose

$$\mathscr{A} = \begin{bmatrix} -1 & 1 \\ 0 & -2 \end{bmatrix}, \quad \mathscr{B} = \begin{bmatrix} 2 \\ -1 \end{bmatrix}, \quad \mathbb{C} = \begin{bmatrix} -2 & 0 \\ 1 & -1 \\ 0 & 3 \end{bmatrix}, \quad \mathscr{D} = \mathbf{0}, \quad \widehat{\mathscr{D}} = \mathbf{0}.$$

Using the procedures of Chapter 4, we find simply that

$$\epsilon^{\mathscr{A}T} = \begin{bmatrix} \epsilon^{-T} & \epsilon^{-T} - \epsilon^{-2T} \\ 0 & \epsilon^{-2T} \end{bmatrix}.$$

Suppose the interval is chosen to be $T = 0.2$; then

$$\epsilon^{0 \cdot 2\mathscr{A}} = \begin{bmatrix} 0.819 & 0.148 \\ 0.000 & 0.670 \end{bmatrix}.$$

Using this, (75) and (76) become

$$\begin{bmatrix} x_{\hat{e}1}(nT) \\ x_{\hat{e}2}(nT) \end{bmatrix} = \begin{bmatrix} 0.819 & 0.148 \\ 0.000 & 0.670 \end{bmatrix} \begin{bmatrix} x_{\hat{e}1}\{(n-1)T\} \\ x_{\hat{e}2}\{(n-1)T\} \end{bmatrix} + \begin{bmatrix} 2 \\ -1 \end{bmatrix} [e(nT)]T$$

$$\begin{bmatrix} w_{\hat{e}1}(nT) \\ w_{\hat{e}2}(nT) \\ w_{\hat{e}3}(nT) \end{bmatrix} = \begin{bmatrix} -2 & 0 \\ 1 & -1 \\ 0 & 3 \end{bmatrix} \begin{bmatrix} x_{\hat{e}1}(nT) \\ x_{\hat{e}2}(nT) \end{bmatrix}.$$

Let the network be initially relaxed and let the excitation be the function $e(t)$ given in Fig. 10. The values for $e(nT)$ are given in the top line of Table 2. Using the recursion formula and numerical values above, computations for $n = 1, \ldots, 5$ yield the results given in Table 2.

Table 2

n	1	2	3	4	5
$e(nT)$	2.0	2.7	4.0	8.0	12.0
$x_{\hat{e}1}(nT)$	0.800	1.68	1.86	4.53	8.29
$x_{\hat{e}2}(nT)$	−0.400	−0.808	−1.34	−1.50	−3.41
$w_{\hat{e}1}(nT)$	−1.60	−3.36	−3.72	−9.06	−16.6
$w_{\hat{e}2}(nT)$	1.20	2.49	3.20	6.03	11.7
$w_{\hat{e}2}(nT)$	−1.20	−2.42	−4.02	−4.50	−10.2

PROPAGATING ERRORS

A particular type of error is created in evaluating $x_{\hat{e}}(nT)$ numerically according to the recursion relation (75). Any error in the numerical value of $e(nT)$ propagates and generates errors in $x_{\hat{e}}(mT)$ with $m \geq n$. Similarly, an error in $x(0)$ propagates and generates errors in $x_{\hat{e}}(mT)$. To get an idea of how such errors propagate, suppose $e(nT)$ is correctly known for all n except $n = n_0$. Let $e(n_0T) = e^a(n_0T) + \zeta$, where $e^a(n_0T)$ is the actual value and ζ is an error vector. Suppose also that $x(0)$, $\epsilon^{\mathscr{A}T}$, $\mathscr{B}$, and T are correctly known. Further, let $x_{\hat{e}}{}^a(nT)$ denote the value of $x_{\hat{e}}(nT)$ for $n \geq n_0$ when $e^a(n_0T)$ replaces $e(n_0T)$. Then, by applying the recursion relation successively for $n = n_0, n_0 + 1, \ldots, n$, we find that

$$x_{\hat{e}}(n_0T) = x_{\hat{e}}{}^a(n_0T) + \mathscr{B}\zeta T$$

and, for $n \geq n_0$,

$$x_{\hat{e}}(nT) = x_{\hat{e}}{}^a(nT) + (\epsilon^{\mathscr{A}T})^{(n-n_0)}\mathscr{B}\zeta T. \tag{77}$$

Note that the error $\mathscr{B}\zeta T$ in the state response at n_0T propagates and causes an error in the state response at nT, equal to the error in $x_{\hat{e}}(n_0T)$ premultiplied by $(\epsilon^{\mathscr{A}T})^{(n-n_0)}$.

In a similar manner, let $x(0) = x^a(0) + \xi$, where $x^a(0)$ is the actual value and ξ is an error vector. The recursion relation yields

$$x_{\hat{e}}(nT) = x_{\hat{e}}{}^a(nT) + (\epsilon^{\mathbf{M}T})^n\xi, \tag{78}$$

where $\mathbf{x}_{\hat{e}}{}^{a}(nT)$ denotes the value of $\mathbf{x}_{\hat{e}}(nT)$ computed when $\mathbf{x}(0)$ is replaced by $\mathbf{x}^{a}(0)$. Again, the error $\boldsymbol{\xi}$ in $\mathbf{x}_{\hat{e}}(0)$ propagates and causes an error in $\mathbf{x}_{\hat{e}}(nT)$ equal to the error in $\mathbf{x}_{\hat{e}}(0)$ premultiplied by $(\epsilon^{\mathscr{A}T})^{n}$.

Errors in the numerical evaluation of $\epsilon^{\mathscr{A}T}$ also propagate. However, we shall defer consideration of these errors till we examine the problem of calculating $\epsilon^{\mathscr{A}T}$ in the next section. Although we shall not consider them here, errors in $\mathscr{B}$ and T also propagate. Using the recursion relationship, you should work out how these errors propagate.

It is clear that the value of $(\epsilon^{\mathscr{A}T})^{(n-m)}$ determines to what extent an error in $\mathbf{x}_{\hat{e}}(mT)$, due to errors in the excitation and in the initial state, affects the accuracy of $\mathbf{x}_{\hat{e}}(nT)$ at a later time. We shall lump the state-response errors contributed by errors in excitation and initial state. Let $\mathbf{x}_{\hat{e}}(mT)$ be in error by $\boldsymbol{\varepsilon}_m$ and let

$$\boldsymbol{\varepsilon}_n = (\epsilon^{\mathscr{A}T})^{(n-m)}\boldsymbol{\varepsilon}_m \tag{79}$$

be the propagated error in $\mathbf{x}_{\hat{e}}(nT)$, assuming $n > m$. A question that is of great significance is: How does the " size " of the error behave as n is increased? Here we are dealing with an error vector; we specify the " size " of a vector in terms of its norm.

For a refresher about norms of vectors and matrices, see Chapter 1. There, the norm of a vector $\boldsymbol{\varepsilon}$ was written as $\|\boldsymbol{\varepsilon}\|$ and defined as a non-negative number having the following properties:

1. $\|\boldsymbol{\varepsilon}\| = 0$ if and only if $\boldsymbol{\varepsilon} = \mathbf{0}$;

2. $\|\alpha\boldsymbol{\varepsilon}\| = |\alpha|\,\|\boldsymbol{\varepsilon}\|$, where α is real or complex number;

3. $\|\boldsymbol{\varepsilon}_1 + \boldsymbol{\varepsilon}_2\| \leq \|\boldsymbol{\varepsilon}_1\| + \|\boldsymbol{\varepsilon}_2\|$, where $\boldsymbol{\varepsilon}_1$ and $\boldsymbol{\varepsilon}_2$ are vectors. (This is the

    triangle inequality.)

A vector may have a number of different norms satisfying these properties. The following three were discussed in Chapter 1:

1. $\|\boldsymbol{\varepsilon}\|_1 = \sum\limits_{i=1}^{n} |\boldsymbol{\varepsilon}_i|$, the sum-magnitude norm; $\hspace{2cm}$ (80a)

2. $\|\boldsymbol{\varepsilon}\|_2 = (\boldsymbol{\varepsilon}'\boldsymbol{\varepsilon})^{1/2} = \left(\sum\limits_{i=1}^{n} |\boldsymbol{\varepsilon}_i|^2 \right)^{1/2}$, the Euclidean norm; $\hspace{1cm}$ (80b)

3. $\|\boldsymbol{\varepsilon}\|_\infty = \max\limits_{i}|\boldsymbol{\varepsilon}_i|$, the max-magnitude norm. $\hspace{2cm}$ (80c)

Of these, the most common and familiar one is the Euclidean norm, which corresponds to the length of a space vector. However, as a practical matter, the numerical evaluation of the other two norms is often significantly easier.

As discussed in Chapter 1, the norm of a matrix as a transformation, transforming one vector into another, satisfies the three properties of a norm listed above. Further, the norm of the vector $\mathbf{K}\boldsymbol{\varepsilon}$, that is, the transformation of $\boldsymbol{\varepsilon}$ by $\mathbf{K}$, satisfies the inequality

$$\|\mathbf{K}\boldsymbol{\varepsilon}\| \leq \|\mathbf{K}\| \, \|\boldsymbol{\varepsilon}\|. \tag{81}$$

In addition and as a direct consequence of (81), the matrix norm has the following property:

$$\|\mathbf{K}_1 \mathbf{K}_2\| \leq \|\mathbf{K}_1\| \, \|\mathbf{K}_2\|. \tag{82}$$

The matrix norms corresponding to the three vector norms in (80) are the following. Let $\mathbf{K} = [k_{ij}]$; then

$$\|\mathbf{K}\|_1 = \max_j \sum_{i=1}^{n} |k_{ij}|, \quad \text{the value of the norm of the column vector with the greatest sum-magnitude norm;} \tag{83a}$$

$$\|\mathbf{K}\|_2 = \lambda_m, \quad \text{where } \lambda_m{}^2 \text{ is the eigenvalue of } \mathbf{K}^*\mathbf{K} \text{ with the largest magnitude;} \tag{83b}$$

$$\|\mathbf{K}\|_\infty = \max_i \sum_{j=1}^{n} |k_{ij}|, \quad \text{the value of the norm of the row vector with greatest sum-magnitude norm;} \tag{83c}$$

Our interest is in determining the norm of the error in the state response. In particular, we want to know in (79) if $\|\boldsymbol{\varepsilon}_n\| < \|\boldsymbol{\varepsilon}_m\|$ when $n > m$. In this equation the two errors are related through the matrix $(\epsilon^{\mathscr{A}T})^{(n-m)}$. Taking the norm of both sides of (79), and using (81) and (82) leads to

$$\|\boldsymbol{\varepsilon}_n\| \leq \|\epsilon^{\mathscr{A}T}\|^{n-m} \|\boldsymbol{\varepsilon}_m\|. \tag{84}$$

We conclude that, if the norm of $\epsilon^{\mathscr{A}T}$ is less than 1, then the norm of the error vector will be a decreasing function of n. Any of the matrix norms in (83) may be used to evaluate $\|\epsilon^{\mathscr{A}T}\|$.

Thus, for the previous example in which

$$\epsilon^{0.2\mathscr{A}} = \begin{bmatrix} 0.819 & 0.148 \\ 0.000 & 0.670 \end{bmatrix}$$

we find

$$\|\epsilon^{0.2\mathscr{A}}\|_1 = \max\,(0.819 + 0.000;\, 0.148 + 0.670) = 0.819;$$

$$\|\epsilon^{0.2\mathscr{A}}\|_2 = \text{root max eigenvalue of } (\epsilon^{0.2\mathscr{A}})'\, \epsilon^{0.2\mathscr{A}} = \max\,(0.819,\, 0.670)$$
$$= 0.819;$$

$$\|\epsilon^{0.2\mathscr{A}}\|_\infty = \max\,(0.819 + 0.148,\, 0.000 + 0.670) = 0.967.$$

For this example, any of the norms shows that the norm of the error vector decreases as n increases.

5.6 NUMERICAL EVALUATION OF $\epsilon^{\mathscr{A}T}$

In the numerical example of the last section we had an exact analytic expression for $\epsilon^{\mathscr{A}T}$; it was

$$\epsilon^{\mathscr{A}T} = \begin{bmatrix} \epsilon^{-T} & \epsilon^{-T} - \epsilon^{-2T} \\ 0 & \epsilon^{-2T} \end{bmatrix}.$$

For $T = 0.2$, this becomes

$$\epsilon^{0.2\mathscr{A}} = \begin{bmatrix} 0.819 & 0.148 \\ 0.000 & 0.670 \end{bmatrix}.$$

Is is evident, however, that this is only an approximate value of $\epsilon^{0.2\mathscr{A}}$, accurate to three digits. The actual $\epsilon^{0.2\mathscr{A}}$ has elements with an infinite number of digits, because $\epsilon^{-0.2}$ and $\epsilon^{-0.9}$ are irrational numbers. Thus, in any finite numerical process, only an approximate value of $\epsilon^{0.2\mathscr{A}}$ can be obtained.

What is true for this example is true for the general case; that is, it is generally true that for any $n \times n$ matrix $\mathscr{A}$ and any nonzero real constant T, only an approximate value of $\epsilon^{\mathscr{A}T}$ is known at the end of a finite numerical process. This becomes even more evident when it is realized that an exact analytic expression for $\epsilon^{\mathscr{A}T}$ is not usually known and that it might be necessary to evaluate $\epsilon^{\mathscr{A}T}$ by evaluating the terms in its defining power series

$$\epsilon^{\mathscr{A}T} = \sum_{k=0}^{\infty} \frac{\mathscr{A}^k T^k}{k!}. \tag{85}$$

It is clear that only a finite number of these terms can possibly be
evaluated in a finite numerical process. We shall now discuss the error
resulting from the truncation of the series in (85) and a criterion for
selecting the number of terms required to achieve a specified accuracy.

Let us write

$$\epsilon^{\mathscr{A}T} = \mathbf{A} + \mathbf{R}, \tag{86}$$

where

$$\mathbf{A} = \sum_{k=0}^{K} \frac{\mathscr{A}^k T^k}{k!} \tag{87}$$

is the truncated series approximation of $\epsilon^{\mathscr{A}T}$ and

$$\mathbf{R} = \sum_{k=K+1}^{\infty} \frac{\mathscr{A}^k T^k}{k!} \tag{88}$$

is the remainder or error matrix. Now if $\mathbf{A}$ is a good approximation of
$\epsilon^{\mathscr{A}T}$, then $\mathbf{A}\boldsymbol{\varepsilon}$ should be a good approximation of $\epsilon^{\mathscr{A}T}\boldsymbol{\varepsilon}$, where $\boldsymbol{\varepsilon}$ is an arbitrary n-vector .A quantitative measure of the quality of the latter approxi-
mation is the norm of the error vector

$$\epsilon^{\mathscr{A}T}\boldsymbol{\varepsilon} - \mathbf{A}\boldsymbol{\varepsilon} = (\epsilon^{\mathscr{A}T} - \mathbf{A})\boldsymbol{\varepsilon} = \mathbf{R}\boldsymbol{\varepsilon}$$

relative to the norm of $\epsilon^{\mathscr{A}T}\boldsymbol{\varepsilon}$; that is,

$$\delta = \frac{\|\mathbf{R}\boldsymbol{\varepsilon}\|}{\|\epsilon^{\mathscr{A}T}\boldsymbol{\varepsilon}\|} \tag{89}$$

is a measure of how well $\mathbf{A}\boldsymbol{\varepsilon}$ approximates $\epsilon^{\mathscr{A}T}\boldsymbol{\varepsilon}$. Obviously, the approximation gets better as δ gets smaller. Thus a reasonable criterion for select-
ing K, the maximum value of the index of summation in the truncated
power series, is as follows: K should be chosen such that δ is less than some
prescribed, positive number Δ. The value assigned to Δ, the prescribed
upper bound on δ, is selected to insure a desired level of accuracy in the
knowledge of $\epsilon^{\mathscr{A}T}\boldsymbol{\varepsilon}$ through evaluation of $\mathbf{A}\boldsymbol{\varepsilon}$.

It must be kept in mind that the accuracy we are talking about is
that achieved through calculations with numbers from the set of *all* real
numbers. When the set of numbers is finite, as it is in doing calculations

on a digital computer, there is a bound on the achievable accuracy. This computer-imposed bound on accuracy is related to the number of significant digits in floating-point numbers used in doing arithmetic on the computer. The implication for the problem we are examining is this: The accuracy of machine-performed calculations is determined by Δ only for Δ greater than some value established by the number of significant digits in numbers used by the computer in performing arithmetic operations.

We shall obtain, first, an upper bound on δ that is independent of $\mathbf{\varepsilon}$. A simple manipulation gives

$$\|\mathbf{\varepsilon}\| = \|\epsilon^{-\mathscr{A}T}\epsilon^{\mathscr{A}T}\mathbf{\varepsilon}\| \leq \|\epsilon^{-\mathscr{A}T}\| \, \|\epsilon^{\mathscr{A}T}\mathbf{\varepsilon}\|,$$

which implies that

$$\frac{\|\mathbf{\varepsilon}\|}{\|\epsilon^{-\mathscr{A}T}\|} = \|\epsilon^{-\mathscr{A}T}\|^{-1} \|\mathbf{\varepsilon}\| \leq \|\epsilon^{\mathscr{A}T}\mathbf{\varepsilon}\|. \tag{90}$$

Thus $\|\epsilon^{\mathscr{A}T}\mathbf{\varepsilon}\|$ is bounded from below by $\|\epsilon^{-\mathscr{A}T}\|^{-1} \|\mathbf{\varepsilon}\|$. Then, since $\|\mathbf{R}\mathbf{\varepsilon}\|$ is bounded from above by $\|\mathbf{R}\| \, \|\mathbf{\varepsilon}\|$, we find in conjunction with (89), the defining equation for δ, that

$$\delta \leq \frac{\|\mathbf{R}\| \, \|\mathbf{\varepsilon}\|}{\|\epsilon^{-\mathscr{A}T}\|^{-1} \|\mathbf{\varepsilon}\|} = \|\mathbf{R}\| \, \|\epsilon^{-\mathscr{A}T}\|. \tag{91}$$

To continue, we would like to evaluate the norm of $\mathbf{R}$ and of $\epsilon^{-\mathscr{A}T}$; but each of these is defined by an infinite series, and evaluation of the norms is not possible. However, we can calculate a bound on $\|\mathbf{R}\|$ as follows: The several inequality relations applied to the norm of the series (88) for $\mathbf{R}$ yield

$$\|\mathbf{R}\| \leq \sum_{k=K+1}^{\infty} \frac{\|\mathscr{A}\|^k T^k}{k!}.$$

Upon setting $l = k - K - 1$, we get, after a simple rearrangement of factors,

$$\|\mathbf{R}\| \leq \frac{\|\mathscr{A}\|^{K+1} T^{K+1}}{(K+1)!} \sum_{l=0}^{\infty} \frac{(K+1)!(K+2)^l}{(K+1+l)!} \left(\frac{\|\mathscr{A}\| T}{K+2}\right)^l$$

$$\leq \frac{\|\mathscr{A}^{K+1}\| T^{K+1}}{(K+1)!} \sum_{l=0}^{\infty} \left(\frac{\|\mathscr{A}\| T}{K+2}\right)^l.$$

The last line follows from the fact that $(K+1)!(K+2)^l/(K+1+l)! \le 1$ for all $l \ge 0$. Now, let K_0 be the least non-negative value of K such that $K+2 > \|\mathscr{A}\|T$; then for all $K \ge K_0$, we have $\|\mathscr{A}\|T/(K+2) < 1$ and

$$\sum_{l=0}^{\infty} \left(\frac{\|\mathscr{A}\|T}{K+2} \right)^l = \frac{1}{1 - \dfrac{\|\mathscr{A}\|T}{K+2}}.$$

Substituting this result in the preceding inequality, we find, for $K \ge K_0$, that

$$\|\mathbf{R}\| \le \frac{\|\mathscr{A}\|^{K+1} T^{K+1}}{(K+1)!} \frac{1}{1 - \dfrac{\|\mathscr{A}\|T}{K+2}}. \tag{92}$$

The right side is an upper bound on the norm of $\mathbf{R}$.

Now let us turn to the norm of $\epsilon^{-\mathscr{A}T}$. If T is replaced by $-T$ in (85), we obtain the defining power series for $\epsilon^{-\mathscr{A}T}$; thus

$$\epsilon^{-\mathscr{A}T} = \sum_{k=0}^{\infty} (-1)^k \frac{\mathscr{A}^k T^k}{k!}.$$

An upper bound on $\|\epsilon^{-\mathscr{A}T}\|$ is easily calculated. We find that

$$\|\epsilon^{-\mathscr{A}T}\| \le \sum_{k=0}^{\infty} \frac{\|\mathscr{A}\|^k T^k}{k!} = \epsilon^{\|\mathscr{A}\|T} \tag{93}$$

When the upper bounds on $\|\mathbf{R}\|$ and $\|\epsilon^{-\mathscr{A}T}\|$ in (92) and (93) are substituted into (91), we get

$$\delta \le \frac{\|\mathscr{A}\|^{K+1} T^{K+1}}{(K+1)!} \frac{1}{1 - \dfrac{\|\mathscr{A}\|T}{K+2}} e^{\|\mathscr{A}\|T} \tag{94}$$

for $K \ge K_0$. The right-hand side of (94) is a decreasing function of K, which tends to zero as K tends to infinity. Thus, for any given value of Δ, there are values of K, in fact a least value, such that

$$\frac{\|\mathscr{A}\|^{K+1} T^{K+1}}{(K+1)!} \frac{1}{1 - \dfrac{\|\mathscr{A}\|T}{K+2}} \epsilon^{\|\mathscr{A}\|T} < \Delta \tag{95}$$

and hence $\delta < \Delta$. Thus the values of $K \geq K_0$ that satisfy (95) also satisfy the previously stated criterion for selecting K. Of all the satisfactory values of K, it is suggested that K should be set equal to the least value that satisfies (95), since the number of arithmetic operations then needed to evaluate the truncated power series (87) will be minimized.

COMPUTATIONAL ERRORS

In the previous section we decomposed the state response $\mathbf{x}(t)$ into the sum of the free response $\mathbf{x}^f(t)$ and the controlled response $\mathbf{x}^c(t)$. As shown in (74), we found that

$$\mathbf{x}^f(nT) = (\epsilon^{\mathscr{A}T})^n \, \mathbf{x}(0). \tag{96}$$

Furthermore, when $\mathbf{e}(t)$ is approximated by the time series

$$\hat{e}(t) = \sum \mathbf{e}(kT) \, \delta(t - kT) T,$$

we found, as shown in (73b), that

$$\mathbf{x}_{\hat{e}}^c(nT) = \sum_{k=1}^{n} (\epsilon^{\mathscr{A}T})^{n-k} \, \mathscr{B} \, \mathbf{e}(kT) T. \tag{97}$$

We are now interested in computing the error resulting from replacing $\epsilon^{\mathscr{A}T}$ by its approximation in these two equations.

ERRORS IN FREE-STATE RESPONSE

Let us first consider the free-state response. Let $\boldsymbol{\varepsilon}^f(nT)$ denote the difference between the actual free state response $(\epsilon^{\mathscr{A}T})^n \mathbf{x}(0)$ and the approximate free-state response $\mathbf{A}^n \mathbf{x}(0)$; that is,

$$\boldsymbol{\varepsilon}^f(nT) = [(\epsilon^{\mathscr{A}T})^n - \mathbf{A}^n]\mathbf{x}(0). \tag{98}$$

After taking the norm of both sides of this equation, we get

$$\|\boldsymbol{\varepsilon}^f(nT)\| \leq \|(\epsilon^{\mathscr{A}T})^n - \mathbf{A}^n\| \, \|\mathbf{x}(0)\|. \tag{99}$$

Thus, in bounding $\|\boldsymbol{\varepsilon}^f(nT)\|$, we must establish an upper bound on terms of the form $\|(\epsilon^{\mathscr{A}T})^l - \mathbf{A}^l\|$. We have used l, rather than n, to denote the power to which $\epsilon^{\mathscr{A}T}$ and $\mathbf{A}$ are raised, since the bound we shall establish will be applied to relations other than just (99). In those other cases n will not be the exponent.

By the properties of matrix norms, we show that

$$\|[\epsilon^{\mathscr{A}T}]^l - \mathbf{A}^l\| = \|[\mathbf{A} + \mathbf{R}]^l - \mathbf{A}^l\|$$

$$= \|\sum_{k=1}^{l} \binom{l}{k} \mathbf{A}^{l-k} \mathbf{R}^k\|$$

$$\leq \sum_{k=1}^{l} \binom{l}{k} \|\mathbf{A}\|^{l-k} \|\mathbf{R}\|^k \tag{100}$$

$$\leq [\|\mathbf{A}\| + \|\mathbf{R}\|]^l - \|\mathbf{A}\|^l.$$

The second line follows from the binomial theorem. The next line results from the triangle inequality and from (82). Finally, the last line is a result of the binomial theorem applied backwards. If we require that K satisfy the inequality relation (95), we know from that equation that $\|\mathbf{R}\| < \Delta \|\epsilon^{-\mathscr{A}T}\|^{-1}$. Furthermore, it is easily shown (Problem 15) that

$$\|\epsilon^{-\mathscr{A}T}\|^{-1} \leq \|\epsilon^{\mathscr{A}T}\|$$

and, therefore, that $\|\mathbf{R}\| < \Delta \|\epsilon^{\mathscr{A}T}\|$. With the added observation that $\|\epsilon^{\mathscr{A}T}\| \leq \|\mathbf{A}\| + \|\mathbf{R}\|$ and the reasonable assumption that $\Delta < 1$, we get

$$\|\mathbf{R}\| < \frac{\Delta}{1 - \Delta} \|\mathbf{A}\|. \tag{101}$$

Combining (100) and (101) results in

$$\|[\epsilon^{\mathscr{A}T}]^l - \mathbf{A}^l\| < \|\mathbf{A}\|^l \left[\left(1 + \frac{\Delta}{1 - \Delta} \right)^l - 1 \right]$$

$$= \|\mathbf{A}\|^l \left[\left(\frac{1}{1 - \Delta} \right)^l - 1 \right]. \tag{102}$$

Observe that this bound on $\|[\epsilon^{\mathscr{A}T}]^l - \mathbf{A}^l\|$ is a decreasing function of l if and only if $\|\mathbf{A}\|/(1 - \Delta) < 1$; therefore we shall assume $\|\mathbf{A}\|/(1 - \Delta) < 1$ in the rest of this section.

Let us return to the task of bounding $\|\boldsymbol{\varepsilon}^f(nT)\|$. The result of substituting (102) in (99) and of some simple manipulations is

$$\|\boldsymbol{\varepsilon}^f(nT)\| < \|\mathbf{A}\|^n \left[\left(\frac{1}{1-\Delta}\right)^n - 1 \right] \|\mathbf{x}(0)\|$$

$$= \left(\frac{\|\mathbf{A}\|}{1-\Delta}\right)^n [1 - (1-\Delta)^n] \|\mathbf{x}(0)\| \tag{103}$$

$$< n \Delta \left(\frac{\|\mathbf{A}\|}{1-\Delta}\right)^n \|\mathbf{x}(0)\|.$$

The second line is just the first line with the terms grouped differently. The last line follows from the fact that $[1 - (1-\Delta)^n]$ is a bounded, increasing function of n that never exceeds $n\Delta$. It is an elementary calculus problem to find the maximum of the right-hand side of (103) with respect to n. In doing so, we obtain

$$\|\boldsymbol{\varepsilon}^f(nT)\| < \|\mathbf{A}\|^n \left[\left(\frac{1}{1-\Delta}\right)^n - 1 \right] \|\mathbf{x}(0)\| \leq -\frac{\epsilon^{-1}\|\mathbf{x}(0)\|\Delta}{\ln[\|\mathbf{A}\|/(1-\Delta)]}. \tag{104}$$

We see, by (104), that $\|\boldsymbol{\varepsilon}^f(nT)\|$ is bounded and, by (103) that $\|\boldsymbol{\varepsilon}^f(nT)\|$ is a decreasing function of n, tending toward zero as n tends to infinity. Furthermore, $\|\boldsymbol{\varepsilon}^f(nT)\|$ tends to zero with Δ. It should be kept in mind that computers that might be used to compute $\mathbf{x}^f(nT)$ recursively use numbers with a limited number of significant digits. Thus, as discussed previously, the error in numerical evaluation of $\mathbf{x}^f(nT)$ will be bounded in accord with (103) and (104) only if Δ is greater than some value dependent upon the computer accuracy.

ERRORS IN CONTROLLED-STATE RESPONSE

Next, let us consider the controlled-state response $\mathbf{x}_{\hat{e}}^c(nT)$ given by (97). Let $\boldsymbol{\varepsilon}^c(nT)$ denote the difference between the actual controlled-state response $\sum_{k=1}^n (\epsilon^{\mathscr{A}T})^{n-k} \mathscr{B} \mathbf{e}(kT) T$ and the approximate controlled-state response $\sum_{k=1}^n \mathbf{A}^{n-k} \mathscr{B} \mathbf{e}(kT) T$; that is,

$$\boldsymbol{\varepsilon}^c(nT) = \sum_{k=1}^n \left[(\epsilon^{\mathscr{A}T})^{n-k} - \mathbf{A}^{n-k} \right] \mathscr{B} \mathbf{e}(kT) T. \tag{105}$$

After taking the norm of both sides of this equation and applying the inequality in (102), we establish that

$$\|\boldsymbol{\varepsilon}^c(nT)\| < \sum_{k=1}^{n} \|\mathbf{A}\|^{n-k} \left[\left(\frac{1}{1-\Delta}\right)^{n-k} - 1\right] \|\mathscr{B}\,\mathbf{e}(kT)\|. \qquad (106)$$

Suppose, for $k = 1, 2, \ldots$, that $\|\mathscr{B}\,\mathbf{e}(kT)\| \leq E$, where E is a constant; that is, suppose $\|\mathscr{B}\,\mathbf{e}(kT)\|$ is a bounded function of k. Then it is found that

$$\begin{aligned}
\|\boldsymbol{\varepsilon}^c(nT)\| &< T\left\{\sum_{k=1}^{n} \|\mathbf{A}\|^{n-k}\left[\left(\frac{1}{1-\Delta}\right)^{n-k} - 1\right]\right\}E \\
&< T\left[\sum_{k=1}^{n} (n-k)\,\Delta\left(\frac{\|\mathbf{A}\|}{1-\Delta}\right)^{n-k}\right]E \qquad (107) \\
&< T\left[\sum_{l=0}^{n-1} l\,\Delta\left(\frac{\|\mathbf{A}\|}{1-\Delta}\right)^{l}\right]E.
\end{aligned}$$

The second line follows from the inequality $[1 - (1-\Delta)^{n-k}] \leq (n-k)\Delta$ for all $k \leq n$. The third line results from replacing $n - k$ by l. Under our assumption that $\|\mathbf{A}\|/(1-\Delta) < 1$, the right-hand side of (107) is a bounded increasing function of n; therefore

$$\|\boldsymbol{\varepsilon}^c(nT)\| < T\left[\Delta\sum_{l=0}^{\infty} l\left(\frac{\|\mathbf{A}\|}{1-\Delta}\right)^{l}\right]E = T\Delta\,\frac{\|\mathbf{A}\|/(1-\Delta)}{[1 - \|\mathbf{A}\|/(1-\Delta)]^2}\,E. \quad (108)$$

The equality at the right follows from the fact that a series of the type $\sum_{m=0}^{\infty} m\alpha^m$ is equal to $\alpha/(1-\alpha)^2$ for $\alpha < 1$. We see by this inequality that $\|\boldsymbol{\varepsilon}^c(nT)\|$ is bounded and tends to zero with Δ. However, for reasons discussed previously, when computing a numerical value for $\mathbf{x}^f(nT)$, the accuracy limitations of a computer preclude the error in numerical evaluation of $\mathbf{x}_{\hat{e}}{}^c(nT)$ being bounded in accord with (108) if Δ is too small.

Example

Let us illustrate the many ideas we have developed in this section with an example. Let us use the same $\mathscr{A}$ and T as in the last example; thus,

$$\mathscr{A} = \begin{bmatrix} -1 & 1 \\ 0 & -2 \end{bmatrix}$$

and $T = 0.2$. For the norm of $\mathscr{A}$ we shall arbitrarily select the least of the two norms (83a) and (83c). We have $\|\mathscr{A}\|_1 = 3$ and $\|\mathscr{A}\|_\infty = 2$; therefore we shall use the latter norm. To simplify the example, certainly not for accuracy, we shall require only $\Delta \leq 0.001$.

Recall that K_0 is the least non-negative (integer) value of K such that $K + 2 > \|\mathscr{A}\|T$. Since $\|\mathscr{A}\|T = 0.4$, we find that $K_0 = 0$. Thus, by (95), we must find a $K \geq 0$ such that

$$\frac{0.4^{K+1}}{(K+1)!} \frac{1}{1 - 0.4/(K+2)} \epsilon^{0.4} < 0.001.$$

We easily find that $K = 4$. The approximation $\mathbf{A}$ is evaluated as follows:

$$\mathbf{A} = \sum_{k=0}^{4} \frac{(0.2)^k}{k!} \mathscr{A}^k$$

$$= \mathbf{U} + 0.2\mathscr{A} + 0.02\mathscr{A}^2 + 0.00133\mathscr{A}^3 + 0.0000666\mathscr{A}^4$$

and

$$\mathscr{A}^2 = \begin{bmatrix} 1 & -3 \\ 0 & 4 \end{bmatrix}, \qquad \mathscr{A}^3 = \begin{bmatrix} -1 & 7 \\ 0 & -8 \end{bmatrix}, \qquad \mathscr{A}^4 = \begin{bmatrix} 1 & -15 \\ 0 & 16 \end{bmatrix}.$$

Hence, to an accuracy of five digits,

$$\mathbf{A} = \begin{bmatrix} 0.81873 & 0.14833 \\ 0.00000 & 0.67040 \end{bmatrix}.$$

Observe that $\|\mathbf{A}\|_\infty = 0.967$. Therefore, by (104), $\|\boldsymbol{\varepsilon}^f(nT)\|_\infty$ is bounded as

$$\|\boldsymbol{\varepsilon}^f(nT)\|_\infty < \frac{0.001 \times 0.368}{\ln\left(\dfrac{0.967}{0.999}\right)} \|\mathbf{x}(0)\|_\infty$$

$$< 0.0117 \|\mathbf{x}(0)\|_\infty$$

and, by (108), $\|\boldsymbol{\varepsilon}^c(nT)\|_\infty$ is bounded as

$$\|\boldsymbol{\varepsilon}^c(nT)\|_\infty < \frac{0.2 \times 0.001 \times 0.967/0.999}{1 - \left(\dfrac{0.967}{0.999}\right)^2} E$$

$$< 0.189E.$$

Let us conclude this section with an observation on one type of error not considered in this or the previous section. Let $e(t)$ be the network excitation and $x(t)$ the corresponding state response. The values of $x(t)$ at $t = nT$ will be $x(nT)$, and, unless $e(t) = \hat{e}(t)$, $x(nT) \neq x_{\hat{e}}(nT)$. The difference between $x(nT)$ and $x_{\hat{e}}(nT)$ is an error that stems from representing $e(t)$ by $\hat{e}(t)$. Even though we have not treated this type of error, we do know that, if we can convert a set of nonhomogeneous state equations into a set of homogeneous state equations, as shown in Chapter 4, then this type of error will not arise, since for the homogeneous equations $e(t) \equiv 0$.

PROBLEMS

1. In the text, the concept of the convolution of two matrix functions is introduced; extend the concept to more than two matrix functions.

2. Prove that the convolution of scalar functions shares the following algebraic properties with ordinary multiplication: If f_1, f_2, and f_3 are integrable functions (so that $f_1{}^*f_2 = \int_0^t f_1(x) f_2(t - x)\, dx$ is defined, as is $f_2{}^*f_3$ and $f_1{}^*f_3$), then

   (a) $f_1{}^*f_2 = f_2{}^*f_1$ (commutative law)

   (b) $f_1{}^*(f_2{}^*f_3) = (f_1{}^*f_2){}^*f_3$ (associative law)

   (c) $u{}^*f = f{}^*u = f$, where u is the unit step function (identity)

   (d) $f_1{}^*(f_2 + f_3) = f_1{}^*f_2 + f_1{}^*f_3$ (distributive law)

   Which of these properties hold when considering the convolution of matrix functions?

3. In the alternative derivation of the convolution integral from the state equations, only one form of the integral is obtained, namely, that given in (21). Show that the other form is also valid.

4. Find the impulse response and the step response of the networks given in Fig. P4 assuming initially relaxed conditions. The desired responses are indicated in the figures. Demonstrate that (31) is satisfied.

5. Find the indicated response of the same networks to the excitation functions in Fig. P5 by using the impulse response or the step response and a superposition integral. (The ordinate shows multiple labels to correspond to the several different source designations in Fig. P4.)

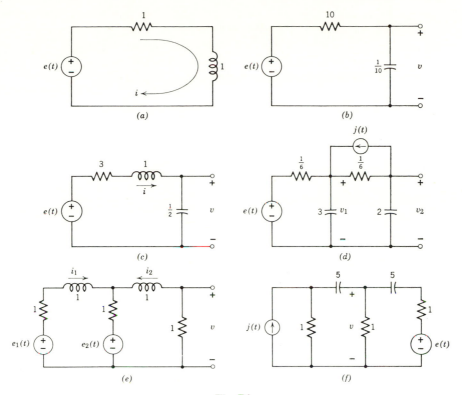

Fig. P4

6. The equivalent circuit of a two-stage RC-coupled amplifier is shown in Fig. P6. Find the response of the amplifier to the excitations shown in Fig. P5 using a superposition integral with (1) the impulse response and (2) the step response.

7. Solve the following integral equation of the convolution type:

$$\mathcal{F}(t) + \int_0^t \mathcal{F}(x)\, \mathcal{G}_1(t - \tau)\, d\tau = \mathcal{G}_2(t).$$

The unknown matrix function is $\mathcal{F}(t)$, and $\mathcal{G}_1(t)$ and $\mathcal{G}_2(t)$ are square matrices and are known (integrable) matrix functions.

8. Obtain a solution of the following scalar integral equations:

(a) $f(t) + \displaystyle\int_0^t f(\tau)\, \epsilon^{-2(t-\tau)}\, d\tau = 2t;$

(b) $f(t) + \displaystyle\int_0^t f(\tau)\, \epsilon^{-(t-\tau)} \sin (t - \tau)\, d\tau = 5;$

(c) $\dfrac{df}{dt} + 2f(t) + 9 \displaystyle\int_0^t f(\tau)\, (t - \tau)\, d\tau = 1 - \epsilon^{-2t}, \qquad f(0) = 0.$

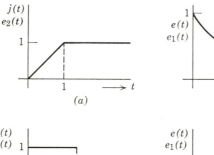

(a)   (b)

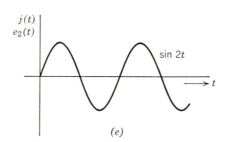

(c)   (d)

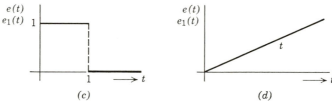

(e)

Fig. P5

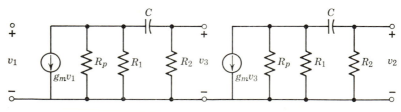

Fig. P6

9. Obtain a solution of the following matrix integral equations:

(a) $\mathscr{F}(t) + \int_0^t \mathscr{F}(\tau) \begin{bmatrix} \epsilon^{-(t-\tau)} & 0 \\ 2 & t+\tau \end{bmatrix} d\tau = \begin{bmatrix} \epsilon^{-2t} & t \\ 0 & t\epsilon^{-t} \end{bmatrix}$

(b) $\mathscr{F}(t) + \int_0^t \mathscr{F}(\tau) \begin{bmatrix} \epsilon^{-2(t-\tau)} & 1 & 0 \\ 0 & \epsilon^{-2(t-\tau)} & 1 \\ 0 & 0 & \epsilon^{-2(t-\tau)} \end{bmatrix} d\tau = \begin{bmatrix} \sin t & 0 & 1 \\ 0 & \cos t & -1 \\ 0 & 0 & \epsilon^{-t} \end{bmatrix}.$

10. Given the differential equation

$$\frac{d^2y}{dt^2} + 2\frac{dy}{dt} + y = \frac{df}{dt} + 3f$$

$$y(0+) = \dot{y}(0+) = 0$$

get an explicit formula for $y(t)$ (the solution) by finding first

$$\mathcal{L}^{-1}\left\{\frac{s+3}{s^2+2s+1}\right\}.$$

Use this formula to find the solution when $f(t)$ is

(a) $f(t) = \begin{cases} 1, & 0 \le t < 1 \\ 0, & 1 \le t \end{cases}$

(b) $f(t) = \begin{cases} t, & 0 \le t \le 1 \\ 0, & 1 \le t \end{cases}$

(c) $f(t) = \begin{cases} t^2, & 0 \le t \le 1 \\ 0, & 1 \le t \end{cases}$

11. The triangular voltage pulse shown in Fig. P11a is applied to the network of Fig. P11b. Find the output-voltage response for all time by using the convolution theorem.

12. Use the convolution-integral theorem to prove the translation (shifting) theorem of Laplace-transform theory.

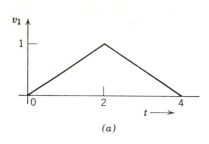

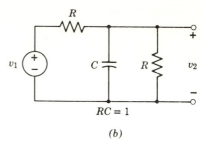

(a)

(b)

Fig. P11

13. The trapezoidal pulse shown in Fig. P13a is applied to the network of Fig. P13b. Find the output-voltage response for all time by using the convolution theorem.

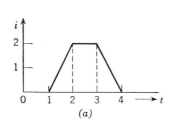

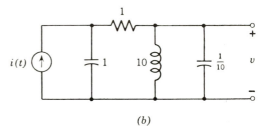

(a)

(b)

Fig. P13

14. The network of Fig. P14a is excited by two voltage sources. The voltages as functions of time are shown in Fig. P14b. Using the convolution theorem, compute the indicated network responses v, i_1, and i_2 for all time.

15. Prove that $\|\epsilon^{\mathscr{A}T}\| \geq \|\epsilon^{-\mathscr{A}T}\|^{-1}$.

16. The network of Fig. P11b is excited by the function

$$
v_1(t) = \begin{cases} \tan t, & 0 \leq t \leq \dfrac{\pi}{4} \\[2ex] \tan\left(\dfrac{\pi}{2} - t\right), & \dfrac{\pi}{4} \leq t < \dfrac{\pi}{2} \\[2ex] 0, & \dfrac{\pi}{2} \leq t \end{cases}
$$

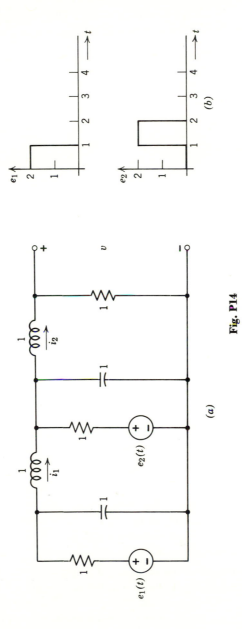

Fig. P14

Find the approximate response of the network for $0 \leq t \leq 2$, using time-series representations. Estimate the maximum error in the solution for the chosen interval.

17. Repeat Problem 16, but use the excitation

$$v_1(t) = \begin{cases} \arcsin t, & 0 \le t < 1 \\ 0, & 1 \le t \end{cases} \qquad \left(0 \le v_1 \le \frac{\pi}{2}\right)$$

Use the staircase approximation and the step response. Estimate the error.

18. The network of Fig. P18a is excited by the function of Fig. P18b. Find the approximate response $v_2(t)$ for $0 \le t \le 5$.

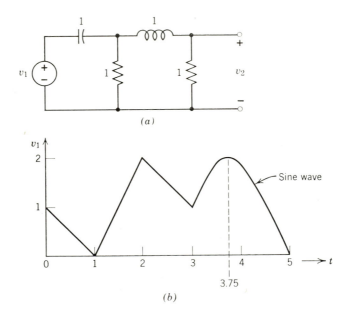

(a)

(b)

Fig. P18

19. For each of the networks of Fig. P4, compute the impulse-response time sequence with $T = 0.1$ and for $n = 0, 1, \ldots, 15$. Then compute the network-response time sequence for $n = 0, 1, \ldots, 15$, when the network excitations depicted in Fig. P5 are approximated by their time sequences.

20. The measured impulse response $w_\delta(t)$ of a single-input, single-output network is shown in Fig. P20a. Using the time sequence for the impulse response, with $T = 0.1$, compute the network-response time sequence for $n = 0, 1, 2, \ldots, 10$, when the network excitations $e(t)$ depicted in Fig. P20b are approximated by their time sequences.

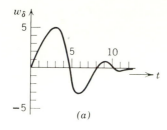

(a)

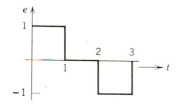

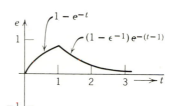

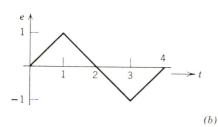

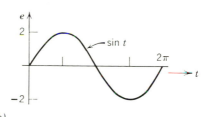

(b)

Fig. P20

21. Consider a network for which

$$\mathcal{A} = \begin{bmatrix} -1 & 1 \\ 0 & -1 \end{bmatrix}, \quad \mathcal{B} = \begin{bmatrix} -2 \\ 3 \end{bmatrix}, \quad \mathbf{C} = \begin{bmatrix} 1 & 0 \\ -1 & 2 \\ 1 & -1 \end{bmatrix}, \quad \mathcal{D} = \begin{bmatrix} 1 \\ 0 \\ -1 \end{bmatrix} \quad \widehat{\mathcal{D}} = \mathbf{0}.$$

Evaluate $\mathbf{x}_{\hat{e}}(nT)$ according to the recursion relation (75) and $\mathbf{w}(nT)$ by (76) for $n = 1, \ldots, 15$, when $\mathbf{e}(t) = [u(t) - (t-1)u(t-1) + (t-2)u(t-2)]$, $T = 0.2$, and $\mathbf{x}(0) = [-1 \quad 2]'$.

22. Suppose that an arithmetic mistake introduces the error vector $\varepsilon_4 = [0.001 \quad -0.012]'$ into the computed value of $\mathbf{x}_{\hat{e}}(4T)$ of Problem 21. Will the norm of the propagated error decrease for increasing time? What is the value of the bound on the propagated error in the computed value of $\mathbf{x}_{\hat{e}}(14T)$ as evaluated by (84)? Answer the questions for each of the norms of (80).

23. Repeat Problems 21 and 22 after replacing $\mathscr{A}$ by each of the following matrices:

$$(a) \quad \begin{bmatrix} 0 & 1 \\ -1 & 0 \end{bmatrix}, \qquad (b) \quad \begin{bmatrix} -4 & 1 \\ -2 & -1 \end{bmatrix}, \qquad (c) \quad \begin{bmatrix} 2 & 0 \\ 1 & -1 \end{bmatrix}.$$

24. Repeat Problems 21 and 22 after replacing $\mathscr{A}$, $\mathscr{B}$, C, $\widehat{\mathscr{D}}$, $\mathbf{x}(0)$, and ε_4 by

$$\mathscr{A} = \begin{bmatrix} -2 & 1 & 3 \\ 0 & -1 & 0 \\ 0 & 2 & -3 \end{bmatrix}, \quad \mathscr{B} = \begin{bmatrix} 0 \\ 2 \\ -1 \end{bmatrix}, \quad C = \begin{bmatrix} 1 & 0 & 2 \\ 1 & -1 & 1 \end{bmatrix}, \quad \mathscr{D} = \mathbf{0},$$

$$\mathbf{x}(0) = [1 \quad -1 \quad 1]', \qquad \varepsilon_4 = [0.001 \quad 0.020 \quad -0.001]'.$$

25. Repeat Problem 21 with $e(t)$ and $\mathbf{x}(0)$ replaced by each of the following:

$(a) \quad e(t) = [te^{-t}], \qquad\qquad \mathbf{x}(0) = [0 \quad 0]'$
$(b) \quad e(t) = [\sin t], \qquad\qquad \mathbf{x}(0) = [1 \quad -1]'$
$(c) \quad e(t) = [2u(t) - u(t-1)], \qquad \mathbf{x}(0) = [0 \quad 1]'$
$(d) \quad e(t) = [u(t) - \epsilon^{-2t}], \qquad \mathbf{x}(0) = [0 \quad 0]'$
$(e) \quad e(t) = [2\epsilon^{-t} - 3\epsilon^{-2t}], \qquad \mathbf{x}(0) = [1 \quad 0]'$

26. Repeat the calculations of the example in Section 5.6 using the other matrix norms in (83).

27. For each of the following $\mathscr{A}$ matrices

$$(a) \quad \begin{bmatrix} -1 & 0 \\ 0 & -1 \end{bmatrix}, \qquad (b) \quad \begin{bmatrix} -6 & 3 \\ -4 & 1 \end{bmatrix}, \qquad (c) \quad \begin{bmatrix} -1 & 0 & 1 \\ 1 & -2 & 1 \\ 0 & 0 & -1 \end{bmatrix}$$

Determine K in the truncation (87) and the error bounds for (104) and (108) when $T = 0.1$ and $\Delta = 0.001$. Use the three different matrix norms in (83).

28. Repeat the calculations of Problem 21 after replacing $\epsilon^{0.2\mathscr{A}}$ by its truncated-power-series approximation $\mathbf{A}$ computed such that, in the truncation criterion, $(a) \Delta = 0.01$, and $(b) \Delta = 0.001$. In each case use the three different norms in (83). Compare the results of these calculations with those obtained in doing Problem 21.

29. Repeat Problem 28 after replacing $\mathscr{A}$ by the matrices in Problem 23.

30. If $\mathbf{r}(t)$ and $\mathbf{e}(t)$ are the response and excitation of a linear time-invariant network N, then, as illustrated in Fig. P30, $\mathbf{r}^{(k)}(t)$ is the response to

$$\mathbf{e}(t) \circ\!\!-\!\!\boxed{\quad N \quad}\!\!-\!\!\circ \mathbf{r}(t) \;\Longleftrightarrow\; \mathbf{e}^{(k)}(t)\circ\!\!-\!\!\boxed{\quad N \quad}\!\!-\!\!\circ \mathbf{r}^{(k)}(t)$$

Fig. P30

$\mathbf{e}^{(k)}(t)$. This suggests a method for computing an approximation of $\mathbf{r}(t)$: (**a**) Approximate $\mathbf{e}$ by straight-line segments, (**b**) differentiate once to get a staircase approximation of $\dot{\mathbf{e}}$. (or twice to get impulse-train approximation of $\ddot{\mathbf{e}}$), (**c**) find the response of N to the approximation of $\dot{\mathbf{e}}$ (or $\ddot{\mathbf{e}}$), and (**d**) integrate once (or twice) to get the approximate response. Using this method, with either the step or impulse response, compute the approximate response of the networks in Fig. P4 for the excitations in Fig. P5.

The next four problems involve the preparation of a computer program to help in implementing the solution to some problems. In each case, prepare a program flow chart and a set of program instructions, in some user language such as FORTRAN IV, for a digital computer program to carry out the job specified in the problem. Include a set of user instructions for the program.

31*. Prepare a program to evaluate $w_{\hat{e}}(nT)$ for $n = 1, \ldots, N$ according to (65). The values of N, $e(kT)$, and $w_\delta(kT)$ are to be supplied as input data.

32*. Prepare a program to evaluate $\mathbf{A}$, the approximation to $\epsilon^{\mathscr{A}T}$, in (87). The program should lead to a selection of K for which (95) is valid. The matrix $\mathscr{A}$ and the scalars T and Δ are to be given as input data. Use the matrix norm of (83a) or (83c).

33*. Prepare a program to evaluate $\mathbf{x}_{\hat{e}}(nT)$ for $n = 1, \ldots, N$ according to (75) when $\mathbf{e}(nT) \equiv \mathbf{0}$ for all n. The values of N, $\mathbf{x}(0)$, and $\epsilon^{\mathscr{A}T}$ are to be supplied as input data.

34*. Combine the program of Problems 32 and 33 to create a single program that, starting with N, $\mathbf{x}(0)$, $\mathscr{A}$, T, and Δ as input data, evaluates $\mathbf{x}_{\hat{e}}(nT)$ for $n = 1, \ldots, N$ when $\mathbf{e}(nT) \equiv \mathbf{0}$.

.6.

REPRESENTATIONS OF NETWORK FUNCTIONS

It is our purpose in this chapter to discuss ways in which network functions are represented and to begin the study of properties of network functions as analytic functions of a complex variable. We shall here concentrate largely on those properties that apply generally to network functions, without regard to their specific nature as driving-point or transfer functions. We shall also study the relationships that exist between parts of a network function—real and imaginary parts, magnitude and angle—and observe how the function is represented by any one of its component parts.

6.1 POLES, ZEROS, AND NATURAL FREQUENCIES

Recall that a *network function* is defined as the ratio of the Laplace transform of a *response* to that of an *excitation* when the network is initially relaxed. Let us begin by observing a few elementary properties of network functions that should have become clear by now, even though some of them may not have been stated explicitly.

We are dealing with lumped, linear, time-invariant networks. Network functions of such networks are *rational functions*, the ratios of two polynomials. In Chapter 3 we related network functions to the determinanst and cofactors of the node admittance or loop impedance matrices. Let us here relate them to the state equations. Recall that the state equations for a network can be written as follows:

$$\frac{d}{dt}\mathbf{x} = \mathscr{A}\mathbf{x} + \mathscr{B}\mathbf{e}, \tag{1a}$$

$$\mathbf{w} = \mathbb{C}\mathbf{x} + \mathscr{D}\mathbf{e} + \widehat{\mathscr{D}}\frac{d\mathbf{e}}{dt}, \tag{1b}$$

where $\mathbf{x}$, $\mathbf{e}$, and $\mathbf{w}$ are the state, excitation, and output vectors, respectively. The last term in the second equation can appear only when an element of the output vector is a capacitor current in a capacitor *and* independent voltage-source loop or an inductor voltage in an inductor *and* independent current-source cut-set.

Assuming initially relaxed conditions, let us take the Laplace transforms of these equations, solve the first one for $\mathbf{X}(s)$ and substitute into the second. The result will be*

$$\mathbf{X}(s) = (s\mathbf{U} - \mathscr{A})^{-1}\mathscr{B}\mathbf{E}(s), \tag{2a}$$

$$\mathbf{W}(s) = \{\mathbb{C}(s\mathbf{U} - \mathscr{A})^{-1}\mathscr{B} + \mathscr{D} + s\widehat{\mathscr{D}}\}\mathbf{E}(s). \tag{2b}$$

The quantity in braces is the transfer matrix $\mathbf{H}(s)$, each of whose elements is a network function. Thus

$$\mathbf{H}(s) = \mathbb{C}(s\mathbf{U} - \mathscr{A})^{-1}\mathscr{B} + \mathscr{D} + s\widehat{\mathscr{D}}. \tag{3}$$

Examine this expression carefully. The last two terms indicate a direct relationship between excitation and response without the mediation of the state vector. These terms control the behavior of the response as s approaches infinity. In fact, as observed in the last chapter, $\widehat{\mathscr{D}}$ is the matrix of residues of $\mathbf{H}(s)$ at infinity.

In the first term of (3), $\mathbb{C}$ and $\mathscr{B}$ are matrices of real numbers. The

---

* Note that $\mathbf{e}(0)$ does not appear in (2b) even though the derivative of $\mathbf{e}$ appears in (1b). This is dictated by the requirement that the network be initially relaxed. The initial value of $\mathbf{e}$ could appear only if the derivative of the excitation term is present in the equation. Since it will be present only when there are all-capacitor loops or all-inductor cut-sets, setting initial capacitor voltages and initial inductor currents to zero will require initial excitation values also to be zero.

complex variable s appears only in $(s\mathbf{U} - \mathscr{A})^{-1}$. Letting $d(s)$ be the characteristic polynomial of $\mathscr{A}$, as before, this term can be written

$$\mathcal{C}(s\mathbf{U} - \mathscr{A})^{-1}\mathscr{B} = \frac{1}{d(s)}\,\mathcal{C}[\mathrm{adj}\,(s\mathbf{U} - \mathscr{A})]\mathscr{B}. \tag{4}$$

Now the elements of $\mathrm{adj}\,(s\mathbf{U} - \mathscr{A})$ are simply cofactors of $\det\,(s\mathbf{U} - \mathscr{A})$ and hence are polynomials. This fact is not modified when $\mathrm{adj}\,(s\mathbf{U} - \mathscr{A})$ is premultiplied by $\mathcal{C}$ and postmultiplied by $\mathscr{B}$. Hence the whole term is a matrix whose elements are polynomials divided by $d(s)$. We have thus verified that network functions are rational functions of s.

Something more can be established from the preceding. In preceding chapters reference has been made to the *natural frequencies* of a network. In Chapter 3 we considered these to be the zeros of the determinant of the loop impedance matrix or the node admittance matrix. There we showed that these two determinants could differ at most by a multiplier Ks^p, and hence their nonzero zeros were the same. However, in Chapter 4 we treated the natural frequencies as the eigenvalues of the matrix $\mathscr{A}$; namely, the zeros of $d(s)$. We now see that the zeros of $d(s)$ are the same as those of the loop impedance determinant and the node admittance determinant. This follows from the fact that $\mathbf{W}(s)$ refers to *any* output. Thus, if we choose all the node voltages, and only these, as the outputs, $\mathbf{W}(s)$ is the matrix of node-voltage transforms. Since network solutions are unique, $(2b)$ must give the same results as the solution of the node equations. In the latter case the denominator of the solution will be Δ_y. Hence Δ_y and $d(s)$ have the same nonzero zeros. A similar conclusion follows concerning Δ_z. We shall state this result as a theorem for ease of reference.

Theorem 1. *The nonzero zeros of* $\det\,(s\mathbf{U} - \mathscr{A})$ *are the same as the nonzero zeros of* $\det\,(\mathbf{AYA}')$ *and* $\det\,(\mathbf{BZB}')$.

LOCATIONS OF POLES

Let $F(s)$ be the generic symbol for a network function. Since it is a rational function, it can be written in the following forms:

$$F(s) = \frac{a_m s^m + a_{m-1} s^{m-1} + \cdots + a_1 s + a_0}{b_n s^n + b_{n-1} s^{n-1} + \cdots + b_1 s + b_0}, \tag{5a}$$

$$F(s) = K\,\frac{(s - s_{01})(s - s_{02}) \cdots (s - s_{0m})}{(s - s_{p1})(s - s_{p2}) \cdots (s - s_{pn})}. \tag{5b}$$

Since each of the matrices $\mathscr{A}$, $\mathscr{B}$, C, $\mathscr{D}$, and $\hat{\mathscr{D}}$, on the right of (3) are matrices of real numbers and $F(s)$ stands for any element of **H**, then all the coefficients of s in (5a) must be real.* Now if s takes on only real values in (5a), then $F(s)$ will be real. A function of a complex variable that is real when the variable is real is called a *real function*. So network functions are real functions of s. From this the *reflection property* immediately follows; namely,

$$F(\bar{s}) = \overline{F}(s); \qquad (6)$$

that is, network functions take on conjugate values at conjugate points in the complex plane.

Now look at the second form in (5) in which the poles s_{pk} and the zeros s_{0k} are placed in evidence. Aside from a scale factor K, the network function is completely specified in terms of its poles and zeros, which determine its analytic properties. In fact, the poles and zeros provide a representation of a network function, as illustrated in Fig. 1. The zeros

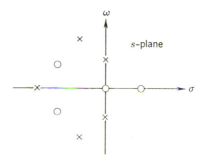

Fig. 1.   Pole-zero pattern.

are shown by circles; and the poles, by crosses. We refer to such diagrams as *pole-zero patterns* or *configurations*. Because of the reflection property (6), the poles and zeros of a network function are either real or occur in complex-conjugate pairs.

Another simple property possessed by network functions follows from a consideration of stability. We know that the free response is governed

---

* This statement should be qualified in a trivial way, since it is possible to multiply every coefficient in the numerator and denominator by an arbitrary complex number without changing the function. This difficulty is overcome by fixing, say, the coefficient of the highest power in the denominator to be 1.

by the poles of the network function. Since

$$\mathcal{L}^{-1}\left[\frac{a_{-k}}{(s - s_p)^k}\right] = \frac{a_{-k}\,t^{k-1}\epsilon^{s_p t}}{(k - 1)!}\,, \tag{7}$$

we immediately conclude that the network function of a stable network cannot have any poles in the right half-plane, and any poles on the $j\omega$-axis must be simple. Otherwise, the free response will be unbounded and the network will be unstable.

This conclusion can be strengthened in the case of a driving-point function. Both driving-point impedance and admittance exhibit this property, and, since one is the reciprocal of the other, the driving-point functions can have neither poles nor zeros in the right half-plane. Furthermore, both poles and zeros on the $j\omega$-axis must be simple.

In the case of a transfer function the reciprocal is not a network function. Hence we can say nothing about its zeros. They may lie anywhere in the complex plane, subject only to the reflection property.

EVEN AND ODD PARTS OF A FUNCTION

Generally speaking, $F(s)$ will have both even and odd powers of s; it will be neither an even function nor an odd function of s. Hence we can write

$$F(s) = \text{Ev } F(s) + \text{Od } F(s), \tag{8}$$

where $\text{Ev } F(s)$ means "even part of $F(s)$," and $\text{Od } F(s)$ means "odd part of $F(s)$." Now an even function $g(s)$ is characterized by the property $g(-s) = g(s)$; and an odd function, by the property $g(-s) = -g(s)$. Using these properties together with (8), we can express the even and odd parts of a function as follows:

$$\text{Ev } F(s) = \tfrac{1}{2}[F(s) + F(-s)], \tag{9a}$$

$$\text{Od } F(s) = \tfrac{1}{2}[F(s) - F(-s)]. \tag{9b}$$

Alternative forms can be obtained if the even and odd powers of s are grouped in both the numerator and denominator of $F(s)$. Thus, write

$$F(s) = \frac{m_1(s) + n_1(s)}{m_2(s) + n_2(s)}\,, \tag{10}$$

where m_1 and m_2 are even polynomials, and n_1 and n_2 are odd polynomials. Then, using this in (9), we get

$$\text{Ev } F(s) = \frac{m_1(s)\, m_2(s) - n_1(s)\, n_2(s)}{m_2{}^2(s) - n_2{}^2(s)}, \tag{11a}$$

$$\text{Od } F(s) = \frac{n_1(s)\, m_2(s) - n_2(s)\, m_1(s)}{m_2{}^2(s) - n_2{}^2(s)}. \tag{11b}$$

Note that the denominator is the same for both the even and the odd part of $F(s)$, and it is an even polynomial. The numerator of Ev $F(s)$ is even, and that of Od $F(s)$ is odd, as they should be.

It is of interest to observe where the poles of Ev $F(s)$ [and Od $F(s)$ also] lie. From (9) it is clear that Ev $F(s)$ has poles where $F(s)$ has poles and also where $F(-s)$ has poles. But the poles of $F(-s)$ are the mirror images about the imaginary axis of the poles of $F(s)$. This can be illustrated by the following $F(s)$ and $F(-s)$:

$$F(s) = \frac{m_1 + n_1}{(s+1)(s^2 + 2s + 2)},$$

$$F(-s) = \frac{m_1 - n_1}{(-s+1)(s^2 - 2s + 2)}.$$

$F(s)$ has a real negative pole and a complex pair in the left half-plane. The poles of $F(-s)$ are the mirror images of these, as shown in Fig. 2. Now Ev $F(s)$ has all the poles in Fig. 2, both left half-plane (lhp) and right half-plane (rhp).

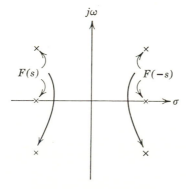

Fig. 2.   Poles of Ev $F(s)$.

The pole pattern in Fig. 2 possesses a certain symmetry. A pole configuration that has symmetry with respect to both the real and imaginary axes is said to have *quadrantal symmetry*. So we say that the poles of Ev $F(s)$ and Od $F(s)$ have quadrantal symmetry.

The function $F(s)$ specifies a value of F for all complex values of s. Of all the values of s, of particular significance are those on the $j\omega$-axis. For $s = j\omega$, we are often interested in the behavior of one of the following quantities: real part, imaginary part, angle, and magnitude (or log magnitude). These are the quantities involved in the steady-state response to sinusoidal excitation. Any one of these quantities can be referred to as a *frequency response*. These components of a function are interrelated by

$$F(j\omega) = R(\omega) + jX(\omega) = |F(j\omega)|\, \epsilon^{j\phi(\omega)}, \tag{12}$$

where the meanings of the symbols are obvious.

Look at (9) under the assumption that $s = j\omega$; what becomes of the even and odd parts of $F(j\omega)$? Since $F(-j\omega) = F(\overline{j\omega}) = \overline{F}(j\omega)$ from (6), we see that

$$\text{Ev } F(j\omega) = \tfrac{1}{2}[F(j\omega) + \overline{F}(j\omega)] = \text{Re } F(j\omega) = R(\omega), \tag{13a}$$

$$\text{Od } F(j\omega) = \tfrac{1}{2}[F(j\omega) - \overline{F}(j\omega)] = j \text{ Im } F(j\omega) = jX(\omega). \tag{13b}$$

That is to say, the real part of a function on the $j\omega$-axis is its even part; the imaginary part on the $j\omega$-axis is its odd part divided by j. Another way of stating this is to say that the real part of $F(j\omega)$ is an even function of angular frequency ω, and the imaginary part is an odd function of ω.

MAGNITUDE AND ANGLE OF A FUNCTION

Similar statements can be made about the magnitude and angle. Thus, using the notation of (12), we can write the square of $F(j\omega)$ as follows:

$$F^2(j\omega) = F(j\omega)\, F(-j\omega)\, \frac{F(j\omega)}{F(-j\omega)} = |F(j\omega)|^2\, \epsilon^{j2\phi(\omega)}$$

Hence

$$|F(j\omega)|^2 = F(j\omega)\, F(-j\omega), \tag{14a}$$

$$\phi(\omega) = \frac{1}{2j} \ln \frac{F(j\omega)}{F(-j\omega)}. \tag{14b}$$

Now replacing ω by $-\omega$, it is seen that the *magnitude-square function* is an even rational function of ω. Observe that $|F(j\omega)|^2$ is the value of the even rational function $G(s) = F(s)\,F(-s)$ on the $j\omega$-axis. It is of further interest to note that both the poles and zeros of $G(s)$ occur in quadrantal symmetry, a fact that is true of any even rational function.

For the angle, we get

$$\phi(-\omega) = \frac{1}{2j} \ln \frac{F(-j\omega)}{F(j\omega)} = -\frac{1}{2j} \ln \frac{F(j\omega)}{F(-j\omega)} = -\phi(\omega). \qquad (15)$$

Hence we are tempted to say that the angle is an odd function of ω. However, the angle is a multivalued function. Only by remaining on the appropriate Riemann surface can the claim about the angle function be made.

THE DELAY FUNCTION

A transfer function will be called ideal if it is of the form $F(s) = \epsilon^{-s\tau}$. For $s = j\omega$, the magnitude identically equals 1, and the angle is proportional to ω. If a network having this transfer function is excited by a signal $e(t)$, the response of the network, in view of the shifting theorem of Laplace transform theory, will be $w(t) = e(t - \tau)$. The response signal is the same as the excitation except that it is delayed in time by an amount τ, called the time delay. Since $\phi(\omega) = -\omega\tau$ for the ideal function, the time delay is the negative derivative of the angle function.

On the basis of the preceding, a function called the delay is defined for an arbitrary transfer function as the negative derivative of the phase function. Thus

$$\tau(\omega) = -\frac{d}{d\omega}\,\phi(\omega) \qquad (16)$$

is the *delay function*. In contrast with the angle function, the delay is a rational function.

6.2 MINIMUM-PHASE FUNCTIONS

As we observed earlier in this chapter, the zeros of transfer functions can occur in any part of the complex plane. However, those functions that have no zeros in the right half-plane have certain properties that are quite important. For this reason we give these functions a distinctive name for ease of identification. We define a *minimum-phase transfer function as one that has no zeros in the right half-plane.* Conversely, any transfer function that has zeros (even one zero) in the right half-plane is labeled *nonminimum-phase.* The reason for these names will become apparent below.

In order to determine the effect of right half-plane zeros on the magni-

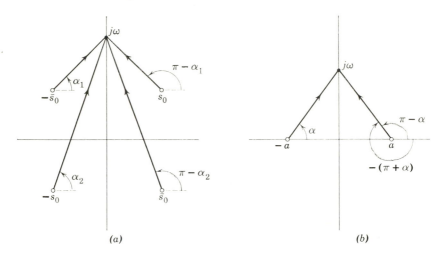

(a)   *(b)*

Fig. 3.   Complex and real zeros in quadrantal symmetry.

tude and angle of a transfer function, consider Fig. 3a. This shows a pair of conjugate zeros in the right half-plane and the left half-plane image of this pair. Let $P_r(s)$ and $P_l(s)$ be quadratics that have the right half-plane pair of factors and the left half-plane pair of factors, respectively; that is,

$$P_r(s) = (s - s_0)(s - \bar{s}_0), \tag{17a}$$

$$P_l(s) = (s + s_0)(s + \bar{s}_0). \tag{17b}$$

It is clear that $P_r(s) = P_l(-s)$. The geometrical construction in the figure indicates that the magnitudes of P_r and P_l are the same when $s = j\omega$. As for the angles, we find

$$\arg P_r(j\omega) = \pi - \alpha_1 - [2\pi - (\pi - \alpha_2)] = -(\alpha_1 + \alpha_2), \tag{18a}$$

$$\arg P_l(j\omega) = \alpha_1 + \alpha_2 = -\arg P_a(j\omega). \tag{18b}$$

Note that in order for the angle of P_r to be zero at $\omega = 0$, as it must be if the angle is to be an odd function, we have written the angle of $(s - \bar{s}_0)$ as $-(\pi + \alpha_2)$, rather than $\pi - \alpha_2$. The difference of 2π corresponds to specifying the angle of one Riemann surface rather than another. This desire to have the angle function an odd function of ω is quite deep-seated in network theorists. The main reason for this desire is that it simplifies the statement of many theorems that we shall state later in the chapter.

It is clear from Fig. 3 that $\alpha_1 + \alpha_2$, the angle contributed by the left-half-plane zeros, is positive for all positive ω. It runs from 0 at $\omega = 0$ to π at infinity. This is illustrated in Fig. 4. It follows, then, that the angle

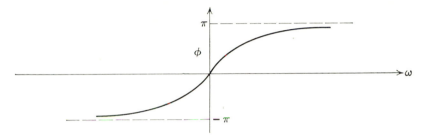

Fig. 4. Angle of a pair of complex left half-plane zeros.

of a pair of conjugate right-half-plane zeros is always negative for positive values of ω, running from 0 at $\omega = 0$ to $-\pi$ at infinity.

Let us now consider the situation in Fig. 3b, which shows a real zero on the positive real axis and its left half-plane image. Again, the magnitudes of the two factors $(j\omega - a)$ and $(j\omega + a)$ are equal. The angle of the left half-plane factor $(j\omega + a)$ is α for positive ω. (It will be $-\alpha$ for negative ω.) We shall choose the angle of the right half-plane factor $(j\omega - a)$ to be $-(\pi + \alpha)$ for positive ω and $\pi - \alpha$ for negative ω in order to make the angle an odd function. Sketches of these angles are shown in Fig. 5.

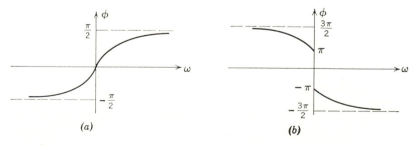

Fig. 5. Examples of angle functions: (a) arg $(j\omega + a)$; (b) arg $(j\omega - a)$.

Note that there is a discontinuity of 2π in the second figure that is introduced simply by our desire to make the angle an odd function. This discontinuity corresponds to jumping from one Riemann surface to another.

Now, if we consider two real right half-plane zeros, we can define the angles in such a way that this discontinuity is eliminated. The situation becomes similar to the case of a pair of conjugate right half-plane zeros. Thus a jump occurs at the origin only when there is an odd number of right-half-plane zeros.

ALL-PASS AND MINIMUM-PHASE FUNCTIONS

With this discussion as background, let us now consider the following two transfer functions:

$$F_1(s) = (s - s_0)(s - \bar{s}_0) \; F(s) = P_l(-s) \; F(s), \qquad (19a)$$

$$F_2(s) = (s + s_0)(s + \bar{s}_0) \; F(s) = P_l(s) \; F(s), \qquad (19b)$$

where s_0 and its conjugate lie in the right half-plane. These two functions are identical except that $F_1(s)$ has a pair of right half-plane zeros, whereas in $F_2(s)$ these are replaced by their left half-plane images. The common function $F(s)$ may have additional right half-plane factors. Suppose we multiply numerator and denominator of $F_1(s)$ by the left half-plane factors $(s + s_0)(s + \bar{s}_0) = P_l(s)$. The result will be

$$F_1(s) = \frac{P_l(s)}{P_l(s)} \; P_l(-s) \; F(s) = F_2(s) \; \frac{P_l(-s)}{P_l(s)} = F_2(s) \; F_0(s), \qquad (20)$$

where

$$F_0(s) = \frac{P_l(-s)}{P_l(s)} = \frac{(s - s_0)(s - \bar{s}_0)}{(s + s_0)(s + \bar{s}_0)}. \qquad (21)$$

Let us define an *all-pass function* as a transfer function all of whose zeros are in the right half-plane and whose poles are the left half-plane images of its zeros. It is clear, therefore, that an all-pass function has a unit magnitude for all values of $s = j\omega$. (This is the reason for its name.) A consideration of the last equation now shows that $F_0(s)$ is an all-pass function. It is a *second-order* all-pass function, the order referring to the number of poles. From (18), the angle of $F_0(j\omega)$ is found to be

$$\arg F_0(j\omega) = \arg P_l(-j\omega) - \arg P_l(j\omega) = -2(\alpha_1 + \alpha_2). \qquad (22)$$

For positive frequencies this is a negative angle. Thus the angle of an all-pass function is negative for all positive frequencies.

Using this equation and (20), we can now write

$$\arg F_1(j\omega) = \arg F_2(j\omega) + \arg F_0(j\omega) < \arg F_2(j\omega), \qquad (\omega > 0). \quad (23)$$

This result tells us that, at all positive frequencies the angle of a function having right-half-plane zeros is less than that of the function obtained when a pair of these zeros is replaced by its left half-plane image.

This procedure of expressing a transfer function as the product of two others may be repeated. At each step a pair of complex zeros or a real zero from the right half-plane may be replaced by their left-half-plane images. A sequence of functions, of which F_1 and F_2 are the first two, will be obtained. Each member of the sequence will have fewer right half-plane zeros than the preceding one. The last member in this sequence will have no right half-plane zeros. Let us label it $F_m(s)$. By definition, $F_m(s)$ is a minimum-phase function (as the subscript is meant to imply). Using (23), and similar results for the other functions, we can write

$$\arg F_1(j\omega) < \arg F_2(j\omega) < \cdots < F_m(j\omega), \qquad (\omega > 0). \qquad (24)$$

Each of the functions in this sequence will have the same j-axis magnitude, but the angles get progressively larger. Paradoxically, the minimum-phase function will have the largest angle of all (algebraically, but not necessarily in magnitude). The reason for this apparent inconsistency is the following. We have defined transfer functions as ratios of output transform to imput transform. When the minimum-phase concept was first introduced by Bode, he defined transfer functions in the opposite way. With such a definition the inequalities in (24) will be reversed, and the minimum-phase function will have the smallest angle algebraically.

At each step in the above procedure a second-order or first-order all-pass function is obtained. The product of any number of all-pass functions is again an all-pass function. It follows that *any non-minimum-phase transfer function can be written as the product of a minimum-phase function and an all-pass function*; that is,

$$F(s) = F_m(s)\, F_a(s), \qquad (25)$$

where F_m is a minimum phase and F_a is an all-pass function.

NET CHANGE IN ANGLE

We can establish one other result from a consideration of the variation of the angle of an all-pass function as ω increases from zero to infinity. Equation 22, together with Fig. 3a, shows that the change in angle $\Delta\phi$, defined as the angle at plus infinity minus the angle at $\omega = 0$, for a second-order all-pass function is -2π. Similarly, for a first-order all-pass function, we can find from Fig. 5 that this change is $\Delta\phi = -\pi$, not counting the discontinuity at $\omega = 0$. It is easy to appreciate that for an nth-order all-pass function the change in angle is $-n\pi$, not counting any discontinuity at $\omega = 0$. If n is even, there will be no discontinuity; however, if n is odd, there will be a discontinuity of $-\pi$, and the total change in angle will become $-n\pi - \pi$.

Consider now a non-minimum-phase function that has n zeros in the right half-plane. This can be expressed as the product of a minimum-phase function and an nth-order all-pass function. The net change in angle of the non-minimum-phase function as ω varies from zero to plus infinity will be the net change in angle of the corresponding minimum-phase function, plus the net change in angle of the all-pass function. Since this latter is a negative quantity, it follows that a non-minimum-phase function has a smaller net change in angle (again, only algebraically), as ω varies from zero to infinity, than the corresponding minimum-phase function, the difference being $n\pi$ or $n\pi + \pi$, where n is the number of right-half-plane zeros.

It is also of interest to determine what the net change in the angle of a minimum-phase function will be as ω varies from zero to plus infinity. The angle contributed by each zero to this net change is $\pi/2$, whereas that contributed by each pole is $-\pi/2$. Hence the net change in angle will be $\pi/2$ times the number of finite zeros, minus the number of finite poles. Thus, if the transfer function is regular at $s = \infty$, *the minimum-phase function will have a smaller* $|\Delta\phi|$ *than the corresponding non-minimum-phase function*, since both angles are nonpositive.

HURWITZ POLYNOMIALS

Let us now consider another aspect of minimum-phase and non-minimum-phase functions; namely, some relationships between the coefficients of a polynomial and the locations of its zeros. Polynomials with no zeros in the open right half-plane are called *Hurwitz polynomials*. If, in addition, there are no zeros on the $j\omega$-axis, the polynomial is called *strictly Hurwitz*, for emphasis. Thus the numerator polynomial of a minimum-phase function is Hurwitz.

Now a necessary condition for a polynomial to be Hurwitz is that all its coefficients have the same sign; however, this is not a sufficient condition. (You can easily make up a counterexample to demonstrate this.) That is to say, some polynomials with zeros in the right half-plane can have all positive or all negative coefficients. However, if a polynomial has coefficients of only one sign, there will be a limitation on the permissible locations of its zeros. The limitation is given by the following theorem:

Theorem 2.   _If a real polynomial $P(s)$ of degree n has coefficients of only one sign, it will have no zeros in the open s-plane sector given by_ $|\arg s| < \pi/n$.

The excluded region is shown in Fig. 6.* In the limiting case, if the only

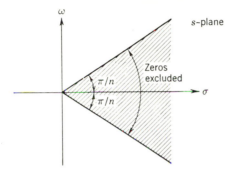

Fig. 6.   Forbidden region for zeros of a polynomial with no negative coefficients.

nonzero coefficients of a polynomial are the first and last (i.e., if $P(s) = s^n + a_0$), then there will be a zero on the boundary $|\arg s| = \pi/n$. Note that the converse of the theorem is not generally true; that is, if the zeros of a polynomial are excluded from the sector $|\arg s| < \pi/n$, the coefficients need not all have the same sign. Thus the polynomials

$$P_1(s) = s^3 + 0.2s^2 + 0.2s + 1,$$

$$P_2(s) = s^3 + s^2 - 0.44s + 1.8,$$

$$P_3(s) = s^3 - 0.3s^2 + 0.6s + 0.5$$

all have a real negative zero and the same right-half-plane factor $s^2 - 0.8s + 1$ whose zeros do not lie in the sector $|\arg s| < \pi/3$; yet two of the polynomials have coefficients with different signs.

* For a proof using the principle of the argument see Norman Balabanian, _Network Synthesis_, Prentice-Hall, Englewood Cliffs, N.J., 1958.

6.3 MINIMUM-PHASE AND NON-MINIMUM-PHASE NETWORKS

Up to this point the discussion has been carried out in terms of functions. We shall now switch to a consideration of the networks of which these functions are transfer functions. The locations of transmission zeros of a network depend both on the types of elements contained in the network and on the structure of the network. Nothing definitive can be stated as to restrictions on the locations of transmission zeros due to element types. Thus RC networks, which have only one type of reactive component, can have complex transmission zeros, as well as real ones, and can even have zeros in the right half-plane. But structure alone does place restrictions.

LADDER NETWORKS

The most notable restriction is given by the following theorem:

Theorem 3.   *The transfer function of a passive, reciprocal ladder network without mutual coupling between branches is minimum-phase.*

The graph of a general ladder network is shown in Fig. 7. The series

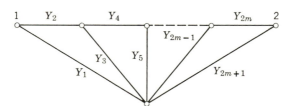

Fig. 7.   General ladder network.

and shunt branches need not be single elements but can be arbitrary two-terminal networks with no coupling between branches. The first and last shunt branches may or may not be present. Using topological formulas for network functions from Chapter 3, the open-circuit transfer impedance can be written as

$$z_{21}(s) = \frac{\sum{}^2 T_{12,0}(y)}{\sum T(y)} = \frac{Y_2 Y_4 \cdots Y_{2m}}{\sum T(y)}. \qquad (26)$$

The right-hand side follows because no shunt branch can appear in a two-tree that includes both nodes 1 and 2 but excludes node 0. The zeros of $z_{21}(s)$ will occur where the numerator has zeros and where the denominator has poles that do not cancel with poles of the numerator. Every tree must contain node 0 and, hence, every tree product must contain at least one of the shunt-branch admittances, Y_1, $Y_3 \cdots Y_{2m+1}$. Hence the poles of these admittances must be the poles of the denominator of $z_{21}(s)$. Some of the series branches may also be in a tree, but the poles of the admittances Y_2, Y_4, etc., in these tree products cancel with the poles of the numerator of z_{21}. The conclusion is that *the zeros $z_{21}(s)$ occur at the zeros of series-branch admittances Y_2, Y_4, etc., and at the poles of shunt-branch admittances Y_1, Y_3, etc.* But the poles and the zeros of the admittance of a passive, reciprocal network cannot lie in the right half-plane. Hence, the result follows.

Although the development was carried out for the z_{21} function, the result is true for other transfer functions also, as discussed in Problem 48 of Chapter 3.

It was shown above that the transmission zeros of a ladder network occur when a shunt-branch admittance has a pole (the branch is a short circuit) or a series-branch admittance has a zero (the branch is an open circuit). It is not true, however, that a transmission zero *must* occur at such points—only that these are the only points at which it *can* occur. Examples where a series-branch admittance zero and a shunt-branch admittance pole are not zeros of transmission are given in Problem 6.

The foregoing becomes very useful in the synthesis of ladder networks. We shall not pursue it here, but the realization of a driving-point function as the open-circuit impedance or short-circuit admittance of a ladder network with certain prescribed transmission zeros utilizes this knowledge. However, if the job is to design a filter, say, with transmission zeros in the right half-plane, we at least know that a ladder network cannot realize such zeros; thus other structures must be sought.

The simplest structures whose transfer functions can be non-minimum-phase are the bridged tee, the twin tee, and the lattice, shown in Fig. 8. Whether they are actually non-minimum-phase will depend on the types of elements present and on their numerical values; for example, the twin-tee network in Fig. 9 will be minimum phase for some values of the resistors and non-minimum-phase for other values, as shown by two particular cases in the figure.

CONSTANT-RESISTANCE NETWORKS

We saw in (25) that a non-minimum-phase transfer function can be written as the product of a minimum-phase function and an all-pass function. This has significance in synthesis; if $F_m(s)$ and $F_a(s)$ can be

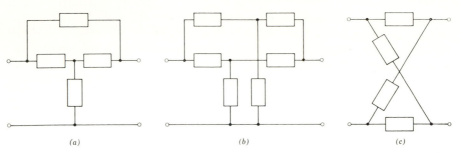

Fig. 8. Potentially non-minimum-phase networks: (a) bridged-tee; (b) twin-tee; (c) lattice.

Non-minimum-phase network Transmission zeros at	Minimum-phase network Transmission zeros at
$s = 1 \pm j3, -\frac{9}{2}$	$s = -1 \pm j3, -\frac{9}{2}$
for $G_1 = 2.28$	for $G_1 = 4.28$
$G_2 = 0.22$	$G_2 = 2.22$
$G_3 = 90$	$G_3 = 1.43$

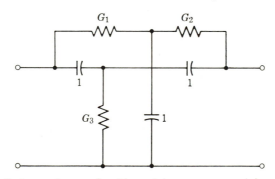

Fig. 9. Twin-tee that can be either minimum- or non-minimum-phase.

realized separately, an interconnection of the two realizations will give the desired network. Consider, for example, the cascade of 2 two-ports shown in Fig. 10, each one realizing one of the two types of functions.

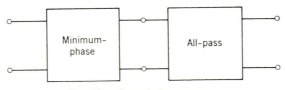

Fig. 10. Cascaded two-ports.

Unfortunately, this interconnection is not necessarily an appropriate realization, because the loading of the second one on the first one causes its transfer function to be changed. If it could be arranged that the two-ports do not load each other in ways that are as yet unaccounted for, then the cascade realization can be used.

One way of eliminating the loading is to make the two-ports *constant-resistance* networks, as shown in Fig. 11. A constant-resistance network

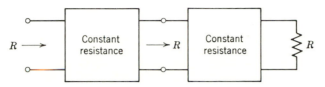

Fig. 11.   Cascade of constant-resistance networks.

is defined as a *two-port whose input impedance at one port is R when the other port is terminated in a resistor R*. Thus whatever the transfer function of the second two-port in Fig. 11 may be, the load it presents at the output port of the first two-port is a resistance R. If the transfer function of each two-port is realized with a termination R, they can then be cascaded without introducing any loading. What has just been described applies to any number of cascaded constant-resistance networks, and it applies whether or not the two-ports are minimum phase.

By direct evaluation of the input impedance of various simple networks when terminated in R, it is found that the two-ports shown in Fig. 12 are constant resistance under the conditions $Z_a Z_b = R^2$; that is, when the impedances Z_a and Z_b are inverses of each other with respect to R^2. When each of these two-ports is terminated in R, the transfer function can be calculated. For concreteness we shall deal with the voltage gain $G_{21}(s) = V_2/V_1$. It is found to be

$$G_{21}(s) = \frac{R - Z_a}{R + Z_a} = \frac{1 - \dfrac{Z_a}{R}}{1 + \dfrac{Z_a}{R}} \qquad \text{(for the lattice),} \qquad (27a)$$

$$G_{21}(s) = \frac{R}{R + Z_a} = \frac{1}{1 + \dfrac{Z_a}{R}} \qquad \text{(for the bridged-tee and ells).} \quad (27b)$$

You should verify these.

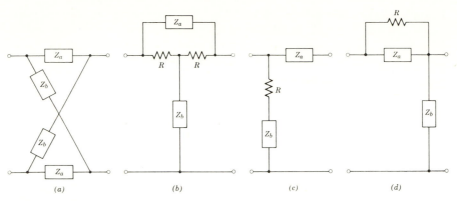

Fig. 12. Some constant-resistance two-ports for $Z_a Z_b = R^2$: (a) lattice; (b) bridged-tee; (c) right-ell; (b) left-ell.

Constant-resistance networks provide a means of realizing a given transfer function of any order. The function can be decomposed into a product of any number of simple transfer functions, each one can be realized separately as a constant-resistance network, and the results can be connected in cascade. We shall now discuss this problem in some detail.

Let us start with an all-pass function. Any all-pass function can be written as the product of the following first-order and second-order all-pass functions.*

$$F_{a1} = \frac{a-s}{a+s} = \frac{1 - \dfrac{s}{a}}{1 + \dfrac{s}{a}}, \tag{28a}$$

$$F_{a2} = \frac{(s^2 - a_1 s + a_0)}{(s^2 + a_1 s + a_0)} = \frac{1 - \dfrac{a_1 s}{s^2 + a_0}}{1 + \dfrac{a_1 s}{s^2 + a_0}}. \tag{28b}$$

When these expressions are compared with (27a), we see that the forms are the same. Hence we can identify Z_a directly and then find Z_b from the relation $Z_a Z_b = R^2$. Thus for the first-order lattice we get

---

*Notice that the numerator factor in the first-order case is written as $a - s$ rather than $s - a$. This amounts to changing the sign of the transfer function or inverting the polarity of the output voltage. Doing this will avoid the discontinuity of π radians in the angle.

$$Z_a(s) = \frac{R}{a}\, s, \qquad Z_b(s) = \frac{1}{\dfrac{s}{aR}} \tag{29}$$

and for the second-order lattice,

$$Z_a(s) = \frac{a_1 R s}{s^2 + a_0}, \qquad Z_b(s) = \frac{R}{a_1}\, s + \frac{1}{\dfrac{a_1}{R a_0}\, s}. \tag{30}$$

Thus the two-ports that realize first- and second-order all-pass functions take the form shown in Fig. 13.

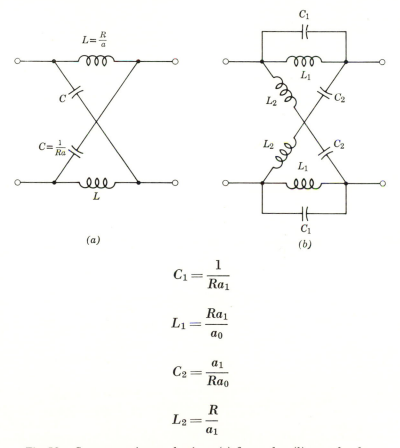

(a)   (b)

$$C_1 = \frac{1}{R a_1}$$

$$L_1 = \frac{R a_1}{a_0}$$

$$C_2 = \frac{a_1}{R a_0}$$

$$L_2 = \frac{R}{a_1}$$

Fig. 13. Constant-resistance lattices: (a) first order; (b) second order.

The lattice network has the disadvantage of not having a common ground for its two-ports. This cannot be avoided in the first-order case because no common-terminal two-port can realize a zero on the positive real axis, as this lattice does. However, for the second-order all-pass lattice a common-terminal equivalent may exist. This will depend on the locations of the zeros. Bridged-tee and twin-tee equivalents of this lattice are discussed in Problem 10.

With the all-pass functions accounted for, there remains the realization of minimum-phase functions. We shall illustrate the realization of a transfer-voltage-ratio function by means of constant-resistance networks in terms of an example. The realization of a minimum-phase function in the general case will become evident from this example. Let the load resistance R be 1 and let

$$G_{21}(s) = K \frac{s^2 - \dfrac{s}{2} + 1}{s^2 + 6s + 1} = \frac{s^2 - \dfrac{s}{2} + 1}{s^2 + \dfrac{s}{2} + 1} K \frac{\left(s^2 + \dfrac{s}{2} + 1\right)}{s^2 + 6s + 1},$$

where K is a gain constant that we shall take to be 1 in this example. The given function has been multiplied and divided by the quadratic *surplus factor* $s^2 + s/2 + 1$ in order to put the result in the form of an all-pass function times a minimum-phase function; $G_{21} = F_a F_m$. The all-pass function is immediately realized by the second-order lattice in Fig. 13b. The angle of the transmission zero is 69°. Hence conditions for both the bridged-tee and twin-tee equivalents of the lattice discussed in Problem 10 are satisfied.

To realize F_m by means of one of the other networks in Fig. 12, we must write this function in the form of (27b). Thus

$$F_m(s) = \frac{s^2 + \dfrac{s}{2} + 1}{s^2 + 6s + 1} = \frac{1}{1 + \dfrac{11s/2}{s^2 + s/2 + 1}}.$$

From this we get

$$Y_a = \frac{1}{Z_a} = \frac{s^2 + \dfrac{s}{2} + 1}{\dfrac{11s}{2}} = \frac{2}{11}s + \frac{1}{11} + \frac{1}{(11/2)s}.$$

With $Z_a Z_b = R^2$ and $R = 1$, we get $Z_b = Y_a$. We see that the $Z_a = 1/Y_a$

branch is the parallel connection of a capacitor, an inductor, and a resistor. If we use Fig. 12d for the realization, the resistor R can be combined with the parallel resistor in the Z_a branch. The final realization is shown in Fig. 14, where one of the bridged-tee equivalents of the lattice has been used.

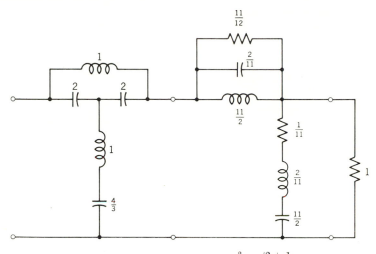

Fig. 14. Realization of $G_{21} = \dfrac{s^2 - s/2 + 1}{s^2 + 6s + 1}$.

The minimum-phase part of the transfer function in the preceding example was realized by a left-ell network, which is a simple ladder. Although we shall not do so here, it can be proved that any minimum-phase transfer function can be realized by a constant-resistance ladder (a cascade of ell networks) by choosing a sufficiently small value of the constant K.* In some cases this may require the introduction of surplus factors by which both numerator and denominator are multiplied, leaving the function unchanged but permitting the identification in the form of (27b). Thus a transfer function $G_{21} = 1/(s^2 + 2s + 2)$ can be written as

$$G_{21} = \frac{s+1}{s^2+2s+2} \cdot \frac{1}{s+1} = \frac{1}{1+\dfrac{s^2+2s+2}{s+1}} \cdot \frac{1}{s+1}$$

$$= \frac{1}{1+\left(s+1+\dfrac{1}{s+1}\right)} \cdot \frac{1}{1+s} = G_{21a}\,G_{21b}.$$

$$(31)$$

* For a proof, see Norman Balabanian, *Network Synthesis*, Prentice-Hall, Englewood Cliffs, N.J., 1958.

The surplus factor $(s + 1)$ converts the transfer function into a product of two functions, each of which can be put in the form of (27b).

The fact that all-pass functions have unit magnitude for all values of $s = j\omega$ is utilized in the design of transmission systems by permitting an independent design for the magnitude and for the angle of the transfer function. Thus a network is designed to realize a desired magnitude function without concern for the angle. From the designed network, an phase function is then determined. Finally, a number of constant-resistance all-pass lattices are cascaded with this network to correct for the angle. Further development of this idea will be left to books on synthesis.

6.4 DETERMINING A NETWORK FUNCTION FROM ITS MAGNITUDE

The preceding sections have been largely concerned with the determination of the properties of a network function. Given a rational function, it is possible to determine, among other things, its real and imaginary parts, and its magnitude and angle. We shall now consider the inverse operation: that of reconstructing a network function when only its real or imaginary part—or its magnitude or angle—is known.

We start first with a consideration of the magnitude-square function. (This is simpler to talk about than the magnitude function.) The necessary conditions for a rational function $G(j\omega)$ to be the magnitude square of a network function on the $j\omega$-axis are quite simple: $G(j\omega)$ must be an even function of ω, and the degree of the numerator should not exceed that of the denominator by more than 2. This is because the network function cannot have more than a simple pole at infinity. In addition, any finite poles of $G(s)$ on the $j\omega$-axis must be double, since the poles of a network function itself on the $j\omega$-axis must be simple.

Given such a $G(j\omega)$, we replace $j\omega$ by s; then all that is left is to identify the poles and zeros of the network function $F(s)$ from those of $G(s)$. The poles and zeros of $G(s)$ occur in quadrantal symmetry. Since $F(s)$ must be regular in the right half-plane, we see, by looking at (15), that all the left-half-plane poles of $G(s)$ must be assigned to the network function $F(s)$. [The right-half-plane poles of $G(s)$ will be mirror images of these and will automatically become poles of $F(-s)$.] Any poles of $G(s)$ on the $j\omega$-axis will be double and will be assigned as simple poles of $F(s)$. As for the zeros, the answer is not as clear-cut. There is generally no limitation on the locations of zeros of a network function, unless it is a driving-point function, in which case the zeros must lie in the left half-plane. For transfer functions, we need not assign to $F(s)$ all the left-half-plane zeros of $G(s)$. Thus the zeros of $F(s)$ are not uniquely determined from a given $G(s)$, unless the transfer function is specified to be minimum-phase. In this case the zeros of $F(s)$, as well as the poles, must lie in the left half-plane.

Let us now consider some examples that illustrate this procedure and are of practical interest. The requirements on most common electrical filters involve transfer functions whose j-axis magnitudes are ideally constant over a given frequency interval, which is referred to as the *pass band*, and are ideally zero over the rest of the $j\omega$-axis, which is referred to as the *stop band*. It is not possible for the j-axis magnitude of a rational function to behave in this ideal manner. (Why?) However, it is possible to find transfer functions whose j-axis magnitudes approximate the desired magnitude in some fashion or other.

Consider the ideal low-pass filter function shown in Fig. 15*a*. Two

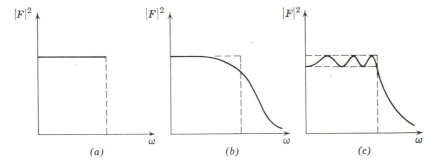

Fig. 15. Approximations of low-pass filter: (*a*) ideal; (*b*) Butterworth; (*c*) Chebyshev.

possible ways of approximating this ideal function are shown in parts (*b*) and (*c*) of the figure. The first of these is called a *maximally flat*, or *Butterworth, approximation,* whereas the second one is called an *equal-ripple*, or *Chebyshev, approximation.* The maximally flat approximation is monotonic in both the pass band and the stop band, the maximum error occurring near the edge of the band. On the other hand, the Chebyshev approximation is oscillatory in the pass band, the peaks of the ripples being equal. In this way the error is distributed more uniformly over the pass band.

The analytical forms of these functions, aside from a scale factor, are given by

$$|F(j\omega)|^2 = \frac{1}{1 + \omega^{2n}}, \qquad \text{(maximally flat)} \qquad (32)$$

and

$$|F(j\omega)|^2 = \frac{1}{1 + \delta^2\, T_n{}^2(\omega)}, \qquad \text{(equal ripple)}, \qquad (33)$$

where δ is a small number that controls the ripple amplitude, $\omega = 1$ corresponds to the edge of the passband, and $T_n(\omega)$ is a Chebyshev polynomial* defined by

$$T_n(s/j) = \cosh\,(n\,\cosh^{-1} s/j), \qquad (34)$$

which reduces on substituting $s = j\omega$ to

$$T_n(\omega) = \cos\,(n\,\cos^{-1}\omega) \quad \text{for} \quad |\omega| \le 1. \qquad (35)$$

Our problem now is to find the transfer function $F(s)$ when its j-axis squared magnitude is known

MAXIMALLY FLAT RESPONSE

Let us first consider the Butterworth response. According to the preceding discussion, we first replace ω^2 by $-s^2$ in (32). The result is

$$G(s) = F(s)\ F(-s) = \frac{1}{1 + (-1)^n s^{2n}}. \qquad (36)$$

This function has no finite zeros, so we need only factor the denominator. In the present case this is a relatively simple task. The zeros of the denominator are found by writing

$$s^{2n} = \epsilon^{j(2k-1+n)\pi} \qquad (37a)$$

which is simply

$$s^{2n} = \pm 1, \qquad (37b)$$

where the minus sign applies for n even. Taking the $2n$th root in (37a), we find the poles of $G(s)$ to be

$$s_k = \epsilon^{j(2k-1+n)\pi/2n}; \qquad k = 1, 2, \cdots, 2n. \qquad (38)$$

* The use of the letter T for the Chebyshev polynomial is a legacy of the past. Some of Chebyshev's work was first published in French, leading to the use of the French spelling "Tschebyscheff," or its variation "Tchebycheff." This spelling of the name has now been discarded in the American literature.

Thus, there are $2n$ poles, each of which has unit magnitude. The poles are uniformly distributed on the unit circle, as shown in Fig. 16 for the case

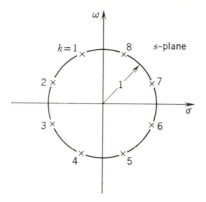

Fig. 16.   Butterworth pole distribution for $n = 4$.

$n = 4$. Notice that the imaginary parts of the poles lie in the pass band—in the range $\omega^2 < 1$.

To form $F(s)$ we simply take the n left half-plane poles of $G(s)$. These are the ones given by values of k from 1 to n. For $n = 4$ these will be

$$s_1 = \epsilon^{j(5\pi/8)}, \qquad s_3 = \epsilon^{j(9\pi/8)},$$

$$s_2 = \epsilon^{j(7\pi/8)}, \qquad s_4 = \epsilon^{j(11\pi/8)}.$$

Finally, for the case $n = 4$,

$$F(s) = \frac{1}{(s - s_1)(s - s_2)(s - s_3)(s - s_4)}$$

$$= \frac{1}{1 + 2.613s + 3.414s^2 + 2.613s^3 + s^4}. \tag{39}$$

The coefficients of the Butterworth polynomials up to order 10 and the factors of these polynomials are given in Tables 1 and 2, respectively, for reference purposes.

The designation " maximally flat " is a result of the following consideration. Suppose a Taylor series of the function in (32) is written (by long division, say) for the region $\omega^2 < 1$. The error between the desired magnitude function in this range, namely 1, and the series will be

$$1 - |F(j\omega)|^2 = \omega^{2n} - \omega^{4n} + \cdots. \tag{40}$$

Table 1. Coefficients of Butterworth Polynomials: $b_0 + b_1 s + \cdots + b_n s^n$

Order n	b_1	b_2	b_3	b_4	b_5	b_6	b_7	b_8	b_9
2	1.4142								
3	2.0000	2.0000							
4	2.6131	3.4142	2.6131						
5	3.2361	5.2361	5.2361	3.2361					
6	3.8637	7.4641	9.1416	7.4641	3.8637				
7	4.4940	10.0978	14.5920	14.5920	10.0978	4.4940			
8	5.1528	13.1371	21.8462	25.6884	21.8462	13.1371	5.1258		
9	5.7588	16.5817	31.1634	41.9864	41.9864	31.1634	16.5817	5.7588	
10	6.3925	20.4317	42.8021	64.8824	74.2334	64.8824	42.8021	20.4317	6.3925

Note: b_0 and b_n are always unity.

Table 2. Factors of Butterworth Polynomials

Order n	Factors
1	$s + 1$
2	$s^2 + 1.4142s + 1$
3	$(s + 1)(s^2 + s + 1)$
4	$(s^2 + 0.7654s + 1)(s^2 + 1.8478s + 1)$
5	$(s + 1)(s^2 + 0.6180s + 1)(s^2 + 1.6180s + 1)$
6	$(s^2 + 0.5176s + 1)(s^2 + 1.4142s + 1)(s^2 + 1.9319s + 1)$
7	$(s + 1)(s^2 + 0.4450s + 1)(s^2 + 1.2470s + 1)(s^2 + 1.8019s + 1)$
8	$(s^2 + 0.3002s + 1)(s^2 + 1.1111s + 1)(s^2 + 1.1663s + 1)(s^2 + 1.9616s + 1)$
9	$(s + 1)(s^2 + 0.3473s + 1)(s^2 + s + 1)(s^2 + 1.5321s + 1)(s^2 + 1.8794s + 1)$
10	$(s^2 + 0.3129s + 1)(s^2 + 0.9080s + 1)(s^2 + 1.4142s + 1)(s^2 + 1.7820s + 1)(s^2 + 1.9754s + 1)$

Since the Taylor series for the error starts with the ω^{2n} power, this means the first $n-1$ derivatives with respect to ω^2 are 0 at $\omega = 0$. Hence the name maximally flat.

The Butterworth function just illustrated is particularly simple, since all of its zeros are at infinity. It is possible to introduce a modified maximally flat function that will have some finite zeros. A glance back at the magnitude-squared functions in (32) and (33) shows that they are both of the form

$$|F(j\omega)|^2 = \frac{1}{1 + f(\omega^2)}, \tag{41}$$

where $f(\omega^2)$ is an even function of ω; the argument is expressed as ω^2 to accent this fact. In the case of the Butterworth function, $f(\omega^2)$ is a power of ω^2; in the equal-ripple case, it is a polynomial.

Now suppose that $f(\omega^2)$ is a rational function,

$$f(\omega^2) = \frac{\omega^{2n}}{P(\omega^2)}, \tag{42}$$

where

$$P(\omega^2) = 1 + a_2\omega^2 + a_4\omega^4 + \cdots + a_{2k}\omega^{2k}$$

is a polynomial whose order, $2k$ in ω, is less than $2n$. Then $|F(j\omega)|^2$ and the difference between the desired function in the pass band, namely 1, and this function will become

$$|F(j\omega)|^2 = \frac{P(\omega^2)}{P(\omega^2) + \omega^{2n}}, \tag{43}$$

and

$$1 - |F(j\omega)|^2 = \frac{\omega^{2n}}{P(\omega^2) + \omega^{2n}}$$
$$= \omega^{2n}[1 - a_2\omega^2 + (a_4 + a_2{}^2)\omega^4 - \cdots]. \tag{44}$$

In the last step a power series is obtained by long division. Again the series starts with the ω^{2n} power, and so the first $n-1$ derivatives of the error with respect to ω^2 are 0 at $\omega = 0$. The magnitude-square function in (43) is therefore also maximally flat. In contrast with the Butterworth function, it has some finite zeros.

As an illustration consider the following magnitude-square function:

$$|F(j\omega)|^2 = \frac{1.838 - 1.346\omega^2 + 0.246\omega^4}{1.838 - 1.346\omega^2 + 0.246\omega^4 + \omega^8} = \frac{(1.355 - 0.496\omega^2)^2}{(1.355 - 0.496\omega^2)^2 + \omega^8}.$$

Note that corresponding coefficients of numerator and denominator are equal up to the highest power of the numerator, as required by (43). Setting $\omega^2 = -s^2$ leads to

$$F(s)\ F(-s) = \frac{(1.355 + 0.496s^2)^2}{(1.355 + 0.496s^2)^2 + s^8}$$

$$= \frac{(1.355 + 0.496s^2)}{(s^2 + 1.9s + 1)(s^2 + 1.05s + 1.355)}$$

$$\times \frac{(1.355 + 0.496s^2)}{(s^2 - 1.9s + 1)(s^2 - 1.05s + 1.355)}$$

or

$$F(s) = \frac{0.496(s^2 + 2.73)}{(s^2 + 1.9s + 1)(s^2 + 1.05s + 1.355)}.$$

Note that the double zeros of $F(s)\ F(-s)$ are assigned equally to $F(s)$ and to $F(-s)$. The locations of the poles and zeros of $F(s)$ are shown in Fig. 17 and are compared with the poles of the fourth-order Butterworth function.

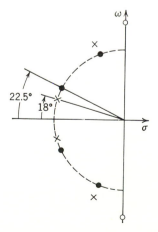

Fig. 17.   Pole locations for example. Dots correspond to poles of Butterworth response.

CHEBYSHEV RESPONSE

Next let us consider the Chebyshev response in (33). The first step is to replace $j\omega$ by s. We then set the denominator equal to zero in order to locate the poles. The result using (34) will be

$$T_n\left(\frac{s}{j}\right) = \cosh\left(n \cosh^{-1}\frac{s}{j}\right) = \frac{\pm j}{\delta}. \qquad (45)$$

In order to solve this equation let us define a new variable $w = x + jy$ and write

$$s = j \cosh w = j \cosh(x + jy) \qquad (46a)$$

and, consequently,

$$T_n\left(\frac{s}{j}\right) = \cosh nw = \cosh n(x + jy) = \frac{\pm j}{\delta}. \qquad (46b)$$

If we now expand $\cosh nw$ in the last equation and set reals and imaginaries equal on both sides of the equation, we will find the values of x and y that satisfy the equation. When these values are substituted in (46a) we find the corresponding values of s. These are the pole locations. If we designate them by $s_k = \sigma_k + j\omega_k$, the result of the indicated operations will be

$$\sigma_k = \sinh\left(\frac{1}{n}\sinh^{-1}\frac{1}{\delta}\right)\sin\frac{2k-1}{n}\frac{\pi}{2}, \qquad (47a)$$

$$\omega_k = \cosh\left(\frac{1}{n}\sinh^{-1}\frac{1}{\delta}\right)\cos\frac{2k-1}{n}\frac{\pi}{2}. \qquad (47b)$$

You should verify these equations.

In order to get some interpretation for these seemingly monstrous expressions, divide each of them by the hyperbolic function, square both sides, and add; the result will be

$$\frac{\sigma_k{}^2}{\sinh^2\left(\frac{1}{n}\sinh^{-1}\frac{1}{\delta}\right)} + \frac{\omega_k{}^2}{\cosh^2\left(\frac{1}{n}\sinh^{-1}\frac{1}{\delta}\right)} = 1. \qquad (48)$$

This is the equation of an ellipse in the s-plane. The major axis of the ellipse will lie along the $j\omega$-axis, since the hyperbolic cosine of a real variable is always greater than the hyperbolic sine. The pole locations for $n = 4$ are shown in Fig. 18.

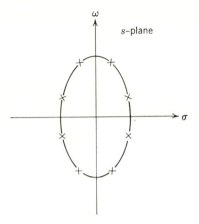

Fig. 18.   Chebyshev pole distribution for $n = 4$.

Finally, the left half-plane poles of $G(s)$ are alloted to $F(s)$, and the task is again complete.

For a typical case, if the permissible ripple is given to be $\delta = 0.1$ and the order $n = 4$, the pole locations are found from (47), and we get the transfer function

$$F(s) = \frac{1}{(s^2 + 0.644s + 1.534)(s^2 + 1.519s + 0.823)}$$

$$= \frac{1}{s^4 + 2.16s^3 + 3.31s^2 + 2.86s + 1.26} \cdot$$

6.5   CALCULATION OF A NETWORK FUNCTION FROM A GIVEN ANGLE

In the last section we found that—starting with an even rational function that satisfies necessary conditions for realizability as the square of the magnitude of a network function—we can determine a rational function $F(s)$ (often more than one) such that the square of the j-axis magnitude of $F(s)$ is equal to the given function; the function becomes unique when it is required to be minimum-phase.

In the present section we shall discuss the possibility of a similar procedure for determining a rational function from a given function of frequency that is claimed to be an angle function. An expression for the angle of a transfer function was given in (14*b*) and is repeated here:

$$\phi(\omega) = \frac{1}{j2} \ln \frac{F(j\omega)}{F(-j\omega)}, \tag{49a}$$

$$\frac{F(j\omega)}{F(-j\omega)} = \epsilon^{j2\phi(\omega)}. \tag{49b}$$

In the following discussion we shall assume that the function which is given is the tangent of $\phi(\omega)$, to which we shall refer as the *tangent function*. In addition, since we shall be using the ratio on the left side of (49*b*) quite often, let us denote it with a single symbol. Let

$$A(s) = \frac{F(s)}{F(-s)}. \tag{50}$$

We shall refer to this function simply as the A-function.

With these preliminaries disposed of, note that for tan $\phi(\omega)$ we can write

$$j \tan \phi(\omega) = \frac{\epsilon^{j\phi} - \epsilon^{-j\phi}}{\epsilon^{j\phi} + \epsilon^{-j\phi}} = \frac{\epsilon^{j2\phi} - 1}{\epsilon^{j2\phi} + 1} = \frac{A(j\omega) - 1}{A(j\omega) + 1}. \tag{51}$$

The last step follows from (49*b*) and (50). If we now invert this last equation and solve for $A(j\omega)$, we get

$$A(j\omega) = \frac{1 + j \tan \phi(\omega)}{1 - j \tan \phi(\omega)}. \tag{52}$$

Let us now inquire into the conditions that the tangent function must satisfy if it is to be a realizable function. Note that

$$\tan \phi(\omega) = \frac{X(\omega)}{R(\omega)}, \tag{53}$$

where R and X are the real and imaginary parts of the network function. We know that these are, respectively, even and odd functions of ω.

Hence tan $\phi(\omega)$ must necessarily be an odd rational function. There are no other requirements that we can place on this function unless we specify whether the desired $F(s)$ is to be a driving-point or a transfer function.

If an odd rational function is prescribed as the tangent function, the first step will be to form $A(j\omega)$ according to (52). If we now replace $j\omega$ by s, we get the ratio of $F(s)$ to $F(-s)$, according to (50). The question now is: How do we extract $F(s)$ from this ratio? The situation here is not as simple as it was in the case of the magnitude function.

In order to carry on, let us write $F(s)$ as the ratio of two polynomials:

$$F(s) = \frac{P_1(s)}{P_2(s)}. \tag{54}$$

Then $A(s)$ can be written

$$A(s) = \frac{F(s)}{F(-s)} = \frac{P_1(s)}{P_1(-s)} \frac{P_2(-s)}{P_2(s)}. \tag{55}$$

Our problem can now be restated as the problem of finding $P_1(s)$ and $P_2(s)$ when the function on the right side of the last equation is known. Note that $A(s)$ will always have zeros in the right half-plane, and it will usually have poles in the right half-plane also. It differs from an all-pass function in that it may have poles in the right half-plane as well as zeros. On the other hand, it is similar to an all-pass function in that each zero is the negative of a pole. As a matter of fact, it can be expressed as the ratio of two all-pass functions, but this has no utility for our present purpose. It can have neither zeros nor poles on the $j\omega$-axis, since, if $P_1(s)$ has a pair of such zeros, so also will $P_1(-s)$, so that they will cancel in the ratio; similarly if $P_2(s)$ has j-axis zeros.

Let us now consider assigning the poles of $A(s)$ to $P_1(-s)$ or $P_2(s)$. If $A(s)$ has any right half-plane poles these must belong to $P_1(-s)$, since $P_2(s)$ cannot have right half-plane zeros. On the other hand, the left half-plane poles cannot uniquely be assigned to either $P_2(s)$ or $P_1(-s)$. If we assign one of the left half-plane poles of $A(s)$ to $P_1(-s)$, then $P_1(s)$ will have the corresponding right half-plane factor, indicating that the transfer function is a non-minimum-phase one. Of course, the distribution of poles and zeros will be dictated by the permissible degrees of numerator and denominator of $F(s)$.

Once $P_2(s)$ and $P_1(-s)$ have been established from the denominator of $A(s)$, it is not necessary to examine the numerator, since the transfer function will now be known; it is only necessary to replace $-s$ by s in $P_1(-s)$ to get $P_1(s)$.

Let us now illustrate this procedure with an example. Suppose we are given

$$\tan \phi(\omega) = \frac{\omega^3 - 4\omega}{2 - 3\omega^2}.$$

(56)

The first step is to substitute this into (52) to obtain $A(j\omega)$. The result is

$$A(j\omega) = \frac{2 - 3\omega^2 + j\omega^3 - j4\omega}{2 - 3\omega^2 - j\omega^3 + j4\omega}.$$

If we now replace $j\omega$ by s, we get

$$A(s) = \frac{-s^3 + 3s^2 - 4s + 2}{s^3 + 3s^2 + 4s + 2} = \frac{(1-s)(s^2 - 2s + 2)}{(s+1)(s^2 + 2s + 2)}.$$

We find that all the poles of $A(s)$ are in the left half-plane, whereas all the zeros are in the right. Hence there is no unique way to assign the zeros and poles of $F(s)$. Any one of the following functions will be suitable:

$$F_1(s) = \frac{1}{(s+1)(s^2 + 2s + 2)},$$

(57a)

$$F_2(s) = \frac{1-s}{s^2 + 2s + 2},$$

(57b)

$$F_3(s) = \frac{(s^2 - 2s + 1)}{s + 1}.$$

(57c)

Notice that the last two have right half-plane zeros. Each of these functions will have the same angle for all values of ω, but their magnitudes will be quite different. If $F(s)$ is required to be minimum-phase, the answer is once again unique—in this case the first function of (57).[*]

In our computations so far we have assumed that $\phi(\omega)$ is specified to be a continuous function of ω. If, however, a function $F(s)$ has either poles or zeros on the $j\omega$-axis, the corresponding phase function $\phi(\omega)$ will have discontinuities of $\pm\pi$ at each pole and zero. In such cases we consider the

---

[*] Even this uniqueness is only to within a constant multiplier. The angle is obviously independent of a real positive gain constant.

discontinuities separately, applying the procedure above to the "continuous part" of the function; that is, we write

$$\phi(\omega) = \phi_c(\omega) + \sum_j \pm \pi u(\omega - \omega_j), \qquad (58)$$

where $\phi_c(\omega)$ is a continuous function. The index j runs over all the zeros and poles on the $j\omega$-axis, and the minus sign applies to the poles.

We now have to identify the step discontinuities. For this we examine a typical factor in $F(s)$ (pole or zero factor):

$$(s - j\omega_0)|_{s=j\omega} = j(\omega - \omega_0).$$

Obviously this factor changes from $-j$ to $+j$ as ω increases through ω_0. Therefore, as we go through a zero on the $j\omega$-axis, in the direction of increasing ω, the angle of $F(s)$ increases abruptly by π; and as we go through a pole, $\phi(\omega)$ decreases by π. Thus we can restore all the poles and zeros of $F(s)$ on the $j\omega$-axis by observing the discontinuities in the given function.

6.6 CALCULATION OF NETWORK FUNCTION FROM A GIVEN REAL PART

In the last two sections we discussed the possibility of determining a network function from a specified rational function of ω which is to be the j-axis magnitude of the function or the tangent of its angle on the $j\omega$-axis. We found that in most cases it is not possible to obtain a unique answer unless the function is a minimum-phase one. Nevertheless, it is possible to calculate a number of functions that will satisfy the requirements. In the case of a specified magnitude we are able to find a number of transfer functions that have the given j-axis magnitude but differ from each other in their angles. Similarly, from a given tangent function, we are able to find a number of transfer functions that have the same angle on the $j\omega$-axis but differ in magnitude. In the present section we shall discuss some computational procedures that will permit us to calculate a network function from its j-axis real part.

Again the question of uniqueness must be answered. Is a network function uniquely determined if its j-axis real part is known? We can very quickly think of several different networks whose network functions have the same real part, so that the question must be answered in the negative. As an example, suppose the desired function is a driving-point admittance function. Consider the network shown in Fig. 19a. In part (b)

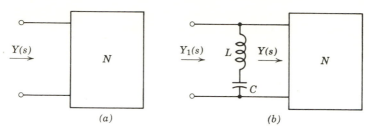

Fig. 19. Two networks whose admittances have the same real part.

of the figure an additional branch is connected at the input terminals. The admittance of the second network is

$$Y_1(s) = Y(s) + \frac{sC}{s^2 LC + 1}.$$

Its j-axis real part is

$$\text{Re}\,[Y_1(j\omega)] = \text{Re}\,[Y(j\omega)] + \text{Re}\,\left[\frac{j\omega C}{1 - \omega^2 LC}\right] = \text{Re}\,[Y(j\omega)];$$

that is, the real parts of both admittances are the same, yet the admittances themselves are different. The function $Y_1(s)$ differs from $Y(s)$ by having a pair of poles on the $j\omega$-axis. If the real part is given, we cannot tell whether to choose $Y(s)$ or $Y_1(s)$ corresponding to this real part. As a matter of fact, an infinite number of functions that differ from $Y(s)$ by having additional poles on the $j\omega$-axis will have the same real part on the $j\omega$-axis. What we can hope to do from a given real part, then, is to find the particular function that has no poles on the $j\omega$-axis.*

THE BODE METHOD

Let us turn back to Section 1 and look at the discussion of the real part of a function starting at (8) and ending at (13). If an even rational function of ω with no finite or infinite poles for real ω is specified to be the real part of a network function, we replace $j\omega$ by s, and the result will be the even part of $F(s)$. Thus

$$R(\omega) \underset{j\omega=s}{\to} \text{Ev}\,F(s) = \tfrac{1}{2}\,[F(s) + F(-s)]. \tag{59}$$

---

* Such a function is a *minimum-susceptance function* if it is an admittance, and a *minimum-reactance function* if it is an impedance. This condition on driving-point functions is the analogue of the minimum-phase condition on the transfer function.

The question is: How can we find $F(s)$ from its even part? As discussed in Section 6.1, the poles of Ev $F(s)$ have quadrantal symmetry. Its left half-plane poles belong to $F(s)$; and its right half-plane poles, to $F(-s)$. If $F(s)$ has a nonzero value at infinity, then $F(-s)$ will have this same value. It is, therefore, clear how to find $F(s)$ from Ev $F(s)$: Expand Ev $F(s)$ in partial fractions and group all the terms contributed by poles in the left half-plane; if there is a constant term in the expansion, we add half of this to the group; finally, we multiply by 2; the result is $F(s)$.

To illustrate, let

$$R(\omega) = \frac{1}{1 + \omega^6}$$

be a specified real part. The first step is to replace ω^2 by $-s^2$, which leads to

$$\text{Ev } F(s) = \frac{1}{1 - s^6} = \frac{1}{(s+1)(s^2+s+1)(1-s)(s^2-s+1)}$$

$$= \tfrac{1}{6}\left[\left(\frac{1}{s+1} + \frac{s+2}{s^2+s+1}\right) + \left(\frac{1}{1-s} + \frac{2-s}{s^2-s+1}\right)\right]. \tag{60}$$

We have already discussed the pole locations of this particular function in connection with the Butterworth response. The denominator can be easily factored, as shown, and the partial-fraction expansion obtained: $F(s)$ is easily identified from the left half-plane poles. It is

$$F(s) = \tfrac{1}{3}\left(\frac{1}{s+1} + \frac{s+2}{s^2+s+1}\right) = \tfrac{1}{3}\frac{2s^2+4s+3}{s^3+2s^2+2s+1}. \tag{61}$$

The procedure described here was first proposed by Bode, so we shall refer to it as the *Bode method*.

THE GEWERTZ METHOD

An alternative approach was first described by Gewertz. To outline this procedure, let us write $F(s)$ as the ratio of two polynomials. Thus

$$F(s) = \frac{a_0 + a_1 s + a_2 s^2 + \cdots + a_m s^m}{b_0 + b_1 s + b_2 s^2 + \cdots + b_n s^n} = \frac{m_1(s) + n_1(s)}{m_2(s) + n_2(s)} \tag{62}$$

where the m's refer to the even parts of the numerator and denominator and the n's refer to the odd parts. The even part of $F(s)$ can now be written as in (11a). Thus

$$\text{Ev } F(s) = \frac{m_1 m_2 - n_1 n_2}{m_2{}^2 - n_2{}^2} = \frac{A_0 + A_1 s^2 + A_2 s^4 + \cdots + A_m s^{2m}}{B_0 + B_1 s^2 + B_2 s^4 + \cdots + B_n s^{2n}}, \quad (63)$$

where the right side has been written in expanded form. When a real-part function is specified as an even rational function in ω, the right side of (63) is obtained when ω^2 is replaced by $-s^2$. We first go to work on the denominator. Since the poles of Ev $F(s)$ are those of both $F(s)$ and $F(-s)$, the ones belonging to $F(s)$ are those that lie in the left half-plane. Hence, when we factor the denominator of (63), we assign all the left half-plane factors to $F(s)$. In this manner the denominator of $F(s)$ in (62) becomes known.

Turn now to the numerator. Suppose we write $F(s)$ as a rational function as in (62) with unknown literal coefficients in the numerator but with known denominator coefficients. We then form the expression $m_1 m_2 - n_1 n_2$ and set it equal to the numerator of the given function in (63). Equating coefficients of like powers of s on the two sides of this equation will permit us to solve for the unknowns. Note that three sets of coefficients are involved: the small a's the capital A's, and the small b's. Of these, the last two sets are known at this point; only the small a's are unknown.

Let us carry out the process just indicated. Identifying m_1, m_2, n_1, and n_2 from (62), we can write

$$m_1 m_2 - n_1 n_2 = (a_0 + a_2 s^2 + \cdots)(b_0 + b_2 s^2 + \cdots)$$
$$-(a_1 s + a_3 s^3 + \cdots)(b_1 s + b_3 s^3 + \cdots) \quad (64)$$
$$= A_0 + A_1 s^2 + \cdots + A_m s^{2m}.$$

Equating the coefficients yields

$$A_0 = a_0 b_0,$$
$$A_1 = a_0 b_2 + b_0 a_2 - a_1 b_1,$$
$$A_2 = a_0 b_4 + a_2 b_2 + a_4 b_0 - a_1 b_3 - a_3 b_1,$$
$$\cdots \cdots \cdots \cdots \cdots \cdots \cdots \cdots \cdots \cdots \cdots,$$
$$A_k = \sum_{j=-k}^{k} (-1)^{k+j} a_{k+j} b_{k-j}. \quad (65)$$

To find the unknown a's, we must solve this set of linear equations simultaneously.

We shall now illustrate, using the function in (60) already treated by Bode's method. The left-half-plane factors in the denominator of that expression are

$$m_2 + n_2 = (s + 1)(s^2 + s + 1) = s^3 + 2s^2 + 2s + 1.$$

Since the given $R(\omega)$ is zero at infinity, so also must $F(s)$ be zero at infinity. (Why?) Hence the numerator of $F(s)$ must be of the form

$$m_1 + n_1 = a_2 s^2 + a_1 s + a_0.$$

By inserting the last two equations into (65) and utilizing the fact that all the capital "A" coefficients are zero except A_0, which is unity, we get

$$1 = a_0,$$

$$0 = 2a_0 + a_2 - 2a_1,$$

$$0 = 2a_2 - a_1.$$

These equations are then solved to yield $a_0 = 1$, $a_1 = \frac{4}{3}$, and $a_2 = \frac{2}{3}$. The network function thus obtained verifies the one previously found in (61).

THE MIYATA METHOD

A variation of these methods is due to Miyata. With $F(s)$ given by (62), the even part is given by (63). Now consider a new function $F_0(s)$ whose even part is

$$\text{Ev } F_0(s) = \frac{1}{m_2{}^2 - n_2{}^2}, \tag{66}$$

where $m_2 + n_2$ is the same denominator as that of $F(s)$. Using either the Bode or the Gewertz method, find the function $F_0(s)$ of which Ev F_0 is the even part. Let it be written as

$$F_0(s) = \frac{m_0 + n_0}{m_2 + n_2}. \tag{67}$$

The numerator of the even part of this expression is $m_0 m_2 - n_0 n_2$ and, according to (66), equals 1. Next consider a new function $\hat{F}(s)$ formed by multiplying $F_0(s)$ by $m_1 m_2 - n_1 n_2$, which is the numerator of the even part of $F(s)$, and form its even part:

$$\hat{F}(s) = \frac{(m_1 m_2 - n_1 n_2)(m_0 + n_0)}{m_2 + n_2}, \tag{68a}$$

$$\mathrm{Ev}\ \hat{F}(s) = \frac{(m_0 m_2 - n_0 n_2)(m_1 m_2 - n_1 n_2)}{m_2{}^2 - n_2{}^2} = \frac{m_1 m_2 - n_1 n_2}{m_2{}^2 - n_2{}^2} \tag{68b}$$

$$= \mathrm{Ev}\ F(s).$$

The next-to-last step follows from the fact that $m_0 m_2 - n_0 n_2 = 1$. Thus $\hat{F}(s)$ and $F(s)$ have the same even part; but $\hat{F}(s)$ in (68a) may have a high-order pole at infinity (because the order of the numerator may be higher than that of the denominator). Suppose the denominator is divided into the numerator yielding a polynomial $q(s)$ as a quotient and a remainder of order no higher than that of the denominator. Thus,

$$\hat{F}(s) = q(s) + F_r(s) \tag{69a}$$

and

$$\mathrm{Ev}\ \hat{F}(s) = \mathrm{Ev}\ q(s) + \mathrm{Ev}\ F_r(s). \tag{69b}$$

The even part of the polynomial $q(s)$ is simply the sum of all its even powers, if any. If $q(s)$ has any even powers, then the right side of the last equation will become infinite as s approaches infinity, whereas we know, by (68b), that the left side does not. The conclusion is that $q(s)$ is an odd polynomial and has no even powers, so that $\mathrm{Ev}\ \hat{F} = \mathrm{Ev}\ F_r$ and, hence, $\mathrm{Ev}\ F_r = \mathrm{Ev}\ F$ from (68b). Furthermore, this remainder function has the same poles as the specified function; consequently, it is the desired function—namely, $F_r(s) = F(s)$.

In summary, we may state that when an even rational function $(m_1 m_2 - n_1 n_2)/(m_2{}^2 - n_2{}^2)$ is specified, a network function $F_0(s)$ whose even part is $1/(m_2{}^2 - n_2{}^2)$ is determined. This function is then multiplied by $(m_1 m_2 - n_1 n_2)$, following which a long division is carried out, yielding a remainder function with no pole at infinity. This is the desired function whose even part is the specified function.

To illustrate, let

$$\mathrm{Ev}\ F(s) = \frac{3s^4 + 6s^2 + 6}{1 - s^6}.$$

Then

$$\mathrm{Ev}\ F_0(s) = \frac{1}{1 - s^6}.$$

But this is the same function as previously considered in (60) and (61). Thus

$$F_0(s) = \tfrac{1}{3}\ \frac{2s^2 + 4s + 3}{s^3 + 2s^2 + 2s + 1}$$

and

$$(3s^4 + 6s^2 + 6)\ F_0(s) = \frac{2s^6 + 4s^5 + 7s^4 + 8s^3 + 10s^2 + 8s + 6}{s^3 + 2s^2 + 2s + 1}$$

$$= 2s^3 + 3s + \frac{4s^2 + 5s + 6}{s^3 + 2s^2 + 2s + 1}.$$

Hence

$$F(s) = \frac{4s^2 + 5s + 6}{s^3 + 2s^2 + 2s + 1}.$$

6.7 INTEGRAL RELATIONSHIPS BETWEEN REAL AND IMAGINARY PARTS

In the preceding several sections, algebraic procedures were discussed for determining a network function as a rational function of s, given one of the components of the function as a rational function, where by "a component of a function" we mean one of the quantities: real part, imaginary part, angle (or tangent function), or magnitude (or log magnitude). One drawback of these procedures is that the given component must already be in a realizable rational form. If, say, the real part is specified graphically or even analytically but not as a rational function, it is necessary first to find a realizable rational approximation to the given function before proceeding to find the network function and, from that, any of the other components.

Network functions are analytic functions of a complex variable, and hence their real and imaginary parts are related by the Cauchy-Riemann equations. However, these equations are implicit relationships and do not provide explicit formulas for computing one component from the other. In this section we shall present a number of relationships between the parts of a network function. These are well known in mathematics as *Hilbert transforms*. However, since they were first used in network theory by Bode, we shall refer to them as the *Bode formulas*. One immediate advantage of these relationships is that the specified component of a function can be given merely as a graph; and beyond this, the Bode formulas have many useful implications and applications, some of which we shall discuss.

Since we are dealing with analytic functions of a complex variable, one point of departure for relating components of a function could be Cauchy's integral formula (see Appendix 2), which states that

$$F(s) = \frac{1}{2\pi j} \oint_C \frac{F(z)}{z - s} \, dz. \tag{70}$$

In this expression C is a closed contour within and on which $F(s)$ is regular; z represents points on the contour, whereas s is any point inside. If we let the contour be a circle and express both z and s in polar coordinates, we shall be able to express the real and imaginary parts of $F(s)$ in terms of either its real or its imaginary part on the circle. Finally, by means of a transformation, the circle is mapped into the imaginary axis. The resulting expressions relating the real and imaginary parts are referred to as *Hilbert transforms*.

An alternative approach, which we shall adopt, is to start with Cauchy's integral theorem. (See Appendix 2.) This theorem states that the contour integral of a function around a path within and on which the function is regular will vanish. In order to apply this theorem, it is necessary to know (1) the integration contour and (2) the function to be integrated. In the present problem the contour of integration should include the $j\omega$-axis, since we want the final result to involve the j-axis real and imaginary parts of a network function. Consequently, since the functions we are dealing with are regular in the entire right half-plane, the contour of integration we shall choose will consist of the $j\omega$-axis and an infinite semicircular arc in the right half-plane. By Cauchy's theorem, the complete contour integral will be zero. Hence it remains only to calculate the contributions of each part of the contour.

Let $F(s)$ be a network function of either the driving-point or the transfer type; in the usual way write

$$F(j\omega) = R(\omega) + jX(\omega), \tag{71a}$$

$$\ln F(j\omega) = \alpha(\omega) + j\phi(\omega), \tag{71b}$$

where $\alpha(\omega) = \ln |F(j\omega)|$ is the gain function and $\phi(\omega)$ is the angle function. If $F(s)$ is a driving-point function, it will have neither zeros nor poles in the right half-plane. Hence $\ln F(s)$ will be regular there. If $F(s)$ is a transfer function, then $\ln F(s)$ will be regular in the right half-plane only if $F(s)$ is a minimum-phase function. Hence the results we develop will apply both to $F(s)$ and to $\ln F(s)$ so long as $F(s)$ is a minimum-phase function.

Let us now consider possible poles of $F(s)$ on the $j\omega$-axis. We know that any such poles must be simple. In carrying out the contour integration such poles must be bypassed by a small indentation to the right. The contribution of this indentation to the total integral is $2\pi j$ times half the residue of the integrand at the pole. (See Appendix 2.) Our objective is to obtain expressions relating the real part of a network function to the imaginary part, so that when one of these is given, the other can be calculated. Thus we are not likely to know the residues at the j-axis poles. Hence we shall assume that $F(s)$ has no poles on the $j\omega$-axis; this includes the points zero and infinity as well, so that $F(s)$ *is assumed regular at zero and infinity*.

If $F(s)$ has a pole on the $j\omega$-axis, then $\ln F(s)$ will have a logarithmic singularity there. If the integrand in question involves $\ln F(s)$, we shall again indent the contour about this singularity. But because the singularity is logarithmic, this indentation will contribute nothing to the contour integral. (See Appendix 2.) Hence, in case the integrand we choose involves $\ln F(s)$, we can permit $F(s)$ to have simple poles on the $j\omega$-axis In the following discussion we shall always take the function in the integrand to be $F(s)$. However, identical results apply if we replace $F(s)$ by $\ln F(s)$. In the formulas, $R(\omega)$ can be replaced by $\alpha(\omega)$, and $X(\omega)$ by $\phi(\omega)$.

Let us now consider integrating a network function $F(s)$, which is regular on the $j\omega$-axis including zero and infinity, around the contour shown in Fig. 20a, which consists of the entire $j\omega$-axis and an infinite semicircular arc to the right. By Cauchy's theorem, the integral of $F(s)$

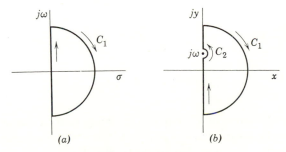

Fig. 20.   Path of integration.

will be zero. Our procedure will be to evaluate the contributions of those parts of the contour that we can evaluate and then to express the remaining parts in terms of these. With these ideas, it is obvious that we shall not be able to obtain the type of relationship we are looking for with $F(s)$ alone as the integrand. No particular point on the $j\omega$-axis is singled out and attention directed thereto.

Suppose we divide $F(s)$ by $s - j\omega_0$ before integrating, where ω_0 is any value of ω. This will put a pole of the integrand on the $j\omega$-axis. In order to apply Cauchy's theorem, we shall have to bypass this pole with a small semicircular arc C_2, as shown in Fig. 20b. The complete contour now consists of three parts, and the contribution of arc C_2 will have to be evaluated. This will focus attention on the value of $F(s)$ at $s = j\omega_0$. Note that the result of the integration will not be a function of s, which is only a dummy variable of integration, but of ω_0, which is an arbitrary point on the $j\omega$-axis. It will be convenient to use a different symbol for the dummy variable; let us use $z = x + jy$. Then the point $j\omega_0$ can be relabeled $j\omega$.

If $F(s)$ is a network function that is regular on the entire $j\omega$-axis as well as on the right half-plane, application of Cauchy's theorem leads to the following result:

$$\oint_C \frac{F(z)}{z - j\omega} \, dz = 0, \tag{72}$$

where the closed contour is the one shown in Fig. 20b.

The complete contour consists of three parts: the large semicircle C_1, the small semicircular indentation C_2 about the point $z = j\omega$, and the imaginary axis. The contribution of the small indentation to the overall integral is $2\pi j$ times half the residue of the integrand at $z = j\omega$, which is simply $F(j\omega)$. To compute the contribution of the infinite semicircle, let us initially assume it to be of finite radius, with $z = R_0 \epsilon^{j\theta}$. Then

$$\int_{C_1} \frac{F(s)}{z - j\omega} \, dz = \int_{C_1} \frac{F(R_0 \epsilon^{j\theta})}{R_0 \epsilon^{j\theta} - j\omega} jR_0 \epsilon^{j\theta} \, d\theta \xrightarrow[R_0 \to \infty]{} jF(\infty) \int_{\pi/2}^{-\pi/2} d\theta$$

$$= -j\pi F(\infty), \tag{73}$$

where $F(\infty)$ is the value of $F(s)$ at $s = \infty$. Thus as R_0 approaches infinity, the integral on C_1 approaches $-j\pi F(\infty)$. Since the imaginary part must be zero at infinity, $F(\infty)$ is also equal to $R(\infty)$.

Now it remains to consider the remainder of the contour. This can be written

$$\lim_{\substack{R_0 \to \infty \\ r \to 0}} \left[\int_{-R_0}^{\omega - r} \frac{F(jy)}{y - \omega} \, dy + \int_{\omega + r}^{R_0} \frac{F(jy)}{y - \omega} \, dy \right] = \int_{-\infty}^{\infty} \frac{F(jy)}{y - \omega} \, dy. \qquad (74)$$

Note that the integration along the imaginary axis must avoid the pole at $z = j\omega$ in a symmetrical manner. This will yield the *principal value* of the integral on the right. In all the subsequent integrals we must keep this point in mind. Now collecting all these results and substituting into (72) we can write

$$\int_{-\infty}^{\infty} \frac{F(jy)}{y - \omega} \, dy = j\pi[F(\infty) - F(j\omega)]. \qquad (75)$$

If we next write $F(j\omega)$ and $F(jy)$ in terms of real and imaginary parts, and equate reals and imaginaries, we get, finally,

$$R(\omega) = R(\infty) - \frac{1}{\pi} \int_{-\infty}^{\infty} \frac{X(y)}{y - \omega} \, dy, \qquad (76a)$$

$$X(\omega) = \frac{1}{\pi} \int_{-\infty}^{\infty} \frac{R(y)}{y - \omega} \, dy. \qquad (76b)$$

We are leaving the algebraic details of these steps for you to work out.

The message carried by these two expressions is very important. The second one states that when a function is specified to be the real part of a network function over all frequencies, the imaginary part of the function is completely determined, assuming the network function has no poles on the $j\omega$-axis. Similarly, if the imaginary part is specified over all frequencies, the real part is completely determined to within an additive constant.

Remember that the same results apply if $F(s)$ is replaced by its logarithm. However, now we must require that $F(s)$ be minimum-phase (if it represents a transfer function). On the other hand, we can relax the requirement of regularity of $F(s)$ on the $j\omega$-axis. A simple pole of $F(s)$ on the $j\omega$-axis becomes a logarithmic singularity of $\ln F(s)$, and such a singularity will contribute nothing to the integral, as mentioned earlier. Thus, for minimum-phase transfer functions, (76) with R and X replaced by α and ϕ, relate the gain and phase functions over all frequencies.

Let us now obtain alternative forms for the two basic expressions in (76) that will throw additional light on the relationships and will bring out points that are not at once apparent from these expressions. Remember that the real and imaginary parts are even and odd functions of frequency, respectively. Let us use this fact and write (76b) as follows:

$$X(\omega) = \frac{1}{\pi} \int_{-\infty}^{0} \frac{R(y)}{y - \omega} + \frac{1}{\pi} \int_{0}^{\infty} \frac{R(y)}{y - \omega} \, dy. \tag{77}$$

In the first of these integrals, replace y and $-y$ and change the limits accordingly. The result is

$$\int_{-\infty}^{0} \frac{R(y)}{y - \omega} \, dy = \int_{\infty}^{0} \frac{R(-y)}{-(y + \omega)} (-dy) = -\int_{0}^{\infty} \frac{R(y)}{y + \omega} \, dy. \tag{78}$$

The last step follows from the fact that $R(y) = R(-y)$. Substituting this into (77), we get

$$X(\omega) = \frac{1}{\pi} \int_{0}^{\infty} R(y) \left[\frac{1}{y - \omega} - \frac{1}{y + \omega} \right] dy = \frac{2\omega}{\pi} \int_{0}^{\infty} \frac{R(y)}{y^2 - \omega^2} \, dy. \tag{79}$$

In a completely similar way, starting with (76a) we get

$$R(\omega) = R(\infty) - \frac{2}{\pi} \int_{0}^{\infty} \frac{y \, X(y)}{y^2 - \omega^2} \, dy. \tag{80}$$

In the last two expressions it still appears that the integrand goes to infinity on the path of integration at the point $y = \omega$. This is really illusory, since we must understand the integral as the principal value. Even this illusory difficulty can be removed if we note by direct integration that

$$\int_{0}^{\infty} \frac{dy}{y^2 - \omega^2} = 0, \tag{81}$$

again using the principal value of the integral. Hence we can subtract $R(\omega)/(y^2 - \omega^2)$ from the integrand in (79) and $\omega \, X(\omega)/(y^2 - \omega^2)$ from the integrand in (80) without changing the values of these integrals. The results of these steps will be

$$R(\omega) = R(\infty) - \frac{2}{\pi} \int_{0}^{\infty} \frac{y \, X(y) - \omega \, X(\omega)}{y^2 - \omega^2} \, dy, \tag{82a}$$

$$X(\omega) = \frac{2\omega}{\pi} \int_{0}^{\infty} \frac{R(y) - R(\omega)}{y^2 - \omega^2} \, dy. \tag{82b}$$

A very important feature of the results that we have established is the fact that it is not necessary to have the real part (or the imaginary part) as a realizable rational function. Corresponding to any given real part, whether in analytical or in graphical form, an imaginary part can be computed from the integral. As a matter of fact, the expressions are quite useful when a desired real part is specified in a vague sort of way and it is desired to obtain an approximate behavior of the imaginary part.

For example, suppose it is desired to know the approximate behavior of the angle function in the pass band of a low-pass filter. In this discussion we shall interpret R and X to represent the gain α and the angle ϕ, respectively. In the pass band the gain is approximately zero up to some frequency ω_0. Hence in (82b) the lower limit becomes ω_0. Furthermore, the point ω, which lies in the pass band, is less than ω_0; thus in the integrand we can neglect ω compared with y, since y varies from ω_0 to infinity. Thus an approximate value is given by

$$\phi(\omega) \simeq \frac{2\omega}{\pi} \int_{\omega_0}^{\infty} \frac{\alpha(y)}{y^2}\, dy. \tag{83}$$

Now let us make the change of variable $y = 1/p$; then $dy/y^2 = -dp$. After appropriately modifying the limits of integration as well, this equation becomes

$$\phi(\omega) \simeq \frac{2\omega}{\pi} \int_0^{1/\omega_0} \alpha\!\left(\frac{1}{p}\right) dp. \tag{84}$$

Note that the integral in (83) or (84) is not a function of ω and that, for a given value of the band edge ω_0, it will be simply a constant. Thus the angle will be approximately a linear function of ω within the pass band.* Of course, the approximation will get progressively worse as we approach the band edge, since then ω can no longer be neglected in comparison to y in the integrand.

REACTANCE AND RESISTANCE-INTEGRAL THEOREMS

The two pairs of expressions obtained so far in (76) and (82) relate the imaginary part at any frequency to the real part at all frequencies; or the real part at any frequency to the imaginary part at all frequencies.

* Such a linear phase characteristic corresponds to a constant time delay in the transmission of sinusoidal functions over this range of frequencies. Therefore for signals that have essentially only this frequency range we get a distortionless transmission. For this reason a linear phase characteristic is desirable.

We should be able to find limiting forms for these expressions when frequency approaches zero or infinity.

First consider (82a) when ω approaches zero. This leads immediately to the result

$$\int_0^\infty \frac{X(y)}{y}\,dy = \frac{\pi}{2}\,[R(\infty) - R(0)]. \tag{85}$$

This expression is referred to as the *reactance-integral theorem*. It states that the integral of the imaginary part over all frequencies, weighted by the reciprocal of frequency, is proportional to the difference of the real part at the two extreme frequencies. It is also called the *phase-area theorem*, since the result remains valid when $F(s)$ is replaced by its logarithm, R by α, and X by ϕ.

A more convenient expression is obtained if a change to logarithmic frequency is made. Define

$$u = \ln \frac{y}{\omega}, \quad \text{or} \quad \frac{y}{\omega} = \epsilon^u, \tag{86}$$

where ω is some arbitrary reference frequency. Then dy/y becomes du, and (85) can be written as follows:

$$\int_{-\infty}^\infty X(y)\,du = \frac{\pi}{2}\,[R(\infty) - R(0)]. \tag{87}$$

Note the change in the lower limit, since $u = -\infty$ when $y = 0$. The argument of $X(y)$ has been retained as y for simplicity, although the integrand should more accurately be written as $X(\omega\epsilon^u)$. Alternatively, a new function $X_1(u) = X(\omega\epsilon^u)$ can be defined. However, this introduces additional new notation to complicate matters. In subsequent equations we shall retain y as the argument of the integrands and write $X(y)$ or $R(y)$, as the case may be, with the understanding that we mean to convert to a function of u by the substitution $y = \omega\epsilon^u$ before performing any operations. Thus we see that *the area under the curve of the imaginary part, when plotted against logarithmic frequency, is proportional to the net change in the real part between zero and infinite frequency.*

Next let us multiply both sides of (82b) by ω and then take the limit as ω approaches infinity. Remember that the upper limit on the integral means that we integrate up to R_0 and then let R_0 approach infinity. Thus (82b) becomes

$$\lim_{\omega \to \infty} \omega X(\omega) = \frac{2}{\pi} \lim_{\omega \to \infty} \left[\lim_{R_0 \to \infty} \int_0^{R_0} \frac{R(y) - R(\omega)}{\dfrac{y^2}{\omega^2} - 1}\,dy \right]. \tag{88}$$

There are two limiting operations involved on the right-hand side. If we interchange these two operations, the expression can be evaluated readily; but we must inquire whether this interchange is permissible. The answer is affirmative if the integral is uniformly convergent for all values of ω, which it is. Hence interchanging the two operations and taking the limits leads to

$$\int_0^\infty [R(y) - R(\infty)] \, dy = -\frac{\pi}{2} \lim_{\omega \to \infty} \omega X(\omega). \qquad (89)$$

The result expressed by this equation is referred to as the *resistance-integral theorem*. (It is also called the *attenuation-integral theorem*, since the result remains valid if $F(s)$ is replaced by its logarithm.) If the asymptotic behavior of the imaginary part of a network function is specified, then—no matter how the j-axis real part behaves with frequency—the area under the curve of the real part, with the horizontal axis shifted upward by an amount $R(\infty)$, must remain constant. Looking at it from the opposite viewpoint, when the integral of the real part of a function over all frequencies is specified, then the infinite-frequency behavior of the imaginary part is fixed.

Consider the special case in which $F(s)$ has a simple zero at infinity; then $F(\infty) = R(\infty) = 0$. Hence

$$- \lim_{\omega \to \infty} \omega X(\omega) = \lim_{s \to \infty} s \, F(s). \qquad (90)$$

However, according to the initial-value theorem, the limit on the right-hand side is simply the initial value of the impulse response of the network represented by $F(s)$. In this case, then, (89) becomes

$$\int_0^\infty R(\omega) \, d\omega = \frac{\pi}{2} \lim_{s \to \infty} s \, F(s) = \frac{\pi}{2} f(0), \qquad (91)$$

where $f(t) = \mathcal{L}^{-1}\{F(s)\}$ is the impulse response. Note that the dummy variable has been changed to ω to suggest the physical meaning.

LIMITATIONS ON CONSTRAINED NETWORKS

What has just been developed can be used to determine some basic limitations on the behavior of networks, when allowance is made for certain inevitable parasitic effects. Consider the situation depicted in Fig. 21a. The capacitance C accounts for parasitic effects that almost inevitably occur, such as junction capacitances in a transistor or just

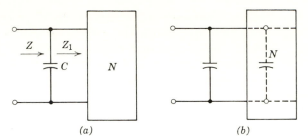

Fig. 21. Network constrained to have a shunt capacitance across its input terminals.

plain wiring capacitance. The presence of such a capacitance imposes some limitations that we shall now discuss.

Let $Z_1(s)$ be the impedance of the network N beyond the capacitance. The total impedance $Z(s)$ is given by

$$Z(s) = \frac{Z_1(s)}{Cs\,Z_1(s) + 1} = \frac{1}{Cs + \dfrac{1}{Z_1(s)}}. \tag{92}$$

Whatever the behavior of $Z_1(s)$ may be at infinity, we observe that the total impedance $Z(s)$ will have a simple zero at infinity. We shall initially assume that the network N does not start with a shunt capacitor as in Fig. 21b, meaning that $Z_1(s)$ has no zero at infinity. If it does, in fact, the result is an effective increase in the value of C.

With these stipulations, (90) is valid with $F(s) = Z(s)$. Inserting (92) into the right side of (90) and evaluating the limit yields

$$\lim_{s \to \infty} s\,Z(s) = \lim_{s \to \infty} \frac{s\,Z_1(s)}{Cs\,Z_1(s) + 1} = \frac{1}{C}.$$

Finally, when this is inserted into (91), the result becomes

$$\int_0^\infty R(\omega)\,d\omega = \frac{\pi}{2C}. \tag{93}$$

We see that the shunt capacitance imposes an effective limit on the area under the curve of the real part. Although this *resistance integral* evolved as the limiting value of the general expression relating the real and imaginary parts of a network function, it appears to provide a figure of merit of some sort for network capability.

Since the resistance-integral theorem applies to functions having no pole on the $j\omega$-axis, (93) is valid for such a function. If a function does have such poles on the $j\omega$-axis, the contour of integration must be indented around these poles and the contributions of these indentations must be taken into account. If one goes through the preceding development carefully, one finds that additional terms are subtracted from the right side of (93) in this case, these terms being proportional to the residues at the poles on the $j\omega$-axis. In the next chapter we shall show that all such residues of driving-point functions are real and positive. Hence, when $Z(s)$ has poles on the $j\omega$-axis, the right side of (93) is reduced in value. For all cases, then, whether $Z(s)$ is regular on the $j\omega$-axis or not, the result can be written as follows:

$$\int_0^\infty R(\omega)\, d\omega \le \frac{\pi}{2C}. \tag{94}$$

Further interpretation of this important result can be obtained from a consideration of Fig. 22, where a two-port is terminated in a resistor R_2.

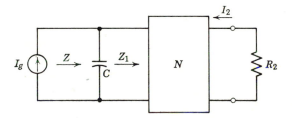

Fig. 22. Resistance-terminated two-port.

Assuming sinusoidal excitation, a calculation of the real power delivered to the input terminals by the source and the power delivered by the network to the load will be

$$\text{power from source} = \tfrac{1}{2}|I_g(j\omega)|^2 \operatorname{Re} Z(j\omega), \tag{95a}$$

$$\text{power to load} = \tfrac{1}{2}|I_2(j\omega)|^2 R_2. \tag{95b}$$

Clearly, the load power cannot exceed the power from the source for a passive two-port. Hence the second expression can be no greater than the first; so

$$\left|\frac{I_2(j\omega)}{I_g(j\omega)}\right|^2 \le \frac{1}{R_2} \operatorname{Re} Z(j\omega). \tag{96}$$

The equality is valid when the two-port is lossless. Thus the squared magnitude of the current gain of a lossless two-port is proportional to the real part of the impedance at the input terminals of the two-port when the output is terminated in R_2. Thus, with (96) inserted into (94), and with $R(\omega)$ interpreted as Re $Z(j\omega)$, there results

$$\int_0^\infty \left| \frac{I_2(j\omega)}{I_g(j\omega)} \right|^2 d\omega \leq \frac{\pi}{2R_2 C}. \tag{97}$$

Suppose the two-port in Fig. 22 is to be a filter with constant power gain over a given band of frequency and zero outside this band. Then the integral in (97) will simply equal the constant-power gain times the bandwidth. In the more general case, even though the transfer function may not be an ideal-filter function, the area under the curve represented by this integral is dimensionally power gain times bandwidth. For this reason the integral in (97) is generally called the *gain-bandwidth* integral. Thus we find a basic limitation on the gain-bandwidth product introduced by the presence of the shunt capacitor C.

ALTERNATIVE FORM OF RELATIONSHIPS

In the preceding discussion two sets of equivalent integral expressions relating the real and imaginary parts of network functions at all frequencies were found in (76) and (82). Still other forms are also possible, one of which is especially convenient for computation and leads to a simple evaluation of stability in closed-loop control systems. This form is most relevant when ln F (the gain and angle) is involved, rather than the network function itself. The expression utilizes the logarithmic frequency defined in (86).

Let us start with (82b) and perform some preliminary manipulations utilizing the change-of-frequency variable. We shall also use α and ϕ instead of R and X. Thus

$$\begin{aligned}
\phi(\omega) &= \frac{2}{\pi} \int_0^\infty \frac{\alpha(y) - \alpha(\omega)}{\left(\dfrac{y}{\omega}\right) - \left(\dfrac{\omega}{y}\right)} \frac{dy}{y} \\[2ex]
&= \frac{2}{\pi} \int_{-\infty}^\infty \frac{\alpha(y) - \alpha(\omega)}{\epsilon^u - \epsilon^{-u}} \, du \tag{98} \\[2ex]
&= \frac{1}{\pi} \int_{-\infty}^\infty \frac{\alpha(y) - \alpha(\omega)}{\sinh u} \, du.
\end{aligned}$$

Note the change in the lower limit, since $u = -\infty$ when $y = 0$. The argument of $\alpha(y)$ has been retained as y, as discussed earlier.

As the next step, we integrate the last form by parts. Using the general formula

$$\int a\,db = ab - \int b\,da$$

with

$$a = \alpha(y) - \alpha(\omega), \qquad db = \frac{du}{\sinh u},$$

$$da = \frac{d\alpha(y)}{du}\,du, \qquad b = -\ln \coth \frac{u}{2}.$$

Hence (98) becomes

$$\phi(\omega) = -\frac{1}{\pi}\left\{[\alpha(y) - \alpha(\omega)]\ln\coth\frac{u}{2}\right\}\Bigg|_{-\infty}^{\infty} + \frac{1}{\pi}\int_{-\infty}^{\infty}\frac{d\alpha(y)}{du}\ln\coth\frac{u}{2}\,du. \tag{99}$$

Note that $\coth u/2$ is an odd function of u, being strictly positive when u is positive and strictly negative when u is negative. Hence its logarithm for negative u will be complex, the imaginary part being simply π. For negative u it can be written

$$\ln\coth\frac{u}{2} = \ln\coth\frac{|u|}{2} + j\pi, \qquad u < 0. \tag{100}$$

When $u = +\infty$, $\ln\coth u/2 = 0$; and when $u = -\infty$, $\ln\coth u/2 = j\pi$. Hence the integrated part of the last equation becomes simply $j[\alpha(0) - \alpha(\omega)]$.

Now consider the remaining integral. If we use (100) for negative values of u, the result will be

$$\int_{-\infty}^{\infty}\frac{d\alpha(y)}{du}\ln\coth\frac{u}{2}\,du = \int_{-\infty}^{\infty}\frac{d\alpha(y)}{du}\ln\coth\frac{|u|}{2}\,du + j\pi\int_{-\infty}^{0}\frac{d\alpha(y)}{du}\,du$$

$$= \int_{-\infty}^{\infty}\frac{d\alpha(y)}{du}\ln\coth\frac{|u|}{2}\,du + j\pi\alpha(y)\Bigg|_{u=-\infty}^{u=0}$$

Finally, using all of these results in (99), we get

$$\phi(\omega) = \frac{1}{\pi} \int_{-\infty}^{\infty} \frac{d\alpha(y)}{du} \ln \coth \frac{|u|}{2} \, du. \tag{101}$$

This equation is quite easy to interpret even though it looks somewhat complicated. Note that the gain α is not an even function of the logarithmic frequency u, and so it is not possible to integrate over only half the range. The equation states that the angle at any frequency depends on the slope of the gain at all frequencies (when plotted against logarithmic frequency), the relative importance of different frequencies being determined by the weighting factor

$$\ln \coth \frac{|u|}{2} = \ln \left| \frac{y+\omega}{y-\omega} \right|. \tag{102}$$

This function is shown plotted in Fig. 23. It rises sharply in the vicinity of $u = 0 (y = \omega)$, falling off to very small values on both sides of this point. This means that most of the contribution to the angle at a frequency ω comes from the slope of the gain in the immediate vicinity of ω.

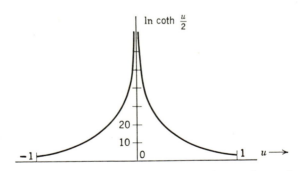

Fig. 23. Plot of weighting factor $\ln \coth \dfrac{|u|}{2} = \ln \left| \dfrac{y+\omega}{y-\omega} \right|$.

Another useful form can be obtained by simply adding and subtracting the slope evaluated at $u = 0 (y = \omega)$ under the integral in (101). We shall leave the details of this operation to you. The result will be

$$\phi(\omega) = \frac{\pi}{2} \frac{d\alpha(\omega)}{du} + \frac{1}{\pi} \int_{-\infty}^{\infty} \left[\frac{d\alpha(y)}{du} - \frac{d\alpha(\omega)}{du} \right] \ln \coth \frac{|u|}{2} \, du. \tag{103}$$

Note that by $d\alpha(\omega)/du$ we mean the slope of the gain as a function of u, evaluated when $u = 0(y = \omega)$. The slope $d\alpha(\omega)/du$ is measured in nepers per unit change of u. A unit change of u means a change in frequency by a factor ϵ.

We see that the angle at any frequency is $\pi/2$ times the gain slope at the same frequency plus another term given by the integral. If the gain is a continuous function, then the difference in the integrand will be small in the vicinity of $y = \omega$, just where the weighting factor has large values. Hence, in this case, the contribution of the integral to the angle will always be small. As a first approximation, then, we can say that the angle will have a value of $\pi/2$ radians whenever the gain slope is 1, a value of π radians whenever the gain slope is 2, etc.

Now suppose a gain function is given in graphical form. We can first approximate the curve by a series of straight-line segments having slopes of n, where n is an integer. An approximation to the (minimum-phase) angle function corresponding to the given gain function can now be quickly sketched according to the discussion of the last paragraph.

As an example of this procedure, suppose the gain plot* shown in Fig. 24 is given. The straight line approximation is superimposed. Now an approximate sketch of the angle, using only the approximate gain plot and completely neglecting the integral in (103), is the discontinuous function shown by the solid lines in the figure. The actual angle function might have the form shown by the dotted curve.

RELATIONS OBTAINED WITH DIFFERENT WEIGHTING FUNCTIONS

In deriving the integral relationships of this section, we started with the integrand in (72) and the closed contour shown in Fig. 20. The function $1/(z - j\omega)$ multiplying the network function $F(z)$ in (72) is a *weighting function*. The same relationships derived here can also be derived by using different weighting functions with integration around the same basic contour. Of course, if the weighting functions introduce additional poles on the $j\omega$-axis, we must avoid these poles by small indentations; for example, the resistance-integral theorem can be derived in short order by integrating the function $[F(z) - R(\infty)]$ around the basic contour. The weighting function here is 1. Similarly, the reactance-integral theorem follows readily when we integrate the function $F(z)/z$ around the basic contour with an indentation around the origin. The weighting function is $1/z$. You should verify these claims.

---

* This is, with a scale change on both axes, the Bode diagram of $|F(j\omega)|$, which finds extensive use in control theory. The Bode diagram, which is just 20 log $|F(j\omega)|$ versus log ω, is discussed in most basic texts on control systems analysis.

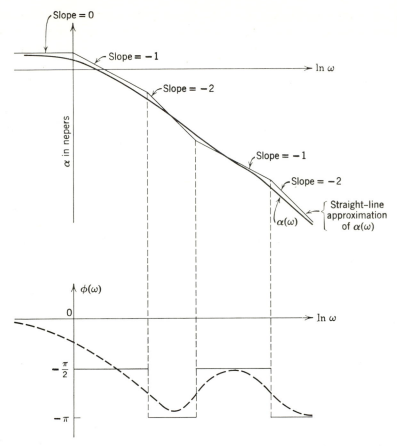

Fig. 24.   Approximate angle corresponding to a given gain function.

From this discussion it seems likely that additional relationships between the real and imaginary parts can be established by using different weighting functions. In fact, a great variety of relationships can be derived, but we have already presented the most important and useful ones. If we consider the two cases mentioned in the preceding paragraph, the criterion for choosing a weighting function appears to be to choose it in such a way that the term in which the known component of the network function appears is an even function of frequency, whereas the term in which the unknown component appears is an odd function. In this way the unknown component will disappear from the integration along the $j\omega$-axis and will appear only in the contributions from the indentations and from the infinite arc. It seems that this consideration in choosing a weighting function will apply quite generally.

So far in this section we have found that for a suitably restricted network function, when the real part is specified over all frequencies, the imaginary part is completely determined. Similarly, when the imaginary part is specified over all frequencies, the real part is completely determined (to within a constant). The question may be asked: Suppose the real part is specified over some frequency intervals and the imaginary part over the remainder of the entire frequency spectrum; is the function completely determined?

Instead of considering this problem in a completely general form, let us suppose that the real part is known for all frequencies less than ω_0 and the imaginary part for all frequencies greater than ω_0. We wish to find an expression that will give the unknown parts of the two components. The discussion concerning the choice of weighting functions suggests that if we can choose a weighting function that changes character at ω_0—so that below ω_0 the term involving the real part is even while above ω_0 the term involving the imaginary part is even—our problem will be solved. What we need is a multivalued weighting function.

Suppose we choose the following weighting function:

$$\frac{1}{(z^2 + \omega^2)\sqrt{1 + \dfrac{z^2}{\omega_0{}^2}}} \cdot$$

Again $z = x + jy$ is taken as a dummy variable. The irrational factor in the denominator is multivalued, with branch points at $z = \pm j\omega_0$. We must choose the branch cut in such a way that the integration along the j-axis will stay on a single sheet of the Riemann surface. This will be the case if, when $z = jy$, we take

$$\sqrt{1 - \frac{y^2}{\omega_0{}^2}} \quad \text{real and positive for} \quad -\omega_0 < y < \omega_0,$$

$$\sqrt{1 - \frac{y^2}{\omega_0{}^2}} \quad \text{imaginary and positive for} \quad y > \omega_0,$$

$$\sqrt{1 - \frac{y^2}{\omega_0{}^2}} \quad \text{imaginary and negative for} \quad y < -\omega_0.$$

With this choice, $\sqrt{1 - y^2/\omega_0{}^2}$ is an even function in the interval $-\omega_0 < y < \omega_0$, whereas over the remainder of the axis it is odd.

The contour of integration consists of the basic contour shown in Fig. 20 but with indentations at $z = \pm j\omega$. In the present case the infinite arc contributes nothing, since the integrand goes down at least as fast as $1/z^3$ at infinity. The contributions of the indentations are $j\pi$ times the residue of the integrand at the corresponding pole, which is easily evaluated. There remains the integration along the $j\omega$-axis. This is broken up into two parts, one beween zero and ω_0, the other between ω_0 and infinity. The details will be left for you to work out. The result will be

$$\frac{2\omega}{\pi} \int_0^{\omega_0} \frac{R(y)}{\sqrt{1 - \dfrac{y^2}{\omega_0{}^2}}} \frac{dy}{(y^2 - \omega^2)} + \frac{2\omega}{\pi} \int_{\omega_0}^{\infty} \frac{X(y)}{\sqrt{\dfrac{y^2}{\omega_0{}^2} - 1}} \frac{dy}{(y^2 - \omega^2)}$$

$$= \frac{X(\omega)}{\sqrt{1 - \dfrac{\omega^2}{\omega_0{}^2}}}, \qquad \omega < \omega_0 \tag{104}$$

$$= \frac{-R(\omega)}{\sqrt{\dfrac{\omega^2}{\omega_0{}^2} - 1}}, \qquad \omega > \omega_0.$$

We have now answered the question posed at the start of this discussion, insofar as the present problem is concerned. If we are given the real part of a function over part of the imaginary axis and the imaginary part over the rest of the axis, then the function is completely defined. Our method of obtaining the result in the last equation can be extended if there are more than two intervals over which one or the other of the two components are known. Additional irrational factors are introduced giving additional branch points at appropriate points on the axis. The resulting expressions, however, become rather complicated and hence limited in usefulness.

Let us now summarize the results of this section. Our objective is to obtain relationships between the real and imaginary parts of a network function $F(s)$ (or between the gain and the phase), so that when one of these is prescribed the other can be calculated. The point of departure is Cauchy's integral theorem, the contour of integration consisting of the imaginary axis with an infinite semicircular arc joining the ends. An integrand is chosen involving $F(s)$ or $\ln F(s)$, multiplied by a weighting function. The contour is indented to bypass poles of the integrand introduced by this function.

If the integrand involves a network function $F(s)$, then the only restriction is that $F(s)$ be regular on the $j\omega$-axis, including the points at zero and infinity. If the integrand involves ln $F(s)$, then $F(s)$ need not be regular on the $j\omega$-axis, but now it must have no zeros in the right half-plane; it must be a minimum-phase function.

The overall contour is divided into the straight segment consisting of the imaginary axis; the semicircular curves bypassing j-axis singularities deliberately introduced into the integrand; and the semicircular arc at infinity. The contributions of the semicircular contours can be computed, leaving only the integral along the imaginary axis.

A very useful feature of these expressions is the fact that the prescribed function need not be given in a realizable analytical form. An approximate graphical form is sufficient. Furthermore, the integrations themselves can be performed graphically.

6.8 FREQUENCY AND TIME-RESPONSE RELATIONSHIPS

The preceding sections have been concerned with the frequency properties of network functions and the relationships among the components of such functions in the frequency domain. Since a network function is the ratio of Laplace transforms of a response function to an excitation function, we might expect relationships to exist between the components of a network function and the time response. In this section we shall examine such relationships.

Let us refer to the notation in Chapter 5 and let $w_u(t)$ be the (scalar) response to a unit step excitation and $w_\delta(t)$ be the response to a unit impulse excitation. The corresponding network function $F(s)$ is related to these as

$$\mathcal{L}[w_\delta(t)] = F(s), \tag{105a}$$

$$\mathcal{L}[w_u(t)] = \frac{F(s)}{s}. \tag{105b}$$

We shall restrict ourselves to network functions having no poles on the $j\omega$-axis.

STEP RESPONSE

Now, from the definition of the Laplace integral, we get

$$\frac{F(s)}{s} = \int_0^\infty w_u(t) \, \epsilon^{-st} \, dt. \tag{106}$$

If we set $s = j\omega$, the exponential does not die out as t goes to infinity, but the integral will converge if $w_u(t) \to 0$ as $t \to \infty$. Next, from the final-value theorem, we find that

$$\lim_{t \to \infty} w_u(t) = \lim_{s \to 0} s \, \frac{F(s)}{s} = \lim_{s \to 0} F(s).$$

Thus, requiring that $w_u(t) \to 0$ as $t \to \infty$ means that $F(s)$ *must have a zero at $s = 0$*. With this stipulation we can now write (106) as

$$F(j\omega) = R(\omega) + jX(\omega) = j\omega \int_0^\infty w_u(t)[\cos \omega t - j \sin \omega t] \, dt.$$

From this it follows that

$$R(\omega) = \int_0^\infty \omega w_u(t) \sin \omega t \, dt, \qquad (107a)$$

$$X(\omega) = \int_0^\infty \omega w_u(t) \cos \omega t \, dt. \qquad (107b)$$

Thus the real and imaginary parts of a network function can be obtained directly from the step response.

Converse relationships giving the step response in terms of the real or imaginary part also exist. These can be obtained starting with the inversion integral for the step response. Since $w_u(t) = \mathcal{L}^{-1}\{F(s)/s\}$, we get

$$w_u(t) = \frac{1}{2\pi j} \int_B \frac{F(s)}{s} \, \epsilon^{st} \, dt. \qquad (108)$$

We are still assuming that $F(s)$ has no poles on the $j\omega$-axis, but let us not restrict it to have a zero at the origin for this development. Then the integrand in the last expression might have a pole at the origin. If it were not for this pole, the Bromwich path could be taken as the $j\omega$-axis. Instead, let us take the path shown in Fig. 25, which consists of the $j\omega$-axis except for a semicircular arc that bypasses the origin. As the radius of the semicircle approaches zero, the path approaches the entire $j\omega$-axis. The three parts of the path have been labeled C_1, C_2, and C_3. Equation 108 can now be written as follows:

$$w_u(t) = \frac{1}{2\pi j} \int_{C_1} \frac{F(s)}{s} \, \epsilon^{st} \, ds + \frac{1}{2\pi j} \int_{C_2} \frac{F(s)}{s} \, \epsilon^{st} \, ds + \frac{1}{2\pi j} \int_{C_3} \frac{F(s)}{s} \, \epsilon^{st} \, ds.$$

$$(109)$$

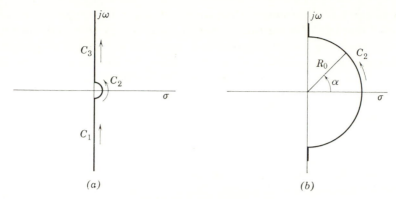

Fig. 25.   Contours of integration.

On the parts C_1 and C_3, $s = j\omega$ and $ds = jd\omega$. On the part C_2, which is shown expanded in part (b) of Fig. 25, we can write

$$s = R_0\, \epsilon^{j\alpha} = R_0 \cos \alpha + jR_0 \sin \alpha,$$

$$ds = jR_0\, \epsilon^{j\alpha}\, d\alpha.$$

Hence (109) becomes

$$w_u(t) = \frac{1}{2\pi j}\left\{\int_{-\infty}^{-R_0} \frac{F(j\omega)}{j\omega}\, \epsilon^{j\omega t}\, j\, d\omega + \int_{R_0}^{\infty} \frac{F(j\omega)}{j\omega}\, \epsilon^{j\omega t}\, j\, d\omega\right\}$$

$$+ \frac{1}{2\pi j}\int_{-\pi/2}^{\pi/2} F(s)\, \exp[t(R_0 \cos \alpha + jR_0 \sin \alpha)]j\, d\alpha. \qquad (110)$$

The last integral on the right involves the radius R_0 in a complicated way. However, we intend to let R_0 approach zero, in which case this term reduces to $F(0)/2$. You should verify this. Note that if we place the additional restriction that $F(s)$ has a zero at $s = 0$, then this term will disappear. When we let $R_0 \to 0$, the remaining two integrals in (110) combine to give the principal value of the integral running from $-\infty$ to $+\infty$. Hence, finally,

$$w_u(t) = \frac{F(0)}{2} + \frac{1}{2\pi j}\int_{-\infty}^{\infty} \frac{F(j\omega)}{\omega}\, \epsilon^{j\omega t}\, d\omega. \qquad (111)$$

(Note that, although it is not explicitly shown, we are to understand the last integral as representing the principal value.) This expression can be further simplified by writing $F(j\omega)$ in terms of its real and imaginary parts, expanding the exponential and using the odd and even properties of the resulting functions to change the range of integration to the positive ω-axis. The details will be left to you. The result is

$$w_u(t) = \frac{R(0)}{2} + \frac{1}{\pi}\int_0^\infty \frac{R(\omega)}{\omega}\sin \omega t \, d\omega + \frac{1}{\pi}\int_0^\infty \frac{X(\omega)}{\omega}\cos \omega t \, d\omega. \quad (112)$$

We have replaced $F(0)$ by $R(0)$, since $X(0) = 0$. Observe that this expression is defined for negative as well as positive values of t. However, $w_u(t) = 0$ for negative values of t. Hence in the two ranges of t we get

$$0 = \frac{R(0)}{2} - \frac{1}{\pi}\int_0^\infty \frac{R(\omega)}{\omega}\sin \omega t \, d\omega + \frac{1}{\pi}\int_0^\infty \frac{X(\omega)}{\omega}\cos \omega t \, d\omega$$

or

$$\frac{R(0)}{2} + \frac{1}{\pi}\int_0^\infty \frac{X(\omega)}{\omega}\cos \omega t \, d\omega = \frac{1}{\pi}\int_0^\infty \frac{R(\omega)}{\omega}\sin \omega t \, d\omega.$$

When we substitute the last equation into (112), we obtain the final result:

$$w_u(t) = R(0) + \frac{2}{\pi}\int_0^\infty \frac{X(\omega)}{\omega}\cos \omega t \, d\omega, \qquad (113a)$$

$$w_u(t) = \frac{2}{\pi}\int_0^\infty \frac{R(\omega)}{\omega}\sin \omega t \, d\omega. \qquad (113b)$$

Up till now we have performed various mathematical manipulations to put the relationships between $F(j\omega)$ and $w_u(t)$ in various equivalent forms. But now we have something new. The last equation shows that the step response of the network can be computed when only the real part of the network function along the $j\omega$-axis is known. Note that this relationship does not require that $R(0) = F(0)$ be zero. With the step response determined, (107b) can be used to compute the imaginary part of $F(j\omega)$. However, from the derivation of (107b) we know that the asymptotic value of the step response that is to be used in (107b) must be zero. Hence, before using $w_u(t)$ as computed from (113b), we first subtract its asymptotic value, $R(0)$, in case it is not equal to zero. In this way $F(j\omega)$ is completely determined from a knowledge of its real part alone.

Similarly, starting with the imaginary part $X(\omega)$, we can compute the step response from the integral in (113a). The portion of the step response computed from this integral will approach zero as t approaches infinity. To the value thus computed we can add any constant, which will become the zero-frequency value of $R(j\omega)$, denoted by $R(0)$ in (113a). However, omitting this step, we can now compute the real part $R(\omega)$ from (107a). Thus $F(j\omega)$ will be completely determined, except for an additive constant, from a knowledge of the imaginary part alone.

The procedures just discussed for finding a network function from its real or imaginary part are quite different from those discussed in earlier sections. They are also apparently more complicated, since they involve evaluating two integrals. However, it should be noted that the real or imaginary part need not be given as a rational function; a graph is sufficient.

IMPULSE RESPONSE

Let us now turn to the impulse response. Everything that was done starting from (106) can be duplicated (with appropriate changes) in terms of the impulse response. We shall list the results and leave the details of the development to you. It will still be *required that $F(s)$ be regular on the $j\omega$-axis*, but now it need not have a zero at $s = 0$. Instead, application of the inversion integral to $F(s)$ will require that $F(s)$ *have a zero at infinity*. If we retrace the steps starting at (106), we shall get the following equations:

$$R(\omega) = \int_0^\infty w_\delta(t) \cos \omega t \, dt, \qquad (114a)$$

$$X(\omega) = -\int_0^\infty w_\delta(t) \sin \omega t \, dt, \qquad (114b)$$

$$w_\delta(t) = \frac{2}{\pi} \int_0^\infty R(\omega) \cos \omega t \, d\omega, \qquad (114c)$$

$$w_\delta(t) = -\frac{2}{\pi} \int_0^\infty X(\omega) \sin \omega t \, d\omega. \qquad (114d)$$

The first two of these are the counterparts of (107), whereas the last two are to be compared with (113). As a matter of fact, the last two equations can be obtained from (113), in view of the fact that the impulse response is the derivative of the step response. (No impulses will be involved, since we assumed $F(\infty) = 0$.)

Equation (114d) shows that the impulse response of the network can be computed even if only the imaginary part $X(\omega)$ is known. Note that $X(\omega)$ will approach zero as ω approaches infinity, even though $F(\infty)$ may not be zero. With the impulse response computed, the real part of $R(\omega)$—or $R(\omega) - R(\infty)$ if $R(\infty) = F(\infty) \neq 0$—can now be found from (114a). Thus $F(j\omega)$ is determined to within the additive constant $F(\infty) = R(\infty)$ by its imaginary part alone.

Similarly, starting from a knowledge of only the real part of $R(\omega)$—or $R(\omega) - R(\infty)$ if $R(\infty) = F(\infty) \neq 0$—the impulse response can be computed from (114c). Having found the impulse response, the imaginary part $X(\omega)$ is now calculated from (114b). Thus we find that a transfer function is completely determined from a knowledge of its real part alone.

In each of the above cases, once the step response or impulse response has been calculated from a given $R(\omega)$ or $X(\omega)$, it is then necessary to find only the Laplace transform, since $\mathcal{L}\{w_u(t)\} = F(s)/s$, and $\mathcal{L}\{w_\delta(t)\} = F(s)$. In this way one of the integrations of (107) and (114) can be avoided.

Examples

Suppose the following is specified to be the real part of a network function on the $j\omega$-axis:

$$R(\omega) = \frac{\omega^4 + 2\omega^2 + 4}{(1 + \omega^2)(4 + \omega^2)}. \tag{115}$$

We see that this has a nonzero value at infinity, and so (114b) cannot be used directly. If we subtract its infinite-frequency value, we get

$$R_1(\omega) = R(\omega) - 1 = \frac{-3\omega^2}{(1 + \omega^2)(4 + \omega^2)}.$$

We can now apply (114b), which leads to

$$w_{\delta 1}(t) = -\frac{6}{\pi} \int_0^\infty \frac{\omega^2 \cos \omega t}{(1 + \omega^2)(4 + \omega^2)} \, d\omega$$

$$= -\frac{3}{\pi} \int_0^\infty \frac{\omega^2 \epsilon^{j\omega t} \, d\omega}{(1 + \omega^2)(4 + \omega^2)} + \int_0^\infty \frac{\omega^2 \epsilon^{-j\omega t} \, d\omega}{(1 + \omega^2)(4 + \omega^2)} \tag{116}$$

$$= -\frac{3}{\pi} \int_{-\infty}^\infty \frac{\omega^2 \epsilon^{j\omega t} \, d\omega}{(1 + \omega^2)(4 + \omega^2)}.$$

The second line follows from the use of the exponential form of $\cos \omega t$. If in the second integral in this line we replace ω by $-\omega$ and appropriately change the limits, the last line will follow.

Now consider the following contour integral in the complex s-plane:

$$I = \oint_C \frac{s^2 \epsilon^{st}\, ds}{(s^2 - 1)(s^2 - 4)},$$

where the contour consists of the entire $j\omega$-axis and an infinite semicircle to the left. The integrand satisfies the conditions of Jordan's lemma, since the rational function in the integrand vanishes at infinity as $1/s^2$. Hence the contribution of the infinite arc will be zero, and the complete integral reduces to its value along the $j\omega$-axis. By the residue theorem, the value of the integral equals $2\pi j$ times the sum of the residues at the left half-plane poles. In the present case there are only two simple poles, at $s = -1$ and $s = -2$, in the left half-plane and their residues are easily computed. Hence we get

$$I = -j \int_{-\infty}^{\infty} \frac{\omega^2 \epsilon^{j\omega t}\, d\omega}{(1 + \omega^2)(4 + \omega^2)} = 2\pi j \left(\frac{1}{6} \epsilon^{-t} - \frac{1}{3} \epsilon^{-2t} \right).$$

When this expression is substituted into (116), we get

$$w_{\delta 1}(t) = \epsilon^{-t} - 2\epsilon^{-2t}.$$

The transfer function can now be found by taking the Laplace transform. The result will be

$$F_1(s) = \frac{1}{s + 1} - \frac{2}{s + 2} = \frac{-s}{(s + 1)(s + 2)}.$$

This function has a zero at infinity. To this we should add the infinite-frequency value of $R(\omega)$, which is $F(\infty)$ and which we subtracted from the original function at the start. Thus

$$F(s) = F_1(s) + F(\infty) = \frac{-s}{(s + 1)(s + 2)} + 1$$

$$= \frac{s^2 + 2s + 2}{(s + 1)(s + 2)}.$$

This is the desired network function.

As a second example, let the real part of a network function be specified by the ideal curve shown in Fig. 26. By using (114c), the impulse response

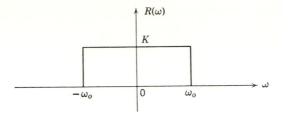

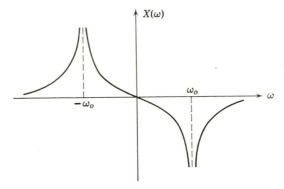

Fig. 26. Specified real part.

is computed to be

$$w_\delta(t) = \frac{2}{\pi} \int_0^{\omega_0} K \cos \omega t \; dt = \frac{2K}{\pi t} \sin \omega_0 t.$$

This is then inserted into (114b) to yield*

$$X(\omega) = -\int_0^\infty \frac{2K}{\pi t} \sin \omega_0 t \sin \omega t \; dt$$

$$= -\frac{K}{\pi} \int_0^\infty \left[\frac{\cos(\omega - \omega_0)t - \cos(\omega + \omega_0)t}{t} \right] dt$$

$$= -\frac{K}{\pi} \ln \left| \frac{\omega_0 + \omega}{\omega_0 - \omega} \right|, \quad |\omega| < \omega_0$$

$$= -\frac{K}{\pi} \ln \left| \frac{\omega + \omega_0}{\omega - \omega_0} \right|, \quad |\omega| > \omega_0.$$

* The last two lines may be obtained from integral 412, in R. S. Burington, *Handbook of Mathematical Tables and Formulas*, 2nd ed., Handbook Publishers, 1940.

PROBLEMS

1. For the networks shown in Fig. P1, verify that the nonzero eigenvalues of the $\mathscr{A}$ matrix in the state equation are the same as the nonzero zeros of the loop impedance matrix and of the node admittance matrix.

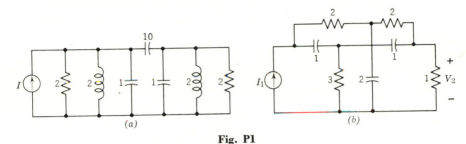

Fig. P1

2. Find the even part, Ev $F(s)$, and the odd part, Od $F(s)$, of the following functions from the even and odd parts of the numerator and denominator:

(a) $F(s) = \dfrac{s^3 + 2s^2 + 3s + 4}{s^2 + 3s + 3}$,   (b) $F(s) = \dfrac{s^3 + s^2 + 2s + 2}{s^4 + 5s^3 + 6s^2 + 4s + 1}$.

3. The polynomial $P_1(s) = s^2 - 6s + 12$ has a pair of zeros in the right half-plane. It is to be multiplied by another polynomial, $P_2(s)$, of degree n so that the resulting polynomial has no negative coefficients. What is the minimum value of n?

4. Let a transfer function be given by

$$F(s) = \frac{1 + a_1 s + a_2 s^2 + \cdots + a_n s^n}{1 + b_1 s + b_2 s^2 + \cdots + b_m s^m}$$

The *angle function* is defined as

$$\hat{\phi}(s) = \frac{1}{2} \ln \frac{F(s)}{F(-s)}$$

which is consistent with (49) when $s = j\omega$. The *delay function* is defined as

$$\tau(s) = -\frac{d}{ds}\hat{\phi}(s) = -\frac{1}{2}\frac{d}{ds}\left(\ln \frac{F(s)}{F(-s)}\right).$$

(**a**) In $F(s)$ let $a_i = 0$ for all i and insert into the expression for the delay function. Find the value of the b_i coefficients if the delay is to be a maximally flat function for the cases $m = 3$, $m = 4$, and $m = 5$.
(**b**) Repeat if $a_i \neq 0$ and $n = m - 1$.

5. Prove that if all the coefficients of a real polynomial $P(s)$ of degree n have the same sign, then $P(s)$ will have no zeros in the sector $|\arg s| < \pi/n$.

6. In Fig. P6 find z_{21} and verify that there is no transmission zero at $s = -1$ even though the left hand shunt branch has an admittance pole at $s = -1$.

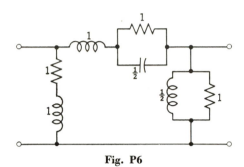

Fig. P6

7. The shunt branch admittances in the diagrams in Fig. P7 have a pole

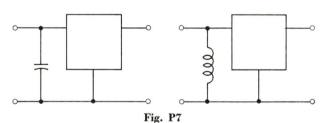

Fig. P7

at infinity and zero, respectively. Show that the overall two-ports *must* have transmission zeros at these frequencies (unlike the case in the previous problem), no matter what the rest of the network contains.

8. A symmetrical lattice has series and cross branch impedances Z_a and Z_b, respectively. Show that when it is terminated in a resistance R, the input impedance equals R if $Z_a Z_b = R^2$. Verify that for this constant-resistance case the voltage gain function is

$$G_{21}(s) = \frac{1 - Z_a/R}{1 + Z_a/R}.$$

9. Verify that the bridged-tee in Fig. 12*b* and the ell-networks in Figs. 12*c* and 12*d* are each constant-resistance networks when terminated in a resistance R if $Z_a Z_b = R^2$. Verify also that under this condition the voltage gain function is

$$G_{21}(s) = \frac{V_2}{V_1} = \frac{1}{1 + Z_a/R}.$$

10. Figure P10*a* shows a symmetrical lattice.
    (*a*) Find the y parameters (possibly using topological formulas) and show that the Fialkow condition (defined in Problem 47 of Chapter 3) will be satisfied under any one of the three conditions listed.

    (1) $\dfrac{L_2}{L_1} \geq 1;$ (2) $\dfrac{C_1}{C_2} \geq 1$ (3) $\dfrac{L_2}{L_1} + \dfrac{C_1}{C_2} \geq 1$

    (*b*) Figures P10*b* and 10*c* show two bridged-tee networks. Show that the first has the same y parameters as the lattice under condition (1) above, and hence is equivalent to the lattice. Show also that the second one has the same y parameters as the lattice under condition (2) above.
    (*c*) If the y parameters of the lattice are expanded in partial fractions, the result will have the form:

    $$y_{11} = y_{22} = k_\infty s + \frac{k_0}{s} + \frac{ks}{s^2 + \omega_0{}^2} = \left(k_\infty s + \frac{\alpha ks}{s^2 + \omega_0{}^2} \right) + \left(\frac{k_0}{s} + \frac{(1 - \alpha)ks}{s^2 + \omega_0{}^2} \right),$$

    $$y_{12} = y_{21} = k_\infty s + \frac{k_0}{s} - \frac{ks}{s^2 + \omega_0{}^2} = \left(k_\infty s - \frac{\alpha ks}{s^2 + \omega_0{}^2} \right) + \left(\frac{k_0}{s} - \frac{(1 - \alpha)ks}{s^2 + \omega_0{}^2} \right).$$

    On the right side a fraction of the finite pole has been combined with the pole at infinity and the rest of it is combined with the pole at the origin. Show that each of the tee networks in the twin-tee shown in Fig. P10*d* has one of the sets of y parameters within the above parentheses. Determine the range of values of α and show that this range of values exists if condition (3) above is satisfied. Thus under this condition the twin-tee is equivalent to the lattice.
    (*d*) Determine the angle of the transmission zeros of the bridged-tees and twin-tee determined by the three conditions in part (*a*).

11. Find a two-port network terminated in a 100-ohm resistor whose voltage gain function is given by each of the following all-pass functions.

In each case involving a lattice determine if a common-terminal equivalent exists, and convert to it if it does.

$$(a)\ G_{21} = \frac{s^2 - 2s + 2}{s^2 + 2s + 2}, \qquad (b)\ G_{21} = \frac{s^2 - 3s + 5}{s^2 + 3s + 5},$$

$$(c)\ G_{21} = \frac{(s^2 - s + 1)(s - 5)}{(s^2 + s + 1)(s + 5)}.$$

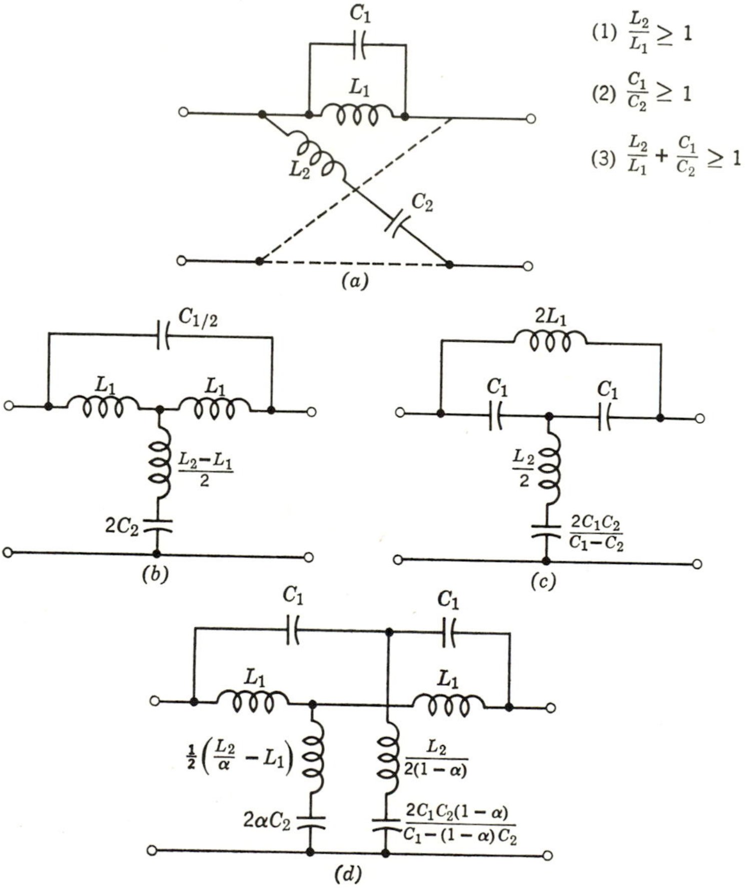
C1
L1
L2
C2
(1) L2/L1 ≥ 1
(2) C1/C2 ≥ 1
(3) L2/L1 + C1/C2 ≥ 1
(a)
C1/2
L1
L1
(L2−L1)/2
2C2
(b)
2L1
C1
C1
L2/2
2C1C2/(C1−C2)
(c)
C1
C1
L1
L1
½(L2/α − L1)
L2/2(1 − α)
2αC2
2C1C2(1 − α)/(C1 − (1 − α)C2)
(d)

Fig. P10

12. Find a two-port network terminated in a 50-ohm resistor whose voltage gain function is given by each of the following nonminimum-phase functions. Select any convenient value of K. Convert any lattices to common-terminal equivalent networks where possible.

$$(a)\ G_{21} = K\frac{s^2 - s + 2}{(s+1)^2}, \qquad (b)\ G_{21} = K\frac{s-2}{s^2 + 4s + 3},$$

$$(c)\ G_{21} = K\frac{s^2 - 3s + 5}{s^2 + 7s + 5}.$$

13. The following functions are specified as the tangent functions of a transfer function. Find the corresponding transfer function $F(s)$. If the answer is not unique, give all possibilities.

$$(a)\ \tan\phi = \frac{2\omega}{1 - 6\omega^2}, \qquad (b)\ \tan\phi = -\omega^3 - \pi u(\omega - 2) - \pi u(\omega + 2),$$

$$(c)\ \tan\phi = \frac{-\omega}{\omega^2 + 2}, \qquad (d)\ \tan\phi = \frac{\omega^3 - 3\omega}{-3\omega^2 + 2},$$

$$(e)\ \tan\phi = \frac{-\omega^3}{\omega^4 - 4\omega^2 + 24}, \qquad (f)\ \tan\phi = \frac{\omega^3 + 2\omega}{\omega^4 - 3\omega^2 - 6}.$$

14. The following functions are specified as the real part of an impedance function $F(s)$. Use any one of the methods of Bode, Gewertz or Miyata to find the function $F(s)$.

$$(a)\ R(\omega) = \frac{16 - 8\omega^2 + \omega^4}{1 + \omega^8}, \qquad (b)\ R(\omega) = \frac{2 + 4\omega^2 + 3\omega^4 + \omega^6}{1 + \omega^8},$$

$$(c)\ R(\omega) = \frac{-\omega^2 + 2\omega^4 - \omega^8}{1 + \omega^8}, \qquad (d)\ R(\omega) = \frac{1 - 2\omega^2 + \omega^4}{1 - 2\omega^2 + \omega^4 + 4\omega^6},$$

$$(e)\ R(\omega) = \frac{(1 - \omega^2 + \omega^4)^2}{(\omega^4 - 4\omega^2 + 3)^2 + \omega^2(\omega^4 - 6\omega^2 + 8)^2}.$$

15. Suppose each function in Problem 14 is the j-axis magnitude squared of a network function. Find the function. If there is more than one possibility, find them all.

16. Give a derivation for (103) in the text starting with (101).

17. Give a derivation for (104) in the text.

18. Derive the reactance integral theorem in (85) by integrating the function $F(s)/s$ around the basic contour with a small indentation around the origin.

19. Derive the resistance integral theorem in (89) by integrating the function $F(s) - R(\infty)$ around the basic contour consisting of the $j\omega$-axis and an infinite semicircle to the right.

20. Derive (79) by integrating the function $F(z)/(z^2 + \omega^2)$ around the basic contour with indentations at $z = \pm j\omega$.

21. Derive (80) by integrating the function $z[F(z) - R(\infty)]/(z^2 + \omega^2)$ around the basic contour with indentations at $z = \pm j\omega$.

22. By integrating the function $[F(z) - R(0)]/z(z^2 + \omega^2)$ around the basic contour with indentations at $z = 0$ and at $z = \pm j\omega$, derive the following relationship:

$$R(\omega) = R(0) - \frac{2\omega^2}{\pi} \int_0^\infty \frac{X(y)/y - X(\omega)/\omega}{y^2 - \omega^2}\, dy.$$

Compare this with (82a) in the text.

23. Each of the curves in Fig. P23 is the magnitude $|F(j\omega)|$ of a transfer function for $\omega > 0$. Assuming the function is minimum-phase, find the corresponding angle function making appropriate approximations.

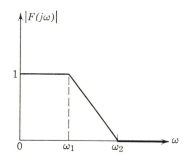

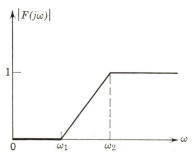
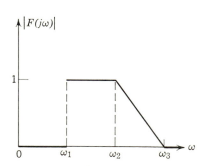

Fig. P23

24. Let the magnitude of a function be the maximally flat function in (32). Show that the angle is given by the following expression:

$$\phi(\omega) = -\frac{n}{\pi} \int_0^\infty \frac{1/y}{1+y^{-2n}} \ln \left| \frac{y+\omega}{y-\omega} \right| dy.$$

25. Let the real part of a function on the $j\omega$ axis be given by the following functions. Find the corresponding step response using (113b) and the impulse response using (114c).

(a) $R(\omega) = \dfrac{1}{4\omega^6 + 12\omega^4 + 11\omega^2 + 3}$,   (b) $R(\omega) = \dfrac{(1-\omega^2)^3}{1+\omega^6}$,

(c) $R(\omega) = \dfrac{1+2\omega^2+\omega^4}{1-2\omega^2+\omega^4+4\omega^6}$,   (d) $R(\omega) = \dfrac{1}{1+\omega^{2n}}$.

26. Suppose the imaginary part of a network function is as shown in Fig. P26. Use (113a) to compute $w_u(t)$ and then use (107a) to determine the real part of the network function.

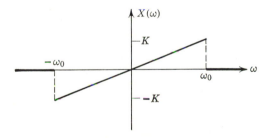

Fig. P26

27. Suppose the step response $w_u(t)$ of a network is as shown in Fig. P27. Use (107a) to determine $R(\omega)$ and the use (114c) to determine the impulse response $w_\delta(t)$. Verify this result by finding $w_\delta(t)$ directly from $w_u(t)$.

The next 3 problems marked with an asterisk involve the preparation of a computer program to help in implementing the solution of some problems. In each case, prepare a program flow chart and a set of program instructions, in some user language like FORTRAN IV, for a digital computer program to carry out the job specified in the problem. Include a set of user instructions for the program.

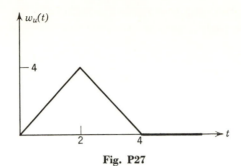

Fig. P27

28.* Suppose $f(\omega^2)$ in (41) is a rational function of ω^2. Let $f(\omega^2)$ be characterized by a list of numbers: The first two numbers will be the numerator degree and the denominator degree; these will be followed by a stream of numbers which are the numerator and denominator coefficients. Prepare a program to accept these data as a description of $f(\omega^2)$ and compute a stable minimum phase $F(s)$, where $F(j\omega)$ satisfies (41). Assume a subroutine exists which will determine all the zeros of a polynomial.†

29.* Prepare a program to determine a stable minimum phase $F(s)$ from $\tan \phi(\omega)$ which is assumed to be a rational function in ω. Use an input data format similar to that in Problem 28. Assume a subroutine exists which will determine all the zeros of a polynomial.

30.* Prepare a program to determine $F(s)$ from Ev $F(s)$ using: (*a*) Bode's method; (*b*) Gewertz's method; (*c*) Miyata's method. Use an input data format similar to that in Problem 28. Assume a subroutine exists which will determine all the zeros of a polynomial.

† One algorithm which might be the basis of such a subroutine program is the quotient-difference algorithm. This algorithm is described in H. Henrici, *Elements of Numerical Analysis*, John Wiley and Sons, New York, 1964, Chap. 8.

.7.

FUNDAMENTALS OF
NETWORK SYNTHESIS

Network synthesis is the process of designing and constructing a network to provide a prescribed response to a specified excitation. This is the converse of the analysis problem where a response is to be calculated when a prescribed excitation is applied to a given network. In contrast with the latter, the synthesis problem may not have a unique solution. In fact, it may not have any solution, since there may be no network that has the desired response to the given excitation. At the outset, then, one may be faced with the necessity of approximating the desired response with one that is obtainable.

Specification of the response and its approximation may be either in the time domain or in the frequency domain. In the frequency domain the end result of the approximation process is the specification of one or more network functions that characterize the desired network. From these functions it is then necessary to realize the network. The realization is guided by the fact that there are various classes of networks. These classes can be characterized by the number of external terminals, by the type of components (lossless, active, *RC*, etc.), by the structure (ladder, grounded, etc.), and so forth.

The first task in the realization process is to determine the properties of network functions that are appropriate for each class of network. These properties include permissible locations of poles and zeros, the signs of

467

residues and real parts, and the relative size of coefficients. This is the task on which we shall largely concentrate in this chapter.

In order to establish the analytical properties of network functions, it will be necessary to introduce some additional mathematical topics. The first two sections will be devoted to this effort. We shall not strive for completeness of exposition but shall often be content simply with the statement of a result with some discussion of plausibility.

7.1 TRANSFORMATION OF MATRICES

Given a square matrix **A**, a number of operations can be performed on it to yield another matrix **B**. This matrix, of course, will be related to the original matrix **A**, the specific relationship depending on the operations that are performed. The matrix **A** is said to be *transformed* in some way.

ELEMENTARY TRANSFORMATIONS

A number of specific operations have properties that are of great importance. They are very simple operations and are collectively called *elementary transformations*. Given a matrix **A**, the elementary transformations of **A** are the following:

1. *The interchange of any two rows or two columns of* **A**.
2. *The addition of the elements of one row or column of* **A** *to the corresponding elements of a second row or column.*
3. *The multiplication of each element of a row or column of* **A** *by a scalar constant.*

Clearly these transformations do not change the order of **A**. From the properties of determinants discussed in Chapter 1, the first transformation simply changes the sign of the determinant of **A**; the second one leaves the determinant unchanged; the third one multiplies the determinant by a constant. Hence, if matrix **A** is nonsingular, then the matrix obtained after an elementary transformation is also nonsingular. In fact, an elementary transformation of a matrix cannot change its rank, even when the rank is less than the order. (See Problem 5.)

The operations on a matrix **A** represented by the elementary transformations can be carried out by multiplying **A** by certain simple, nonsingular matrices. These matrices, called *elementary matrices*, are themselves obtained by carrying out the corresponding operation on a unit matrix. Thus, adding the third row to the second row of a third-order unit matrix leads to the elementary matrix on the left below; similarly, adding the third column to the second column of a third-order unit matrix leads to the elementary matrix on the right:

$$\begin{bmatrix} 1 & 0 & 0 \\ 0 & 1 & 1 \\ 0 & 0 & 1 \end{bmatrix} \quad \begin{bmatrix} 1 & 0 & 0 \\ 0 & 1 & 0 \\ 0 & 1 & 1 \end{bmatrix}$$

When a matrix **A** having three rows is *pre*multiplied by the elementary matrix on the left, the effect is to add the third row of **A** to the second row. When a matrix **A** having three columns is *post*multiplied by the elementary matrix on the right, the effect is to add the third column of **A** to its second column. Thus

$$\begin{bmatrix} 1 & 0 & 0 \\ 0 & 1 & 1 \\ 0 & 0 & 1 \end{bmatrix} \begin{bmatrix} a_{11} & a_{12} & a_{13} & a_{14} \\ a_{21} & a_{22} & a_{23} & a_{24} \\ a_{31} & a_{32} & a_{33} & a_{34} \end{bmatrix}$$

$$= \begin{bmatrix} a_{11} & a_{12} & a_{13} & a_{14} \\ a_{21} + a_{31} & a_{22} + a_{32} & a_{23} + a_{33} & a_{24} + a_{34} \\ a_{31} & a_{32} & a_{33} & a_{34} \end{bmatrix},$$

$$\begin{bmatrix} a_{11} & a_{12} & a_{13} \\ a_{21} & a_{22} & a_{23} \\ a_{31} & a_{32} & a_{33} \end{bmatrix} \begin{bmatrix} 1 & 0 & 0 \\ 0 & 1 & 0 \\ 0 & 1 & 1 \end{bmatrix} = \begin{bmatrix} a_{11} & a_{12} + a_{13} & a_{13} \\ a_{21} & a_{22} + a_{23} & a_{23} \\ a_{31} & a_{32} + a_{33} & a_{33} \end{bmatrix}.$$

Note that **A** need not be square; it must, of course, be conformable with the elementary matrix.

Since an elementary transformation of a unit matrix will not change its rank, *any elementary matrix is nonsingular*. Since the product of two nonsingular matrices is nonsingular, this leads to the conclusion that *the product of any number of elementary matrices is nonsingular*. A question of greater significance is, Does it work backwards? Can any nonsingular matrix be decomposed into elementary matrices? The answer is yes. It can be shown that *any nonsingular matrix can be written as the product of a finite number of elementary matrices*.

As further illustrations, suppose it is desired (1) to add the first row of a (4×3) matrix **A** to its third row after multiplying the first row by 5, and

(2) to interchange the third column and the second column after the second has been multiplied by 3. The two elementary matrices that will accomplish this are the following:

$$
\mathbf{E}_1 = \begin{bmatrix} 1 & 0 & 0 & 0 \\ 0 & 1 & 0 & 0 \\ 5 & 0 & 1 & 0 \\ 0 & 0 & 0 & 1 \end{bmatrix}, \qquad \mathbf{E}_2 = \begin{bmatrix} 1 & 0 & 0 \\ 0 & 0 & 3 \\ 0 & 1 & 0 \end{bmatrix}.
$$

The first one should premultiply $\mathbf{A}$; the second one, postmultiply $\mathbf{A}$. You should confirm this result.

The further exposition of the details of elementary matrices will be left to an extended set of problems. In the following development it will be assumed that the results of these problems are available.

EQUIVALENT MATRICES

Let $\mathbf{A}$ and $\mathbf{B}$ be two matrices of the same order. We say that $\mathbf{B}$ is *equivalent* to $\mathbf{A}$ if it can be obtained from $\mathbf{A}$ by a finite number of elementary transformations. If all the transformations are carried out on the rows, $\mathbf{B}$ is *row equivalent* to $\mathbf{A}$; if all the transformations are carried out on the columns, it is *column equivalent*. Carrying out a number of consecutive elementary transformations means multiplying $\mathbf{A}$ by the product of a number of elementary matrices. Such a product can be represented by a single matrix that is necessarily nonsingular, since each of the elementary matrices is nonsingular. Hence the general definition of equivalence can be restated as follows.

Theorem 1. *Let $\mathbf{A}$ and $\mathbf{B}$ be two matrices of the same order. Matrix $\mathbf{B}$ is equivalent to $\mathbf{A}$ if and only if*

$$\mathbf{B} = \mathbf{PAQ}, \tag{1}$$

where $\mathbf{P}$ and $\mathbf{Q}$ are nonsingular.

Since $\mathbf{P}$ and $\mathbf{Q}$ are nonsingular, then $\mathbf{A} = \mathbf{P}^{-1}\mathbf{B}\mathbf{Q}^{-1}$. This is of the same form as (1); hence, if $\mathbf{B}$ is equivalent to $\mathbf{A}$, then $\mathbf{A}$ is equivalent to $\mathbf{B}$; that is, the equivalence of two matrices is a mutual property.

Since an elementary transformation of a matrix does not change its rank, a sequence of elementary transformations leaves the rank of a matrix unchanged. Hence two equivalent matrices have the same rank.

In particular, if a square matrix $\mathbf{A}$ is nonsingular, a matrix equivalent to $\mathbf{A}$ is also nonsingular.

In fact, if $\mathbf{A}$ is a nonsingular matrix, it can always be reduced to a unit matrix by successive elementary transformations; that is, there will always be nonsingular matrices $\mathbf{P}$ and $\mathbf{Q}$ (each one being the product of elementary matrices) such that

$$\mathbf{PAQ} = \mathbf{U}. \tag{2}$$

Then

$$\mathbf{A} = \mathbf{P}^{-1}(\mathbf{PAQ})\mathbf{Q}^{-1} = \mathbf{P}^{-1}\mathbf{U}\mathbf{Q}^{-1} = \mathbf{P}^{-1}\mathbf{Q}^{-1}. \tag{3}$$

Thus, if $\mathbf{A}$ is nonsingular, it can always be factored into the product of two nonsingular matrices $\mathbf{P}^{-1}$ and $\mathbf{Q}^{-1}$. This, of course, is an "existence" statement; it does not specify an algorithm for carrying out the factoring.

The fact that a nonsingular matrix is equivalent to a unit matrix, as expressed in (2), is a special case of a more general case. Let $\mathbf{A}$ be a matrix of order $(m \times n)$ and of rank r. Then by elementary transformations it can always be reduced to a matrix $\mathbf{B}$ of the form

$$\mathbf{B} = \mathbf{PAQ} = \begin{array}{c} \\ r \\ m-r \end{array} \overset{\begin{array}{cc} r & n-r \end{array}}{\begin{bmatrix} \mathbf{U}_r & \mathbf{0} \\ \mathbf{0} & \mathbf{0} \end{bmatrix}}. \tag{4}$$

The upper-left-hand submatrix is a unit matrix of order r. When $\mathbf{A}$ is square and nonsingular, $n = m = r$, and (4) reduces to (2). The matrix on the right side of (4) is called the *normal form* of matrix $\mathbf{A}$.

In (1), suppose that $\mathbf{Q} = \mathbf{U}$; the result is $\mathbf{B} = \mathbf{PA}$. The nonsingular matrix $\mathbf{P}$ is the product of elementary matrices. Multiplying $\mathbf{A}$ by $\mathbf{P}$ means carrying out elementary transformations on the rows of $\mathbf{A}$. In the product matrix $\mathbf{B}$, then the rows are simply linear combinations of the rows of $\mathbf{A}$. Consequently, if two matrices are row equivalent, the rows of one are linear combinations of the rows of the other, and vice versa. In a similar way, if two matrices are column equivalent, the columns of one are linear combinations of the columns of the other; for example, in Chapter 2 it was found that the fundamental cut-set matrix $\mathbf{Q}_f$ of a network is obtained by premultiplying the incidence matrix $\mathbf{A}$ by the nonsingular matrix $\mathbf{A}_t$. So we should expect the rows of $\mathbf{Q}_f$ to be linear combinations of the rows of $\mathbf{A}$ or, equivalently, the cut-set equations to be linear combinations of Kirchhoff's-current-law equations at nodes, which we know to be true.

SIMILARITY TRANSFORMATION

In the equivalence relation (1), there need be no relationship between the matrices $\mathbf{P}$ and $\mathbf{Q}$. However, when there are certain specific relationships, the equivalence takes on such useful properties that it becomes convenient to classify and to name the corresponding transformations.

Suppose that in (1), $\mathbf{A}$ is a square matrix and $\mathbf{P} = \mathbf{Q}^{-1}$. Then

$$\mathbf{B} = \mathbf{Q}^{-1}\mathbf{A}\mathbf{Q} \tag{5a}$$

or

$$\mathbf{Q}\mathbf{B} = \mathbf{A}\mathbf{Q}. \tag{5b}$$

This transformation is a *similarity transformation*; $\mathbf{A}$ and $\mathbf{B}$ are called *similar* matrices. This transformation has already been discussed in Chapter 1, where we saw that two similar matrices have the same eigenvalues. It is included here for completeness.

CONGRUENT TRANSFORMATION

Another special kind of equivalence is the following. In (1) suppose $\mathbf{P} = \mathbf{Q}'$. Then the transformation

$$\mathbf{B} = \mathbf{Q}'\mathbf{A}\mathbf{Q} \tag{6}$$

is called a *congruent transformation*; $\mathbf{B}$ is said to be *congruent to* $\mathbf{A}$.

Since $\mathbf{Q}$ can be written as a product of elementary matrices, $\mathbf{Q}'$ will equal the product of the transposes of these elementary matrices, but in reverse order. Hence $\mathbf{Q}'\mathbf{A}\mathbf{Q}$ is obtained from $\mathbf{A}$ by carrying out pairs of elementary transformations, one transformation on the rows and a corresponding one on the columns.

A comparison of the similar transformation in (5) and the congruent transformation in (6) shows that the two will be identical if $\mathbf{Q}^{-1} = \mathbf{Q}'$. This property is given a special name. A matrix having the property

$$\mathbf{Q}^{-1} = \mathbf{Q}' \tag{7}$$

is called an *orthogonal* matrix.

If $\mathbf{A}$ is a real symmetric matrix of rank r, we can prove by means of a sequence of elementary transformations that it is congruent to a diagonal matrix $\mathbf{D}$ in the form

$$\mathbf{D} = \mathbf{Q}'\mathbf{A}\mathbf{Q} = \begin{bmatrix} \mathbf{D}_r & \mathbf{0} \\ \mathbf{0} & \mathbf{0} \end{bmatrix}, \tag{8}$$

where $\mathbf{D}_r$ is a diagonal matrix of order r and rank r, and $\mathbf{Q}$ is nonsingular. This resembles the normal form in (4), but there are a number of differences. In the general case of (4), $\mathbf{A}$ need not be square, and the two matrices $\mathbf{P}$ and $\mathbf{Q}$ need not be related.

The nonzero elements on the diagonal of $\mathbf{D}$ may be positive or negative. The rows and columns can always be interchanged to place the positive elements first. The corresponding product of elementary matrices can be lumped into $\mathbf{Q}$. When the positive and negative terms are shown explicitly, the result can be written as follows:

$$\mathbf{D} = \mathbf{Q}'\mathbf{A}\mathbf{Q} = \begin{bmatrix} \mathbf{D}_p & 0 & 0 \\ 0 & -\mathbf{D}_{r-p} & 0 \\ 0 & 0 & 0 \end{bmatrix}, \qquad \mathbf{D}_r = \begin{bmatrix} \mathbf{D}_p & 0 \\ 0 & -\mathbf{D}_{r-p} \end{bmatrix}, \qquad (9)$$

where both $\mathbf{D}_p$ and $\mathbf{D}_{r-p}$ are diagonal matrices with positive diagonal elements, the order and rank of $\mathbf{D}_p$ being p, and the order and rank of $\mathbf{D}_{r-p}$ being $r - p$.

Let us now define a matrix

$$\mathbf{D}^{-\frac{1}{2}} = \begin{bmatrix} \mathbf{D}_p^{-\frac{1}{2}} & 0 & 0 \\ 0 & -\mathbf{D}_{r-p}^{-\frac{1}{2}} & 0 \\ 0 & 0 & \mathbf{U} \end{bmatrix}, \qquad (10)$$

where

$$\mathbf{D}_p^{-\frac{1}{2}} = \begin{bmatrix} d_1^{-\frac{1}{2}} & & \bigcirc \\ & d_2^{-\frac{1}{2}} & \\ & & \ddots \\ \bigcirc & & d_p^{-\frac{1}{2}} \end{bmatrix}, \qquad \mathbf{D}_{r-p}^{-\frac{1}{2}} = \begin{bmatrix} d_{p+1}^{-\frac{1}{2}} & & \bigcirc \\ & \ddots & \\ \bigcirc & & d_r^{-\frac{1}{2}} \end{bmatrix}. \quad (11)$$

Then after a further congruent transformation of $\mathbf{D}$ by the matrix $\mathbf{D}^{-\frac{1}{2}}$, (9) may be written as

$$(\mathbf{D}^{-\frac{1}{2}})'\mathbf{D}\mathbf{D}^{-\frac{1}{2}} = (\mathbf{Q}\mathbf{D}^{-\frac{1}{2}})'\mathbf{A}(\mathbf{Q}\mathbf{D}^{-\frac{1}{2}}) = \begin{bmatrix} \mathbf{U}_p & 0 & 0 \\ 0 & -\mathbf{U}_{r-p} & 0 \\ 0 & 0 & 0 \end{bmatrix} = \mathbf{C}. \quad (12)$$

Since $\mathbf{D}^{-\frac{1}{2}}$ is nonsingular, so also is $\mathbf{Q}\mathbf{D}^{-\frac{1}{2}}$. Hence the right-hand side is simply a congruent transformation of $\mathbf{A}$. It is said to be a *canonical matrix*; the congruent transformation of **A** in (12) is said to put **A** in *canonical form*. The integer p in this expression is called the *index of the matrix*.

7.2   QUADRATIC AND HERMITIAN FORMS

The subject of this section is a mathematical form that arises in networks from a consideration of power dissipated or energy stored. In order to see how it arises, before we delve into its mathematical properties, let us consider a purely resistive network with a branch-resistance matrix **R**; the branch voltage and current vectors at any time are $\mathbf{v}(t)$ and $\mathbf{i}(t)$. The power dissipated in the network at any time is $p(t) = \mathbf{i}(t)'\mathbf{v}(t)$. When the branch relation $\mathbf{v} = \mathbf{Ri}$ is introduced, the power becomes

$$p = \mathbf{i}(t)'\mathbf{v}(t) = \mathbf{i'Ri}. \tag{13}$$

For a network with three branches, for example, the quantity on the right is

$$\mathbf{i'Ri} = \begin{bmatrix} i_1 & i_2 & i_3 \end{bmatrix} \begin{bmatrix} R_1 & 0 & 0 \\ 0 & R_2 & 0 \\ 0 & 0 & \mathbf{R_3} \end{bmatrix} \begin{bmatrix} i_1 \\ i_2 \\ i_3 \end{bmatrix}$$

$$= R_1 i_1{}^2 + R_2 i_2{}^2 + R_3 i_3{}^2.$$

This expression is quadratic in the currents and illustrates what we shall now describe. To indicate that the results are general we shall use a general notation.

DEFINITIONS

Let $\mathbf{A} = [a_{ij}]$ be a real, square matrix and $\mathbf{x} = [x_i]$ be a column vector, real or complex. The expression

$$\mathbf{x'Ax} = \begin{bmatrix} x_1 & x_2 \cdots x_n \end{bmatrix} \begin{bmatrix} a_{11} & a_{12} \cdots a_{1n} \\ a_{21} & a_{22} \cdots a_{2n} \\ \vdots & \vdots \quad \vdots \\ a_{n1} & a_{n2} \cdots a_{nn} \end{bmatrix} \begin{bmatrix} x_1 \\ x_2 \\ \vdots \\ x_n \end{bmatrix} \tag{14}$$

when $\mathbf{x}$ is a real vector (i.e., the elements of $\mathbf{x}$ are real), and the expression

$$\mathbf{x^*Ax} = [\bar{x}_1 \quad \bar{x}_2 \cdots \bar{x}_n] \begin{bmatrix} a_{11} & a_{12} \cdots a_{1n} \\ a_{21} & a_{22} \cdots a_{2n} \\ \cdot & \cdot \quad\quad \cdot \\ \cdot & \cdot \quad\quad \cdot \\ \cdot & \cdot \quad\quad \cdot \\ a_{n1} & a_{n2} \cdots a_{nn} \end{bmatrix} \begin{bmatrix} x_1 \\ x_2 \\ \cdot \\ \cdot \\ \cdot \\ x_n \end{bmatrix} \tag{15}$$

when $\mathbf{x}$ is a complex vector, are called *quadratic forms*. The reason for the name becomes clear when we perform the indicated matrix multiplications and get

$$\mathbf{x'Ax} = \sum_{i=1}^{n} \sum_{j=1}^{n} a_{ij} x_i x_j \tag{16}$$

when the x's are real, and

$$\mathbf{x^*Ax} = \sum_{i=1}^{n} \sum_{j=1}^{n} a_{ij} \bar{x}_i x_j \tag{17}$$

when the x's are complex. We see that these are homogeneous expressions of degree 2 in the variables $x_1, x_2, \cdots, x_n$.

The matrix $\mathbf{A}$ in (14) through (17) is called the *matrix of the quadratic form*. We consider the x's to be variables, so that the matrix essentially defines the quadratic form. We shall concern ourselves with quadratic forms in which the matrix $\mathbf{A}$ is real and symmetric. Actually, *any* real quadratic form with a real matrix can be converted into a quadratic form with a symmetric matrix, because, if the x's and the a_{ij}'s are real, we can write

$$a_{ij} x_i x_j + a_{ji} x_j x_i = 2 \left(\frac{a_{ij} + a_{ji}}{2} \right) x_i x_j. \tag{18}$$

We see that the contribution to the quadratic form of the two terms on the left of this equation will remain unchanged if we replace both a_{ij} and a_{ji} in the matrix by half their sum. Thus, if $\mathbf{A}$ is not symmetric, we define the symmetric matrix $\mathbf{B}$ as

$$\mathbf{B} = \tfrac{1}{2}(\mathbf{A} + \mathbf{A'}). \tag{19}$$

The matrix $\mathbf{B}$ is called the *symmetric part* of $\mathbf{A}$. This operation leaves the diagonal elements of $\mathbf{A}$ unchanged, whereas the off-diagonal elements are

modified in the manner just described. From the preceding discussion it
follows that

$$\mathbf{x}'\mathbf{A}\mathbf{x} = \mathbf{x}'\mathbf{B}\mathbf{x}. \tag{20}$$

Let us now turn our attention to a quadratic form in which the vector
$\mathbf{x}$ is complex. As long as the matrix $\mathbf{A}$ of the quadratic form is real and
symmetric, the quadratic form $\mathbf{x}^*\mathbf{A}\mathbf{x}$ will be real. To prove this result,
observe that

$$\sum_{i=1}^{n}\sum_{j=1}^{n} a_{ij}\,\bar{x}_i\,x_j = \sum_{i=1}^{n} a_{ii}\,\bar{x}_i\,x_i + \sum_{\substack{i=1\\j\neq i}}^{n}\sum_{j=1}^{n} a_{ij}\,\bar{x}_i\,x_j$$

$$= \sum_{i=1}^{n} a_{ii}|x_i|^2 + \tfrac{1}{2}\sum_{\substack{i=1\\j\neq i}}^{n}\sum_{j=1}^{n} a_{ij}(\bar{x}_i\,x_j + \bar{x}_j\,x_i) \tag{21}$$

$$= \sum_{i=1}^{n} a_{ii}|x_i|^2 + \sum_{\substack{i=1\\j\neq i}}^{n}\sum_{j=1}^{n} a_{ij}\,\mathrm{Re}\,(\bar{x}_i\,x_j).$$

The second line is a consequence of $\mathbf{A}$ being symmetric, whereas the last
term in the last line is a result of the fact that $\bar{x}_j\,x_i$ is the conjugate of
$\bar{x}_i\,x_j$. Everything in the last line is now real, thus proving the result.

TRANSFORMATION OF A QUADRATIC FORM

Let us now observe what happens to a quadratic form when the vector
$\mathbf{x}$ is subjected to a real, nonsingular linear transformation. Let $\mathbf{x} = \mathbf{Q}\mathbf{y}$,
where $\mathbf{Q}$ is nonsingular and $\mathbf{y}$ is a column vector. The quadratic form
becomes

$$\mathbf{x}^*\mathbf{A}\mathbf{x} = (\mathbf{Q}\mathbf{y})^*\mathbf{A}(\mathbf{Q}\mathbf{y}) = \mathbf{y}^*(\mathbf{Q}'\mathbf{A}\mathbf{Q})\mathbf{y}, \tag{22}$$

where we used the fact that $\mathbf{Q}$ is real to write $\mathbf{Q}^* = \mathbf{Q}'$. Within the parentheses we find a congruent transformation of $\mathbf{A}$. It was observed earlier
that a real, symmetric matrix $\mathbf{A}$ can always be reduced to the canonical
form of (12) by means of a nonsingular, congruent transformation. Hence
the quadratic form can be reduced to

$$\mathbf{x}^*\mathbf{A}\mathbf{x} = \mathbf{y}^*\mathbf{C}\mathbf{y} = |y_1|^2 + |y_2|^2 + \cdots + |y_p|^2 - |y_{p+1}|^2 - \cdots - |y_r|^2.$$
$$\tag{23}$$

We can state this result as the following theorem:

Theorem 2. *Every real quadratic form* x*Ax *in which* A *is real and symmetric can be reduced by means of a real, nonsingular, linear transformation* x = Qy *to the canonical form given in (23) in which* r *is the rank of* A *and* p *is the index.*

This theorem is, of course, an existence theorem, it does not give any guidance as to how one goes about finding the appropriate linear transformation. One procedure for doing this is called the *Lagrange reduction,* which consists of repeatedly carrying out a process similar to completing the square. Let us illustrate this with a number of examples.

Examples

1. For simplicity, suppose x is a real vector. Let

$$\mathbf{x'Ax} = [x_1 \quad x_2]\begin{bmatrix} 1 & 2 \\ 2 & 2 \end{bmatrix}\begin{bmatrix} x_1 \\ x_2 \end{bmatrix}$$

$$= x_1^2 + 4x_1 x_2 + 2x_2^2$$

$$= x_1^2 + 4x_1 x_2 + (4x_2^2 - 4x_2^2) + 2x_2^2$$

$$= (x_1 + 2x_2)^2 - 2x_2^2.$$

In this process $4x_2^2$ was added and subtracted to complete the square. Now set

$$\begin{array}{ll} y_1 = x_1 + 2x_2 \\ y_2 = \sqrt{2}\, x_2 \end{array} \quad \text{or} \quad \begin{bmatrix} x_1 \\ x_2 \end{bmatrix}\begin{bmatrix} 1 & -\sqrt{2} \\ 0 & 1/\sqrt{2} \end{bmatrix}\begin{bmatrix} y_1 \\ y_2 \end{bmatrix}.$$

Then

$$\mathbf{x'Ax} = y_1^2 - y_2^2 = [y_1 \quad y_2]\begin{bmatrix} 1 & 0 \\ 0 & -1 \end{bmatrix}\begin{bmatrix} y_1 \\ y_2 \end{bmatrix}.$$

In this case the rank of A equals its order (2) and the index is 1.

2. This time let x be a complex vector and

$$\mathbf{x^*Ax} = [\bar{x}_1 \quad \bar{x}_2 \quad \bar{x}_3]\begin{bmatrix} 1 & -1 & 3 \\ -1 & 2 & 0 \\ 3 & 0 & 14 \end{bmatrix}\begin{bmatrix} x_1 \\ x_2 \\ x_3 \end{bmatrix}$$

$$= x_1 \bar{x}_1 - \bar{x}_1 x_2 + 3\bar{x}_1 x_3 - \bar{x}_2 x_1 + 2x_2 \bar{x}_2 + 3\bar{x}_3 x_1 + 14x_3 \bar{x}_3$$

$$= [x_1 \bar{x}_1 - x_1(\bar{x}_2 - 3\bar{x}_3) - \bar{x}_1(x_2 - 3x_3)] + (2x_2 \bar{x}_2 + 14x_3 \bar{x}_3)$$

The first set of terms can be written as a magnitude square by adding $(x_2 + 3x_3)(\bar{x}_2 + 3\bar{x}_3)$, which means subtracting the same quantity from the second set of terms. The result of this operation is

$$\mathbf{x^*Ax} = (x_1 - x_2 + 3x_3)(\bar{x}_1 - \bar{x}_2 + 3\bar{x}_3) + (x_2 \bar{x}_2 + 3x_2 \bar{x}_3 + 3\bar{x}_2 x_3) + 5x_3 \bar{x}_3$$

$$= (x_1 - x_2 + 3x_3)(\bar{x}_1 - \bar{x}_2 + 3\bar{x}_3) + (x_2 + 3x_3)(\bar{x}_2 + 3\bar{x}_3) - 4x_3 \bar{x}_3.$$

In the last step, $9x_3 \bar{x}_3$ was added and subtracted in order to "complete the square" in the preceding step. Now let

$$
\begin{aligned}
y_1 &= x_1 - x_2 + 3x_3 \\
y_2 &= \quad\;\; x_2 + 3x_3 \\
y_3 &= \qquad\qquad 2x_3
\end{aligned}
\quad \text{or} \quad
\begin{bmatrix} x_1 \\ x_2 \\ x_3 \end{bmatrix} =
\begin{bmatrix} 1 & 1 & -3 \\ 0 & 1 & -\frac{3}{2} \\ 0 & 0 & \frac{1}{2} \end{bmatrix}
\begin{bmatrix} y_1 \\ y_2 \\ y_3 \end{bmatrix}.
$$

Then the quadratic form finally becomes

$$\mathbf{x^*Ax} = |y_1|^2 + |y_2|^2 - |y_3|^2. \tag{24}$$

DEFINITE AND SEMIDEFINITE FORMS

It can be observed from (23) that the value of the quadratic form will normally depend on the values of the y-variables. However, it may happen that the value of the quadratic form will remain of one sign independent of the values of the variables. Such forms are called *definite*. In particular, a real, quadratic form $\mathbf{x^*Ax}$ is called *positive definite* if for any set of complex or real numbers $x_1, x_2, \cdots, x_n$, not all zero, the value of the quadratic form is strictly positive. Similarly, we say the quadratic form is *positive semidefinite* if

$$\mathbf{x^*Ax} \geq 0 \tag{25}$$

for all $\mathbf{x} \neq 0$, provided there is at least one set of values of the variables for which the equality holds. Since the positive property of such a quadratic form is not dependent on the values of the variables, it must be associated with the matrix $\mathbf{A}$ of the quadratic form. The following terminology, then, appears quite natural. *A real, symmetric matrix $\mathbf{A}$ is said to be positive definite or semidefinite according as the quadratic form $\mathbf{x^*Ax}$ is positive definite or semidefinite.*

We need to find means for determining whether or not a quadratic form is positive definite. An approach to this is obtained by considering the canonic form in (23). The matrix $\mathbf{A}$ of the form is characterized by three integers: the order n, the rank r, and the index p. If the index is less then the rank (but greater than zero), the matrix can be neither positive definite nor positive semidefinite.

Suppose that the index equals the rank: $p = r$. Then all the signs in (23) will be positive. There are two possibilities: (1) the rank is equal to the order, $r = n$, so that $\mathbf{A}$ is nonsingular; or (2) $r < n$, so that $\mathbf{A}$ is singular. Suppose $r < n$. Then choose y_1 up to $y_r = 0$ and y_{r+1} to $y_n \neq 0$. This will cause the quadratic form to vanish but, with $\mathbf{x} = \mathbf{Q}\mathbf{y}$, not all the x's will be zero. For any other choice of y-variables, the quadratic form will be positive. Hence the quadratic form satisfies (25) and is positive semidefinite. The converse is, clearly, also true.

On the other hand, if $r = n$ (with p still equal to r), so that $\mathbf{A}$ is nonsingular, then every nonzero choice of the y's (and hence of the x's) will lead to a positive value of the quadratic form. The conclusion is the following theorem:

Theorem 3. *A quadratic form having a real, symmetric matrix $\mathbf{A}$ of order* n, *rank* r *and index* p *is positive definite if and only if $\mathbf{A}$ is nonsingular and the index equals the rank:* p $=$ r $=$ n. *It is positive semidefinite if $\mathbf{A}$ is singular and* p $=$ r.

If a quadratic form is positive definite, then it can be seen from (23) that its canonical matrix will be a unit matrix; that is, the nonsingular linear transformation $\mathbf{x} = \mathbf{Q}\mathbf{y}$ leads to

$$\mathbf{Q'AQ} = \mathbf{U}. \tag{26}$$

It is possible to find the determinant of $\mathbf{A}$ by taking the determinant of both sides of this expression. Since det $\mathbf{U} = 1$, and the determinant of a product of matrices of the same order equals the product of determinants, we get

$$(\det \mathbf{Q'})(\det \mathbf{A})(\det \mathbf{Q}) = 1. \tag{27}$$

Since $\mathbf{Q}$ and its transpose have the same determinant, which is nonzero since $\mathbf{Q}$ is nonsingular, we obtain,

$$\det \mathbf{A} = \frac{1}{(\det \mathbf{Q})^2}. \tag{28}$$

This result expresses the fact that the determinant of a positive definite matrix is positive. Furthermore, suppose we set the last variable x_n in the quadratic form equal to zero. Then none of the coefficients a_{ni} or a_{in} of the matrix $\mathbf{A}$ will appear in the quadratic form. This is most easily seen from (16) with $x_n = 0$. Hence we might as well remove the nth row and column of $\mathbf{A}$ and consider it to be of the $(n-1)$th order. For this new matrix (28) still applies. But the determinant of the new matrix is the principal cofactor of the old matrix obtained by removing the last row and column. Since permuting the variables has no effect on the quadratic form, it is immaterial which one of the variables we call x_n. It follows that all the first principal cofactors of a positive definite matrix will be positive.

This argument can now be repeated by setting two of the variables equal to 0, then 3, and so on, up to all but one. We shall find that all the principal cofactors of $\mathbf{A}$ will be positive. In the last case, with all but one of the variables equal to zero, we find that all the elements of $\mathbf{A}$ on the principal diagonal must be positive. (These elements are the $(n-1)$th principal cofactors of $\mathbf{A}$).

What we have succeeded in proving is that, if a matrix is known to be positive definite, then its determinant and all its principal cofactors will be positive. Actually, what we need for testing a given matrix is the converse of this result. It happens that this is also true. The proof, however, is quite lengthy and will not be given. For future reference we shall list this result as a theorem.

Theorem 4. *A real symmetric matrix $\mathbf{A}$ is positive definite if and only if its determinant and principal cofactors are all positive. It is positive semidefinite if and only if its determinant is zero and all its principal cofactors are non-negative.*

As an example, consider the real matrix previously used to illustrate the Lagrange reduction. It is required to form the determinant and principal cofactors.

$$\mathbf{A} = \begin{bmatrix} 1 & -1 & 3 \\ -1 & 2 & 0 \\ 3 & 0 & 14 \end{bmatrix}.$$

We observe that the diagonal elements are all positive. Since this is a third-order matrix, the diagonal elements are the second cofactors. The first principal cofactors are easily formed:

$$\Delta_{11} = 28, \qquad \Delta_{22} = 5, \qquad \Delta_{33} = 1.$$

They are all positive. This leaves the determinant, which is found to be -4. This is less than zero, and hence $\mathbf{A}$ is not positive definite, or semidefinite.

HERMITIAN FORMS

Up to this point we have been dealing with quadratic forms having real matrices that are symmetric. If the matrix of a quadratic form is complex, it is possible to replace the matrix by its Hermitian part without changing the value of the form, just as a real matrix was replaced by its symmetric part. Let $\mathbf{H}$ be a Hermitian matrix ($h_{ji} = \bar{h}_{ij}$). The expression

$$\mathbf{x}^*\mathbf{H}\mathbf{x} \tag{29}$$

is called a *Hermitian form*. When $\mathbf{H}$ is real, the Hermitian form reduces to a quadratic form. It should be expected, then, that the properties of Hermitian forms are analogous to those of quadratic forms. We shall merely list a few, without extensive comment.

By carrying out an expansion as we did in (21) for a quadratic form, it can be shown that *the value of a Hermitian form is real*.

A Hermitian form of rank r can be reduced to the canonical form given on the right side of (23) by a nonsingular linear transformation $\mathbf{x} = \mathbf{Q}\mathbf{y}$, where $\mathbf{Q}$ is generally complex.

The terms " positive definite " and " semidefinite " apply to Hermitian forms and are defined in the same way as for quadratic forms. The theorem relating to the determinant and principal cofactors of positive definite and semidefinite matrices applies to a Hermitian matrix also.

7.3 ENERGY FUNCTIONS

Now that a mathematical background has been presented, we are ready to turn to a consideration of network functions. Specifically, we shall relate certain network functions to the energy stored and dissipated in the network. Then, from a physical knowledge of the nature of this energy, we can draw some conclusions about the properties of network functions.

Consider a multiport network excited by voltage sources at each port. Figure 1 shows the use of a two-port network, but the discussion will be

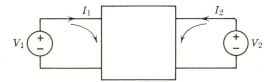

Fig. 1. Excited two-port.

carried out in terms of a general multiport. The network, which is linear and time invariant, is assumed to be initially relaxed. Now consider writing a set of loop equations for this network. Referring back to Chapter 2, we find that it will have the form

$$\mathbf{Z}_m(s)\,\mathbf{I}_m(s) = \mathbf{E}(s), \qquad (30a)$$

$$\left(\mathbf{R}_m + s\mathbf{L}_m + \frac{1}{s}\,\mathbf{D}_m\right)\mathbf{I}_m(s) = \mathbf{E}(s), \qquad (30b)$$

where $\mathbf{Z}_m$ is the loop impedance matrix, and $\mathbf{R}_m$, $\mathbf{L}_m$, and $\mathbf{D}_m$ are the loop-parameter matrices; $\mathbf{E}$ is the equivalent loop source voltage vector. Since there are no other sources except the ones at the ports, and the loops are chosen such that each voltage source lies on only one loop, and the loop orientation relative to the source is as shown in Fig. 1, then an element of $\mathbf{E}$ will be nonzero for the loops that include a port and 0 for the internal loops.

In (30), the variables are Laplace transforms of voltages and currents. When excitations are sinusoids of the same frequency, the same expressions, (30), are valid, with the transform variables replaced by phasors and s replaced by $j\omega$. (See Problem 50.) Let us use the subscript p to designate phasors; for example, I_p is a complex number whose magnitude is the rms value of a sinusoidal current and whose angle is its phase. Then, assuming an n-port so that $\mathbf{E}_p$ has n nonzero components, and assuming m loops so that $\mathbf{I}_{mp}$ has m components, (30) becomes

$$\left\{\mathbf{R}_m + j\omega\left(\mathbf{L}_m - \frac{1}{\omega^2}\,\mathbf{D}_m\right)\right\}
\begin{bmatrix} I_{mp1} \\ I_{mp2} \\ \\ \\ \\ \vdots \\ \\ I_{mpm} \end{bmatrix}
=
\begin{bmatrix} V_{p1} \\ V_{p2} \\ \vdots \\ V_{pn} \\ 0 \\ \vdots \\ 0 \end{bmatrix}. \qquad (31)$$

Let us now compute the power supplied to the network. The complex power supplied at port k is $\bar{I}_{mpk}\,V_{pk}$. The total complex power delivered at all the ports is therefore $\mathbf{I}_{mp}^*\,\mathbf{E}_p$. The real part of this is the real, average power. The imaginary part is proportional to the net average energy stored in the network, the difference between the average energy stored in the inductors and in the capacitors. Thus,

$$\text{Re } (\mathbf{I}_{mp}^* \mathbf{E}_p) = P, \tag{32a}$$

$$\text{Im } (\mathbf{I}_{mp}^* \mathbf{E}_p) = 2\omega(W_L - W_C). \tag{32b}$$

The complex power input to the network can be obtained by premultiplying both sides of (31) by $\mathbf{I}_{mp}^*$. The result becomes

$$\mathbf{I}_{mp}^* \mathbf{R}_m \mathbf{I}_{mp} + j\omega\left(\mathbf{I}_{mp}^* \mathbf{L}_m \mathbf{I}_{mp} - \frac{1}{\omega^2} \mathbf{I}_{mp}^* \mathbf{D}_m \mathbf{I}_{mp}\right) = \mathbf{I}_{mp}^* \mathbf{E}_p$$

$$= \bar{I}_{p1} V_{p1} + \bar{I}_{p2} V_{p2} + \cdots + \bar{I}_{pn} V_{pn}. \tag{33}$$

(On the right side appear only those loop currents that are port currents, so that the m subscript can be omitted.) We find that the complex power input on the right equals the sum of three terms on the left. We recognize each of these terms as a quadratic form.

For a nonreciprocal network the loop-parameter matrices are not symmetric. However, as discussed in the last section, the value of the quadratic form is unchanged if the matrix of the form is replaced by its symmetric part. We shall assume that this has been done. Each of the quadratic forms on the left side of (33) is real. Hence comparing (32) and (33) leads to the conclusion that

$$\mathbf{I}_{mp}^* \mathbf{R}_m \mathbf{I}_{mp} = \text{real power supplied to network}, \tag{34a}$$

$$\tfrac{1}{2}\mathbf{I}_{mp}^* \mathbf{L}_m \mathbf{I}_{mp} = \text{average energy stored in inductors}, \tag{34b}$$

$$\frac{1}{2\omega^2} \mathbf{I}_{mp}^* \mathbf{D}_m \mathbf{I}_{mp} = \text{average energy stored in capacitors}. \tag{34c}$$

It is possible to obtain equivalent expressions for each of these quadratic forms. The matrix of each form is one of the loop-parameter matrices. Returning to Chapter 2, we find that the loop-parameter matrices can be written in terms of the branch-parameter matrices, as follows:

$$\mathbf{R}_m = \mathbf{B}\mathbf{R}\mathbf{B}', \tag{35a}$$

$$\mathbf{L}_m = \mathbf{B}\mathbf{L}\mathbf{B}' \tag{35b}$$

$$\mathbf{D}_m = \mathbf{B}\mathbf{D}\mathbf{B}'. \tag{35c}$$

where $\mathbf{R}$, $\mathbf{L}$, and $\mathbf{D}$ are the branch-parameter matrices, and $\mathbf{B}$ is the loop matrix.

Let us consider the quadratic form that involves $\mathbf{R}_m$. Using (35a). results in

$$\mathbf{I}_{mp}^* \mathbf{R}_m \mathbf{I}_{mp} = \mathbf{I}_{mp}^* \mathbf{BRB'I}_{mp} = (\mathbf{B'I}_{mp})^* \mathbf{R}(\mathbf{B'I}_{mp}). \tag{36}$$

Recall from (56) in Chapter 2 that $\mathbf{B'I}_{mp} = \mathbf{I}_p$ is the loop transformation that specifies the branch currents $\mathbf{I}_p$ in terms of loop currents. Thus

$$\mathbf{I}_{mp}^* \mathbf{R}_m \mathbf{I}_{mp} = \mathbf{I}_p^* \mathbf{R} \mathbf{I}_p = \sum_{k=1}^{b} \sum_{j=1}^{b} \mathbf{R}_{jk} \bar{I}_{pj} I_{pk}, \tag{37}$$

where b is the number of branches in the network. For a general non-passive, nonreciprocal network, nothing specific can be stated about this quadratic form.

PASSIVE, RECIPROCAL NETWORKS

Let us now restrict consideration to passive, reciprocal networks. In this case, the branch-resistance matrix is diagonal. Then (37) becomes

$$\mathbf{I}_{mp}^* \mathbf{R}_m \mathbf{I}_{mp} = \mathbf{I}_p^* \mathbf{R} \mathbf{I}_p = \sum_{k=1}^{b} R_k |I_{pk}|^2. \tag{38}$$

We know that the real power supplied cannot be negative for such networks. Hence the quadratic form must be at least positive semidefinite. It will be positive definite if none of the branch resistances is zero. This is true because in that case the diagonal matrix $\mathbf{R}$ will be nonsingular. The same conclusion follows from the right side of (38), as it should, since R_k is non-negative for all k.

Tracing through identical arguments for the other two quadratic forms involving the loop-inductance and inverse-capacitance parameters leads to

$$\frac{1}{2\omega^2} \mathbf{I}_{mp}^* \mathbf{D}_m \mathbf{I}_{mp} = \frac{1}{2\omega^2} \mathbf{I}_p^* \mathbf{D} \mathbf{I}_p = \frac{1}{2\omega^2} \sum_{k=1}^{b} D_k |I_{pk}|^2, \tag{39a}$$

$$\mathbf{I}_{mp}^* \mathbf{L}_m \mathbf{I}_{mp} = \mathbf{I}_p^* \mathbf{L} \mathbf{I}_p = \sum_{k=1}^{b} \sum_{j=1}^{b} L_{jk} \bar{I}_{pj} I_{pk}. \tag{39b}$$

Note the differences on the right-hand sides of these two expressions. The branch-inverse-capacitance matrix is diagonal, whereas the branch-inductance matrix is not necessarily diagonal. When there is no mutual inductance, **L** is also diagonal. Again, from the interpretation in (34) as average energy stored, these quadratic forms must be positive semidefinite.

For purposes of convenient reference we define the following notation:

$$F(j\omega) = \mathbf{I}^*_{mp} R_m \mathbf{I}_{mp}, \tag{40a}$$

$$T(j\omega) = \tfrac{1}{2}\mathbf{I}^*_{mp} \mathbf{L}_m \mathbf{I}_{mp}, \tag{40b}$$

$$V(j\omega) = \frac{1}{2\omega^2} \mathbf{I}^*_{mp} \mathbf{D}_m \mathbf{I}_{mp}. \tag{40c}$$

From their physical interpretations, these quantities are collectively called *energy functions*, even though the first one is dimensionally not energy. The choice of symbols for these functions is an unfortunate one, since they can be confused with other quantities having similar symbols; but they have become quite standard in the literature, so we shall continue to use them.

The positive semidefinite condition on $T(j\omega)$ imposes conditions on the sizes of mutual inductances. If mutual coupling in a network is always between pairs of branches only, the semidefinite condition is equivalent to the usual constraint that the coupling coefficient cannot exceed unity. If more than two branches are mutually coupled, the restriction of the coupling coefficient to values less than unity is not sufficiently strong to insure positive semidefiniteness; that is, in this case positive definiteness is a stronger condition than unity coupling. (See Problem 17.)

As an illustration, consider the network in Fig. 2. Both sources are

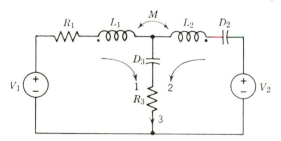

Fig. 2.   Illustrative example for energy functions.

sinusoidal at an angular frequency ω. The loop-parameter matrices are

$$\mathbf{R}_m = \begin{bmatrix} R_1 + R_3 & R_3 \\ R_3 & R_3 \end{bmatrix}, \quad \mathbf{L}_m = \begin{bmatrix} L_1 & -M \\ -M & L_2 \end{bmatrix}, \quad \mathbf{D}_m = \begin{bmatrix} D_3 & D_3 \\ D_3 & D_2 + D_3 \end{bmatrix}$$

The energy functions are

$$F(j\omega) = [\bar{I}_{p1} \quad \bar{I}_{p2}] \begin{bmatrix} R_1 + R_3 & R_3 \\ R_3 & R_3 \end{bmatrix} \begin{bmatrix} I_{p1} \\ I_{p2} \end{bmatrix} = R_1 |I_{p1}|^2 + R_3 |I_{p1} + I_{p2}|^2,$$

$$T(j\omega) = \tfrac{1}{2}[\bar{I}_{p1} \quad \bar{I}_{p2}] \begin{bmatrix} L_1 & -M \\ -M & L_2 \end{bmatrix} \begin{bmatrix} I_{p1} \\ I_{p2} \end{bmatrix}$$

$$= \frac{L_1}{2}|I_{p1}|^2 - M \operatorname{Re}(\bar{I}_{p1} I_{p2}) + \frac{L_2}{2}|I_{p2}|^2,$$

$$V(j\omega) = \frac{1}{2\omega^2}[\bar{I}_{p1} \quad \bar{I}_{p2}] \begin{bmatrix} D_3 & D_3 \\ D_3 & D_2 + D_3 \end{bmatrix} \begin{bmatrix} I_{p1} \\ I_{p2} \end{bmatrix}$$

$$= \frac{D_2}{2\omega^2}|I_{p2}|^2 + \frac{D_3}{2\omega^2}|I_{p1} + I_{p2}|^2.$$

Since $I_{p1} + I_{p2}$ equals branch current 3, the term $R_3 |I_{p1} + I_{p2}|^2$ is the power dissipated in R_3 and $|I_{p1} + I_{p2}|^2 D_3/2\omega^2$ is the energy stored in D_3. The positive semidefinite nature of $T(j\omega)$ is not evident from the right-hand side. Observe, however, that the matrix $\mathbf{L}_m$ is singular only for $L_1 L_2 - M^2 = 0$, which is the condition of unity coupling.

To summarize the result so far obtained, the *loop resistance, inductance, and reciprocal-capacitance matrices* $\mathbf{R}_m$, $\mathbf{L}_m$, *and* $\mathbf{D}_m$ *of a passive, reciprocal network are positive semidefinite*. This result was established by giving physical interpretations to certain quadratic forms based on a sinusoidal steady-state analysis.

Let us now return to the original loop equations in (30) in which the variables are Laplace transforms. Without any concern for physical interpretation, let us premultiply both sides by $\mathbf{I}_m^*(s)$. The result will be

$$\mathbf{I}_m^*(s) \, \mathbf{R}_m \mathbf{I}_m(s) + s\mathbf{I}_m^*(s) \, \mathbf{L}_m \mathbf{I}_m(s) + \frac{1}{s}\mathbf{I}_m^*(s) \, \mathbf{D}_m \mathbf{I}_m(s) = \mathbf{I}_m^*(s) \, \mathbf{E}(s). \quad (41)$$

Again we find the same quadratic forms we had before, only now the variables are loop-current transforms rather than phasors. The quadratic forms in this equation do not have an energy interpretation like those of (33). However, *the matrices of these quadratic forms are identical with the former ones*. Hence these quadratic forms are positive semidefinite. We therefore give them symbols similar to those of (40) and continue to call them energy functions, although even dimensionally they do not represent energy.

$$F_0(s) = \mathbf{I}_m^*(s)\mathbf{R}_m\mathbf{I}_m(s), \tag{42a}$$

$$T_0(s) = \mathbf{I}_m^*(s)\mathbf{L}_m\mathbf{I}_m(s). \tag{42b}$$

$$V_0(s) = \mathbf{I}_m^*(s)\mathbf{D}_m\mathbf{I}_m(s). \tag{42c}$$

When this notation is used, (41) becomes

$$F_0(s) + sT_0(s) + \frac{1}{s}V_0(s) = \mathbf{I}^*\mathbf{V}. \tag{43}$$

(The notation on the right side has been modified in two ways. The m subscripts have been dropped because the only loop currents that remain in the product $\mathbf{I}_m^*\mathbf{E}$ are those that are the same as port currents. Also, the only nonzero components in $\mathbf{E}$ are the port voltages. Hence $\mathbf{I}_m^*\mathbf{E}$ can be replaced by $\mathbf{I}^*\mathbf{V}$, where $\mathbf{I}$ and $\mathbf{V}$ are the port vectors.)

Let us digress here for a moment. This entire development started from the loop equations. Alternatively, a completely dual development can proceed on the basis of the node equations. Instead of the loop-parameter matrices $\mathbf{R}_m$, $\mathbf{L}_m$, and $\mathbf{D}_m$, the conductance, inverse-inductance, and capacitance node-parameter matrices $\mathbf{G}_n$, $\mathbf{\Gamma}_n$, and $\mathbf{C}_n$, respectively, will appear. Energy functions can now be defined in terms of these parameter matrices and the node-voltage vector $\mathbf{V}_n$. These will have the same form as (42) with $\mathbf{V}_n$ in place of $\mathbf{I}_m$ and with the node-parameter matrices in place of the loop-parameter ones. From these it is concluded that *the node conductance, capacitance, and inverse-inductance matrices $\mathbf{G}_n$, $\mathbf{C}_n$, and $\mathbf{\Gamma}_n$ of a passive, reciprocal network are positive semidefinite.* An equation similar to (43) can now be written with these new energy functions, with $\mathbf{V}$ and $\mathbf{I}$ interchanged. This alternative development is not needed to carry on the subsequent discussion, just as the node system of equations itself is really superfluous. However, just as node equations provide helpful viewpoints and often simplify computation, so also this alternative approach may sometimes be useful. You should work out the details of the procedure just outlined if you are interested.

Look again at (43). The quantities appearing on the left side are defined in terms of loop-current variables (or branch-current variables through the loop transformation). But on the right we find port variables. Of course, the port voltage and current variables are related to each other.

If this relationship is used on the right side of (43), an extremely important result follows. As a relation between the vectors $\mathbf{V}$ and $\mathbf{I}$, we can use

$$\mathbf{V}(s) = \mathbf{Z}_{oc}(s)\mathbf{I}(s) \tag{44a}$$

or

$$\mathbf{I}(s) = \mathbf{Y}_{sc}(s)\mathbf{V}(s). \tag{44b}$$

The first of these can be inserted directly into (43); the second can be inserted after taking the conjugate transpose of (43) which leads to

$$\overline{\left(F_0 + sT_0 + \frac{1}{s}V_0\right)} = (\mathbf{I}*\mathbf{V})*$$

or
$$\tag{45}$$

$$F_0 + \bar{s}T_0 + \frac{1}{\bar{s}}V_0 = \mathbf{V}*\mathbf{I}.$$

This follows because the quadratic forms are real scalars. The result of inserting (44a) into (43) and (44b) into (45) is

$$F_0 + sT_0 + \frac{1}{s}V_0 = \mathbf{I}*\mathbf{Z}_{oc}(s)\mathbf{I}, \tag{46a}$$

$$F_0 + \bar{s}T_0 + \frac{1}{\bar{s}}V_0 = \mathbf{V}*\mathbf{Y}_{sc}(s)\mathbf{V}. \tag{46b}$$

It is from these expressions that some of the most fundamental properties of network functions originate. We shall now embark on a study of these properties.

THE IMPEDANCE FUNCTION

Let us consider first the simplest multiport; namely, a one-port. In this case $\mathbf{Z}_{oc}$ is the scalar $Z(s)$, the impedance of the one-port, and $\mathbf{I}$ reduces to the scalar input current. From (46a) the expression for $Z(s)$ becomes

$$Z(s) = \frac{1}{|I(s)|^2}\{F_0(s) + sT_0(s) + \frac{1}{s}V_0(s)\}. \tag{47}$$

Note that the quadratic forms are functions of s only through the fact that the loop currents are functions of s. The real, positive semidefinite nature of the quadratic forms does not depend on the current variables, but only on the loop-parameter matrices, which are constant matrices.

The preceding expression can be separated into real and imaginary parts after replacing s by $\sigma + j\omega$. Thus

$$\operatorname{Re}[Z(s)] = \frac{1}{|I(s)|^2} \left[F_0(s) + \sigma T_0(s) + \frac{\sigma}{\sigma^2 + \omega^2} V_0(s) \right] \qquad (48a)$$

$$\operatorname{Im} Z(s) = \frac{\omega}{|I(s)|^2} \left[T_0(s) - \frac{V_0(s)}{\sigma^2 + \omega^2} \right]. \qquad (48b)$$

Notice that these equations apply no matter what the value of s may be, except at zeros of $I(s)$. These two are extremely important equations, from which we can draw some interesting conclusions. For later reference let us state these results as a theorem.

Theorem 5.   Let $Z(s)$ be the driving-point impedance of a linear time-invariant, passive, reciprocal network N. Then the following statements are true.

(a) *Whenever* $\sigma \geq 0$, $\operatorname{Re}[Z(s)] \geq 0$.

(b) *If* N *contains no resistances* $(F_0(s) = 0)$, *then*

$$\sigma > 0 \ implies \ \operatorname{Re}[Z(s)] > 0,$$
$$\sigma = 0 \ implies \ \operatorname{Re}[Z(s)] = 0,$$
$$\sigma < 0 \ implies \ \operatorname{Re}[Z(s)] < 0.$$

(c) *If* N *contains no capacitances* $(V_0(s) = 0)$, *then*

$$\omega > 0 \ implies \ \operatorname{Im}[Z(s)] > 0,$$
$$\omega = 0 \ implies \ \operatorname{Im}[Z(s)] = 0,$$
$$\omega < 0 \ implies \ \operatorname{Im}[Z(s)] < 0.$$

(d) *If* N *contains no inductances* $(T_0(s) = 0)$, *then*

$$\omega > 0 \ implies \ \operatorname{Im}[Z(s)] < 0,$$
$$\omega = 0 \ implies \ \operatorname{Im}[Z(s)] = 0,$$
$$\omega < 0 \ implies \ \operatorname{Im}[Z(s)] > 0.$$

These results follow immediately from (48). Part (a) states that the value of $Z(s)$ corresponding to a value of s lying in the right half-plane must itself lie in the right half-plane. It leads to the discussion of positive real functions, which we shall take up next. Part (b) leads to the historic-ally important reactance theorem of Foster. Parts (c) and (d) lead to Cauer's results on RL and RC networks.

CONDITION ON ANGLE

Another property of the impedance function can be determined from (47). Note that $|I|^2$, F_0, T_0, and V_0 are all positive constants for any value of s. Hence $Z(s)$ can be written as

$$Z(s) = a_0 + a_1 s + \frac{a_2}{s}, \tag{49}$$

where each of the coefficients is positive. Let $s_0 = \sigma_0 + j\omega_0$ be a point in the right half-plane; that is, $\sigma_0 > 0$, as shown in Fig. 3. Each of the

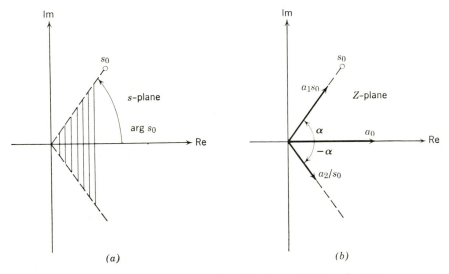

Fig. 3. Demonstration that $|\arg Z| \leq |\arg s|$ for $|\arg s| \leq \pi/2$.

terms on the right of (49) can be represented by a directed line in the complex plane, as shown in Fig. 3b for the corresponding value of s. Whatever the legnths of these lines may be, the sum cannot lie outside the cross-hatched sector shown in Fig. 3a. By observing from the diagram what happens for a number of possible angles of s, including 0 and $\pi/2$ radians, the following result is obtained:

$$|\arg Z(s)| \leq |\arg s| \quad \text{for} \quad 0 \leq |\arg s| \leq \pi/2. \tag{50}$$

This seems to be a stronger condition on the impedance than the condition $\text{Re}\,[Z(s)] \geq 0$ for $\text{Re}\,s \geq 0$. Not only must $Z(s)$ be in the right half-plane when s is but its location there is limited by a condition on its angle. It is not, however, a stronger condition, since it followed from the previous one.

What has been done in this section is to start with a class of networks and to derive some properties that the driving-point impedance of such networks necessarily satisfies. This was done from considerations of energy in the frequency domain. An alternative approach would be to start from the definition of a passive network given in Chapter 1 and repeated here for a one-port:

$$\int_{-\infty}^{t} v(\tau)i(\tau)\,d\tau \geq 0. \tag{51}$$

Suppose the current and voltage at the terminals of a passive network are

$$i(t) = 2|I_0|\,\epsilon^{\sigma_0 t}\cos(\omega_0 t + \alpha) = I_0\epsilon^{s_0 t} + \bar{I}_0\epsilon^{\bar{s}_0 t}, \tag{52a}$$

$$v(t) = Z(s_0)I_0\epsilon^{s_0 t} + Z(\bar{s}_0)\bar{I}_0\epsilon^{\bar{s}_0 t}, \tag{52b}$$

where $s_0 = \sigma_0 + j\omega_0$ with $\sigma_0 > 0$ and $I_0 = |I_0|\epsilon^{j\alpha}$. We assume these signals were initiated at $t = -\infty$, at which time there was no initial energy stored in the network. Because $\epsilon^{\sigma_0 t} = 0$ for $t = -\infty$, both signals start from 0. There is, then, no question of a transient, and the given expressions for current and voltage represent the excitation and total response.

Inserting these expressions for v and i into (51) leads, after some manipulation, to

$$|I_0|^2\,\frac{\text{Re}\,[Z(s_0)]}{\sigma_0}\,\epsilon^{2\sigma_0 t} + \text{Re}\left[\frac{Z(s_0)I_0{}^2}{s_0}\,\epsilon^{2s_0 t}\right] \geq 0. \tag{53}$$

Now express the multiplier of the exponential in the last brackets in terms of its magnitude and angle:

$$\frac{Z(s_0)I_0{}^2}{s_0} = \frac{|Z(s_0)|\,|I_0|^2}{|s_0|}\,\epsilon^{j\theta_0}. \tag{54}$$

When this is inserted into the preceding equation, the result will be

$$|I_0|^2\epsilon^{2\sigma_0 t}\left[\frac{\text{Re}\,[Z(s_0)]}{\sigma_0} + \frac{|Z(s_0)|}{|s_0|}\cos\,(2\omega_0 t + \theta_0)\right] \geq 0 \tag{55}$$

The worst case occurs when the cosine equals -1. For this case the condition reduces to

$$\frac{\text{Re}\,[Z(s_0)]}{\sigma_0} - \frac{|Z(s_0)|}{|s_0|} \geq 0,$$

or

$$\frac{\text{Re}\,[Z(s_0)]}{|Z(s_0)|} \geq \frac{\sigma_0}{|s_0|}. \tag{56}$$

Each side of this expresssion is the real part divided by the magnitude of a complex quantity, which equals the cosine of the corresponding angle. Hence

$$\cos\,[\arg\,Z(s_0)] \geq \cos\,[\arg\,s_0], \tag{57}$$

from which it follows that

$$|\arg\,Z(s_0)| \leq |\arg\,s_0|. \tag{58}$$

Since Re $s_0 = \sigma_0 > 0$, this is identical with (50).

This completes the development of the general necessary properties of impedance functions of passive networks. Completely similar properties could have been developed for the admittance function by starting from (46b) instead of (46a). Thus Theorem 5 and Eq. (50) are true when $Z(s)$ is replaced by $Y(s)$. We shall now define a class of mathematical functions having these same properties and shall investigate the detailed behavior of this class of functions.

7.4   POSITIVE REAL FUNCTIONS

A *positive real function* $F(s)$ is an analytic function of the complex variable $s = \sigma + j\omega$, which has the following properties:

1. $F(s)$ is regular for $\sigma > 0$.
2. $F(\sigma)$ is real.
3. $\sigma \geq 0$ implies Re $[F(s)] \geq 0$.

This is a mathematical definition for a class of mathematical functions. Our motivation in making this definition is the fact that a network function of interest—namely, a driving-point impedance (or admittance)—possesses these properties. By making a mathematical study of positive real functions we can perhaps determine things about impedances that we could not establish from physical reasoning alone. The concept of a positive real function, as well as many of the properties of positive real functions that we shall consider, are due to Otto Brune.

We shall now show that, if a function is rational and satisfies the last two of the above conditions, it will automatically satisfy condition 1. We shall do this by showing that a pole of order n of a real rational function is surrounded by $2n$ sectors in which the real part of the function is alternately positive and negative. Let s_0 be a pole of order n of the rational function $F(s)$. The case $n = 3$ is illustrated in Fig. 4. In the neighborhood

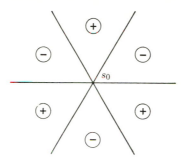

Fig. 4.   Pole of order 3.

of the pole of order n, the function has a Laurent expansion of the form

$$F(s) = \frac{a_{-n}}{(s-s_0)^n} + \frac{a_{-n+1}}{(s-s_0)^{n-1}} + \cdots + \frac{a_{-1}}{s-s_0} + \sum_{j=0}^{\infty} a_j(s-s_0)^j. \quad (59)$$

If a sufficiently small neighborhood of s_0 is chosen, the first term of the Laurent expansion can be made much larger in magnitude than the rest; hence the real part of $F(s)$ in this neighborhood will take on both positive and negative values. We show this as follows: If we write

$$a_{-n} = k\epsilon^{j\theta}, \tag{60a}$$

$$(s-s_0) = \rho\epsilon^{j\phi}, \tag{60b}$$

then

$$\text{Re}\left(\frac{a_{-n}}{(s-s_0)^n}\right) = \frac{k}{\rho^n}\cos(\theta - n\phi). \tag{61}$$

Since θ is a fixed angle and ϕ can vary from 0 to 2π in this neighborhood, we see that the real part of the dominant term changes sign $2n$ times as ϕ varies from 0 to 2π. Therefore the real part of $F(s)$ also changes sign $2n$ times (although not necessarily at exactly the same values of ϕ, due to the other terms in the Laurent expansion).

Now suppose that the function $F(s)$ satisfies the last two conditions in the definition of a positive real function, but it has a pole in the interior of the right half-plane. According to what we have just proved, the real part of $F(s)$ will then take on both negative and positive values in the right half-plane, which contradicts condition 3.

We conclude that in the case of rational functions, whose only singular points are poles, condition 1 of the definition of a positive real function is a consequence of the other two conditions and hence is unnecessary.

As a further aid to the understanding, the definition of a positive real function can be interpreted as a *conformal mapping*. A positive real function $W = F(s)$ maps the real s-axis into the real W-axis, and maps the right half s-plane into the right half W-plane. This is illustrated in Fig. 5.

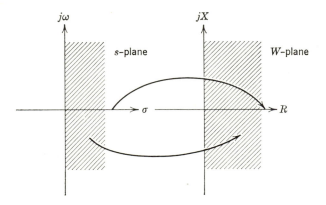

Fig..5.   Mapping by positive real functions.

An immediate consequence of this interpretation is the fact that *a positive real function of a positive real function is itself positive real*; that is, if $F_1(s)$ and $F_2(s)$ are pr (this is used as an abbreviation for positive real), then

$$F_3(s) = F_1[F_2(s)] \tag{62}$$

is also pr; because the right half s-plane goes into the right half F_2-plane since $F_2(s)$ is positive real. Also, the right half F_2-plane goes into the right half F_1-plane since F_1 is positive real. The composite mapping therefore maps the right half s-plane into the right half F_3-plane. The real axis is preserved throughout.

This is a useful result. We can use it to show immediately that, if $F(s)$ is pr, so are $1/F(s)$ and $F(1/s)$. To prove this result we merely observe that

$$\frac{1}{s} = \frac{\sigma}{\sigma^2 + \omega^2} - j\,\frac{\omega}{\sigma^2 + \omega^2} \tag{63}$$

is a pr function. Now we use $1/s$ and $F(s)$ as $F_1(s)$ and $F_2(s)$ in (62), in both possible ways, and the result follows immediately.

From the fact that the reciprocal of a pr function is itself pr, it follows that a positive real function can have no zeros in the right half-plane; because if it did, then its reciprocal would have poles in the right half-plane, which is impossible. Since the impedance of a passive reciprocal network is a pr function, its reciprocal—the admittance—is also a pr function.

From a conformal-mapping point of view, the points $F(s) = 0$ and ∞ (these are the zeros and poles of the function), which are on the boundary of the right half F-plane, cannot be images of any interior points of the right half s-plane. Let us now inquire into the properties resulting when other boundary points of the right half F-plane are images of boundary points of the right half s-plane; that is, let a point on the $j\omega$-axis be mapped by a pr function F into a point on the imaginary axis of the F-plane. If $j\omega_0$ is the point in question, then

$$F(j\omega_0) = jX_0, \tag{64}$$

where X_0 is real (positive, negative, or zero).

Consider a neighborhood of $j\omega_0$ in the s-plane and the corresponding neighborhood of jX_0 in the F-plane, as shown in Fig. 6. Let s_1 denote a point in the right half-plane, in this neighborhood of $j\omega_0$. Let us now

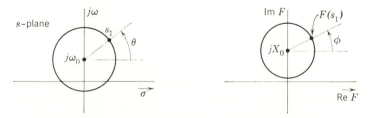

Fig. 6.　Conformal mapping by positive real functions.

expand $F(s)$ in a Taylor series about $j\omega_0$ and evaluate it at $s = s_1$. The result is

$$F(s_1) - jX_0 = F^{(n)}(j\omega_0)(s_1 - j\omega_0)^n + F^{(n+1)}(j\omega_0)(s_1 - j\omega_0)^{n+1} + \cdots, \tag{65}$$

where $F^{(n)}(j\omega_0)$ is the first nonvanishing derivative of $F(s)$ at $j\omega_0$.

As s_1 approaches $j\omega_0$, the dominant term on the right will be the first term. Let us define.

$$\phi = \arg\,[F(s_1) - jX_0], \tag{66a}$$

$$\theta = \arg\,(s_1 - j\omega_0), \tag{66b}$$

$$\beta = \arg\,[F^{(n)}(j\omega_0)]. \tag{66c}$$

Then, in the limit, we shall find from (65) that

$$\lim_{s_1 \to j\omega_0} \phi = \beta + n \lim_{s_1 \to j\omega_0} \theta. \tag{67}$$

But the positive real condition requires that $|\phi| \leq \pi/2$ as long as $|\theta| \leq \pi/2$. Therefore we conclude from (67) that

$$n = 1, \tag{68a}$$

$$\beta = 0. \tag{68b}$$

Thus the first nonvanishing derivative is the first one, and its angle is zero at $s = j\omega_0$. This is a very important result. For future reference we shall state it as a theorem:

Theorem 6. *If any point on the* $j\omega$*-axis is mapped by a positive real function* F *into a point on the imaginary axis in the* F*-plane, then at this point the derivative* dF/ds *is real and positive.*

A number of other results follow from this important theorem. Note that if $F(s)$ has a zero or a pole on the $j\omega$-axis, the conditions of the theorem are satisfied. In the case of a zero $(X_0 = 0)$, a point on the $j\omega$-axis is mapped into the origin of the F-plane, which is on the imaginary axis. Hence the derivative dF/ds is real and positive. This also implies that the zero is a simple one, since at a higher order zero the first derivative will be zero. If $F(s)$ has a pole on the $j\omega$-axis, its reciprocal will have a zero there and the theorem will apply to the reciprocal. However, $d(1/F)/ds$ evaluated at a pole of $F(s)$ is the reciprocal of the residue of $F(s)$ at the pole. (See

Appendix 2.) These considerations can now be stated as the following theorem.

Theorem 7. *If a positive real function has any poles or zeros on the* $j\omega$-*axis (including* $s = 0$, ∞), *such poles or zeros must be simple. At a simple zero on the* $j\omega$-*axis the derivative is real and positive. At a simple pole on the* $j\omega$-*axis, the residue is real and positive.*

NECESSARY AND SUFFICIENT CONDITIONS

We have up to this point collected quite a number of necessary conditions that a positive real function satisfies. What we would like to do is to find a set from among these necessary conditions which proves to be sufficient as well. The result is contained in the following theorem:

Theorem 8. *A rational function* F(s) *with real coefficients is positive real if and only if*

(a)   F(s) *is regular for* $\sigma > 0$;
(b)   *Poles on the* $j\omega$-*axis (including* $s = 0$, ∞) *are simple, with real positive residues;*
(c)   *Re* $[F(j\omega)] \geq 0$ *for all* ω, *except at the poles.*

That these conditions are necessary is obvious from the definition of a pr function and from Theorem 7. Therefore only the sufficiency needs to be proved; that is, let us assume that a function $F(s)$ satisfies these conditions and show that the function must be positive real. Let ω_1, ω_2, $\cdots$, ω_k be the poles on the $j\omega$-axis and let us examine the principal parts at these poles. If there is a pole at the origin, the principal part is

$$F_0(s) = k_0/s,$$

where k_0 is real and positive. It is evident that $F_0(s)$ is itself pr and, in addition, that

$$\text{Re}\,[F_0(j\omega)] = 0.$$

Similarly, the principal part at a possible simple pole of $F(s)$ at infinity is

$$F_\infty(s) = k_\infty s,$$

where k_∞ is real and positive; $F_\infty(s)$ is also pr and, in addition, its real part on the $j\omega$-axis is zero; that is,

$$\text{Re}\,[F_\infty(j\omega)] = 0.$$

Any other poles on the $j\omega$-axis must occur in conjugate pairs and with conjugate residues, since $F(s)$ is a real function. Since the residues are real by hypothesis, the two residues are equal. Taking the principal parts at the conjugate poles $j\omega_i$ and $-j\omega_i$ together, we get

$$F_i(s) = \frac{k_i}{s - j\omega_i} + \frac{k_i}{s + j\omega_i} = \frac{2k_i s}{s^2 + \omega_i{}^2},$$

where k_i is real and positive. This function is also positive real, and, in addition, has the property

$$\mathrm{Re}\,[F_i(j\omega)] = 0.$$

(We may note that $F_0(s)$ is the impedance of a capacitance, $F_\infty(s)$ that of an inductance, and $F_i(s)$ that of a parallel tuned circuit.)

Thus we can subtract from the given function $F(s)$ the principal parts at all of its poles on the $j\omega$-axis. The remainder function $F_r(s)$ still has property (c) of the theorem; that is,

$$\mathrm{Re}\,[F_r(j\omega)] = \mathrm{Re}\,[F(j\omega)] \geq 0. \tag{69}$$

The remainder function $F_r(s)$ is a function that is regular in the right half-plane and its entire boundary, the $j\omega$-axis, including the point at infinity. For such a function the minimum value of the real part throughout its region of regularity lies on the boundary. This can be proved by using the maximum-modulus theorem (see Appendix 2) in the following way. Let $G(s) = \epsilon^{-F_r(s)}$ This function will have the same region of regularity as $F_r(s)$. Hence, according to the maximum-modulus theorem, the maximum magnitude of $G(s)$ for all $\sigma \geq 0$ lies on the $j\omega$-axis. Since

$$|G(s)| = \epsilon^{-\mathrm{Re}\,[F_r(s)]} \tag{70}$$

the maximum magnitude of $G(s)$ will correspond to the smallest value of $\mathrm{Re}\,[F_r(s)]$. This proves the desired result that the minimum value of $\mathrm{Re}\,[F_r(s)]$ for all $\sigma \geq 0$ occurs on the $j\omega$-axis. Since according to (69) this value is nonnegative, the real part of $F_r(s)$ must be non-negative everywhere in the right half-plane; that is,

$$\mathrm{Re}\,[F_r(s)] \geq 0 \qquad (\sigma \geq 0).$$

Since, in addition, $F_r(\sigma)$ is real, we conclude that $F_r(s)$ is a positive real function. Now we can write

$$F(s) = F_r(s) + k_\infty s + \frac{k_0}{s} + \sum_i \frac{2k_i s}{s^2 + \omega_i{}^2}. \tag{71}$$

We have shown that each term on the right is pr. You can easily show that the sum of two (or more) pr functions is itself pr. Hence, $F(s)$ is positive real. This completes the proof of the sufficiency of the stated conditions.

Since the reciprocal of a pr function is also pr, we can restate these necessary and sufficient conditions in terms of the zeros of $F(s)$.

Theorem 9. *A real rational function* $F(s)$ *is positive real if and only if—*

(a) $F(s)$ *has no zeros in* $\sigma > 0$.
(b)   *Zeros on the $j\omega$-axis (including* $s = \infty$*) are simple, with real positive derivatives.*
(c)   *Re* $[F(j\omega)] \geq 0$ *for all* ω *(except at poles).*

This theorem follows directly from the preceding one if we remember that the residue of a function at a simple pole is the reciprocal of the derivative of the reciprocal of the function.

In testing a given function to determine positive realness, it may not always be necessary to use the necessary and sufficient conditions listed in the preceding two theorems. It may be possible to eliminate some functions from consideration by inspection because they violate certain simple necessary conditions. Let us now discuss some of these conditions.

We have seen that a rational positive real function has neither zeros nor poles in the right half s-plane. We previously defined a *Hurwitz polynomial* as one that has no zeros in the right half-s-plane. This definition permits zeros on the $j\omega$-axis. With this terminology, we see that a positive real function is the ratio of two Hurwitz polynomials.

The factors that constitute a Hurwitz polynomial must have one of the following two forms: $(s + a)$ for real zeros or $(s^2 + as + b)$ for a pair of complex zeros, with a being non-negative and b being positive. If any number of such factors are multiplied, the result must be a polynomial all of whose coefficients are non-negative. Furthermore, unless *all* the factors correspond to zeros on the $j\omega$-axis, *all* the coefficients of the polynomial will be strictly positive. If we introduce the added condition that zeros on the $j\omega$-axis be simple, then it is found that, when all the zeros are on the $j\omega$-axis, every other coefficient will be zero and the remaining coefficients will be strictly postive. Even though this is a necessary condition for a Hurwitz polynomial, it is not sufficient, as the following counter-example readily demonstrates:

$$(s^2 - s + 4)(s + 2) = s^3 + s^2 + 2s + 8. \qquad (72)$$

The polynomial on the right has no missing powers of s and all coefficients are positive, yet it has a pair of zeros in the right half-plane.

Hence, if a rational function is presented as a candidate for positive realness, this criterion can serve as a negative type of test. If the numerator or denominator polynomials have any negative coefficients, or missing coefficients (other than all alternate coefficients as described above), the function can be discarded. On the other hand, if this test is passed, nothing definite can be said about the function.

Another simple test follows from the fact that a positive real function can have no more than a simple pole or a simple zero at zero or infinity (which are on the $j\omega$-axis). This requires that the highest powers of s in numerator and denominator not differ by more than unity; and similarly for the lowest powers.

To illustrate we shall list some rational functions and see if they can be ruled out rapidly as not satisfying certain necessary conditions for positive real functions.

Function	Remarks
$F(s) = \dfrac{s^2 + 2s + 2}{s + 1}$	No more than simple pole at infinity; positive coefficients. Might be positive real.
$F(s) = \dfrac{s^4 + 3s^2 + 2s + 2}{s^3 + 4s^2 + s + 2}$	Coefficient of cubic term missing. Not positive real.
$F(s) = \dfrac{s + 2}{s^3 + 5s^2 + 3s + 1}$	Double zero at infinity. Not positive real.
$F(s) = \dfrac{s^3 + 2s}{s^3 + 5s^2 + 2s + 5}$	Coefficients missing in numerator looks bad, but all even powers missing. Might still be positive real. (In fact, it is.)
$F(s) = \dfrac{(s + 1)(s + 3)(s + 4)}{(s + 2)^3}$	No negative or missing coefficients, but triple pole at $s = -2$ might seem peculiar. Not ruled out. (In fact, it is pr.)

THE ANGLE PROPERTY OF POSITIVE REAL FUNCTIONS

An important property of the impedance of a passive network that was found earlier was the angle property given in (50). This property can be proved mathematically, without recourse to energy considerations, simply from the definition of a positive real function. However, the proof is somewhat lengthy even though it is straightforward. It will therefore not be given but will be outlined in a problem. Furthermore, assuming the truth of (50) for a real function $F(s)$, it follows that $F(s)$ is positive real.

$$\text{Re}\,[F(s)] = |F(s)|\,\cos\,[\arg\,F(s)]$$

$$\geq |F(s)|\,\cos\,[\arg\,s] \geq 0 \qquad \left(0 \leq |\arg\,s| \leq \frac{\pi}{2}\right).$$

Thus this angle property is not only necessary but sufficient as well. We shall therefore state it here as a theorem.

Theorem 10. *A real rational function* F(s) *is positive real if and only if*

$$|\arg\,F(s)| \leq |\arg\,s| \qquad \left(0 \leq |\arg\,s| \leq \frac{\pi}{2}\right). \tag{73}$$

BOUNDED REAL FUNCTIONS

It is possible to relate another function to a positive real function through a bilinear transformation. The function so obtained possesses some interesting properties. Consider the following bilinear transformation:

$$W(s) = \frac{1 - F(s)}{1 + F(s)} \quad \text{or} \quad F(s) = \frac{1 - W(s)}{1 + W(s)}. \tag{74}$$

The mapping between the F- and W-plane is shown in Fig. 7. The right half of the F-plane is mapped into the inside of the unit circle in the W-plane. The $j\omega$-axis becomes the contour of the unit circle.

If $F(s)$ is a pr function, right-half-plane values of s map into right-half-plane values of F and so are mapped inside the W-plane unit circle. When

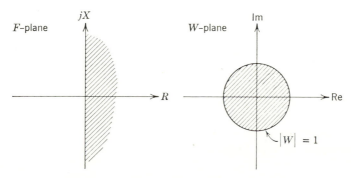

Fig. 7.   Mapping of bilinear transformation.

$s = j\omega$, $F(j\omega)$ takes on values in the right-half-plane or on the jX-axis, and these fall inside or on the W-plane unit circle. Hence, if $F(s)$ is pr and $W(s)$ is related to F through (74), then

$$|W(j\omega)| \leq 1. \tag{75}$$

Now consider the poles of $W(s)$. From (74), these occur where $F(s) = -1$. Values of s for which this is true cannot lie in the closed right half-plane if $F(s)$ is pr, since that would require $\operatorname{Re} F < 0$ for $\operatorname{Re} s \geq 0$. Hence $W(s)$ must be regular both in the right half-plane and on the $j\omega$-axis.

A function $W(s)$ having these properties—namely, $W(s)$ regular in the closed right half-plane and $|W(j\omega)| \leq 1$—is called *bounded real*. Thus *a bilinear transformation of a positive real function is bounded real*. The converse of this is also true; that is *a bilinear transformation of a bounded real function is positive real*. Sketching the few steps in the proof is left to you.

This relationship of bounded real and positive real functions leads to an interesting conclusion. Suppose a pr function $F(s)$ is written as

$$F(s) = \frac{m_1(s) + n_1(s)}{m_2(s) + n_2(s)}, \tag{76}$$

where m_1 and m_2 are even polynomials, and n_1 and n_2 are odd. Then the bilinear transformation (74) gives

$$W(s) = \frac{(m_2 - m_1) + (n_2 - n_1)}{(m_2 + m_1) + (n_2 + n_1)}. \tag{77}$$

The magnitude squared $|W(j\omega)|^2$ becomes

$$|W(j\omega)|^2 = \frac{(m_2 - m_1)^2 - (n_2 - n_1)^2}{(m_2 + m_1)^2 - (n_2 + n_1)^2}\bigg|_{s=j\omega} \leq 1. \tag{78}$$

Now suppose m_1 and m_2 are interchanged; the value of $|W(j\omega)|$ will clearly not be affected, as observed from (78), neither will the poles of $W(s)$, as observed from (77), although the zeros of $W(s)$ will. Let the new W-function obtained by interchanging m_1 and m_2 be called $\hat{W}(s)$. It is clearly a bounded real function. From (77), $\hat{W}$ is found to be

$$\hat{W}(s) = \frac{(m_1 - m_2) + (n_2 - n_1)}{(m_2 + m_1) + (n_1 + n_2)}. \tag{79}$$

The corresponding pr function $\hat{F}(s)$ is found from the bilinear transformation to be

$$\hat{F}(s) = \frac{m_2(s) + n_1(s)}{m_1(s) + n_2(s)}. \tag{80}$$

This is simply the original $F(s)$ with the even powers of the numerator and denominator interchanged. Since $\hat{F}(s)$ is a bilinear transformation of a bounded real function, it is pr. The reciprocal of $\hat{F}$ is also pr. But the reciprocal of $\hat{F}$ is the same as the original $F(s)$, with $n_1(s)$ and $n_2(s)$ interchanged. The conclusion is given as the following theorem:

Theorem 11. *If in a positive real function the even powers of the numerator and denominator, or the odd powers, are interchanged, the result is a positive real function.*

THE REAL PART FUNCTION

Since the real part of a pr function plays such a central role in its properties, we should examine the behavior of the real part of such a function on the $j\omega$-axis. Remember that the j-axis real part of F is equal to the even part evaluated at $s = j\omega$; that is,

$$R(\omega) = \text{Re}\,[F(j\omega)] = \text{Ev}\,F(s)|_{s=j\omega} = \tfrac{1}{2}[F(j\omega) + F(-j\omega)], \tag{81}$$

so that statements made about the even part can easily be interpreted in terms of the real part on the $j\omega$-axis.

We already know that $R(\omega)$ is necessarily an even function of ω and non-negative for all ω. It is also easy to establish that the even part of $F(s)$ can have no poles on the $j\omega$-axis. Any poles of the even part would also have to be poles of $F(s)$; but on the $j\omega$-axis, these are simple. If we consider $F(s)$ expanded in partial fractions as in (71), the function $F(-s)$ will contain the same terms, but all those involving the poles on the $j\omega$-axis will have a negative sign. Hence, in forming the even part, $F(s) + F(-s)$, these will all cancel, leaving the function with no poles on the $j\omega$-axis. Interpreted in terms of the real part, this means that $R(\omega)$ *must be bounded for all* ω.

Now let us consider a possible zero of $R(\omega)$. Figure 8 shows a sketch of $R(\omega)$ versus ω in the vicinity of a zero. Because of the positive real requirement, $R(\omega)$ must remain positive on both sides of the zero. It follows that a zero of $R(\omega)$ on the ω-axis cannot be of odd multiplicity; it must be of even multiplicity.

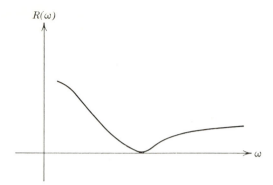

Fig. 8.   Sketch of real part of positive real function.

We have here determined certain necessary conditions for the *j*-axis real part of a positive real function. Let us now list a set of necessary and sufficient conditions as a theorem.

Theorem 12. *A real function* R(ω) *of a real variable* ω *is the* j-*axis real part of a rational positive real function* F(s) *if and only if—*

(a)   R(ω) *is an even rational function with real coefficients.*
(b)   R(ω) *is bounded for all* ω.
(c)   R(ω) ≥ 0 *for all* ω.

We have already seen that these conditions are necessary. As a matter of fact, it has already been demonstrated in Chapter 6, by actual construction, that conditions (a) and (b) are sufficient to find a real rational function $F(s)$ from a given $R(\omega)$. If condition (c) is also satisfied by $R(\omega)$, this is sufficient to make the rational function in question a positive real function.

7.5   REACTANCE FUNCTIONS

Let us now turn our attention to some special types of positive real functions. These arise from a consideration of networks containing only two types of elements (*LC, RC, RL*). Historically, such networks were studied before the more general ones, starting with the work done by Foster in 1924.

We shall initially consider networks that have no resistance. Such networks are referred to as *lossless*, or *reactance*, networks. In Theorem 5 we noted that the driving-point impedance of a lossless network is purely imaginary on the $j\omega$-axis; that is, $\mathrm{Re}\,[Z(j\omega)] = 0$. Stated in terms of a transformation, the impedance of a lossless network maps the imaginary axis of the s-plane into the imaginary axis of the Z-plane. Having observed this property of the impedance of a network, we shall now revert to mathematics and make this property the basis of a definition. We shall make the following definition: *A reactance function is a positive real function that maps the imaginary axis into the imaginary axis.* In this terminology, the driving-point impedance of a lossless network is a reactance function.

Let us now establish some properties of reactance functions. In the first place we shall show that the poles and zeros of a reactance function all lie on the $j\omega$-axis.

To prove this theorem note that, just as a function that maps the real axis into the real axis has symmetry about the real axis [i.e., $F(\bar{s}) = \bar{F}(s)$], so a function that maps the imaginary axis into the imaginary axis has symmetry about the imaginary axis. To see this clearly let us rotate the two planes (the s-plane and the F-plane) clockwise by $\pi/2$ radians. We do this by defining

$$z = \frac{s}{j}, \tag{82a}$$

$$\psi(z) = \frac{1}{j}\,F(jz). \tag{82b}$$

These transformations are shown in Fig. 9. Note that the real s-axis

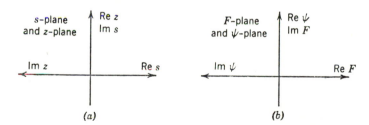

Fig. 9.   Transformation that rotates axes by $\pi/2$ radians.

becomes the imaginary z-axis, and vice versa. A similar case obtains for the other transformation. When z is real, the argument of $F(jz)$ is imaginary, so by hypothesis $F(jz)$ will also be imaginary. Hence $\psi(z)$ will be real when z is real. It follows from the reflection property given in (6) of Chapter 6 that

$$\psi(\bar{z}) = \bar{\psi}(z). \tag{83}$$

If we now translate back through the transformations in (82) this relation becomes

$$F(-\bar{s}) = -\bar{F}(s). \tag{84}$$

Note that the point $-\bar{s}$ is the image of the point s with respect to the imaginary axis. A similar case obtains for the points $-\bar{F}$ and F. Hence the result in (84) states that image points with respect to the imaginary axis in the s-plane go into image points with respect to the imaginary axis in the F-plane.

It follows that, if $F(s)$ has a pole or a zero in the left half-plane, then the image point in the right half-plane is also a pole or a zero, which is not possible for a pr function. Hence *the poles and zeros of a reactance function must all lie on the $j\omega$-axis*.

Let us turn back to Theorem 6 for a moment. There we saw that if a pr function maps a point on the $j\omega$-axis into a point on the imaginary axis, then the derivative of the function at that point is real and positive. But according to Theorem 5, a reactance function maps the entire $j\omega$-axis into the imaginary axis in the F-plane. Hence for such a function the derivative property will hold at all points on the $j\omega$-axis (except at poles). This is the basis of another very important property; namely, *the poles and zeros of a reactance function alternate on the $j\omega$-axis*; that is, between any two poles is a zero and between any two zeros is a pole.

As already noted, Theorem 6 applies at all points on the $j\omega$-axis except at poles. Hence the derivative dF/ds evaluated at $s = j\omega$ is real and positive. Let us compute the derivative along the $j\omega$-axis, which we are permitted to do since the derivative exists. The result will be

$$\left.\frac{dF}{ds}\right|_{s=j\omega} = \frac{dF(j\omega)}{d(j\omega)} = \frac{d[jX(\omega)]}{d(j\omega)} = \frac{j\,dX(\omega)}{j\,d\omega} = \frac{dX(\omega)}{d\omega} > 0. \tag{85}$$

We have used the usual notation $F(j\omega) = R(\omega) + jX(\omega)$, and, since F is here a reactance function, $R(\omega)$ is zero. Notice that $X(\omega)$ is a real function of a real variable. Therefore, if there is no pole between two zeros of $X(\omega)$, the derivative will become negative somewhere in between, which, as we have just shown, is impossible. A similar conclusion applies to successive poles. Figure 10 illustrates the form that $X(\omega)$ should have for successive zeros or poles without intervening poles or zeros, respectively.

The property that we have just proved is referred to as the alternation property of the poles and zeros. From this property it is clear that the plot of $X(\omega)$ against ω must have the general shape shown in Fig. 11.

Since $X(\omega)$ is an odd function of ω and the alternation of poles and

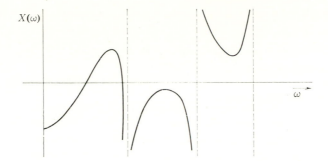

Fig. 10.   Impossible behavior of a reactance function.

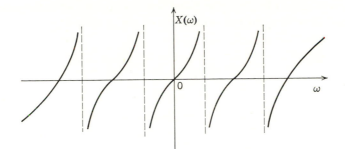

Fig. 11.   Behavior of a reactance function.

zeros must hold on the entire imaginary axis (positive and negative values of ω), we conclude that the point $s = 0$ is either a zero or a pole of a reactance function.

Note that, if $F(s)$ is a pr function mapping the imaginary axis into the imaginary axis, so is the function $F(1/s)$. With the transformation $s \rightarrow 1/s$, the point ∞ in the s-plane goes into the origin in the $1/s$-plane. Hence, by using the immediately preceding result, we find that *the point* $s = \infty$ *is either a zero or a pole of a reactance function*.

We have now discussed several properties of reactance functions. We should also note that certain properties of general pr functions apply in particular to reactance functions. Thus, since we have shown that poles and zeros of a pr function that lie on the $j\omega$-axis are simple and that residues at such poles are real and positive, we conclude that *all* poles and zeros of reactance functions are simple and that residues at all poles are real and positive.

We are now in a position to consolidate our results about reactance functions and to state necessary and sufficient conditions for a rational function of s to be a reactance function.

Theorem 13. *A real rational function* $\psi(s)$ *is a reactance function if and only if —*

1. *All of its poles are simple and lie on the* $j\omega$-*axis;*
2. *The residues are all real and positive;*
3. *The function has either a pole or a zero at* $s = 0$ *and at* $s = \infty$; *and*
4. *Re* $\psi(j\omega) = 0$ *for some* ω.

Notice that this statement involves only the poles and the residues, not the zeros. We have already shown these conditions to be necessary; it remains to prove that they are sufficient; that is, assuming a rational function to satisfy the stated conditions, we must show that the function is a reactance function. This is most easily done by considering the partial-fraction expansion of such a function. If we combine the two terms due to conjugate poles, the most general form of the partial-fraction expansion will be*

$$\psi(s) = \frac{k_0}{s} + k_\infty s + \sum_{i=1}^{n} \frac{2k_i s}{s^2 + \omega_i^2}, \qquad (86)$$

where the summation runs over all the poles, and all the k's are positive. Of course, the pole at the origin or at infinity, or both, may be absent. This expression is consistent with (71) with $F_r(s) = 0$, since in the present case there are no other poles except these on the $j\omega$-axis. The desired result follows immediately. Each term in this expansion is imaginary for imaginary values of s, so that $\psi(s)$ maps the imaginary axis into the imaginary axis, which makes $\psi(s)$ a reactance function by definition.

The alternation property of the poles and zeros forms the basis of an alternate set of necessary and sufficient conditions, stated as follows:

Theorem 14. *A real rational function of* s *is a reactance function if and only if all of its poles and zeros are simple, lie on the* $j\omega$-*axis, and alternate with each other.*

Again, we have already proved that a reactance function necessarily satisfies these conditions. It remains to show that the conditions are sufficient. A rational function that satisfies the given conditions must have the following form:

$$\psi(s) = K \frac{s(s^2 + \omega_1^2)(s^2 + \omega_3^2) \cdots (s^2 + \omega_{2n^2-1})}{(s^2 + \omega_0^2)(s^2 + \omega_2^2) \cdots (s^2 + \omega_k^2)}, \qquad (87)$$

---

* In the absence of condition 4 of Theorem (13) a constant term would be permitted in the expansion of (86). Condition 4 is required to eliminate this constant, which cannot be part of a reactance function.

where

$$0 \le \omega_0 < \omega_1 < \omega_2 < \omega_3 < \cdots < \omega_{2n-2} < \omega_{2n-1} < \omega_{2n} < \infty. \qquad (88)$$

In (87) K is a positive constant, and $k = 2n - 2$ or $2n$ according as $\psi(s)$ has a zero or a pole at infinity. If $\psi(s)$ has a pole at $s = 0$, we take ω_0 to be zero. A factor s will then cancel. The desired result now follows immediately. Each of the quadratic pole and zero factors in (87) is real when s is imaginary. This means that, due to the factor s, $\psi(s)$ is imaginary when s is imaginary. Hence, $\psi(s)$ is a reactance function, by definition.*

REALIZATION OF REACTANCE FUNCTIONS

At the start of this discussion we showed that the driving-point impedance of a lossless network is necessarily a reactance function. Note that the driving-point admittance of a lossless network is also a reactance function; that is,

$$Y(s) = 1/Z(s) \qquad (89)$$

is also imaginary for imaginary values of s if $Z(s)$ is.

The question now arises as to whether the converse of this condition is also true; that is, given a reactance function, is this the driving-point impedance (or admittance) of some lossless network? In order to answer this question in the affirmative, we shall have to construct a lossless network that has the given reactance function as its impedance or admittance. The question was answered in 1924 by Foster in his famous reactance theorem (although not in the form given here).

Theorem 15. *A rational function of s is a reactance function if and only if it is the driving-point impedance or admittance of a lossless network.*

We have already established the sufficiency. It remains to show that, given a reactance function, it is necessarily the impedance or the admittance of a lossless network. To show this, turn back to the partial-fraction expansion of a reactance function given in (86). We can recognize each of the summands of the partial-fraction expansion to be the impedance or admittance of a very simple reactance structure. The structures are shown in Fig. 12. Thus, if $\psi(s)$ is to be an impedance, we can represent it as a

---

* It appears that this argument requires only that $\psi(s)$ be an odd rational function: ratio of two even polynomials with an additional factor of s in numerator or denominator. But, in addition to mapping the imaginary axis into the imaginary axis, a reactance function has to be pr. Without the alternation property, an odd rational function will not be pr.

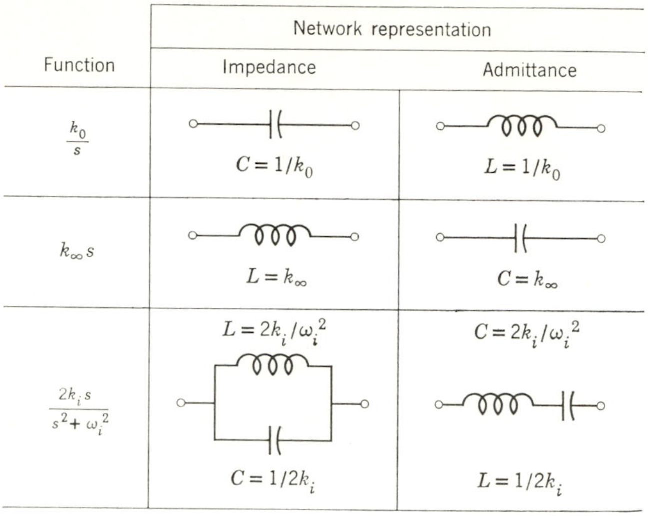

	Network representation	
Function	Impedance	Admittance
$\dfrac{k_0}{s}$	$C = 1/k_0$	$L = 1/k_0$
$k_\infty s$	$L = k_\infty$	$C = k_\infty$
$\dfrac{2k_i s}{s^2 + \omega_i^2}$	$L = 2k_i/\omega_i^2$ $C = 1/2k_i$	$C = 2k_i/\omega_i^2$ $L = 1/2k_i$

Fig. 12.   Representation of partial-fraction summands.

series combination of the elementary one-port networks in column 2 of Fig. 12. Or, if $\psi(s)$ is to be an admittance, we can represent it as a parallel combination of the elementary one-port networks in column 3. The forms of the resulting networks are shown in Fig. 13. They are referred to as Foster's first and second form.

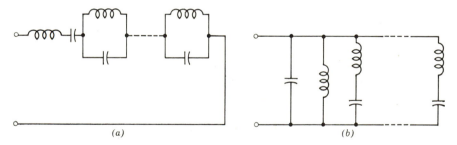

Fig. 13.   Foster's forms of lossless one-ports.

We have now proved the theorem with a vengeance. We found that a given reactance function can be both the impedance and the admittance of some lossless network (not the same network, of course).

Let us illustrate this result with the following function. Let

$$Z(s) = \frac{4(s^2+1)(s^2+9)}{s(s^2+4)} = 4s + \frac{9}{s} + \frac{15s}{s^2+4} \tag{90a}$$

or

$$Y(s) = \frac{s(s^2+4)}{4(s^2+1)(s^2+9)} = \frac{1}{32}\left(\frac{3s}{s^2+1} + \frac{5s}{s^2+9}\right). \qquad (90b)$$

In the first of these, the term $4s$ is recognized as the impedance of a four-unit inductor. Similarly, $9/s$ is recognized as the impedance of a $\frac{1}{9}$-unit capacitor. (The units are not henry or farad because this is presumably a normalized function.) The impedance of a parallel LC branch is

$$Z = \frac{s/C}{s^2 + \dfrac{1}{LC}}.$$

Hence, by direct comparison, the values of L and C are found to be $C = \frac{1}{15}$, $L = \frac{15}{4}$. The network takes the form shown in Fig. 14a.

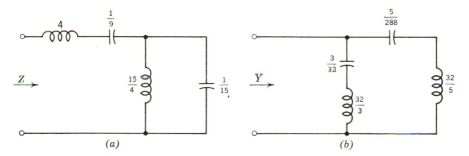

Fig. 14.   Reactive network realizations.

The admittance in $(90b)$ is seen to consist of two terms, each of which can be realized by an inductor and a capacitor in series. The admittance of such a series-tuned circuit is

$$Y = \frac{s/L}{s^2 + \dfrac{1}{LC}}.$$

Hence the values of L and C for each of the two branches can be found by comparison of this expansion with each of the numerical terms in the given function. The result is shown in Fig. 14b.

The two networks obtained are entirely equivalent at their terminals. No measurements made there could distinguish one from the other.

LADDER FORM OF NETWORK

The Foster forms are not the only possible networks that realize a given function. (There are, in fact, an infinite number of alternative structures.) Let us illustrate one possible alternative with the example already treated, before generalizing. The impedance in (90) has a pole at infinity. If this entire pole is subtracted from $Z(s)$, the remaining function will no longer have a pole at infinity and so it must have a zero there. Thus

$$Z_1(s) = Z(s) - 4s = \frac{4(s^2 + 1)(s^2 + 9)}{s(s^2 + 4)} - 4s = \frac{24s^2 + 36}{s^3 + 4s}$$

$$= \left(\frac{s^3 + 4s}{24s^2 + 36}\right)^{-1} = \left(\frac{s}{24} + \frac{5s/2}{24s^2 + 36}\right)^{-1}.$$

The result of subtracting $4s$ from the impedance means removing a four-unit inductor from the network as illustrated in Fig. 15a. leaving a

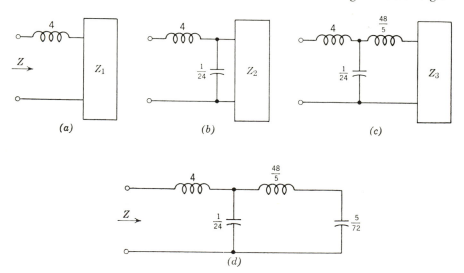

Fig. 15.   Ladder-network realization of impedance.

network whose impedance is Z_1. Now the reciprocal of Z_1 will have a pole at infinity. This pole can be totally removed by subtracting $s/24$, leaving

$$Y_2 = Y_1 - \frac{s}{24} = \frac{5s/2}{24s^2 + 36} = \left(\frac{24s^2 + 36}{5s/2}\right)^{-1} = \left(\frac{48s}{5} + \frac{72}{5s}\right)^{-1}.$$

The equivalent of subtracting $s/24$ from the admittance is to remove a $\frac{1}{24}$-unit capacitor from across the input terminals of Z_1, as shown in Fig. 15b. The admittance Y_2 remaining after this removal has no pole at infinity, but its reciprocal, Z_2, does. This can be removed, leaving

$$Z_3 = Z_2 - \frac{48s}{5} = \frac{72}{5s}$$

$$Y_3 = \frac{5}{72}s.$$

The network equivalent of subtracting an impedance of $48s/5$ is to remove a $\frac{48}{5}$ unit inductor, as shown in Fig. 15c. The remaining impedance Z_3 is simple enough to identify as a capacitor. The final network is shown in Fig. 15d. It is in the form of a ladder network.

A ladder network having arbitrary branches is shown in Fig. 16. Its

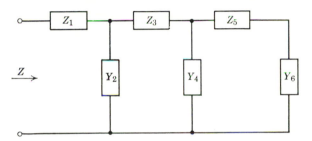

Fig. 16.   Ladder network with arbitrary branch impedances.

impedance can be written in the following *continued-fraction* form:

$$Z = Z_1 + \cfrac{1}{Y_2 + \cfrac{1}{Z_3 + \cfrac{1}{Y_4 + \cfrac{1}{Z_5 + \cfrac{1}{Y_6}}}}} \tag{91}$$

In the example just treated the process carried out step by step is actually a continued-fraction expansion, where each of the Z_i and Y_i functions in (91) is of the form ks, in which k is inductance or capacitance. The expansion deals exclusively with the pole at infinity, removing this pole alternately from the impedance and then the remaining admittance. The result of this process for an arbitrary reactance function will have the network structure shown in Fig. 17a.

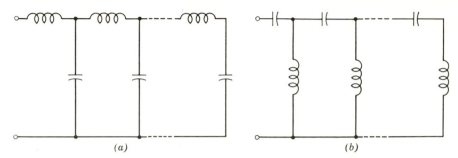

Fig. 17.   First and second Cauer forms of lossless ladders.

An alternate continued fraction expansion can be made by dealing with the pole at the origin, $s = 0$. The expansion is obtained by removing the pole at $s = 0$ alternately from the impedance and then the remaining admittance until the function is exhausted. For the example previously treated the expansion is

$$Z(s) = \frac{4(s^2 + 1)(s^2 + 9)}{s(s^2 + 4)} = 9/s + \frac{4s^3 + 31s}{s^2 + 4} = 9/s + \cfrac{1}{4/31s + \cfrac{15s/31}{4s^2 + 31}}$$

$$= 9/s + \cfrac{1}{4/31s + \cfrac{1}{961/15s + \cfrac{1}{15/124s}}}$$

The result is like Fig. 17b with the first four elements being $C_1 = \frac{1}{9}$, $L_2 = \frac{31}{4}$, $C_3 = \frac{15}{961}$, and $L_4 = \frac{124}{15}$.

To summarize, we have shown that the impedance and admittance of a lossless network are reactance functions; and conversely, given any reactance function, a number of networks can be found whose impedance or admittance is equal to the given reactance function.

HURWITZ POLYNOMIALS AND REACTANCE FUNCTIONS

We have found that a reactance function is an odd rational function, the ratio of an odd to an even polynomial, or vice versa, as observed in

(87). If we denote even and odd polynomials by $m(s)$ and $n(s)$, respectively. then a reactance function $\psi(s)$ can be written as

$$\psi(s) = \frac{m(s)}{n(s)} \quad \text{or} \quad \frac{n(s)}{m(s)}. \tag{92}$$

where m and n have no common factors.

Now consider the parallel combination of a lossless network and a one-ohm resistance. Taking $\psi(s)$ to be the admittance of the lossless network, the impedance of this combination will be

$$Z(s) = \frac{1}{1 + \psi(s)} = \frac{n(s)}{m(s) + n(s)} \quad \text{or} \quad \frac{m(s)}{m(s) + n(s)}, \tag{93}$$

where (92) is used for $\psi(s)$. The impedance of this RLC network will be pr and regular on the $j\omega$-axis; hence its poles cannot lie in the closed right half-plane. The polynomial $m + n$ in (93) is therefore a strictly Hurwitz polynomial. This is a very useful result. We shall state this result and its converse as a theorem.

Theorem 16. *If* $P(s) = m(s) + n(s)$ *is a Hurwitz polynomial, then the ratio* m/n *is a reactance function. Conversely, if the ratio of the even and odd parts of a polynomial* $P(s)$ *is found to be a reactance function, then* $P(s)$ *will differ from a strictly Hurwitz polynomial by at most a multiplicative even polynomial.**

This theorem provides us with a means for easily determining whether a given rational function is regular in the right half-plane, as positive realness requires. We take the ratio of the even and odd parts (or its reciprocal) of the denominator polynomial, then expand in a continued fraction or a partial fraction. To illustrate, let

$$P(s) = 2s^4 + 5s^3 + 6s^2 + 3s + 1,$$

$$m(s) = 2s^4 + 6s^2 + 1,$$

$$n(s) = 5s^3 + 3s.$$

---

* A proof is outlined in problem 43 for you to work out. An alternate proof is given in Norman Balabanian, *Network Synthesis*, Prentice-Hall, Englewood, Cliffs, N.J., 1958, pp. 77–81.

Now form the ratio m/n and expand in a continued fraction. The result will be

$$\frac{m}{n} = \frac{2s^4 + 6s^2 + 1}{5s^3 + 3s} = \frac{2s}{5} + \cfrac{1}{\cfrac{25s}{24} + \cfrac{1}{\cfrac{576s}{235} + \cfrac{1}{\cfrac{47s}{24}}}}$$

The elements in a lossless-network realization of this continued fraction will all be positive. Hence m/n is a reactance function and $P(s)$ is a strictly Hurwitz polynomial. In this example we assumed without detailed investigation the absence of a multiplicative even polynomial. A criterion for verifying this will evolve from the next example

As another illustration consider

$$P(s) = s^5 + 2s^4 + 3s^3 + 6s^2 + 4s + 8,$$

$$m = 2s^4 + 6s^2 + 8,$$

$$n = s^5 + 3s^3 + 4s.$$

Then

$$\frac{n}{m} = \frac{s^5 + 3s^3 + 4s}{2s^4 + 6s^2 + 8} = \frac{s}{2} + 0.$$

Observe that $n/m = s/2$ is a reactance function; but the continued-fraction expansion terminates prematurely because a polynomial

$$s^4 + 3s^2 + 4 = (s^2 - s + 2)(s^2 + s + 2)$$

is a factor of both even and odd parts. This is an even polynomial that has two zeros in the right half-plane and two in the left. The original polynomial is

$$P(s) = (s + 2)(s^4 + 3s^2 + 4)$$

and is exactly a Hurwitz polynomial times an even polynomial, in accordance with the theorem.

To conclude: Given a polynomial $P(s) = m(s) + n(s)$, in order to detect the presence of an even polynomial factor, we note that this must be a factor of both the even and odd parts. Hence, when the ratio is formed and a continued-fraction expansion is carried out, the premature termination of the expansion will signal the presence of such an even factor. The final divisor, just before the termination, will be the even polynomial in question.

7.6   IMPEDANCES AND ADMITTANCES OF *RC* NETWORKS

Let us now turn to another type of two-element network: namely, *RC* networks. We can, if we like, carry out a complete discussion of this case, without referring to the discussion of *LC* networks. However, this would be a waste of time, since it is possible to interrelate the driving-point functions by means of suitable transformations. The procedure we shall follow was first used by Cauer in extending Foster's work to *RC* and *RL* networks.

Let $Z(s)$ be the driving-point impedance of an *RC* network N. With the usual choice of loops, let the loop impedance matrix of N be

$$\mathbf{Z}_m(s) = [c_{ij}(s)], \tag{94}$$

where the elements of the matrix are

$$c_{ij}(s) = R_{ij} + \frac{1}{sC_{ij}}. \tag{95}$$

Let us replace each resistance in N by an inductance of equal value (R ohms becomes R henrys). Then the loop impedance matrix of the new network N' becomes

$$\boldsymbol{\zeta}_m(s) = \left[sR_{ij} + \frac{1}{sC_{ij}} \right] = \left[s\left(R_{ij} + \frac{1}{s^2 C_{ij}} \right) \right] = s\mathbf{Z}_m(s^2). \tag{96}$$

The driving-point impedance of network N' is found from a solution of the corresponding loop equations; it will be given by the ratio of det $(\boldsymbol{\zeta}_m)$ and one of its first principal cofactors. The impedance of network N will equal the ratio of det $(\mathbf{Z}_m)$ and one of its first principal cofactors. But $\boldsymbol{\zeta}_m$ and $\mathbf{Z}_m$ are related through (96). Hence, remembering the effect on the determinant of multiplication of a matrix by a scalar s, we find the driving-point impedance of the network N' to be

$$\psi(s) = sZ(s^2).$$

The network N' contains only capacitance and inductance, so that $\psi(s)$ in the last equation is a reactance function. Thus we have found that the impedance of an RC network can be transformed to a reactance function by replacing s by s^2 and then multiplying by s.

It would be of interest to see if the converse is also true; that is, given a reactance function $\psi(s)$, can we convert to the impedance of an RC network with the opposite transformation? To do this, consider the reactance function to be expanded in partial fractions, as shown in (86). Now divide the entire result by s and replace s by $\sqrt{s}$. (This is the opposite of the transformation just used.) The result will be

$$\frac{1}{\sqrt{s}}\,\psi(\sqrt{s}) = \frac{1}{\sqrt{s}}\left(\frac{k_0}{\sqrt{s}} + k_\infty\sqrt{s} + \sum_i \frac{2k_i\sqrt{s}}{s + \omega_i^2}\right)$$

$$= \frac{k_0}{s} + k_\infty + \sum_i \frac{2k_i}{s + \omega_i^2}.$$

Each term on the right can be recognized as the impedance of a simple RC structure. The term k_0/s is a capacitor; and k_∞ a resistor. Each of the other terms represents the impedance of a branch consisting of R and C in parallel, given by $(1/C)/(s + 1/RC)$. The values of R and C are obtained by comparing the two expressions. As a matter of fact, the representations of column 2 in Fig. 12 will apply, with inductances replaced by resistances. For convenient reference let us state this result as follows:

Theorem 17. *If $Z_{RC}(s)$ is the driving point impedance of an* RC *network, then*

$$\psi(s) = sZ_{RC}(s^2) \tag{97}$$

is a reactance function. Conversely, if $\psi(s)$ is a reactance function, then

$$Z_{RC}(s) = \frac{1}{\sqrt{s}}\,\psi(\sqrt{s}) \tag{98}$$

is the driving-point impedance of an RC *network.*

Let us now consider the admittance of an RC network. By using (98), it can be expressed as

$$Y_{RC}(s) = \sqrt{s}\,\frac{1}{\psi(\sqrt{s})}.$$

But the reciprocal of a reactance function ψ is itself a reactance function. Hence, given a reactance function $\psi(s)$, to obtain an *RC* admittance we replace s by $\sqrt{s}$, then multiply by $\sqrt{s}$. For convenient reference we shall state this as follows:

Theorem 18. *If* $Y_{RC}(s)$ *is the admittance of an* RC *network, then*

$$\psi(s) = \frac{1}{s}\, Y_{RC}(s^2) \qquad (99)$$

is a reactance function. Conversely, if $\psi(s)$ *is a reactance function, then*

$$Y_{RC}(s) = \sqrt{s}\, \psi\left(\sqrt{s}\right) \qquad (100)$$

is the driving-point admittance of an RC *network.*

Here we find a basic distinction between reactance functions and *RC* impedance-and-admittance functions. Whereas the reciprocal of a reactance function is again a member of the same class of functions, the reciprocal of an *RC* impedance is a member of the class of *RC* admittances, and vice versa.

With the preceding transformations we are in a position to translate all the properties of reactance functions into properties of *RC* impedances and admittances. The procedure for establishing these results is quite straightforward. To start with, let us apply (98) and (100) to the partial-fraction expansion of a reactance function given in (86). With appropriate changes in notation for the poles and residues, the results will be

$$Z_{RC}(s) = k_\infty + \frac{k_0}{s} + \sum_i \frac{k_i}{s + \sigma_i}, \qquad (101)$$

$$Y_{RC}(s) = k_\infty s + k_0 + \sum_i \frac{k_i s}{s + \sigma_i}, \qquad (102)$$

where the k's and σ's are all real and positive. Note that we have used the same symbols for the residues and poles in both cases, but these are general expressions for classes of functions and the two are not supposed to be related.

Equation (102) is not a partial-fraction expansion of $Y_{RC}(s)$. It is, rather an expansion of $Y_{RC}(s)/s$, after which the result is multiplied through by s. If we divide (102) by s, we find that the form is identical with (101). This shows that *an* RC *admittance function divided by* s *is an* RC *im-*

pedance function. We see that the poles of both these functions are negative real, and the residues of Z_{RC} and Y_{RC}/s are all positive.

By differentiating the last two equations along the real axis $(s = \sigma)$, we obtain a result that is the counterpart of the positive-slope property of a reactance function; that is,

$$\frac{dZ_{RC}(\sigma)}{d\sigma} < 0, \tag{103a}$$

$$\frac{dY_{RC}(\sigma)}{d\sigma} > 0. \tag{103b}$$

Thus the curves of RC driving-point functions plotted for real values of s are monotonic; $Z_{RC}(\sigma)$ is strictly decreasing, whereas $Y_{RC}(\sigma)$ is strictly increasing. Just as in the case of reactance functions, this implies that the zeros and poles of both must alternate, in this case along the real axis.

Sketches of typical RC driving-point functions for real values of s are shown in Figs. 18 and 19. In Fig. 18 note that the first pole near the origin may in fact move into the origin, making $F(0)$ infinite. Also, the last zero on the negative real axis may move out to infinity, causing $F(\infty)$ to become zero. Similarly, in Fig. 19 the first zero may be at the origin, causing $F(0)$ to be zero. Also, the final pole may move out to infinity, causing $F(\infty)$ to become infinite.

Let us now collect all of these results and state them in the form of theorems. Theorems 19 and 20 are for RC impedances; Theorems 21 and 22, for RC admittances.

Theorem 19. *A rational function* F(s) *is the driving-point impedance of an* RC *network if and only if all of its poles are simple and restricted to the finite negative real axis (including* s $= 0$*), with real positive residues at all poles and with* F(∞) *real and non-negative.* (This is the counterpart of Theorem 13 for reactance functions.)

Theorem 20. *A rational function* F(s) *is the driving-point impedance of an* RC *network if and only if all the poles and zeros are simple, lie on the negative real axis, and alternate with each other, the first critical point (pole or zero), starting at the origin and moving down the negative real axis, being a pole.* (This is the counterpart of Theorem 14 for reactance functions.)

Theorem 21. *A rational function* F(s) *is the driving-point admittance of an* RC *network if and only if all of its poles are simple and restricted to the negative real axis (excluding the point* s $= 0$*, but including infinity), with*

F(0) *real and non-negative, and with all the residues of* F(s)/s *real and positive.*

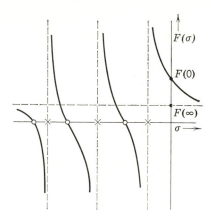

Fig. 18.   Typical $Z_{RC}(\sigma)$.

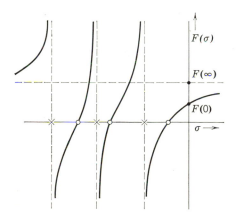

Fig. 19.   Typical $Y_{RC}(\sigma)$.

Theorem 22. *A rational function* F(s) *is the driving-point admittance of an* RC *network if and only if all the poles and zeros are simple, lie on the negative real axis, and alternate with each other, the first critical point (pole or zero), starting at the origin and moving down the negative real axis, being a zero.* (The only difference between this theorem for admittances and Theorem 20 for impedances is the last word.)

We have already sketched the proofs of all these theorems in the preceding discussion. You may organize the proofs as an exercise.

We have now stated several sets of necessary and sufficient conditions for a rational function to be the driving-point impedance or admittance of an *RC* one-port network. Generally, when it is desired to prove the sufficiency of a set of conditions for a given function to be the driving-point (or transfer) function of a network from a class of networks, it is done by showing that a network (at least one) of the given class can be realized from the given function. In the present case we tied up the proof with reactance functions by showing that the given function can always be transformed to a reactance function. This function can then be realized as an *LC* network. The desired *RC* network is then obtained by performing the inverse transformation. This step amounts to replacing each *L* in the *LC* network with an *R* of equal value.

Alternatively, we can work on the given function itself, expanding it in partial fractions just as we did for reactance functions. We have already obtained the desired forms in (101) and (102). Each term in these expressions can be recognized as the impedance or admittance of a simple *RC* structure. The series or parallel connection of these structures (depending on whether the function is to be impedance or admittance) gives the desired result. The networks have the same form as the Foster forms of lossless networks shown in Fig. 12. Hence they are referred to as Foster realizations of *RC* networks, although it was Cauer who first gave these results. Figure 20 shows the realizations of the terms in (101) and (102).

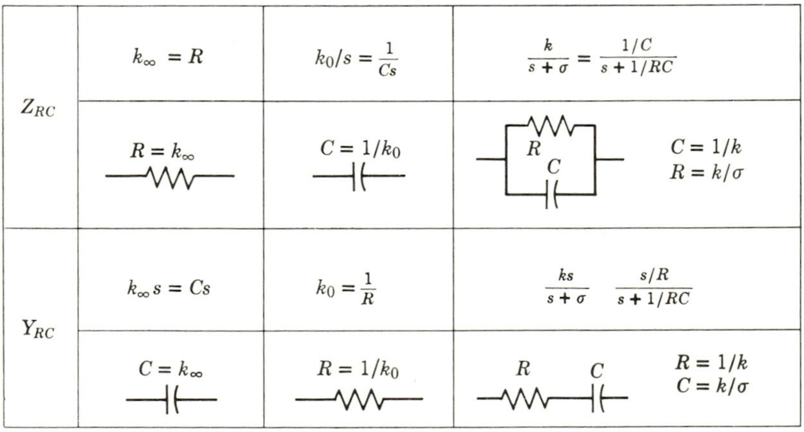

Fig. 20. Foster realizations of *RC* components.

To illustrate, let the given function be

$$Z(s) = \frac{2(s+2)(s+4)}{(s+1)(s+3)} = 2 + \frac{3}{s+1} + \frac{1}{s+3}.$$

The poles are real, negative, and simple; and the residues are positive, as the partial-fraction expansion shows. The constant term can be recognized as a two-unit resistor. The term $3/(s+1)$ represents a parallel *RC* branch. By reference to Fig. 20, the values of *C* and *R* are found to be $C=\frac{1}{3}$, $R=3$. The last term has the same form as the second one and can be realized by the same kind of branch. The complete realization is given in Fig. 21*a*.

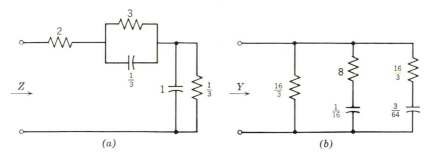

Fig. 21. *RC* networks realizing a given function.

Now consider the reciprocal of the function under consideration:

$$Y(s) = \frac{(s+1)(s+3)}{2(s+2)(s+4)} = \frac{1}{16}\left(3 + \frac{2s}{s+2} + \frac{3s}{s+4}\right).$$

The right side is obtained by first expanding $Y(s)/s$ in partial fractions and then multiplying by *s*. By reference to Fig. 20, the realization shown in Fig. 21*b* is obtained. You should verify that both networks have the same impedance.

LADDER-NETWORK REALIZATION

Since a one-to-one correspondence has been established between reactance functions and *RC* driving-point functions through appropriate transformations, we should expect that whatever procedures are used to obtain a network realization of a reactance function can also be used to obtain realizations of *RC* functions. The Foster forms have already been obtained. The ladder (Cauer) forms can also be obtained by expanding Z_{RC} and Y_{RC} in continued fractions. We shall not give the detailed development in the general case, since it should be quite obvious. Instead, an illustrative example will be given, with the same function for which Foster-form realizations were obtained above. There will be some characteristic differences in obtaining the continued-fraction expansion of Z_{RC} or Y_{RC} compared with that of a reactance function, because Z_{RC} cannot have a pole at infinity and Y_{RC} cannot have a pole at zero. Also, (see Problem 31) the smallest value of the real part occurs for different values of s for Z_{RC} and Y_{RC}.

Starting with the previously given $Z(s)$, a continued-fraction expansion dealing with the infinite-frequency behavior at each step is obtained as follows:

$$Z(s) = \frac{2s^2 + 12s + 16}{s^2 + 4s + 3} = 2 + \cfrac{1}{\cfrac{s^2 + 4s + 3}{4s + 10}}$$

$$= 2 + \cfrac{1}{\cfrac{s}{4} + \cfrac{1}{\cfrac{8s + 20}{3s + 1}}} = 2 + \cfrac{1}{\cfrac{s}{4} + \cfrac{1}{\frac{8}{3} + \cfrac{1}{\cfrac{3s + 1}{5\frac{2}{3}}}}}$$

$$= 2 + \cfrac{1}{\cfrac{s}{4} + \cfrac{1}{\frac{8}{3} + \cfrac{1}{\cfrac{6s}{52} + \cfrac{1}{53\frac{1}{3}}}}}$$

The corresponding network is shown in Fig. 22a.

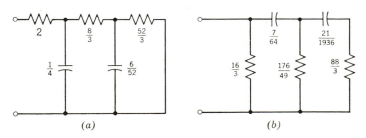

Fig. 22.   Ladder-network realizations.

An alternative realization is obtained by starting with the admittance and dealing with the zero-frequency behavior. Thus

$$Y(s) = \frac{3 + 4s + s^2}{16 + 12s + 2s^2} = \frac{3}{16} + \frac{1}{\dfrac{16 + 12s + 2s^2}{\dfrac{7}{4}s + \dfrac{5}{8}s^2}} = \frac{3}{16} + \cfrac{1}{\dfrac{64}{7s} + \cfrac{1}{\dfrac{14 + 5s}{352\tfrac{2}{7} + 16s}}}$$

$$= \frac{3}{16} + \cfrac{1}{\dfrac{64}{7s} + \cfrac{1}{49\!/\!176 + \cfrac{3s\!/\!11}{176\tfrac{2}{7} + 8s}}} = \frac{3}{16} + \cfrac{1}{\dfrac{64}{7s} + \cfrac{1}{49\!/\!176 + \cfrac{1}{\dfrac{1936}{21s} + \dfrac{1}{3\!/\!88}}}}$$

The corresponding network is shown in Fig. 22*b*.

RESISTANCE-INDUCTANCE NETWORKS

What has been done for *RC* networks can be duplicated for *RL* networks. The starting point is again a transformation that will take a reactance function into an *RL* impedance or admittance. It is immediately found that the class of *RL* impedance functions is identical with the class of *RC* admittance functions, and vice versa. Hence there is no need to duplicate the detailed development. In any theorem involving *RC* networks it is only necessary to replace the word " impedance " with the word " admittance " (or " admittance " with " impedance ") to arrive at a valid theorem for *RL* networks. We shall not pursue this subject here but will suggest some of the results as problems.

7.7   TWO-PORT PARAMETERS

In the last three sections we studied some of the most important properties of driving-point functions of linear, time-invariant, passive, reciprocal one-port networks. We shall now go on to a consideration of multiport networks; in particular, two-ports. The groundwork was already laid in (46) for a consideration of the open-circuit impedance and short-circuit admittance matrices. These expressions are repeated here for convenience.

$$F_0 + sT_0 + \frac{1}{s}\,V_0 = \mathbf{I}^*\mathbf{Z}_{\mathrm{oc}}\,\mathbf{I}, \tag{104a}$$

$$F_0 + \bar{s}T_0 + \frac{1}{\bar{s}}\,V_0 = \mathbf{V}^*\mathbf{Y}_{\mathrm{sc}}\,\mathbf{V}. \tag{104b}$$

The left-hand side of each of these equations is a positive real function. (The left-hand side of the second equation is the conjugate of that of the first). The right-hand side of each equation is a quadratic form that is now seen to equal a positive real function. Just as we say that the matrix of a positive definite quadratic form is positive definite, so also we say that the matrix of a positive real quadratic form is positive real. The conclusion is as follows:

Theorem 23. *The open-circuit impedance and short-circuit admittance matrices of a linear, time-invariant, passive, reciprocal multiport are positive real matrices.*

The same result can be demonstrated in a different way. It will be illustrated for a two-port by the network shown in Fig. 23. The two pairs

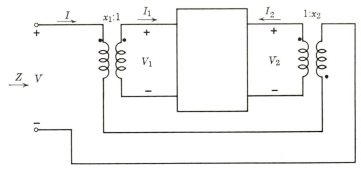

Fig. 23. Brune's demonstration that the z- and y-matrices are positive real.

of terminals of the two-port are connected in series through ideal transformers whose turns ratios are $x_1 : 1$ and $x_2 : 1$, respectively. The voltage and current at the input terminals will be given by

$$V = x_1 V_1 + x_2 V_2,$$ (105a)

$$I = \frac{I_1}{x_1} = \frac{I_2}{x_2}.$$ (105b)

If we now compute the driving-point impedance $Z(s) = V/I$ at the input terminals, we shall get

$$Z(s) = x_1{}^2 z_{11} + 2 x_1 x_2 z_{21} + x_2{}^2 z_{22}$$ (106)

$$= [x_1 \quad x_2] \begin{bmatrix} z_{11} & z_{12} \\ z_{21} & z_{22} \end{bmatrix} \begin{bmatrix} x_1 \\ x_2 \end{bmatrix}.$$

Since the impedance is positive real, this proves that the quadratic form on the right is also positive real. To prove the condition for the y-matrix, the two pairs of terminals can be connected in parallel through ideal transformers and the overall input admittance calculated. This is left for you as an exercise.

Let us now restrict ourselves to two-ports; the extension of the subsequent results to higher order multiports is simple and will become evident. The fact that $\mathbf{Z}_{oc}$ and $\mathbf{Y}_{sc}$ are positive real matrices has some interesting consequences. Let x_1 and x_2 be two arbitrary real numbers. Since the quadratic forms

$$Q_1 = \begin{bmatrix} x_1 & x_2 \end{bmatrix} \begin{bmatrix} z_{11} & z_{12} \\ z_{21} & z_{22} \end{bmatrix} \begin{bmatrix} x_1 \\ x_2 \end{bmatrix} \qquad (107a)$$

$$Q_2 = \begin{bmatrix} x_1 & x_2 \end{bmatrix} \begin{bmatrix} y_{11} & y_{12} \\ y_{21} & y_{22} \end{bmatrix} \begin{bmatrix} x_1 \\ x_2 \end{bmatrix} \qquad (107b)$$

are positive real functions, it follows that any pole of these functions on the $j\omega$-axis must be simple, and the residue at such a pole must be real and positive, Suppose, for instance, that the z-parameters have a pole at $s = j\omega_i$. Since this pole is a simple pole of the quadratic form, the residue of Q_1 is

$$\text{residue of } Q_1 = \lim_{s \to j\omega_i} (s - j\omega_i)\begin{bmatrix} x_1 & x_2 \end{bmatrix} \begin{bmatrix} z_{11} & z_{12} \\ z_{21} & z_{22} \end{bmatrix} \begin{bmatrix} x_1 \\ x_2 \end{bmatrix}$$

$$\qquad (108)$$

$$= \lim_{s \to j\omega_i} \begin{bmatrix} x_1 & x_2 \end{bmatrix} \begin{bmatrix} (s - j\omega_i)z_{11} & (s - j\omega_i)z_{12} \\ (s - j\omega_i)z_{21} & (s - j\omega_i)z_{22} \end{bmatrix} \begin{bmatrix} x_1 \\ x_2 \end{bmatrix}.$$

If we give the residues of z_{11}, $z_{21}(= z_{12})$, and z_{22} at the pole $s = j\omega_i$ the labels $k_{11}{}^{(i)}$, $k_{21}{}^{(i)}$, and $k_{22}{}^{(i)}$, respectively, then the residue of the quadratic form will become

$$\text{residue of } Q_1 = \begin{bmatrix} x_1 & x_2 \end{bmatrix} \begin{bmatrix} k_{11}{}^{(i)} & k_{21}{}^{(i)} \\ k_{21}{}^{(i)} & k_{22}{}^{(i)} \end{bmatrix} \begin{bmatrix} x_1 \\ x_2 \end{bmatrix}. \qquad (109)$$

Thus the residue itself is a quadratic form whose matrix is the matrix of residues of the z-parameters. However, this residue must be real and non-negative for all values of x_1 and x_2. Hence the matrix of residues of the z-parameters at any poles on the $j\omega$-axis must be positive definite or semidefinite. As discussed in Section 7.2, this requires that the determinant of the matrix and all of its principal cofactors be non-negative; that is,

$$k_{11}{}^{(i)} \geq 0, \qquad k_{22}{}^{(i)} \geq 0, \tag{110a}$$

$$k_{11}{}^{(i)} k_{22}{}^{(i)} - (k_{21}{}^{(i)})^2 \geq 0. \tag{110b}$$

The first line in these expressions is already known, since z_{11} and z_{22} are driving-point functions and therefore positive real. The second line, however, is a new and important result. It is known as the *residue condition*.

What was done for quadratic form Q_1 is also valid for Q_2/s. Thus the same conclusions follow for residues of $\mathbf{Y}_{sc}/s$. We shall state this result as a theorem.

Theorem 24.   *At any pole on the $j\omega$-axis of $\mathbf{Z}_{oc}$ or $\mathbf{Y}_{sc}/s$ of a linear, time-invariant, passive, reciprocal two-port network, the residues of the parameters satisfy the condition*

$$k_{11} k_{22} - k_{21}^2 \geq 0, \tag{111}$$

where k_{ij} is the residue of z_{ij} or y_{ij}/s at the $j\omega$-axis pole.
(The superscript has been omitted for simplicity.)

In particular, if the network is lossless, all the poles z_{ij} and y_{ij} are on the $j\omega$-axis, and hence the residue condition (111) applies at all the poles. One of the implications of this fact for a lossless network is that it is impossible for z_{21} to have a pole that is not also a pole of z_{11} and z_{22}, nor for y_{21} to have a pole that is not also a pole of y_{11} and y_{22}. For if either k_{11} or k_{22} is zero when k_{21} is not, the residue condition will be violated. On the other hand, it is possible for z_{11} or z_{22} (or both) to have a pole not shared by the other parameters. We refer to such poles as *private poles* of z_{11} or z_{22}. A similar statement applies to the y-parameters.

Let us now turn to another consequence of the positive real nature of $\mathbf{Z}_{oc}$ and $\mathbf{Y}_{sc}$. By definition, positive realness is linked with the real part of a function. Hence we should expect to obtain some relationship among the real parts of the z- and y-parameters. Let us denote these real parts

r_{11}, $r_{21}(= r_{12})$, and r_{22} for the z-parameters and g_{11}, $g_{21}(= g_{12})$, and g_{22} for the y-parameters. The real part of the quadratic forms Q_1 and Q_2 in (107) can then be written as follows:

$$\text{Re}\,(Q_1) = [x_1 \quad x_2] \begin{bmatrix} r_{11} & r_{12} \\ r_{21} & r_{22} \end{bmatrix} \begin{bmatrix} x_1 \\ x_2 \end{bmatrix}, \tag{112a}$$

$$\text{Re}\,(Q_2) = [x_1 \quad x_2] \begin{bmatrix} g_{11} & g_{12} \\ g_{21} & g_{22} \end{bmatrix} \begin{bmatrix} x_1 \\ x_2 \end{bmatrix}. \tag{112b}$$

Whenever s lies in the right half-plane or on the $j\omega$-axis, these quadratic forms must be positive semidefinite or definite, since Q_1 and Q_2 are positive real functions. As in the case of the matrix of residues, it follows that

$$r_{11} \geq 0 \qquad r_{22} \geq 0,$$
$$r_{11} r_{22} - r_{21}^2 \geq 0, \qquad \text{Re } s \geq 0 \tag{113}$$

for the real parts of the z-parameters, and

$$g_{11} \geq 0, \qquad g_{22} \geq 0,$$
$$g_{11} g_{22} - g_{21}^2 \geq 0, \qquad \text{Re } s \geq 0 \tag{114}$$

for the real parts of the y-parameters. Again the first lines in each set carry no surprises, since z_{11}, z_{22}, y_{11}, and y_{22} are driving-point functions and hence positive real. The second lines, however, express a new result, which is called the *real-part condition*. In fact, the real part condition alone is a sufficient condition that $\mathbf{Z}_{oc}$ or $\mathbf{Y}_{sc}$ be positive real matrices. (Verify this statement.)

RESISTANCE-CAPACITANCE TWO-PORTS

As a final note, observe that what was said about lossless two-ports is also true for RC two-ports, by virtue of the transformations previously discussed and with appropriate and obvious modifications. Thus for RC two-ports, the z- and y-parameters will have all their poles on the negative real axis, and the residue condition will apply at these poles. No pole of z_{21} can fail to be a pole of both z_{11} and z_{22}, but z_{11} and z_{22} can have private poles; and similarly for the y-parameters.

As an illustration, consider the RC two-port shown in Fig. 24. The short-

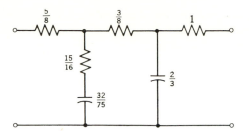

Fig. 24. Resistance-capacitance two-port.

circuit admittance parameters of this two-port as are follows:

$$y_{11}(s) = \tfrac{1}{25}\,\frac{(28s^2 + 124s + 75)}{(s + \tfrac{3}{2})(s + 4)} = \tfrac{1}{2} + \frac{192\!\big/\!375}{s + \tfrac{3}{2}} + \frac{27\!\big/\!250}{s + 4},$$

$$-y_{21}(s) = \tfrac{6}{5}\,\frac{(s + \tfrac{5}{2})}{(s + \tfrac{3}{2})(s + 4)} = \tfrac{1}{2} - \frac{8\!\big/\!25}{s + \tfrac{3}{2}} - \frac{9\!\big/\!50}{s + 4},$$

$$y_{22}(s) = \frac{(s + 1)(s + 3)}{(s + \tfrac{3}{2})(s + 4)} = \tfrac{1}{2} + \frac{\tfrac{8}{5}}{s + \tfrac{3}{2}} + \frac{3\tfrac{8}{10}}{s + 4}.$$

Note, first, that all three functions have the same poles. The zeros of y_{11} are at (approximately) $-\tfrac{5}{7}$ and $-\tfrac{26}{7}$. Thus the zeros and poles of both y_{11} and y_{22} alternate on the negative real axis, and they have the appropriate behavior at infinity (which is?), so that both are *RC* admittance functions. On the other hand, y_{21} does not have all the properties of an *RC* admittance; among other things, the residues of y_{21}/s are not all positive.

A test for the residue condition shows that it is satisfied at all the poles; in fact, the residue condition is satisfied with the equals sign. To distinguish when the residue condition is satisfied with the equals sign and when it is not, we say the pole is *compact* if the residue condition at the pole is satisfied with the equals sign. Thus, for the illustration, all the poles (including the constant term) are compact. Verification that the real-part condition is satisfied will be left to you.

For driving-point functions of networks containing two kinds of elements only we found sets of necessary conditions that were also proved to be sufficient. In fact, actual procedures for realizing one or more networks from a given function were obtained. The case for a two-port is not as simple, since a set of three parameters is involved. Although necessary conditions on these parameters have been obtained, these conditions turn out to be generally sufficient for reali zability only if we admit the inclusion of ideal transformers. If transform ers are not permitted, a general set of sufficient conditions is n ot available. We shall not pursue this subject any further in this b ook.

7.8   LOSSLESS TWO-PORT TERMINATED IN A RESISTANCE

Up to this point we have accomplished the following. For a linear, time-invariant, passive, reciprocal network with two kinds of elements, we have established necessary and sufficient conditions for the driving-point functions. Given a function satisfying these conditions, procedures have been discussed for finding a network having a given function as its impedance or admittance. More generally, we have seen that positive realness is a necessary condition for the impedance or admittance of the most general network of the class under discussion. We have not, however, shown that this is a sufficient condition. This was shown initially by Brune in 1932, but we shall here discuss an alternative structure first developed by Darlington in 1939. He showed that any positive real function can be realized as the impedance of a lossless network terminated in a single resistance.

Consider the network of Fig. 25. It consists of a lossless network

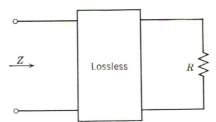

Fig. 25.   Resistance-terminated lossless two-port.

terminated in a resistance R. The impedance at the input terminals can be written in terms of R and the two-port parameters as

$$Z(s) = \frac{Rz_{11} + z_{11}z_{22} - z_{12}z_{21}}{R + z_{22}} = z_{11}\frac{R + 1/y_{22}}{R + z_{22}} \tag{115a}$$

or

$$Z(s) = z_{11}\frac{1 + 1/y_{22}}{1 + z_{22}}. \tag{115b}$$

(See Chapter 3.) In the final form all impedances have been normalized to R, which is equivalent to taking the value of R to equal 1.

Now suppose a rational positive real function $Z(s)$ is given. The even and odd parts of its numerator and denominator can be separated, and the function can be written in the usual form. Then this expression can be placed in the same form as (115b) in two possible ways, as follows:

$$Z(s) = \frac{m_1(s) + n_1(s)}{m_2(s) + n_2(s)}, \tag{116}$$

$$Z(s) = \frac{m_1}{n_2} \frac{1 + n_1/m_1}{1 + m_2/n_2} \qquad \text{(case A)} \tag{117}$$

$$Z(s) = \frac{n_1}{m_2} \frac{1 + m_1/n_1}{1 + n_2/m_2} \qquad \text{(case B)} \tag{118}$$

For each of these two cases, formal identifications can be made by comparing these expressions with (115b). Thus

$$
\begin{array}{cc}
\textit{Case A} & \textit{Case B} \\[2mm]
z_{11} = \dfrac{m_1}{n_2} & z_{11} = \dfrac{n_1}{m_2} \\[4mm]
z_{22} = \dfrac{m_2}{n_2} & z_{22} = \dfrac{n_2}{m_2} \\[4mm]
y_{22} = \dfrac{m_1}{n_1} & y_{22} = \dfrac{n_1}{m_1}
\end{array} \tag{119}
$$

Since $Z(s)$ is positive real, both $m_1 + n_1$ and $m_2 + n_2$ are Hurwitz polynomials. Hence the ratios m_1/n_1 and m_2/n_2, and their reciprocals, are reactance functions. Also, using Theorem 11 (concerning the interchange of the even or odd parts of numerator and denominator of a pr function), the ratios m_1/n_2 and n_1/m_2 are also reactance functions. Thus all the functions in (119) are reactance functions.

There remains the task of determining z_{21} from the expressions in (119), so that a complete set of parameters of the lossless two-port in Fig. 25 will be available. To do this, observe that

$$z_{11}z_{22} - z_{21}^2 = \frac{z_{11}}{y_{22}} = \frac{n_1}{n_2} \qquad \text{(case A)},$$

$$= \frac{m_1}{m_2} \qquad \text{(case B)}. \tag{120}$$

Since

$$z_{21} = \sqrt{z_{11} z_{22} - (z_{11} z_{22} - z_{21}^2)}$$

we get for z_{21}, by using (120) and (119):

$$z_{21} = \frac{\sqrt{m_1 m_2 - n_1 n_2}}{n_2} \qquad \text{(case A)},$$

$$\tag{121}$$

$$z_{21} = \frac{\sqrt{-(m_1 m_2 - n_1 n_2)}}{m_2} \qquad \text{(case B)}.$$

Of course, once the z-parameters are known, the y-parameters can also be found. (See Table 1 in Chapter 3.) The complete results are tabulated in Table 1.

The question is, do these open-circuit or short-circuit parameters satisfy realizability conditions for a passive, reciprocal, and lossless two-port? The first difficulty appears to be that z_{21} is not a rational function because of the indicated square root. However, if $m_1 m_2 - n_1 n_2$ is a perfect square, the apparent difficulty will disappear. Observe that $m_1 m_2 - n_1 n_2$ is the numerator of the even part of $Z(s)$. Because $Z(s)$ is a positive real function, zeros of its even part on the $j\omega$-axis must necessarily have even multiplicity. There is no such requirement, however, on any other zeros. Hence, unless some remedy can be found, it appears that z_{21} will generally be irrational.

A remedy has been found in the following way. Suppose the given $Z(s)$ is *augmented* by multiplying its numerator and denominator by a strictly Hurwitz polynomial $m_0 + n_0$, which certainly does not change the function. Thus

$$Z(s) = \frac{m_1 + n_1}{m_2 + n_2} \cdot \frac{m_0 + n_0}{m_0 + n_0} = \frac{\hat{m}_1 + \hat{n}_1}{\hat{m}_2 + \hat{n}_2}. \tag{122}$$

The new even part of $Z(s)$ will be

$$\text{Ev}\,[Z(s)] = \frac{\hat{m}_1 \hat{m}_2 - \hat{n}_1 \hat{n}_2}{\hat{m}_2{}^2 - \hat{n}_2{}^2} = \frac{m_1 m_2 - n_1 n_2}{m_2{}^2 - n_2{}^2} \cdot \frac{m_0{}^2 - n_0{}^2}{m_0{}^2 - n_0{}^2}. \tag{123}$$

The new z_{21} for case A will be

$$z_{21} = \frac{\sqrt{\hat{m}_1 \hat{m}_2 - \hat{n}_1 \hat{n}_2}}{\hat{n}_2} = \frac{\sqrt{(m_1 m_2 - n_1 n_2)(m_0{}^2 - n_0{}^2)}}{m_0 n_2 + n_0 m_2}. \tag{124}$$

Table 1.

Condition	z-Parameters			y-Parameters		
	z_{11}	z_{22}	z_{21}	y_{11}	y_{22}	y_{21}
Case A: *No* pole or zero of $Z(s)$ at $s = 0$	$\dfrac{m_1}{n_2}$	$\dfrac{m_2}{n_2}$	$\dfrac{\sqrt{m_1 m_2 - n_1 n_2}}{n_2}$	$\dfrac{m_2}{n_1}$	$\dfrac{m_1}{n_1}$	$\dfrac{\sqrt{m_1 m_2 - n_1 n_2}}{n_1}$
Case B: Pole or zero of $Z(s)$ at $s = 0$	$\dfrac{n_1}{m_2}$	$\dfrac{n_2}{m_2}$	$\dfrac{\sqrt{-(m_1 m_2 - n_1 n_2)}}{m_2}$	$\dfrac{n_2}{m_1}$	$\dfrac{n_1}{m_1}$	$\dfrac{\sqrt{-(m_1 m_2 - n_1 n_2)}}{m_1}$

It is now clear how to make z_{21} a rational function: we set $m_0{}^2 - n_0{}^2$ equal to the product of all factors in $m_1 m_2 - n_1 n_2$ that are of odd multiplicity and thus cause z_{21} to be irrational in the first place. Since $m_0{}^2 - n_0{}^2 = (m_0 + n_0)(m_0 - n_0)$, the augmenting polynomial $m_0 + n_0$ is found by taking the left-half-plane zeros of $m_0{}^2 - n_0{}^2$.

The question arises as to the significance of cases A and B. When is one appropriate and when is the other? We observe from Table 1 that the denominator of z_{21} (or y_{21}) is odd for case A and even for case B. Since z_{21} should be an odd rational function, the numerator of z_{21} should be even for case A and odd for case B. If $m_1 m_2 - n_1 n_2$ has s^2 as a factor, taking the square root will make the numerator of z_{21} odd, and so case B is appropriate. On the other hand, if $m_1 m_2 - n_1 n_2$ does not have s^2 as a factor, case A will be appropriate. The only way in which s^2 can be a factor of $m_1 m_2 - n_1 n_2$ is for either m_1 or m_2 to have its constant term missing. This means that *case B applies when* $Z(s)$ *has a pole or a zero at* $s = 0$, *and case A applies if* $Z(s)$ *has neither a pole nor a zero at* $s = 0$.

To illustrate the preceding, let

$$Z(s) = \frac{s^3 + 4s^2 + 4s}{s^3 + 5s^2 + 8s + 4}.$$

This function has a zero at $s = 0$; hence case B is appropriate. We form

$$m_1 m_2 - n_1 n_2 = 4s^2(5s^2 + 4) - (s^3 + 4s)(s^3 + 8s) = -s^2(s^2 - 4)^2.$$

This is a complete square, so that augmentation is not needed. Hence, from Table 1.

$$z_{21} = \frac{\sqrt{-(m_1 m_2 - n_1 n_2)}}{m_2} = \frac{s(s^2 - 4)}{5s^2 + 4} = \frac{s}{5} - \frac{0.96s}{s^2 + \frac{4}{5}}.$$

Also,

$$z_{11} = \frac{n_1}{m_2} = \frac{s(s^2 + 4)}{5s^2 + 4} = \frac{s}{5} + \frac{0.64s}{s^2 + \frac{4}{5}},$$

$$z_{22} = \frac{n_2}{m_2} = \frac{s(s^2 + 8)}{5s^2 + 4} = \frac{s}{5} + \frac{1.44s}{s^2 + \frac{4}{5}}.$$

Whatever the given $Z(s)$, it is possible to make z_{21} an odd rational function by augmenting the original function if necessary. Now let us consider the other realizability conditions. The real-part condition of (113) and (114) is satisfied identically with the equals sign when $s = j\omega$. Since the functions are regular in the right half-plane, the maximum modulus theorem can be used to show that the real-part condition will be satisfied anywhere in the right half-plane.

There remains the residue condition. The residue of a function can be calculated as the numerator divided by the derivative of its denominator, evaluated at a zero of the denominator. For the z-parameters the residues at finite nonzero poles are given in Table 2, in which the primes indicate

Table 2.

Condition	k_{11}	k_{22}	k_{21}
Case A	$\dfrac{m_1}{n_2'}\bigg\|_{n_2 = 0}$	$\dfrac{m_2}{n_2'}\bigg\|_{n_2 = 0}$	$\dfrac{\sqrt{m_1 m_2 - n_1 n_2}}{n_2'}\bigg\|_{n_2 = 0}$
Case B	$\dfrac{n_1}{m_2'}\bigg\|_{m_2 = 0}$	$\dfrac{n_2}{m_2'}\bigg\|_{m_2 = 0}$	$\dfrac{\sqrt{-(m_1 m_2 - n_1 n_2)}}{m_2'}\bigg\|_{m_2 = 0}$

the derivative with respect to s. Thus, by forming $k_{11} k_{22} - k_{21}^2$, it is found that at all the finite nonzero poles the residue condition is satisfied, and, furthermore, it is satisfied with the equality sign. Hence *all the finite, nonzero poles of the z-parameters are compact.*

It is also true that the residue condition is satisfied at a possible pole at infinity or at zero, but not always with an equality sign. (See problem 53.) A similar development can be carried out for the y-parameters divided by s, leading to similar results. The conclusion is that the residue condition is satisfied by both the z-parameters and the y-parameters divided by s at all their poles.

We have now established that, given a positive real function, it is possible to find a set of open- or short-circuit parameters that satisfy realizability conditions of a lossless two-port terminated in a unit resistance. There remains the task of actually realizing (constructing) the two-port. One procedure for doing this was developed by Cauer. It starts by expanding, say, the z-parameters into partial fractions. The terms in each z-parameter corresponding to a particular pole are lumped together; they are simple enough so that a lossless two-port realizing this set of parameters can be recognized. The component two-ports so obtained are then connected in series. However, as the discussion of interconnecting two-ports in Chapter 3 described, in order to permit a series connection it may be necessary to use ideal transformers. The series structure of two-ports is not a very desirable one; one objection is that all but one of the two-ports will be floating above ground.

A more desirable structure is a cascade structure. Darlington's contribution was to show that such a realization is possible. We shall do no more here than outline his procedure. Observe, first, that if the given impedance function has poles and zeros on the $j\omega$-axis, these can be removed as the branches of a ladder network. Such branches are shown in Fig. 26 as type

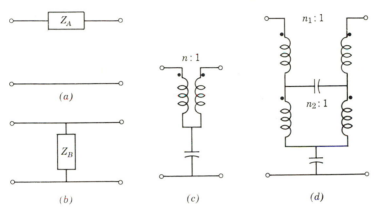

Fig. 26. Canonic sections for cascade realization: (*a*) Type A; (*b*) type B; (*c*) type C; (*d*) type D.

A and type B. These branches may consist of a single inductance or capacitance, a series-tuned circuit, or a parallel-tuned circuit.

After all poles and zeros of $Z(s)$ on the $j\omega$-axis have been removed by branches of type A and B, the even part of the remaining impedance will have three types of zeros: real, imaginary, and complex. A typical even part will be

$$\text{Ev}\left[Z(s)\right] = \frac{(s^2 + \omega_0{}^2)^2(s^2 - a^2)^2(s^4 + b_1 s^2 + b_0)^2}{m_2{}^2 - n_2{}^2}. \tag{125}$$

All zeros are shown as double zeros. For a positive real function, those on the $j\omega$-axis must necessarily be double. If the other zeros are not initially double, the function is augmented to make them so. *The transmission zeros of the lossless two-port are exactly the zeros of the even part of* $Z(s)$.

It turns out that a type C section, as shown in Fig. 26, has a pair of transmission zeros that are either real or imaginary (depending on the dot arrangement of the ideal transformer), and a type D section has a quadruplet of complex transmission zeros. Removal of a type C section is found to leave a reduced impedance whose even part has all the other zeros except the pair realized by the section. Similarly, removal of a type D section leaves a reduced impedance whose even part has all the other zeros except the quadruplet realized by the section. Hence a cascade connection of these four types of two-ports will realize the desired lossless two-port. We observe that, generally, ideal transformers are required in the realization.

It is not our purpose here to examine the details of the realization of the lossless two-port but simply to observe the result. Thus it is now possible to state the following (Darlington) theorem:

Theorem 25. *A positive real function can be realized as the impedance of a lossless two-port terminated in a unit resistance.*

This actually constitutes an existence theorem; it proves the sufficiency of the positive real condition for the realizability of a passive, reciprocal network.

For specific functions other, more useful, structures of the two-port may be found than the type C and type D sections with their ideal transformers. To illustrate this point, let

$$Z(s) = \frac{s^4 + s^3 + 3s^2 + 2s + 1}{5s^3 + 5s^2 + 2s + 5},$$

$$\mathrm{Ev}\,[Z(s)] = \frac{(2s^2 + 1)^2}{(5s^2 + 1)^2 - (5s^3 + 2s)^2}.$$

It appears that we shall need a type C section in the lossless two-port N to realize the pair of imaginary transmission zeros, as well as something for the transmission zero at infinity. Suppose the z-parameters are now formed by using Table 1. The result will be as follows:

$$z_{11} = \frac{s^4 + 3s^2 + 1}{5s^3 + 2s} = \frac{\frac{13}{5}s^2 + 1}{5s^3 + 2s} + \frac{s}{5},$$

$$z_{22} = \frac{5s^2 + 1}{5s^3 + 2s},$$

$$z_{21} = \frac{2s^2 + 1}{5s^3 + 2s}.$$

Observe that z_{11} has a private pole at infinity; this can be removed as a

$\frac{1}{5}$ unit inductance, leaving a set of z-parameters having the same poles. The partial result is shown in Fig. 27a. The z-parameters remaining after

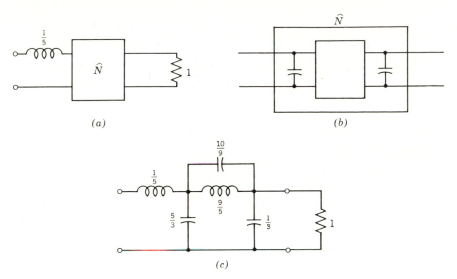

(a)   (b)

(c)

Fig. 27. Realization of impedance as lossless ladder termination in R.

the series inductor is removed are pertinent to the two-port $\hat{N}$.

Let us now invert these and find the y-parameters of $\hat{N}$. The result will be as follows:

$$y_{11} = \frac{z_{22}}{z_{11}z_{22} - z_{21}^2} = \frac{5(5s^2 + 1)}{9s} = \frac{25s}{9} + \frac{5}{9s},$$

$$y_{22} = \frac{z_{11}}{z_{11}z_{22} - z_{21}^2} = \frac{(13s^2 + 5)}{9s} = \frac{13s}{9} + \frac{5}{9s},$$

$$-y_{21} = \frac{z_{21}}{z_{11}z_{22} - z_{21}^2} = \frac{5(2s^2 + 1)}{9s} = \frac{10s}{9} + \frac{5}{9s}.$$

This can be rewritten as

$$y_{11} = Y + \frac{5s}{3},$$

$$y_{22} = Y + \frac{s}{3},$$

$$-y_{21} = Y,$$

where

$$Y = \frac{10}{9} s + \frac{5}{9s}.$$

Notice that, in addition to a common term Y, y_{11} and y_{22} have an extra term. These terms can each be realized by a capacitor, one across the input and one across the output of $\hat{N}$, as shown in Fig. 27b. That leaves Y, which is simple to realize; a type A section, as in Fig. 26a, with $Z_A = 1/Y$, has exactly the y-parameters $y_{11} = y_{22} = -y_{21} = Y$. The complete network, then, is shown in Fig. 27c. It is in the form of a lossless ladder network terminated in a unit resistance. No ideal transformer, as required by the type C section, appears.

It was earlier found that the impedance of a passive, reciprocal network is a positive real function. What has now been established is the converse; namely, that a positive real function can always be realized as a passive reciprocal network. In particular, the network will be a lossless two-port terminated in a resistance. In the most general case, the lossless two-port may require the use of type C and type D sections. In some cases, these complicated sections can be avoided. Certain sufficient conditions exist under which a realization avoiding type C and D sections can be found but we shall not pursue it further here.

7.9 PASSIVE AND ACTIVE RC TWO-PORTS

Let us now look back over the procedure followed in the last section to observe if there are features of a general nature that can serve as guides to other kinds of networks. The first step was to consider a network structure representative of a particular class. In Fig. 25 this was a lossless two-port terminated in a resistance. An expression was then written for the function of interest in terms of the component parts of the structure, as in (115). Then a rational function of the class under consideration was manipulated to make it take on this same form, as in (117) and (118). Finally, the pertinent functions of the component parts of the network were identified. Of course, it must be verified that these functions satisfy realizability conditions for the class of networks under consideration. Furthermore, procedures must be established for realizing these functions.

In this section we shall carry out similar procedures for networks consisting of resistance, capacitance, and possibly active devices. A number of considerations make such networks of practical importance. These include the relatively large size and weight of inductors, the low cost of transistors and other active devices, and the fact that realization of resistance, capacitance, and active devices is obtainable in integrated circuits.

CASCADE CONNECTION

Consider first the network shown in Fig. 28, in which two *RC* two-ports

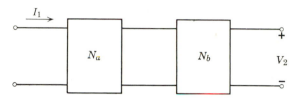

Fig. 28.   Cascaded *RC* two-ports.

are connected in cascade. The function of interest is the open-circuit transfer impedance $z_{21} = V_2/I_1$. This function can be written in terms of the open-circuit parameters of the individual two-ports as follows:*

$$z_{21}(s) = \frac{z_{21a}\, z_{21b}}{z_{22a} + z_{11b}}. \tag{126}$$

If a rational function is specified as the transfer impedance of an *RC* two-port, this function must be manipulated into the form of (126). In the first place, for an *RC* network we know that the poles of z_{21} must be real and negative (possibly including $s = 0$), but there are no restrictions on the locations of the zeros. There should, however, be no more finite zeros than poles; for example, two possible functions are the following:

$$z_{21} = \frac{P(s)}{Q(s)} = K\,\frac{(s+5)}{(s+2)(s+4)(s+6)}, \tag{127a}$$

$$z_{21} = \frac{P(s)}{Q(s)} = K\,\frac{(s^2+s+1)(s^2+2s+2)}{(s+1)(s+2)(s+3)(s+4)}. \tag{127b}$$

The first has a single finite zero, which is real. The second has two pairs of complex zeros. Each of these is a realizable *RC* transfer impedance for any value of the constant K.

* See Chapter 3.

Observe from (126) that the denominator is the sum of two RC impedances—which is, again, an RC impedance; but the denominator of any given rational function will be a polynomial. The situation is remedied by dividing numerator and denominator of the given function by an auxiliary polynomial $D(s)$. The degree and the zeros of this polynomial must be chosen to make Q/D an RC impedance. This is easy to do: the degree of $D(s)$ must be equal to, or be one greater than, the degree of $Q(s)$; and its zeros must alternate with those of $Q(s)$. Thus, if the first function of (127) is used as an example,

$$z_{21} = \frac{K(s+5)/D(s)}{(s+2)(s+4)(s+6)/D(s)}. \tag{128}$$

The choices

$$D(s) = (s + \sigma_1)(s + \sigma_2)(s + \sigma_3) \quad \text{or} \quad (s + \sigma_1)(s + \sigma_2)(s + \sigma_3)(s + \sigma_4)$$

with

$$0 \leq \sigma_1 < 2 < \sigma_2 < 4 < \sigma_3 < 6 < \sigma_4$$

are both acceptable. Let us choose $D(s) = (s+1)(s+3)(s+5)$. Then, by comparison with (126), we can write

$$z_{22a} + z_{11b} = \frac{(s+2)(s+4)(s+6)}{(s+1)(s+3)(s+5)} = \frac{\frac{15}{8}}{s+1} + \frac{\frac{3}{4}}{s+3} + \frac{\frac{3}{8}}{s+5} + \frac{16}{5},$$

$$z_{21a}z_{21b} = \frac{K(s+5)}{(s+1)(s+3)(s+5)} = \frac{K}{(s+1)(s+3)} = \left(\frac{\frac{15}{8}}{s+1}\right)\left(\frac{\frac{3}{4}}{s+3}\right). \tag{129}$$

Observe that choosing one of the zeros of $D(s)$ at $s = -5$ has the collateral advantage of a cancellation in $z_{21a}z_{21b}$. This is not a general feature for all cases.

The identification of the individual parameters from the last set of equations is not unique. Choosing the constant K as $\frac{45}{32}$, the following identifications can be made:

$$z_{21a} = \frac{\frac{15}{8}}{s+1}; \qquad z_{21b} = \frac{\frac{3}{4}}{s+3},$$

$$z_{22a} = \frac{\frac{15}{8}}{s+1}; \qquad z_{11b} = \frac{\frac{3}{4}}{s+3} + \frac{\frac{3}{8}}{s+5} + \frac{16}{5}.$$

Each of these two sets is now easily recognized. Two port N_a is easily seen to be a degenerate tee network with a shunt branch only (like the type B section in Fig. 26*b*), as shown in Fig. 29*a*. As for N_b, in addition to a

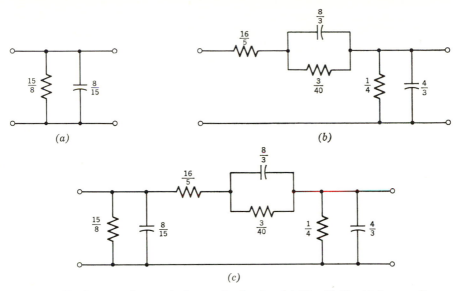

Fig. 29. Realization of numerical example showing: (*a*) N_a ; (*b*) N_b ; (*c*) the overall network with a cascade connection.

common term between z_{21b} and z_{11b}, which will lead to a degenerate tee network, there are private terms in z_{11b} that will appear in series with the input. The network N_b is shown in Fig. 29*b*, and the overall network is shown in Fig. 29*c*. You should verify that this network has the given function as its transfer impedance.

To review the procedure, the first step when presented with a realizable transfer function $P(s)/Q(s)$ (having real and negative poles) is to divide both numerator and denominator by the auxiliary polynomial $D(s)$, so chosen as to make Q/D an *RC* impedance. Then Q/D is expanded in partial fractions. An assignment of terms from this expansion is made to z_{22b} and z_{11b}. This assignment is guided by the decomposition that is to be made of P/D into the product of z_{21a} and z_{21b}. Clearly, the realization is not unique.

CASCADING A NEGATIVE CONVERTER

In the synthesis procedure just discussed, when a function $z_{21} = P(s)/Q(s)$ is given, it is required that a polynomial $D(s)$ be selected to make Q/D an *RC* impedance. Thus the poles of the original function must be real and negative. This is a limited class of functions. If this restriction is removed and the poles are permitted to lie anywhere in the left half-plane, the function will not be realizable as an RC network. The function Q/D will have complex zeros (some), and the residues at the poles will not all be positive. This latter observation provides a clue as to a remedy.

Suppose a negative converter of the voltage-conversion variety is cascaded between the two component two-ports, as shown in Fig. 30.

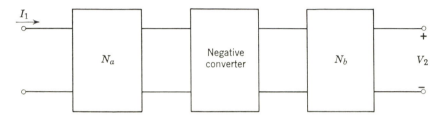

Fig. 30. RC network with a negative converter cascaded between two-ports.

The load on N_a is not simply N_b, as it was before, but N_b preceded by a negative converter. Assuming a conversion ratio of one, the impedance at the input terminals of the negative converter is $-z_{11b}$, and this replaces $+z_{11b}$ in (126). The overall transfer impedance now becomes

$$z_{21}(s) = \frac{z_{21a}\,z_{21b}}{z_{22a} - z_{11b}}. \tag{130}$$

Now suppose a rational function $P(s)/Q(s)$ with complex poles is given. We divide numerator and denominator by an auxiliary polynomial $D(s)$ with an appropriate number of real, negative zeros. Next we expand Q/D in partial fractions. Some of the residues will be positive and some negative. If these are collected together, the result will be

$$z_{21} = \frac{P(s)}{Q(s)} = \frac{P/D}{Q/D} = \frac{P/D}{k_0 + \sum \dfrac{k_{ip}}{s + \sigma_{ip}} - \sum \dfrac{k_{in}}{s + \sigma_{in}}}, \tag{131}$$

where the subscripts p and n stand for "positive" and "negative." The identification of z_{22a} and z_{11b} is now immediate: all those terms in the partial-fraction expansion of Q/D having positive residue belong to z_{22a}; those with negative residue, to z_{11b}. The factors of the numerator polynomial $P(s)$ are assigned to z_{21a} or z_{21b} with the requirement that z_{21a} have no pole that z_{22a} does not have and that z_{21b} have no pole that z_{11b}

does not have. (The term " pole " includes the behavior at infinity. Thus, if z_{21b} is nonzero at infinity, z_{11b} must also be nonzero.)

As an illustration take a fourth-order Chebyshev function with two pairs of $j\omega$-axis transmission zeros thrown in:

$$z_{21} = \frac{P}{Q} = \frac{(s^2 + 4)(s^2 + 9)}{6(s^4 + 2.16s^3 + 3.31s^2 + 2.86s + 1.26)} .$$

The auxiliary polynomial must be at least of the fourth degree; suppose we choose $D(s) = s(s + 1)(s + 2)(s + 3)$. Then

$$z_{22a} - z_{11b} = 6 \frac{s^4 + 2.16s^3 + 3.31s^2 + 2.86s + 1.26}{s(s + 1)(s + 2)(s + 3)}$$

$$= 6 + \frac{1.26}{s} + \frac{22.5}{s + 2} - \left(\frac{1.65}{s + 1} + \frac{45.15}{s + 3} \right),$$

$$z_{21a} z_{21b} = \frac{(s^2 + 4)(s^2 + 9)}{s(s + 1)(s + 2)(s + 3)} .$$

Clearly, each of the quadratic factors in $P(s)$ must be assigned to one of the component two-ports. It is clear that two poles (at 0 and -2) belong to N_a, and two poles (at -1 and -3) belong to N_b. Also, one of the pairs of imaginary zeros must be assigned to N_a; the other pair, to N_b. Thus

$$z_{21a} = \frac{s^2 + 4}{s(s + 2)},$$

$$z_{21b} = \frac{s^2 + 9}{(s + 1)(s + 3)} .$$

Both of these are nonzero at infinity, and so both z_{22a} and z_{11b} must be nonzero at infinity. However, in the partial-fraction expansion of Q/D, no constant term appears among the negative-residue terms. This is remedied by adding and subtracting a constant, say 1. Hence

$$z_{22a} = 7 + \frac{1.26}{s} + \frac{22.5}{s + 2},$$

$$z_{11b} = 1 + \frac{1.65}{s + 1} + \frac{45.15}{s + 3} .$$

Each of the pairs of functions z_{22a}, z_{21a} and z_{11b}, z_{21b} is realizable as an RC two-port. There remains the job of carrying out the realization. Because the transmission zeros are not real, the problem is somewhat more difficult than the preceding example. Nevertheless, routine procedures are available for doing the job, but we shall not pursue the details here. (See Bibliography on Synthesis.)

PARALLEL CONNECTION

As another configuration, consider the network in Fig. 31a. An RC

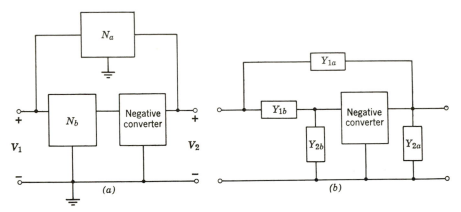

Fig. 31. Parallel-connected two-ports with a negative converter.

two-port N_b is cascaded with a negative converter of the current-conversion variety, and the combination is connected in parallel with another RC two-port N_a. The function of interest is the voltage gain $V_2(s)/V_1(s)$, which is given in terms of the overall y-parameters as $V_2/V_1 = -y_{21}/y_{22}$. For parallel-connected two-ports the overall y-parameters equal the sum of the corresponding component y-parameters. However, the presence of the negative converter reverses the sign of y_{21b} and y_{22b} (assuming a unity conversion ratio), and hence the desired function becomes

$$\frac{V_2}{V_1} = \frac{-(y_{21a} - y_{21b})}{y_{22a} - y_{22b}}, \tag{132}$$

Now when a rational function P/Q is given, the usual auxiliary polynomial $D(s)$ can be used to divide numerator and denominator, and the result manipulated to the form of (132). We shall not carry out this general development but shall replace the general two-ports in Fig. 31a with a specific configuration, as shown in Fig. 31b. This does not limit the class of transfer functions that can be handled.

When the y-parameters of these simple networks are calculated and substituted into (132), the result becomes

$$\frac{V_2}{V_1} = \frac{Y_{1a} - Y_{1b}}{(Y_{1a} + Y_{2a}) - (Y_{1b} + Y_{2b})} = \frac{Y_{1a} - Y_{1b}}{(Y_{1a} - Y_{1b}) + (Y_{2a} - Y_{2b})}. \quad (133)$$

The denominator of this expression contains the numerator added to another term. Now when a transfer function is given as $P(s)/Q(s)$, it must be manipulated to the form of (133). Thus

$$\frac{V_2}{V_1} = \frac{KP(s)}{Q(s)} = \frac{KP(s)}{KP(s) + [Q(s) - KP(s)]} = \frac{\dfrac{KP}{D}}{\dfrac{KP}{D} + \dfrac{Q - KP}{D}}. \quad (134)$$

In this expression a multiplying constant K has been shown explicitly. The polynomial $KP(s)$ was added and subtracted in the denominator, and then the usual auxiliary polynomial $D(s)$ with real, negative zeros was introduced. Comparison of the last two equations leads to

$$Y_{1a} - Y_{1b} = \frac{KP(s)}{D(s)} = k_\infty s + k_0 + \sum \frac{k_{ip}s}{s + \sigma_{ip}} - \sum \frac{k_{in}s}{s + \sigma_{in}}, \quad (135a)$$

$$Y_{2a} - Y_{2b} = \frac{Q(s) - KP(s)}{D(s)} = \hat{k}_\infty s + \hat{k}_0 + \sum \frac{\hat{k}_{ip}s}{s + \hat{\sigma}_{ip}} - \sum \frac{\hat{k}_{in}s}{s + \hat{\sigma}_{in}}. \quad (135b)$$

After a partial-fraction expansion of Y/s is made, the terms with positive residues and those with negative residues are grouped together. The terms with positive residues in the two expressions are identified with Y_{1a} and Y_{2a}, respectively; and those with negative residues, with Y_{1b} and Y_{2b}, respectively. From their manner of construction each of these quantities will be a realizable *RC* admittance function, and a realization can be easily obtained.

As an illustration, take the Chebyshev function treated in the last subsection:

$$\frac{V_2}{V_1} = \frac{KP(s)}{Q(s)} = \frac{K(s^2 + 4)(s^2 + 9)}{s^4 + 2.16s^3 + 3.31s^2 + 2.86s + 1.26}.$$

For convenience K had earlier been taken as $\frac{1}{6}$. The auxiliary polynomial chosen earlier was of degree 4 and had a factor s. An RC admittance cannot have a pole at $s = 0$, and therefore s cannot be a factor of the auxiliary polynomial in the present case. Also, since an RC admittance can have a pole at infinity, let us choose $D(s) = (s + 1)(s + 2)(s + 3)$. Then

$$\frac{KP(s)}{D(s)} = \frac{K(s^2 + 4)(s^2 + 9)}{(s + 1)(s + 2)(s + 3)} = K\left(s + 6 + \frac{52s}{s + 2} - \frac{25s}{s + 1} - \frac{39s}{s + 3}\right),$$

$$\frac{Q(s)}{D(s)} = \frac{s^4 + 2.16s^3 + 3.31s^2 + 2.86s + 1.26}{(s + 1)(s + 2)(s + 3)}$$

$$= s + 0.21 + \frac{3.75s}{s + 2} - \left(\frac{0.275s}{s + 1} + \frac{7.52s}{s + 3}\right).$$

The value of K can be chosen with a view toward simplifying the network. Since KP/D is to be subtracted from Q/D, K can be chosen to cancel one of the terms. Thus, with $K = 3.75/52 = 0.072$, we get

$$Y_{1a} - Y_{1b} = \frac{KP(s)}{D(s)} = 0.072s + 0.432 + \frac{3.75s}{s + 2} - \left(\frac{1.8s}{s + 1} + \frac{2.81s}{s + 3}\right),$$

$$Y_{2a} - Y_{2b} = \frac{Q - KP}{D} = 0.928s + \frac{1.53s}{s + 1} - \left(0.222 + \frac{4.71s}{s + 3}\right).$$

Finally,

$$Y_{1a} = 0.072s + 0.432 + \frac{3.75s}{s + 2},$$

$$Y_{2a} = 0.928s + \frac{1.53s}{s + 1},$$

$$Y_{1b} = \frac{1.8s}{s + 1} + \frac{2.81s}{s + 3},$$

$$Y_{2b} = 0.222 + \frac{4.71s}{s + 3}.$$

The complete network is shown in Fig. 32.

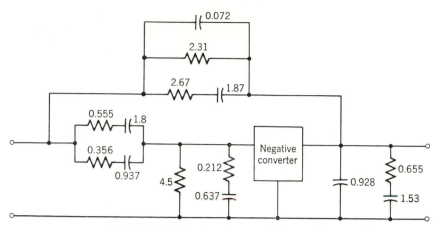

Fig. 32.   Realization of numerical example.

For the structure under consideration, it is necessary to realize only two-terminal *RC* networks. This circumstance makes the design effort relatively small.

THE *RC*-AMPLIFIER CONFIGURATION

In the preceding subsections the active device used in conjunction with *RC* networks has been a negative converter. Obviously, other network configurations can be found that utilize different active devices. The simplest possibility would be an amplifier in a feedback arrangement. The amplifier can be represented in its simplest form as a controlled source (voltage or current). Although many different configurations are possible, we shall treat only one and suggest others as problems.

Consider the configuration in Fig. 33. This contains two voltage ampli-

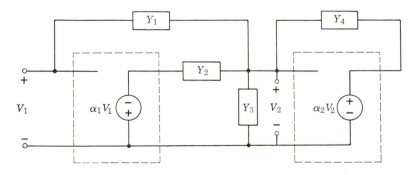

Fig. 33.   Feedback-amplifier configuration.

fiers (voltage-controlled voltage sources), each having a different polarity. Note that the overall output voltage is taken as the input voltage of the second amplifier. An evaluation of the voltage gain function for this configuration leads to

$$\frac{V_2}{V_1} = \frac{Y_1 - \alpha_1 Y_2}{Y_1 + Y_2 + Y_3 - (\alpha_2 - 1) Y_4}. \tag{136}$$

We shall assume that the second amplifier gain α_2 is greater than 1.

Now, given a rational function $P(s)/Q(s)$ as the voltage gain, we divide numerator and denominator by the usual auxiliary polynomial $D(s)$ having an appropriate degree (equal to, or one less than, the degree of Q or P, whichever is greater) and distinct real, negative zeros. Each of the functions P/sD and Q/sD is expanded in partial fractions, and the positive-residue and negative-residue terms are collected. Finally, identification of the admittances in (136) is made by comparison with this expression. Thus

$$\frac{V_2}{V_1} = \frac{P(s)}{Q(s)} = \frac{P(s)/D(s)}{Q(s)/D(s)},$$

$$Y_1 - \alpha_1 Y_2 = \frac{P(s)}{D(s)} = k_\infty s + k_0 + \sum \frac{k_{ip} s}{s + \sigma_{ip}} - \sum \frac{k_{in} s}{s + \sigma_{in}}, \tag{137a}$$

$$Y_1 + Y_2 + Y_3 - (\alpha_2 - 1) Y_4 = \frac{Q(s)}{D(s)} = \hat{k}_\infty s + \hat{k}_0 + \sum \frac{\hat{k}_{ip} s}{s + \hat{\sigma}_{ip}} - \sum \frac{\hat{k}_{in} s}{s + \hat{\sigma}_{in}}. \tag{137b}$$

From the first of these we get

$$Y_1(s) = k_\infty s + k_0 + \sum \frac{k_{ip} s}{s + \sigma_{ip}}, \tag{138a}$$

$$Y_2(s) = \frac{1}{\alpha_1} \sum \frac{k_{in} s}{s + \sigma_{in}}. \tag{138b}$$

When these are inserted into (137b), Y_3 and Y_4 can be identified. Thus

$$Y_3 - (\alpha_2 - 1) Y_4 = (\hat{k}_\infty - k_\infty)s + (\hat{k}_0 - k_0) + \sum \frac{\hat{k}_{ip} s}{s + \sigma_{ip}}.$$

$$- \sum \left(\frac{\hat{k}_{in} s}{s + \hat{\sigma}_{in}} + \frac{k_{ip} s}{s + \sigma_{ip}} + \frac{1}{\alpha_1} \frac{k_{in} s}{s + \sigma_{in}} \right). \tag{139}$$

General expressions for Y_3 and Y_4 cannot be written, because there will be uncertainties in the right-hand side; for example, the constant term $\hat{k}_0 - k_0$ may be positive or negative. In the former case, the constant term will go to Y_3; in the latter case, to Y_4. Furthermore, the set of poles $-\hat{\sigma}_{ip}$ is contained in the set $\{-\sigma_{ip}, -\sigma_{in}\}$, since the latter is the set of all zeros of $D(s)$. Hence there will be some cancellations of terms, with uncertainty in the general case as to the resultant sign. These uncertainties disappear, of course, when dealing with a specific case. This will be clarified with an example below. In any case, it is required only to realize two-terminal *RC* networks. Also, it will be necessary to specify the amplifier gains.

In an actual realization it would be useful for elements to appear at appropriate places at the input and output of the amplifier to account for parameters of an actual amplifier other than the gain. An equivalent circuit of an amplifier is shown in Fig. 34. The actual amplifier that is used to realize the controlled source will have input, output, and feedback impedances, as represented by G_a, G_b, and G_c in Fig. 34. Look back at

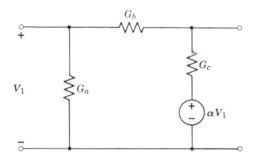

Fig. 34.   Amplifier equivalent circuit.

Fig. 33. Because of the position of Y_2, it is possible for this branch to account for the output impedance of the first amplifier. Similarly, Y_3 and Y_4 can be made to account for the input, output and feedback impedances of the second amplifier. It is therefore useful to have a constant term in the expansion of Y_3. If it does not appear there by virtue of the fact that $k_0 \geq \hat{k}_0$ in (139), then an appropriate constant can be added and subtracted on the right side of (139).

As an illustration, take the following all-pass function:

$$\frac{V_2}{V_1} = \frac{s^2 - 2s + 2}{s^2 + 2s + 2}.$$

In the present case, the degree of $D(s)$ need be no more than 1. Let $D(s) = s + 1$. Then

$$Y_1 - \alpha_1 Y_2 = \frac{s^2 - 2s + 2}{s + 1} = s + 2 - \frac{5s}{s + 1},$$

from which

$$Y_1 = s + 2, \qquad Y_2 = \frac{1}{\alpha_1} \frac{5s}{s + 1},$$

$$Y_3 - (\alpha_2 - 1) Y_4 = \frac{s^2 + 2s + 2}{s + 1} - Y_1 - Y_2 = -\frac{(\alpha_1 + 5)s}{\alpha_1(s + 1)}.$$

We observe that there is no constant term on the right side. Hence, in order to account for the input impedance of the second amplifier, we add and subtract a constant G. Identifications for Y_3 and Y_4 become

$$Y_3 = G, \qquad Y_4 = \frac{1}{\alpha_2 - 1} \left(G + \frac{(\alpha_1 + 5)s}{\alpha_1(s + 1)} \right).$$

For concreteness let $\alpha_1 = 5$, $\alpha_2 = 2$, and $G = 1$; Y_4 then becomes

$$Y_4 = 1 + \frac{2s}{s + 1} = \frac{3s + 1}{s + 1} = \frac{1}{\frac{1}{3} + \frac{\frac{2}{3}}{3s + 1}}.$$

Thus Y_4 is realized as a $\frac{1}{3}$-unit resistance in series with a parallel RC branch. The complete realization is shown in Fig. 35. Note that the realization does not account for the input impedance of the first amplifier. If the input impedance of the actual amplifier is not to have any influence, then the network should be driven by a voltage source, or at least a source whose output impedance is small in magnitude compared with the input impedance of the amplifier.

To summarize this section: we have illustrated the classical pattern of formal network synthesis in terms of RC passive and active two-ports. The first step is to seek a specific structure incorporating components of the class of networks under consideration and to write the appropriate transfer function in terms of the parameters of the components. Then a rational function is manipulated into the same form, after introducing an auxiliary polynomial by which the numerator and denominator are divided. This

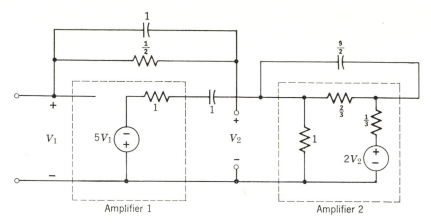

Fig. 35.   Realization of RC amplifier.

permits an identification of the parameters of the component networks, care being taken to ensure that they are realizable. The final step is to realize the components.

PROBLEMS

Elementary Matrices.   Elementary matrices are of three types, corresponding to the three types of elementary transformation. Use the following notation:

Type 1: $^rE_{ij}$ is the elementary matrix which interchanges rows j and i.
        $^cE_{ij}$ is the elementary matrix which interchanges columns j and i.

Type 2: $^rE_{i+j}$ is the elementary matrix which adds row j to row i.
        $^cE_{i+j}$ is the elementary matrix which adds column j to column i.

Type 3: $^rE_{\alpha i}$ is the elementary matrix which multiplies row i by α.
        $^cE_{\alpha i}$ is the elementary matrix which multiplies column i by α.

1. Construct the single matrix that will perform the following operations on a matrix **A**:

(*a*) Interchange the third and second rows after the third row is multiplied by 2. Order of **A**: (3,5).

(*b*) Add the first column to the third after multiplying the first by 3 and the third by 4. Order of **A** $= (2,3)$.

(*c*) Add the second row to the first after multiplying the first by -1. Order of **A**: (3,3).

2. Find the single matrix which will perform the following operations on a matrix A:

(a) Multiply the first row by k and add it to the third row, then multiply the second row by 3 and interchange this row with the first. Order of A: (4,4).

(b) Add the second column to the first, then multiply the second column by -2 and add it to the third, then interchange the second and third columns. Order of $A = (4,3)$.

3. Write the single matrix which performs each of the following operations on the rows of a matrix A:

(a) Multiply row 3 by -2, then add to row 1, then interchange row 1 with row 3. Order of A: (3,3).

(b) Interchange row 4 with row 2; then multiply row 1 by 5 and add to row 3; then subtract row 3 from row 2. Order of A: (4,4).

4. Repeat Problem 3 if the operations are performed on the columns.

5. Let a matrix A of order (m,n) with $m \leq n$ have rank r. Show that the product of any elementary matrix with A will have the same rank, r. Do this by multiplying A by each type of elementary matrix and determining the effect of this on the determinant of appropriate sub-matrices of A.

6. Show that the inverse of an elementary matrix is an elementary matrix of the same type.

7. Prove that a nonsingular matrix A can always be written as the product of a finite number of elementary matrices.

8. Prove that two matrices A and B of order (m,n) are equivalent if and only if they have the same rank.

9. For each matrix A given below find matrices P and Q such that PAQ is the normal form of A.

$$(a) \ A = \begin{bmatrix} 1 & 2 & -1 & 4 \\ 2 & 4 & 3 & 5 \\ -1 & -2 & 6 & -7 \end{bmatrix} \qquad (b) \ A = \begin{bmatrix} 3 & -3 & 1 & 2 \\ 2 & 1 & -3 & -6 \\ 1 & 1 & 1 & 2 \end{bmatrix}$$

$$(c) \ A = \begin{bmatrix} 1 & 2 & 1 & 0 \\ 1 & 2 & 3 & 2 \\ 2 & -1 & 2 & 5 \\ 3 & 6 & 5 & 2 \\ -1 & 3 & 1 & -3 \end{bmatrix}.$$

10. Reduce the following matrices to normal form:

(a)
$$\begin{bmatrix} 2 & 0 & 1 & 1 \\ 3 & 5 & -2 & 0 \\ 3 & 2 & 0 & 1 \\ 4 & 0 & 2 & 2 \end{bmatrix}$$

(b)
$$\begin{bmatrix} 6 & 8 & -2 & 4 \\ 3 & 4 & -1 & 2 \\ 12 & 16 & -4 & 8 \\ -6 & -8 & 2 & -4 \end{bmatrix}$$

(c)
$$\begin{bmatrix} 1 & 0 & 2 & -1 \\ 0 & 6 & 2 & 2 \\ 1 & -6 & 0 & -3 \\ -2 & 0 & -4 & 2 \end{bmatrix}$$

(d)
$$\begin{bmatrix} 1 & 1 & 1 & 1 \\ -1 & 1 & 1 & 1 \\ -1 & -1 & 1 & 1 \\ -1 & -1 & -1 & 1 \end{bmatrix}.$$

11. Reduce the following matrices to canonic form using the Lagrange reduction procedure:

(a) $\mathbf{A} = \begin{bmatrix} 1 & 1 & 1 \\ 1 & 0 & 0 \\ 1 & 0 & -1 \end{bmatrix}$

(b) $\mathbf{A} = \frac{1}{4} \begin{bmatrix} 1 & -1 & 3 & -7 \\ -1 & 5 & 1 & 3 \\ 3 & 1 & 29 & -49 \\ -7 & 3 & -49 & 113 \end{bmatrix}$

(c) $\mathbf{A} = \frac{1}{4} \begin{bmatrix} 1 & -1 & 3 & -7 \\ -1 & 5 & 1 & 3 \\ 3 & 1 & -3 & 7 \\ -7 & 3 & 7 & 15 \end{bmatrix}.$

12. Determine whether the following matrices are positive definite:

$$A = \begin{bmatrix} 2 & 1 \\ 1 & 0 \end{bmatrix}$$

$$B = \begin{bmatrix} 1 & 0 & \frac{1}{2} & \frac{1}{2} \\ 0 & 2 & 1 & -1 \\ \frac{1}{2} & 1 & 4 & 1 \\ \frac{1}{2} & -1 & 1 & 5 \end{bmatrix}$$

$$C = \begin{bmatrix} 14 & -2 & 4 & -3 \\ -2 & 5 & 1 & 0 \\ 4 & 1 & 5 & -2 \\ -3 & 0 & -2 & 6 \end{bmatrix}$$

$$D = \begin{bmatrix} 4 & 1 \\ 1 & -1 \end{bmatrix} \qquad E = \begin{bmatrix} 1 & -1 & 1 \\ -1 & 2 & 0 \\ 1 & 0 & 3 \end{bmatrix}.$$

13. Prove that a symmetric matrix $\mathbf{A}$ is positive definite if and only if $\mathbf{A}^{-1}$ is symmetric and positive definite.

14. (*a*) Prove that a real symmetric matrix $\mathbf{A}$ is positive definite if and only if there exists a nonsingular real matrix $\mathbf{B}$ such that $\mathbf{A} = \mathbf{BB}'$. (*b*) Prove that a real symmetric matrix $\mathbf{A}$ of order n and rank r is positive semidefinite if and only if there exists a matrix $\mathbf{B}$ of rank r such that $\mathbf{A} = \mathbf{B}'\mathbf{B}$.

15. Let $\mathbf{A}$ be a real symmetric matrix of order n and $\mathbf{B}$ a real matrix of order (r,n) and rank r. Prove that if $\mathbf{A}$ is positive definite, then $\mathbf{BAB}'$ is positive definite.

16. Let $\mathbf{A}$ be a real nonsymmetric matrix and $\mathbf{x}$ a complex vector. $\mathbf{A}$ can be written as the sum of its symmetric and skew symmetric parts, $\mathbf{A}_s$ and $\mathbf{A}_{ss}$, respectively, where

$$\mathbf{A}_s = \tfrac{1}{2}(\mathbf{A} + \mathbf{A}'),$$

$$\mathbf{A}_{ss} = \tfrac{1}{2}(\mathbf{A} - \mathbf{A}').$$

Show that

$$(a)\ \ Re(\mathbf{x}'\mathbf{Ax}) = \mathbf{x}'A_s\mathbf{x},$$

$$(b)\ \ Im(\mathbf{x}'\mathbf{Ax}) = \mathbf{x}'A_{ss}\mathbf{x}.$$

17. In the network of Fig. P17 all three inductances are mutually coupled. Suppose it is possible to have the given inductance matrix in which all mutual inductances are less than unity. Verify that this inductance matrix is not positive definite or semidefinite. Setting $R_1 = R_2 = R_3 = 1$ for convenience, compute the natural frequencies.

$$\mathbf{L} = \begin{bmatrix} 1 & .9 & .2 \\ .9 & 1 & .9 \\ .2 & .9 & 1 \end{bmatrix}.$$

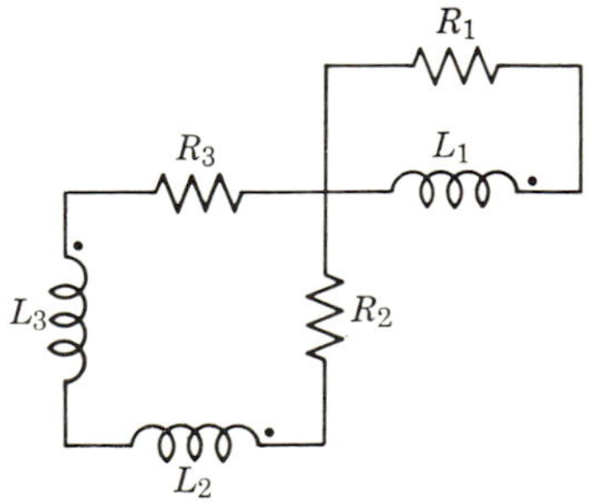

Fig. P17

18. Let $\mathbf{A} = [a_{ij}]$. The *ascending principal cofactors* of a square matrix $\mathbf{A}$ are defined as:

$$p_1 = a_{11}, \quad p_2 = \begin{vmatrix} a_{11} & a_{12} \\ a_{21} & a_{22} \end{vmatrix}, \quad p_3 = \begin{vmatrix} a_{11} & a_{12} & a_{13} \\ a_{21} & a_{22} & a_{23} \\ a_{31} & a_{32} & a_{33} \end{vmatrix}, \dots p_n = \det \mathbf{A}.$$

Prove the following stronger condition for positive definiteness than Theorem 4 in the text. *A real matrix symmetric $\mathbf{A}$ is positive definite if and only if all ascending principal cofactors are positive.*

19. (*a*) Find the rank of each of the following matrices. (They will be different.)

$$\mathbf{A} = \begin{bmatrix} 1 & -6 & 4 \\ 0 & 0 & 0 \\ 0 & 0 & 2 \end{bmatrix} \qquad \mathbf{B} = \begin{bmatrix} 1 & -3 & 2 \\ -3 & 3 & 0 \\ 2 & 0 & 2 \end{bmatrix}.$$

(*b*) Verify that the quadratic forms $\mathbf{X'AX}$ and $\mathbf{X'BX}$ are equal.

20. Let $\mathbf{A}$ be a real, symmetric, positive definite matrix. Prove that $\mathbf{A}^2$ is also positive definite.

21. Let $F(s)$ be a positive real function. Prove the angle property, namely that $|\arg F(s)| \le |\arg s|$ for $0 \le |\arg s| \le \pi/2$. Hint: make the following bilinear transformations:

$$p = \frac{s-1}{s+1}, \quad W = \frac{F-1}{F+1}.$$

Observe how the right half F and s-planes are mapped into the W and p planes. Show that $W(p)$ satisfies Schwarz's lemma (see Appendix 2), then substitute the above transformations in the lemma.

22. It is claimed that the functions below are not positive real for $n > N_0$. Verify this claim and determine the value of N_0 for each function.

(*a*) $F(s) = \left(\dfrac{s+1}{s+2}\right)^n$,   (*b*) $F(s) = \left(\dfrac{s+10}{s+11}\right)^n$.

23. Let $F(s)$ be a reactance function and let $F(j\omega)=jX(\omega)$. Prove that

$$\frac{dX(\omega)}{d\omega} \geq \left|\frac{X}{\omega}\right|.$$

(Hint: use energy functions.)

24. Prove that a bilinear transformation of a bounded real function is positive real.

25. Let

$$Z(s) = \frac{s^2 + s + 1}{s^2 + s + 4}.$$

The even part of this positive real function has a zero (double) on the $j\omega$ axis. Hence, this function maps this point on the $j\omega$ axis to a point on the jX axis in the Z-plane. Verify that the derivative dZ/ds at this point is real and positive.

26. Let $Z(s) = (m_1 + n_1)/(m_2 + n_2)$ be a positive real function. Prove that

$$P(s) = m_1 + an_1 + bm_2 + cn_2$$

is a Hurwitz polynomial, where a, b, and c are positive constants.

27. Let $P(s)$ be a Hurwitz polynomial. Prove that $F(s)$ is a pr function, where

$$F(s) = \frac{1}{P(s)} \frac{dP(s)}{ds}.$$

28. Let $Z(s) = P(s)/Q(s)$ be a positive real function. Prove that the following function is also positive real:

$$Z_1(s) = \frac{dP(s)/ds}{dQ(s)/ds}.$$

29. Let $Y_{RC} = \frac{(s+1)(s+3)(s+8)}{(s+2)(s+4)(s+10)}.$

To obtain a realization for Y_{RC} it is suggested that the first branch across the terminals be a resistor of $\frac{1}{2}$-ohm. The remaining admittance will be $Y_{RC} - 2$. Is this a pr function?

30. Using the partial fraction expansions in (99) and (100), show that

(a) $\text{Im}\,[Z_{RC}(j\omega)] \lessgtr 0$ for $\omega \gtrless 0$, respectively.
(b) $\text{Im}\,[Y_{RC}(j\omega)] \gtrless 0$ for $\omega \gtrless 0$, respectively.
(c) $\text{Re}\,[Z_{RC}(j\omega)]$ is a monotonically *decreasing* function of ω for $\omega \geq 0$.
(d) $\text{Re}\,[Y_{RC}(j\omega)]$ is a monotonically *increasing* function of ω for $\omega \geq 0$.

31. From the result of Problem 30, show that:

(a) $\text{Re}\,[Z_{RC}(0)] > \text{Re}\,[Z_{RC}(\infty)]$ or $Z_{RC}(0) > Z_{RC}(\infty)$,
(b) $\text{Re}\,[Y_{RC}(0)] < \text{Re}\,[Y_{RC}(\infty)]$ or $Y_{RC}(0) < Y_{RC}(\infty)$.

32. The results of Problem 30 can be described as a mapping of the upper half s-plane into the lower half Z-plane and a mapping of the lower half s-plane to the upper half Z-plane. Use this to obtain an alternative proof that $dZ_{RC}(s)/ds$ is negative on the real axis.

33. Using the approach of Problem 32 show that, for the impedance $Z_{RL}(s)$ of a passive RL network, $dZ_{RL}(s)/ds$ is real and positive on the real axis.

34. Let $Z_1(s)$ and $Z_2(s)$ be RC impedance functions. Prove that Z_1/Z_2 is a positive real function.

35. Find one or more realizations of the following reactance functions:

(a) $Z(s) = \dfrac{10s(s^2+1)(s^2+4)}{(s^2+1/4)(s^2+2)}$,   (b) $Z(s) = \dfrac{2(s^2+1)(s^2+25/4)}{s(s^2+9/4)(s^2+9)}$.

36. The following rational function is given:

$$Y(s) = \frac{K(s^2+1)(s^2+\alpha^2)}{s(s^2+4)}.$$

It is also known that

(a) $\left.\dfrac{dY}{ds}\right|_{s=j1} = \tfrac{1}{2}$,   (b) $\left.\dfrac{dZ}{ds}\right|_{s=j2} = 2.$

Find a realization as a ladder network and as one of the Foster forms.

37. Let $P(s)$ be a polynomial whose zeros are all imaginary. Prove that

$$F(s) = \frac{P(s)}{dP/ds}$$

is a reactance function.

38. The admittance of a passive RC network has zeros at $s = 0$, -1 and -3. At each of these points the slope of $Y(\sigma)$ equals 2. For large values of

$s = \sigma$, $Y(\sigma)$ approaches 4. Find a ladder network and a network in one of the Foster forms to realize this function.

39. Let $Y_{RC}(s)$ be the admittance of a passive RC network. Prove that the residues of Y_{RC} are negative at all poles except the pole at infinity.

40. Let $P(s)$ be a polynomial whose zeros are all real and negative. Prove that each of the following two functions is the admittance of a passive RC network:

$$F_1(s) = \frac{P(s)}{dP/ds}, \qquad F_2(s) = \frac{s\, dP/ds}{P(s)}.$$

41. Let $P(s)$ be a polynomial whose zeros are all real and negative. If K is a real, positive constant, prove that all the zeros of $F(s)$ are real and negative, where

$$F(s) = \frac{s\, dP}{ds} + KP$$

42. (a) Let $F(s) = P(s)/Q(s)$ be an RC impedance function. Prove that $F(s)$ is also an RC impedance function, where

$$F(s) = \frac{dP(s)/ds}{dQ(s)/ds}$$

(b) Repeat, replacing the word admittance for impedance.

43. It is desired to prove Theorem 16 on Hurwitz polynomials. In the first place, to prove that if $P(s) = m(s) + n(s)$ is Hurwitz, then m/n is a reactance function, note that the factors of a Hurwitz polynomial are of the form $(s^2 + as + b)$ or $(s + c)$, where a, b, and c are real and positive. Write:

$$P(s) = (s^2 + as + b) P_1(s) = (s^2 + as + b)(m_1 + n_1)$$
$$= [(s^2 + b)m_1 + asn_1] + [(s^2 + b)n_1 + asm_1].$$

Then, the ratio of the even and odd parts is:

$$\frac{m}{n} = \frac{(s^2 + b)m_1 + asn_1}{(s^2 + b)n_1 + asm_1} = \left(\frac{n_1}{m_1} + \frac{as}{s^2 + b}\right)^{-1} + \left(\frac{s^2 + b}{as} + \frac{m_1}{n_1}\right)^{-1}$$

Show that m/n will be a reactance function provided m_1/n_1 is also. Repeat this procedure, with $P_1 = m_1 + n_1$ until all the factors of $P(s)$, both quadratic and linear, are exhausted.

In the second place, it must be proved that, if $P(s) = m(s) + n(s)$ and m/n is a reactance function then $P(s)$ is a Hurwitz polynomial (except for a possible even factor). The zeros of $P(s)$ occur when

$$P(s) = m + n = n\left(\frac{m}{n} + 1\right) = 0.$$

Prove that, since m/n is a reactance function, $m/n = -1$ cannot occur in the right half s-plane.

44. Verify whether the following polynomials are Hurwitz:

(a) $P(s) = s^4 + 7s^3 + 6s^2 + 4s + 1$;
(b) $P(s) = 2s^4 + 4s^3 + 7s^2 + 7s + 3$;
(c) $P(s) = s^7 + 3s^6 + 6s^5 + 6s^4 + 6s^3 + 6s^2 + 5s + 3$;
(d) $P(s) = 2s^5 + 9s^4 + 16s^3 + 15s^2 + 7s + 2$.

45. Suppose that a network having a driving-point impedance $Z_1 = F(s)$ is given. It is desired to find a second network whose driving-point admittance Y_2 is equal to $F(s)$. Such networks are called inverse. Discuss the conditions under which the inverse of a given network may be found by the method of duality discussed in Chapter 2.

46. For each of the one-ports shown in Fig. P46 find the inverse network.

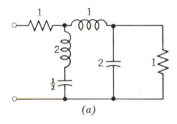 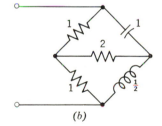

(a)

(b)

Fig. P46

Verify that the driving-point admittance of the inverse is the same as the driving-point impedance of the given network.

47. Show that the residue condition is satisfied with the "equals" sign (compact poles) at all the finite, nonzero poles of the y parameters of a lossless two-port terminated in a unit resistance, as given in Table 1.

48. Show that the symmetric matrix of rational functions

$$\begin{bmatrix} z_{11}(s) & z_{12}(s) \\ z_{12}(s) & z_{22}(s) \end{bmatrix}$$

is a positive real matrix if and only if the matrix of real parts

$$\begin{bmatrix} r_{11}(\sigma, \omega) & r_{12}(\sigma, \omega) \\ r_{12}(\sigma, \omega) & r_{22}(\sigma, \omega) \end{bmatrix}$$

is positive definite or semidefinite in $\sigma \geq 0$.

49. Let $y_{11} = y_{22}$ and $y_{21} = y_{12}$ be two real rational functions. Suppose the lattice shown in Fig. P49 is to have these functions as its short-circuit

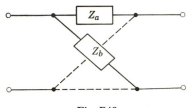

Fig. P49

parameters. Show that the branch impedances Z_a and Z_b will be positive real if

(a) y_{11} is positive real;
(b) the real part condition $(\text{Re } y_{11})^2 - (\text{Re } y_{21})^2 \geq 0$ is satisfied for $\text{Re } s \geq 0$.

If in (b) it is only known that the real part condition is satisfied on the $j\omega$-axis, what additional conditions must be placed on the given functions y_{11} and y_{12} before the theorem will again be true?

50. Show that at a zero of z_{11} on the $j\omega$-axis, z_{21} is imaginary. Hence show that any $j\omega$-axis poles of the open-circuit voltage gain

$$g_{21}(s) = \frac{V_2(s)}{V_1(s)}\bigg|_{I_2=0}$$

are simple and with imaginary residues. Repeat for the short-circuit current gain $h_{21}(s)$.

51. Figure P51a shows a lossless two-port terminated in a unit resistance at one port. In Fig. P51b the two ports have been interchanged. In the first case, the impedance is

$$Z_1(s) = \frac{m_1 + n_1}{m_2 + n_2}.$$

(a) Find expressions for $Z_2(s)$ appropriate to the Darlington case A and case B.

(b) Observe the relationship between Z_1 and Z_2 when (1) $m_1 = m_2$, and (2) $n_1 = n_2$.

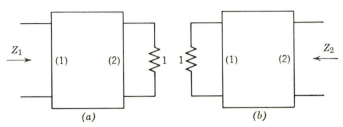

Fig. P51

52. Find the z or y parameters of the lossless two-port which, when terminated in a resistance, realizes the following impedance functions. Specify whether the Darlington case A or case B applies.

(a) $Z(s) = \dfrac{s^2 + s + 4}{s^2 + s + 1}$,

(b) $Z(s) = \dfrac{s^3 + s^2 + 2s + 1}{s^2 + s + 1}$,

(c) $Z(s) = \dfrac{s^2 + 5s + 1}{s^3 + 5s^2 + 2s + 1}$,

(d) $Z(s) = \dfrac{s^4 + 2s^3 + 3s^2 + s + 1}{s^4 + s^3 + 3s^2 + 2s + 1}$,

(e) $Z(s) = \dfrac{s^3 + 2s^2 + 5s}{s^3 + 3s^2 + 4s + 35/8}$,

(f) $Z(s) = \dfrac{s^4 + s^3 + 3s^2 + 2s + 6}{s^3 + 3s^2 + 7s}$.

53. Let $Z(s)$ be the impedance of a lossless two-port terminated in a 1 unit resistance. Show that:

(a) if $Z(s)$ has a pole at $s = 0$, the z parameters of the lossless two-port have a noncompact pole there;

(b) if $Z(s)$ has a zero at $s = 0$, the y parameters have a noncompact pole there;

(c) if $Z(s)$ has neither a pole nor a zero at $s = 0$, then both the z's and the y's will have a compact pole there;

(d) the conclusions of the previous parts are true if the point $s = 0$ is replaced by the point $s = \infty$.

54. Darlington's theorem shows that $Z(s)$ can be realized as a reactive two-port terminated in a resistance at one-port. However, the two-port will generally require transformers in the realization. An alternative which

does not is given in the following. Let a positive real function be written as:

$$Z(s) = \frac{m_1(s) + n_1(s)}{m_2(s) + n_2(s)}$$

where m_1, m_2 are even polynomials and n_1, n_2 are odd. The zeros of the polynomial $m_1 m_2 - n_1 n_2$ have quadrantal symmetry.

Let $H(s)$ be a Hurwitz polynomial formed from the left-half-plane zeros of $m_1 m_2 - n_1 n_2$; that is

$$H(s)H(-s) = m_1 m_2 - n_1 n_2$$

and

$$H(s) = m(s) + n(s).$$

(*a*) Prove that

$$\mathbf{Z}_{oc} = \begin{bmatrix} \dfrac{n^2 + m_1 m_2}{m_2 n_2} & \sqrt{R_2}\,\dfrac{m}{n_2} & \sqrt{R_3}\,\dfrac{n}{m_2} \\[3ex] \sqrt{R_2}\,\dfrac{m}{n_2} & R_2\,\dfrac{m_2}{n_2} & 0 \\[3ex] \sqrt{R_3}\,\dfrac{n}{m_2} & 0 & R_3\,\dfrac{n_2}{m_2} \end{bmatrix}$$

is the open circuit impedance matrix of a reactive three-port which, when terminated in resistances at two of its ports, as shown in Fig. P54 realizes $Z(s)$ as the impedance at the remaining port. (The major element of the proof is in proving the residue condition.)

(*b*) Prove that

$$\mathbf{Y}_{sc} = \begin{bmatrix} \dfrac{n^2 + m_1 m_2}{m_1 n_1} & \dfrac{m}{\sqrt{R_2}\,n_1} & \dfrac{n}{\sqrt{R_3}\,m_1} \\[3ex] \dfrac{m}{\sqrt{R_2}\,n_1} & \dfrac{m_1}{R_2 n_1} & 0 \\[3ex] \dfrac{n}{\sqrt{R_3}\,m_1} & 0 & \dfrac{n_1}{R_3 m_1} \end{bmatrix}$$

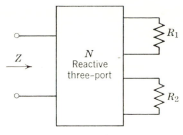

Fig. P54

is the short-circuit admittance matrix of the above reactive three-port. (c) Realize the following impedances in the above form. (The Darlington realization requires transformers.)

$$\text{(1)} \ Z(s) = \frac{s^3 + s^2 + 2s + 1}{s^3 + 2s^2 + 3s/2 + 2}, \qquad \text{(2)} \ Z(s) = \frac{3s^3 + 5s^2 + 4s + 6}{s^3 + 2s^2 + 2s + 1}.$$

55. Prove that the inverse of a positive real matrix is positive real.

56. Let $Q(s)$ be a Hurwitz polynomial whose zeros are all complex. Let $P(p)$ be a *Hurwitz* polynomial *associated with* $Q(s)$ in the following way, after the transformation $s = p^2$:

$$Q(s) = Q(p^2) = P(p)P(-p).$$

That is, $P(p)$ is the polynomial containing all the left-half-plane zeros of $Q(p^2)$.

Write
$$P(p) = A(p^2) + pB(p^2)$$

where A and B are even polynomials, as emphasized by the manner of writing their arguments.

(a) Prove that $Q(s)$ can always be written as the difference of two polynomials in the form

$$Q(s) = A^2(s) - sB^2(s)$$

where $A(s)$ and $B(s)$ have real negative zeros only and where A/sB and B/A are both RC impedance functions. This form of $Q(s)$ is called the *Horowitz decomposition* of Q (not to be confused with Hurwitz).

(b) Apply the result of (a) to the following polynomials:

(1) $Q(s) = (s^2 + 2s + 2)(s^2 + s + 10)$,
(2) $Q(s) = (s^2 + s + 1)(s^2 + 2s + 4)$,
(3) Fourth order Chebyshev, with ripple $\delta = 0.1$,
(4) Fourth order Butterworth,
(5) $Q(s) = (s^2 + s + 2)(s^2 + 2s + 5)(s^2 + s + 8)$,
(6) $Q(s) = (s^2 + s + 1)(s^2 + 2s + 3)(s^2 + 3s + 5)$,
(7) Sixth order Chebyshev, with ripple $\delta = 0.1$,
(8) Sixth order Butterworth.

57. Let Q_1 and Q_2 be polynomials of the same degree, with associated Hurwitz polynomials P_1 and P_2, and let their Horowitz decompositions be

$$Q_1 = A_1{}^2 - sB_1{}^2$$

$$Q_2 = A_2{}^2 - sB_2{}^2$$

(a) Prove that, if P_1/P_2 is positive real, then

$$A_1{}^2 - sB_2{}^2 \quad \text{and} \quad A_2{}^2 - sB_1{}^2$$

are also Horowitz decompositions; that is,

$$\frac{A_1}{sB_2}, \frac{B_2}{A_1}, \frac{A_2}{sB_1}, \frac{B_1}{A_2}$$

are all RC impedance functions.

(b) Apply this result to pairs of applicable polynomials from Problem 56.

58. Let $Q_1(s) = A_1{}^2(s) - sB_1{}^2(s)$ be the Horowitz decomposition of a polynomial $Q_1(s)$. Let $Q_2(s)$ be another polynomial of degree no greater than that of Q_1 and with no real zeros.

(a) Prove that Q_2 can always be written as

$$Q_2(s) = A_1(s)A_2(s) - sB_1(s)B_2(s)$$

where

$$\frac{A_1}{sB_2}, \frac{B_2}{A_1}, \frac{A_2}{sB_1} \quad \text{and} \quad \frac{B_1}{A_2}$$

are all RC impedance functions.

[With the transformation $s = p^2$, $Q_2(p^2)/Q_1(p^2)$ is the even part of a pr function. From this form the pr function itself, say $F(p)$, and write numerator and denominator in terms of even and odd polynomials. Finally, from $F(p)$ form the even part; its numerator should be $Q_2(p^2)$.]

(b) Apply this result to the appropriate polynomials in Problem 56.

59. If in Problem 58 the polynomial Q_2 is allowed to have real negative zeros as well, show that Q_2 can be decomposed into

$$Q_2 = A_1 B_2 - A_2 B_1$$

where

$$\frac{A_1}{sB_2}, \frac{B_2}{A_1}, \frac{A_2}{sB_1}, \frac{B_1}{A_2}, \frac{A_1}{sB_1}, \frac{B_1}{A_1}, \frac{A_2}{sB_2}, \frac{B_2}{A_2}$$

are all RC impedance functions.

[This time $pQ_2(p^2)/Q_1(p^2)$ is the odd part of a pr function. Construct the pr function from its odd part, adding enough of a constant so that the result has no zeros on the $j\omega$ axis. Reconstruct the odd part from the pr function; its numerator should be $pQ_2(p^2)$.]

60. Let $F(s) = \dfrac{Q_1(s)}{Q_2(s)}$ be any real rational function.

(a) Prove that it can be decomposed into

$$F(s) = \frac{B_1(s)A_2(s) - A_1(s)B_2(s)}{A_1(s)A_2(s) - sB_1(s)B_2(s)}$$

where

$$\frac{A_1}{sB_1}, \frac{B_1}{A_1}, \frac{A_2}{sB_2} \text{ and } \frac{B_2}{A_2}$$

are all RC impedance functions.

(b) Apply this result to the following functions:

(1) $F(s) = \dfrac{s^2 + s + 1}{(s^2 + 2s + 2)(s^2 + 3s + 5)}$,

(2) $F(s) = \dfrac{(s^2 + s + 1)(s + 2)(s^2 + 2s + 8)}{(s^2 + 2s + 6)(s^2 + 4s + 10)}$,

(3) $F(s) = \dfrac{(s^2 + s + 1)(s^2 + 3s + 5)}{(s^2 + 4s + 8)(s^2 + 5s + 10)}$.

61. Let the rational function $F(s)$ in Problem 60 be further restricted as follows: (a) $F(s)$ has no pole at ∞ and $F(\infty) \geq 0$. (b) Any real negative poles of $F(s)$ are simple with positive residues.

(a) Prove that $F(s)$ can be decomposed as

$$F(s) = \frac{A_1(s)B_2(s) - A_2(s)B_1(s)}{A_1^2(s) - sB_1^2(s)}$$

where

$$\frac{A_2}{sB_2}, \frac{B_2}{A_2}, \frac{A_1}{sB_1}, \frac{B_1}{A_1}$$

are all RC impedances.

(b) Apply this result to the appropriate functions in Problem 60.

62. The following transfer impedance functions are to be realized by the cascade RC-negative converter network shown in Fig. 30. Find the Horowitz decomposition of the denominators (as in Problem 56) and from this the z parameters of the RC two-ports.

(a) $$\frac{K(s^2+1)}{s^4 + 4s^3 + 9s^2 + 8s + 5},$$

(b) $$\frac{K(2s^2 - s + 2)}{s^2 + 2s + 6},$$

(c) $$\frac{K(s^2 - 2s + 3)}{(s^2 + s + 3)(s^2 + s + 5)},$$

(d) $$\frac{K(s^2 + 1)(s^2 + 4)}{(s^2 + 2s + 2)(s^2 + 3s + 6)}.$$

63. Let the functions in Problem 62 be transfer voltage ratios which are to be realized in the parallel RC-negative converter network shown in Fig. 31. Use an appropriate auxiliary polynomial and obtain a network.

64. The functions in Problem 62 are to be realized in the amplifier network in Fig. 33. Use an appropriate auxiliary polynomial and obtain a realization.

65. Suppose a given rational function is augmented by an auxiliary polynomial with real negative zeros, and a partial fraction expansion is carried out as in (131) in the text. The function can be expressed as in (130). The result is to be realized by the cascade connection shown in Fig. P65a, were N_a is RC. Two-port N_b is shown in dashed lines. Its input impedance z_{11b} is $-z$. This two-port, in turn, is made up of the components shown in Fig. P65b. Show that the parameters of N_c are given in terms of those of N_b by

$$z_{11c} = \frac{1}{g}\frac{1/G - z}{1/g - z},$$

$$z_{21c} = \left(\frac{1}{g} - \frac{1}{G}\right)\frac{z_{21b}}{1/g - z}.$$

Note that $-z$ is the negative of an RC impedance. From a sketch of $-z(s)$ specify conditions on the parameters g and G for which N_c will be an RC two-port.

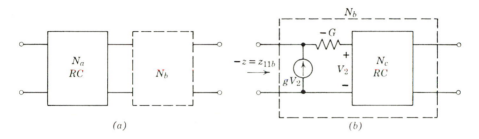

(a)   (b)

Fig. P65

66. (a) Develop a procedure for the realization of a voltage transfer ratio in the form of Fig. P22 in Chapter 3, where the two-port and the admittance Y are RC. (b) Realize each of the functions in Problem 62 in this configuration.

67. A transfer impedance function is to be realized by a cascade of the units shown in Fig. P26 (Chapter 3) in which each of the two-ports N_b and N_a are to be RC. (a) Develop a procedure which will permit the identification of RC-realizable functions z_{21b} and y_{21a}, and similar functions for the other units in the cascade. (b) Illustrate the procedure with the functions a and d in Problem 62.

68. Let $Z(s)$ be a rational positive real function.
(a) Prove that

$$F(s) = \frac{kZ(s) - sZ(k)}{kZ(k) - sZ(s)}$$

is also positive real, where k is a real, positive constant.

(b) Show that the numerator and denominator have a common factor $(s + k)$ if and only if $Z(-k) = -Z(k)$.

(c) When $Z(s)$ is expressed in terms of $F(s)$, the result is

$$Z(s) = Z(k)\frac{s + kF(s)}{k + sF(s)}.$$

Show that if $Z(-k) \neq -Z(k)$, then the numerator and denominator of the expression for $Z(s)$ will have a common factor $(s + k)$.

(d) Illustrate the above with the following pr functions:

$$(1) \quad Z(s) = \frac{s^2 + s + 1}{s^2 + s + 4}, \qquad (2) \quad Z(s) = \frac{s^3 + 2s^2 + s + 1}{s^3 + s^2 + 2s + 1}.$$

69. Two positive real functions $F_1(s)$ and $F_2(s)$ are said to be *complementary* if their sum is equal to a positive constant K. Suppose that $F_1(s)$ and K are given. Determine the restrictions on $F_1(s)$ and K such that $F_1(s)$ will have a complementary function. If $F_1(s)$ and $F_2(s)$ are complementary and represent driving-point impedance functions, this means that the series connection of the two corresponding networks has a constant input impedance. In case $F_1(s)$ and $F_2(s)$ are admittance functions, then the parallel connection of the corresponding networks will have a constant input admittance. We refer to such pairs of networks as being complementary.

70. Refer to Problem 69. Let $Z_1(s)$ be the driving-point impedance function of an RC network and assume that it is regular at the origin. Show that its complementary function $Z_2(s)$ will be an RL impedance function regular at infinity.

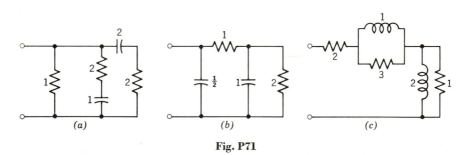

Fig. P71

71. Find complementary networks for each of the networks shown in Fig. P71.

.8.

THE SCATTERING
PARAMETERS

The properties and behavior of multiport networks can be described in terms of *impedance*, *admittance*, or *hybrid matrices*, as discussed in Chapter 3. These matrices are defined in terms of either open-circuit port voltages or short-circuit port currents. In actual operation the multiport may not have open- or short-circuit conditions at any of its ports. Nevertheless, such open- and short-circuit parameters can adequately describe the operation of the multiport under any terminating condition. Of course, some networks may not have a z-matrix representation, some may not have a y-matrix representation, and some (such as ideal transformers) may have neither.

It would be of value to have another representation of a multiport network, one that describes network operation with port-loading conditions other than open- or short-circuit. If a set of parameters are defined with some finite loading at each port, this set should be more convenient to use when describing transmission (of power) from a physical generator (with internal impedance) at one port to some loads at the other ports. The scattering parameters are such a set.

Scattering parameters originated in the theory of transmission lines. They are of particular value in microwave network theory where the concept of power is much more important than the concepts of voltage and current; in fact, the latter become somewhat artificial. Scattering parameters should be defined in such a way, then, that the quantities of interest in power transmission take on very simple expressions.

In the development here we shall freely use concepts such as incidence and reflection from transmission-line theory, purely for motivational purposes. The resulting mathematical expressions, however, do not depend for their validity on such interpretation. We shall start by treating the simpler one-port network before graduating to two-ports and multiports.

8.1 THE SCATTERING RELATIONS OF A ONE-PORT

We begin by considering the situation shown in Fig. 1. A one-port net-

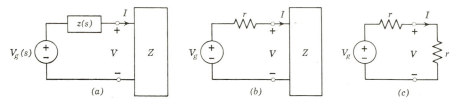

Fig. 1. A one-port network terminating a second one represented by its Thévenin equivalent.

work is shown terminating a voltage source in series with an impedance $z(s)$, which can be considered as the Thévenin equivalent of another network, a source network, to which Z is connected. The lower case letter z stands for the *source impedance*. If it is real, the situation will be as shown in Fig. 1b. The one-port will absorb power from the source network. Optimal matching will occur when $Z(s) = z(-s)$, in which case maximum power is transferred. (When $s = j\omega$, $z(-s)$ becomes $z(-j\omega) = z(\overline{j\omega}) = \bar{z}(j\omega)$. Thus $Z(s) = z(-s)$ reduces to $Z(j\omega) = \bar{z}(j\omega)$, which is the usual form for maximum power transfer.)

When z is real (equal to r), matching will occur when $Z = r$. Using the terminology of wave propagation, we say that if the one-port is matched to the source (network), there will be no reflection at the terminals.

Under unmatched conditions the voltage transform V at the terminals is pictured as having contributions from the "incident wave" arriving from the left and the "reflected wave" coming back from the one-port. A similar case obtains for the current transform I. Thus we can write

$$V = V_i + V_r \tag{1a}$$

and

$$I = I_i - I_r, \tag{1b}$$

$$V_r = r I_r = r(I_i - I)$$
$$V = V_i + r I_i - r I \,, \quad r I_i = V_i$$
$$\therefore \quad V_i = \tfrac{1}{2}(V + r I)$$

where the subscripts i and r refer to "incident" and "reflected," respectively. The negative sign in the second equation is a result of the reference chosen for the reflected current. Suppose we think of a real quantity r as the "characteristic impedance" of the transmission system to the left of the one-port terminals. Then the incident and reflected quantities are related by

$$\frac{V_i}{I_i} = \frac{V_r}{I_r} = r, \tag{2}$$

which are well-known relations for a transmission line. By using this result, (1) can be inverted to give

$$V_i = \tfrac{1}{2}(V + rI), \qquad I_i = \tfrac{1}{2}(gV + I), \tag{3a}$$

$$V_r = \tfrac{1}{2}(V - rI), \qquad I_r = \tfrac{1}{2}(gV - I), \tag{3b}$$

where $g = 1/r$. It is now possible to define a *voltage reflection coefficient* ρ as the ratio between reflected and incident voltage transforms, and a *current reflection coefficient* as the ratio between reflected and incident current transforms. Thus, using (3) for the incident and reflected variables, we get

$$\rho = \frac{V_r}{V_i} = \frac{V - rI}{V + rI} = \frac{Z - r}{Z + r} = \frac{Z/r - 1}{Z/r + 1} = \frac{gZ - 1}{gZ + 1} = \frac{gV - I}{gV + I} = \frac{I_r}{I_i}. \tag{4}$$

Some of the steps in this sequence used $V = ZI$. Just as the impedance Z can characterize the behavior of the one-port network, so also the reflection coefficient can characterize it completely. There is a one-to-one correspondence between Z and ρ given by the bilinear transformation $\rho = (Z - r)(Z + r)^{-1}$. We observe that the current and voltage reflection coefficients are the same. It must be emphasized, however, that this is true only for the case under consideration; namely, a real source impedance. When we consider the general case later, we will find that the two reflection coefficients are different.

The wave-propagation concepts that were used in the preceding discussion are artificial in the case of lumped networks. Nevertheless, it is possible to regard (3) as formal definitions of the variables V_i, V_r and I_i, I_r without attaching any interpretive significance to these quantities that reflect their intuitive origin. In the development we used r as the characteristic impedance. However, this idea is not necessary in the definitions expressed by (3) or (4); r is simply an arbitrary real positive number that has the dimensions of impedance.

It is, in fact, possible to introduce the incident and reflected voltages in an alternative way. Consider again the one-port in Fig. 1; it is characterized by the two variables V and I. Instead of these, a linear combination of these variables can be used as an equally adequate set. Thus the transformation

$$V_i = a_{11} V + a_{12} I,$$
$$V_r = a_{21} V + a_{22} I \tag{5}$$

defines two new variables V_i and V_r in terms of the old ones, V and I. The coefficients of the transformation should be chosen in such a way that the new variables become convenient to use. The choice $a_{11} = a_{21} = \frac{1}{2}$ and $a_{12} = -a_{22} = r/2$ will make (5) reduce to (3). Other choices could lead to additional formulations, which may or may not be useful for different applications.

It is possible to interpret the incident and reflected variables by reference to the situation shown in Fig. 1c, in which the one-port is matched to the real source impedance. In this case $V = rI$. Hence, from (3a) we find that

$$V_i = V, \quad I_i = I \tag{6}$$

when matched. This tells us that *when the one-port is matched to its terminations, the voltage at the port is V_i and the current is I_i*. Furthermore, under matched conditions, (3b) tells us that $V_r = 0$ and $I_r = 0$; and from Fig. 1c we observe that

$$V_i = \tfrac{1}{2} V_g. \checkmark \tag{7}$$

From (4) we see that, under matched conditions, the reflection coefficient is zero.

When the one-port is not matched, V_r and ρ are not zero. In fact, (1) can be rewritten as

$$V_r = V - V_i, \tag{8a}$$

$$I_r = I_i - I; \quad = -(I - I_i) \tag{8b}$$

that is, the reflected voltage V_r is a measure of the deviation of the one-port voltage, when under actual operation, from its value when matched.

Similarly, I_r is a measure of the deviation of the current, when under actual operation, from its value when matched. Note the slight asymmetry, in that one deviation is positive and the other negative.

NORMALIZED VARIABLES—REAL NORMALIZATION

The preceding discussion has been carried out by using two pairs of variables: the incident and reflected voltages, and the incident and reflected currents. Since these quantities are proportional in pairs, from (2), it should be sufficient to talk about one incident variable and one reflected variable. However, rather than select either the voltage or current, we use normalized variables related to both.

The *normalized* incident and reflected variables are defined as follows:

$$a(s) = \sqrt{r}\, I_i(s) = \frac{V_i(s)}{\sqrt{r}}, \tag{9a}$$

$$b(s) = \sqrt{r}\, I_r(s) = \frac{V_r(s)}{\sqrt{r}}. \tag{9b}$$

We shall refer to a and b as the *scattering variables*.

By using (3) these new variables can be expressed in terms of the voltage and current as

$$a = \tfrac{1}{2}\left(r^{-1/2}V + r^{1/2}I\right),$$
$$b = \tfrac{1}{2}\left(r^{-1/2}V - r^{1/2}I\right). \tag{10}$$

The square root of r appearing on the right of these expressions is disconcerting. It could be eliminated by defining a normalized voltage and current,

$$V_n = r^{-1/2}V, \tag{11a}$$

$$I_n = r^{1/2}I. \tag{11b}$$

Then the scattering variables become

$$a = \tfrac{1}{2}\left(V_n + I_n\right), \tag{12a}$$

$$b = \tfrac{1}{2}\left(V_n - I_n\right). \tag{12b}$$

A glance at (4) will show that the reflection coefficient is invariant to the normalization. Thus

$$\rho = \frac{V_r}{V_i} = \frac{b}{a} = \frac{V_n - I_n}{V_n + I_n} = \frac{Z_n - 1}{Z_n + 1} = \frac{Z - r}{Z + r}, \tag{13}$$

where $Z_n = Z/r$ is the normalized impedance.

Conversely, the normalized voltage, current, and impedance can be expressed in terms of the scattering variables and the reflection coefficient by inverting (12) and (13). Thus

$$V_n = a + b, \tag{14a}$$

$$I_n = a - b, \tag{14b}$$

$$Z_n = \frac{1 + \rho}{1 - \rho} = (1 + \rho)(1 - \rho)^{-1}. \tag{14c}$$

AUGMENTED NETWORK

The normalization just carried out can be interpreted by reference to the network shown in Fig. 2. The normalized value of the source resistance

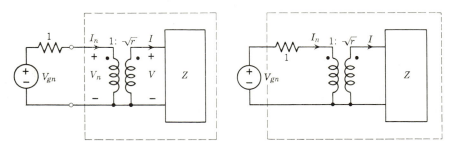

Fig. 2.   (a) Normalized network; (b) augmented normalized network.

is 1. The ideal transformer of turns ratio $1 : \sqrt{r}$ gives the appropriate equations relating the actual voltage and current on its secondary side to the normalized voltage and current on the primary side. The original one-port in cascade with the ideal transformer can be called the normalized one-port network. When the original one-port is matched to r, this is equivalent to the normalized one-port being matched to unity. Because of its normalizing function, r is called the *normalizing number*, or the *reference resistance*. In the event that the normalizing number is unity, the resulting $1:1$ ideal transformer in Fig. 2 need not be included.

From the normalized network it is clear that when the one-port is matched, the input impedance at the transformer primary is a unity resistance. Hence $V_n = I_n$; from (12), this means $a = V_n$ and $b = 0$. Thus, under matched conditions, the incident scattering variable equals the normalized voltage, and the reflected scattering variable is zero. Furthermore, with the normalized source voltage defined as $V_{gn} = r^{-1/2}V_g$, we see that the relationship of V_g to V_i given in (7) is invariant to normalization. Thus

$$a = \tfrac{1}{2} V_{gn}. \tag{15}$$

We can go one more step and include the series unity resistance in with the normalized network, as illustrated in Fig. 2b. The result is then called the (series) *augmented normalized network*. Of course, it is possible to think of the original network as being augmented by the series resistance r, without reference to normalization. The reflection coefficient of the original network can be expressed in terms of the input admittance Y_a of the augmented network or Y_{an} of the augmented normalized network. It is clear from Fig. 2 that we can write

$$Y_a = \frac{I}{V_g} = \frac{1}{Z+r}, \qquad Y_{an} = \frac{I_n}{V_{gn}} = \frac{r}{Z+r} = rY_a.$$

Then

$$\rho = \frac{Z-r}{Z+r} = \frac{Z+r-2r}{Z+r} = 1 - \frac{2r}{Z+r}$$

or

$$\rho = 1 - 2rY_a = 1 - 2Y_{an}. \tag{16}$$

(In the derivation, r was added and subtracted in the numerator.) This is a useful expression. It can often be the simplest form for computing the reflection coefficient.

As an example, consider the situation shown in Fig. 3. Two one-ports having impedances given by $Z_a = f_a(s)$ and $Z_b = f_b(s)$ have reflection coefficients ρ_1 and ρ_2, respectively. It is desired to find the reflection coefficient of the one-port shown in Fig. 3b in terms of ρ_1 and ρ_2. Since ρ is invariant to normalization, let the normalizing number be unity and con-

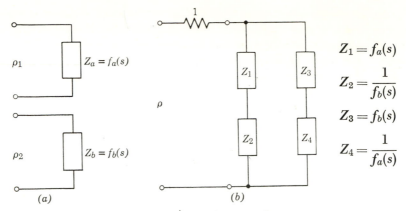

Fig. 3. Illustrative example.

sider the augmented network. The augmented input admittance can be readily written as follows:

$$Y_a = \cfrac{1}{1 + \cfrac{1}{\cfrac{1}{f_a(s) + \cfrac{1}{f_b(s)}} + \cfrac{1}{f_b(s) + \cfrac{1}{f_a(s)}}}} = \cfrac{1}{1 + \cfrac{1 + f_a f_b}{f_a + f_b}}$$

$$= \frac{f_a + f_b}{f_a f_b + f_a + f_b + 1} = \frac{f_a + f_b}{(f_a + 1)(f_b + 1)}.$$

Hence, by using (16), we obtain

$$\rho = 1 - 2Y_a = 1 - \frac{2(f_a + f_b)}{(f_a + 1)(f_b + 1)} = \frac{(f_a - 1)(f_b - 1)}{(f_a + 1)(f_b + 1)},$$

$$\rho = \rho_1 \rho_2.$$

For this example it is clear that the reflection coefficient of the overall network is more simply expressed in terms of the reflection coefficients of the components than is the overall admittance in terms of the component admittances.

REFLECTION COEFFICIENT FOR TIME-INVARIANT, PASSIVE, RECIPROCAL NETWORK

The analytic properties of $\rho(s)$ for a time-invariant, passive, reciprocal network can be obtained by reference to (13). For such a network $Z_n(s)$ is a positive real function of s. A positive real Z_n maps the $j\omega$-axis of the

s-plane into the closed right half Z_n-plane. Equation 13 is a bilinear transformation. (See Appendix 2.) The bilinear transformation maps the closed right half of the Z_n-plane into the interior or boundary of the ρ-plane unit circle. Hence, when $s = j\omega$, the corresponding point lies inside or on the boundary of the ρ-plane unit circle. Thus

$$|\rho(j\omega)| \leq 1. \tag{17}$$

As for the poles of $\rho(s)$, they are given by the zeros of $Z_n(s) + 1$. They cannot lie in the closed right half-plane since this would require Re $Z_n = -1$ for a point in the closed right half-plane, which is impossible for a positive real function. Hence $\rho(s)$ is regular in the closed right half-plane. We see that the positive real condition on the impedance of a one-port can be translated into equivalent conditions on the reflection coefficient.

A *bounded real function* $\rho(s)$ was defined in Chapter 7 as a function that is (1) real when s is real, (2) regular in the closed right half-plane, and (3) $|\rho(j\omega)| \leq 1$ for all ω.

The above discussion has shown that *for a time-invariant, passive, reciprocal network the reflection coefficient is a bounded real function.*

POWER RELATIONS

We have seen that the scattering variables have special significance in describing power transfer from a source to a load. We shall here discuss the power relation in the network of Fig. 1 in terms of the scattering variables. Let us assume sinusoidal steady-state conditions. The complex power delivered to the one-port is $W = V(j\omega) \bar{I}(j\omega)$. What happens to this expression if the voltage and current are normalized as in (11)? With the stipulation that r is a positive real number, the answer is: nothing. When the normalization is inserted, we still have

$$W = V_n(j\omega) \bar{I}_n(j\omega). \tag{18}$$

We can now use (14) to express this result in terms of the scattering variables. The real power will be

$$P = \text{Re } W = \text{Re } (a + b)\overline{(a - b)} = |a(j\omega)|^2 - |b(j\omega)|^2$$
$$= |a|^2(1 - |\rho(j\omega)|^2). \tag{19}$$

The last step follows from $\rho = b/a$.

A number of observations can be made from here. The magnitude square of both a and b has the dimensions of power. Thus the *dimensions of the scattering variables are the square root of power*, (voltage $\times$ current)$^{1/2}$. We can think of the net power delivered to the one-port as being made up of the power in the incident wave, P_i, less the power returned to the source by the reflected wave, P_r. Of course, under matched conditions there is no reflection. The power delivered under these conditions is the maximum available power, say P_m, from the source in series with resistance r. This power is easily found from Fig. 1c to be

$$P_m = \frac{|V_g|^2}{4r} = \frac{|r^{-1/2}V_g|^2}{4} = \frac{|V_{gn}|^2}{4} = |a|^2. \tag{20}$$

The last step follows from (15). With these ideas, (19) can be rewritten as

$$P = P_i - P_r$$

and

$$\frac{P}{P_m} = 1 - |\rho(j\omega)|^2, \tag{21}$$

where $P_i = |a|^2$ and $P_r = |b|^2$. This is an extremely important result. The right side specifies the fraction of the maximum available power that is actually delivered to the one-port. If there is no reflection ($\rho = 0$), this ratio is unity.

For a passive one-port, the power delivered cannot exceed the maximum available; that is, $P/P_m \leq 1$. Hence

$$|\rho(j\omega)| \leq 1 \qquad \text{(for passive one-port).} \tag{22}$$

8.2 MULTIPORT SCATTERING RELATIONS

As discussed in the last section, the scattering parameters are particularly useful in the description of power transfer. The simplest of such situations is the transfer of power from a source with an internal impedance to a load, which we have already discussed. More typical is the transfer from a source to a load through a coupling network N, as shown in Fig. 4. The two-port network N may be a filter or an equalizer matching network. The load may be passive, either real or complex, or it may be active (e.g., a tunnel diode). More generally, the coupling network is a multiport with transmission of power from one or more ports to one or more other ports.

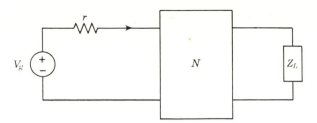

Fig. 4. Two-port filter or matching network.

We shall deal in detail with the situation shown in Fig. 5. A multiport is terminated at each port with a *real* positive resistance and a source. A special case is the two-port shown in Fig. 8.5b. The development is simply a generalization of the one-port case except that scalar relationships will now be replaced by matrices. We shall treat the general n-port case but will illustrate the details with the two-port for ease of visualization.

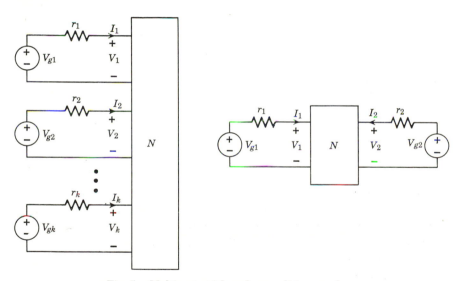

Fig. 5. Multiports with real normalizing numbers.

To begin, we define the vector variables as follows:

$$\mathbf{V} = \begin{bmatrix} V_1 \\ V_2 \\ \vdots \\ V_k \end{bmatrix}, \quad \mathbf{I} = \begin{bmatrix} I_1 \\ I_2 \\ \vdots \\ I_k \end{bmatrix}, \quad \mathbf{V}_g = \begin{bmatrix} V_{g1} \\ V_{g2} \\ \vdots \\ V_{gk} \end{bmatrix} \tag{23}$$

and the diagonal matrix

$$\mathbf{r} = \begin{bmatrix} r_1 & 0 & 0 \cdots 0 \\ 0 & r_2 & 0 \cdots 0 \\ 0 & 0 & r_3 \cdots 0 \\ \vdots & \vdots & \vdots & \vdots \\ 0 & 0 & 0 \cdots r_k \end{bmatrix}. \tag{24}$$

The matrix $\mathbf{r}$ is nonsingular and positive definite, since all r_j's are assumed positive. From Fig. 5 we can write

$$\mathbf{V}_g = \mathbf{V} + \mathbf{rI} = (\mathbf{Z}_{oc} + \mathbf{r})\mathbf{I}, \tag{25}$$

where $\mathbf{Z}_{oc}$ is the open-circuit impedance matrix of multiport N. Suppose now that at each of the ports the multiport is matched to the source resistance. This means the ratio V_j/I_j is to equal the resistance r_j at the jth port. In matrix form this becomes $\mathbf{V} = \mathbf{rI}$, or $\mathbf{Z}_{oc} = \mathbf{r}$ when the multiport is matched. By analogy with the one-port case, we introduce the incident-voltage vector $\mathbf{V}_i$ and the incident-current vector $\mathbf{I}_i$ as equal to the port-voltage vector and the port-current vector, respectively, when the ports are all matched; that is,

$$\mathbf{V}_i = \mathbf{V}, \tag{26a}$$

$$\mathbf{I}_i = \mathbf{I}, \tag{26b}$$

and

$$\mathbf{V}_i = \mathbf{rI}_i. \tag{26c}$$

when matched.

Similarly, we introduce the *reflected-voltage* vector $\mathbf{V}_r$ and the *reflected-current* vector $\mathbf{I}_r$, as the deviation of the port-voltage vector and the port-current vector, respectively, from their matched values. In analogy with (8) for the one-port case, they are written as

$$\mathbf{V}_r = \mathbf{V} - \mathbf{V}_i, \tag{27a}$$

$$\mathbf{I}_r = \mathbf{I}_i - \mathbf{I}. \tag{27b}$$

When the last two pairs of equations are used with (25), the incident and reflected variables can be written as

$$\mathbf{V}_i = \tfrac{1}{2}\mathbf{V}_g, \qquad \mathbf{I}_i = \tfrac{1}{2}\mathbf{r}^{-1}\mathbf{V}_g \tag{28}$$

and

$$\mathbf{V}_i = \tfrac{1}{2}(\mathbf{V} + \mathbf{rI}), \tag{29a}$$

$$\mathbf{V}_r = \tfrac{1}{2}(\mathbf{V} - \mathbf{rI}), \tag{29b}$$

$$\mathbf{I}_i = \tfrac{1}{2}\mathbf{r}^{-1}(\mathbf{V} + \mathbf{rI}), \tag{29c}$$

$$\mathbf{I}_r = \tfrac{1}{2}\mathbf{r}^{-1}(\mathbf{V} - \mathbf{rI}). \tag{29d}$$

These expressions should be compared with (7) and (3) for the one-port.

THE SCATTERING MATRIX

For the one-port a reflection coefficient was defined relating reflected to incident voltage and another one relating reflected to incident current. These two turned out to be the same reflection coefficient whose value was invariant to normalization. In the multiport case the relationship of reflected to incident variables is a matrix relationship. We define two such relationships—one for the voltages and one for the currents—as follows:

$$\mathbf{I}_r = \mathbf{S}_I\mathbf{I}_i, \tag{30a}$$

$$\mathbf{V}_r = \mathbf{S}_V\mathbf{V}_i, \tag{30b}$$

where $\mathbf{S}_I$ is the *current-scattering matrix* and $\mathbf{S}_V$ is the *voltage-scattering matrix*. These matrices can be expressed in terms of $\mathbf{Z}_{oc}$ and the terminating impedances by using (29) for the incident and reflected variables, and by using $\mathbf{V} = \mathbf{Z}_{oc}\mathbf{I}$. The details will be left to you; the result is as follows:

$$\mathbf{S}_I = \mathbf{r}^{-1}(\mathbf{Z}_{oc} - \mathbf{r})(\mathbf{Z}_{oc} + \mathbf{r})^{-1}\mathbf{r} = (\mathbf{Z}_{oc} + \mathbf{r})^{-1}(\mathbf{Z}_{oc} - \mathbf{r}), \tag{31a}$$

$$\mathbf{S}_V = (\mathbf{Z}_{oc} - \mathbf{r})(\mathbf{Z}_{oc} + \mathbf{r})^{-1} = \mathbf{r}(\mathbf{Z}_{oc} + \mathbf{r})^{-1}(\mathbf{Z}_{oc} - \mathbf{r})\mathbf{r}^{-1}. \tag{31b}$$

Study these expressions carefully, note how $\mathbf{S}_I$ is relatively simple when $(\mathbf{Z}_{oc} + \mathbf{r})^{-1}$ premultiplies $(\mathbf{Z}_{oc} - \mathbf{r})$, and $\mathbf{S}_V$ is relatively simple when it *postmultiplies* $(\mathbf{Z}_{oc} - \mathbf{r})$.

What is $(\mathbf{Z}_{oc} + \mathbf{r})$? It is, in fact, the open-circuit impedance matrix of the *augmented* network; that is, the multiport in Fig. 5, which includes the series resistance at each port as part of the multiport. The inverse, $(\mathbf{Z}_{oc} + \mathbf{r})^{-1}$, is the short-circuit admittance matrix of the augmented multiport, which we shall label $\mathbf{Y}_a$. In terms of $\mathbf{Y}_a$, the two reflection coefficients are found from (31), after some manipulation, to be

$$\mathbf{S}_I = \mathbf{U} - 2\mathbf{Y}_a\mathbf{r}, \tag{32a}$$

$$\mathbf{S}_V = \mathbf{U} - 2\mathbf{r}\mathbf{Y}_a. \tag{32b}$$

The only difference seems to be that in one case $\mathbf{r}$ postmultiplies $\mathbf{Y}_a$ and in the other case it premultiplies $\mathbf{Y}_a$. Each of these equations can be solved for $\mathbf{Y}_a$. When these two expressions for $\mathbf{Y}_a$ are equated, a relationship between $\mathbf{S}_I$ and $\mathbf{S}_V$ is found to be

$$\mathbf{S}_I \mathbf{r}^{-1} = \mathbf{r}^{-1} \mathbf{S}_V \qquad (33a)$$

$$\mathbf{r}\mathbf{S}_I = \mathbf{S}_V \mathbf{r} \qquad (33b)$$

We seem to be at an impasse; the matrix $\mathbf{r}$ or $\mathbf{r}^{-1}$ seems to crop up and spoil things. Perhaps normalization will help. Look back at (9) to see how the incident and reflected voltages and currents were normalized. Suppose we carry out a similar normalization, but in matrix form. To normalize currents, we multiply by the matrix $\mathbf{r}^{1/2}$; and to normalize voltages, we multiply by $\mathbf{r}^{-1/2}$, where

$$\mathbf{r}^{1/2} = \begin{bmatrix} r_1^{1/2} & & & & \\ & r_2^{1/2} & & \bigcirc & \\ & & r_3^{1/2} & & \\ & & & \ddots & \\ & \bigcirc & & & \ddots \\ & & & & r_k^{1/2} \end{bmatrix} \qquad (34)$$

is a real diagonal matrix, each diagonal entry of which is the square root of the corresponding entry of matrix $\mathbf{r}$. To see what will happen, let us multiply both sides of (30a) by $\mathbf{r}^{1/2}$ and both sides of (30b) by $\mathbf{r}^{-1/2}$. Then

$$\mathbf{r}^{1/2}\mathbf{I}_r = \mathbf{r}^{1/2}\mathbf{S}_I(\mathbf{r}^{-1/2}\mathbf{r}^{1/2})\mathbf{I}_i = (\mathbf{r}^{1/2}\mathbf{S}_I\mathbf{r}^{-1/2})(\mathbf{r}^{1/2}\mathbf{I}_i), \qquad (35a)$$

$$\mathbf{r}^{-1/2}\mathbf{V}_r = \mathbf{r}^{-1/2}\mathbf{S}_V(\mathbf{r}^{1/2}\mathbf{r}^{-1/2})\mathbf{V}_i = (\mathbf{r}^{-1/2}\mathbf{S}_V\mathbf{r}^{1/2})(\mathbf{r}^{-1/2}\mathbf{V}_i). \qquad (35b)$$

But observe from (33)—through premultiplying and postmultiplying both sides by $\mathbf{r}^{1/2}$—that

$$\mathbf{r}^{1/2}\mathbf{S}_I\mathbf{r}^{-1/2} = \mathbf{r}^{-1/2}\mathbf{S}_V\mathbf{r}^{1/2} = \mathbf{S}, \qquad (36)$$

where the matrix $\mathbf{S}$ is introduced for convenience.

Since $\mathbf{V}_i = \mathbf{r}\mathbf{I}_i$, it follows that the two normalized variables on the right side of (35) are equal; that is, $\mathbf{r}^{-1/2}\mathbf{V}_i = \mathbf{r}^{1/2}\mathbf{I}_i$. In view of (36), it follows that the two normalized variables on the left side of (35) are also equal.

With the preceding discussion as justification, we now define the *normalized vector scattering variables* as

$$\mathbf{a} = \mathbf{r}^{1/2}\mathbf{I}_i = \mathbf{r}^{-1/2}\mathbf{V}_i \,, \tag{37a}$$

$$\mathbf{b} = \mathbf{r}^{1/2}\mathbf{I}_r = \mathbf{r}^{-1/2}\mathbf{V}_r \,. \tag{37b}$$

These scattering variables are related by a *scattering matrix* $\mathbf{S}$, which is related to the current- and voltage-scattering matrices through (36). Thus

$$\mathbf{b} = \mathbf{S}\mathbf{a}. \tag{38}$$

RELATIONSHIP TO IMPEDANCE AND ADMITTANCE MATRICES

The relationship between this scattering matrix and the matrices $\mathbf{Z}_{oc}$ and $\mathbf{Y}_a$ can be found by appropriately pre- and post-multiplying (32) and (31) by $\mathbf{r}^{1/2}$ and $\mathbf{r}^{-1/2}$ consistent with (36). If we define

$$\mathbf{Y}_{an} = \mathbf{r}^{1/2}\mathbf{Y}_a\mathbf{r}^{1/2}, \tag{39a}$$

$$\mathbf{Z}_n = \mathbf{r}^{-1/2}\mathbf{Z}_{oc}\,\mathbf{r}^{-1/2}, \tag{39b}$$

$$\mathbf{Y}_n = \mathbf{r}^{1/2}\mathbf{Y}_{sc}\mathbf{r}^{1/2}, \tag{39c}$$

where $\mathbf{Y}_{sc}$, the short-circuited admittance matrix of the multiport, is $\mathbf{Z}_{oc}^{-1}$, then

$$\mathbf{S} = \mathbf{U} - 2(\mathbf{Z}_n + \mathbf{U})^{-1} = \mathbf{U} - 2\mathbf{Y}_{an}\,, \tag{40}$$

$$\begin{aligned} \mathbf{S} &= (\mathbf{Z}_n - \mathbf{U})(\mathbf{Z}_n + \mathbf{U})^{-1} = (\mathbf{Z}_n + \mathbf{U})^{-1}(\mathbf{Z}_n - \mathbf{U}) \\ &= (\mathbf{U} - \mathbf{Y}_n)(\mathbf{U} + \mathbf{Y}_n)^{-1} = (\mathbf{U} + \mathbf{Y}_n)^{-1}(\mathbf{U} - \mathbf{Y}_n). \end{aligned} \tag{41}$$

We leave the details for you to work out. Compare (40) with (16), which is the corresponding scalar result for a one-port.

The relationship between $\mathbf{S}$ and $\mathbf{Y}_{an}$ points up an important property that is not evident from the manner in which it was obtained. Because of the series resistances in the augmented network, this network may have an admittance matrix even though the original multiport has neither an impedance nor an admittance matrix. This will be true for any passive network. Thus an advantage of *scattering parameters* (as we call the elements of the scattering matrix) is that they exist for all passive networks, even those for which impedance or admittance parameters do not exist. An illustration is provided by the ideal transformer shown in Fig. 6.

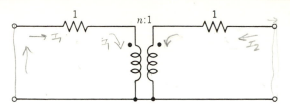

Fig. 6.   Ideal transformer.

The augmented network for unity reference resistances at both ports is shown. The ideal transformer has neither an impedance nor an admittance matrix. Nevertheless, the augmented network has an admittance representation. The short-circuit admittance matrix can be calculated directly, and then (40) will give the scattering matrix. The details will be left to you; the result will be

$$\mathbf{Y}_{an}=\begin{bmatrix} \dfrac{1}{n^2+1} & \dfrac{-n}{n^2+1} \\[2ex] \dfrac{-n}{n^2+1} & \dfrac{n^2}{n^2+1} \end{bmatrix}, \qquad \mathbf{S}=\begin{bmatrix} \dfrac{n^2-1}{n^2+1} & \dfrac{2n}{n^2+1} \\[2ex] \dfrac{2n}{n^2+1} & -\dfrac{n^2-1}{n^2+1} \end{bmatrix}.$$

This reduces to an especially simple matrix for a turns ratio of $n=1$.

Note that S_{22} is the negative of S_{11}. Two-ports that satisfy the conditions $S_{22}=-S_{11}$ are said to be *antimetric*, in contrast to symmetric two-ports, for which $S_{22}=S_{11}$. (See Problems 10 and 11.)

NORMALIZATION AND THE AUGMENTED MULTIPORT

The normalized scattering variables **a** and **b** can be expressed in terms of voltages and currents by applying the normalization to (29). Defining the normalized voltage and current as $\mathbf{V}_n=\mathbf{r}^{-1/2}\mathbf{V}$ and $\mathbf{I}_n=\mathbf{r}^{1/2}\mathbf{I}$, these equations and their inverses take the relatively simple forms

$$\begin{aligned} \mathbf{a}&=\tfrac{1}{2}\left(\mathbf{V}_n+\mathbf{I}_n\right) & \mathbf{V}_n&=\mathbf{a}+\mathbf{b}, \\ \mathbf{b}&=\tfrac{1}{2}\left(\mathbf{V}_n-\mathbf{I}_n\right) & \mathbf{I}_n&=\mathbf{a}-\mathbf{b}. \end{aligned} \tag{42}$$

Comparing these with (12) and (14) shows that the expressions for the multiport scattering variables are identical with those for the one-port, except that they are matrix expressions in the present case.

Finally, note that the two expressions in (28) relating incident voltage and current to $\mathbf{V}_g$, reduce to a single equation under normalization:

$$\mathbf{a} = \tfrac{1}{2}\,\mathbf{V}_{gn}. \tag{43}$$

The normalization can again be interpreted by appending ideal transformers to the ports. This is illustrated for the two-port in Fig. 7. The

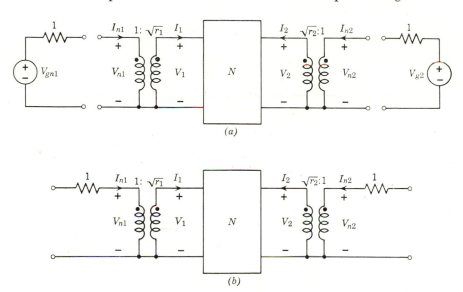

(a)

(b)

Fig. 7. Normalized and augmented two-port.

turns ratios of the transformers are $1 : \sqrt{r_1}$ and $1 : \sqrt{r_2}$. They provide the appropriate equations relating the actual voltages and currents to the normalized values. The total network, including the ideal transformers, is called the normalized network. Even though the reference resistances at the ports may be different, the matched condition corresponds to an input resistance of unity at each port. If we include a series unit resistance at each port, the resulting overall network is the augmented (normalized) network, as shown in Fig. 7b.

How are the port normalizing numbers r_j chosen? In the ideal-transformer example, for instance, what caused us to choose unity normalizing numbers for both ports? The answer to these questions is simply convenience. If a network will actually operate with certain terminating resistances, it would clearly be convenient, and would simplify the resulting expressions, if these resistances were chosen as the normalizing numbers.

In other cases choice of some parameter in the network as the normalizing number leads to simplification. Consider, for example, the gyrator shown in Fig. 8. It has an impedance representation containing the gyration

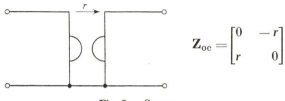

$$\mathbf{Z}_{oc} = \begin{bmatrix} 0 & -r \\ r & 0 \end{bmatrix}$$

Fig. 8.   Gyrator.

resistance r. If the port normalizing numbers r_1 and r_2 for both ports are chosen equal to r, the matrix $\mathbf{Z}_n$ can be written very simply. Equation 40 then leads to the scattering matrix. Thus

$$\mathbf{Z}_n = \begin{bmatrix} 0 & \dfrac{-r}{\sqrt{r_1 r_2}} \\ \dfrac{r}{\sqrt{r_1 r_2}} & 0 \end{bmatrix} = \begin{bmatrix} 0 & -1 \\ 1 & 0 \end{bmatrix},$$

$$\mathbf{S} = \mathbf{U} - 2(\mathbf{U} + \mathbf{Z}_n)^{-1} = \begin{bmatrix} 1 & 0 \\ 0 & 1 \end{bmatrix} - 2\begin{bmatrix} 1 & -1 \\ 1 & 1 \end{bmatrix}^{-1} = \begin{bmatrix} 0 & -1 \\ 1 & 0 \end{bmatrix}.$$

8.3   THE SCATTERING MATRIX AND POWER TRANSFER

The preceding discussion provides means for finding the scattering parameters of a multiport from either its impedance or admittance matrix—or the admittance matrix of the augmented network. But the scattering parameters appear in the relationships between incident and reflected scattering variables. We shall now consider in greater detail what these relationships are. For simplicity we shall treat the two-port, for which the equations are

$$b_1 = S_{11}a_1 + S_{12}a_2, \tag{44a}$$

$$b_2 = S_{21}a_1 + S_{22}a_2. \tag{44b}$$

We shall assume arbitrary loading of the two-port, with possible signals applied to each port, as shown in Fig. 9.

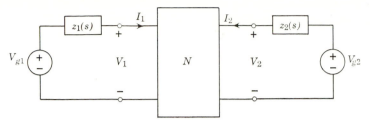

Fig. 9. Two-port with arbitrary loading.

INTERPRETATION OF SCATTERING PARAMETERS

From (44) the scattering parameters can be interpreted as follows:

$$S_{11} = \frac{b_1}{a_1}\bigg|_{a_2=0}, \qquad S_{12} = \frac{b_1}{a_2}\bigg|_{a_1=0},$$

$$(45)$$

$$S_{21} = \frac{b_2}{a_1}\bigg|_{a_2=0}, \qquad S_{22} = \frac{b_2}{a_2}\bigg|_{a_j=0}.$$

We see that each parameter is the ratio of a reflected to an incident variable, under the condition of zero incident variable at the other port. What does it mean for an incident variable, say a_2, to be zero? This is easily answered by reference to (42) or, equivalently, (29), which relate scattering variables to voltages and currents. Thus $a_2 = 0$ means $V_{n2} = -I_{n2}$, or $V_2 = -r_2 I_2$. Now look at Fig. 9. The condition $V_2 = -r_2 I_2$ means that port 2 is terminated in r_2 (rather than in impedance z_2) and that there is no voltage source; that is, *port 2 is matched*. Furthermore, if there is no incident variable and $V_2 = -r_2 I_2$, (29b), shows that the reflected voltage is the total voltage at the port ($V_{r2} = V_2$) and (29d) shows that the reflected current is the total current ($I_{r2} = I_2$). Similar meanings attach to the condition $a_1 = 0$.

   Thus the scattering parameters of a two-port are defined as the ratio of a reflected to an incident variable when the other incident variable is zero—meaning, when the other port is *match-terminated*. More specifically, consider S_{11}. This is the ratio of reflected to incident variables at port 1 (b_1/a_1) when port 2 is matched. Let us now substitute into b_1/a_1 the scalar equation for port 1 resulting from (42). Then

$$S_{11} = \frac{b_1}{a_1} = \frac{V_{n1} - I_{n1}}{V_{n1} + I_{n1}} = \frac{Z_{n1} - 1}{Z_{n1} + 1} = \frac{Z_1 - r_1}{Z_1 + r_1}, \qquad (46)$$

where Z_{n1} is the normalized input impedance at port 1. Comparing this with (13), which gives the reflection coefficient for a one-port, we see that S_{11} *is the reflection coefficient at port 1 when the other port is matched.* A similar conclusion is reached about S_{22}.

Now look at the off-diagonal terms, specifically S_{21}. When $a_2 = 0$, as we have already mentioned, $b_2 = V_{n2}$, or $V_{r2} = V_2$. Furthermore, $a_1 = V_{gn1}/2$, from (43). Hence

$$S_{21} = \frac{b_2}{a_1}\bigg|_{a_2=0} = \frac{V_{n2}}{V_{gn1}/2} = 2\sqrt{\frac{r_1}{r_2}}\frac{V_2}{V_{g1}}. \tag{47}$$

Thus S_{21} is seen to be proportional to a voltage gain, a *forward* voltage gain. A further clarification is given in terms of Fig. 10. Here the two-port

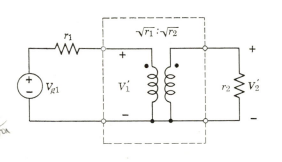

Resistance looking into transformer primary $= r_1$

$$V_1' = \frac{V_{g1}}{2}$$

$$V_2' = \sqrt{\frac{r_2}{r_1}}\,V_1' = \frac{1}{2}\sqrt{\frac{r_2}{r_1}}\,V_{g1}$$

$$\frac{V_2}{V_2'} = 2\sqrt{\frac{r_1}{r_2}}\frac{V_2}{V_{g1}} = S_{21}$$

Fig. 10.   Transducer voltage ratio.

has been replaced by an ideal transformer of such turns ratio that the resistance r_2 becomes matched to r_1. The output voltage under this condition is called V_2'. Now if we take the ratio of the actual output voltage V_2 to V_2', the result is called the *transducer voltage ratio.* The calculation given in the figure shows this quantity to be the same as the right-hand side of (47). Hence S_{21} is the *forward transducer voltage ratio.* When the source is matched as in Fig. 10, it will deliver to the network at its terminals (namely, the primary side of the transformer) the maximum available power. This is seen to be $P_{m1} = |V_{g1}|^2/4r_1$.

Now return to the original setup. The power transferred to the load with network N in place, when it is matched and when the port 2 voltage is V_2, will be $P_2 = |V_2|^2/r_2$. Hence the magnitude square of S_{21} in (47) will be

$$|S_{21}(j\omega)|^2 = 4\,\frac{r_1}{r_2}\frac{|V_2(j\omega)|^2}{|V_{g1}(j\omega)|^2} = \frac{P_2}{P_{m1}} = \mathscr{G}(\omega^2). \tag{48}$$

Thus the magnitude square of S_{21}, which is often called the *transducer power gain*, $\mathscr{G}(\omega^2)$, is simply the ratio of actual load power to the maximum power available from the generator when both ports are matched. A completely similar discussion can be carried out for S_{12}, the *reverse transducer voltage ratio*. (The argument is written ω^2 because $\mathscr{G}$ is an even function of frequency.)

A slight manipulation of (48) leads to $|S_{21}|^2 = |V_{n2}|^2/|V_{gn1}/2|^2$. This is, in fact, the transducer power gain for the normalized network. The conclusion is that the transducer power gain is unchanged by normalization. When a two-port is to be inserted between two resistive terminations, this result gives added reason for using the terminations as the reference resistances; the transmission coefficient then directly describes the power gain property of the network.

Without detailed exposition, it should be clear that the scattering parameters for a multiport can be interpreted in a similar fashion. The parameters on the main diagonal of **S** will be reflection coefficients. Thus

$$S_{jj} = \frac{b_j}{a_j}\bigg|_{\substack{\text{all other}\\ a\text{'s}=0}} = \frac{Z_j - r_j}{Z_j + r_j} \tag{49}$$

is the reflection coefficient at port j when all other ports are matched; that is, terminated in their reference resistances.

The parameters off the main diagonal are called *transmission coefficients*, in contrast with the reflection coefficients. They are given by

$$S_{ij} = \frac{b_i}{a_j}\bigg|_{\substack{\text{all other}\\ a\text{'s}=0}} = 2\sqrt{\frac{r_j}{r_i}}\,\frac{V_i}{V_{gj}}. \tag{50}$$

Following an analysis like that leading to (48), we find the magnitude square of a transmission coefficient to be the ratio of load power at one port to maximum available power from a source at another port when all ports are matched. Thus

$$|S_{ij}|^2 = \frac{4r_j}{r_i}\frac{|V_i|^2}{|V_{gj}|^2} = \frac{P_i}{P_{mj}}. \tag{51}$$

As an illustration, consider the network of Fig. 11. A three-port is formed with a gyrator by making an extra port, as shown. (The gyrator symbol shown was mentioned in Problem 3 in Chapter 3.) The objective is to find the scattering matrix of the three-port and to interpret its elements. Let the gyration resistance be taken as unity. This amounts to

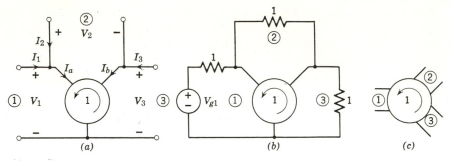

Fig. 11. Gyrator as three-port circulator.

choosing the normalizing numbers of the ports as the gyration resistance. Then the gyrator relationship is

$$\begin{bmatrix} V_1 \\ V_3 \end{bmatrix} = \begin{bmatrix} 0 & -1 \\ 1 & 0 \end{bmatrix} \begin{bmatrix} I_a \\ I_b \end{bmatrix}.$$

But $I_a = I_1 + I_2$, $I_b = I_3 - I_2$, and $V_2 = V_1 - V_3$, as observed from Fig. 11. With these, we get

$$\begin{bmatrix} V_1 \\ V_2 \\ V_3 \end{bmatrix} = \begin{bmatrix} 0 & 1 & -1 \\ -1 & 0 & -1 \\ 1 & 1 & 0 \end{bmatrix} \begin{bmatrix} I_1 \\ I_2 \\ I_3 \end{bmatrix}.$$

The coefficient matrix here is $\mathbf{Z}_n$. The scattering matrix can now be computed from (41). The result is

$$\mathbf{S} = \begin{bmatrix} 0 & 1 & 0 \\ 0 & 0 & -1 \\ 1 & 0 & 0 \end{bmatrix} \quad \text{or} \quad \begin{matrix} b_1 = a_2, \\ b_2 = -a_3, \\ b_3 = a_1. \end{matrix} \tag{52}$$

This is a very interesting result. Note first that the diagonal elements are all zero. Hence the reflection coefficients at all ports are zero, which means all the ports are matched. As for the transmission coefficients, consider Fig. 11b, in which only port 1 is excited and all 3 ports are matched. Thus, $V_2 = -I_2$ and $V_3 = -I_3$. Hence, from (28) and (37), we find $a_1 = V_{i1} = V_{g1}/2$ and $a_2 = a_3 = 0$; from (52), $b_1 = b_2 = 0$ and $b_3 = V_{g1}/2$. But $b_3 = V_{r3} = V_3$; hence the power in the matched load at port 3 is

$$P_3 = |V_3|^2 = \frac{|V_{g1}|^2}{4} = P_{m1}.$$

The conclusion is that when a signal is incident at port 1, with all ports match-terminated, none is reflected ($b_1 = 0$), none is transmitted to port 2 ($b_2 = 0$), but all of it is transmitted, without loss, to port 3, the power there being the maximum available from the source at port 1.

Similar conclusions follow from (52) for the other transmissions; namely, that a signal incident at port 2 is all transmitted to port 1, and a signal incident at port 3 is all transmitted to port 2. In this latter case, because of the minus sign in $b_2 = -a_3$, there is a reversal in voltage phase, but the power transmitted is not affected.

The three-port in Fig. 11 seems to have a cyclic power-transmission property. Power entering one port is transmitted to an adjacent port in a cyclic order, as shown by the circular arrow. A multiport device having this property is called a *circulator*, the symbol for which is shown in Fig. 11c for a three-port. For our circulator in Fig. 11, the cyclic order is 132. Clearly, a circulator of the opposite cyclic order (123) is also possible. Its scattering matrix must clearly be of the form

$$
\mathbf{S} = \begin{bmatrix} 0 & 0 & S_{13} \\ S_{21} & 0 & 0 \\ 0 & S_{32} & 0 \end{bmatrix}
$$

$$
= \begin{bmatrix} 0 & 0 & e^{j\theta_1} \\ e^{j\theta_2} & 0 & 0 \\ 0 & e^{j\theta_3} & 0 \end{bmatrix}
$$

where $|S_{13}| = |S_{21}| = |S_{32}| = 1$. At a single frequency the angles of the nonzero parameters can have any values without influencing the power transmitted—and we have so indicated on the right. For the simplest case all the angles are zero and the nonzero scattering parameters are all unity. More generally, each of the nonzero transmission coefficients can be an all-pass function.

The particularly happy way in which the scattering parameters are related to power transmission and reflection gives a further, though belated, justification for the normalization that we carried out in (37). It was done there for the purpose of arriving at a single scattering matrix rather than the two based on current and voltage. These simple interpretations of the scattering parameters in terms of power transmission and reflection would not have been possible had we continued to deal with either the current or the voltage scattering matrix.

8.4 PROPERTIES OF THE SCATTERING MATRIX

Since the scattering parameters have meanings intimately connected with power, the processes of power transmission are quite conveniently expressed in terms of scattering parameters. Assume a multiport with arbitrary terminating impedances in series with voltage sources, as in Fig 5, except that the terminations are arbitrary. The complex power input to the multiport in the sinusoidal steady state will be $W = \mathbf{V}^*\mathbf{I}$, where $\mathbf{V}^*$ is the conjugate transpose of $\mathbf{V}$. If you go through the details o substituting the normalized variables for the actual ones, you will find that the expression for power is invariant to normalization. Thus

$$W = \mathbf{V}_* \mathbf{I}_n = (\mathbf{a}^* + \mathbf{b}^*)(\mathbf{a} - \mathbf{b})$$
$$= (\mathbf{a}^*\mathbf{a} - \mathbf{b}^*\mathbf{b}) + (\mathbf{b}^*\mathbf{a} - \mathbf{a}^*\mathbf{b}). \tag{53}$$

Here (42) was used to replace the normalized voltage and current vectors by the scattering variables. The last term in parentheses on the far right is the difference of two conjugate quantities, since

$$(\mathbf{a}^*\mathbf{b}) = (\mathbf{b}^*\mathbf{a})^* = \overline{(\mathbf{b}^*\mathbf{a})}.$$

The last step here follows because these matrix products are scalars and the transpose of a scalar is itself. But the difference of two conjugates is imaginary. Hence the real power will be

$$P = \mathbf{a}^*\mathbf{a} - \mathbf{b}^*\mathbf{b} = \mathbf{a}^*\mathbf{a} - \mathbf{a}^*\mathbf{S}^*\mathbf{S}\mathbf{a}$$
$$= \mathbf{a}^*(\mathbf{U} - \mathbf{S}^*\mathbf{S})\mathbf{a}. \tag{54}$$

This was obtained by substituting $\mathbf{S}\mathbf{a}$ for $\mathbf{b}$. This equation should be compared with (19) for the one-port case.

The properties of the scattering matrix for different classes of networks can be established from (54). First observe that the right side is a quadratic form. For convenience, define

$$\mathbf{Q} = \mathbf{U} - \mathbf{S}^*\mathbf{S}, \tag{55}$$

so

$$P = \mathbf{a}^*\mathbf{Q}\mathbf{a}. \tag{56}$$

Without placing any restrictions on the type of network, let us take the conjugate transpose of $\mathbf{Q}$:

$$\mathbf{Q}^* = (\mathbf{U} - \mathbf{S}^*\mathbf{S})^* = \mathbf{U} - \mathbf{S}^*\mathbf{S} = \mathbf{Q}. \tag{57}$$

As discussed in Chapter 1, a matrix that is equal to its own conjugate transpose is said to be *Hermitian*. Thus $\mathbf{Q}$ is a Hermitian matrix. Its elements are related by $q_{ij} = \bar{q}_{ji}$, which requires the diagonal elements to be real.

We are interested in particular classes of networks: active and passive, reciprocal and nonreciprocal, lossless and lossy. There is not much of a specific nature that can be said about active networks. Hence we shall concentrate mainly on passive networks, which may be reciprocal or nonreciprocal. We shall also focus on the lossless subclass of passive networks; these also may be reciprocal or nonreciprocal.

First, for passive networks in general, the real power delivered to the multiport from sinusoidal sources at the ports must never be negative. Hence

$$\mathbf{U} - \mathbf{S}^*\mathbf{S} \text{ is positive semidefinite} \tag{58}$$

This is the fundamental limitation on the scattering matrix of a passive multiport. It should be compared with (22) for the one-port case.

A necessary and sufficient condition for a matrix to be positive semi-definite is that the principal cofactors be no less than zero, as discussed in Chapter 7. The diagonal elements of $\mathbf{Q}$ are also principal cofactors and must be non-negative. In terms of the elements of $\mathbf{S}$, this means

$$q_{jj} = 1 - \sum_i \bar{S}_{ij} S_{ij} = 1 - \sum_i |S_{ij}|^2 \geq 0. \tag{59}$$

Each term in this summation is positive. The expression tells us that a sum of positive terms cannot exceed unity. This requires, *a fortiori*, that each term not exceed unity, or

$$|S_{ij}(j\omega)| \leq 1. \tag{60}$$

This is a fundamental limitation imposed on the scattering parameters as a consequence of passivity. It tells us that *for a passive network, the magnitude of a reflection coefficient cannot exceed unity, nor can the magnitude of a transmission coefficient.*

Next, consider a lossless multiport, whether reciprocal or nonreciprocal.

In this case no power is dissipated within the multiport. Hence the real-power input shown in (56) must be identically zero for any possible vector **a**. This is possible only if the matrix of the quadratic form vanishes; that is,

$$\mathbf{Q} = \mathbf{U} - \mathbf{S^*S} = 0$$

or

$$\mathbf{S^*S} = \mathbf{U} = \mathbf{SS^*}.$$

(61)

By definition of the inverse, we see that $\mathbf{S}^{-1} = \mathbf{S^*}$. A matrix whose inverse equals its conjugate transpose is called a *unitary matrix*. Thus *the scattering matrix of a lossless multiport is unitary*. (The last step in the equation follows because a matrix commutes with its inverse.)

The unitary property imposes some constraints on the elements of the scattering matrix that can be established by expanding the products in (61). The result will be

$$|S_{1j}|^2 + |S_{2j}|^2 + |S_{3j}|^2 + \cdots + |S_{nj}|^2 = 1$$

$$\bar{S}_{1j}S_{1k} + \bar{S}_{2j}S_{2k} + \cdots + \bar{S}_{nj}S_{nk} = 0$$

from which

$$\sum_{i=1}^{n} \bar{S}_{ij}S_{ik} = \delta_{jk},$$

(62)

or

$$|S_{j1}|^2 + |S_{j2}|^2 + |S_{j3}|^2 + \cdots + |S_{jn}|^2 = 1$$

$$S_{j1}\bar{S}_{k1} + S_{j2}\bar{S}_{k2} + \cdots + S_{jn}\bar{S}_{kn} = 0$$

from which

$$\sum_{i=1}^{n} S_{ji}\bar{S}_{ki} = \delta_{jk},$$

(63)

where δ_{jk} is the Kronecker delta.

TWO-PORT NETWORK PROPERTIES

The immediately preceding equations specify properties of the scattering parameters of multiports. We shall examine these in detail for the two-port network, specifically limiting ourselves to lossless networks, both reciprocal and nonreciprocal

First, with $n = 2$, set $j = k = 1$ in (62); then $j = k = 2$ in (63). The results will be

$$|S_{11}|^2 + |S_{21}|^2 = 1, \tag{64a}$$

$$|S_{21}|^2 + |S_{22}|^2 = 1. \tag{64b}$$

Subtracting one from the other gives

$$|S_{11}(j\omega)|^2 = |S_{22}(j\omega)|^2. \tag{65}$$

Thus, *for a lossless two-port, whether reciprocal or nonreciprocal, the magnitude of the reflection coefficients at the two ports is equal.* This result can be extended to complex frequencies by analytic continuation. Using the symbol ρ for reflection coefficient, the result can be written as follows:

$$\rho_1(s)\, \rho_1(-s) = \rho_2(s)\, \rho_2(-s). \tag{66}$$

In terms of poles and zeros, we conclude the following. The poles and zeros of $\rho_1(s)\, \rho_1(-s)$ and $\rho_2(s)\, \rho_2(-s)$ are identical, and they occur in quadrantal symmetry. In making the assignment of poles and zeros of $\rho_1(s)$ from those of $\rho_1(s)\, \rho_1(-s)$, the only consideration is stability. No poles of $\rho_1(s)$ can lie in the right half-plane. Hence the left-half-plane poles of $\rho_1(s)\, \rho_1(-s)$ must be poles of $\rho_1(s)$. As for the zeros, no limitation is imposed by stability; zeros of $\rho_1(s)\, \rho_1(-s)$ can be assigned to $\rho_1(s)$ from either the left or the right half-plane, subject to the limitation that these zeros, plus their images in the $j\omega$-axis, must account for all the zeros of $\rho_1(s)\, \rho_1(-s)$. Similar statements can be made about the poles and zeros of $\rho_2(s)$.

Let us return again to (62) and this time set $j = k = 2$. The result will be

$$|S_{12}|^2 + |S_{22}|^2 = 1. \tag{67}$$

When this is compared with (64b), we see that

$$|S_{12}(j\omega)|^2 = |S_{21}(j\omega)|^2; \tag{68}$$

that is, the magnitude of the forward-transmission coefficient equals that of the reverse-transmission coefficient. This is not surprising for reciprocal networks, since then $S_{12}(s) = S_{21}(s)$, but it is true for nonreciprocal networks also. In fact, even more detailed relationships can be found by setting $j = 2$, $k = 1$ in (62). The result will be

$$S_{11}(j\omega) = -\frac{S_{21}(j\omega)}{\bar{S}_{12}(j\omega)}\, \bar{S}_{22}(j\omega) \tag{69a}$$

or

$$\rho_1(s) = -\frac{S_{21}(s)}{S_{12}(-s)}\, \rho_2(-s). \tag{69b}$$

This applies to both reciprocal and nonreciprocal lossless two-ports. For the reciprocal case $S_{12} = S_{21}$; hence the ratio of $S_{21}(s)$ to $S_{12}(-s)$ will then be an all-pass function. Since $\rho_1(s)$ can have no poles in the right half-plane, the zeros of this all-pass function must cancel the right-half-plane poles of $\rho_2(-s)$.

AN APPLICATION—FILTERING OR EQUALIZING

A number of different applications can be handled by the configuration shown in Fig. 12a. A lossless coupling network N is to be designed for

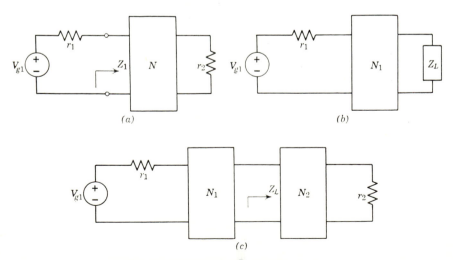

Fig. 12.   Filter, equalizer or matching network.

insertion between a source with a real internal impedance and a real load. The network may be required to perform the function of filtering or equalizing; that is, shaping the frequency response in a prescribed way.

Alternatively, the load may be an impedance Z_L as shown in Fig. 12b, and the coupling network is to be designed to provide a match to the resistive source over a band of frequencies. This case can be related to the former one by using Darlington's theorem. That is to say, Z_L can be realized as a lossless two-port N_2 terminated in a resistance r_2, as shown in Fig. 12c. The cascade combination of N_1 and N_2 plays the role of the lossless network N in Fig. 12a. A limited discussion of this matching problem will be given in the next section.

Let us concentrate on the filter or equalizer problem. What is specified is the transducer power gain as a function of frequency. We shall label this function $\mathscr{G}(\omega^2)$. According to (48), $\mathscr{G}(\omega^2)$ is simply the magnitude square of S_{21}. But $|S_{21}|^2$ is related to the magnitude of the input reflection coefficient by (64a). If this expression is analytically continued, it can be written as

$$\rho_1(s)\,\rho_1(-s) = 1 - S_{21}(s)\,S_{21}(-s), \tag{70}$$

where we have again replaced S_{11} by ρ_1. If $\mathscr{G}(\omega^2)$ is specified, the right-hand side is a known even function. It is now only necessary to assign the poles and zeros of the right side appropriately to $\rho_1(s)$.

Furthermore, the reflection coefficient ρ_1 and the impedance Z_1 looking into the input terminals of N with the output terminated in r_2 are related by (46), which can be solved for Z_1 as follows:

$$\frac{Z_1(s)}{r_1} = \frac{1 + \rho_1(s)}{1 - \rho_1(s)}. \tag{71}$$

Hence, once $\rho_1(s)$ has been determined from (70), the input impedance Z_1 becomes known as a function of s. The task is then reduced to an application of Darlington's theorem; namely, realizing $Z_1(s)$ as a lossless two-port terminated in a resistance.

To illustrate this discussion let the transducer power gain be given as

$$\mathscr{G}(\omega^2) = \frac{1}{1 + \omega^6}.$$

This is a third-order Butterworth filter function. The continuation is obtained through replacing $(-\omega^2)$ by s^2. When this is inserted into (70), the result is

$$\rho_1(s)\,\rho_1(-s) = 1 - \frac{1}{1-s^6} = \frac{-s^6}{1-s^6}$$

$$= \frac{-s^6}{(1+s)(1+s+s^2)(1-s)(1-s+s^2)}.$$

In the last step the denominator has been factored, putting into evidence the left- and right-half-plane poles. The left-half-plane poles must belong to $\rho_1(s)$, as opposed to $\rho_1(-s)$. In the example the zeros also are uniquely assignable: three zeros at the origin to $\rho_1(s)$ and three to $\rho_1(-s)$. The only ambiguity arises in the appropriate sign. There is no *a priori* reason why $\rho_1(s)$ must have a positive sign. The conclusion is that $\rho_1(s)$ must be the following:

$$\rho_1(s) = \frac{\pm s^3}{(s+1)(s^2+s+1)}.$$

When this is inserted into (71), the impedance is found to be

$$\frac{Z_1(s)}{r_1} = \frac{2s^2+2s+1}{2s^3+2s^2+2s+1} \quad \text{or} \quad \frac{2s^3+2s^2+2s+1}{2s^2+2s+1},$$

depending on the sign chosen for $\rho_1(s)$. But these are inverse impedances, and their realizations will be duals. In the present case the realization is rather simply obtained by expanding in a continued fraction. Thus, by using the second function, we obtain

$$\frac{Z_1(s)}{r_1} = s + \cfrac{1}{2s + \cfrac{1}{s+1}}.$$

The network realizing this function, and its dual, are shown in Fig. 13. These are the normalized realizations. Recall that the normalized open-circuit impedance matrix is obtained by dividing all elements of $\mathbf{Z}_{oc}$ by $\sqrt{r_1 r_2}$. To undo this normalization, all branch impedances must, therefore, be multiplied by $\sqrt{r_1 r_2}$. One denormalized network, with the actual source and load resistances, is shown in Fig. 13c.

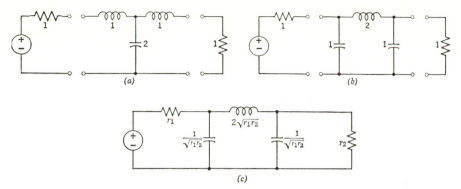

Fig. 13. Illustrative example.

LIMITATIONS INTRODUCED BY PARASITIC CAPACITANCE

The general matching problem illustrated in Fig. 12b and c takes on some additional significance for the special case shown in Fig. 14a, where

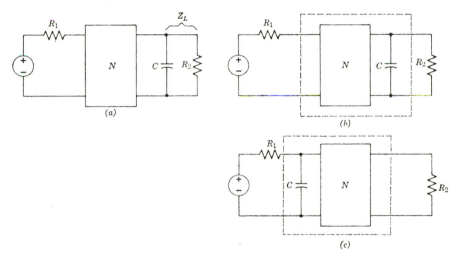

Fig. 14. Two-port constrained by parasitic capacitance.

the load impedance is the parallel combination of a capacitor and a resistor. This configuration is identical with that of a network working between two resistive terminations R_1 and R_2, but constrained by a parasitic shunt capacitance across the output terminals, which can be treated as part of the two-port, as suggested in Fig. 14b. This is equivalent to the situation illustrated in Fig. 14c, where a parasitic shunt capacitance occurs across the input port.

We have already seen in Chapter 6 that such a parasitic capacitance leads to some integral constraints on the real part of the driving-point impedance. We shall here derive a similar integral constraint on the reflection coefficient and use it to find a limitation on the transducer power gain.

The two situations in Fig. 14b and c are similar and can be treated simultaneously by considering Fig. 15, where R is either R_1 or R_2. In the

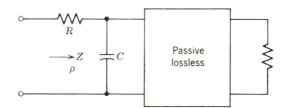

Fig. 15.   Shunt capacitance constraint.

first case, Z is the impedance Z_1 looking into the left-hand port of the dashed two-port in Fig. 14c, with R_2 in place; and ρ is the corresponding reflection coefficient ρ_1. For $R = R_2$, Z is the impedance Z_2 looking into the right-hand port of the dashed two-port in Fig. 14b, with the other port terminated in R_1; and ρ is the corresponding reflection coefficient ρ_2. In either case,

$$\rho = \frac{Z - R}{Z + R}.$$

Ideally, under matched conditions, $\rho = 0$. This requires $Z = R$ independent of frequency, at least over the frequency band of interest. But this cannot be achieved exactly; it might only be approximated. If ρ cannot be identically zero, at least we would like to make its magnitude as close to zero as possible.

It is customary to define another quantity related to ρ as follows:

$$\text{return loss} = \ln \frac{1}{|\rho(j\omega)|}. \tag{72}$$

When $\rho = 0$, the return loss is infinite; under totally mismatched conditions, (when $\rho = 1$), the return loss is zero. Thus maximizing the return loss over a band of frequencies is a measure of the optimization of matching over this band.

An integral constraint on the return loss can be obtained by taking the contour integral of $\ln (1/\rho)$ around the standard contour—the $j\omega$-axis and an infinite semicircle to the right, as described in Chapter 6 and shown again in Fig. 16. To apply Cauchy's theorem, the integrand must be regular in the right half-plane. However, although ρ is regular in the right

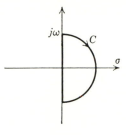

Fig. 16

half-plane, $1/\rho$ need not be. Hence we first multiply $1/\rho$ by an all-pass function, as follows:

$$\frac{1}{\rho} A(s) = \frac{Z+R}{Z-R} \cdot \frac{(s-s_1)(s-s_2)\cdots(s-s_n)}{(s+s_1)(s+s_2)\cdots(s+s_n)}. \tag{73}$$

where each s_k is a pole of $1/\rho$ in the right half-plane. The resulting function is regular in the right half-plane, and the contour integration can be carried out. There will be contributions to the contour integral from the $j\omega$-axis and from the infinite semicircle. The evaluate the latter, observe that as $s \to \infty$, $Z \to 1/sC$ because of the shunt capacitor. Hence

$$\frac{Z+R}{Z-R} \xrightarrow[s \to \infty]{} \frac{\dfrac{1}{sC}+R}{\dfrac{1}{sC}-R} \to -\left(1+\frac{2}{RCs}\right). \tag{74}$$

Also, the all-pass function approaches

$$A(s) = \frac{s^n - s^{n-1}\sum s_k + \cdots}{s^n + s^{n-1}\sum s_k + \cdots} \xrightarrow[s \to \infty]{} 1 - \frac{2\sum s_k}{s}. \tag{75}$$

As a consequence

$$\frac{1}{\rho} A(s) \xrightarrow[s \to \infty]{} -\left[1+\frac{2}{s}\left(\frac{1}{RC}-\sum s_k\right)\right]. \tag{76}$$

The negative sign before the right-hand side suggests that we take the logarithm of $-A(s)/\rho$ instead of $+A(s)/\rho$. Thus, since $\ln(1+x) \to x$ for $|x| \ll 1$,

$$\int_C \ln\left[-\frac{A(s)}{\rho}\right] ds \to \int_C \frac{2}{s}\left(\frac{1}{RC} - \sum s_k\right) ds = -j2\pi\left(\frac{1}{RC} - \sum s_k\right), \quad (77)$$

where C is the infinite semicircular part of the standard contour and $\int_C ds/s = -j\pi$.

Along the $j\omega$-axis,

$$\int_{-\infty}^{\infty} \ln\left[-\frac{A(j\omega)}{\rho(j\omega)}\right] jd\omega = \int_{-\infty}^{\infty} \ln\left|\frac{1}{\rho}\right| jd\omega + \int_{-\infty}^{\infty} j \arg\left[-\frac{A(j\omega)}{\rho(j\omega)}\right] jd\omega$$

$$= j2\int_0^{\infty} \ln\frac{1}{|\rho|} d\omega. \quad (78)$$

The first step follows from the fact that the magnitude of an all-pass function is unity on the $j\omega$-axis, so $\ln|A(j\omega)/\rho(j\omega)| = \ln|1/\rho|$. The last step follows because $\ln|1/\rho|$ is an even function of ω, whereas the angle is an odd function.

From Cauchy's theorem, the sum of the integrals on the left sides of the last two equations should equal zero. Hence

$$\int_0^{\infty} \ln\frac{1}{|\rho(j\omega)|} d\omega = \pi\left(\frac{1}{RC} - \sum s_k\right). \quad (79)$$

Recall that s_k is a pole of $1/\rho$ in the right half-plane, so that its real part is positive. The sum of all such poles will, therefore, be real and positive. If $1/\rho$ has no poles in the right half-plane, this sum will vanish. The final result will therefore be

$$\int_0^{\infty} \ln\frac{1}{|\rho(j\omega)|} d\omega \le \frac{\pi}{RC}. \quad (80)$$

This is a fundamental limitation on the return loss (or the reflection coefficient) when a two-port matching network is constrained by a shunt capacitor across one port.

This constraint places a limitation on the achievable transducer power gain also. To illustrate this, suppose the band of interest is the low-frequency region $0 \le \omega \le \omega_c$, which is the passband, the remainder of the frequency range being the stopband. There will be contributions to the integral in (80) from both bands. The most favorable condition will occur when $1/\rho$ has no poles in the right half-plane and the magnitude of ρ is constant, say $|\rho_0|$, in the passband. Then

$$\int_0^\infty \ln \frac{1}{|\rho|} \, d\omega = \int_0^{\omega_c} \ln \frac{1}{|\rho_0|} \, d\omega + \int_{\omega_c}^\infty \ln \frac{1}{|\rho|} \, d\omega$$

$$= \omega_c \ln \frac{1}{|\rho_0|} + \int_{\omega_c}^\infty \ln \frac{1}{|\rho|} \, d\omega = \frac{\pi}{RC}. \tag{81}$$

Outside the passband, there should be a total mismatch and $|\rho|$ should ideally equal 1. More practically, although its value will be less than 1, it should be close to 1. Hence $\ln 1/|\rho|$ will be a small positive number, ideally zero. Therefore the integral from ω_c to ∞ will be positive, and (81) will yield

$$\ln \frac{1}{|\rho_0|} \leq \frac{\pi}{\omega_c \, RC} \quad \text{or} \quad \frac{1}{|\rho_0|} \leq \epsilon^{\pi/\omega_c \, RC}$$

$$|\rho_0| \geq \epsilon^{-\pi/\omega_c RC}. \tag{82}$$

When this expression is combined with (64), the magnitude squared of S_{21} becomes

$$|S_{21}(j\omega)|^2 \leq 1 - \epsilon^{-2\pi/\omega_c \, RC}. \tag{83}$$

This puts an upper limit on the achievable transducer power gain, $|S_{21}(j\omega)|$, over a wide frequency band, even if we assume a constant value is possible over the passband. The wider the frequency band of interest, the more stringent will this limitation become for a fixed shunt capacitance. Note that the immediately preceding result is valid for Fig. 14c; it also applies to Fig. 14b through (65) and (68).

8.5   COMPLEX NORMALIZATION

The scattering parameters treated up to this point were defined for a multiport operating with resistive terminations. Normalization was carried out with these real numbers. Suppose the terminations of a multiport are not resistances but impedances; how do we normalize them? We shall now turn our attention to this general problem. For simplicity, we shall illustrate with a two-port, but the results will apply in matrix form to the general multiport.

The situation to be treated is illustrated with a two-port in Fig. 17.

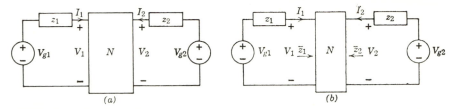

Fig. 17.   Two-port with general terminations: (*a*) general case; (*b*) matched case.

The terminating impedances are strictly passive with positive real parts:

$$r_k = \operatorname{Re} z_k > 0. \tag{84}$$

These real parts play the same normalizing role as the resistive terminations before, and they are represented by the positive definite diagonal matrix in (24). From Fig. 17*a*, we can write

$$\begin{bmatrix} V_{g1} \\ V_{g2} \end{bmatrix} = \begin{bmatrix} V_1 \\ V_2 \end{bmatrix} + \begin{bmatrix} z_1 & 0 \\ 0 & z_2 \end{bmatrix} \begin{bmatrix} I_1 \\ I_2 \end{bmatrix}$$

which can be generalized for the multiport as

$$\mathbf{V}_g = \mathbf{V} + \mathbf{z}\mathbf{I} = (\mathbf{Z}_{\mathrm{oc}} + \mathbf{z})\mathbf{I}. \tag{85}$$

The right-hand side results from the substitution of $\mathbf{V} = \mathbf{Z}_{\mathrm{oc}}\,\mathbf{I}$. This should be compared with (25) for resistive terminations.

Now suppose that the multiport is simultaneously matched to the terminating impedances at all the ports; that is, the impedance looking into port j of the multiport is $\bar{z}_j$. This is shown in Fig. 17*b* for the two-port. From Fig. 17,

$$\begin{bmatrix} V_1 \\ V_2 \end{bmatrix} = \begin{bmatrix} \bar{z}_1 & 0 \\ 0 & \bar{z}_2 \end{bmatrix} \begin{bmatrix} I_1 \\ I_2 \end{bmatrix} \quad \text{or} \quad \mathbf{V} = \bar{\mathbf{z}}\mathbf{I} \quad \textit{when matched.} \tag{86}$$

As before, the incident voltages and currents are defined as the port voltages and currents under matched conditions; that is, $\mathbf{V}_i = \mathbf{V}$ and $\mathbf{I}_i = \mathbf{I}$, when matched. Hence, generalizing the last equation and inserting into (85), we obtain

$$\mathbf{I}_i = \tfrac{1}{2}\,\mathbf{r}^{-1}\mathbf{V}_g\,, \tag{87a}$$

$$\mathbf{V}_i = \overline{\mathbf{z}}\mathbf{I}_i = \tfrac{1}{2}\,\overline{\mathbf{z}}\mathbf{r}^{-1}\mathbf{V}_g\,, \tag{87b}$$

since $\mathbf{z} + \overline{\mathbf{z}} = 2\mathbf{r}$. These should be compared with (28) for real terminations. Note that the expression for $\mathbf{V}_i$ in terms of $\mathbf{V}_g$ is not as simple as it is in the case of real terminations.

The condition of simultaneous match at all ports, requiring as it does that $\mathbf{Z}_{oc}$ be the conjugate of the terminating impedance $\mathbf{z}$, is not possible of attainment at all frequencies, but only at a single frequency. Hence the procedure being described here is strictly applicable only at a single frequency, which may be at any point on the $j\omega$-axis. It may also be used in narrow-band applications without excessive error.

Again we define the reflected variables as deviations from their matched values, according to (27). When (85) and (87) are inserted there, and after some manipulation whose details you should supply, the incident and reflected variables become

$$\mathbf{V}_i = \tfrac{1}{2}\,\overline{\mathbf{z}}\mathbf{r}^{-1}(\mathbf{V} + \mathbf{z}\mathbf{I}), \tag{88a}$$

$$\mathbf{V}_r = \tfrac{1}{2}\,\mathbf{z}\mathbf{r}^{-1}(\mathbf{V} - \overline{\mathbf{z}}\mathbf{I}), \tag{88b}$$

$$\mathbf{I}_i = \tfrac{1}{2}\,\mathbf{r}^{-1}(\mathbf{V} + \mathbf{z}\mathbf{I}), \tag{88c}$$

$$\mathbf{I}_r = \tfrac{1}{2}\,\mathbf{r}^{-1}(\mathbf{V} - \overline{\mathbf{z}}\mathbf{I}). \tag{88d}$$

Again we note that the expressions for the voltages are somewhat complicated compared with those for the currents.

Let us now introduce voltage, current, and impedance normalizations. To normalize currents we multiply by $\mathbf{r}^{1/2}$, and to normalize voltages we multiply by $\mathbf{r}^{-1/2}$. Impedances are normalized according to (39b). The normalized incident and reflected variables become

$$\mathbf{r}^{1/2}\mathbf{I}_i = \tfrac{1}{2}\,(\mathbf{V}_n + \mathbf{z}_n\,\mathbf{I}_n) = \tfrac{1}{2}\,(\mathbf{Z}_n + \mathbf{z}_n)\mathbf{I}_n\,, \tag{89a}$$

$$\mathbf{r}^{-1/2}\mathbf{V}_i = \tfrac{1}{2}\,\overline{\mathbf{z}}_n(\mathbf{V}_n + \mathbf{z}_n\,\mathbf{I}_n) = \tfrac{1}{2}\,\overline{\mathbf{z}}_n(\mathbf{Z}_n + \mathbf{z}_n)\mathbf{I}_n \tag{89b}$$

and

$$\mathbf{r}^{1/2}\mathbf{I}_r = \tfrac{1}{2}\,(\mathbf{V}_n - \overline{\mathbf{z}}_n\,\mathbf{I}_n) = \tfrac{1}{2}\,(\mathbf{Z}_n - \overline{\mathbf{z}}_n)\mathbf{I}_n\,, \tag{90a}$$

$$\mathbf{r}^{-1/2}\mathbf{V}_r = \tfrac{1}{2}\,\mathbf{z}_n(\mathbf{V}_n - \overline{\mathbf{z}}_n\,\mathbf{I}_n) = \tfrac{1}{2}\,\mathbf{z}_n(\mathbf{Z}_n - \overline{\mathbf{z}}_n)\mathbf{I}_n\,, \tag{90b}$$

where $\mathbf{z}_n = \mathbf{r}^{-1/2}\mathbf{z}\mathbf{r}^{-1/2} = [z_j/r_j]$. Examine these expressions carefully. For the case of real terminations, both quantities in (89) are the same; they were together labeled $\mathbf{a}$ in (37). Likewise, both quantities in (90) are the same for real terminations; they were collectively labeled $\mathbf{b}$ before. Clearly, this is no longer true for impedance terminations. Observe that if $\mathbf{z}$ is made real, the two expressions in (89) reduce to $\mathbf{a}$ and the two in (90) reduce to $\mathbf{b}$ in (42).

Two different scattering matrices can be defined, one for the currents and one for the voltages, even for the normalized variables. We shall arbitrarily define $\mathbf{a}$ and $\mathbf{b}$ as the *normalized* incident and reflected *currents*. Thus from (89) and (90) we get

$$\mathbf{a} = \mathbf{r}^{1/2}\mathbf{I}_i = \tfrac{1}{2}\,(\mathbf{V}_n + \mathbf{z}_n\mathbf{I}_n) = \tfrac{1}{2}\,(\mathbf{Z}_n + \mathbf{z}_n)\mathbf{I}_n\,, \tag{91a}$$

$$\mathbf{b} = \mathbf{r}^{1/2}\mathbf{I}_r = \tfrac{1}{2}\,(\mathbf{V}_n - \bar{\mathbf{z}}_n\mathbf{I}_n) = \tfrac{1}{2}\,(\mathbf{Z}_n - \bar{\mathbf{z}}_n)\mathbf{I}_n\,. \tag{91b}$$

We now define the scattering matrix $\mathbf{S}$, as before, by the relationship $\mathbf{b} = \mathbf{S}\mathbf{a}$. When (91) are inserted for $\mathbf{a}$ and $\mathbf{b}$, we can solve for the scattering matrix:

$$\tfrac{1}{2}\,(\mathbf{Z}_n - \bar{\mathbf{z}}_n)\mathbf{I}_n = \mathbf{S}\tfrac{1}{2}\,(\mathbf{Z}_n + \mathbf{z}_n)\mathbf{I}_n\,,$$

$$\mathbf{S} = (\mathbf{Z}_n - \bar{\mathbf{z}}_n)(\mathbf{Z}_n + \mathbf{z}_n)^{-1}. \tag{92}$$

This should be compared with (41) for the case of real terminations. Note that $(\mathbf{Z}_n + \mathbf{z}_n)^{-1}$ is the normalized admittance matrix of the augmented network, $\mathbf{Y}_{an}$. Hence, by adding and subtracting $\mathbf{z}_n$ within the first parentheses in (92), this expression can be rewritten as

$$\mathbf{S} = \mathbf{U} - 2\mathbf{Y}_{an}\,, \tag{93}$$

since $\mathbf{z}_n + \bar{\mathbf{z}}_n = 2\mathbf{U}$. This is the same expression as (40) for real terminations.

Another matrix, say $\hat{\mathbf{S}}$, can be defined for the normalized voltage variables by writing $\mathbf{r}^{-1/2}\mathbf{V}_r = \hat{\mathbf{S}}\mathbf{r}^{-1/2}\mathbf{V}_i$. Inserting from (89) and (90) leads to

$$\hat{\mathbf{S}} = \mathbf{z}_n(\mathbf{Z}_n - \bar{\mathbf{z}}_n)(\mathbf{Z}_n + \mathbf{z}_n)^{-1}\bar{\mathbf{z}}_n^{-1}. \tag{94}$$

Comparing this with (92) gives the relationship between the two matrices as

$$\hat{\mathbf{S}}\bar{\mathbf{z}}_n = \mathbf{z}_n\mathbf{S}. \tag{95}$$

The relative complexity of the voltage scattering matrix $\hat{\mathbf{S}}$ is adequate reason for our having chosen the definitions of $\mathbf{a}$ and $\mathbf{b}$, and the scattering matrix $\mathbf{S}$ as we did. But even more pertinent is their relation to power. To see this, first solve for $\mathbf{V}_n$ and $\mathbf{I}_n$ in terms of $\mathbf{a}$ and $\mathbf{b}$ from (91). Thus

$$\mathbf{V}_n = \bar{\mathbf{z}}_n \mathbf{a} + \mathbf{z}_n \mathbf{b}, \qquad (96a)$$

$$\mathbf{I}_n = \mathbf{a} - \mathbf{b}, \qquad (96b)$$

since $\mathbf{z}_n + \bar{\mathbf{z}}_n = 2\mathbf{U}$. These are not quite the same as (42) for the case of real terminations. However, if the expression for $\mathbf{V}_n^* \mathbf{I}_n$ is formed similar to (53), the expression for the real part will turn out to be exactly the same in terms of the scattering matrix defined here as it was before in (54).

Hence the properties of the scattering matrix as discussed in Section 84 apply also to the scattering matrix $\mathbf{S}$ given in (92). This is, then, the appropriate extension of the scattering variables and the scattering matrix to complex normalization.

FREQUENCY-INDEPENDENT NORMALIZATION

The preceding discussion concerning complex normalization, being based on optimal matching in the sinusoidal steady state, is limited to a single frequency. We shall now extend the discussion to arbitrary signals and to all values of s. In Fig. 18 is shown a two-port that is representative

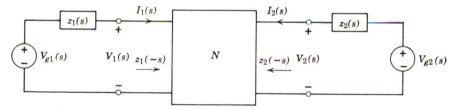

Fig. 18. General matching with arbitrary signals and passive impedance terminations.

of any multiport, excited by arbitrary voltage signals through passive impedance terminations. For the general case,

$$\mathbf{V}_g(s) = \mathbf{V}(s) + \mathbf{z}(s)\mathbf{I}(s) \qquad (97a)$$

$$= [\mathbf{Z}_{oc}(s) + \mathbf{z}(s)]\, \mathbf{I}(s). \qquad (97b)$$

These are identical with (85) except that they apply for all s. Now let

$$\mathbf{r}(s) = \tfrac{1}{2}\left\{\mathbf{z}(s) + \mathbf{z}(-s)\right\}; \qquad (98)$$

that is, $r_1(s)$ and $r_2(s)$ are the even parts of $z_1(s)$ and $z_2(s)$. Since $\mathbf{r}(s)$ is even, $\mathbf{r}(-s) = \mathbf{r}(s)$; so we shall not write the argument of $\mathbf{r}$ except for emphasis. Without making any claims about real power matching, let us define a condition of the two-port in which

$$\mathbf{Z}_{oc}(s) = \begin{bmatrix} z_1(-s) & 0 \\ 0 & z_2(-s) \end{bmatrix} \quad \text{or} \quad \mathbf{Z}_{oc}(s) = \mathbf{z}(-s). \tag{99}$$

We take this as a reference condition and call it the *matched condition**. When $s = j\omega$, $\mathbf{r}$ is simply the real part of $\mathbf{z}(s)$.

As before, the incident voltage and current are defined as the port voltages and currents under matched conditions; that is, $\mathbf{V}_i(s) = \mathbf{V}(s)$ and $\mathbf{I}_i(s) = \mathbf{I}(s)$ when $\mathbf{Z}_{oc} = \mathbf{z}(-s)$. From Fig. 18 and from (97) there follows that

$$\mathbf{I}_i(s) = \tfrac{1}{2}\,\mathbf{r}^{-1}\mathbf{V}_g(s) = \tfrac{1}{2}\,\mathbf{r}^{-1}\left[\mathbf{V}(s) + \mathbf{z}(s)\,\mathbf{I}(s)\right], \tag{100a}$$

$$\mathbf{V}_i(s) = \mathbf{z}(-s)\,\mathbf{I}_i(s) = \tfrac{1}{2}\,\mathbf{z}(-s)\mathbf{r}^{-1}\mathbf{V}_g(s) \tag{100b}$$

$$= \tfrac{1}{2}\,\mathbf{z}(-s)\mathbf{r}^{-1}\left[\mathbf{V}(s) + \mathbf{z}(s)\,\mathbf{I}(s)\right].$$

The reflected variables are again defined as deviations from the reference (matched) condition, just as before. By inserting (100) and (97) into $\mathbf{I}_r = \mathbf{I}_i - \mathbf{I}$ and $\mathbf{V}_r = \mathbf{V} - \mathbf{V}_i$, we obtain

$$\mathbf{I}_r(s) = \tfrac{1}{2}\,\mathbf{r}^{-1}\left[\mathbf{V}(s) - \mathbf{z}(-s)\,\mathbf{I}(s)\right] \tag{101a}$$

$$\mathbf{V}_r(s) = \tfrac{1}{2}\,\mathbf{z}(s)\mathbf{r}^{-1}\left[\mathbf{V}(s) - \mathbf{z}(-s)\,\mathbf{I}(s)\right]. \tag{101b}$$

Compare these with (88); note that $\bar{\mathbf{z}}$ has been replaced by $\mathbf{z}(-s)$. When $s = j\omega$, $\mathbf{z}(-s) = \mathbf{z}(-j\omega) = \mathbf{z}[\overline{(j\omega)}] = \bar{\mathbf{z}}(j\omega)$.

Another useful form is obtained by inserting $\mathbf{V} = \mathbf{Z}_{oc}\,\mathbf{I}$. Then

$$\mathbf{I}_i = \tfrac{1}{2}\,\mathbf{r}^{-1}[\mathbf{Z}_{oc}(s) + \mathbf{z}(s)]\,\mathbf{I}(s) \qquad \mathbf{V}_i(s) = \mathbf{z}(-s)\,\mathbf{I}_i(s), \tag{102a}$$

$$\mathbf{I}_r = \tfrac{1}{2}\,\mathbf{r}^{-1}[\mathbf{Z}_{oc}(s) - \mathbf{z}(-s)]\,\mathbf{I}(s) \qquad \mathbf{V}_r(s) = \mathbf{z}(s)\,\mathbf{I}_r(s). \tag{102b}$$

The next step, as before, is normalization. However, this process is now more complicated. Previously the matrix $\mathbf{r}$ was a matrix of real numbers,

---

* A more complete treatment in the time domain is given in E. S. Kuh and R. A. Rohrer, *Theory of Linear Active Networks*, Holden-Day, San Francisco, 1967, pp. 287–300.

and normalization amounted to dividing or multiplying the components of vectors to be normalized by the square roots of the scalar elements of **r**. Now **r**(s) is an even rational function of s. The process of taking the square root is not that simple. Let us digress here a moment in order to examine the properties of **r**(s).

Let us start with the impedance **z**, whose elements are rational positive real functions, hence regular in the right half-plane. We write the numerator and denominator of the jth element of **z** as $n_j(s)$ and $d_j(s)$. Thus

$$z_j(s) = \frac{n_j(s)}{d_j(s)}. \tag{103}$$

Then

$$r_j(s) = \tfrac{1}{2}\left[\frac{n_j(s)}{d_j(s)} + \frac{n_j(-s)}{d_j(-s)}\right] = \frac{n_j(s)\,d_j(-s) + n_j(-s)\,d_j(s)}{2d_j(s)\,d_j(-s)}. \tag{104}$$

Both the numerator and denominator of the right side are even polynomials of s. Hence their zeros occur in quadrantal symmetry. We now define a rational function $f_j(s)$ whose zeros and poles are all the zeros and poles of $r_j(s)$ that lie in the left half-plane. The function containing all the right-half-plane poles and zeros of $r_j(s)$ will then be $f_j(-s)$. Thus

$$r_j(s) = f_j(s)\,f_j(-s). \tag{105}$$

Each of the function $f_j(s)$ and $f_j(-s)$ is something like a " square root " of $r_j(s)$, since their product gives $r_j(s)$.

Before we proceed, here is a simple example. Suppose z_j is the positive real function

$$z_j(s) = \frac{s+4}{s+1}.$$

Then

$$r_j(s) = \tfrac{1}{2}\left(\frac{s+4}{s+1} + \frac{-s+4}{-s+1}\right) = \frac{(s+2)(s-2)}{(s+1)(s-1)}.$$

Clearly,

$$f_j(s) = \frac{s+2}{s+1}, \qquad f_j(-s) = \frac{s-2}{s-1}.$$

Observe that the real parts of these two functions for $s = j\omega$ are equal; thus for complex normalization at a single frequency $s = j\omega$, either one will do.

We dealt here with a single element of the $\mathbf{r}(s)$ matrix. In matrix form what was just described becomes

$$\mathbf{r}(s) = \mathbf{f}(s)\,\mathbf{f}(-s), \tag{106}$$

where $\mathbf{f}(s)$ is a diagonal matrix each of whose nonzero elements is a rational function whose numerator and denominator are both Hurwitz polynomials.

Let us now return to the main discussion. Our impulse is to normalize by multiplying currents by the " square root " of $\mathbf{r}(s)$, but which " square root," $\mathbf{f}(s)$ or $\mathbf{f}(-s)$? Our guide must be the desire that the scattering variables to be defined must lead to a scattering matrix that satisfies the fundamental relationship relative to power given in (54). In the present case the conjugate of a vector is generalized by making the argument of the vector $(-s)$. Thus $\bar{V}$ becomes $\mathbf{V}(-s)$, which reduces to $\bar{V}$ when $s = j\omega$ What should be done is to form an expression for the power, and from it to determine the required normalization.

The first step is to solve (100a) and (101a) for the voltage and current in terms of the incident and reflected currents. The result will be

$$\begin{aligned}
\mathbf{V}(s) &= \mathbf{r}(s)\,[\mathbf{I}_i(s) + \mathbf{I}_r(s)], \\
\mathbf{I}(s) &= \qquad \mathbf{I}_i(s) - \mathbf{I}_r(s).
\end{aligned} \tag{107}$$

Next we form $\mathbf{V}'(-s)\,\mathbf{I}(s)$ and take the even part. Calling this P, we get

$$\begin{aligned}
P &= \text{even part of } \mathbf{V}'(-s)\,\mathbf{I}(s) \\
&= \text{even part of } [\mathbf{I}_i'(-s)\,\mathbf{r} + \mathbf{I}_i'(-s)\mathbf{r}][\mathbf{I}_i(s) - \mathbf{I}_r(s)] \\
&= \text{even part of } \mathbf{X}(s) + \text{even part of } \mathbf{Y}(s),
\end{aligned}$$

where

$$\begin{aligned}
\mathbf{X}(s) &= \mathbf{I}_i'(-s)\,\mathbf{r}\mathbf{I}_i(s) - \mathbf{I}_r'(-s)\,\mathbf{r}\mathbf{I}_r(s), \\
\mathbf{Y}(s) &= \mathbf{I}_r'(-s)\,\mathbf{r}\mathbf{I}_i(s) - \mathbf{I}_i'(-s)\,\mathbf{r}\mathbf{I}_r(s).
\end{aligned}$$

Note that $\mathbf{r}$ is diagonal and hence $\mathbf{r}' = \mathbf{r}$. By direct evaluation it is found that

$$\begin{aligned}
\mathbf{X}(-s) &= \mathbf{X}'(s) = \mathbf{X}(s), \\
\mathbf{Y}(-s) &= -\mathbf{Y}'(s) = -\mathbf{Y}(s),
\end{aligned}$$

since $\mathbf{X}$ and $\mathbf{Y}$ are diagonal matrices. Thus $\mathbf{X}(s)$ is even, and $\mathbf{Y}(s)$ is odd. Therefore

$$P = \mathbf{I}'_i(-s)\,\mathbf{r}\mathbf{I}_i(s) - \mathbf{I}'_r(-s)\,\mathbf{r}\mathbf{I}_r(s). \tag{108}$$

Now let us define the current scattering matrix as

$$\mathbf{I}_r(s) = \mathbf{S}_I\,\mathbf{I}_i(s) \tag{109}$$

and insert it into (108), with the result

$$P = \mathbf{I}'_i(-s)\,[\mathbf{r} - \mathbf{S}_I(-s)\,\mathbf{r}\mathbf{S}_I(s)]\,\mathbf{I}_i(s).$$

Recalling that $\mathbf{r} = \mathbf{f}(s)\,\mathbf{f}(-s)$, this expression can be put in the form of (54) *uniquely* as follows:

$$P = [\mathbf{f}(s)\,\mathbf{I}_i(-s)]'\{\mathbf{U} - [\mathbf{f}^{-1}(s)\,\mathbf{S}'_I(-s)\,\mathbf{f}(-s)][\mathbf{f}(s)\,\mathbf{S}_I(s)\mathbf{f}^{-1}(-s)]\}[\mathbf{f}(-s)\,\mathbf{I}_i(s)]. \tag{110}$$

Now it is clear! We must define the scattering variables $\mathbf{a}$ and $\mathbf{b}$ as the normalized incident and reflected *current* variables; and we must define the scattering matrix $\mathbf{S}$ as the *normalized* current scattering matrix in the following way:

$$\mathbf{a}(s) = \mathbf{f}(-s)\,\mathbf{I}_i(s), \tag{111a}$$

$$\mathbf{b}(s) = \mathbf{f}(s)\,\mathbf{I}_r(s), \tag{111b}$$

$$\mathbf{S} = \mathbf{f}(s)\,\mathbf{S}_I(s)\,\mathbf{f}^{-1}(-s). \tag{112}$$

[Equation 111b is justified if (109) is multiplied by $\mathbf{f}(s)$ and the expressions for $\mathbf{a}(s)$ and $\mathbf{S}$ are used.] Then (110) becomes

$$P = \mathbf{a}'(-s)\,[\mathbf{U} - \mathbf{S}'(-s)\,\mathbf{S}(s)]\,\mathbf{a}(s). \tag{113}$$

This expression reduces, for $s = j\omega$, to the power input as given in (54) for the case of real normalization. Hence the scattering matrix defined by (112) and (109) has the same properties as discussed in Section 8.4 for real normalization.

The only thing left to do is to find expressions for $\mathbf{S}$ in terms of $\mathbf{Z}_{oc}$ and the terminating impedances. For this purpose return to (102) and use (109) to find the current scattering matrix

$$
\begin{aligned}
\mathbf{S}_I &= \mathbf{r}^{-1}[\mathbf{Z}_{oc}(s) - \mathbf{z}(-s)]\,[\mathbf{Z}_{oc}(s) + \mathbf{z}(s)]^{-1}\,\mathbf{r} \\
&= [\mathbf{Z}_{oc}(s) + \mathbf{z}(s)]^{-1}\,[\mathbf{Z}_{oc}(s) - \mathbf{z}(-s)] \\
&= \mathbf{U} - 2\mathbf{Y}_a(s)\,\mathbf{r},
\end{aligned}
\tag{114}
$$

where $\mathbf{Y}_a$ is the admittance matrix (not normalized) of the augmented network:

$$
\mathbf{Y}_a(s) = [\mathbf{Z}_{oc}(s) + \mathbf{z}(s)]^{-1}.
\tag{115}
$$

When (114) is inserted into (112), the scattering matrix becomes

$$
\begin{aligned}
\mathbf{S} &= \mathbf{f}(s)\,[\mathbf{Z}_{oc}(s) + \mathbf{z}(s)]^{-1}\,[\mathbf{Z}_{oc}(s) - \mathbf{z}(-s)]\,\mathbf{f}^{-1}(-s) \\
&= \mathbf{f}(s)\,\mathbf{f}^{-1}(-s) - 2\mathbf{f}(s)\,\mathbf{Y}_a(s)\,\mathbf{f}(s).
\end{aligned}
\tag{116}
$$

Compare this with (92) and (93) for the single-frequency complex normalization.

We see that in the case of frequency-independent normalization the expression for $\mathbf{S}$ is somewhat more complicated than the expression for the single-frequency scattering matrix.

For a one-port having an impedance $Z(s)$, the *current* reflection coefficient ρ_I is obtained from (114) by noting that the matrices in that expression are scalars in the one-port case. Thus

$$
\rho_I(s) = \frac{Z(s) - z(-s)}{Z(s) + z(s)}.
\tag{117}
$$

The reflection coefficient itself is seen from (116) to differ from this expression by the function $f(s)/f(-s)$, which is an all-pass function. Thus

$$
\rho(s) = \frac{Z(s) - z(-s)}{Z(s) + z(s)} \cdot \frac{f(s)}{f(-s)},
\tag{118}
$$

where $A(s) = f(s)/f(-s)$ is an all-pass function.

Examples

Consider, for example, the one-port in Fig. 19 terminated in a parallel

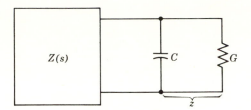

Fig. 19. Reflection coefficient of a one-port.

R and C. Then

$$z(s) = \frac{1}{G + sC}, \qquad r(s) = \tfrac{1}{2}\left[z(s) + z(-s)\right] = \frac{G}{(G + sC)(G - sC)},$$

$$f(s) = \frac{\sqrt{G}}{G + sC}, \qquad f(-s) = \frac{\sqrt{G}}{G - sC}.$$

The current reflection coefficient from (117) is

$$\rho_I(s) = \frac{Z(s) - \dfrac{1}{(G - sC)}}{Z(s) + \dfrac{1}{(G + sC)}} = \frac{(G - sC)Z(s) - 1}{(G + sC)Z(s) + 1} \cdot \frac{G + sC}{G - sC}.$$

Finally, from (118) the reflection coefficient is

$$\rho(s) = \rho_I(s)\frac{f(s)}{f(-s)} = \frac{(G - sC)}{(G + sC)}\rho_I(s) = \frac{(G - sC)Z(s) - 1}{(G + sC)Z(s) + 1}.$$

As another example of frequency-independent complex normalization, consider the network in Fig. 20. A gyrator is terminated at one port by a

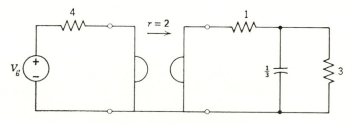

Fig. 20. Illustrative example.

resistor and at the other port by an impedance

$$z_2 = \frac{s+4}{s+1}.$$

The gyrator has an open-circuit impedance matrix; thus the following can be written:

$$\mathbf{Z}_{oc} = \begin{bmatrix} 0 & -2 \\ 2 & 0 \end{bmatrix}, \qquad \mathbf{z} = \begin{bmatrix} 4 & 0 \\ 0 & \dfrac{s+4}{s+1} \end{bmatrix}$$

and

$$\mathbf{Y}_a = (\mathbf{Z}_{oc} + \mathbf{z})^{-1} = \begin{bmatrix} 4 & -2 \\ 2 & \dfrac{s+4}{s+1} \end{bmatrix}^{-1} = \begin{bmatrix} \dfrac{s+4}{8(s+\frac{5}{2})} & \dfrac{s+1}{4(s+\frac{5}{2})} \\ \dfrac{-(s+1)}{4(s+\frac{5}{2})} & \dfrac{s+1}{2(s+\frac{5}{2})} \end{bmatrix}.$$

The terminating impedance z_2 is the same as the one treated earlier just below (105). Thus the functions $f(s)$ and $f^{-1}(-s)$ are

$$f(s) = \begin{bmatrix} 2 & 0 \\ 0 & \dfrac{s+2}{s+1} \end{bmatrix}, \qquad f^{-1}(-s) = \begin{bmatrix} \frac{1}{2} & 0 \\ 0 & \dfrac{-s+1}{-s+2} \end{bmatrix}.$$

When all the above is inserted into (116), the scattering matrix becomes

$$\mathbf{S} = \begin{bmatrix} 2 & 0 \\ 0 & \dfrac{s+2}{s+1} \end{bmatrix} \begin{bmatrix} \frac{1}{2} & 0 \\ 0 & \dfrac{-s+1}{-s+2} \end{bmatrix} - \frac{2}{8(s+\frac{5}{2})} \begin{bmatrix} 2 & 0 \\ 0 & \dfrac{s+2}{s+1} \end{bmatrix}$$

$$\begin{bmatrix} (s+4) & 2(s+1) \\ -2(s+1) & 4(s+1) \end{bmatrix} \begin{bmatrix} 2 & 0 \\ 0 & \dfrac{s+2}{s+1} \end{bmatrix}$$

$$\mathbf{S} = \begin{bmatrix} \dfrac{-\frac{3}{2}}{s+\frac{5}{2}} & \dfrac{-(s+2)}{s+\frac{5}{2}} \\ \dfrac{s+2}{s+\frac{5}{2}} & \dfrac{-\frac{3}{2}}{s+\frac{5}{2}} \cdot \dfrac{s+2}{-s+2} \end{bmatrix}.$$

Observe that S_{22} equals S_{11} multiplied by an all-pass function; their magnitudes on the $j\omega$-axis are, thus, the same. Equation 65 and, more generally, (66), are satisfied. You can verify that (64), (67), (68), and (69) are also satisfied.

NEGATIVE-RESISTANCE AMPLIFIER

As a further illustration of the application of scattering parameters, we shall discuss a network whose analysis and design are greatly simplified when scattering parameters are used. An impetus to the development of the negative-resistance amplifier came from the advent of the tunnel diode, a simple linear model for which is shown in Fig. 21a. A more complete model is that shown in Fig. 21b, but for many purposes the simpler model is adequate.

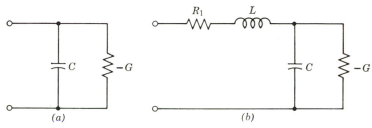

Fig. 21.   Models of tunnel diode.

The impedance of the simple tunnel-diode model is

$$Z_d(s) = \frac{1}{-G + sC}.$$

This is clearly not a positive real function. Let us now form the function $-Z_d(-s)$, as follows:

$$-Z_d(-s) = \frac{-1}{-G - sC} = \frac{1}{G + sC} = Z(s),$$

which *is* a positive real function, the impedance of a parallel G and C. This is, in fact, an example of a more general result that can be stated as follows. Let $Z_d(s)$ be the impedance of an *active* network consisting of positive inductances and capacitances, and *negative* resistances. Let $Z(s)$ be the impedance of the *passive* network obtained when the sign of each resistance is changed. Then $Z(s) = -Z_d(-s)$. The proof of this result is left to you. (See Problems 30 and 31.)

Now suppose a tunnel diode represented by the model in Fig. 21a is the termination at one port of a lossless three-port network, as shown in Fig. 22a. Let the normalizing impedances at the three ports be $z_1 = R_1$, $z_2 = R_2$, and $z_3 = 1/(G + sC)$. Thus ports 1 and 2 are terminated in their

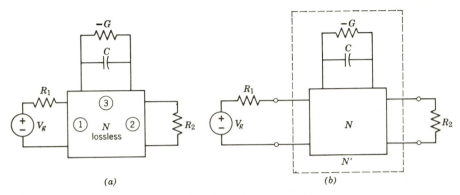

(a)   (b)

Fig. 22. Negative resistance amplifier.

normalizing impedances, whereas port 3 is terminated in an impedance $z_a(s)$ related to its normalizing impedance by $z_a(s) = -z_3(-s)$. The scattering relations of the three-port can be written as follows:

$$b_1 = S_{11}a_1 + S_{12}\,a_2 + S_{13}\,a_3, \tag{119a}$$

$$b_2 = S_{21}a_1 + S_{22}\,a_2 + S_{23}\,a_3, \tag{119b}$$

$$b_3 = S_{31}a_1 + S_{32}\,a_2 + S_{33}\,a_3. \tag{119c}$$

Alternatively, if the termination at port 3 is included in the network, the structure can be regarded as a two-port N', as shown in Fig. 22b, with the real terminations R_1 and R_2. Let the scattering parameters of this two-port be labeled with a prime. By definition,

$$S_{11}'(j\omega) = \frac{b_1}{a_1}\bigg|_{a_2=0} = S_{11}(j\omega) + S_{13}(j\omega)\frac{a_3}{a_1}$$

$$= S_{11}(j\omega) - \frac{S_{13}(j\omega)S_{31}(j\omega)}{S_{33}(j\omega)}. \tag{120}$$

The first step follows from (119a) with $a_2 = 0$. Since port 3 in Fig. 22a is terminated in an impedance that is the j-axis negative conjugate of the normalizing impedance, then $b_3 = 0$, according to Problem 27. Hence the second step in the last equation follows from (119c).

By a similar approach the remaining scattering parameters of Fig. 22b can be formed. The results are given here with the details left for you to work out.

$$S'_{11} = \frac{S_{11}S_{33} - S_{13}S_{31}}{S_{33}}, \qquad S'_{12} = \frac{S_{12}S_{33} - S_{13}S_{32}}{S_{33}},$$

$$S'_{21} = \frac{S_{21}S_{33} + S_{23}S_{31}}{S_{33}}, \qquad S'_{22} = \frac{S_{22}S_{33} + S_{23}S_{32}}{S_{33}}. \tag{121}$$

Now since the three-port in Fig. 22 is lossless, its scattering matrix is unitary. This property imposes certain conditions among the scattering parameters. (See problem 35.) Under these conditions the above equations become

$$S'_{11}(j\omega) = \frac{S_{22}(-j\omega)}{S_{33}(j\omega)}, \qquad S'_{12}(j\omega) = \frac{-S_{21}(-j\omega)}{S_{33}(j\omega)},$$

$$S'_{21}(j\omega) = \frac{-S_{12}(-j\omega)}{S_{33}(j\omega)}, \qquad S'_{22}(j\omega) = \frac{S_{11}(-j\omega)}{S_{33}(j\omega)}. \tag{122}$$

These equations provide relationships for the reflection and transmission coefficients of the two-port negative-resistance amplifier in terms of the scattering parameters of the lossless three-port. The transducer power gain of the amplifier is $\mathcal{G}(\omega^2) = |S'_{21}(j\omega)|^2$, which, from the above,

$$\mathcal{G}(\omega^2) = |S'_{21}(j\omega)|^2 = \frac{|S_{12}(j\omega)|^2}{|S_{33}(j\omega)|^2}, \tag{123}$$

since $|S_{12}(-j\omega)| = |S_{12}(j\omega)|$. Because N is a lossless three-port, $|S_{12}(j\omega)|^2 \leq 1$. Then

$$\mathcal{G}(\omega^2) \leq \frac{1}{|S_{33}(j\omega)|^2}. \tag{124}$$

But S_{33} is the reflection coefficient at port 3, across which there appears a capacitor. Hence, as discussed in the last section, there is a fundamental limitation on this function as given in (80), with $S_{33} = \rho$.

The optimum design over a given frequency band $0 - \omega_c$ is to give $|S_{12}(j\omega)|$ its largest constant value over the band and to make $|S_{33}(j\omega)|$ a constant. The maximum of $|S_{12}(j\omega)|$ is 1. With the limitation given in (82), we find the transducer power gain to be limited by

$$\mathscr{G}(\omega^2) \leq \epsilon^{2\pi/\omega_c RC} \tag{125}$$

where $R = 1/G$ in Fig. 22. This is a basic "gain-bandwidth" limitation dependent only on the parameters of the tunnel diode.

We turn now to a consideration of the design of the three-port network. Suppose the three-port in Fig. 22a is to be a reciprocal network. If $|S_{12}(j\omega)| = 1$ over a band of frequencies, all the other scattering parameters of the three-port (S_{11}, S_{23}, etc.) will vanish over that frequency band. (See Problem 36.) Thus a nontrivial *reciprocal* three-port with the above optimum design is not possible.

Consider the three-port circulator shown in Fig. 23a. Its scattering

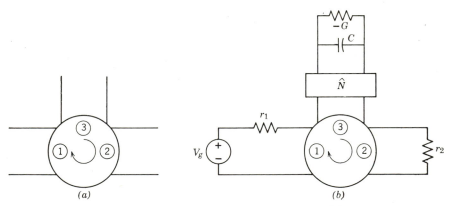

Fig. 23.   Nonreciprocal negative resistance amplifier.

matrix, normalized with respect to resistances r_1, r_2, and r_3, where r_3 is an arbitrary real number, is

$$\mathbf{S}_c = \begin{bmatrix} 0 & 1 & 0 \\ 0 & 0 & 1 \\ 1 & 0 & 0 \end{bmatrix}. \tag{126}$$

Thus S_{12} identically equals 1; but the remaining parameters of the circulator are not adequate to make just the circulator the desired three-port. If something else is attached to port 2, the resulting S_{12} will no longer equal 1. This leaves port 3.

Consider the structure shown in Fig. 23b. It consists of the circulator with a reciprocal two-port $\hat{N}$ in cascade at one of its ports. Let the scattering matrix $\hat{S}$ of the two-port $\hat{N}$ be

$$\hat{S} = \begin{bmatrix} \hat{S}_{11} & \hat{S}_{12} \\ \hat{S}_{21} & \hat{S}_{22} \end{bmatrix} \tag{127}$$

normalized with respect to r_3 and $z_3 = 1/(G + sC)$, at its input and output, respectively; r_3 is the same as the port 3 normalizing resistance of the circulator. It remains to express the scattering matrix of the overall three-port in terms of the parameters of S_c and $\hat{S}$ in (126) and (127). This can be done by using the results of Problem 24. The details are left for you; the result is

$$S = \begin{bmatrix} 0 & 1 & 0 \\ \hat{S}_{11} & 0 & \hat{S}_{12} \\ \hat{S}_{21} & 0 & \hat{S}_{22} \end{bmatrix}. \tag{128}$$

Finally, the scattering matrix of the overall negative-resistance amplifier (the one denoted by S') is obtained by using (122). Thus

$$S' = \begin{bmatrix} 0 & \dfrac{\hat{S}_{11}(-j\omega)}{\hat{S}_{22}(j\omega)} \\ \dfrac{-1}{\hat{S}_{22}(j\omega)} & 0 \end{bmatrix}. \tag{129}$$

The reflection coefficients are both zero, indicating that the amplifier is matched at both input and output. If (65) is used, it is observed that the reverse transmission S'_{12} has a unit magnitude. The forward transmission of the amplifier is related only to the output reflection coefficient of the two-port $\hat{N}$. Thus

$$\mathcal{G}(\omega^2) = \frac{1}{|\hat{S}_{22}(j\omega)|^2} \tag{130}$$

The design problem now becomes the following. A gain function $\mathcal{G}(\omega^2)$ is to be selected, subject to the gain-bandwidth limitation of (125), so as to maximize $1/|\hat{S}_{22}(j\omega)|^2$. Now, from $|\hat{S}_{22}(j\omega)|^2$, it is a problem of determining the two-port $\hat{N}$, shown in Fig. 24 terminated in r_3 at one port and

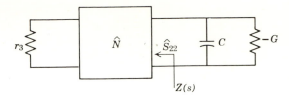

Fig. 24. Two port $\hat{N}$ to be designed.

in the tunnel diode at the other. The problem is not unlike that of the filter problem illustrated in Section 8.4. From $|\hat{S}_{22}(j\omega)|^2$ it is necessary to determine $\hat{S}_{22}(s)$. This is related to the impedance Z looking into port 2 of $\hat{N}$ by (118), in which ρ plays the role of $\hat{S}_{22}$. The two-port $\hat{N}$ can then be designed from Z.

This quick description has ignored a number of problems having to do with the proper selection of $\hat{S}_{22}(s)$ from its magnitude squared, since this is not a unique process. The details of this process would take us far afield and will not be pursued any further here.

PROBLEMS

1. Two one-ports having impedances $Z_a = f_a(s)$ and $Z_b = f_b(s)$ have reflection coefficients ρ_1 and ρ_2. Find the reflection coefficient ρ of the one-port shown in Fig. P1 in terms of ρ_1 and ρ_2.

$$Z_1 = f_a(s), \qquad Z_3 = f_b(s)$$

$$Z_2 = \frac{1}{f_b(s)}, \qquad Z_4 = \frac{1}{f_a(s)}$$

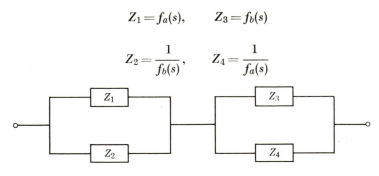

Fig. P1

2. The sense of incident and reflected waves is related to what is taken to be the direction of power flow. Figure P2 shows a one-port in part (a) with

the usual references of voltage and current. The reflection coefficient for this one-port is ρ_1. In Fig. P2b the current in the one-port is reversed.

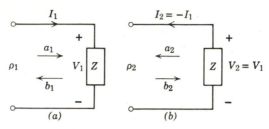

Fig. P2

The one-port is thus considered as supplying power to the network to the left of the terminals. Find the new reflection coefficient ρ_2 in terms of Z and in terms of ρ_1.

3. It was shown in the text that $\mathbf{S} = \mathbf{U} - 2\mathbf{Y}_{an}$, where $\mathbf{Y}_{an}$ is the y-matrix of the normalized augmented network, by assuming that a multiport has a short-circuit admittance matrix. Prove this result from the augmented network without making this assumption.

4. Show that $\mathbf{S} = 2(\mathbf{U} + \mathbf{Y}_n)^{-1} - \mathbf{U}$.

5. Consider a *matched, nonreciprocal, lossless* three-port. Starting with the general form of the scattering matrix, and using the appropriate properties of matched and lossless multiports, determine the elements of the scattering matrix. Can you identify the class of multiport from this scattering matrix?

6. (a) Write out a scattering matrix to represent a four-port circulator.
(b) Take the transpose of this matrix and identify the kind of four-port it represents.

7. In Fig. P7 are shown the two controlled sources that have no impedance or admittance representations. Find the scattering matrices for these two-ports.

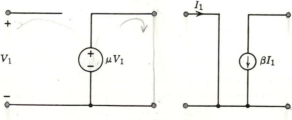

Fig. P7

8. In Fig. P8 are shown two controlled sources. One has no impedance representation, the other no admittance representation; but they both have a scattering matrix. Find them.

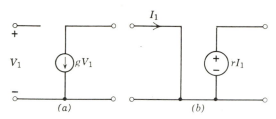

Fig. P8

9. Use (41) in the text to find expressions for S_{11}, S_{12}, S_{21}, and S_{22} for a two-port in terms of the z-parameters.

10. *Symmetrical* passive two-ports are defined as two-ports for which $z_{11} = z_{22}$. Using the results of Problem 9, show that symmetrical two-ports have $S_{11} = S_{22}$.

11. *Antimetrical* passive two-ports are defined as two-ports for which $z_{11} = y_{22}$, $z_{22} = y_{11}$ and $z_{21} = -y_{21}$. Show that antimetrical two-ports are characterized by $S_{22} = -S_{11}$.

12. (*a*) Let the circulator of Fig. 11 in the text be terminated in a resistance $-r$ at port 1 and r at port 3. Find the relationship between voltage and current at port 2. A one-port device having this *v-i* relationship has been called a *norator*.

(*b*) Repeat (a) but with $-r$ and r interchanged. A one-port device having this *v-i* relationship has been called a *nullator*.

13. Show that

(*a*) $(\mathbf{Z}_n + \mathbf{U})^{-1}(\mathbf{Z}_n - \mathbf{U}) = (\mathbf{Z}_n - \mathbf{U})(\mathbf{Z}_n + \mathbf{U})^{-1}$.

(*b*) $\mathbf{r}^{-1}\{\mathbf{Z}_{oc}(s) - \mathbf{z}(-s)\}\{\mathbf{Z}_{oc}(s) + \mathbf{z}(s)\}^{-1}\mathbf{r}$

$$= \{\mathbf{Z}_{oc}(s) + \mathbf{z}(s)\}^{-1}\{\mathbf{Z}_{oc}(s) - \mathbf{z}(-s)\},$$

where $\mathbf{Z}_n$ is the impedance matrix of a multiport normalized to real numbers.

14. Each of the multiports in Fig. P14 is an ideal junction consisting of direct connections between the ports. Find the scattering matrix of each for 1-ohm normalizing resistances at each port. In each case suppose that power is supplied by a voltage source in series with the terminating resistance at one port. Find the fraction of the power reflected at that

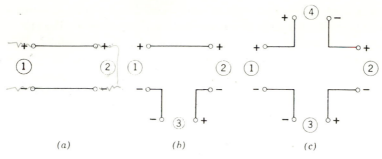

Fig. P14

port and transmitted to each of the other ports. Is this what you would have expected without finding S?

15. The structure in Fig. P15 is a hybrid coil. It consists of a three-winding ideal transformer from which a four-port is formed. The two transformer

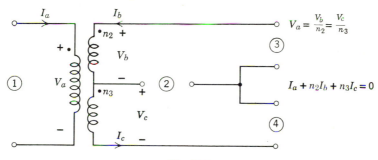

Fig. P15

secondary windings have turns ratios n_2 and n_3 relative to the primary. The equations characterizing the transformer are given in the figure. Assume each port is terminated in its real normalizing resistances r_1, r_2, r_3, and r_4. The turns ratios and the normalizing resistances are to be chosen so that—

(a) When port 1 is excited (by a voltage source in series with its terminating resistance), there is no transmission to port 2, and vice versa;

(b) When port 3 is excited, there is to be no transmission to port 4, and vice versa;

(c) All ports are matched (no reflections).

Find the scattering matrix of this four-port in terms of n_2 and n_3 only.

16. It is desired to investigate the possible existence of a lossless, reciprocal three-port that is match terminated with real impedances. Use the properties of the scattering matrix to determine the realizability of such a device. If realizable, find S_{12} and S_{13}.

17. Figure P17 shows a lossless, reciprocal three-port network that is assumed to be symmetrical. The three-port is *not* match terminated.

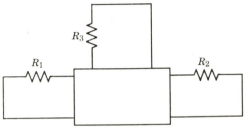

Fig. P17

When one of the ports is excited (by a voltage source in series with its termination), it is assumed that equal power is delivered to the other two-ports. Find the maximum fraction of available power that is delivered to each port under these conditions and find the fraction of power that is reflected.

18. The network in Fig. P18 is a lattice operating between two resistive terminations. Compute the transducer power gain. Determine conditions

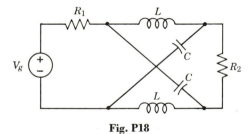

Fig. P18

on the lattice elements that will make the gain identically unity, independent of frequency. Under these conditions find the reflection and transmission coefficients.

19. Figure P19 shows a two-port terminated at the output by an impedance that is not the normalizing impedance. Let ρ_2 be the reflection coefficient of Z_2 normalized to r_2, which is the output normalizing resistance of the two-port. The input is match terminated; that is, r_1 is the normalizing

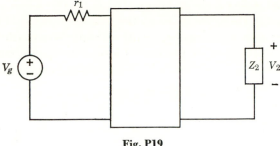

Fig. P19

resistance. Find the input reflection coefficient ρ and the voltage gain V_{2n}/V_{gn} in terms of ρ_2 and the scattering parameters of the two-port.

20. This is a generalization of the last problem. An n-port network is match terminated at m of its ports and terminated in arbitrary loads at the remaining $n-m$ ports, as illustrated in Fig. P20. the scattering equations of the multiport are partitioned, as shown. The reflection coefficient at the port terminated by Z_k is ρ_k. Let $\boldsymbol{\rho}$ be the diagonal matrix

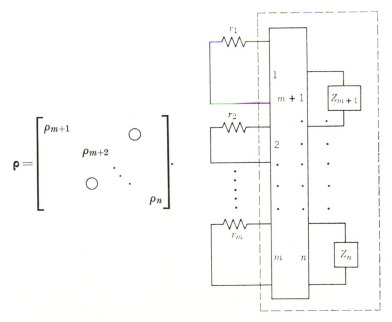

Fig. P20

It is desired to find the scattering matrix $\mathbf{S}'$ for the m-port within the dashed lines in the figure, given by the relation $\mathbf{b}_1 = \mathbf{S}'\mathbf{a}_1$. Using the relation between a_k, b_k, and ρ_k, write an expression relating $\mathbf{a}_2$, $\mathbf{b}_2$, and $\boldsymbol{\rho}$. Insert this into the partitioned form of the scattering relations and show that

$$\mathbf{S}' = \mathbf{S}_1 + \mathbf{S}_2\,\boldsymbol{\rho}(\mathbf{U} - \mathbf{S}_4\,\boldsymbol{\rho})^{-1}\mathbf{S}_3.$$

$$\begin{bmatrix} b_1 \\ b_2 \\ \vdots \\ b_m \\ \cdots \\ b_{m+1} \\ \vdots \\ b_n \end{bmatrix} = \begin{bmatrix} S_{11} & S_{12}\cdots S_{1m} & \vdots & S_{1m+1} & \cdots S_{1n} \\ S_{21} & S_{22}\cdots S_{2m} & \vdots & & \cdots S_{2n} \\ \vdots & \vdots & \vdots & & \\ S_{m1} & S_{m2}\cdots S_{mm} & \vdots & S_{mm+1}\cdots S_{mn} \\ \cdots & \cdots & \vdots & \\ S_{m+1_1} & & \vdots & \\ \vdots & & \vdots & \\ S_{n1} & S_{n2}\cdots S_{nm} & \vdots & S_{nn} \end{bmatrix} \begin{bmatrix} a_1 \\ a_2 \\ \vdots \\ a_m \\ \cdots \\ a_{m+1} \\ \vdots \\ a_n \end{bmatrix}$$

$$\begin{bmatrix} \mathbf{b}_1 \\ \mathbf{b}_2 \end{bmatrix} = \begin{bmatrix} \mathbf{S}_1 & \mathbf{S}_2 \\ \mathbf{S}_3 & \mathbf{S}_4 \end{bmatrix} \begin{bmatrix} \mathbf{a}_1 \\ \mathbf{a}_2 \end{bmatrix}$$

21. One of the ports of the circulator in Fig. P21 is not match terminated. Use the result of the previous problem to find the scattering matrix $\mathbf{S}'$ of the two port within the dashed line.

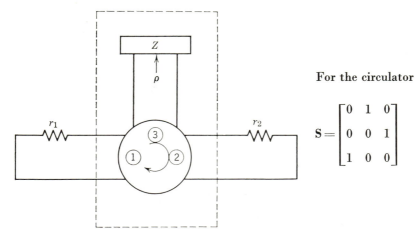

For the circulator

$$\mathbf{S} = \begin{bmatrix} 0 & 1 & 0 \\ 0 & 0 & 1 \\ 1 & 0 & 0 \end{bmatrix}$$

Fig. P21

22. The four-port network in Fig. P22 is a hybrid that is match terminated at two of the ports but not at the other two. Find the scattering matrix of the two-port within the dashed line in terms of ρ_3 and ρ_4. Under what condition will the two-port have no reflections at either port?

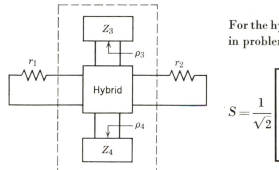

For the hybrid, with $n_2 = n_3$ in problem 8–15,

$$S = \frac{1}{\sqrt{2}} \begin{bmatrix} 0 & 0 & 1 & 1 \\ 0 & 0 & 1 & -1 \\ 1 & 1 & 0 & 0 \\ 1 & -1 & 0 & 0 \end{bmatrix}$$

Fig. P22

23. In order to find the relationships among the port voltages and currents of a two-port that consists of the cascade connection of two sub-two-ports, it is convenient to express the port relationships of the sub-two-ports in terms of the chain matrix. It is desired to find a matrix $\mathbf{T}$ that can play a similar role for two cascaded two-ports; but this time the variables are scattering variables, rather than actual currents and voltages. For the network shown in Fig. P23, let the desired overall relationship be written $\mathbf{x} = \mathbf{T}\mathbf{y}$.

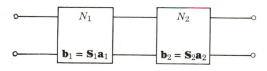

Fig. P23

(*a*) Determine the elements of the vectors **x** and **y** (from among the scattering variables) such that the overall **T** matrix equals $\mathbf{T}_1\mathbf{T}_2$, where $\mathbf{x}_1 = \mathbf{T}_1\mathbf{y}_1$ and $\mathbf{x}_2 = \mathbf{T}_2\mathbf{y}_2$. Specify any condition on the normalizing impedances that may be required for complex normalization and also for real normalization. (This can be referred to as a *compatibility condition*.)

(b) Express the elements of the matrix $\mathbf{T}$ for a two-port in terms of the scattering parameters of that two-port.

(c) Determine a condition that the elements of $\mathbf{T}$ satisfy if the two-port is reciprocal.

24. Instead of terminating some ports in Problem 20 in individual loads, suppose these ports are connected to other ports of a multiport, as shown in Fig. P24. The multiport N has $m + k$ ports, and multiport $\hat{N}$ has

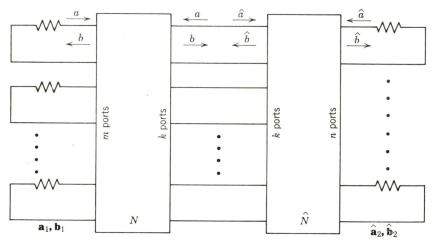

Fig. P24

$n + k$ ports. A total of k ports of one are connected to k ports of the other, leaving an $(m + n)$ port network. Let $\mathbf{S}$ and $\hat{\mathbf{S}}$ be the scattering matrices of the two multiports. The scattering equations can be partitioned as follows:

$$\begin{bmatrix} \mathbf{b}_1 \\ \mathbf{b}_2 \end{bmatrix} = \begin{bmatrix} \mathbf{S}_{11} & \mathbf{S}_{12} \\ \mathbf{S}_{21} & \mathbf{S}_{22} \end{bmatrix} \begin{bmatrix} \mathbf{a}_1 \\ \mathbf{a}_2 \end{bmatrix}, \qquad \begin{bmatrix} \hat{\mathbf{b}}_1 \\ \hat{\mathbf{b}}_2 \end{bmatrix} = \begin{bmatrix} \hat{\mathbf{S}}_{11} & \hat{\mathbf{S}}_{12} \\ \hat{\mathbf{S}}_{21} & \hat{\mathbf{S}}_{22} \end{bmatrix} \begin{bmatrix} \hat{\mathbf{a}}_1 \\ \hat{\mathbf{a}}_2 \end{bmatrix}$$

where $\mathbf{b}_1$ and $\mathbf{a}_1$ are m-vectors, $\mathbf{b}_2$, $\mathbf{a}_2$, $\hat{\mathbf{b}}_1$, and $\hat{\mathbf{a}}_1$ are k-vectors, and $\hat{\mathbf{b}}_2$ and $\hat{\mathbf{a}}_2$ and n-vectors. The normalizing impedance matrix (frequency independent) of each multiport is also partitioned conformally, as follows:

$$\begin{array}{cc} m & k \\ \begin{array}{c} m \\ k \end{array} \begin{bmatrix} \mathbf{z}_1(s) & 0 \\ 0 & \mathbf{z}_2(s) \end{bmatrix}, & \begin{array}{cc} k & n \\ \begin{array}{c} k \\ n \end{array} \begin{bmatrix} \bar{\mathbf{z}}_1(s) & 0 \\ 0 & \bar{\mathbf{z}}_2(s) \end{bmatrix}. \end{array}$$

Let S' be the scattering matrix of the overall $(m + n)$ port and write the scattering equations as

$$\begin{bmatrix} \mathbf{b}_1 \\ \hat{\mathbf{b}}_2 \end{bmatrix} = \begin{bmatrix} S'_{11} & S'_{12} \\ S'_{21} & S'_{22} \end{bmatrix} \begin{bmatrix} \mathbf{a}_1 \\ \hat{\mathbf{a}}_2 \end{bmatrix}.$$

(a) Find the *compatibility condition* that will permit the result $\hat{\mathbf{b}}_1 = \mathbf{a}_2$, $\hat{\mathbf{a}}_1 = \mathbf{b}_2$. (See Problem 23.)

(b) Show that the overall scattering parameters are given by

$$S'_{11} = S_{11} + S_{12}\,\hat{S}_{11}(U - S_{22}\,\hat{S}_{11})^{-1}\,S_{21} = S_{11} + S_{12}(U - \hat{S}_{11}S_{22})^{-1}\,\hat{S}_{11}S_{21}$$

$$S'_{12} = S_{12}(U - \hat{S}_{11}S_{22})^{-1}\,\hat{S}_{12}$$

$$S'_{21} = \hat{S}_{21}(U - S_{22}\,\hat{S}_{11})^{-1}\,S_{21}$$

$$S'_{22} = \hat{S}_{22} + \hat{S}_{21}(U - S_{22}\,\hat{S}_{11})^{-1}\,S_{22}\,\hat{S}_{12} = \hat{S}_{22} + \hat{S}_{21}S_{22}(U - \hat{S}_{11}S_{22})^{-1}\,\hat{S}_{21}$$

(c) Compare with the more special result of Problem 20 and verify that this result reduces to that one.

25. The four-port network within the dashed lines in Fig. P25 represents a telephone repeater. The two-ports labeled L are low-pass filters, and those labeled H are high-pass filters. They are reciprocal and symmetric two-ports and hence are characterized by only two parameters each.

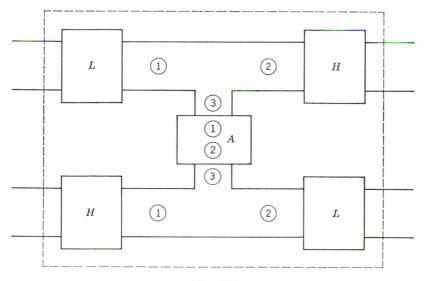

Fig. P25

The two-port labeled A is an amplifier that has transmission in one direction only (down). There are also two ideal three-port junctions in the repeater. The scattering matrices of each component are the following with all normalizing resistances being 1 ohm:

$$\mathbf{S}_L = \begin{bmatrix} \rho_L & t_L \\ t_L & \rho_L \end{bmatrix}, \qquad \mathbf{S}_H = \begin{bmatrix} \rho_H & t_H \\ t_H & \rho_H \end{bmatrix}, \qquad \mathbf{S}_A = \begin{bmatrix} \rho_1 & 0 \\ t & \rho_2 \end{bmatrix},$$

$$\mathbf{S}_j = \tfrac{1}{3} \begin{bmatrix} 1 & 2 & 2 \\ 2 & 1 & -2 \\ 2 & -2 & 1 \end{bmatrix}$$

Find the scattering matrix of the four-port repeater. By examining the elements of this matrix, describe the reflection and transmission of low- and high-frequency signals at each of the ports.

26. For single-frequency complex normalization, current and voltage scattering matrices were not defined in the text. Define $\mathbf{S}_I$ and $\mathbf{S}_V$ as

$$\mathbf{I}_r = \mathbf{S}_I \mathbf{I}_i \quad \text{and} \quad \mathbf{V}_r = \mathbf{S}_V \mathbf{V}_i.$$

Starting from (88) in the text, show that

$$\mathbf{S}_I = \mathbf{U} - 2\mathbf{Y}_a\,\mathbf{r}, \quad \text{where} \quad \mathbf{Y}_a = (\mathbf{Z}_{\text{oc}} + \mathbf{z})^{-1}$$

and

$$\mathbf{z}\mathbf{S}_I = \mathbf{S}_V\,\bar{\mathbf{z}}.$$

The first of these agrees with (114) derived for frequency-independent normalization. Similarly, for frequency-independent normalization, show that

$$\mathbf{z}(s)\mathbf{S}_I = \mathbf{S}_V\,\mathbf{z}(-s).$$

27. Let $z_k(s)$ be the complex normalizing impedance of the kth port of a multiport.

(a) If this port is terminated in an impedance $z_k(-s)$, show that the incident variable at that port is zero: $a_k = 0$.

(b) If the port is terminated in an impedance $-z_k(-s)$, show that the corresponding condition is $b_k = 0$.

28. Find the scattering matrix of the ideal transformer shown in Fig. P28 normalized to the terminating impedances. (Frequency-independent normalization.)

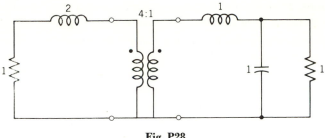

Fig. P28

29. Find the scattering matrix of the gyrator in Fig. P29 normalized (frequency independent) to is terminating impedances. Verify the properties of reflection and transmission coefficients of a two-port.

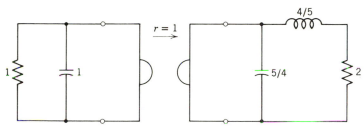

Fig. P29

30. In Fig. P30a a passive, lossless two-port is terminated in a -1-ohm resistance. In Fig. P30b the same two-port is terminated in a $+1$-ohm

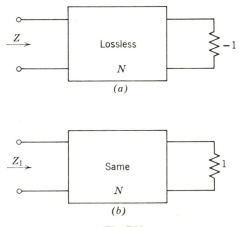

Fig. P30

resistance. Show, by writing expressions for Z and Z_1 in terms of the z- or y-parameters of the two-port, that $Z_1(s) = -Z(-s)$.

31. In Fig. P31a the network contains positive inductances and capacitances, and $n-1$ negative resistances, all taken to be -1 ohm for convenience. It can be represented by the n-port network shown in Fig. P31b with all but one port terminated in -1 ohm. Let the same n-port be terminated in $+1$-ohm resistors, as shown in Fig. P31c and let the corresponding impedance be Z. By writing the open-circuit impedance equations and solving for Z, show that $Z(s) = -Z_a(-s)$.

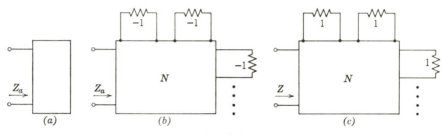

Fig. P31

32. Let $z_a(s)$ be the impedance of an active network containing inductances, capacitances, and negative resistances. Let $z(s)$ be the corresponding impedance when the signs of all resistors are reversed. Each of these networks is to be the terminating impedance of a one-port $Z(s)$. Let the two frequency-independent, complex normalized reflection coefficients be ρ_a and ρ, respectively. Show that $\rho_a(s) = 1/\rho(s)$.

33. Derive the results given in (121).

34. Prove that the determinant of a unitary matrix is unity.

35. Let S be the scattering matrix of a passive, lossless three-port. From the unitary property of S, prove the following. [Where the argument of a parameter is not given, it is $j\omega$; thus S_{12} means $S_{12}(j\omega)$.]

$$S_{11}(-j\omega) = S_{22}S_{33} - S_{23}S_{32} \qquad S_{21}(-j\omega) = S_{13}S_{32} - S_{12}S_{33}$$

$$S_{12}(-j\omega) = S_{23}S_{31} - S_{21}S_{33} \qquad S_{22}(-j\omega) = S_{11}S_{33} - S_{13}S_{31}$$

$$S_{13}(-j\omega) = S_{21}S_{32} - S_{22}S_{31} \qquad S_{23}(-j\omega) = S_{12}S_{31} - S_{11}S_{32}$$

$$S_{31}(-j\omega) = S_{12}S_{23} - S_{13}S_{22}$$

$$S_{32}(-j\omega) = S_{13}S_{21} - S_{11}S_{23}$$

$$S_{33}(-j\omega) = S_{11}S_{22} - S_{12}S_{21}$$

36. In a lossless reciprocal three-port, suppose the j-axis magnitude of one of the transmission coefficients is identically 1. Show from the unitary and symmetric properties of $\mathbf{S}$ that all the other scattering parameters must be identically zero.

37. A negative-resistance amplifier is to have the structure shown in Fig. P37. This is obtained from the one discussed in Fig. 23 by appending another two-port at port 2 of the circulator. Following the procedure in the text, obtain expressions for the overall amplifier scattering matrix $\mathbf{S}'$ in terms of the scattering parameters of each of the subnetworks in the figure.

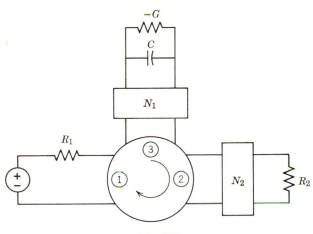

Fig. P37

. 9 .

SIGNAL-FLOW GRAPHS
AND FEEDBACK

The linear network model that has been developed up to this point includes a number of components: resistors, capacitors, gyrators, transformers, etc. Each component of the network model has been characterized by a parameter, such as R, and by a graphical symbol, such as this -⋀⋀⋀- for the resistor. Networks are made up of interconnections of these components. In each voltage-current relationship, our interest often has been directed at the scalar parameter; for example, in writing $v = L \, di/dt$, attention has been focused on the L.

But each voltage-current relationship defines a *mathematical operation*. Instead of focussing on the parameter, one could emphasize the mathematical operations and the signals on which the operations are performed. An operational symbol could be used to represent the mathematical operation, and these operational symbols could be interconnected into an *operational diagram*. The analysis of this operational diagram would provide an alternative means for determining transfer functions of networks. The *signal-flow graph* is just such an operational diagram.

Most methods of analysis discussed in preceding chapters apply to linear networks in general, whether passive or active, reciprocal or nonreciprocal. No special attention was directed at active, nonreciprocal networks. Signal-flow graph analysis is particularly appropriate for such networks. Furthermore, for active nonreciprocal networks, a number of

636

topics, such as feedback and stability, become quite important. This chapter will be concerned with the ideas of signal-flow graphs, feedback, and stability.

9.1 AN OPERATIONAL DIAGRAM

Each element, or interconnected group of elements in a network, operates on an excitation signal in some way to produce a response. The element or system of elements can be considered to be an operator. The

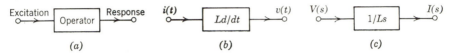

Fig. 1. Operational symbols.

process can be represented as in Fig. 1*a*. The excitation and response signals are each represented by a node between which lies a block labeled "operator." The arrows leading to and from the block indicate a "signal flow."

The special case of an inductor is shown in Fig. 1*b* and *c*. In the first of these the signals are in the time domain. The second figure, which gives the signals in the transform domain, shows that the variable taken as excitation, or as response, is arbitrary, in the initially relaxed case. This flexibility is not universal, however, since it is not available for controlled sources. Henceforth we shall deal with Laplace transforms, and so all operators will be functions of s.

The operational symbol in Fig. 1 is quite simple, but it can be simplified even further. There is really no need for the rectangle; it can be replaced by a line segment and the operator can be shown alongside the line. Only one arrow will then be necessary to show the signal flow. This is illustrated in Fig. 2*a* for the symbol of Fig. 1*c*. Dimensionally, the operator may be one of the number of things: impedance, admittance, current gain, etc. the word "transmittance" is used to stand for the function by which the excitation is to be multiplied in order to give the response. The transmittance in Fig. 2*a* is $1/Ls$.

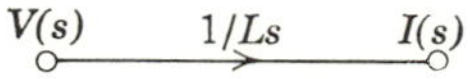

$$V(s) \qquad 1/Ls \qquad I(s)$$

Fig. 2. Branch of diagram.

There are two types of relationships among variables in a network or system. In one of these relationships one variable is expressed in terms of another one by means of an operator. The second type of relationship expresses an equilibrium; Kirchhoff's laws are of this nature. In all such equilibrium relationships, one variable can be expressed as a linear combination of a number of other variables. Thus the expression

$$V_1 = 3V_2 - 2sI + \frac{s+1}{s+2} V_3$$

can be drawn as an operational diagram by representing each variable by a node and directing a branch from all other nodes to the one corresponding to V_1, with appropriate transmittances. This is illustrated in Fig. 3.

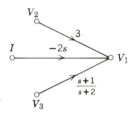

Fig. 3.   Diagram of equation.

Given a network, then, an operational diagram can be drawn by expressing the equations that describe the network performance in operational form. Before formalizing this process, some simple illustrations will be given. Consider the series-parallel connection of two two-ports shown in Fig. 4. A complete description of the terminal behavior can be given in

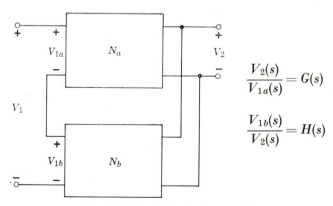

$$\frac{V_2(s)}{V_{1a}(s)} = G(s)$$

$$\frac{V_{1b}(s)}{V_2(s)} = H(s)$$

Fig. 4.   Simple feedback network.

terms of the g-parameters if we know the g-parameters of the individual two-ports. However, we may not be interested in such a complete description; we may want to know only the ratio of the output to input voltage transforms, assuming we know the corresponding ratios for the individual two-ports. The appropriate functions are defined in the diagram. We assume that there is no " loading " of one two-port on another when the series-parallel connection is made; that is, we assume that the port relationships of each two-port remain the same after they are interconnected.

The equations describing this network can be written as follows:

$$V_2 = G(s)V_{1a}$$
$$V_{1a} = V_1 - V_{1b} \tag{1}$$
$$V_{1b} = H(s)V_2.$$

The first and last equations can be represented by the operational branches shown in Fig. 5a. The second equation expresses an equilibrium of voltages

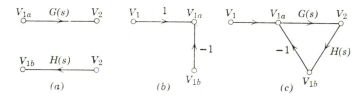

Fig. 5.   Development of operational diagram of example.

and can be represented by the branches in Fig. 5b. All three of these can now be combined to give the final result shown in Fig. 5c. This operational diagram represents the network in Fig. 4 to the same extent that the equations (1) do.

As a second example, consider the network in Fig. 6. The total output resistance is R_3. The following equations can be written:

$$I = \frac{1}{R_3} V_2$$

$$V_2 = V_b - \frac{1}{sC} I \tag{2}$$

$$V_b = R_2(-gV_a - I) = -gR_2 V_a - R_2 I$$

$$V_a = V_1 + aV_2.$$

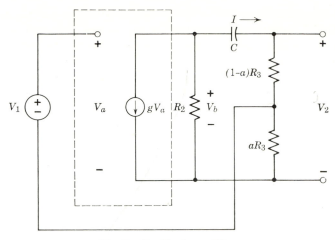

Fig. 6. Feedback amplifier.

An operational diagram can now be drawn for each equation, as shown successively in Fig. 7a to d. When these are all superimposed the final result is shown in Fig. 7e.

By comparing the operational diagram with the original amplifier

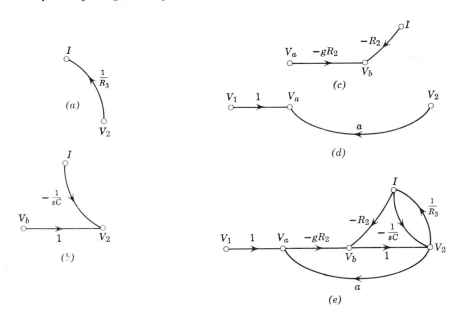

Fig. 7. Operational diagram of amplifier.

diagram, it is clear that the two bear no structural resemblance to each other. The operational diagram shows the manner in which parts of the network operate on signals within the network to yield other signals. However, for passive components, which signal is excitation and which one response is not uniquely fixed. This means that the operational diagram will take on a different structure depending on the manner in which the equations are written.

Suppose, for example, that the first three equations in (2) are rearranged as follows

$$V_2 = R_3 I$$

$$V_b = V_2 + \frac{1}{sC} I$$

$$I = -g V_a - \frac{1}{R_2} V_b.$$

The resulting operational diagram will take the form shown in Fig. 8,

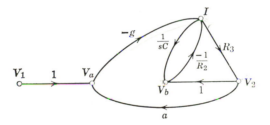

Fig. 8.   Alternate operational diagram of amplifier.

as you should verify. A comparison with the previous diagram shows the two to be quite different. This example illustrates the interesting fact that there is not a unique operational diagram representing a given network.

The preceding discussion has skirted a number of fundamental questions. Drawing an operational diagram for a network requires, first, the writing of equations describing the behavior of the network. This poses the questions of what variables to select for describing the network performance, how many of them are required, and how many equations are required. For a systematic development, such questions must be answered. The answers will be postponed, however, for a later section, in order to turn to a more detailed discussion of operational diagrams.

9.2 SIGNAL-FLOW GRAPHS

Look back at the operational diagram in Fig. 8 to observe its abstract features. It consists of a number of branches joined together at nodes. It is therefore a topological graph. The branches are *directed*, or oriented, and carry weights, or are *weighted*.

These observations will be made the basis of a definition. We shall define a *signal-flow graph* as a representation of a system of equations by a *weighted, directed* graph. It is clear, then, that a signal-flow graph is related to a network only through the network equations. It is the equations that are represented by the graph. To emphasize the fact that signal-flow-graph analysis is general and not limited only to electric networks, a general notation will be used.

Consider a set of linear algebraic equations of the form

$$\mathbf{AX} = \mathbf{Y}. \tag{3}$$

In typical applications the entries of $\mathbf{Y}$ are transforms of excitations and entries of $\mathbf{X}$ are transforms of response functions. This is the standard form for writing a set of linear equations, but it is not an appropriate form for drawing a signal-flow graph. A look back at the preceding examples shows that the desired form has each variable expressed explicitly in terms of other variables. This can be achieved by adding vector $\mathbf{X}$ to both sides in (3) and rearranging terms to get

$$\mathbf{X} = -\mathbf{Y} + (\mathbf{A} + \mathbf{U})\mathbf{X}. \tag{4}$$

In many applications there is a single driving function. This function may, of course, appear in more than one of the scalar equations in (4). If a single driving function, y_0, is assumed, then $\mathbf{Y}$ can be written as

$$\mathbf{Y} = \mathbf{K}y_0, \tag{5}$$

where $\mathbf{K}$ is a column vector. With this expression (4) can be rewritten as

$$\mathbf{X} = [-\mathbf{K} \quad (\mathbf{A} + \mathbf{U})]\begin{bmatrix} y_0 \\ \mathbf{X} \end{bmatrix}. \tag{6}$$

If there are additional driving functions, the scalar y_0 will become the vector $\mathbf{Y}_0$ and $\mathbf{K}$ will have several columns instead of only one.

If $\mathbf{X}$ is a vector of order n, then the coefficient matrix on the right side of (6) is of order $(n, n + 1)$. For future convenience, let us augment this matrix, by adding a row of zeros, so as to create a square matrix and define a matrix $\mathbf{C}$ as

$$\mathbf{C} = \begin{bmatrix} 0 & 0 \\ -\mathbf{K} & \mathbf{A}+\mathbf{U} \end{bmatrix}. \tag{7}$$

This matrix can be associated with a directed, weighted graph as follows. For each column of $\mathbf{C}$ a node is labeled with the symbol for the associated variable. For each nonzero entry c_{ij} of matrix $\mathbf{C}$ a directed branch is drawn from node j to node i and labeled with the value of c_{ij} as its weight. This weight is called the *transmittance* of the branch. If $c_{ij} = 0$, there will be no branch from node j to node i. The resulting directed, weighted graph is the signal-flow graph of the set of equations (6). Matrix $\mathbf{C}$ is called the *connection matrix* of the graph.

As an illustration, consider the following set of equations:

$$\begin{bmatrix} -1 & 0 & 2 \\ -3 & 0 & 1 \\ 2 & -1 & -1 \end{bmatrix} \begin{bmatrix} x_1 \\ x_2 \\ x_3 \end{bmatrix} = \begin{bmatrix} y_0 \\ -y_0 \\ 0 \end{bmatrix}.$$

Then

$$\mathbf{K} = \begin{bmatrix} 1 \\ -1 \\ 0 \end{bmatrix}, \quad \mathbf{A} = \begin{bmatrix} -1 & 0 & 2 \\ -3 & 0 & 1 \\ 2 & -1 & -1 \end{bmatrix}, \quad \mathbf{A}+\mathbf{U} = \begin{bmatrix} 0 & 0 & 2 \\ -3 & 1 & 1 \\ 2 & -1 & 0 \end{bmatrix}$$

and

$$\begin{bmatrix} x_1 \\ x_2 \\ x_3 \end{bmatrix} = \begin{bmatrix} -1 & 0 & 0 & 2 \\ 1 & -3 & 1 & 1 \\ 0 & 2 & -1 & 0 \end{bmatrix} \begin{bmatrix} y_0 \\ x_1 \\ x_2 \\ x_3 \end{bmatrix}.$$

The resulting connection matrix is

$$\mathbf{C} = \begin{bmatrix} 0 & 0 & 0 & 0 \\ -1 & 0 & 0 & 2 \\ 1 & -3 & 1 & 1 \\ 0 & 2 & -1 & 0 \end{bmatrix}.$$

The signal-flow graph of this set of equations will have four nodes, labeled y_0, x_1, x_2, and x_3. These are first placed in a convenient arrangement, as in Fig. 9a. The branches are then inserted in accordance with the connection matrix. Thus in the third row (corresponding to x_2, since the first row containing zeros was added to create the square matrix C) and in the fourth column (corresponding to x_3) there is a nonzero entry of value 1. Hence there will be a branch from x_3 to x_2 with a weight of 1. All other branches are similarly inserted; the final result is shown in Fig. 9b.

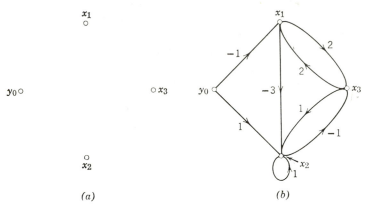

Fig. 9.   Setting up signal-flow graph.

GRAPH PROPERTIES

From a consideration of the previous examples and from additional illustrations, a number of properties of signal-flow graphs can be observed. Observe the node labeled y_0 in Fig. 9. There are two branches leaving this node but none entering it. The signal y_0 is thus not caused by other signals, it is a source. A node at which only outgoing branches are incident is called a *source node*. Similarly, a node at which only incoming branches are incident is called a *sink node*. None of the nodes in Fig. 9 satisfies this condition. However, a sink node can always be trivially introduced in a signal-flow graph. In Fig. 9, for example, an added equation $x_3 = x_3$ will add a new node x_3 with a branch having unity transmittance entering it from the old node x_3. Any node but a source node can, therefore, be considered a sink node.

Another observation from Fig. 9 is that there are sequences of branches that can be traced from a node back to itself. Such a sequence is called a *feedback loop*, and each branch in a feedback loop is a *feedback branch*. Thus one can leave node x_1 along branch -3 to x_2; then from x_2 to x_3

along branch -1; then from x_3 back to x_1 along the outgoing branch 2. There is, however, no feedback loop with the opposite orientation starting at x_1. One can go from x_1 to x_3 and then to x_2 but cannot return to x_1 since the orientation of branch -3 is wrong for this. Some feedback loops consist of a single branch, such as the one having unity transmittance at node x_2. Such a loop is called a *self-loop*. Any node that lies on a feedback loop is called a *feedback node*. In Fig. 9 all nodes but y_0 are feedback nodes.

Not all branches of a graph are in feedback loops. Any branch that is not a feedback branch is called a *cascade branch*. In Fig. 9b the two branches leaving node y_0 are cascade branches. All other branches in that graph are feedback branches.

Those nodes that are neither source nor sink nodes will have both incoming and outgoing branches incident on them. The variable corresponding to the node plays two roles: it is the signal caused by all incoming branches and also the signal which is carried by all outgoing branches. These two roles can be separated by *splitting* the node. As an illustration, consider the graph in Fig. 10a. Node x_1 has both incoming and outgoing

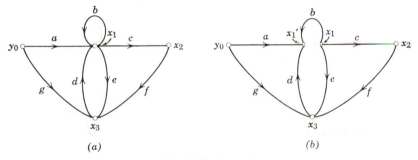

Fig. 10.   Splitting node x_1.

branches. In Fig. 10b node x_1 has been split into two nodes, labeled x_1 and x_1'. One of these is a source, from which all outgoing branches leave, the other is a sink to which all incoming branches come. Splitting a node clearly interrupts all feedback loops that pass through that node. By splitting a sufficient number of nodes all feedback loops of a graph can be interrupted. The *index* of a signal-flow graph is the smallest number of nodes that must be split to interrupt all feedback loops in the graph. In the original graph of Fig. 10a there are three feedback loops; in the modified graph, after node x_1 is split, there are no feedback loops. Hence the index of the graph of Fig. 10a is one. A set of nodes, equal in number to the index of a graph, that must be split in order to interrupt all feedback loops is called a set of *essential nodes*. In Fig. 10 this set (only one node in this case) is unique: only x_1 is an essential node. In other cases there **may** be more than one set of essential nodes.

INVERTING A GRAPH

The fact that a particular node in a signal-flow graph is a source node **is** a result of the manner in which the equations represented by the graph **are** written. By rearranging the equations, what was a source node may become a nonsource node and what was a nonsource node may become a source node. Specifically, consider an equation for x_2 which is then rearranged to give x_1 explicitly as follows:

$$x_2 = ax_1 + bx_3 + cx_4$$

$$x_1 = \frac{1}{a}x_2 - \frac{1}{a}(bx_3 + cx_4).$$

The graph corresponding to the original is shown in Fig. 11a. Similarly, the graph corresponding to the modified equation is shown in Fig. 11b.

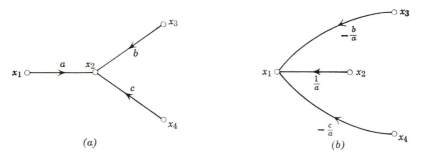

Fig. 11. Inverting a branch: (a) original; (b) inverted.

Focus attention on the original branch with transmittance a from x_1 to x_2. In the modified graph the direction has been reversed and the transmittance has been inverted. At the same time observe what has happened to the other branches. The branch originally incoming to x_2 from x_3 has been redirected to x_1, and its transmittance has been divided by the negative of the transmittance of the branch that was inverted. The same thing has happed to the branch originally incoming to x_2 from x_4. From the equation it is clear that any other branch incoming to x_2 would have undergone the same change.

The result of this process is called *inverting a branch*. The inversion of a branch can be carried out for any branch that leaves a source node with the result that the node is converted to a sink node. The same process can be carried out for a path consisting of any number of branches from a source node to another node. The inversion is carried out one branch at a time starting at a source node. This is illustrated in Fig. 12. (The graph

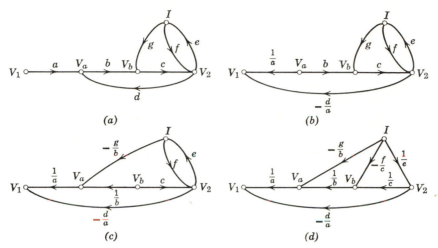

Fig. 12. Inverting a path: (*a*) original; (*b*) after inversion of branch *a*; (*c*) after inversion of branch *b*; (*d*) final.

is the same as that in Fig. 7 with general symbols for the transmittances.) It is desired to invert the path from V_1 to I. Branch *a* is inverted first, leading to Fig. 12*b*. Now node V_a is a source node; so branch *b* can be inverted, leading to Fig. 12*c*. Similarly, branches *c* and *e* can be inverted, in order. The final graph with the inverted path is shown in Fig. 12*d*.

Note that the original graph in Fig. 12*a* has three feedback loops and is of index 1; that is, one node (V_2) must be split before all loops are interrupted. However, the graph with the path inverted is a cascade graph, having no feedback loops. This is inherently a simpler graph.

Besides inverting an open path, it is also possible to invert a loop. In this case, the process is started by splitting a node on the loop, thus creating a source and a sink node. The open path between these two is then inverted, following which the split node is recombined. Details will be left for you to work out.

REDUCTION OF A GRAPH

A signal-flow graph is a representation of a set of equations. Just as the set of equations can be solved for any of the variables in terms of the excitations, the graph can also be "solved." One method of solving the equations is by a process of successive elimination of variables. The analogous procedure for the graph is to successively eliminate the nodes of the graph until only source nodes and sink nodes remain. This process will now be discussed.

Consider the signal flow graph in Fig. 10a and suppose that node x_3 is to be eliminated. The proper relationships among the remaining variables will be maintained if branches are added to the graph between pairs of nodes. The transmittances of these branches must be such as to maintain the path transmittances of the original graph from one of the nodes of the pair to the other one, through the node to be eliminated. To illustrate, in the original graph of Fig. 10, which is reproduced in Fig. 13a,

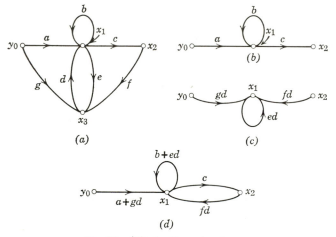

Fig. 13. Elimination of node x_3.

there is a path from y_0 to x_1 through x_3, with a path transmittance gd. In the reduced graph in which node x_3 is eliminated there must be added a branch from y_0 to x_1 with a transmittance gd. In Fig. 13b node x_3, as well as all incoming and outgoing branches incident at x_3 have been removed, leaving transmittances that go directly from one node to another without passing through the intermediate node x_3. Figure 13c shows the same nodes with branches that must be added to account for the transmittances between them for paths that originally went through node x_3. Finally, the reduced graph shown in Fig. 13d combines the last two parts. This graph is said to be *equivalent* to the original graph since the signals at any node are the same as the signals at the corresponding nodes in the original graph. (You should verify this.) Note that the node removed in this illustration did not have a self-loop. Clearly, the same process cannot be used if a self-loop is present.

What was illustrated by means of an example will now be discussed in more general terms. Consider the equations defining a signal-flow graph that are the scalar components of (6). Let x_n be a node without a self-loop. The nth node is chosen, rather than an arbitrary node, strictly for convenience. Removing node x_n from the graph is equivalent to eliminating variable x_n from the equations. The equation for x_n is

$$x_n = c_{n0} y_0 + \sum_{k=1}^{n-1} c_{nk} x_k . \tag{8}$$

To eliminate x_n this expression is substituted into all other equations; then (8) is disregarded; for instance, if the original equation for x_p is

$$x_p = c_{p0} y_0 + \sum_{k=1}^{n-1} c_{pk} x_k + c_{pn} x_n , \tag{9}$$

the modified equation, after substituting (8) for x_n, becomes

$$x_p = c_{p0} y_0 + \sum_{k=1}^{n-1} c_{pk} x_k + c_{pn} c_{n0} y_0 + \sum_{k=1}^{n-1} c_{pn} c_{nk} x_k \tag{10}$$

$$= (c_{p0} + c_{pn} c_{n0}) y_0 + \sum_{k=1}^{n-1} (c_{pk} + c_{pn} c_{nk}) x_k .$$

To interpret this expression in terms of the graph, observe that $c_{pn} c_{n0}$ is the transmittance from the source node y_0 to the node x_p *through* the intermediate node x_n. In the first term on the right this transmittance is added to the direct transmittance c_{p0} from y_0 to x_p. Similarly, $c_{pn} c_{nk}$ is the transmittance from node x_k to node x_p through node x_n. The equation shows that this term is added to the direct transmittance from x_k to x_p. If every transmittance is modified in this way, x_n can be eliminated from the graph, and transmittances between all other nodes will remain invariant.

Now consider the case when there is a self-loop at node x_n, which means that $c_{nn} \neq 0$. In this case the equation for x_n is

$$x_n = c_{n0} y_0 + \sum_{k=1}^{n-1} c_{nk} x_k + c_{nn} x_n . \tag{11}$$

The last term can be transposed to the left side and the result solved for x_n.

$$x_n = \frac{c_{n0}}{1 - c_{nn}} y_0 + \sum_{k=1}^{n-1} \frac{c_{nk}}{1 - c_{nn}} x_k . \tag{12}$$

This is the equation for a node without a self-loop. In terms of the graph the equation shows that if the transmittance of each incoming branch to x_n is divided by $1 - c_{nn}$, the self-loop can be eliminated. Following this, node x_n can be eliminated as before.

Division by $1 - c_{nn}$ is possible only if $c_{nn} \neq 1$. We are clearly in trouble of $c_{nn} = 1$. A glance back at Fig. 9 shows that there is in that graph a self-loop with a transmittance of 1. Nevertheless, the set of equations represented by that graph are linearly independent and so can be solved. In such cases, it is always possible to rearrange the equations in such a way that a unity entry on the main diagonal of the connection matrix can be avoided. Thus an unavoidable such entry can occur only if the equations are not independent—in which case we should not expect a solution anyway.

The preceding operations can also be interpreted in terms of the connection matrix. This matrix has the following form, assuming node x_n has no self-loop.

$$
\mathbf{C} = \left[
\begin{array}{ccccc:c}
0 & 0 & 0 & \cdots & 0 & 0 \\
c_{10} & c_{11} & c_{12} & \cdots & c_{1,n-1} & c_{1n} \\
c_{20} & c_{21} & c_{22} & \cdots & c_{2,n-1} & c_{2n} \\
\vdots & \vdots & \vdots & \vdots & \vdots & \vdots \\
c_{n-1,0} & c_{n-1,1} & c_{n-1,2} & \cdots & c_{n-1,\,n-1} & c_{n-1,\,n} \\
\hdashline
c_{n0} & c_{n1} & c_{n2} & \cdots & c_{n,\,n-1} & 0
\end{array}
\right].
\tag{13}
$$

The partition indicates that the last row and column are to be eliminated. Note that $c_{nn} = 0$ since x_n has no self-loop.

Now consider the typical equation (10) of the set, after the modification introduced by eliminating node x_n. The corresponding connection matrix will have the form

$$
\mathbf{C}_1 = \left[
\begin{array}{cccc}
0 & 0 & \cdots & 0 \\
c_{10} + c_{1n}c_{n0} & c_{11} + c_{1n}c_{n1} & \cdots & c_{1,\,n-1} + c_{1n}c_{n,\,n-1} \\
c_{20} + c_{2n}c_{n0} & c_{21} + c_{2n}c_{n1} & \cdots & c_{2,\,n-1} + c_{2n}c_{n,\,n-1} \\
\vdots & \vdots & & \vdots \\
c_{n-1,\,0} + c_{n-1,\,n}c_{n0} & c_{n-1,\,1} + c_{n-1,\,n}c_{n1} & \cdots & c_{n-1,\,n-1} + c_{n-1,\,n}c_{n,\,n-1}
\end{array}
\right]
\begin{array}{c}
0 \\
c_{1n} \\
c_{2n} \\
\vdots \\
c_{n-1,\,n}
\end{array}
$$

$$
\begin{array}{cccc}
c_{n0} & c_{n1} & \cdots & c_{n,\,n-1}
\end{array} \quad 0
$$

$$
\tag{14}
$$

where the external row and column have been added for comparison with the preceding equations.

Now observe the manner in which the modified connection matrix is obtained from the original one in (13). An entry in the last column of $\mathbf{C}$, say c_{1n}, is multiplied by each entry in the last row. These products are then added to the corresponding entry in the connection matrix. Thus c_{1n} multiplied by c_{n3} is added to c_{13}, which is in the (2, 4) position of the connection matrix. (Note that subscripts of rows and columns start from zero.) This is repeated for each entry in the last column. This process of successive reduction of a connection matrix is called the *node-pulling* algorithm.

If node x_n has a self-loop, the entry c_{nn} in the connection matrix is not zero. The process of dividing all incoming transmittances to node x_n in the graph by $1 - c_{nn}$ corresponds to dividing all entries of the last row in the connection matrix by $1 - c_{nn}$ and then replacing the diagonal entry by zero. The node-pulling algorithm can then be applied.

By repeated applications of these operations, a signal-flow graph can be reduced so that only source and sink nodes remain. If there is only one source and one sink node, the graph reduces to a single branch from the source to the sink. The transmittance of this branch is called the *graph gain*.

As an illustration of the process, consider the set of equations

$$\begin{bmatrix} 2 & 1 & 2 \\ 1 & -1 & -3 \\ 2 & 1 & 1 \end{bmatrix} \begin{bmatrix} x_1 \\ x_2 \\ x_3 \end{bmatrix} = \begin{bmatrix} 1 \\ 0 \\ -1 \end{bmatrix} y_0.$$

The system can be modified as in (6) to get

$$\begin{bmatrix} x_1 \\ x_2 \\ x_3 \end{bmatrix} = \begin{bmatrix} -1 & 3 & 1 & 2 \\ 0 & 1 & 0 & -3 \\ 1 & 2 & 1 & 2 \end{bmatrix} \begin{bmatrix} y_0 \\ x_1 \\ x_2 \\ x_3 \end{bmatrix}. \tag{15}$$

The corresponding signal-flow graph is shown in Fig. 14.

Suppose that a solution for x_1 is desired. Then we go through the following operations, starting with the connection matrix derived from (15).

The corresponding reductions of the flow graph are illustrated in Fig. 15, the corresponding steps being labeled with the same letter.

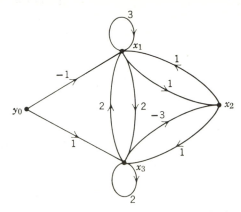

Fig. 14.   Signal-flow graph of example.

1. Remove self-loop at x_3 by dividing all incoming transmittances by $(1 - c_{44}) = 1 - 2 = -1$. The resulting connection matrix is

$$
\mathbf{C}_1 = \begin{array}{c} \\ y_0 \\ x_1 \\ x_2 \\ \\ x_3 \end{array}
\begin{array}{cccc}
y_0 & x_1 & x_2 & x_3 \\
\end{array}
\left[
\begin{array}{ccc:c}
0 & 0 & 0 & 0 \\
-1 & 3 & 1 & 2 \\
0 & 1 & 0 & -3 \\
\hdashline
-1 & -2 & -1 & 0
\end{array}
\right].
$$

2. Remove node x_3 by using the node-pulling algorithm. The resulting connection matrix is

$$
\mathbf{C}_2 = \begin{array}{c} \\ y_0 \\ x_1 \\ x_2 \end{array}
\begin{array}{ccc}
y_0 & x_1 & x_2 \\
\end{array}
\left[
\begin{array}{ccc}
0 & 0 & 0 \\
-3 & -1 & -1 \\
3 & 7 & 3
\end{array}
\right].
$$

Notice that in the process a self-loop is created at x_2 with a transmittance equal to 3 and notice how the other transmittances are altered.

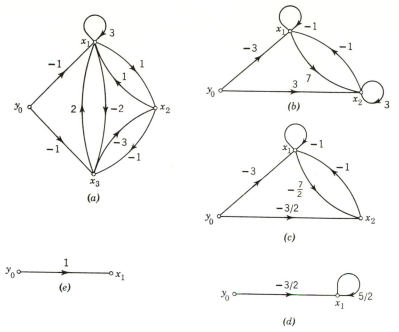

Fig. 15. Reduction of signal-flow graph.

3. Remove self-loop at x_2 by dividing incoming transmittances by $1 - 3 = -2$. Then

$$\mathbf{C}_3 = \begin{array}{c} \\ y_0 \\ x_1 \\ \\ x_2 \end{array} \begin{array}{c} y_0 \quad x_1 \quad x_2 \\ \left[\begin{array}{ccc} 0 & 0 & 0 \\ -3 & -1 & -1 \\ \cdots\cdots\cdots\cdots \\ -\frac{3}{2} & -\frac{7}{2} & 0 \end{array} \right] \end{array}.$$

4. Remove node x_2 to obtain

$$\mathbf{C}_4 = \begin{array}{c} \\ y_0 \\ x_1 \end{array} \begin{array}{c} y_0 \quad x_1 \\ \left[\begin{array}{cc} 0 & 0 \\ -\frac{3}{2} & \frac{5}{2} \end{array} \right] \end{array}.$$

5. Remove self-loop at x_1 by dividing incoming transmittances by $1 - \frac{5}{2} = -\frac{3}{2}$. The connection matrix at this step is

$$\mathbf{C}_5 = \begin{array}{c} \\ y_0 \\ x_1 \end{array} \begin{array}{c} y_0 \quad x_1 \\ \left[\begin{array}{cc} 0 & 0 \\ 1 & 0 \end{array} \right] \end{array}.$$

Thus we find that $x_1 = y_0$. It is clear that the solution can be obtained by working on the matrix only, or the flow graph only.

The process of path inversion can also be useful in the reduction of a signal-flow graph to determine the graph gain. This is illustrated in the graph shown in Fig. 16a. If the path from y_0 to x_3 is inverted, the result

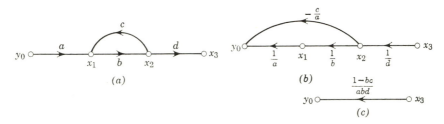

Fig. 16.   Inversion and reduction of a graph.

is as shown in Fig. 16b. The graph can now be rapidly reduced. First, branches $1/a$ and $1/b$ are in cascade. The combination is in parallel with branch $-c/a$. Finally, the combination of these three is in cascade with branch $1/d$. The reduced graph is shown in Fig. 16c. The graph gain of the original graph is thus found to be

$$\frac{x_3}{y_0} = \frac{abd}{1 - bc}.$$

REDUCTION TO AN ESSENTIAL GRAPH

The graph-reduction process described above proceeds by the elimination of one node at a time; however, this is not an absolute requirement. Several nodes can be eliminated simultaneously. The only requirement is that branches be included between pairs of nodes in the reduced graph to account for all the path transmittances between the same pairs of nodes in the original graph.

Thus an alternative viewpoint is to focus on the nodes that are retained in the reduced graph, as opposed to those which are eliminated. A particularly useful reduced graph is obtained by retaining only a set of essential nodes, together with source and sink nodes; such a graph is called an *essential graph*. No matter what structure the original graph may have, the essential graph for a given index will have a fixed structure. Thus for index 2 and a single source node, an essential diagram will have the structure shown in Fig. 17. There will be, in the general case, transmittances from the source node to each of the essential nodes, and from each of the

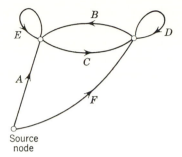

Fig. 17.   Essential graph of index 2.

essential nodes to all other essential nodes. There will also be self-loops at each essential node. It is only necessary to determine what the values of the transmittances will be. In specific cases, of course, one or more of the branches in the essential graph might be missing.

To illustrate, consider the graph of Fig. 20 in the next subsection. This is of index 2, although it has six feedback loops; V_1 and I_3 constitute a set of essential nodes. In Fig. 17, then, the two essential nodes will be V_1 and I_3, and the source node will be V_g. It remains to compute the transmittances; for example, in the original graph there are three paths from V_1 to I_3, with the following path transmittances: $-G_3$, $-\alpha Y_1 R_4 G_3$, and $Y_2 \mu Z_2 G_3 = \mu G_3$. Hence in the essential graph the branch from V_1 to I_3 will have a transmittance $G_3(\mu - 1 - \alpha Y_1 R_4)$. The remaining transmittances are evaluated similarly, with the final result shown in Fig. 18.

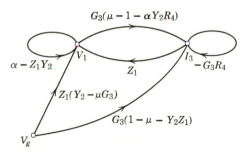

Fig. 18.   Essential graph of example.

GRAPH-GAIN FORMULA

Although simple and straightforward, the graph-reduction process that has been described is somewhat tedious. It would be valuable to have a single formula for the graph gain that could be written directly from the graph without any further manipulations. Such a formula has, in fact, been established.* The derivation of the formula does not require advanced concepts or mathematics, but it is quite long. We shall, therefore, state the result here but omit the proof.

Consider a signal-flow graph with a single source node and a single sink node. Let G be the graph gain and let G_k be the path transmittance of the kth direct path (without loops) from source to sink. In the graph some of the feedback loops may have nodes or branches in common with other loops, and some may not. We say a set of loops is *nontouching* if there is neither a node nor a branch in common between any two loops of the set. The graph gain is given by the following expression:

$$G = \frac{1}{\Delta} \sum_{\substack{\text{all} \\ \text{direct} \\ \text{paths}}} G_k \Delta_k, \tag{16}$$

where the *graph determinant* Δ is given by

$$\Delta = 1 - \sum_j P_{j1} + \sum_j P_{j2} - \sum_j P_{j3} + \cdots, \tag{17}$$

in which

P_{j1} is the *loop gain* (the product of all branch transmittances around a loop) of the jth feedback loop;

P_{j2} is the *product* of loop gains of the jth *pair* of nontouching loops;

P_{j3} is the product of loop gains of the jth *triplet* of nontouching loops; etc.

The second subscript in P_{ji} refers to the number of loops in the nontouching set. The determinant Δ_k is the graph determinant of the subgraph which does not touch the kth direct path from source to sink.

Let us now illustrate the use of this expression. As a first illustration, consider the original signal-flow graph of Fig. 12. A sink node can be trivially created by adding a new node labeled I, with a unity transmittance branch from the old node I. There are three loops in this graph with loop gains bcd, gce, and ef. Since they all have at least one node in common, they are all touching. Hence

$$\Delta = 1 - ef - gce - bcd.$$

---

* S. J. Mason, "Feedback Theory—Further Properties of Signal-Flow Graphs," *Proc. IRE*, Vol. 44, July 1956, pp. 920–926.

There is only one direct path from V_1 to I, and the path gain (or trans-mittance) is $G_1 = abce$. All three loops touch this path; hence $\Delta_1 = 1$. The graph gain therefore equals

$$G = \frac{V_1}{I} = \frac{abce}{1 - ef - gce - bcd}.$$

Now consider the inverted graph in Fig. 12d. This is a strictly cascade graph, without feedback loops. Hence $\Delta = 1$, and all Δ_k's $= 1$. The graph gain from I to V_1 is therefore simply the sum of direct path transmittances. There are a total of four paths from I to V_1. Hence

$$\frac{I}{V_1} = -\frac{d}{ae} + \frac{1}{abce} - \frac{f}{abc} - \frac{g}{ab} = \frac{1 - ef - gce - bcd}{abce}.$$

This agrees with the preceding value of graph gain.

As a second illustration, let us take the signal-flow graph in Fig. 20 in the next subsection. Let V_4 be the desired response. The graph has a total of six loops. Of these there are two pairs of loops that are nontouch-ing; namely, loop $V_4 I_3 V_4$ with loop $V_1 I_2 V_1$ and with loop $V_1 I_5 V_1$. There are no triplets of nontouching loops. The following calculations can now be made:

Loop	Loop gains	
	P_{j1}	P_{j2}
1. $V_1 I_2 V_1$	$-Z_1 Y_2$	$\left.\begin{array}{l} G_3 R_4 Z_1 Y_2 \\ -\alpha G_3 R_4 \end{array}\right\}$
2. $V_4 I_3 V_4$	$-G_3 R_4$	
3. $V_1 I_5 V_1$	$\alpha Y_1 Z_1 = \alpha$	
4. $V_1 I_3 V_1$	$-Z_1 G_3$	
5. $V_1 I_2 V_6 I_3 V_1$	$Y_2 \mu Z_2 G_3 Z_1 = \mu Z_1 G_3$	
6. $V_1 I_5 V_4 I_3 V_1$	$-\alpha Y_1 R_4 G_3 Z_1 = -\alpha G_3 R_4$	

Then

$$\sum P_{j1} = \alpha - G_3 R_4 - \alpha G_3 R_4 - Z_1 Y_2 - (1 - \mu)Z_1 G_3$$

$$\sum P_{j2} = -\alpha G_3 R_4 + G_3 R_4 Z_1 Y_2$$

$$\Delta = 1 - \sum P_{j1} + \sum P_{j2} = 1 - \alpha + G_3 R_4 + \alpha G_3 R_4 + Z_1 Y_2$$
$$+ (1 - \mu)Z_1 G_3 - \alpha G_3 R_4 + Z_1 Y_2 G_3 R_4$$
$$= 1 - \alpha + G_3 R_4 + Z_1 Y_2(1 + G_3 R_4) + (1 - \mu)Z_1 G_3.$$

Notice that in the computation of Δ the term $\alpha G_3 R_4$ from P_{j1} cancelled with a similar term from P_{j2}. This is an illustration of the fact that the gain formula is not a minimum-effort formula, as are the topological formulas in Chapter 3.

Next we determine the numerator of the graph gain. The following calculations are self-explanatory:

Direct paths	Path gains G_k	Δ_k
1. $V_g\, I_3\, V_4$	$G_3\, R_4$	$\Delta_1 = 1 - \alpha + Z_1 Y_2$
2. $V_g\, I_2\, V_1\, I_5\, V_4$	$\alpha Y_2\, R_4$	$\Delta_2 = 1$
3. $V_g\, I_2\, V_6\, I_3\, V_4$	$-\mu G_3\, R_4$	$\Delta_3 = 1 - \alpha$
4. $V_g\, I_2\, V_1\, I_3\, V_4$	$-Z_1 Y_2\, G_3\, R_4$	$\Delta_4 = 1$
5. $V_g\, I_3\, V_1\, I_5\, V_4$	$\alpha G_3\, R_4$	$\Delta_5 = 1$

Then

$$\sum G_k \Delta_k = G_3\, R_4 - \alpha G_3\, R_4 + G_3\, R_4\, Z_1 Y_2 + \alpha Y_2\, R_4$$
$$- (1 - \alpha)\mu G_3\, R_4 + \alpha G_3\, R_4 - G_3\, R_4\, Z_1 Y_2$$
$$= G_3\, R_4[1 - \mu(1 - \alpha)] + \alpha Y_2\, R_4 .$$

Again we find a cancellation taking place. Finally, the graph gain is

$$G = \frac{V_4}{V_g} = \frac{G_3\, R_4[1 - \mu(1 - \alpha)] + \alpha Y_2\, R_4}{(1 + G_3\, R_4)(1 + Z_1\, Y_2) - \alpha + (1 - \mu)Z_1\, G_3} .$$

It is observed that the formula for determining the graph gain is quite simple to apply. A source of error is the possible overlooking of one or more feedback loops and one or more direct paths from input to output. A systematic search procedure, however, should eliminate this possible source of error.

The graph gain expression in (16) is a network function expressing the ratio of a response transform to an excitation transform. Clearly, the same network function should be obtained whether it is calculated with the use of a signal-flow graph or from a solution of the network equations expressed in the form of $\mathbf{AX} = \mathbf{Y}$, as in (3). However, in the latter case the network function will be obtained in terms of det $\mathbf{A}$. Thus the graph determinant must be the same as det $\mathbf{A}$.

DRAWING THE SIGNAL-FLOW GRAPH OF A NETWORK

In the preceding discussion it has been assumed that a set of linear equations is given in the standard form of (3). The more usual problem is that a network is presented for analysis, with the ultimate goal of finding an expression for a network function. Given the network, it is desired to draw the signal-flow graph directly, merely by inspection of the network diagram. An alternative would be to write down the connection matrix directly.

In either case, the first order of business is to choose a set of variables. Different choices of the variables will lead to different flow graphs representing the same network. The choice of variables is guided by many considerations, but in any case must include the independent source variables and the response variables. No matter what variables and equations relating them are chosen, two things must be ensured.

1. The system of equations obtained must be an adequate description of the network;

2. The equations obtained must be linearly independent.

Some choices of the variables and equations will not be very useful if we are to make full use of the signal-flow graph technique; for example, we might simply choose loop currents or node voltages as variables and represent the loop or node equations as a signal-flow graph. This procedure is certainly valid and needs no comment.

A more useful approach is to choose a mixed set of variables, as in Section 2.7. Tree branch voltages and link currents are known to be topologically independent. Hence twig voltages and link currents are selected as variables. There is one slight difference here from the discussion in Chapter 2. There independent sources were not counted as separate branches but were always taken together with their accompanying branches. As we did in Chapter 4, however, we shall now count such sources as separate branches. Independent voltage sources must then be made twigs; and independent current sources, links.

By Kirchhoff's laws and appropriate partitioning, the following equations result:

$$\mathbf{I}_{ta} = -\mathbf{Q}_l \mathbf{I}_{la} \tag{18a}$$

$$\mathbf{V}_{la} = -\mathbf{B}_t \mathbf{V}_{ta}, \tag{18b}$$

where the subscript a stands for " all." The links will include all independent current sources, and the twigs will include all independent voltage sources. Let us partition the current and voltage vectors to put the sources in evidence. When $\mathbf{Q}_l$ and $\mathbf{B}_t$ are conformally partitioned, the above expressions become

$$\begin{bmatrix} \mathbf{I}_t \\ \mathbf{I}_{tg} \end{bmatrix} = - \begin{bmatrix} \mathbf{Q}_1 & \mathbf{Q}_{g1} \\ \mathbf{Q}_2 & \mathbf{Q}_{g2} \end{bmatrix} \begin{bmatrix} \mathbf{I}_l \\ \mathbf{I}_g \end{bmatrix} = - \begin{bmatrix} \mathbf{Q}_1 \mathbf{I}_l + \mathbf{Q}_{g1} \mathbf{I}_g \\ \mathbf{Q}_2 \mathbf{I}_l + \mathbf{Q}_{g2} \mathbf{I}_g \end{bmatrix} \tag{19}$$

$$\begin{bmatrix} \mathbf{V}_l \\ \mathbf{V}_{lg} \end{bmatrix} = - \begin{bmatrix} \mathbf{B}_1 & \mathbf{B}_{g1} \\ \mathbf{B}_2 & \mathbf{B}_{g2} \end{bmatrix} \begin{bmatrix} \mathbf{V}_t \\ \mathbf{V}_g \end{bmatrix} = - \begin{bmatrix} \mathbf{B}_1 \mathbf{V}_t + \mathbf{B}_{g1} \mathbf{V}_g \\ \mathbf{B}_2 \mathbf{V}_t + \mathbf{B}_{g2} \mathbf{V}_g \end{bmatrix}. \tag{20}$$

In these expressions $\mathbf{I}_{tg}$ represents the currents of voltage sources (which are twigs) and $\mathbf{V}_{lg}$ represents the voltages of current sources (which are links).

There remain the voltage-current relationships. Since link currents and twig voltages are to be the variables in the signal-flow graph, the $V\text{-}I$ relationships should express $\mathbf{I}_l$ and $\mathbf{V}_t$ explicitly in terms of $\mathbf{V}_l$ and $\mathbf{I}_t$. This means the following hybrid form:

$$\mathbf{I}_l = \mathbf{Y}_l \mathbf{V}_l + \mathbf{G}_{12} \mathbf{I}_t \tag{21a}$$

$$\mathbf{V}_t = \mathbf{G}_{21} \mathbf{V}_l + \mathbf{Z}_t \mathbf{I}_t. \tag{21b}$$

Note that voltages and currents pertaining to independent sources are not included in these equations.

The first row of (19) and of (20) are now inserted into (21) to yield

$$\mathbf{I}_l = -(\mathbf{Y}_l \mathbf{B}_1 \mathbf{V}_t + \mathbf{Y}_l \mathbf{B}_{g1} \mathbf{V}_g + \mathbf{G}_{12} \mathbf{Q}_1 \mathbf{I}_l + \mathbf{G}_{12} \mathbf{Q}_{g1} \mathbf{I}_g) \tag{22a}$$

$$\mathbf{V}_t = -(\mathbf{G}_{21} \mathbf{B}_1 \mathbf{V}_t + \mathbf{G}_{21} \mathbf{B}_{g1} \mathbf{V}_g + \mathbf{Z}_t \mathbf{Q}_1 \mathbf{I}_l + \mathbf{Z}_t \mathbf{Q}_{g1} \mathbf{I}_g). \tag{22b}$$

These equations are in the form of (4), from which a signal-flow graph can be drawn.

The variables for which equations are written are the currents of all links except independent current sources, and the voltages of all twigs except independent voltage sources. Note that these variables do not depend on the currents in voltage sources ($\mathbf{I}_{tg}$) and voltages across current sources ($\mathbf{V}_{lg}$). However, it may be that these latter variables are response variables for which solutions are required. In the signal flow graph each such variable will constitute a sink node; the corresponding equations, are the second rows of (19) and (20), which are repeated here.

$$\mathbf{I}_{tg} = -(\mathbf{Q}_2 \mathbf{I}_l + \mathbf{Q}_{g2} \mathbf{I}_g) \tag{23a}$$

$$\mathbf{V}_{lg} = -(\mathbf{B}_2 \mathbf{V}_t + \mathbf{B}_{g2} \mathbf{V}_g). \tag{23b}$$

From (22) and (23) a signal-flow graph can be drawn. We are sure that the equations constitute an independent set and therefore a solution can be obtained. We also know that the equations constitute an adequate set since any other variable can be determined once these variables are known. However, if we were really required first to write the $\mathbf{Q}_l$ and $\mathbf{B}_t$ matrices; then to partition them; then to write the V-I relationships and determine all the matrices in (21); and then finally to set up (22) and (23), the value of using a signal-flow graph would be obscured. The preceding development, in fact, constitutes an "existence proof." For a given network we would not go through the same steps in drawing the signal-flow graph as are needed to establish the existence. A much simpler process would be used, illustrated by the examples to follow.

One further point must be clarified. When discussing the mixed-variable equations in Section 2.7, the V-I relationship used was the inverse of that given in (21), as shown in (110) in Chapter 2. For that choice the selection of the branches of a controlled source as twig or link was given in Table 2. With the present V-I relations these selections must be reversed. Thus, for example, the controlled branch of a controlled current source must now be a link, and the controlled branch of a controlled voltage source a twig.

Example 1.

Consider the network in Fig. 19a. There are two controlled sources. The graph of the network is shown in Fig. 19b, with the tree in heavy lines.

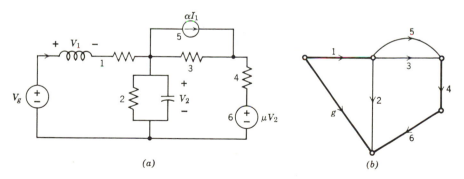

$$(a) \qquad\qquad\qquad\qquad (b)$$

Fig. 19.   Example.

Equations for the link currents and twig voltages must now be written. The current of any link which is not a controlled source can be expressed in terms of its own voltage which, in turn, can be written in terms of twig voltages. Thus for links 2 and 3 the equations are

$$I_2 = Y_2 V_2 = Y_2 V_g - Y_2 V_1$$
$$I_3 = G_3 V_3 = -G_3 V_1 + G_3 V_g - G_3 V_6 - G_3 V_4.$$

The only other link is a controlled current source, $I_5 = \alpha I_1$. The term I_1 is the current in twig 1 and can be expressed simply in terms of its own voltage. Thus

$$I_5 = \alpha Y_1 V_1.$$

As for the voltage of any twig branch that is not a controlled source, it can be expressed in terms of its own current, which, in turn, can be expressed in terms of link currents. Thus for twigs 1 and 4 the equations are

$$V_1 = Z_1 I_1 = Z_1 I_2 + Z_1 I_3 + Z_1 I_5$$
$$V_4 = R_4 I_4 = R_4 I_3 + R_4 I_5.$$

For twig 6, which is a controlled source, the equation is

$$V_6 = \mu V_2 = \mu Z_2 I_2.$$

The signal-flow graph can now be drawn and is shown in Fig. 20. It has a

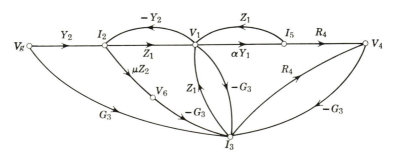

Fig. 20.   Signal-flow graph of example.

large number (six) of feedback loops but its index is only 2, since opening nodes V_1 and I_3 will interrupt all loops. The graph-reduction procedure can now be used to solve for any of the variables.

If it is desired to find the input impedance of the network from the terminals of the independent voltage source, it will be necessary to solve for the current in the voltage source. This variable does not now appear in the graph, but a sink node for I_g can be easily added since $I_g = -I_2 - I_3 - I_5$. The graph gain relating I_g to V_g is the negative of the input impedance.

Example 2.

As a second example, consider the network in Fig. 21. For the link

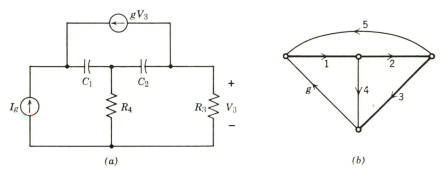

(a)   (b)

Fig. 21.   Network example.

currents the equations are

$$I_4 = G_4 V_4 = G_4 V_2 + G_4 V_3$$

$$I_5 = g V_3 .$$

Similarly, the equations for the twig voltages are

$$V_1 = \frac{1}{sC_1} I_1 = \frac{1}{sC_1} I_g + \frac{1}{sC_1} I_5$$

$$V_2 = \frac{1}{sC_2} I_2 = \frac{1}{sC_2} I_g - \frac{1}{sC_2} I_4 + \frac{1}{sC_2} I_5$$

$$V_3 = R_3 I_3 = R_3 I_g - R_3 I_4 .$$

The resulting signal-flow graph is shown in Fig. 22. This is a simpler graph than that of the previous example. There are three feedback loops, all of which would be interrupted if node I_4 were split. Hence the index is 1.

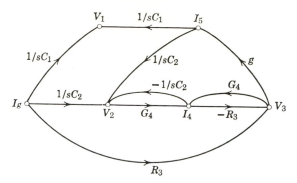

Fig. 22.   Signal-flow graph of example.

9.3   FEEDBACK

An intuitive meaning for the concept of feedback in physical processes has existed for a long time, but it was only in the 1930's that Bode gave the concept a more accurate and mathematical significance. In a qualitative way, we say that a network is a feedback network if some variable, either an output variable or an internal one, is used as the input to some part of the network in such a way that it is able to affect its own value. It is said, again qualitatively, that the output, or part of it, is fed back to the input.

As an illustration, look back at the amplifier diagram in Fig. 6. Here part of the input voltage V_a to the dashed box is a function of the output voltage V_2. Thus whatever the output voltage becomes is influenced by its own value. This feedback effect has a great influence on the behavior of a network. We shall now discuss the concept of feedback in a quantitative way.

RETURN RATIO AND RETURN DIFFERENCE

Consider first the signal-flow graph shown in Fig. 23a. Focus attention on the branch with a transmittance k. This quantity is assumed to be a specific parameter in a network, such as the μ or α of a controlled source.* This assumption implies that the signal-flow graph of a network can be drawn in such a way that the desired parameter appears by itself in only

---

* The assumption that k is a specific network parameter is not essential to the definition of return ratio. However, the assumption is made in order to provide a simple relationship between return ratio and sensitivity of the gain to variation of a network parameter, to be defined later.

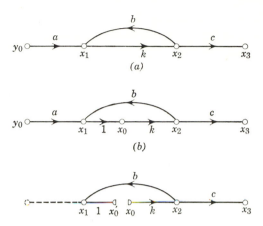

Fig. 23.   Determining the return ratio.

one branch transmittance. It is always possible to modify network equations in such a way (by introducing auxiliary variables or by suitable linear combinations) that this result can be achieved, except in those cases when the same parameter appears twice in the network diagram itself. This is the case, for example, with the turns ratio of a transformer and with the gyration ratio of a gyrator.

We now insert an auxiliary node x_0 in the branch k with a unity transmittance from x_1 to x_0, as in Fig. 23b. This simply introduces an auxiliary equation $x_0 = x_1$ without modifying the remaining equations. The next step is to split the new node, as shown in Fig. 23c, thus creating a source node and a sink node. At the same time, all other source nodes in the graph are removed. (This amounts to shorting independent voltage sources and opening independent current sources in the network.) A measure of the feedback with reference to parameter k is obtained by determining the signal returned to the sink half of the split node per unit of signal sent out from the source half. We define the *return ratio* of k, T_k, to be the ratio of $-x_0'$ to x_0 when a node x_0 is inserted into branch k and then split, as described. In Fig. 23, the return ratio of k is found to be $T_k = -kb$. (The reason for the negative sign in the definition is to conform with standard usage in feedback control theory.)

Another measure of the feedback is obtained by taking the difference between the signal sent from the source half of the split node and the signal received at the sink half, per unit signal sent. This quantity is called the *return difference* and is labeled F_k. The two measures of feedback are related by

$$F_k = 1 + T_k.\tag{24}$$

Let us illustrate the calculation of the return ratio and return difference for a more substantial graph. Look back at the signal-flow graph in Fig. 7 representing the amplifier in Fig. 6. Let the transconductance g be the parameter of interest. This parameter appears only once in the graph, but it is not alone. The graph can be modified as shown in Fig. 24a to put it in

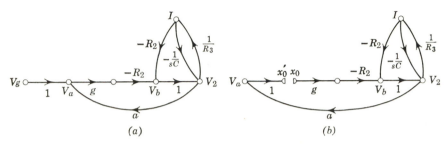

(a)   (b)

Fig. 24.   Calculation of return difference.

the desired form. Next the source node V_g is removed, an auxiliary node is introduced in branch g and then split. The result is shown in Fig. 24b. The negative graph gain from x_0 to x_0' is the return ratio. This is found from the gain formula in (16) to be

$$T_g = \frac{-x_0'}{x_0} = \frac{gR_2 a}{1-(-1/R_3 Cs)-(-R_2/R_3)} = \frac{gaR_2 R_3 Cs}{(R_2 + R_3)Cs + 1}.$$

The return difference is now found from (24) to be

$$F_g = 1 + T_g = \frac{(R_2 + R_3 + gaR_2 R_3)Cs + 1}{(R_2 + R_3)Cs + 1}. \tag{25}$$

An interesting observation can be made by calculating the graph determinant of the original graph in Fig. 24a. Thus

$$\Delta = 1 + \frac{R_2}{R_3} + gaR_2 + \frac{1}{R_3 Cs} = \frac{(R_2 + R_3 + gaR_2 R_3)Cs + 1}{R_3 Cs}. \tag{26}$$

By comparing with the previous expression for F_g, we see that the numerator of F_g and the numerator of Δ are the same. This result is not accidental, but quite general, as we shall now discuss.

Consider the diagram in Fig. 25, which shows a portion of a signal-flow graph containing the branch k between nodes x_a and x_b. A node has been inserted and then split. Before the operation is performed, the equa-

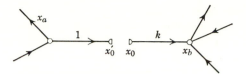

Fig. 25.   Formula for the return difference.

tions for the entire graph are as given previously in (6) and repeated here for convenience.

$$\mathbf{X} = -\mathbf{K}y_0 + (\mathbf{A} + \mathbf{U})\mathbf{X}. \tag{27}$$

For node x_b the equation is

$$x_b = kx_a + a_{b0}y_0 + \sum_{j \neq a} a_{bj}x_j, \tag{28}$$

where it is assumed that the parameter k does not appear in any of the other a_{bj} coefficients and that there is no other direct path between x_a and x_b.

Now suppose that the equation for x_b and for all the other nodes is rewritten in the form

$$\mathbf{AX} = \mathbf{Y}. \tag{29}$$

Equation 28 for x_b becomes

$$-kx_a + x_b - \sum_{j \neq a} a_{bj}x_j = a_{b0}y_0. \tag{30}$$

The coefficient matrix of the set of equation is $\mathbf{A}$. To find an expression for the determinant of $\mathbf{A}$, suppose it is expanded along the bth row and then all terms except the one containing k are collected. The sum of these terms will simply be the value of det $\mathbf{A}$ when $k = 0$. Let this be designated Δ^0. Then

$$\det \mathbf{A} = \Delta = \Delta^0 - k\Delta_{ba}, \tag{31}$$

where Δ_{ba} is a cofactor of $\mathbf{A}$.

Now we turn to a determination of the return difference. We first remove all source nodes. This means equating to zero the right side of (29). Then the insertion and splitting of the node is performed as in Fig. 25. This introduces a new source node x_0, which enters into the equation of node x_b only. This amounts to adding kx_0 to the right side of (29) in the bth row. Furthermore, as Fig. 25 shows, the term kx_a that appeared in the equation for x_b no longer appears. Hence kx_a is removed from the bth row on the left side of (29). All feedback loops in the original graph except the ones containing branch k are included in the modified graph. Hence the graph determinant is obtained from the old graph determinant simply by setting $k = 0$. But this was called Δ^0 above. Hence the graph gain of the new graph, which is just the negative return ratio, is given by

$$- T_k = \frac{x_0'}{x_0} = \frac{k\Delta_{ba}}{\Delta^0}. \tag{32}$$

The return difference can now be formed as follows:

$$F_k = 1 + T_k = 1 - \frac{k\Delta_{ba}}{\Delta^0} = \frac{\Delta^0 - k\Delta_{ba}}{\Delta^0},$$

or, by using (31),

$$F_k = \frac{\Delta}{\Delta^0}. \tag{33}$$

This is a highly significant result. A measure of feedback with reference to a specific parameter k is obtained by taking the ratio of the graph determinant with the parameter in place to its value when the parameter is set equal to zero.

You should demonstrate the validity of this expression for the graph of Fig. 24, for which the return difference was already found in (25), by using the expression for the graph determinant given in (26).

SENSITIVITY

Generally speaking, each parameter in a network has an influence on the response. As that parameter is changed (due to aging, temperature changes, replacement, etc.), the response will change. It is of interest to know by what fraction the response will change when a given parameter changes by a certain fraction. This information is given by what is called the *sensitivity*. We define the sensitivity of a quantity, say the graph gain G, to a parameter k by

$$S_k^G = \frac{\partial G/G}{\partial k/k}. \tag{34}$$

In this definition the changes are considered to be differentially small.

It is possible to relate the sensitivity to the return difference. Suppose that the parameter k is in a branch lying in a direct path between source and sink. Hence the graph gain in (16) can be written as follows:

$$G = \frac{k\Delta_{ba} + R}{\Delta^0 - k\Delta_{ba}}, \qquad (35)$$

where R is the remainder of the numerator after the term containing k is removed. By direct evaluation, by using the definition in (34), and by using (33) for the return difference (and after some manipulation), the following relationship is obtained:

$$S_k^G = \frac{1}{F_k}\left(1 - \frac{G^0}{G}\right), \qquad (36)$$

where $G^0 = R/\Delta^0$ is the graph gain when k is set equal to zero; that is, this is the graph gain due to "leakage" paths not traversing branch k.

In the event that there are no leakage paths, meaning that all direct paths from source to sink pass through branch k, the sensitivity becomes simply the reciprocal of the return difference. This is the case, for example, in the signal-flow graph of Fig. 24. Hence in this example the sensitivity of the gain to the transconductance g is the reciprocal of the return difference already determined in (25).

9.4 STABILITY

In the preceding sections we have discussed a method of analysis that constitutes an alternative approach to the determination of network functions and thus network responses to specified excitations. The signal-flow graph approach can be used for passive, reciprocal networks as well as active, nonreciprocal networks. It is of greatest value for the latter type network, however.

A very important concern in the study of active networks is whether the response remains bounded or increases indefinitely following an excitation. This concern is absent in lossy, passive networks since the poles of network functions for such networks necessarily lie in the left half-plane. In this section we shall investigate this concern. We shall deal with signals in both the time and the frequency domain and we shall *assume that all networks are initially relaxed.* Only single-input, single-output networks will be considered. This is not restrictive, since multi-input, multi-output networks can be handled by the principle of superposition.

Let $w(t)$ denote the network response corresponding to an excitation $e(t)$. We shall say the network is *stable* if, given a constant $0 \leq E < \infty$, there exists another constant $0 \leq W < \infty$ such that $|w(t)| \leq W$ whenever $|e(t)| \leq E$ for $0 \leq t < \infty$. In less precise terms, we designate a network as *stable if to every bounded excitation there corresponds a bounded response.* To distinguish this from other definitions to be introduced later, we refer to this as *bounded-input bounded output* (BIBO) stability.

To find functional criteria for a network to be BIBO stable, the response $w(t)$ must be expressed in terms of the excitation $e(t)$. Start with the convolution integral

$$w(t) = \int_0^t h(t - \tau)\, e(\tau) d\tau, \tag{37}$$

where $h(t)$ is the impulse response. Then the following can be shown:

Theorem 1. *A network is BIBO stable if and only if*

$$\int_0^\infty |h(\tau)|\, d\tau < \infty. \tag{38}$$

That is to say, if the impulse response is absolutely integrable, then the response to any bounded excitation will remain bounded. For the *if* portion of the proof, start by taking the absolute value of both sides of (37). After applying the usual inequalities, the result becomes

$$|w(t)| \leq \int_0^t |h(t - \tau)|\, |e(\tau)|\, d\tau.$$

If $|e(\tau)|$ is replaced by its upper bound E, this inequality can only be strengthened. Thus

$$|w(t)| \leq E \int_0^t |h(t - \tau)|\, d\tau = E \int_0^t |h(\tau)|\, d\tau.$$

The right side follows from the change of variable $(t - \tau) \to \tau$. Now, if the upper limit is increased to infinity, the integral will not be reduced; so the inequality will be further strengthened. Hence

$$|w(t)| \leq E \int_0^\infty |h(\tau)|\, d\tau = W.$$

Because of the condition of the theorem given in (38), $W < \infty$ and $|w(t)|$ is bounded for $0 \leq t < \infty$.

For the *only if* portion of the proof, we start with this observation. If

$$\int_0^\infty |h(\tau)|\, d\tau \not< \infty, \tag{39}$$

then, given any $0 \leq H < \infty$, there exists a $0 \leq t' < \infty$ such that

$$\int_0^{t'} |h(\tau)|\, d\tau > H. \tag{40}$$

This portion of the proof will be by contradiction; that is, we shall assume (38) to be invalid and show, given $0 \leq E < \infty$ and *any* $0 \leq W < \infty$, that there exists a $0 \leq t' < \infty$ such that $|w(t')| > W$ for some $|e(t)| \leq E$. Now, choose an excitation

$$e(t) = E \operatorname{sgn} [h(t' - t)],$$

where $\operatorname{sgn}[x]$ is simply the sign of its argument. Thus $e(t)$ is a function that alternates between $+E$ and $-E$ as the sign of $h(t' - t)$ changes. With this excitation, the convolution integral in (37) yields

$$w(t') = \int_0^{t'} h(t' - \tau)\{E \operatorname{sgn} [h(t' - \tau)]\}\, d\tau$$

$$= E \int_0^{t'} |h(t' - \tau)|\, d\tau.$$

The final result is a consequence of the fact that $x \operatorname{sgn} x = |x|$. Now let $H = W/E$ in (40) and insert the result into the last equation. It follows that $w(t') > W$ and, hence, $|w(t')| > W$. This completes the proof.

This theorem specifies a condition for BIBO stability in the time domain. When $H(s) = \mathcal{L}\{h(t)\}$ is a proper rational fraction, it is possible to give an equivalent frequency domain condition. Thus

Theorem 2.   *If* $H(s)$ *is a proper rational fraction in* s, *then the network will be BIBO stable, which means that*

$$\int_0^\infty |h(\tau)|\, d\tau < \infty \tag{41}$$

if and only if all the poles of H(s) *have negative real part.*

The proof of the *if* statement is initiated with the partial-fraction expansion

$$H(s) = \sum_{i=1}^{l} \sum_{j=1}^{\nu_i} \frac{k_{ij}}{(s + s_i)^j},$$

where $-s_i$ is a pole of $H(s)$ of multiplicity ν_i. The inverse transform of this expression yields

$$h(t) = \sum_{i=1}^{l} \sum_{j=1}^{\nu_i} \frac{k_{ij}}{(j-1)!} t^{j-1} \epsilon^{-s_i t},$$

from which we easily obtain

$$|h(t)| \leq \sum_{i=1}^{l} \sum_{j=1}^{\nu_i} \frac{|k_{ij}|}{(j-1)!} t^{j-1} \left| \epsilon^{-s_i t} \right|.$$

Since $\left| \epsilon^{-s_i t} \right| = \epsilon^{-(\mathrm{Re}\, s_i)t}$, the expression yields

$$|h(t)| \leq \sum_{i=1}^{l} \sum_{j=1}^{\nu_i} \frac{|k_{ij}|}{(j-1)!} t^{j-1}\, \epsilon^{-(\mathrm{Re}\, s_i)t}. \tag{42}$$

If all the poles of $H(s)$ have negative real part, that is, if $\mathrm{Re}\, s_i > 0$ for $i = 1, 2, \cdots, l$, then each of the terms $t^{j-1}\epsilon^{-(\mathrm{Re}\, s_i)t}$ is integrable from 0 to ∞. A finite linear combination of integrable terms, as in (42), is also integrable. Hence (41) is satisfied.

To prove the *only if* statement we turn to the Laplace-transform integral for $H(s)$; that is,

$$H(s) = \int_0^\infty h(t)\, \epsilon^{-st}\, dt.$$

When the usual inequalities are used, this yields

$$|H(s)| \leq \int_0^\infty |h(t)|\, \left| \epsilon^{-st} \right|\, dt.$$

For $\mathrm{Re}\, s \geq 0, \left| \epsilon^{-st} \right| \leq 1$; hence

$$|H(s)| \leq \int_0^\infty |h(t)|\, dt \quad \text{for} \quad \mathrm{Re}\, s \geq 0.$$

Thus, if (41) is valid, then $|H(s)|$ is bounded for all s such that $\text{Re } s \geq 0$. This means that $H(s)$ can have no poles with non-negative real part, and the proof is complete.

If a proper rational fraction

$$H(s) = \frac{N(s)}{D(s)} \tag{43}$$

is given, and it is desired to know if this is the transform of the impulse response of a BIBO-stable network, it will be necessary to locate the zeros of $D(s)$ and to see if they have negative real part. Fortunately, to settle the question of stability, it is not essential to know exactly where the poles of the function are, but only that they lie in the left half-plane. We shall now turn our attention to criteria that provide just such an indication without factoring $D(s)$.

ROUTH CRITERION

If the poles of $H(s)$ are all to lie in the left half-plane, the polynomial $D(s)$ must be *strictly Hurwitz*, as defined in Section 6.2. It was observed there that a necessary condition for a polynomial to be strictly Hurwitz is that *all* the coefficients must be present and must have the same sign. This is a useful bit of knowledge in that it provides a basis for easily eliminating polynomials that cannot possibly be strictly Hurwitz. However, we still need a basis—a sufficiency condition—for selecting a strictly Hurwitz polynomial from amongst the remaining candidate polynomials. The next theorem, which will present necessary and sufficient conditions for a polynomial to be strictly Hurwitz, is an extension of Theorem 16 in Chapter 7 on Hurwitz polynomials.

Suppose $D(s)$ is a polynomial of degree n. For convenience, and certainly without loss of generality, assume that the leading coefficient is positive. Thus

$$D(s) = a_0 s^n + a_1 s^{n-1} + \cdots + a_{n-1} s + a_n \qquad (a_0 > 0). \tag{44}$$

Now let $\alpha(s)$ and $\beta(s)$ be polynomials derived by taking alternate terms from $D(s)$, starting with $a_0 s^n$ and $a_1 s^{n-1}$, respectively. Thus

$$\alpha(s) = a_0 s^n + a_2 s^{n-2} + a_4 s^{n-4} + \cdots \tag{45a}$$

$$\beta(s) = a_1 s^{n-1} + a_3 s^{n-3} + a_5 s^{n-5} + \cdots. \tag{45b}$$

Next form the ratio $\alpha(s)/\beta(s)$ and express it as a continued fraction as follows:

$$\frac{\alpha(s)}{\beta(s)} = \gamma_1 s + \cfrac{1}{\gamma_2 s + \cfrac{1}{\gamma_3 s + \cfrac{1}{\begin{matrix} \cdot \\ \quad \cdot \\ \quad \quad \cdot \ + \cfrac{1}{\gamma_n s} \end{matrix}}}} \tag{46}$$

The desired relationship is the following:*

Theorem 3. *The polynomial* D(s) *is strictly Hurwitz if and only if* $\gamma_i > 0$ *for* i $= 1, 2, \cdots,$ n.

Note that *all* the γ_i numbers must be *positive*; none can be zero, down to the nth one. If all γ_i after the kth one are zero, this is an indication that an even polynomial is a factor of both $\alpha(s)$ and $\beta(s)$, and thus of $D(s)$. This even polynomial can have pairs of j-axis zeros or quadruplets of complex zeros, two of which are in the right half-plane. In either case $D(s)$ cannot be strictly Hurwitz.

HURWITZ CRITERION

The Routh criterion is ideally suited to determining the stability of a network when each of the coefficients of $D(s)$ is known numerically. If one or more of the coefficients depend on one or more unspecified parameters, the Routh criterion becomes difficult to work with. An alternative to the Routh criterion would be useful in this case.

Let Δ_n denote the determinant formed from the coefficients of $D(s)$ as follows: The first row contains a_1 in column 1, a_3 in column 2, and so on until the a_{1+2i} are exhausted. The remaining columns, up to a total of n, are filled with zeros. The second row contains a_0 in column 1, a_2 in column 2, and so on until the a_{2i} are used up. The remaining columns are filled with zeros. The second pair of rows each begins with one zero, after which the first pair of rows is repeated until a total of n columns are filled. The third pair of rows each begins with two zeros, after which the first pair of rows is repeated until a total of n columns are filled. This process is continued until an array of n rows and n columns has been constructed. The array

---

* A particularly lucid proof is given in R. J. Schwarz and B. Friedland, *Linear Systems*, McGraw-Hill, New York, 1965.

illustrated below will always have a_n in the lower right-hand corner, in the (n, n) position, and 0 in all the other rows of the last column. (Why?)

$$\Delta_n = \begin{vmatrix} a_1 & a_3 & a_5 & a_7 & \cdots & 0 & 0 \\ a_0 & a_2 & a_4 & a_6 & \cdots & 0 & 0 \\ 0 & a_1 & a_3 & a_5 & a_7 \cdots & 0 & 0 \\ 0 & a_0 & a_2 & a_4 & a_6 \cdots & 0 & 0 \\ 0 & 0 & a_1 & a_3 & a_5 \cdots & 0 & 0 \\ 0 & 0 & a_0 & a_2 & a_4 \cdots & 0 & 0 \\ \cdot & \cdot & \cdot & \cdot & \cdot & \cdot & \cdot \\ \cdot & \cdot & \cdot & \cdot & \cdot & \cdot & \cdot \\ \cdot & \cdot & \cdot & \cdot & \cdot & \cdot & \cdot \\ 0 & 0 & \cdot & \cdot & \cdot \cdots a_{n-1} & 0 \\ 0 & 0 & \cdot & \cdot & \cdot \cdots a_{n-2} & a_n \end{vmatrix} . \tag{47}$$

Let Δ_{n-1} denote the determinant derived from Δ_n by deleting the last row and column of Δ_n. Continuing in this way, let Δ_{n-k} denote the determinant obtained by deleting the last row and column of Δ_{n-k+1}. By this procedure we construct a total of n determinants Δ_i, known as the *Hurwitz determinants*, which are the basis for the next theorem.

Theorem 4. *The polynomial* D(s) *is strictly Hurwitz if and only if* $\Delta_i > 0$ *for* i $= 1, 2, \cdots,$ n.

This theorem, known as the *Hurwitz criterion*, is reasonably easy to apply to polynomials in which the coefficients are functions of some set of parameters.

LIÉNARD-CHIPART CRITERION

In the application of the Hurwitz criterion, there is clearly the need to evaluate a large number of determinants. Any help in reducing the number of determinants to be evaluated would be greatly appreciated. This is accomplished by the next theorem, known as the *Liénard-Chipart criterion*:

Theorem 5. *The polynomial* D(s) *is strictly Hurwitz if and only if all elements are positive in one of the following:**

1. $a_n, a_{n-2}, a_{n-4}, \cdots$ and $\Delta_n, \Delta_{n-2}, \Delta_{n-4}, \cdots$
2. $a_n, a_{n-2}, a_{n-4}, \cdots$ and $\Delta_{n-1}, \Delta_{n-3}, \Delta_{n-5}, \cdots$
3. $a_n, a_{n-1}, a_{n-3}, \cdots$ and $\Delta_n, \Delta_{n-2}, \Delta_{n-4}, \cdots$
4. $a_n, a_{n-1}, a_{n-3}, \cdots$ and $\Delta_{n-1}, \Delta_{n-3}, \Delta_{n-5}, \cdots$

* A proof of this theorem and the preceding theorem may be found in F. R. Gantmacher, *The Theory of Matrices*, Vol. 2, Chelsea Publishing Co., New York, 1959.

Observe that only every other Hurwitz determinant need be evaluated by this criterion. Since the effort needed to evaluate a determinant increases with the size of the determinant, it is to advantage to select either condition (2) of (4) of the theorem since neither includes Δ_n, the largest Hurwitz determinant.

We will now illustrate the Liénard-Chipart criterion. Consider

$$D(s) = s^4 + 6s^3 + 11s^2 + (6 + k)s + ak$$

and the problem of determining those values of a and k for which $D(s)$ is strictly Hurwitz. We shall approach this task by using condition 2 of the Liénard-Chipart criterion. Note immediately that

$$a_n = a_4 = ak \tag{48a}$$

$$a_{n-2} = a_2 = 11. \tag{48b}$$

The desired Hurwitz determinants are now constructed and evaluated; thus

$$\Delta_{n-1} = \Delta_3 = \begin{vmatrix} 6 & 6+k & 0 \\ 1 & 11 & ak \\ 0 & 6 & 6+k \end{vmatrix} = 360 + 54k - k^2 - 36ak \tag{49a}$$

$$\Delta_{n-3} = \Delta_1 = |6| = 6. \tag{49b}$$

If conditions 2 of the theorem are to be satisfied, then from (48) and (49) we get the following two inequality relations in a and k:

$$ak > 0 \tag{50a}$$

$$360 + 54k - k^2 - 36ak > 0. \tag{50b}$$

The curves $ak = 0$ and $360 + 54k - k^2 - 36ak = 0$ are boundaries of the open regions in which (50a) and (50b), respectively, are satisfied. The region in which (50a) is valid is shown in Fig. 26a; similarly, (50b) is valid in the region shown in Fig. 26b. Both relations in (50) are valid simultaneously in the intersection of these regions, shown in Fig. 26c. Thus, for example, if $k = 30$ is chosen, then a can be no greater than 1.

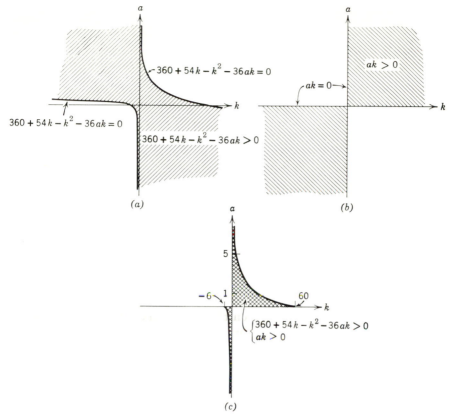

Fig. 26.　Example.

9.5　THE NYQUIST CRITERION

The Routh and Liénard-Chipart criteria are relatively simple criteria. However, in order to use them, we require that the denominator $D(s)$ of the impulse-response transform $H(s)$ be known as a function of s. This is not always available. It would be useful to have another method for stability testing that would use experimental data or only approximate plots of the magnitude and angle of $H(j\omega)$. Such a technique is the *Nyquist criterion*. which we shall discuss now.

The objective is to determine whether or not $H(s)$ has poles in the left half-plane only. The Nyquist criterion uses the principle of the argument from complex-variable theory to decide this issue. We shall focus on determining the number of poles of $H(s)$ outside the left half-plane; therefore, we choose a contour that encloses the right half-plane and contains the imaginary axis. The oriented contour, known as the *Nyquist contour*, is illustrated in Fig. 27.

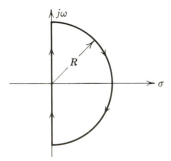

Fig. 27.   Nyquist contour.

For the validity of the principle of the argument, $H(s)$ can have no zeros or poles on the contour. Therefore we first assume that $H(s)$ has no zeros or poles on the imaginary axis. Similarly, since we wish to let R go to infinity, we must make the additional assumption that $H(s)$ is regular and nonzero at infinity. We shall return to this assumption later and show that the assumption relative to the zeros can be relaxed.

Let us consider the mapping of the contour by the function H; that is, the locus of the point $H(s)$ as s traverses the contour of Fig. 27. This may be a curve such as the one shown in Fig. 28a. Since H is a network function, it is real on the real axis, and so the map is symmetric with respect to the real axis. Let N_0 and N_p be the number of zeros and number of poles of $H(s)$, respectively, that lie inside the oriented Nyquist contour C. Now the argument principle states that

$$\Delta \arg H(s)|_C = 2\pi(N_p - N_0); \tag{51}$$

that is, the change in the argument of $H(s)$ as s traverses the contour C, which is *oriented in the negative direction*, is 2π times the number of poles, minus the number of zeros of $H(s)$ within C (taking into account the multiplicity of each).

Let us see what the nature of the locus of $H(s)$ must be if this change of angle is to be nonzero. It is quite evident that the locus must go around the origin in the H-plane if there is to be any change in the argument. This is the case in Fig. 28b. But in Fig. 28a, there is no change in the argument as we traverse the locus once. In other words, the locus must *enclose* the

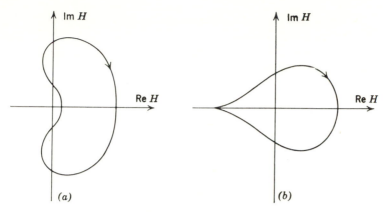

Fig. 28.   Map of the contour of Fig. 27.

origin in the H-plane if there is a net change in the angle of $H(s)$ over the Nyquist contour C. Clearly, if the origin is enclosed by the locus, it is enclosed an integral number of times. Let N_{cw} denote the number of *clockwise* encirclements of the origin by the locus. Then

$$\Delta \arg H(s)|_C = - 2\pi\, N_{\mathrm{cw}}. \tag{52}$$

Substituting this expression into (51) yields

$$N_{\mathrm{cw}} = N_0 - N_p. \tag{53}$$

Thus, if the $H(s)$ locus does not enclose the origin, we can conclude that $H(s)$ has *as many poles as zeros in the right half-plane.* But we really want to know whether it has any poles in the right half-plane. Therefore for this test to be useful we must know, by some other means, how many zeros $H(s)$ has in the right half-plane; for example, that $H(s)$ has no zeros in the right half-plane, which means $H(s)$ is a minimum-phase function. This is by no means an easy task. However, there is no need to abandon the procedure because difficulty is encountered. What we can do is to find another function involving $D(s)$, the denominator polynomial of $H(s)$, and some other factor, this other factor being such that its zero locations are known.

We shall suppose that a flow-graph representation of the network has been reduced to an equivalent graph of the form depicted in Fig. 29. Then by the graph-gain formula in (16), it follows that the transfer function associated with this flow graph is

$$H(s) = \frac{kA(s)}{1 + kA(s)B(s)}. \tag{54}$$

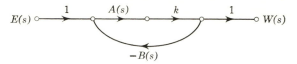

Fig. 29.   Signal-flow graph of feedback network.

It must now be established where the poles of $H(s)$ come from. Observe first that the return ratio T of parameter k, and the return difference F, are given by

$$T(s) = -kA(s)B(s) \tag{55a}$$

$$F(s) = 1 + kA(s)B(s). \tag{55b}$$

Both of these functions have the same poles. Now the poles of the transfer function $H(s)$ are either (1) zeros of the return difference $F(s)$ that are not common with zeros of $A(s)$ or (2) poles of $A(s)$ that are common with zeros of $B(s)$. In the latter case $T(s)$ and $F(s)$ will not have a pole at such a pole of $A(s)$. Hence $H(s) = kA(s)/F(s)$ will have this pole.

Suppose that, even though $B(s)$ may have zeros in common with poles of $A(s)$, none of these lie in the right half-plane or j-axis. Hence any right half-plane or j-axis poles of the transfer function $H(s)$ must be zeros of the return difference $F(s)$. Stated differently: *If all the zeros of* F(s) = 1 *+* kA(s)B(s) *are in the left half-plane, then all the poles of the transfer function* H(s) *are in the left half-plane, provided that* B(s) *has no right half-plane and* j-*axis zeros in common with poles of* A(s). (In this statement we cannot say " *If and only if* all the zeros of $F(s) = 1 + kA(s)B(s)$ are in the left half-plane..." Why?)

Under the stated condition, investigating the location of the poles of the transfer function $H(s)$ can be changed to investigating the location of the zeros of the return difference $F(s) = 1 + kA(s)T(s)$.

In order to apply the argument principle to $F(s)$, we must assume that $F(s)$ has no poles or zeros on the imaginary axis and is regular and non-zero at infinity. Now consider the locus of points $F(s)$ on the Nyquist contour C; a typical plot is shown in Fig. 30a, where the dashed curve corresponds to $\omega < 0$. Let N_{cw} denote the number of clockwise encirclements of the origin by the locus $F(s)$. Then, as before,

$$N_{\text{cw}} = N_0 - N_p, \tag{56}$$

where N_0 and N_p are the numbers of zeros and poles, respectively, of $F(s)$ in the right half-plane.

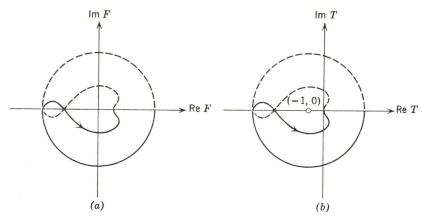

Fig. 30.   Nyquist plots.

Observe that the origin in the F-plane corresponds to the point $(-1, 0)$ in the T-plane; that is, $F(s) = 0$ implies $T(s) = -1$. Thus the number of encirclements of the origin by the locus $F(s)$ is equal to the number of encirclements of the point $(-1, 0)$ by the locus $T(s)$. This is illustrated in Fig. 30b. But, as observed in (54), the return difference and return ratio have the same poles. Hence a new interpretation can be given to (56); namely, N_{cw} is the number of clockwise encirclements of the point $(-1, 0)$ by the locus $T(s)$, and N_p is the number of poles of $T(s)$ in the right half-plane.

DISCUSSION OF ASSUMPTIONS

Now we return to some of the assumptions made earlier about $F(s)$. One of these was that $F(s)$ has no zeros on the imaginary axis and at infinity. This is not a serious restriction, since we can tell from the behavior of the locus whether or not the assumption holds in a given case. If the locus $F(s)$ *intersects* the origin, or, equivalently, if the locus $T(s)$ intersects the point $(-1, 0)$, then it will be known that $F(s)$ has a zero somewhere on the Nyquist locus, either on the j-axis or at infinity, depending on the value of s at the intersection point.

Specifically, if the intersection occurs for s infinite, $F(s)$ will have a zero at infinity. But this is of no interest, since only finite zeros influence stability. Nevertheless, there will be a problem in such a case in counting the number of encirclements of the point $(-1, 0)$. We shall shortly discuss a slight modification of the locus that settles this question.

As for the second possibility, namely, that the locus $T(s)$ intersects the $(-1, 0)$ point at a finite value of $s = j\omega$, then $F(s)$ will have a zero on the

$j\omega$-axis. But this fact gives the information we were seeking; it tells us that not all the zeros of $F(s)$ are in the left half-plane, at least one being on the $j\omega$-axis.

Another earlier assumption on $F(s)$ was that $F(s)$ has no poles on the imaginary axis or at infinity. This same assumption applies to $T(s)$, since $T(s)$ and $F(s)$ have the same poles. Again the validity of the assumption is easily observed from the behavior of the locus. Thus if the locus $T(s)$ becomes unbounded, then $F(s)$ must have a pole at the corresponding value of s.

The locations of such poles are, therefore, known. (Would this be true if $T(j\omega)$ were known from experimental data?) We shall further assume that the multiplicity of such poles is known. The principle of the argument requires the function to which it is applied to have no poles or zeros on the contour. But what should be done if it is discovered that $F(s)$ has such poles or zeros? For finite j-axis zeros, the question has been answered. Let us turn to the case of finite j-axis poles. If the Nyquist contour is to avoid such poles, the contour can be modified by indentations into the right half-plane with vanishingly small semicircular arcs centered at the pole, as shown in Fig. 31a. The corresponding change in the locus of $T(s)$ is shown in Fig. 31b. The solid lines show the locus for values of $s = j\omega$, with ω less than and greater than ω_0, where $s = j\omega_0$ is the location of the pole. As ω approaches ω_0 from below, the locus moves out to infinity at some angle. In Fig. 31b, infinity is approached in the third quadrant. As s takes on values along the vanishingly small semicircular arc, the locus $T(s)$ approaches an infinite-radius circular arc of $m\pi$ radians, where m is the multiplicity of the pole. The orientation on this circular part of the locus is clockwise, as shown by the dashed curve in Fig. 31b. (You should approximate $T(s)$ in the vicinity of the pole by the dominant term in its

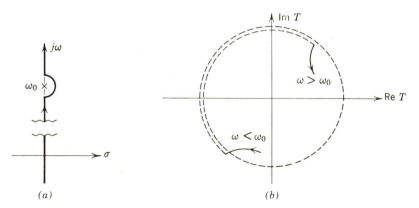

Fig. 31.   Modification of Nyquist contour and Nyquist diagram.

Laurent series and verify the above statements.) The "infinite-radius" circular arc in the T-plane joins the ends of the segments of the locus $T(s)$ that result as s approaches or leaves the vicinity of a pole on the imaginary axis. Now it is possible to count the encirclements of the point $(-1, 0)$ even when $T(s)$ has imaginary-axis poles. Note that these vanishingly small indentations into the right half-plane do not affect the number of zeros of $F(s)$ computed by (56) to be in the left half-plane. Why?

Finally, consider the case where $F(s)$ has a pole or zero at infinity. In the case of a pole, $T(s)$ also has that pole. In this case we must examine $T(s)$ on the arbitrarily large semicircular arc in Fig. 27 used to close the Nyquist contour in the right half-plane. Just as in the case of finite poles, the locus $T(s)$ will go to infinity as s approaches infinity along the imaginary axis. Again, corresponding to the circular arc on the Nyquist locus (this time the infinite arc); the locus $T(s)$ will be an infinite clockwise-oriented circular arc of $n\pi$-radians, where n is the multiplicity of the pole at infinity. This is depicted in Fig. 32a.

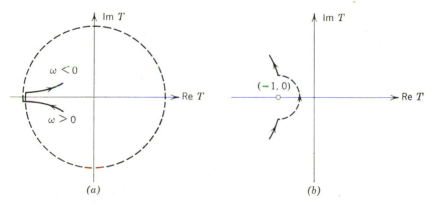

Fig. 32.   Loci corresponding to a pole or zero of $F(s)$ at infinity.

In the case of a zero of $F(s)$ at infinity, $T(s)$ equals -1 there, so that the locus $T(s)$ intersects the $(-1, 0)$ point. In this case consider the limit of the large circular arc of the Nyquist contour as the radius approaches infinity. The locus $T(s)$ will approach a vanishingly small counter-clockwise oriented arc of $n\pi$ radians centered at the point $(-1, 0)$, where n is the multiplicity of the zero at infinity. This is illustrated in Fig. 32b.

NYQUIST THEOREM

The overall locus $T(s)$ in the general case will be a combination of Figs. 30b, 31b, and 32. This modified locus, which accounts for poles of

$T(s)$, or $F(s)$, on the finite $j\omega$-axis and poles or zeros of $F(s) = 1 + T(s)$ at infinity, is called a *Nyquist diagram*. In (56) the number of clockwise encirclements of the point $(-1, 0)$ refers to the number of encirclements by the Nyquist diagram.

The preceding discussion will now be summarized in the form of a theorem, called the *Nyquist criterion*.

Theorem 6.   *A network, having*

$$H(s) = \frac{kA(s)}{1 + kA(s)B(s)}$$

as its transfer function, is BIBO stable (a) if no right half-plane or imaginary-axis pole of A(s) *is also a zero of* B(s) *and (b) if the Nyquist diagram of* T(s) = kA(s)B(s) *does not intersect the point* $(-1, 0)$ *and encircles it* $-N_p$ *times in the clockwise direction,[*] where* N_p *is the number of poles of* T(s) *in the right half-plane.*

Note that this is a sufficiency theorem only. It is not necessary because right half-plane or imaginary-axis zeros of $A(s)$ might cancel all right half-plane or imaginary-axis zeros of $1 + T(s)$, if the latter had any. Of course, if $A(s)$ is known to have left-half-plane zeros only, this cannot happen, and the conditions of the theorem become necessary as well as sufficient. In any case, from a practical point of view, we would view a network as unstable—not just possibly unstable—if the conditions of the theorem were not satisfied. Why?

If this approach is to tell us anything about the zeros of $F(s)$ through (56), then we must know that $T(s)$ has no poles in the right half-plane; or, if it has any there, we must know how many. There is one case in which we can definitely say that $T(s)$ has no right-half-plane poles; that is, when $A(s)B(s)$ may be written as a product of passive network functions.

Lastly, if the Nyquist diagram of $T(s)$ is to be the key to predicting the stability of the network, then we must be certain that no right half-plane or imaginary-axis pole of $A(s)$ is also a zero of $B(s)$. In one case we can say with absolute certainty that this cannot occur; that is, when the numerator of $B(s)$ is a constant.

The preceding discussion of stability was based on the single-loop feedback representation of Fig. 29. Nyquist's criterion can also be extended to multiple-loop flow-graph representations. This involves plotting several Nyquist diagrams. We shall not discuss this extension here.

---

[*] By (56), the number of zeros, N_0, of $1 + T(s)$ in the right half-plane is $N_{cw} + N_p$. The condition of the theorem corresponds to $N_0 = 0$, or $N_{cw} = -N_p$. An equivalent condition would be: "...encircles it N_0 times in the counter-clockwise direction."

Example

Let us now illustrate the Nyquist stability criterion by means of an example. We shall go through this example in some detail and show how approximations can be incorporated in the procedure.

Consider the three-stage RC-coupled amplifier with frequency-sensitive feedback shown in Fig. 33. In this example we shall try to show the ad-

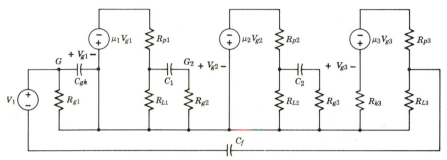

Fig. 33   Example for Nyquist locus.

vantage of the Nyquist criterion by not computing the return ratio $T(s)$. Instead we shall estimate $T(s)$ by making use of a number of approximations. This network is a simplified model of a vacuum-tube network in which many of the interelectrode capacitances have been neglected to simplify the example.

This network can easily be modeled by a single loop flow graph of the type we considered. The value of k will be $\mu_1\mu_2\mu_3$ and $A(s)B(s)$ can be written as a product of passive network functions, though we shall not do so here. Thus we may properly set $N_p = 0$ and proceed in the knowledge that $N_0 = N_{\text{cw}}$. Furthermore, $B(s) \equiv 1$; hence poles of $A(s)$ cannot be common to zeros of $B(s)$. You should convince yourself of these facts.

Interest lies entirely on the $j\omega$-axis; hence we shall deal mostly with steady-state phasors instead of Laplace transforms. Remembering the flow-graph definition of the return ratio, we must open the loop and apply a unit signal at the right-hand node of the pair of nodes thus formed. The signal returned to the left-hand node of that pair of nodes will be $T(j\omega)$. We shall now interpret this step in terms of the actual network rather than the signal-flow graph. Imagine that the loop is opened at the input to the first amplifier stage, and the voltage V_{g1} is set equal to unity. Observe that this is equivalent to replacing the first controlled voltage source by an independent voltage source of value μ_1. Thus the condition shown in Fig. 34 is the appropriate one for computing the return ratio. Here the external source is removed and the first controlled source is assumed to be an independent source having a phasor voltage μ_1.

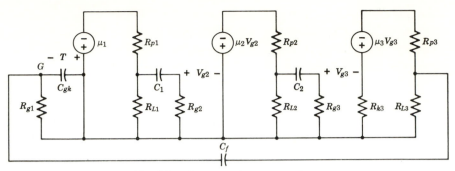

Fig. 34.   Return ratio computation.

Notice that the reference for T is chosen to conform with the definition that it is the negative of the returned signal to the back half of the split node. It is easy to see how this interpretation can be used for experimental measurement of $T(j\omega)$ on the network of which this example is a model.

In order to construct the Nyquist diagram, let us split the frequency range $0 \le \omega < \infty$ into a number of bands and use suitable approximations in each band. At very low frequencies the returned signal will be very small due to the coupling capacitances C_1, C_2, and C_f. The influence of C_{gk} can be neglected in this range. There are three RC coupling networks in the loop. Let us use the notation

$$R_e = \frac{R_L R_g}{R_L + R_g}, \tag{57}$$

with suitable subscripts for each of the coupling networks. Then the voltage ratio of each stage will be

$$\frac{-\mu R_e C s}{(R_e + R_p)C s + (R_p + R_L)/(R_g + R_L)}, \tag{58}$$

with appropriate subscripts. Hence in this range the return ratio will be given by

$$
\begin{aligned}
T(s) = {} & \frac{\mu_1 R_{e1} C_1 s}{(R_{e1} + R_{p1})C_1 s + (R_{p1} + R_{L1})/(R_{g2} + R_{L1})} \\[2mm]
& \times \frac{\mu_2 R_{e2} C_2 s}{(R_{e2} + R_{p2})C_2 s + (R_{p2} + R_{L2})/(R_{g3} + R_{L2})} \\[2mm]
& \times \frac{\mu_3 R_{e3} C_f s}{(R_{e3} + R_{p3})C_f s + (R_{p3} + R_{L3})/(R_{g1} + R_{L3})}.
\end{aligned} \tag{59}
$$

(The negative signs disappear because of the reference for T.) The asymptotic phase of each of the factors in (59) as $\omega \to 0$ will be $\pi/2$ radians. Thus the total asymptotic phase of $T(j\omega)$ will be $3\pi/2$ radians, the magnitude approaching zero. Hence the low-frequency portion of the locus of $T(j\omega)$ looks like the curve in Fig. 35.

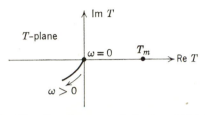

Fig. 35.   Low frequency behavior of $T(j\omega)$.

Let us assume that the upper half-power frequency $1/R_1 C_{gk}$ is considerably higher than the half-power frequencies of the three RC coupling networks. Thus there will be a midband frequency range in which the behavior of the network in Fig. 34 can be approximated by that shown in Fig. 36.

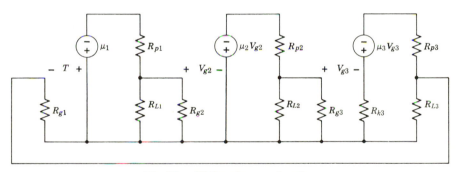

Fig. 36.   Midband approximation.

For this network T is computed quite easily. Alternatively, the desired expression can be obtained from (59) by neglecting the constant terms in the denominators compared with the frequency-dependent terms. In either case the result will be

$$T = \mu_1 \mu_2 \mu_3 \frac{R_{e1}}{R_{p1} + R_{e1}} \frac{R_{e2}}{R_{p2} + R_{e2}} \frac{R_{e3}}{R_{p3} + R_{e3}} = T_m. \qquad (60)$$

This is obviously a real positive number. Thus the midband T-locus is on the positive real axis. The point T_m is marked on Fig. 37.

At high frequencies Fig. 36 can still be used, except that the effect of C_{gk} must now be included. Since C_{gk} is in parallel with R_{e3}, the third factor in (60) should be modified and replaced by the following:

$$\frac{R_{e3}}{R_{e3} + R_{p3} + j\omega C_{gk} R_{e3} R_{p3}}. \tag{61}$$

Hence the angle of T will asymptotically approach $-\pi/2$. The high end of the $T(j\omega)$ locus therefore takes the form shown in Fig. 37.

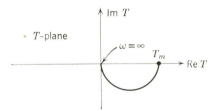

Fig. 37.   High-frequency behavior of $T(j\omega)$.

We can now estimate the $T(j\omega)$ locus for $0 \leq \omega < \infty$ to have roughly the shape shown in Fig. 38. To improve the approximation, we should

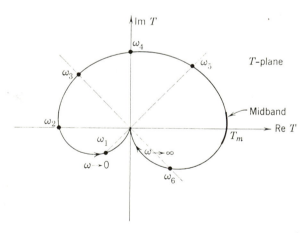

Fig. 38.   Approximate T-locus for example.

estimate a few points on the curve. Suppose, for simplicity, that the three significant half-power frequencies of the interstage circuits are either identical or widely separated. In the first case we know that at the common half-power frequency each circuit contributes an angle of 45° and a 3-dB attenuation. This point, marked ω_3 in Fig. 38, must be 9 dB less than 20 log T_m. Similarly, we can find the frequency at which each circuit contributes a 60° angle. This is the frequency at which each of the denominator factors in (59) contributes 30°, which is easily found to be approximately $\omega_2 = 0.58\omega_3$. At this frequency each factor will be down about 4 dB. Therefore $T(j\omega_2)$ will be down 12 dB from 20 log T_m. The frequency ω_2, marked in Fig. 38, is the point where the locus crosses the negative real axis. The other points ω_1, ω_4, ω_5, ω_6 are similarly computed. The widely separated case is left as a problem. (See Problem 25.)

Once the locus for the positive range of ω is known, the diagram can be completed by symmetry about the real axis. The complete locus for the example is shown in Fig. 39.

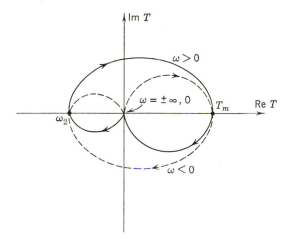

Fig. 39.   Complete T-locus.

It is evident that the stability of the system is determined by the value of $|T(j\omega_2)|$. If this magnitude is greater than or equal to 1, the system is unstable. In such a case the system can be made stable by modifying some of the element values. Even if the point $(-1, 0)$ is not enclosed or intersected by the T-locus, the proximity of the curve to this point gives a measure of the "relative stability"; that is, it gives an indication of the closeness of a pole to the $j\omega$-axis.

This idea can be expressed in a somewhat more quantitative way by defining stability margins, the *gain margin*, and the *phase margin*. As a matter of fact, consideration of the Nyquist locus leads to many other concepts that are useful in system design, such as *conditional stability*. However, we shall arbitrarily terminate the discussion at this point, leaving such extensions to books on control theory, which treat feedback systems in considerable detail.

PROBLEMS

9-4(c)
H-L.
Jan 11, 1971

1. Determine the index of each of the signal-flow graphs in Fig. P1. The numbers on the branches are numerical values of branch transmittances.

2. By reducing each of the signal-flow graphs of Fig. P1, determine the graph gain.

3. Determine the graph gain of each of the signal-flow graphs of Fig. P1 by applying the graph gain formula.

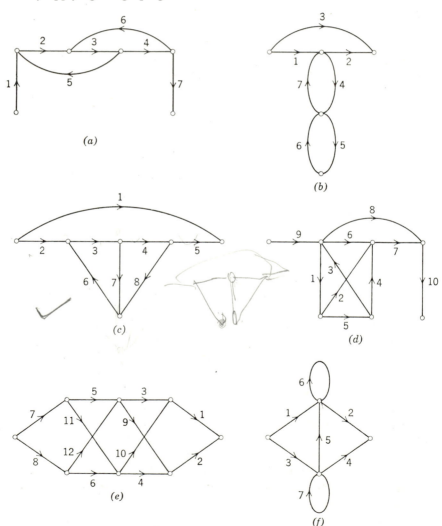

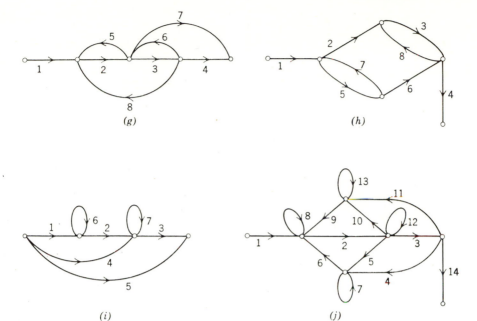

Fig. P1

4. Each of the signal-flow graphs in Fig. P1 has one source node and one sink node. (*a*) Invert the path from source to sink in each graph. (*b*) Determine the graph gain of the inverted graph and from this, that of the original graph.

5. Reduce each of the signal-flow graphs in Fig. P1 to an essential signal-flow graph. Then evaluate the gain using the graph gain formula. Compare with the values from Problem 2 or 3.

6. In the graph of Fig. 12*a*, invert the loop $V_aV_bV_2V_a$. Find the graph gain I/V_1 and verify that it is the same as found in the text.

7. Draw a signal-flow graph for the networks of Fig. P7. Reduce the signal-flow graph to find the transfer function V_2/V_1.

8. Solve the following systems of equations for x_1 using signal-flow graphs. Also find x_1 by Cramer's rule (both as a check and to illustrate the amount of work involved).

(*a*)
$$\begin{bmatrix} 2 & 1 & -1 & 3 \\ 4 & 1 & 0 & -1 \\ 1 & 0 & 1 & 2 \\ -4 & -7 & 2 & 3 \end{bmatrix} \begin{bmatrix} x_1 \\ x_2 \\ x_3 \\ x_4 \end{bmatrix} = \begin{bmatrix} 2 \\ 1 \\ 0 \\ 0 \end{bmatrix}$$

(b)
$$\begin{bmatrix} 4 & 1 & 0 & 2 \\ 2 & 5 & 1 & -1 \\ 1 & 0 & 3 & -1 \\ 0 & 1 & 0 & 2 \end{bmatrix}\begin{bmatrix} x_1 \\ x_2 \\ x_3 \\ x_4 \end{bmatrix} = \begin{bmatrix} 0 \\ 1 \\ -1 \\ 0 \end{bmatrix}$$

(c)
$$\begin{bmatrix} -4 & -1 & 2 & 0 \\ 1 & 3 & 0 & -1 \\ -3 & -1 & 6 & 1 \\ 1 & 0 & 0 & -2 \end{bmatrix}\begin{bmatrix} x_1 \\ x_2 \\ x_3 \\ x_4 \end{bmatrix} = \begin{bmatrix} 2 \\ -1 \\ 1 \\ -2 \end{bmatrix}$$

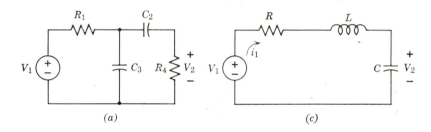

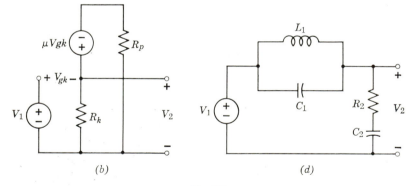

Fig. P7

9. Find the transfer function V_4/V_g for the network of Fig. 19 whose signal-flow graph is given in Fig. 20 by (a) reducing the signal-flow graph alone, (b) operating on the connection matrix alone, (c) applying the graph gain formula alone, and (d) using mixed-variable equations.

10. Find the transfer impedance V_3/I_g for the network of Fig. 21 with signal-flow graph in Fig. 22 by (a) reducing the signal-flow graph alone,

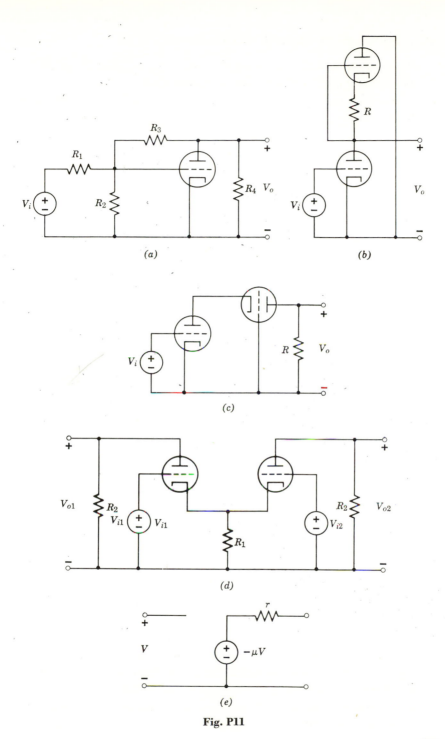

Fig. P11

693

(b) operating on the connection matrix alone, (c) applying the graph gain formula alone, and (d) using node equations.

11. Determine a signal-flow graph for each of the networks of Fig. P11 a to d. Use the triode model shown in Fig. P11e. Then evaluate the graph gain(s).

12. Set up a signal-flow graph for the network of Fig. P12 to find the transfer function $Y_{21}(s) = I_2/V_1$. Find this function by reducing the signal-flow graph. Also, apply the graph gain formula to the original graph, and compare the answers obtained.

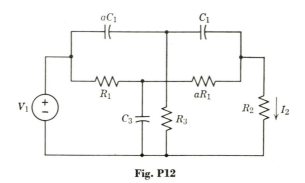

Fig. P12

13. Find the voltage ratio V_2/V_1 for the general ladder network of Fig. P13 by first setting up the signal-flow graph. From the signal-flow graph show that the transmission zeros occur at series impedance poles or shunt impedance zeros.

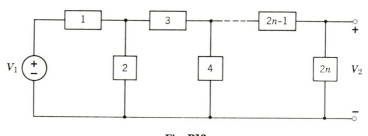

Fig. P13

14. Determine a signal-flow graph for each of the networks of Fig. P14a to c. Use the transistor model shown in Fig. P14d. Then evaluate the graph gain.

15. Find the gain V_2/V_1 for the "pseudo-tuned" amplifier of Fig. P15 using signal-flow graphs.

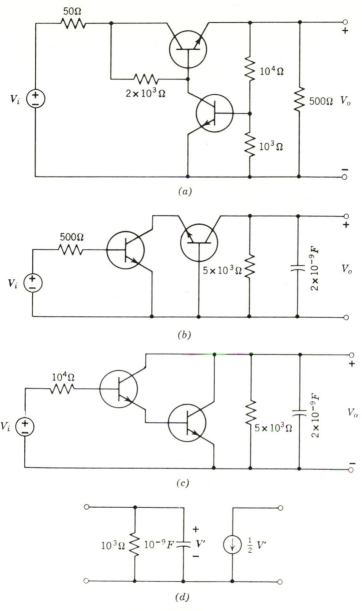

(a)

(b)

(c)

(d)

Fig. P14

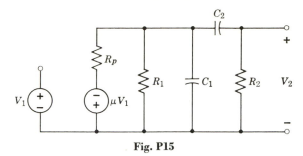

Fig. P15

16. For each of the networks of Fig. P16a to d, use (40) to determine the sensitivity of the transfer function V_0/V_i to (a) c, (b) r, and (c) g. Use the transistor model of Fig. P16e. The nominal values are c = 100 μF, r = $10^3\Omega$, and g = 0.5℧.

17. The return ratio and return difference of a branch are defined in the text. The same quantities can be defined for a node. The *return ratio of node j*, represented by T_j, is the negative graph gain of the graph obtained by splitting the node and removing all other source and sink nodes. The *return difference of node j* is defined as $F_j = 1 + T_j$. The *partial return ratio of node j*, T_j', is defined as the return ratio of node j when all *higher* numbered nodes are removed from the graph. This is obviously dependent on the ordering of the nodes. The *partial return difference* of node j is also defined; it is $F_j' = 1 + T_j'$.

(a) Suppose the nodes of a graph are numbered in a particular order. Now remove all nodes above the one numbered k. Next, draw a reduced graph in general form, retaining only nodes k and $k - 1$. Determine the partial return differences and find the product $F_k' F_{k-1}'$. Finally, interchange the numbers of nodes k and $k - 1$ and again find the product $F_k' F_{k-1}'$. Show that this product is independent of the node numbering.

(b) Show that the product of the partial return differences of all nodes of a graph is a unique property of the graph, independent of the node numbering.

18. Which of the following impulse response functions are associated with BIBO stable networks:

(a) $te^{-2t} \sin ut \ (t)$

(b) $\dfrac{2}{(t - 3)^2} u(t)$

(c) $\dfrac{\sin^2 2t}{t^2} u(t)$

(d) $\dfrac{\epsilon^{-2t} - \epsilon^{-t/2}}{t} u(t)$

(e) $[\cos t + t^2 \epsilon^{-t}]u(t)$

(f) $\dfrac{5}{t} u(t - 3)$

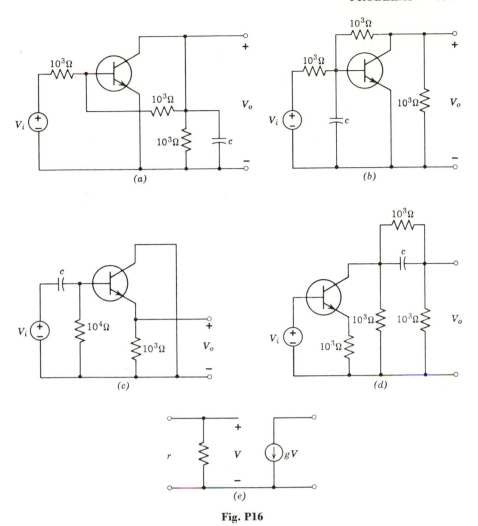

Fig. P16

19. Use (*i*) the Routh criterion and (*ii*) the Liénard-Chipart criterion to determine which of the following polynomials in *s* have zeros only in the left half-plane:

(*a*) $s^3 + s^2 + s + 6$

(*b*) $s^4 + 5s^3 + 9s^2 + 7s + 2$

(*c*) $s^4 + 3s^3 + 7s^2 + 6s + 4$

(*d*) $s^5 + 2s^4 + 3s^3 + 8s^2 + 7s + 6$

(*e*) $2s^5 + 9s^4 + 19s^3 + 25s^2 + 19s + 6$

(*f*) $4s^3 + 7s^2 + 7s + 2$

(*g*) $s^3 + 3s^2 + 4s + 2$

(*h*) $3s^3 + 5s^2 + 4s + 6$

(*i*) $s^6 + 7s^5 + 7s^4 + 5s^3 + 4s^2 + s + 1$

(*j*) $2s^5 + 4s^4 + 3s^3 + 3s^2 + 4s + 2$

(k) $2s^3 + s^2 + 5s + 2$ (l) $s^4 + 3s^3 + 2s^2 + s + 1$

(m) $2s^4 + 4s^3 + 7s^2 + 7s + 3$

20. Suppose $H(s)$ is a rational fraction and is regular at infinity; thus, $H(s)$ equals a constant (which may be zero) plus a proper rational fraction. Then prove that a network having $H(s)$ as its transfer function is BIBO stable if and only if all the poles of $H(s)$ have negative real part.

21. Use the Liénard-Chipart criterion to determine the value of μ for which the networks in Fig. P21 are BIBO stable.

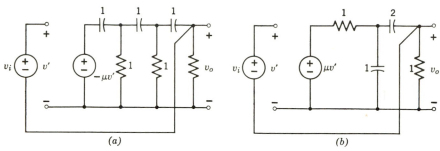

(a)   (b)

Fig. P21

22. Use the Liénard-Chipart criterion to determine the values of ρ and γ for which the networks in Fig. P22 are BIBO stable. In the ρ–γ parameter plane, shade the stability regions.

23. Use the Liénard-Chipart criterion to determine the values of r and g for which the oscillator networks shown in Fig. P23a to c are *not* BIBO stable. Assume a ficticious voltage source input to be in series with the transistor base lead. Use the transistor model shown in Fig. P23d. In the r–g parameter plane, shade the instability regions.

24. Draw the Nyquist diagram for each of the following return ratio functions:

(a) $T(s) = k \dfrac{4}{s(s+1)(s+2)}$ (b) $T(s) = k \dfrac{s+1}{s^2 + 3s + 3}$

(c) $T(s) = k \dfrac{s+3}{s(s+2)(s^2 + 2s + 2)}$ (d) $T(s) = k \dfrac{(s+1)(s+3)}{s^2 + 4s + 4}$

(e) $T(s) = k \dfrac{(s+1)(s+3)}{s(s^2 + 4s + 4)}$ (f) $T(s) = k \dfrac{3}{s(s+2)(s^2 + 2s + 10)}$

(g) $T(s) = k \dfrac{s+2}{(s+4)(s-1)}$ (h) $T(s) = k \dfrac{s-1}{s(s+1)}$

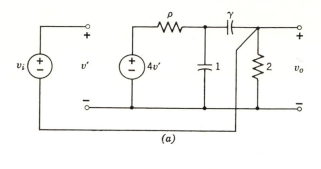

(a)

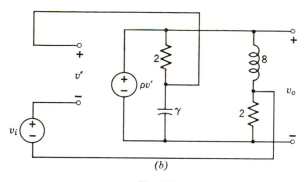

(b)

Fig. P22

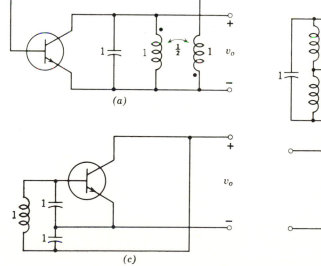

(a)

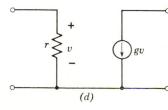

(b)

(c)

(d)

Fig. P23

(i) $T(s) = k \dfrac{(s+1)}{s^2(s+2)}$ (j) $T(s) = k \dfrac{s+4}{s^2(s+2)}$

(k) $T(s) = k \dfrac{(s+1)^2}{s^3(s+8)}$ (l) $T(s) = k \dfrac{s}{(s+1)(s+2)^2}$

(m) $T(s) = k \dfrac{s^2 - s + 1}{s(s+1)}$ (n) $T(s) = k \dfrac{(s+1)(s+2)}{s(s^2+4)}$

(o) $T(s) = k \dfrac{2}{s(s^2+4)}$ (p) $T(s) = k \dfrac{(s+1)}{s(s^2-s+1)}$

For what values of k is the associated feedback network known to be stable by the Nyquist criterion? Assume $A(s)$ has only left half-plane zeros.

25. In the network of Fig. 33 in the text, let the values of the components be such that the break-frequencies of the three interstage networks are:

$$\omega_a = 100, \qquad \omega_b = 1000, \qquad \omega_c = 100{,}000$$

Sketch the Nyquist diagram carefully for this case. Find the maximum value that T_m can be, if the network is to be stable.

26. Sketch the Nyquist diagram for the network of Fig. P26. Find values of $R_f L_f$ for which the network is stable. What is the maximum value of α under this condition, if the network is to remain stable for small variations of parameter values (R_e, G_e, R_f, L_f in particular) from the design value?

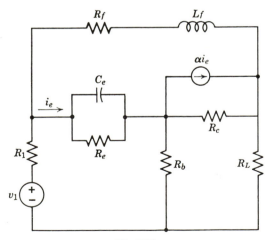

Fig. P26

27. Sketch the Nyquist diagram for the network of Fig. P27 and give the condition for stability.

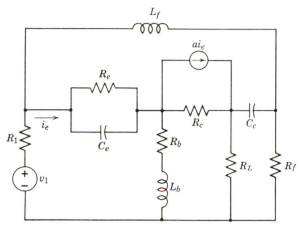

Fig. P27

28. Draw a signal-flow graph for the network given in Fig. P28. By reducing the graph, calculate the transfer voltage ratio.

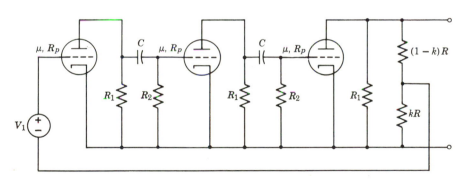

Fig. P28

29. Prove the following theorem, known as the *inverse Nyquist criterion:*
A network, having

$$H(s) = \frac{kA(s)}{1 - kA(s)B(s)}$$

as its transfer function, is BIBO stable, (a) if no RHP or imaginary axis pole of A(s) is also a zero of B(s) and (b) if the Nyquist diagram of

$1/T(s) = 1/kA(s)B(s)$ *does not intersect the point* $(-1, 0)$ *and encircles it* $-N_p$ *times in the clockwise direction, where* N_p *is the number of poles of* $1/T(s)$ *in the RHP.* Note: The Nyquist diagram of $1/T(s)$ is known as the *inverse Nyquist diagram* of $T(s)$.

30. With reference to Problem 29, draw the inverse Nyquist diagram for each of the return ratio functions listed in Problem 24. Then, indicate the values of k for which the associated feedback network is known to be stable by the inverse Nyquist criterion. Assume $A(s)$ has only left half-plane zeros.

31. Consider all possible combinations of $A(s)$ and $B(s)$ derivable from the following list. Apply (*i*) the Nyquist criterion and (*ii*) the inverse Nyquist criterion (of Problem 29) to determine the values of k for which each network is known to be stable.

$$A(s) = \frac{1}{(s+1)(s+2)} \qquad B(s) = 1$$

$$A(s) = \frac{1}{s(s+1)} \qquad B(s) = \frac{1}{s+4}$$

$$A(s) = \frac{(s+1)}{s^2(s+2)} \qquad B(s) = \frac{s}{s+4}$$

32. Consider an active network that can be excited either by a voltage source or a current source, as illustrated in Fig. P32*a* and *b*. The respective transfer functions will be

$$(a)\ \frac{V_2}{V_1} = R\left.\frac{\Delta_{12}}{\Delta}\right|_z, \qquad (b)\ \frac{V_2}{I_1} = \left.\frac{\Delta_{12}}{\Delta}\right|_y.$$

In the first case the determinant is that of the loop impedance matrix and is formed by short-circuiting the input terminals. In the second case the determinant is that of the node admittance matrix and is formed with the input terminals open-circuited. The zeros of these determinants—which are, the poles of the respective transfer functions—will be different. Hence, the stability properties under the two kinds of excitation need not be the same.

If a network is stable when its terminals are short-circuited, it is said to be *short-circuit stable*. Similarly, if a network is stable when its terminals are open-circuited, it is said to be *open-circuit stable*. It is possible for a network to be both open-circuit and short-circuit stable.

It is also possible for it to be stable under one of the terminal conditions but not the other.

(*a*) The network of Fig. P32*c* includes two transistors, represented by linear models, and a passive, reciprocal two-port. By examining the poles of the input impedance, show that this network is open-circuit stable.

(*b*) A second network is shown in Fig. P32*d*. Show that it is short-circuit stable.

Now consider a two-port with terminations on both ends, as shown in Fig. P32*e*. Either port can be short-circuited or open-circuited by giving appropriate values to R_s and R_L. The overall network may be short-circuit stable and/or open-circuit stable at either end.

(*c*) As a specific example take the network shown in Fig. P32*f*. This is

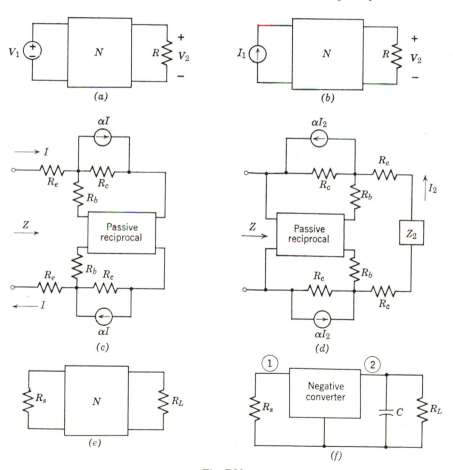

Fig. P32

a negative converter with a capacitor representing inevitable capacitances in the active devices. Determine whether this is short-circuit stable and/or open-circuit stable at each end.

33. In the example of Fig. 21, it is desired to find the input impedance. Place a node (and appropriate branches) in Fig. 22 representing the voltage across the independent current source. Find the input impedance by reducing the graph. Also find the transfer impedance V_3/I_g.

34. In Fig. 20, nodes V_1 and V_4 constitute a set of essential nodes. Reduce this graph to an essential graph with V_1 and V_4 as essential nodes and compare with Fig. 18.

35. Signal-flow graphs for two feedback amplifiers are shown in Fig. P35;

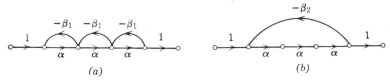

(a)   (b)

Fig. P35

β_2 is to be adjusted so as to make the graph gains of the two graphs equal. Determine the sensitivity S_a^G of the gain of each of the two graphs to the transmittance α. Compare and determine which type of amplifier is less sensitive to changes in α.

. 10 .

LINEAR TIME-VARYING AND NONLINEAR NETWORKS

In the preceding parts of this book, consideration has been limited to linear, time-invariant networks. However, actual networks display characteristics that cannot be completely and adequately described by a linear, time-invariant model. A simple device such as a resistor, for example, will have a time-varying resistance induced by changes in the temperature of its environment; this effect might not be negligible in all cases. Similarly, magnetic-field saturation in the ferromagnetic core of an inductor, large-signal operation of an active device such as a transistor, excessive energy dissipation in a resistor, etc., lead to nonlinearity. Actual networks, then, under many conditions of operation, will be nonlinear.

Quite aside from undesired departures from time-invariance and/or linearity, which we seek to minimize, there are other situations in which such departures are introduced by design in the quest for networks that perform in some prescribed manner; for example, the parametric amplifier is designed as a time-varying, linear network. Oscillators, modulators, and demodulators are only a few of the many networks designed with nonlinear elements.

This chapter will be concerned with the formulation and analytic solution of (1) linear, time-varying networks and (2) nonlinear networks. Since the analytic solution of a nonlinear network cannot, in general, be

705

determined, there is included a section on numerical solution techniques. Finally, there is a section on Liapunov stability theory, which is one of the basic tools available to the analyst studying nonlinear networks. Even though we can present only limited coverage of each of these topics—the existing knowledge is extensive enough to require books devoted to each—the treatment will be careful and not shallow. References for further study will be provided.

10.1   STATE EQUATION FORMULATION FOR TIME-VARYING NETWORKS

In Section 4.5 we presented a method for formulating the state equations of a time-invariant network. By reference to that treatment, you may verify that the notion of time dependence *or* independence was not introduced until after equations 129. Therefore, we shall begin our consideration of time-varying networks with these equations, which for convenience are repeated here:

$$\frac{d}{dt}[\mathscr{C}\mathbf{v}_{Ct}] = [-\mathscr{Y} \quad \mathscr{H}]\begin{bmatrix} \mathbf{v}_{Ct} \\ \mathbf{i}_{Ll} \end{bmatrix} + [-\widehat{\mathscr{Y}} \quad \widehat{\mathscr{H}}]\begin{bmatrix} \mathbf{v}_E \\ \mathbf{i}_J \end{bmatrix} + \frac{d}{dt}[\widehat{\mathscr{C}}\mathbf{v}_E] \tag{1a}$$

$$\frac{d}{dt}[\mathscr{L}\mathbf{i}_{Ll}] = [\mathscr{G} \quad -\mathscr{Z}]\begin{bmatrix} \mathbf{v}_{Ct} \\ \mathbf{i}_{Ll} \end{bmatrix} + [\widehat{\mathscr{G}} \quad -\widehat{\mathscr{Z}}]\begin{bmatrix} \mathbf{v}_E \\ \mathbf{i}_J \end{bmatrix} + \frac{d}{dt}[\widehat{\mathscr{L}}\mathbf{i}_J] \tag{1b}$$

or, after consolidation into a single equation,

$$\frac{d}{dt}\begin{bmatrix} \mathscr{C} & \mathbf{0} \\ \mathbf{0} & \mathscr{L} \end{bmatrix}\begin{bmatrix} \mathbf{v}_{Ct} \\ \mathbf{i}_{Ll} \end{bmatrix} = \begin{bmatrix} -\mathscr{Y} & \mathscr{H} \\ \mathscr{G} & -\mathscr{Z} \end{bmatrix}\begin{bmatrix} \mathbf{v}_{Ct} \\ \mathbf{i}_{Ll} \end{bmatrix} + \begin{bmatrix} -\widehat{\mathscr{Y}} & \widehat{\mathscr{H}} \\ \widehat{\mathscr{G}} & -\widehat{\mathscr{Z}} \end{bmatrix}\begin{bmatrix} \mathbf{v}_E \\ \mathbf{i}_J \end{bmatrix}$$

$$+ \frac{d}{dt}\begin{bmatrix} \widehat{\mathscr{C}} & \mathbf{0} \\ \mathbf{0} & \widehat{\mathscr{L}} \end{bmatrix}\begin{bmatrix} \mathbf{v}_E \\ \mathbf{i}_J \end{bmatrix}. \tag{2}$$

REDUCTION TO NORMAL FORM

This equation can be reduced to the normal form

$$\frac{d}{dt}\mathbf{x} = \mathscr{A}\mathbf{x} + \mathscr{B}\mathbf{e}, \tag{3}$$

provided the matrix

$$\begin{bmatrix} \mathscr{C} & 0 \\ 0 & \mathscr{L} \end{bmatrix} \tag{4}$$

is bounded and nonsingular. Since we are interested in a solution for all $t \geq t_0$, we shall assume that (4) is bounded and nonsingular for $t \geq t_0$; that is, that each of the elements of (4) is bounded and

$$\det \begin{bmatrix} \mathscr{C} & 0 \\ 0 & \mathscr{L} \end{bmatrix} \neq 0$$

for $t \geq t_0$.

Now we can proceed in two ways to put (2) into normal form. For the first way, set

$$\mathbf{x} = \begin{bmatrix} \mathscr{C} & 0 \\ 0 & \mathscr{L} \end{bmatrix} \begin{bmatrix} \mathbf{v}_{Ct} \\ \mathbf{i}_{Ll} \end{bmatrix} - \begin{bmatrix} \widehat{\mathscr{C}} & 0 \\ 0 & \widehat{\mathscr{L}} \end{bmatrix} \begin{bmatrix} \mathbf{v}_E \\ \mathbf{i}_J \end{bmatrix}; \tag{5}$$

then

$$\begin{bmatrix} \mathbf{v}_{Ct} \\ \mathbf{i}_{Ll} \end{bmatrix} = \begin{bmatrix} \mathscr{C} & 0 \\ 0 & \mathscr{L} \end{bmatrix}^{-1} \left(\mathbf{x} + \begin{bmatrix} \widehat{\mathscr{C}} & 0 \\ 0 & \widehat{\mathscr{L}} \end{bmatrix} \begin{bmatrix} \mathbf{v}_E \\ \mathbf{i}_J \end{bmatrix} \right). \tag{6}$$

Upon substituting this expression into (2) and rearranging terms, the desired form in (3) is obtained with

$$\mathscr{A} = \begin{bmatrix} -\mathscr{Y} & \mathscr{H} \\ \mathscr{G} & -\mathscr{Z} \end{bmatrix} \begin{bmatrix} \mathscr{C} & 0 \\ 0 & \mathscr{L} \end{bmatrix}^{-1} \tag{7}$$

$$\mathscr{B} = \begin{bmatrix} -\widehat{\mathscr{Y}} & \widehat{\mathscr{H}} \\ \widehat{\mathscr{G}} & -\widehat{\mathscr{Z}} \end{bmatrix} + \begin{bmatrix} -\mathscr{Y} & \mathscr{H} \\ \mathscr{G} & -\mathscr{Z} \end{bmatrix} \begin{bmatrix} \mathscr{C} & 0 \\ 0 & \mathscr{L} \end{bmatrix}^{-1} \begin{bmatrix} \widehat{\mathscr{C}} & 0 \\ 0 & \widehat{\mathscr{L}} \end{bmatrix} \tag{8}$$

and

$$\mathbf{e} = \begin{bmatrix} \mathbf{v}_E \\ \mathbf{i}_J \end{bmatrix}. \tag{9}$$

For the second way, set

$$\mathbf{x} = \begin{bmatrix} \mathbf{v}_{Ct} \\ \mathbf{i}_{Ll} \end{bmatrix} - \begin{bmatrix} \mathscr{C} & 0 \\ 0 & \mathscr{L} \end{bmatrix}^{-1} \begin{bmatrix} \widehat{\mathscr{C}} & 0 \\ 0 & \widehat{\mathscr{L}} \end{bmatrix} \begin{bmatrix} \mathbf{v}_E \\ \mathbf{i}_J \end{bmatrix}. \tag{10}$$

In this case

$$\begin{bmatrix} \mathbf{v}_{Ct} \\ \mathbf{i}_{Ll} \end{bmatrix} = \mathbf{x} + \begin{bmatrix} \mathscr{C} & \mathbf{0} \\ \mathbf{0} & \mathscr{L} \end{bmatrix}^{-1} \begin{bmatrix} \widehat{\mathscr{C}} & \mathbf{0} \\ \mathbf{0} & \widehat{\mathscr{L}} \end{bmatrix} \begin{bmatrix} \mathbf{v}_E \\ \mathbf{i}_J \end{bmatrix}. \tag{11}$$

After this expression is substituted into (2) and the terms in the resulting equation are regrouped, we obtain the normal-form equation (3), with

$$\mathscr{A} = \begin{bmatrix} \mathscr{C} & \mathbf{0} \\ \mathbf{0} & \mathscr{L} \end{bmatrix}^{-1} \left\{ \begin{bmatrix} -\mathscr{Y} & \mathscr{H} \\ \mathscr{G} & -\mathscr{Z} \end{bmatrix} - \frac{d}{dt} \begin{bmatrix} \mathscr{C} & \mathbf{0} \\ \mathbf{0} & \mathscr{L} \end{bmatrix} \right\} \tag{12}$$

and

$$\mathscr{B} = \begin{bmatrix} \mathscr{C} & \mathbf{0} \\ \mathbf{0} & \mathscr{L} \end{bmatrix}^{-1} \left\{ \begin{bmatrix} -\widehat{\mathscr{Y}} & \widehat{\mathscr{H}} \\ \widehat{\mathscr{G}} & -\widehat{\mathscr{Z}} \end{bmatrix} + \begin{bmatrix} -\mathscr{Y} & \mathscr{H} \\ \mathscr{G} & -\mathscr{Z} \end{bmatrix} \begin{bmatrix} \mathscr{C} & \mathbf{0} \\ \mathbf{0} & \mathscr{L} \end{bmatrix}^{-1} \begin{bmatrix} \widehat{\mathscr{C}} & \mathbf{0} \\ \mathbf{0} & \widehat{\mathscr{L}} \end{bmatrix} \right\}. \tag{13}$$

The vector **e** is again given by (9).

The second way of choosing the state vector leads to an $\mathscr{A}$ matrix, set forth in (12), which requires the derivative of the parameter matrix of (4). Since our interest is in a solution of the state equation for $t \geq t_0$, we must assume that this parameter matrix is differentiable for $t \geq t_0$. This is an assumption that does not have to be made when the state vector is chosen by the first method.

In addition, the second way of choosing the state vector yields $\mathscr{A}$ and $\mathscr{B}$ matrices that, in general, are functionally more complicated than those obtained when the state vector is chosen by the first method. This is most evident when

$$\begin{bmatrix} \widehat{\mathscr{C}} & \mathbf{0} \\ \mathbf{0} & \widehat{\mathscr{L}} \end{bmatrix} = \mathbf{0}.$$

In this instance the first method yields matrices

$$\mathscr{A} = \begin{bmatrix} -\mathscr{Y} & \mathscr{H} \\ \mathscr{G} & -\mathscr{Z} \end{bmatrix} \begin{bmatrix} \mathscr{C} & \mathbf{0} \\ \mathbf{0} & \mathscr{L} \end{bmatrix}^{-1} \quad \text{and} \quad \mathscr{B} = \begin{bmatrix} -\widehat{\mathscr{Y}} & \widehat{\mathscr{H}} \\ \widehat{\mathscr{G}} & -\widehat{\mathscr{Z}} \end{bmatrix},$$

which, clearly, are functionally simpler than the matrices

$$\mathscr{A} = \begin{bmatrix} \mathscr{C} & \mathbf{0} \\ \mathbf{0} & \mathscr{L} \end{bmatrix}^{-1} \left\{ \begin{bmatrix} -\mathscr{Y} & \mathscr{H} \\ \mathscr{G} & -\mathscr{Z} \end{bmatrix} - \frac{d}{dt} \begin{bmatrix} \mathscr{C} & \mathbf{0} \\ \mathbf{0} & \mathscr{L} \end{bmatrix} \right\}$$

and

$$\mathscr{B} = \begin{bmatrix} \mathscr{C} & \mathbf{0} \\ \mathbf{0} & \mathscr{L} \end{bmatrix}^{-1} \begin{bmatrix} -\hat{\mathscr{Y}} & \hat{\mathscr{H}} \\ \hat{\mathscr{G}} & -\hat{\mathscr{Z}} \end{bmatrix}$$

that are established by the second method. From these observations we conclude that the first way of selecting $\mathbf{x}$ is more desirable.

THE COMPONENTS OF THE STATE VECTOR

Let us therefore turn our attention to characterizing the elements of $\mathbf{x}$, as given by (5). We have, after performing the multiplications indicated in (5),

$$\mathbf{x} = \begin{bmatrix} \mathscr{C}\mathbf{v}_{Ct} - \hat{\mathscr{C}}\mathbf{v}_E \\ \mathscr{L}\mathbf{i}_{Ll} - \hat{\mathscr{L}}\mathbf{i}_J \end{bmatrix}. \tag{14}$$

By reference to (117) and (118) in Chapter 4, observe that

$$\mathscr{C} = \mathbf{C}_t + \mathbf{Q}_{CC}\mathbf{C}_l\mathbf{Q}'_{CC}$$

$$\hat{\mathscr{C}} = -\mathbf{Q}_{CC}\mathbf{C}_l\mathbf{Q}'_{EC}.$$

As a consequence, we find that

$$\mathscr{C}\mathbf{v}_{Ct} - \hat{\mathscr{C}}\mathbf{v}_E = \mathbf{C}_t\mathbf{v}_{Ct} + \mathbf{Q}_{CC}\mathbf{C}_l(\mathbf{Q}'_{CC}\mathbf{v}_{Ct} + \mathbf{Q}'_{EC}\mathbf{v}_E)$$
$$= \mathbf{C}_t\mathbf{v}_{Ct} + \mathbf{Q}_{CC}\mathbf{C}_l\mathbf{v}_{Cl}. \tag{15}$$

The last step follows from (113e) in Chapter 4, which is repeated here:

$$\mathbf{Q}'_{CC}\mathbf{v}_{Ct} + \mathbf{Q}'_{EC}\mathbf{v}_E = \mathbf{v}_{Cl}.$$

But the elements of $\mathbf{C}_t\mathbf{v}_{Ct}$ and $\mathbf{C}_l\mathbf{v}_{Cl}$ are simply charges on the capacitors in the network. Thus the elements of $\mathbf{x}$ established by $\mathscr{C}\mathbf{v}_{Ct} - \hat{\mathscr{C}}\mathbf{v}_E$ are seen by (15) to be linear combinations of the capacitor charges.

Next we turn to the second row on the right side of (14). By reference to (122) and (123) of Chapter 4, we determine that

$$\mathscr{L} = \mathbf{L}_{ll} - \mathbf{L}_{lt}\mathbf{Q}_{LL} - \mathbf{Q}'_{LL}\mathbf{L}_{tl} + \mathbf{Q}'_{LL}\mathbf{L}_{tt}\mathbf{Q}_{LL}$$

$$\hat{\mathscr{L}} = -\mathbf{Q}'_{LL}\mathbf{L}_{tt}\mathbf{Q}_{LJ} + \mathbf{L}_{lt}\mathbf{Q}_{LJ}.$$

Thus

$$\mathscr{L}\mathbf{i}_{Ll} - \widehat{\mathscr{L}}\mathbf{i}_J = \mathbf{L}_{ll}\mathbf{i}_{Ll} - \mathbf{Q}'_{LL}\mathbf{L}_{tl}\mathbf{i}_{Ll} + \mathbf{L}_{lt}(-\mathbf{Q}_{LL}\mathbf{i}_{Ll} - \mathbf{Q}_{LJ}\mathbf{i}_J)$$

$$- \mathbf{Q}'_{LL}\mathbf{L}_{tt}(-\mathbf{Q}_{LL}\mathbf{i}_{Ll} - \mathbf{Q}_{LJ}\mathbf{i}_J) \qquad (16)$$

$$= \mathbf{L}_{ll}\mathbf{i}_{Ll} - \mathbf{Q}'_{LL}\mathbf{L}_{tl}\mathbf{i}_{Ll} + \mathbf{L}_{lt}\mathbf{i}_{Lt} - \mathbf{Q}'_{LL}\mathbf{L}_{tt}\mathbf{i}_{Lt}.$$

The last step follows from (113d), in Chapter 4, which is repeated here:

$$-\mathbf{Q}_{LL}\mathbf{i}_{Ll} - \mathbf{Q}_{LJ}\mathbf{i}_J = \mathbf{i}_{Lt}.$$

But the elements of $\mathbf{L}_{ll}\mathbf{i}_{Ll}$, $\mathbf{L}_{tl}\mathbf{i}_{Lt}$, $\mathbf{L}_{lt}\mathbf{i}_{Lt}$, and $\mathbf{L}_{tt}\mathbf{i}_{Lt}$ are flux linkages of the inductors. In particular, the elements of $\mathbf{L}_{tl}\mathbf{i}_{Ll}$ and $\mathbf{L}_{lt}\mathbf{i}_{Lt}$ are mutual flux linkages. Hence the elements of $\mathbf{x}$ established by $\mathscr{L}\mathbf{i}_{Ll} - \widehat{\mathscr{L}}\mathbf{i}_J$ are linear combinations of the inductor flux linkages.

A word is also in order on the network output equation. As in Chapter 4, it may be shown that any of the network voltage and current variables may be specified in terms of $\mathbf{v}_{Ct}$, $\mathbf{i}_{Ll}$, $\mathbf{v}_E$, $\mathbf{i}_J$, $d\mathbf{v}_E/dt$, and $d\mathbf{i}_J/dt$. If $\mathbf{w}$ denotes the output vector—the elements of $\mathbf{w}$ are the voltage and/or current variables that are the desired network outputs—then it can be shown that, for either way of defining the state vector $\mathbf{x}$ given previously, $\mathbf{w}$ is conveniently expressed as

$$\mathbf{w} = \mathbf{C}\mathbf{x} + \mathscr{D}\mathbf{e} + \frac{d}{dt}\widehat{\mathscr{D}}\mathbf{e}. \qquad (17)$$

You should verify this statement.

Example

Let us illustrate the reduction of network equations from the form of (2) to the standard form of (3) by using the network shown in Fig. 1,

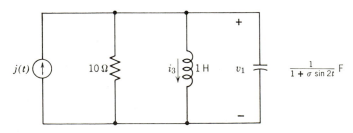

Fig. 1. Periodically time-varying network.

which contains a periodically varying capacitance. Let v_1 denote the voltage across the capacitor and i_3 the current through the inductor. It is easy to verify that the vector $[v_1 \quad i_3]'$ satisfies the differential equation

$$\frac{d}{dt}\begin{bmatrix} \dfrac{1}{1+\sigma \sin 2t} & 0 \\ 0 & 1 \end{bmatrix}\begin{bmatrix} v_1 \\ i_3 \end{bmatrix} = \begin{bmatrix} -\frac{1}{10} & -1 \\ 1 & 0 \end{bmatrix}\begin{bmatrix} v_1 \\ i_3 \end{bmatrix} + \begin{bmatrix} 1 \\ 0 \end{bmatrix}[j(t)].$$

(You should do this.) This is, of course, the result that would be obtained by using the methods of Section 4.5 through (129a) and (129b).

As indicated in Fig. 1, $|\sigma|$ is less than 1. Therefore

$$\begin{bmatrix} \dfrac{1}{1+\sigma \sin 2t} & 0 \\ 0 & 1 \end{bmatrix}$$

is bounded for all t and has an inverse, which is

$$\begin{bmatrix} 1+\sigma \sin 2t & 0 \\ 0 & 1 \end{bmatrix}.$$

In accordance with (5), the state vector is taken as

$$\mathbf{x} = \begin{bmatrix} \dfrac{1}{1+\sigma \sin 2t} & 0 \\ 0 & 1 \end{bmatrix}\begin{bmatrix} v_1 \\ i_3 \end{bmatrix}.$$

Observe that the elements of $\mathbf{x}$ are the capacitor charge and inductor flux linkage. Now, by (7) and (8), we find

$$\mathscr{A} = \begin{bmatrix} -\frac{1}{10} & -1 \\ 1 & 0 \end{bmatrix}\begin{bmatrix} 1+\sigma \sin 2t & 0 \\ 0 & 1 \end{bmatrix} = \begin{bmatrix} -\dfrac{(1+\sigma \sin 2t)}{10} & -1 \\ 1+\sigma \sin 2t & 0 \end{bmatrix}$$

$$\mathscr{B} = \begin{bmatrix} 1 \\ 0 \end{bmatrix}.$$

The alternative choice of the network state vector, indicated in (10), is

$$\mathbf{x} = \begin{bmatrix} v_1 \\ i_3 \end{bmatrix},$$

Associated with this state vector, we have, by (12) and (13),

$$\mathscr{A} = \begin{bmatrix} 1 + \sigma \sin 2t & 0 \\ 0 & 1 \end{bmatrix} \left\{ \begin{bmatrix} -\frac{1}{10} & -1 \\ 1 & 0 \end{bmatrix} - \begin{bmatrix} \dfrac{-2\sigma \cos 2t}{(1 + \sigma \sin 2t)^2} & 0 \\ 0 & 1 \end{bmatrix} \right\}$$

$$= \begin{bmatrix} \dfrac{-(1 + \sigma \sin 2t)^2 + 20\sigma \cos 2t}{10 + 10\sigma \sin 2t} & -1 - \sigma \sin 2t \\ 1 & 0 \end{bmatrix}$$

$$\mathscr{B} = \begin{bmatrix} 1 + \sigma \sin 2t & 0 \\ 0 & 1 \end{bmatrix} \begin{bmatrix} 1 \\ 0 \end{bmatrix} = \begin{bmatrix} 1 + \sigma \sin 2t \\ 0 \end{bmatrix}.$$

Observe that the $\mathscr{A}$ and $\mathscr{B}$ matrices in this case are functionally more complicated than before. Thus by example, as well as by the previous general discussion, we see that the better choice of $\mathbf{x}$ is as given in (5)—linear combinations of charges and flux linkages as elements of the state vector. This will again be found to be true when we turn our attention to nonlinear networks.

10.2 STATE-EQUATION SOLUTION FOR TIME-VARYING NETWORKS

As indicated in the last section, the input and output variables of a time-varying network are related by the state equations

$$\frac{d}{dt}\mathbf{x} = \mathscr{A}\mathbf{x} + \mathscr{B}\mathbf{e} \tag{18a}$$

$$\mathbf{w} = \mathcal{C}\mathbf{x} + \mathscr{D}\mathbf{e} + \frac{d}{dt}\widehat{\mathscr{D}}\mathbf{e}. \tag{18b}$$

As we seek to solve these equations, we must anticipate that any or all of the matrices $\mathscr{A}$, $\mathscr{B}$, C, $\mathscr{D}$, and $\widehat{\mathscr{D}}$ may be functions of time.

We shall solve the state equation 18a for the state vector $\mathbf{x}$ by using the variation-of-parameter method. You should compare the result at each step with the result at the corresponding step for the time-invariant case in Chapter 4. It is assumed that $\mathbf{x}$ is an n-vector and, consequently, that $\mathscr{A}$ is a square matrix of order n. Now let

$$\mathbf{x}(t) = \mathbf{Y}(t)\,\hat{\mathbf{x}}(t), \tag{19}$$

where $\mathbf{Y}(t)$ is a square matrix of order n. Upon substituting this transformation in (18a) and rearranging terms, we get

$$\left(\frac{d}{dt}\mathbf{Y} - \mathscr{A}\,\mathbf{Y}\right)\hat{\mathbf{x}} = -\mathbf{Y}\frac{d}{dt}\hat{\mathbf{x}} + \mathscr{B}\mathbf{e}. \tag{20}$$

It is evident that, if the expression in parentheses is zero, the solution will be simplified. We shall suppose that this is true. To be more precise, we shall suppose that the homogeneous differential equation

$$\frac{d}{dt}\mathbf{Y} = \mathscr{A}\,\mathbf{Y}, \tag{21}$$

with $\mathbf{Y}(t)$ equal to $\mathbf{Y}(t_0)$ at time t_0, possesses a nonsingular solution for all finite $t \geq t_0$. By combining (20) and (21), we find that

$$\mathbf{Y}\frac{d}{dt}\hat{\mathbf{x}} = \mathscr{B}\mathbf{e}.$$

Then, since we have assumed that $\mathbf{Y}$ is nonsingular for $t \geq t_0$, $\mathbf{Y}^{-1}$ exists, and

$$\frac{d}{dt}\hat{\mathbf{x}} = \mathbf{Y}^{-1}\mathscr{B}\mathbf{e}. \tag{22}$$

A solution for $\hat{\mathbf{x}}$ is obtained by integrating. The result is

$$\hat{\mathbf{x}}(t) = \hat{\mathbf{x}}(t_0) + \int_{t_0}^{t} \mathbf{Y}(\tau)^{-1}\,\mathscr{B}(\tau)\,\mathbf{e}(\tau)\,d\tau. \tag{23}$$

By (19) and the assumption that $\mathbf{Y}(t_0)$ is nonsingular, the initial vector $\hat{\mathbf{x}}(t_0)$ is given by the equation

$$\hat{\mathbf{x}}(t_0) = \mathbf{Y}(t_0)^{-1}\mathbf{x}(t_0). \tag{24}$$

To obtain $\mathbf{x}(t)$ we now premultiply (23) by $\mathbf{Y}(t)$. If, at the same time, we introduce $\hat{\mathbf{x}}(t_0)$ as given by (24), we get

$$\mathbf{x}(t) = \mathbf{Y}(t)\,\mathbf{Y}(t_0)^{-1}\,\mathbf{x}(t_0) + \int_{t_0}^{t} \mathbf{Y}(t)\,\mathbf{Y}(\tau)^{-1}\,\mathscr{B}(\tau)\,\mathbf{e}(\tau)\,d\tau. \qquad (25)$$

It is permissible to take $\mathbf{Y}(t)$ inside the integral, as the integration variable is τ, *not* t.

When the network is not subject to external excitation—$\mathbf{e}(t) \equiv 0$ for $t \geq t_0$—it is evident from (25) that the matrix $\mathbf{Y}(t)\,\mathbf{Y}(t_0)^{-1}$ characterizes the transition of the state from $\mathbf{x}(t_0)$ at time t_0 to $\mathbf{x}(t)$ at time t. Thus the matrix $\mathbf{Y}(t)\,\mathbf{Y}(\tau)^{-1}$ is known as the *state-transition matrix* and is denoted by $\mathbf{\Phi}(t, \tau)$; that is,

$$\mathbf{\Phi}(t, \tau) = \mathbf{Y}(t)\,\mathbf{Y}(\tau)^{-1}. \qquad (26)$$

In Chapter 4 it was seen that the state-transition matrix is a function of $t - \tau$ for time-invariant networks; *this is not true in the more general case of time-varying networks.* The solution for $\mathbf{x}$ can be expressed in terms of the state-transition matrix by inserting (26) into (25) to get

$$\mathbf{x}(t) = \mathbf{\Phi}(t, t_0)\,\mathbf{x}(t_0) + \int_{t_0}^{t} \mathbf{\Phi}(t, \tau)\,\mathscr{B}(\tau)\,\mathbf{e}(\tau)\,d\tau. \qquad (27)$$

A SPECIAL CASE OF THE HOMOGENEOUS EQUATION SOLUTION

Before the state vector $\mathbf{x}(t)$ can be determined from (27), it is necessary to find the state-transition matrix $\mathbf{\Phi}(t, \tau)$, or, equivalently, $\mathbf{Y}(t)$. This is the task to which we now turn our attention. We shall consider the solution of the homogeneous matrix equation 21 when, in particular, we set $\mathbf{Y}(t_0) = \mathbf{U}$. This is no real restriction since (24) requires only that $\mathbf{Y}(t_0)$ have an inverse.

First, consider the corresponding scalar equation

$$\frac{d}{dt} y = ay,$$

with $y(t_0) = 1$. The solution of this equation is

$$y(t) = \exp\left[\int_{t_0}^{t} a(\tau)\,d\tau\right].$$

It is tempting, therefore, to suppose that the solution to the homogeneous matrix equation 21 will be of the form

$$\mathbf{Y}(t) = \exp \left[\int_{t_0}^{t} \mathscr{A}(\tau) \, d\tau \right]. \tag{28}$$

We shall now show that this is the solution only for the very special case when the product of $\mathscr{A}(t)$ with $\int_{t_0}^{t} \mathscr{A}(\tau) \, d\tau$ is commutative.

We know from the definition of a matrix exponential that

$$\exp \left[\int_{t_0}^{t} \mathscr{A}(\tau) \, d\tau \right] = \sum_{k=0}^{\infty} \frac{1}{k!} \left[\int_{t_0}^{t} \mathbf{A}(\tau) \, d\tau \right]^{k}. \tag{29}$$

By differentiating both sides, we get

$$\frac{d}{dt} \exp \left[\int_{t_0}^{t} \mathscr{A}(\tau) \, d\tau \right] = \sum_{k=1}^{\infty} \frac{1}{k!} \frac{d}{dt} \left[\int_{t_0}^{t} \mathscr{A}(\tau) \, d\tau \right]^{k}. \tag{30}$$

We are tacitly assuming that each of the indicated infinite series converges. Now let us examine a typical term from the right-hand side of this last equation. It can be shown that

$$\frac{1}{k!} \frac{d}{dt} \left[\int_{t_0}^{t} \mathscr{A}(\tau) \, d\tau \right]^{k} = \frac{1}{k!} \sum_{i=1}^{k} \left[\int_{t_0}^{t} \mathscr{A}(\tau) \, d\tau \right]^{i-1} \mathscr{A}(t) \left[\int_{t_0}^{t} \mathscr{A}(\tau) \, d\tau \right]^{k-i}.$$

$$\tag{31}$$

This follows as a generalization of

$$\frac{d}{dt} \mathbf{A}^{2} = \frac{d}{dt} \mathbf{A}\mathbf{A} = \left(\frac{d}{dt} \mathbf{A} \right) \mathbf{A} + \mathbf{A} \left(\frac{d}{dt} \mathbf{A} \right),$$

where $\mathbf{A}$ is any differentiable matrix.

In general, $\mathscr{A}(t)$ does not commute with $\int_{t_0}^{t} \mathscr{A}(\tau) \, d\tau$; you may demonstrate this to yourself with the simple matrix

$$\mathscr{A}(t) = \begin{bmatrix} 1 & t \\ 1 & 1 \end{bmatrix}.$$

Hence it is only when

$$\mathscr{A}(t) \left[\int_{t_0}^{t} \mathscr{A}(\tau) \, d\tau \right] = \left[\int_{t_0}^{t} \mathscr{A}(\tau) \, d\tau \right] \mathscr{A}(t)$$

that we get

$$\frac{1}{k!} \frac{d}{dt} \left[\int_{t_0}^{t} \mathscr{A}(\tau) \, d\tau \right]^{k} = \frac{1}{k!} \sum_{i=1}^{k} \mathscr{A}(t) \left[\int_{t_0}^{t} \mathscr{A}(\tau) \, d\tau \right]^{k-1}$$

$$= \frac{1}{(k-1)!} \mathscr{A}(t) \left[\int_{t_0}^{t} \mathscr{A}(\tau) \, d\tau \right]^{k-1}. \quad (32)$$

The last step follows from the fact that the summands do not depend on the summation index i. By combining this result with (30), we find, *in this special case*, that

$$\frac{d}{dt} \exp \left[\int_{t_0}^{t} \mathscr{A}(\tau) \, d\tau \right] = \sum_{k=1}^{\infty} \mathscr{A}(t) \frac{1}{(k-1)!} \left[\int_{t_0}^{t} \mathscr{A}(\tau) \, d\tau \right]^{k-1}$$

$$= \mathscr{A}(t) \exp \left[\int_{t_0}^{t} \mathscr{A}(\tau) \, d\tau \right]. \quad (33)$$

Hence (28) is the solution of the homogeneous equation.

When the solution of the homogeneous equation is given by (28), we find that the state-transition matrix $\mathbf{\Phi}(t, \tau)$ has a particularly simple form. We know that $\mathbf{\Phi}(t, \tau) = \mathbf{\Phi}(t, t_0)|_{t_0 = \tau}$ and that $\mathbf{\Phi}(t, t_0) = \mathbf{Y}(t) \mathbf{Y}(t_0)^{-1} = \mathbf{Y}(t)$, since $\mathbf{Y}(t_0) = \mathbf{U}$. Thus, by replacing t_0 with τ in the integral appearing in (28), the relation for $\mathbf{\Phi}(t, t_0) = \mathbf{Y}(t)$, we get

$$\mathbf{\Phi}(t, \tau) = \exp \left[\int_{\tau}^{t} \mathscr{A}(\nu) \, d\nu \right]. \quad (34)$$

Example

Before turning our attention to the solution of (21) under less restrictive conditions, let us consider as an example a network for which the preceding commutativity condition is satisfied. The network illustrated in Fig. 2 with two time-varying resistors is easily shown to satisfy the

following differential equation for the state vector $[q_2 \quad \lambda_5]'$:

$$\frac{d}{dt} \begin{bmatrix} q_2 \\ \lambda_5 \end{bmatrix} = \begin{bmatrix} -2t & -1 \\ 1 & -2t \end{bmatrix} \begin{bmatrix} q_2 \\ \lambda_5 \end{bmatrix} + \begin{bmatrix} 0 \\ -1 \end{bmatrix} [e(t)].$$

You will find it easy to verify that the product of

$$\mathscr{A}(t) = \begin{bmatrix} -2t & -1 \\ 1 & -2t \end{bmatrix}$$

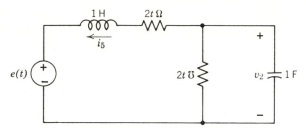

Fig. 2.   Time-varying network.

with

$$\int_{t_0}^{t} \mathscr{A}(\tau)\, d\tau = \begin{bmatrix} -(t+t_0) & -1 \\ 1 & -(t+t_0) \end{bmatrix}(t-t_0)$$

is commutative. Therefore, from (28),

$$\mathbf{Y}(t) = \exp\left\{ \begin{bmatrix} -(t+t_0) & -1 \\ 1 & -(t+t_0) \end{bmatrix}(t-t_0) \right\},$$

and, from (34), the state-transition matrix is

$$\boldsymbol{\Phi}(t, \tau) = \exp\left\{ \begin{bmatrix} -(t+\tau) & -1 \\ 1 & -(t+\tau) \end{bmatrix}(t-\tau) \right\}.$$

By the theory of functions of a matrix that was established in Chapter 4, the matrix exponential for $\boldsymbol{\Phi}(t, \tau)$ can be replaced by a closed-form equivalent. The appropriate function of s is $f(s) = \epsilon^{s(t-\tau)}$. By the procedures in Chapter 4, the equivalent form for $\boldsymbol{\Phi}(t, \tau)$ is found to be

$$\boldsymbol{\Phi}(t, \tau) = \begin{bmatrix} \cos(t-\tau) & -\sin(t-\tau) \\ \sin(t-\tau) & \cos(t-\tau) \end{bmatrix} \epsilon^{-(t^2-\tau^2)}.$$

You should verify this expression. ·

Finally, upon substituting this relation for $\boldsymbol{\Phi}(t, \tau)$ into (27), we find for the state vector

$$\begin{bmatrix} q_2(t) \\ \lambda_5(t) \end{bmatrix} = \epsilon^{-(t^2-t_0^2)} \begin{bmatrix} \cos(t-t_0) & -\sin(t-t_0) \\ \sin(t-t_0) & \cos(t-t_0) \end{bmatrix} \begin{bmatrix} q_2(t_0) \\ \lambda_5(t_0) \end{bmatrix}$$

$$+ \int_{t_0}^{t} \epsilon^{-(t^2-\tau^2)} \begin{bmatrix} \sin(t-\tau) \\ -\cos(t-\tau) \end{bmatrix} [e(\tau)]\, d\tau.$$

This completes the example.

EXISTENCE AND UNIQUENESS OF SOLUTION OF THE HOMOGENEOUS EQUATION

We have now discussed the solution of the homogeneous equation 21 for the special case when the product of $\mathscr{A}(t)$ with $\int_{t_0}^{t} \mathscr{A}(\tau)\, d\tau$ is commutative. We now remove this restriction. In this case no general method of solution is available. In fact, we must usually be satisfied not with a solution but with a knowledge that one or more solutions exist, that there is a unique solution, and/or that the solution has certain definable properties, such as periodicity. The solution itself is often approximated by truncating a perturbation series or an iteration, or by numerical integration. The former is outside the scope of this text.* Numerical integration will be taken up in Section 10.5, where numerical methods applicable to nonlinear, time-varying networks will be treated.

Our objective now is to establish conditions under which (21) possesses a unique solution. So that there shall be no abiguity, we state that $\mathbf{Y}$ is a solution *in the ordinary sense*, if it satisfies the homogeneous equation (21) for all finite $t \geq t_0$, with $d\mathbf{Y}/dt$ at $t = t_0$ being interpreted as a derivative from the right.† The following well-established theorem provides us with a sufficient condition‡:

Theorem 1. *Given any* $\mathbf{Y}(t_0)$, *the homogeneous differential equation* 21 *has a unique solution in the ordinary sense, equal to* $\mathbf{Y}(t_0)$ *at time* t_0, *if* $\mathscr{A}$ *is a continuous function of* t *for* $t_0 \leq t < \infty$.

It is quite possible for $\mathscr{A}$ not to be continuous. Furthermore, it may be possible to find a $\mathbf{Y}$ that satisfies the differential equation for almost all values of $t \geq t_0$. To take care of this situation, integrate both sides of (21) from t_0 to t to obtain the associated integral equation

$$\mathbf{Y}(t) = \mathbf{Y}(t_0) + \int_{t_0}^{t} \mathscr{A}(\tau)\, \mathbf{Y}(\tau)\, d\tau. \tag{35}$$

---

* For a further discussion of this topic see Nicholas Minorsky, *Nonlinear Oscillations*, D. Van Nostrand Co., Princeton, N.J., 1962.

† If the limit exists, then

$$\lim_{\Delta \to 0} \left[\frac{\mathbf{Y}(t_0 + \Delta) - \mathbf{Y}(t_0)}{\Delta} \right] \quad \text{with} \quad \Delta > 0$$

is the derivative of $\mathbf{Y}$ from the right at $t = t_0$.

‡ For further discussion on the topic of existence and uniqueness of solutions to ordinary differential equations, see Earl A. Coddington and Norman Levinson, *Theory of Ordinary Differential Equations*, McGraw-Hill, New York, 1955.

We now introduce a new notion; we say that $\mathbf{Y}$ is a solution of the differential equation 21 *in the extended sense*, if it satisfies the integral equation 35 for all finite $t \geq t_0$. The following theorem provides us with a satisfactory sufficient condition:*

Theorem 2. *Given any* $\mathbf{Y}(t_0)$, *the homogeneous differential equation* 21 *has a unique continuous solution, in the extended sense, equal to* $\mathbf{Y}(t_0)$ *at time* t_0, *if* $\mathscr{A}$ *is a locally integrable function† for* $t \geq t_0$.

The sufficiency condition of this second theorem is considerably weaker than that of the first theorem; the price paid is that the solution may not satisfy (21) in the ordinary sense.

As an illustration of these notions, consider the network with a discontinuous time-varying conductance that is shown in Fig. 3. The state

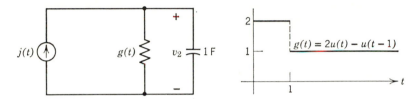

Fig. 3.   Time-varying network with a discontinuous time-varying conductance.

equation for this network is

$$\frac{d}{dt} q_2 = -[2u(t) - u(t-1)]q_2 + j(t).$$

The homogeneous differential equation corresponding to (21) is

$$\frac{d}{dt} y = -[2u(t) - u(t-1)]y,$$

and with $t_0 = 0$ the integral equation corresponding to (35) is

$$y(t) = y(0) - \int_0^t [2u(\tau) - u(\tau - 1)]y(\tau)\, d\tau.$$

---

* See the footnote * on page 718.

† A vector- or matrix-valued function is said to be *locally integrable* if each of its elements (a) is a continuous function except at a finite number of points on any finite interval and (b) possesses a proper or absolutely convergent improper Riemann integral on every finite interval over the interior of which the function is continuous.

Now you may easily verify that

$$y(t) = \epsilon^{-2t} \qquad (0 \leq t \leq 1)$$
$$= \epsilon^{-1}\epsilon^{-t} \qquad (1 \leq t)$$

is a solution of the integral equation and hence of the homogeneous differential equation *in the extended sense.* Observe that it is not a solution of the homogeneous differential equation *in the ordinary sense* at $t = 1$. We anticipated that a solution in the ordinary sense might not exist, since $2u(t) - u(t - 1)$ is not continuous; furthermore, we knew that a solution in the extended sense did exist, since $\int_0^t [2u(\tau) - u(\tau - 1)]d\tau$ exists for all finite t. In addition, the theorem tells us the solution we have found is the unique continuous solution—we need search for no other.

We must still consider one assumption relative to the solution of the homogeneous state equation. This is the assumption that $\mathbf{Y}(t)^{-1}$ exists for all finite $t \geq t_0$. By reference to Chapter 1, you may verify that

$$\frac{d}{dt}|\mathbf{Y}| = \sum_{k=1}^{n} \sum_{l=1}^{n} \frac{dy_{lk}}{dt} \Delta_{lk}, \tag{36}$$

where y_{lk} denotes the (l, k)th element of $\mathbf{Y}$ and Δ_{lk} denotes the (l, k)th cofactor of $\mathbf{Y}$. From (21) we know that

$$\frac{dy_{lk}}{dt} = \sum_{m=1}^{n} a_{lm} y_{mk}, \tag{37}$$

where a_{lm} is the (l, m)th element of $\mathscr{A}$; therefore

$$\frac{d}{dt}|\mathbf{Y}| = \sum_{k=1}^{n} \sum_{l=1}^{n} \sum_{m=1}^{n} a_{lm} y_{mk} \Delta_{lk}. \tag{38}$$

Since $\sum_{k=1}^{n} y_{mk} \Delta_{lk} = \delta_{ml}|\mathbf{Y}|$, we get

$$\frac{d}{dt}|\mathbf{Y}| = \sum_{l=1}^{n} \sum_{m=1}^{n} a_{lm} \delta_{ml}|\mathbf{Y}| = \left(\sum_{l=1}^{n} a_{ll}\right)|\mathbf{Y}|. \tag{39}$$

Recalling the definition of the trace of a matrix, we see that $\sum_{l=1}^{n} a_{ll} =$ tr $\mathscr{A}$. Thus

$$\frac{d}{dt}|\mathbf{Y}| = (\text{tr } \mathscr{A})|\mathbf{Y}|. \tag{40}$$

This is a first-order scalar differential equation for the determinant of $\mathbf{Y}$. The solution is, of course,

$$|\mathbf{Y}(t)| = \exp\left\{\int_0^t [\operatorname{tr}\,\mathscr{A}(\tau)]\,d\tau\right\}|\mathbf{Y}(t_0)|.$$

But since $\mathbf{Y}(t_0) = \mathbf{U}$ implies $|\mathbf{Y}(t_0)| = 1$, then

$$|\mathbf{Y}(t)| = \exp\left\{\int_{t_0}^t [\operatorname{tr}\,\mathscr{A}(\tau)]\,d\tau\right\}. \tag{41}$$

It is now evident that $\mathbf{Y}(t)^{-1}$ exists—$|\mathbf{Y}(t)| \neq 0$—provided $\int_{t_0}^t [\operatorname{tr}\,\mathscr{A}(\tau)]\,d\tau$ is finite for all finite $t \geq t_0$. If either of the two preceding theorems are satisfied, $\int_{t_0}^t \mathscr{A}(\tau)\,d\tau$ is finite for $t \geq t_0$; hence $\int_{t_0}^t [\operatorname{tr}\,\mathscr{A}(\tau)]\,d\tau = \operatorname{tr}\int_{t_0}^t \mathscr{A}(\tau)\,d\tau$ is finite for $t \geq t_0$, The conclusion is that $\mathbf{Y}(t)$ is nonsingular under the conditions of either Theorem 1 or Theorem 2.

SOLUTION OF STATE EQUATION—EXISTENCE AND UNIQUENESS

We turn our attention again to the state equation (nonhomogeneous) itself. What is initially needed is a precise statement of what constitutes a solution. We say that $\mathbf{x}$ is a solution *in the ordinary sense* when the state equation 18a is satisfied for all finite $t \geq t_0$ with $d\mathbf{x}/dt$ at $t = t_0$ being interpreted as a derivative from the right. The following theorem* concerns the existence of a solution in the ordinary sense.

Theorem 3. *Given any* $\mathbf{x}(t_0)$, *the state equation* 18a *has a unique solution, in the ordinary sense, equal to* $\mathbf{x}(t_0)$ *at time* t_0, *if* $\mathscr{A}$ *and* $\mathscr{B}\mathbf{e}$ *are continuous functions of* t *for* $t_0 \leq t < \infty$.

When the above sufficiency condition is violated because $\mathscr{A}$ and/or $\mathscr{B}\mathbf{e}$ are not continuous, it may still be possible to find an $\mathbf{x}$ that satisfies the state equation for almost all values of $t \geq t_0$. To accommodate this possibility we must introduce the associated integral equation, as we did in the case of the homogeneous equation. Upon integrating both sides of (18a) from t_0 to t, we get

$$\mathbf{x}(t) = \mathbf{x}(t_0) + \int_{t_0}^t [\mathscr{A}(\tau)\,\mathbf{x}(\tau) + \mathscr{B}(\tau)\,\mathbf{e}(\tau)]\,d\tau. \tag{42}$$

* See the footnote on page 718.

We say $\mathbf{x}$ is a solution of the state equation 18a *in the extended sense*, if it satisfies the integral equation 42 for all finite $t \geq t_0$. It turns out that $\mathbf{x}$ will satisfy the integral equation if $\mathscr{A}$ and $\mathscr{B}\mathbf{e}$ are locally integrable functions. Thus we can state the following theorem:*

Theorem 4. *Given any* $\mathbf{x}(t_0)$, *the state equation* 18a *has a unique continuous solution, in the extended sense, equal to* $\mathbf{x}(t_0)$ *at time* t_0, *if* $\mathscr{A}$ *and* $\mathscr{B}\mathbf{e}$ *are locally integrable functions for* $t \geq t_0$.

Observe that the sufficiency conditions of these two theorems incorporate those of the corresponding existence theorems for the homogeneous equation. This is comforting to know. It tells us that if the state equation has a unique solution the homogeneous equation also possesses a unique solution. In other words, the variation-of-parameter method is a valid method by which to obtain the solution whenever the state equation has a solution.

You must bear in mind that the conditions of the theorems are merely sufficient conditions, *not* necessary and sufficient conditions. Thus, even though the conditions of the theorem are not satisfied, a solution may exist, and, in fact, a unique solution may exist.

It may appear that the preceding discussion is overly concerned with matters that are self-evident or trivial. After all, does not a network always have a solution?

To illustrate that it is not wise blindly to assume the existence of a unique state response, consider the network illustrated in Fig. 4. The

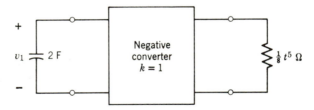

Fig. 4.   Time-varying network that does not have a unique solution.

state equation for this network is easily found to be†

$$\frac{dq_1}{dt} = 4\,\frac{1}{t^5}\,q_1 .$$

---

* See the footnote on page 718.

† Here, and throughout the chapter, if the right hand side of an indicated state equation is not defined at a distinct set of points in t for fixed x, you should make the convenient assumption that the right-hand side of the state equation is zero at these points. In this case, a solution only in the extended sense is being sought, hence the above action will not effect the result.

Suppose $q_1(0) = 0$. Then you may verify that

$$q_1 = \alpha \exp\left(-t^{-4}\right) \tag{43}$$

is a solution for every finite value of α. Several different solutions are shown in Fig. 5. The primary observation to be made is that there is no unique solution.

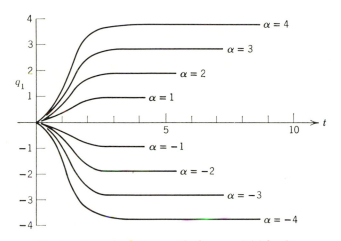

Fig. 5.   Several solutions with the same initial value.

Of course, we had no basis for anticipating a *unique* solution. The sufficiency conditions of our second theorem are not satisfied; that is, $\int_0^t \mathscr{A}(\tau)\, d\tau = \int_0^t [4/\tau^5]\, d\tau$ does not exist for any $t > 0$. Furthermore, we had no basis for anticipating the existence of even any solution. In fact, no solution exists if the initial condition $q_1(0)$ is other than zero.

PERIODIC NETWORKS

The class of time-varying networks is very broad, since no stipulation is made as to the manner of time variation. It should be expected that, for certain special kinds of time variation, additional results can be obtained about the solution that are valid for that class but not for others. A particularly important class of time functions is the class of periodic functions. We shall now consider the special case in which the $\mathscr{A}$-matrix of a time-varying network is periodic. In this case it is possible to establish some additional properties of the solution $\mathbf{Y}$ of the homogeneous equation 21.

Assume that $\mathscr{A}$ is periodic with period T; that is, $\mathscr{A}(t + T) = \mathscr{A}(t)$ for all t.

First of all, we find that, if $\mathbf{Y}(t)$ is a solution of (21), then so is $\mathbf{Y}(t + T)$; that is, if there is a shift of the time variable by an amount equal to the period of $\mathscr{A}$, a solution of (21) will remain a solution. To see this, note that $\mathbf{Y}(\tau)$ satisfies

$$\frac{d}{d\tau} \mathbf{Y}(\tau) = \mathscr{A}(\tau)\, \mathbf{Y}(\tau). \tag{44}$$

Set $\tau = t + T$ and observe that

$$d\mathbf{Y}(\tau)/d\tau = (d\mathbf{Y}(t + T)/dt)(dt/d\tau) = d\mathbf{Y}(t + T)/dt.$$

Therefore (44) becomes,

$$\frac{d}{dt} \mathbf{Y}(t + T) = \mathscr{A}(t + T)\, \mathbf{Y}(t + T) = \mathscr{A}(t)\, \mathbf{Y}(t + T) \tag{45}$$

because $\mathscr{A}$ is periodic, and so $\mathscr{A}(t + T) = \mathscr{A}(t)$. Comparison of this equation with (21) verifies that $\mathbf{Y}(t + T)$ is a solution of the homogeneous equation.

We shall now seek to establish some properties of the solution of (21) when $\mathscr{A}(t)$ is periodic. In particular, we shall be concerned with the stability of the solution. In what follows we shall assume that conditions sufficient to guarantee a unique solution of (21) are satisfied. Then $\mathbf{Y}(t + T)$ is the unique solution with the value $\mathbf{Y}(t_0 + T)$ at time t_0, and $\mathbf{Y}(t)$ is the unique solution with the value $\mathbf{Y}(t_0)$ at time t_0. Since $\mathbf{Y}(t)$ is nonsingular for $t \geq t_0$, $\mathbf{Y}(t_0)^{-1}$ exists. Thus we can define a constant matrix $\mathbf{M}$ as

$$\mathbf{M} = \mathbf{Y}(t_0)^{-1}\mathbf{Y}(t_0 + T) \tag{46}$$

and know that $\mathbf{M}$ exists. It is easy to verify that $\mathbf{Y}(t)\mathbf{M}$ is a solution of (21) and hence the unique solution having the value $\mathbf{Y}(t_0)\mathbf{M}$ at time t_0. Now, after premultiplying (46) by $\mathbf{Y}(t_0)$, we see that $\mathbf{Y}(t_0 + T) = \mathbf{Y}(t_0)\mathbf{M}$. Therefore, since the solution of (21) with the initial value $\mathbf{Y}(t_0 + T) = \mathbf{Y}(t_0)\mathbf{M}$ is unique, the solutions $\mathbf{Y}(t)$ and $\mathbf{Y}(t + T)$ are linearly related as

$$\mathbf{Y}(t + T) = \mathbf{Y}(t)\mathbf{M}. \tag{47}$$

This is an important result, and we shall return to it shortly. But we shall firist show that the matrix $\mathbf{M}$ can itself be expressed as a matrix exponential in the form

$$\mathbf{M} = \epsilon^{\mathbf{P}T}. \tag{48}$$

It is only necessary to show that $\ln \mathbf{M}$ exists and hence

$$\mathbf{P} = \frac{1}{T} \ln \mathbf{M} \tag{49}$$

exists. Observe that $\mathbf{M}$ is nonsingular, since both $\mathbf{Y}(t_0)^{-1}$ and $\mathbf{Y}(t_0 + T)$ in (46) are nonsingular. From $d(s) = \det (s\mathbf{U} - \mathbf{M})$ we see by setting $s = 0$ that the determinant of a matrix is equal in magnitude to the constant term in the characteristic equation of the matrix. Then, since the constant term is equal in magnitude to the product of the eigenvalues of the matrix, we can infer that none of the eigenvalues of the nonsingular matrix $\mathbf{M}$, denoted by s_i, is zero. Let k denote the number of distinct eigenvalues of $\mathbf{M}$ and let r_i denote their multiplicities. Denote the constituent matrices of $\mathbf{M}$ by $\mathbf{K}_{ij}$, $i = 1, \cdots, k$ and $j = 1, \cdots, r_i$. Using the constituent matrix expansion of Section 4.4,

$$\mathbf{P} = \frac{1}{T} \sum_{i=1}^{k} \sum_{j=1}^{r_i} \mathbf{K}_{ij} \left[\frac{1}{(j-1)!} \frac{d^{j-1}}{ds^{j-1}} \ln s \bigg|_{s=s_i} \right], \tag{50}$$

where, for $j > 1$, $d^{j-1} \ln s/ds^{j-1}$ is equal, of course, to $(-1)^{j-2}(j-2)!/s^{j-1}$. Since none of the eigenvalues is zero, these quantities as well as $\ln s_i$ are finite. Thus $\mathbf{P}$ exists.

Observe that the matrix $\mathbf{P}$ is not uniquely defined. This is a consequence of the multiple-valued nature of $\ln s_i$. The various values of $\ln s_i$ differ by an additive factor of $j2\pi n$. In defining $\mathbf{P}$ any one of the values of $\ln s_i$ is satisfactory.

In order to obtain some additional information about the solution, let us assume that $\mathbf{Y}(t)$ can be written as the product of two functions. In particular, let

$$\mathbf{Y}(t) = \mathbf{Q}(t)\epsilon^{\mathbf{P}(t-t_0)}, \tag{51}$$

where $\mathbf{P}$ is given by (49) and $\mathbf{Q}(t)$ is a matrix to be determined. In fact, this expression can be inverted to solve for $\mathbf{Q}(t)$. Thus we have

$$\mathbf{Q}(t) = \mathbf{Y}(t)\epsilon^{-\mathbf{P}(t-t_0)}. \tag{52}$$

Now we can use the relationship in (47) to find out something about $\mathbf{Q}(t)$. Substitution of (51) into (47) yields

$$\mathbf{Q}(t + T)\epsilon^{\mathbf{P}(t+T-t_0)} = \mathbf{Q}(t)\epsilon^{\mathbf{P}(t-t_0)}\mathbf{M} = \mathbf{Q}(t)\epsilon^{\mathbf{P}(t-t_0)}\epsilon^{\mathbf{P}T}.$$

The last step follows from (48). Hence

$$\mathbf{Q}(t + T) = \mathbf{Q}(t); \tag{55}$$

that is, $\mathbf{Q}(t)$ is periodic with the same period T as $\mathscr{A}$.

It is now an established fact that $\mathbf{Y}(t)$ may be expressed as the product of a periodic matrix $\mathbf{Q}(t)$ and a matrix exponential $\epsilon^{\mathbf{P}(t-t_0)}$. Since $\mathbf{Y}(t)$ is known to be continuous, $\mathbf{Q}(t)$ is continuous and hence bounded* on the closed interval t_0 to $t_0 + T$. Then, because $\mathbf{Q}(t)$ is also periodic, it is uniformly bounded for all $t \geq t_0$. Consequently the behavior of $\mathbf{Y}(t)$ as t tends to infinity is governed by the behavior of the exponential $\epsilon^{\mathbf{P}(t-t_0)}$. After making note of the fact that the eigenvalues of $\mathbf{P}$ are known as the *characteristic exponents* of $\mathscr{A}$, we may make the following observations:

1. If all of the characteristic exponents have negative real part, then $\epsilon^{\mathbf{P}(t-t_0)}$ and hence $\mathbf{Y}(t)$ tend to zero as t tends to infinity.

2. If all the characteristic exponents have nonpositive real part, then $\epsilon^{\mathbf{P}(t-t_0)}$ and hence $\mathbf{Y}(t)$ are bounded as t tends to infinity.

3. If one or more of the characteristic exponents have positive real part, then $\epsilon^{\mathbf{P}(t-t_0)}$ and hence $\mathbf{Y}(t)$ are unbounded as t tends to infinity.

You should seek to verify these statements.

Thus the stability of the solution is tied up with the matrix $\mathbf{P}$. This, in turn, is related to $\mathbf{M}$ by (48) or (49), and $\mathbf{M}$ is related to the solution $\mathbf{Y}$ by (46). However, this sequence of relationships is not helpful. We want to know $\mathbf{P}$ in order to say something about the behavior of the solution, without actually finding $\mathbf{Y}(t)$. Unfortunately, there is no general procedure by which $\mathbf{P}$ may be determined without knowing $\mathbf{Y}$. The value of the theory of solutions for periodic $\mathscr{A}$ is therefore not as a computational tool but as a means of reaching other, more useful, theoretical results. We shall see this more clearly shortly.

To conclude, it should be observed that both $\epsilon^{\mathbf{P}(t-t_0)}$. and $\mathbf{Q}$ satisfy certain differential equations. Thus the exponential $\epsilon^{\mathbf{P}(t-t_0)}$ satisfies

$$\frac{d}{dt}\mathbf{X} = \mathbf{PX}, \tag{56}$$

with $\mathbf{X}(t_0) = \mathbf{U}$. As for $\mathbf{Q}$, substituting for $\mathbf{Y}(t)$ from (51) into (21) leads to

$$\frac{d}{dt}\mathbf{Q} = \mathscr{A}\mathbf{Q} - \mathbf{QP}, \tag{57}$$

with $\mathbf{Q}(t_0) = \mathbf{Y}(t_0)$.

---

* A vector or matrix is said to be bounded if each of its elements is bounded. A vector or matrix is unbounded if one or more of its elements are unbounded.

10.3 PROPERTIES OF THE STATE-EQUATION SOLUTION

We are now ready for a study of the properties of the solution of the state equation for a time-varying network. This will be done by comparison with some "reference" network about whose solution we have some knowledge.

THE GRONWALL LEMMA*

In this study we shall make strong use of the following result from mathematical analysis and will therefore discuss it first:

Lemma. If

$$\varphi(t) \leq \gamma + \int_{t_0}^{t} [\psi(\tau)\,\varphi(\tau) + \theta(\tau)]d\tau, \tag{58}$$

where φ, ψ, and θ are non-negative continuous functions of t for $t \geq t_0$ and where γ is a positive constant, then

$$\varphi(t) \leq \gamma \exp\left\{\int_{t_0}^{t} [\psi(\tau) + \theta(\tau)/\gamma]d\tau\right\}. \tag{59}$$

This result is called Gronwall's Lemma; its proof follows. From (58) we see that

$$\frac{\varphi(t)}{\gamma + \int_{t_0}^{t} [\psi(\tau)\,\varphi(\tau) + \theta(\tau)]d\tau} \leq 1. \tag{60}$$

After multiplying both sides by the non-negative function $\psi(t)$ and adding $\theta(t)/\left\{\gamma + \int_{t_0}^{t} [\psi(\tau)\,\varphi(\tau) + \theta(\tau)]d\tau\right\}$ to both sides, it is seen that

$$\frac{\psi(t)\,\varphi(t) + \theta(t)}{\gamma + \int_{t_0}^{t} [\psi(\tau)\,\varphi(\tau) + \theta(\tau)]d\tau} \leq \psi(t) + \frac{\theta(t)}{\gamma + \int_{t_0}^{t} [\psi(\tau)\,\varphi(\tau) + \theta(\tau)]d\tau}. \tag{61}$$

Note that $\psi(\tau)\,\varphi(\tau) + \theta(\tau) \geq 0$ and, therefore, that $\int_{t_0}^{t} [\psi(\tau)\,\varphi(\tau) + \theta(\tau)]d\tau \geq 0$. Thus $\theta(t)/\left\{\gamma + \int_{t_0}^{t} [\psi(\tau)\,\varphi(\tau) + \theta(\tau)]d\tau\right\} \leq \theta(t)/\gamma$. Using this inequality

---

* This lemma is also known as the Gronwall-Bellman lemma in the literature on mathematical analysis.

in (61), we get

$$\frac{\psi(t)\,\varphi(t) + \theta(t)}{\gamma + \int_{t_0}^{t} [\psi(\tau)\,\varphi(\tau) + \theta(\tau)]d\tau} \leq \psi(t) + \frac{\theta(t)}{\gamma}.$$

Integration of both sides from t_0 to t yields

$$\ln\left\{\gamma + \int_{t_0}^{t} [\psi(\tau)\,\varphi(\tau) + \theta(\tau)]d\tau\right\} - \ln\gamma \leq \int_{t_0}^{t}\left[\psi(\tau) + \frac{\theta(\tau)}{\gamma}\right]d\tau,$$

which is equivalent to

$$\gamma + \int_{t_0}^{t} [\psi(\tau)\,\varphi(\tau) + \theta(\tau)]d\tau \leq \gamma\,\exp\left\{\int_{t_0}^{t}\left[\psi(\tau) + \frac{\theta(\tau)}{\gamma}\right]d\tau\right\}. \qquad (62)$$

By combining (58) and (62), we get the indicated inequality (59). This completes the proof.

Now we turn to a consideration of the state equation, which is repeated here:

$$\frac{d}{dt}\,\mathbf{x}(t) = \mathscr{A}(t)\,\mathbf{x}(t) + \mathscr{B}(t)\,\mathbf{e}(t).$$

Suppose that the right-hand side is close, in some sense not yet defined, to the right-hand side of the *homogeneous reference equation*

$$\frac{d}{dt}\,\mathbf{x}(t) = \hat{\mathscr{A}}(t)\,\mathbf{x}(t).$$

Suppose also that all solutions of the reference equation either approach zero as t tends to infinity or are bounded. It would be useful to be able to infer as a consequence that all solutions of the original state equation either approach zero or are bounded. The next several theorems state precise conditions under which such inferences are valid. The conditions imposed by the theorems on the difference between $\mathscr{A}(t)$ and $\hat{\mathscr{A}}(t)$ and on $\mathscr{B}(t)\,\mathbf{e}(t)$ establish the sense in which we view the state equation as being close to the reference homogeneous equation.

ASYMPTOTIC PROPERTIES RELATIVE TO A TIME-INVARIANT REFERENCE

For networks that are close to being described by a time-invariant homogeneous equation, we have the following theorem:

Theorem 5.   *If all solutions of*

$$\frac{d}{dt}\mathbf{y} = \hat{\mathscr{A}}\mathbf{y}, \tag{63}$$

where $\hat{\mathscr{A}}$ is a constant matrix, are bounded as $t \to \infty$, *then all solutions of*

$$\frac{d}{dt}\mathbf{x} = \mathscr{A}(t)\,\mathbf{x} + \mathscr{B}(t)\,\mathbf{e}(t) \tag{64}$$

are bounded as $t \to \infty$, *provided*

$$\int_{t_0}^{\infty} \|\mathscr{A}(t) - \hat{\mathscr{A}}\| \, dt < \infty \tag{65a}$$

$$\int_{t_0}^{\infty} \|\mathscr{B}(t)\,\mathbf{e}(t)\| \, dt < \infty. \tag{65b}$$

The double bars denote the norm of the matrix or vector enclosed. Refer to Chapter 1 to refresh your memory on the properties of norms, since they will be used extensively in the rest of the chapter.

This theorem is quite valuable because it is fairly easy to establish whether or not all solutions of the reference equation 63 are bounded. Recall that $\epsilon^{\hat{\mathscr{A}}(t-t_0)}\mathbf{y}(t_0)$ for any $\mathbf{y}(t_0)$ is a solution of (63). It is then an immediate consequence of the constituent-matrix expansion of $\epsilon^{\hat{\mathscr{A}}(t-t_0)}$ that all solutions will be bounded if none of the eigenvalues of $\mathscr{A}$ has positive real part.

The proof of Theorem 5 is as follows. Rewrite the state equation as

$$\frac{d}{dt}\mathbf{x} = \hat{\mathscr{A}}\mathbf{x} + [\mathscr{A}(t) - \hat{\mathscr{A}}]\mathbf{x} + \mathscr{B}(t)\,\mathbf{e}(t).$$

Now you may easily show that an equivalent integral equation is

$$\mathbf{x}(t) = \mathbf{y}(t) + \int_{t_0}^{t} \epsilon^{\hat{\mathscr{A}}(t-\tau)} \{[\mathscr{A}(\tau) - \hat{\mathscr{A}}]\mathbf{x}(\tau) + \mathscr{B}(\tau)\,\mathbf{e}(\tau)\} d\tau,$$

where $\mathbf{y}(t)$ is the solution of the reference equation 63 with $\mathbf{y}(t_0) = \mathbf{x}(t_0)$, and $\epsilon^{\hat{\mathscr{A}}(t-\tau)}$ is the transition matrix associated with (63). Taking the norm on both sides of the integral equation, applying the triangle inequality,* and using the inequality for a matrix product brings us to the relation

$$\|\mathbf{x}(t)\| \leq \|\mathbf{y}(t)\| + \int_{t_0}^{t} \|\epsilon^{\hat{\mathscr{A}}(t-\tau)}\| \{\|\mathscr{A}(\tau) - \hat{\mathscr{A}}\| \|\mathbf{x}(\tau)\| + \|\mathscr{B}(\tau)\,\mathbf{e}(\tau)\|\}d\tau.$$

Since all solutions of (63) are bounded, $\|\mathbf{y}(t)\|$ and $\|\epsilon^{\hat{\mathscr{A}}(t-\tau)}\|$, with $t_0 \leq \tau \leq t$, must be bounded for $t \geq t_0$. Let $\gamma > 0$ and $\delta > 0$, respectively, be the corresponding bounds. Then

$$\|\mathbf{x}(t)\| \leq \gamma + \int_{t_0}^{t} \{\delta\|\mathscr{A}(\tau) - \hat{\mathscr{A}}\| \|\mathbf{x}(\tau)\| + \delta\|\mathscr{B}(\tau)\,\mathbf{e}(\tau)\|\}d\tau.$$

This expression is in the form of (58), and so Gronwall's lemma can be applied. Associate $\|\mathbf{x}(t)\|$ with $\varphi(t)$, $\delta\|\hat{\mathscr{A}}(t) - \mathscr{A}\|$ with $\psi(t)$, and $\delta\|\mathscr{B}(t)\,\mathbf{e}(t)\|$ with $\theta(t)$. Then, by the lemma,

$$\|\mathbf{x}(t)\| \leq \gamma \exp\left\{\int_{t_0}^{t} [\delta\|\mathscr{A}(\tau) - \hat{\mathscr{A}}\| + (\delta/\gamma)\|\mathscr{B}(\tau)\,\mathbf{e}(\tau)\|]d\tau\right\}$$

$$\leq \gamma \exp\left\{\int_{t_0}^{\infty} [\delta\|\mathscr{A}(\tau) - \hat{\mathscr{A}}\| + (\delta/\gamma)\|\mathscr{B}(\tau)\,\mathbf{e}(\tau)\|]d\tau\right\}.$$

The second inequality is obtained by letting t tend to infinity. The right-hand side of this relation is finite because of the conditions (65) of the theorem. Thus $\|\mathbf{x}(t)\|$ is bounded, and, by implication, $\mathbf{x}(t)$ is bounded as $t \to \infty$. The theorem is thus proved.

Observe that this theorem, like the others to follow, does not provide a constructive method by which the stability properties of the solution can be discovered. Instead, given a time-varying state equation as in (64), the theorem tells us that we must first verify that the norm of $\mathscr{B}\mathbf{e}$ is integrable over infinite time. Then we must look for a constant matrix $\hat{\mathscr{A}}$

---

* The term "triangle inequality," heretofore applied to the relation

$$\|\mathbf{x}_1 + \mathbf{x}_2\| \leq \|\mathbf{x}_1\| + \|\mathbf{x}_2\|,$$

will now also be applied to the integral relation $\|\int_a^b \mathbf{x}(\tau)\,d\tau\| \leq \int_a^b \|\mathbf{x}(\tau)\|\,d\tau$. We have not shown this to be a valid inequality; you should do so. It is a natural extension of the first inequality, if you think of the integral in terms of its defining sum under a limiting process.

with no eigenvalues in the right half-plane, and finally we must verify that the norm of $\mathscr{A}(t) - \hat{\mathscr{A}}$ is integrable. Then we can conclude that the solution of the original equation will remain bounded as $t \to \infty$.

The absence of eigenvalues of $\hat{\mathscr{A}}$ in the right half-plane still permits one or more on the $j\omega$-axis. However, if all eigenvalues of $\hat{\mathscr{A}}$ are in the open left half-plane, they will have negative real part. Then all of the solutions of the reference homogeneous equation 63 will approach zero. Since this is a more restrictive property than boundedness, it might possibly serve to relax the conditions (65). The next theorem shows how this is accomplished.

Theorem 6.   *If all solutions of*

$$\frac{d}{dt}\mathbf{y} = \hat{\mathscr{A}}\mathbf{y}, \tag{63}$$

where $\hat{\mathscr{A}}$ is a constant matrix, approach zero as $t \to \infty$, *then all solutions of*

$$\frac{d}{dt}\mathbf{x} = \mathscr{A}(t)\mathbf{x} + \mathscr{B}(t)\,\mathbf{e}(t) \tag{64}$$

are bounded as $t \to \infty$, *provided*

$$\int_{t_0}^{\infty} \|\mathscr{A}(t) - \hat{\mathscr{A}}\| dt < \infty \tag{66}$$

and $\mathscr{B}(t)\,\mathbf{e}(t)$ is bounded for $t \geq t_0$.

That is, if $\hat{\mathscr{A}}$ has no eigenvalues in the right-hand plane *and* on the $j\omega$-axis; then the norm of $\mathscr{B}\mathbf{e}$ need not be integrable over infinite time. It is only necessary for $\mathscr{B}\mathbf{e}$ to be bounded.

The proof is as follows. We start with the equation

$$\frac{d}{dt}\mathbf{z} = \hat{\mathscr{A}}\mathbf{z} + \mathscr{B}(t)\,\mathbf{e}(t).$$

The solution of this equation when $\mathbf{z}(t_0) = \mathbf{0}$ is

$$\mathbf{z}(t) = \int_{t_0}^{t} \epsilon^{\hat{\mathscr{A}}(t-\tau)}\mathscr{B}(\tau)\,\mathbf{e}(\tau)\,d\tau.$$

Taking the norm of both sides of this equation and using the usual norm inequalities, we obtain

$$\|\mathbf{z}(t)\| \leq \int_{t_0}^{t} \|\epsilon^{\hat{\mathscr{A}}(t-\tau)}\| \|\mathscr{B}(\tau)\,\mathbf{e}(\tau)\|\,d\tau.$$

By hypothesis, all solutions of (63) approach zero at $t \to \infty$; therefore all of the eigenvalues of $\hat{\mathscr{A}}$ have negative real part. Thus positive constants α and δ exist such that $\|\epsilon^{\hat{\mathscr{A}}(t-\tau)}\| \leq \delta\epsilon^{-\alpha(t-\tau)}$. You should verify this statement and indicate how an α and δ may be selected. Furthermore, $\mathscr{B}(t)\,\mathbf{e}(t)$ is bounded for $t \geq t_0$; therefore its norm is bounded. Let β be a bound for $\|\mathscr{B}(t)\,\mathbf{e}(t)\|$. Substituting these bounds in the above inequality leads to

$$\|\mathbf{z}(t)\| \leq \int_{t_0}^{t} \beta\delta\epsilon^{-\alpha(t-\tau)}d\tau = \frac{\beta\delta}{\alpha}[1 - \epsilon^{-\alpha(t-t_0)}].$$

This inequality shows that $\|\mathbf{z}(t)\|$ and hence $\mathbf{z}(t)$ are bounded for $t \geq t_0$.

Now set $\mathbf{w} = \mathbf{x} - \mathbf{z}$. Differentiation yields $d\mathbf{w}/dt = d\mathbf{x}/dt - d\mathbf{z}/dt$. Since the differential equations for $\mathbf{x}$ and $\mathbf{z}$ determine $d\mathbf{x}/dt$ and $d\mathbf{z}/dt$, we easily establish that $\mathbf{w}$ satisfies the differential equation

$$\frac{d}{dt}\mathbf{w} = \mathscr{A}(t)\mathbf{w} + [\mathscr{A}(t) - \hat{\mathscr{A}}]\mathbf{z}(t).$$

This equation is similar to the state equation treated in the previous theorem (Theorem 5), with $[\mathscr{A}(t) - \hat{\mathscr{A}}]\mathbf{z}(t)$ replacing $\mathscr{B}(t)\,\mathbf{e}(t)$. Thus if

$$\int_{t_0}^{\infty} \|[\mathscr{A}(t) - \hat{\mathscr{A}}]\mathbf{z}(t)\|\,dt < \infty,$$

the conclusion of the previous theorem applies, and $\mathbf{w}(t)$ is bounded as $t \to \infty$. The inequality is obviously true by (1) condition (66), (2) the boundedness of $\|\mathbf{z}(t)\|$ for $t \geq t_0$, and (3) the inequality

$$\int_{t_0}^{\infty} \|[\mathscr{A}(t) - \hat{\mathscr{A}}]\mathbf{z}(t)\|\,dt \leq \int_{t_0}^{\infty} \|\mathscr{A}(t) - \hat{\mathscr{A}}\|\,\|\mathbf{z}(t)\|\,dt.$$

We have now established that $\mathbf{z}(t)$ and $\mathbf{w}(t)$ are bounded as $t \to \infty$. Thus $\mathbf{x}(t) = \mathbf{w}(t) + \mathbf{z}(t)$ is bounded as $t \to \infty$, and the theorem is proved.

Things improve even more if there is no external excitation; that is, if $\mathcal{B}(t)\,\mathbf{e}(t) = 0$, then the conclusion of the last theorem can be strengthened.

Theorem 7. *If all solutions of*

$$\frac{d}{dt}\mathbf{y} = \hat{\mathcal{A}}\mathbf{y},$$

where $\hat{\mathcal{A}}$ is a constant matrix, approach zero as $t \to \infty$, then all solutions of

$$\frac{d}{dt}\mathbf{x} = \mathcal{A}(t)\mathbf{x}$$

approach zero as $t \to \infty$, provided $\|\mathcal{A}(t) - \hat{\mathcal{A}}\| < \alpha/\delta$ for $t \geq t_0$.

The constants α and δ have the same meanings as those given to them in the proof of the last theorem. The proof of this theorem is left to you.

As an illustration of the second of the previous three theorems, consider the filter network shown in Fig. 6. The state equation is easily shown to be

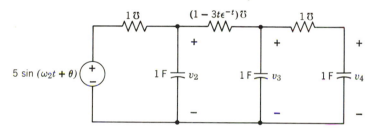

Fig. 6.   Time-varying network.

$$\frac{d}{dt}\begin{bmatrix} q_2 \\ q_3 \\ q_4 \end{bmatrix} = \begin{bmatrix} -2 + 3t\epsilon^{-t} & 1 - 3t\epsilon^{-t} & 0 \\ 1 - 3t\epsilon^{-t} & -2 + 3t\epsilon^{-t} & 1 \\ 0 & 1 & -1 \end{bmatrix}\begin{bmatrix} q_2 \\ q_3 \\ q_4 \end{bmatrix} + \begin{bmatrix} 1 \\ 0 \\ 0 \end{bmatrix}[5\,\sin\,(\omega_2 t + \theta)].$$

If $\|\mathcal{A}(t) - \hat{\mathcal{A}}\|$ is to be integrable from t_0 to ∞, then we must pick $\hat{\mathcal{A}}$ such that $\lim\,(\mathcal{A}(t) - \hat{\mathcal{A}}) = 0$ as $t \to \infty$. As a consequence, an examination of $\mathcal{A}(t)$ reveals that

$$\hat{\mathcal{A}} = \begin{bmatrix} -2 & 1 & 0 \\ 1 & -2 & 1 \\ 0 & 1 & -1 \end{bmatrix}.$$

Next the characteristic polynomial of $\widehat{\mathscr{A}}$ is determined. The resolvent matrix algorithm of Chapter 4 may be used to do this by strictly numerical methods. The characteristic polynomial is thus found to be

$$d(s) = s^3 + 5s^2 + 6s + 1.$$

Using the Routh criterion discussed in Chapter 9, we find, by strictly numerical methods, that all of the zeros of $d(s)$ have negative real part. Therefore all solutions of (63) approach zero as $t \to \infty$.

Now form $\mathscr{A}(t) - \widehat{\mathscr{A}}$:

$$\mathscr{A}(t) - \widehat{\mathscr{A}} = \begin{bmatrix} 3t\epsilon^{-t} & -3t\epsilon^{-t} & 0 \\ -3t\epsilon^{-t} & 3t\epsilon^{-t} & 0 \\ 0 & 0 & 0 \end{bmatrix},$$

then

$$\|\mathscr{A}(t) - \widehat{\mathscr{A}}\|_1 = 6t\epsilon^{-t}.$$

We have arbitrarily chosen to use the sum-magnitude vector norm and the associated matrix norm in this example. Now, since

$$\int_{t_0}^{\infty} \|\mathscr{A}(t) - \widehat{\mathscr{A}}\|_1 dt = \int_{t_0}^{\infty} 6t\epsilon^{-t} dt = 6(1 + t_0)\epsilon^{-t_0},$$

condition (66) of Theorem 6 is satisfied. Finally,

$$\mathscr{B}(t)\,\mathbf{e}(t) = \begin{bmatrix} 1 \\ 0 \\ 0 \end{bmatrix} [5 \sin (\omega_2 t + \theta)]$$

is bounded.

Thus all of the conditions of Theorem 6 are satisfied. Hence, by the theorem, $\mathbf{x}(t) = [q_2(t)\ q_3(t)\ q_4(t)]'$ is bounded as $t \to \infty$. Observe that we have established this fact without explicitly computing the solution to the state equation *or* (63).

ASYMPTOTIC PROPERTIES RELATIVE TO A PERIODIC REFERENCE

We now turn our attention to networks that are close to being described by a homogeneous equation with periodic coefficients. We shall call this a *periodic homogeneous reference equation.* The next theorem is on the boundedness of the state vector.

Theorem 8. *If all solutions of*

$$\frac{d}{dt}\mathbf{y} = \hat{\mathscr{A}}(t)\mathbf{y}, \tag{67}$$

where $\hat{\mathscr{A}}(t)$ is periodic, are bounded as t $\to \infty$, *then all solutions of*

$$\frac{d}{dt}\mathbf{x} = \mathscr{A}(t)\mathbf{x} + \mathscr{B}(t)\,\mathbf{e}(t)$$

are bounded as t $\to \infty$, *provided*

$$\int_{t_0}^{\infty} \|\mathscr{A}(t) - \hat{\mathscr{A}}(t)\|\,dt < \infty \tag{68a}$$

$$\int_{t_0}^{\infty} \|\mathscr{B}(t)\,\mathbf{e}(t)\|\,dt < \infty. \tag{68b}$$

The proof of this theorem rests on the fact that any solution of (67) may be expressed as $\mathbf{Q}(t)\epsilon^{\mathbf{P}(t-t_0)}\mathbf{y}(t_0)$ for some $\mathbf{y}(t_0)$, where $\mathbf{Q}(t)$ is nonsingular and periodic, and $\mathbf{P}$ is a constant matrix. To start, the state equation is first rewritten as

$$\frac{d}{dt}\mathbf{x} = \hat{\mathscr{A}}(t)\mathbf{x} + [\mathscr{A}(t) - \hat{\mathscr{A}}(t)]\mathbf{x} + \mathscr{B}(t)\,\mathbf{e}(t).$$

Let $\mathbf{y}(t)$ denote the solution of the reference equation 67, with $\mathbf{y}(t_0) = \mathbf{x}(t_0)$; then, using the transition matrix $\mathbf{Q}(t)\epsilon^{\mathbf{P}(t-\tau)}\mathbf{Q}(\tau)^{-1}$ associated with (67), the state equation may be put in the equivalent integral equation form

$$\mathbf{x}(t) = \mathbf{y}(t) + \int_{t_0}^{t} \mathbf{Q}(t)\,\epsilon^{\mathbf{P}(t-\tau)}\mathbf{Q}(\tau)^{-1}\{[\mathscr{A}(\tau) - \hat{\mathscr{A}}(\tau)]\mathbf{x}(\tau) + \mathscr{B}(\tau)\,\mathbf{e}(\tau)\}d\tau.$$

Taking the norm of both sides of this equation and applying the usual inequalities associated with norms, we get

$$\|\mathbf{x}(t)\| \leq \|\mathbf{y}(t)\| + \int_{t_0}^{t} \|\mathbf{Q}(t)\|\,\|\epsilon^{\mathbf{P}(t-\tau)}\|\,\|\mathbf{Q}(\tau)^{-1}\|$$

$$\times \{\|\mathscr{A}(\tau) - \hat{\mathscr{A}}(\tau)\|\,\|\mathbf{x}(\tau)\| + \|\mathscr{B}(\tau)\,\mathbf{e}(\tau)\|\}d\tau. \tag{69}$$

Since all solutions of the reference equation 67 are bounded, $\|\mathbf{y}(t)\|$ is bounded for $t \geq t_0$, and none of the characteristic exponents associated with (67) have positive real part. Thus $\|\epsilon^{\mathbf{P}(t-\tau)}\|$, with $t_0 \leq \tau \leq t$, is bounded for $t \geq t_0$. Since $\mathbf{Q}(t)$ is nonsingular and periodic, $\|\mathbf{Q}(t)\|$ and $\|\mathbf{Q}(\tau)^{-1}\|$, with $t_0 \leq \tau \leq t$, are bounded for $t \geq t_0$. Let γ and δ denote bounds on $\|\mathbf{y}(t)\|$ and $\|\mathbf{Q}(t)\| \, \|\epsilon^{\mathbf{P}(t-\tau)}\| \, \|\mathbf{Q}(\tau)^{-1}\|$, respectively. Using these bounds in (69), it is found that

$$\|\mathbf{x}(t)\| \leq \gamma + \int_{t_0}^{\infty} [\delta \|\mathscr{A}(\tau) - \widehat{\mathscr{A}}(\tau)\| \, \|\mathbf{x}(\tau)\| + \delta \|\mathscr{B}(\tau) \, \mathbf{e}(\tau)\|] d\tau.$$

This is a form on which Gronwall's lemma can be applied. If we also let t tend to infinity, the result becomes

$$\|\mathbf{x}(t)\| \leq \gamma \, \exp \left\{ \int_{t_0}^{\infty} [\delta \|\mathscr{A}(\tau) - \widehat{\mathscr{A}}(\tau)\| + (\delta/\gamma) \|\mathscr{B}(\tau) \, \mathbf{e}(\tau)\|] d\tau \right\}.$$

Then, by the conditions (68) of the theorem, $\|\mathbf{x}(t)\|$ is bounded and hence $\mathbf{x}(t)$ is bounded as $t \to \infty$. Thus the theorem is proved.

Again, the conditions of the theorem can be relaxed if $\widehat{\mathscr{A}}(t)$ is further restricted. Instead of asking only that the solutions of the periodic homogeneous reference equation 67 be asymptotically bounded, we ask that they approach zero as $t \to \infty$. Then conditions (68) can be relaxed. This is stated more precisely in the next theorem.

Theorem 9.   *If all solutions of*

$$\frac{d}{dt} \mathbf{y} = \widehat{\mathscr{A}}(t)\mathbf{y},$$

where $\widehat{\mathscr{A}}(t)$ is periodic, approach zero as $t \to \infty$, then all solutions of

$$\frac{d}{dt} \mathbf{x} = \mathscr{A}(t)\mathbf{x} + \mathscr{B}(t) \, \mathbf{e}(t)$$

are bounded as $t \to \infty$, provided

$$\int_{t_0}^{\infty} \|\mathscr{A}(t) - \widehat{\mathscr{A}}(t)\| \, dt < \infty \tag{70}$$

and $\mathscr{B}(t) \, \mathbf{e}(t)$ is bounded for $t \geq t_0$.

As a key item, the proof, the details of which are left for you, requires that all of the characteristic exponents have negative real part. This is implied by the assumption that all solutions of (67) approach zero as $t \to \infty$. Thus positive constants α and δ exist such that $\|\epsilon^{\mathbf{P}(t-\tau)}\| \le \delta \epsilon^{-\alpha(t-\tau)}$.

Finally, as in the previous sequence of theorems, when there is no excitation—$\mathscr{B}(t)\, e(t) = \mathbf{0}$—the conclusion can be greatly strengthened.

Theorem 10. *If all solutions of*

$$\frac{d}{dt}\,\mathbf{y} = \widehat{\mathscr{A}}(t)\mathbf{y},$$

where $\widehat{\mathscr{A}}(t)$ is periodic, approach zero as $t \to \infty$, then all solutions of

$$\frac{d}{dt}\,\mathbf{x} = \mathscr{A}(t)\mathbf{x}$$

approach zero as $t \to \infty$, provided $\|\mathscr{A}(t) - \widehat{\mathscr{A}}(t)\| < \beta$ for $t \ge t_0$, where β is positive constant that depends on $\widehat{\mathscr{A}}(t)$ and is established during the proof.

The proof, which is similar to that for Theorem 7, is left for you.

To illustrate the first of the last sequence of theorems, consider the network in Fig. 7. The state equation is

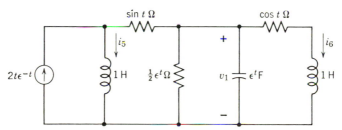

Fig. 7. Time-varying network with periodic homogeneous reference equation.

$$\frac{d}{dt}\begin{bmatrix} q_1 \\ \lambda_5 \\ \lambda_6 \end{bmatrix} = \begin{bmatrix} 2 & -1 & -1 \\ \epsilon^{-t} & -\sin t & 0 \\ \epsilon^{-t} & 0 & -\cos t \end{bmatrix}\begin{bmatrix} q_1 \\ \lambda_5 \\ \lambda_6 \end{bmatrix} + \begin{bmatrix} 1 \\ \sin t \\ 0 \end{bmatrix}[t\epsilon^{-t}].$$

The norm of $\|\mathscr{A}(t) - \widehat{\mathscr{A}}(t)\|$ will be integrable from t_0 to infinity only if $\lim_{t \to \infty}[\mathscr{A}(t) - \widehat{\mathscr{A}}(t)] = 0$. Thus an examination of $\mathscr{A}(t)$ reveals that

$$\widehat{\mathscr{A}}(t) = \begin{bmatrix} 2 & -1 & -1 \\ 0 & -\sin t & 0 \\ 0 & 0 & -\cos t \end{bmatrix}.$$

You may verify that the solution of the reference equation 67 is

$$\mathbf{y}(t) = \begin{bmatrix} \epsilon^{-2(t-t_0)} & -\int_{t_0}^{t} \epsilon^{-2(t-\tau)} \epsilon^{\cos \tau - \cos t_0} \, d\tau & -\int_{t_0}^{t} \epsilon^{-2(t-\tau)} \epsilon^{-\sin \tau - \sin t_0} \, d\tau \\ 0 & \epsilon^{\cos t - \cos t_0} & 0 \\ 0 & 0 & \epsilon^{-\sin t + \sin t_0} \end{bmatrix} \mathbf{y}(t_0).$$

It is seen that for all $\mathbf{y}(t_0)$ the solution is bounded as $t \to \infty$.

Next, examine

$$\mathscr{A}(t) - \hat{\mathscr{A}}(t) = \begin{bmatrix} 0 & 0 & 0 \\ \epsilon^{-t} & 0 & 0 \\ \epsilon^{-t} & 0 & 0 \end{bmatrix}$$

and

$$\mathscr{B}(t)\,\mathbf{e}(t) = \begin{bmatrix} 2t\epsilon^{-t} \\ 2t \sin t\epsilon^{-t} \\ 0 \end{bmatrix}.$$

For the norm of these expressions we obtain

$$\|\mathscr{A}(t) - \hat{\mathscr{A}}(t)\|_{\infty} = \epsilon^{-t}$$

and

$$\|\mathscr{B}(t)\,\mathbf{e}(t)\|_{\infty} = 2t\epsilon^{-t}.$$

Note that the max-magnitude vector norm is a more convenient norm for this problem than the sum-magnitude norm, since the sum-magnitude norm of $\mathscr{B}(t)\,\mathbf{e}(t)$ is a more cumbersome expression. In particular,

$$\|\mathscr{B}(t)\,\mathbf{e}(t)\|_{1} = 2t\epsilon^{-t} + 2t\,|\sin t|\,\epsilon^{-t}.$$

It is clear that conditions (68) are satisfied, since

$$\int_{t_0}^{\infty} \|\mathscr{A}(t) - \hat{\mathscr{A}}(t)\|_{\infty} \, dt = \epsilon^{-t_0} < \infty$$

and

$$\int_{t_0}^{\infty} \|\mathscr{B}(t)\,\mathbf{e}(t)\|_{\infty} \, dt = 2(1 + t_0)\epsilon^{-t_0} < \infty.$$

Thus the conditions of Theorem 8 are satisfied and the state vector $\mathbf{x} = [q_1 \quad \lambda_5 \quad \lambda_6]'$ is bounded as $t \to \infty$. Unlike in the previous example, it was necessary to solve the homogeneous equation $d\mathbf{y}/dt = \hat{\mathscr{A}}(t)\mathbf{y}$. Thus the fact that $\hat{\mathscr{A}}(t)$ is not constant requires more work in the application of the theorem than the corresponding case when $\hat{\mathscr{A}}(t)$ is constant.

ASYMPTOTIC PROPERTIES RELATIVE TO A GENERAL TIME-VARYING REFERENCE

We have now discussed a number of theorems that lay out conditions for the solution of a general state equation, (64), to possess asymptotic boundedness properties. This has been done with reference to the solution of a homogeneous reference equation of the form

$$\frac{d}{dt}\mathbf{y} = \hat{\mathscr{A}}(t)\mathbf{y}. \tag{71}$$

Two cases have been considered: (1) $\hat{\mathscr{A}}(t)$, a constant matrix and (2) $\hat{\mathscr{A}}(t)$, a periodic matrix. When $\hat{\mathscr{A}}(t)$ is not limited to these cases, but assuming a solution of (71) to have the appropriate asymptotic property, it may be tempting to suppose that the general solution of the state equation will have similar properties if similar conditions are satisfied by $\mathscr{A}(t)$ and $\mathscr{B}(t)\,\mathbf{e}(t)$. Such a supposition would be incorrect. Counter examples can be found to demonstrate this; one such counter example follows.

Suppose

$$\mathscr{A}(t) = \begin{bmatrix} -\nu & 0 \\ \epsilon^{-\nu t} & \sin t + t \cos t - 2\nu \end{bmatrix}$$

and

$$\mathscr{B}(t)\,\mathbf{e}(t) = \begin{bmatrix} 0 \\ 0 \end{bmatrix}.$$

Then set

$$\hat{\mathscr{A}}(t) = \begin{bmatrix} -\nu & 0 \\ 0 & \sin t + t \cos t - 2\nu \end{bmatrix}.$$

It is easily verified that the solution of the reference homogeneous equation 71 is

$$y_1(t) = \epsilon^{-\nu t} y_1(0)$$

$$y_2(t) = \epsilon^{t \sin t - 2\nu t} y_2(0).$$

It is clear that $\mathbf{y} = [y_1 \quad y_2]'$ is bounded for $2\nu \geq 1$ and, in fact, approaches zero for $2\nu > 1$. Observe that $\int_0^\infty \|\mathscr{A}(t) - \hat{\mathscr{A}}(t)\|_\infty \, dt = \int_0^\infty \epsilon^{-\nu t} \, dt = 1/\nu < \infty$ and $\int_0^\infty \|\mathscr{B}(t) \, \mathbf{e}(t)\|_\infty \, dt = 0 < \infty$. We arbitrarily chose to use the max-magnitude vector norm and the associated matrix norm in this example. The solution of the state equation is

$$x_1(t) = \epsilon^{-\nu t} x_1(0)$$

$$x_2(t) = \epsilon^{t \sin t - 2\nu t} \left[x_2(0) + x_1(0) \int_0^t \epsilon^{-\tau \sin \tau} \, d\tau \right].$$

You should verify this statement. Let us turn our attention to the integral in this last expression. Set $t_n = 2n\pi + \pi/2$; then, since $\epsilon^{-\tau \sin \tau}$ is positive for $0 \leq \tau \leq t_n$.

$$\int_0^{t_n} \epsilon^{-\tau \sin \tau} d\tau > \int_{t_n - 5\pi/4}^{t_n - 3\pi/4} \epsilon^{-\tau \sin \tau} d\tau = \int_{2n\pi - 3\pi/4}^{2n\pi - \pi/4} \epsilon^{-\tau \sin \tau} d\tau.$$

The least value of the integrand in the interval $2n\pi - 3\pi/4$ to $2n\pi - \pi/4$ occurs at $\tau = 2n\pi - 3\pi/4$ and is $\epsilon^{(2n\pi - 3\pi/4)/\sqrt{2}}$. Therefore

$$\int_0^{t_n} \epsilon^{-\tau \sin \tau} d\tau > \epsilon^{(2n\pi - 3\pi/4)/\sqrt{2}} \int_{2n\pi - 3\pi/4}^{2n\pi - \pi/4} d\tau$$

$$> \frac{\pi}{2} \epsilon^{(2n\pi - 3\pi/4)/\sqrt{2}}.$$

Since $2n\pi - 3\pi/4 = t_n - 5\pi/4$,

$$\int_0^{t_n} \epsilon^{-\tau \sin \tau} d\tau > \frac{\pi}{2} \epsilon^{-5\pi/4\sqrt{2}} \epsilon^{t_n/\sqrt{2}}$$

We now know that at $t = t_n$

$$x_2(t_n) > \epsilon^{t_n \sin t_n - 2\nu t_n} \left[x_2(0) + x_1(0) \frac{\pi}{2} \epsilon^{-5\pi/4\sqrt{2}} \epsilon^{t_n/\sqrt{2}} \right].$$

Since $\sin t_n = \sin(2n\pi + \pi/2) = 1$, this inequality may be written as

$$x_2(t_n) > \epsilon^{(1-2\nu)t_n}\left[x_2(0) + x_1(0)\frac{\pi}{2}\,\epsilon^{-5\pi/4\sqrt{2}}\,\epsilon^{t_n/\sqrt{2}}\right].$$

Obviously, as t_n tends to infinity, $x_2(t_n)$ becomes unbounded when $2\nu < 1 + 1/\sqrt{2}$ and $x_1(0) \neq 0$. This implies that $\mathbf{x} = [x_1 \quad x_2]'$ becomes unbounded as $t \to \infty$. Thus, for $1 \leq 2\nu < 1 + 1/\sqrt{2}$, all solutions of the reference equation 71 are bounded, and some solutions of the state equation are unbounded.

This example makes it clear that the simple modification we contemplated making in our previous theorems will not work. It will be of value, nevertheless, to determine what additional restrictions are needed to still achieve the desired result. The state equation can be rewritten as

$$\frac{d}{dt}\mathbf{x} = \mathscr{A}(t)\mathbf{x} + [\widehat{\mathscr{A}}(t) - \widehat{\mathscr{A}}(t)]\mathbf{x} + \mathscr{B}(t)\,\mathbf{e}(t).$$

Let $\mathbf{y}(t)$ denote the solution of (71), with $\mathbf{y}(t_0) = \mathbf{x}(t_0)$, and let $\mathbf{Y}(t)\mathbf{Y}(\tau)^{-1}$ be the transition matrix associated with (71). Then the state equation has the equivalent integral equation form

$$\mathbf{x}(t) = \mathbf{y}(t) + \int_{t_0}^{t_0}\mathbf{Y}(t)\,\mathbf{Y}(\tau)^{-1}\{[\mathscr{A}(\tau) - \widehat{\mathscr{A}}(\tau)]\mathbf{x}(\tau) + \mathscr{B}(\tau)\,\mathbf{e}(\tau)\}d\tau.$$

Taking the norm of both sides and applying the norm inequalities as previously, we get

$$\|\mathbf{x}(t)\| \leq \|\mathbf{y}(t)\| + \int_{t_0}^{t} \|\mathbf{Y}(t)\|\,\|\mathbf{Y}(\tau)^{-1}\|$$
$$\times \{\|\mathscr{A}(\tau) - \widehat{\mathscr{A}}(\tau)\|\,\|\mathbf{x}(\tau)\| + \|\mathscr{B}(\tau)\,\mathbf{e}(\tau)\|\}d\tau.$$

To apply Gronwall's lemma it is imperative that $\|\mathbf{y}(t)\|$, $\|\mathbf{Y}(t)\|$, and $\|\mathbf{Y}(\tau)^{-1}\|$, with $t_0 \leq \tau \leq t$, be bounded for $t \geq t_0$. If all solutions of (71) are bounded, $\|\mathbf{y}(t)\|$ and $\|\mathbf{Y}(t)\|$ are bounded; however, it cannot be inferred that $\|\mathbf{Y}(\tau)^{-1}\|$ is bounded.

Thus we need a condition that insures the boundedness of $\|\mathbf{Y}(\tau)^{-1}\|$. To this end, note that $\mathbf{Y}(t)^{-1} = \text{adj }[\mathbf{Y}(t)]/\det[\mathbf{Y}(t)]$ exists for all finite $t \geq t_0$ provided $\det[\mathbf{Y}(t)] \neq 0$ and is bounded if $\det[\mathbf{Y}(t)]$ is bounded away from zero. Now, recall from (41) that

$$\det[\mathbf{Y}(t)] = \exp\left\{\int_{t_0}^{t} \text{tr }[\widehat{\mathscr{A}}(\tau)]d\tau\right\}.$$

Therefore if we add the condition

$$\lim_{t \to \infty} \int_{t_0}^{t} \text{tr}\,[\hat{\mathscr{A}}(\tau)]d\tau > -\infty,$$

then $\det[\mathbf{Y}(t)]$ is bounded away from zero, and $\|\mathbf{Y}(t)^{-1}\|$ is bounded for $t \geq t_0$. Equivalently, $\|\mathbf{Y}(\tau)^{-1}\|$ with $t_0 \leq \tau \leq t$ is bounded for $t \geq t_0$.

With this added condition, the proof proceeds as with previous proofs, making use of Gronwall's lemma at the end. Having sketched the proof—the details are for you to fill in—we state the last theorem in this section.

Theorem 11. *If all solutions of*

$$\frac{d}{dt}\mathbf{y} = \hat{\mathscr{A}}(t)\mathbf{y}$$

are bounded as $t \to \infty$, *then all solutions of*

$$\frac{d}{dt}\mathbf{x} = \mathscr{A}(t)\mathbf{x} + \mathscr{B}(t)\,\mathbf{e}(t)$$

are bounded as $t \to \infty$, *provided*

$$\lim_{t \to \infty} \int_{t_0}^{t} \text{tr}\,[\hat{\mathscr{A}}(\tau)]d\tau > -\infty \tag{72a}$$

$$\int_{t_0}^{\infty} \|\mathscr{A}(t) - \hat{\mathscr{A}}(t)\|\,dt < \infty \tag{72b}$$

$$\int_{t_0}^{\infty} \|\mathscr{B}(t)\,\mathbf{e}(t)\|\,dt < \infty. \tag{72c}$$

The last two conditions are the same as before. The modification lies in condition (72a). This is the price we pay for not limiting $\hat{\mathscr{A}}(t)$ to a constant or periodic matrix. Condition (72a) is quite severe; so much so, in fact, that a network which is known to have an asymptotically bounded state vector by one of the previous theorems may fail to satisfy (72a). For example, suppose $\hat{\mathscr{A}}(t)$ is a constant matrix and the conditions of Theorem 5 are satisfied. Then, in addition, suppose all solutions of the reference homogeneous equation 63 approach zero as $t \to \infty$—certainly a more demanding condition than that of boundedness required by Theorem

5 and the above theorem. In this case condition (72a) cannot be fulfilled. You should verify this.

Now let us apply this theorem to an example. Consider the network in Fig. 8. All elements are time-varying. The state equation is found to be

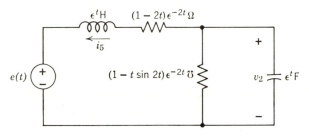

Fig. 8.   Time-varying network.

$$\frac{d}{dt}\begin{bmatrix} q_2 \\ \lambda_5 \end{bmatrix} = \begin{bmatrix} -(1 - t \sin 2t)\epsilon^{-3t} & -\epsilon^{-t} \\ \epsilon^{-t} & -(1 - 2t)\epsilon^{-3t} \end{bmatrix}\begin{bmatrix} q_2 \\ \lambda_5 \end{bmatrix} + \begin{bmatrix} 0 \\ -1 \end{bmatrix}[e(t)].$$

If condition (72b) is to be satisfied, then $\hat{\mathscr{A}}(t)$ must be selected such that $\lim_{t \to \infty} [\mathscr{A}(t) - \hat{\mathscr{A}}(t)] = 0$; one possible choice is

$$\hat{\mathscr{A}}(t) = \begin{bmatrix} -\epsilon^{-3t} & 0 \\ 0 & -\epsilon^{-3t} \end{bmatrix}.$$

Then the reference equation 71 has the solution

$$\mathbf{y}(t) = \begin{bmatrix} \exp\{\tfrac{1}{3}(\epsilon^{-3t} - \epsilon^{-3t_0})\} & 0 \\ 0 & \exp\{\tfrac{1}{3}(\epsilon^{-3t} - \epsilon^{-3t_0})\} \end{bmatrix}\mathbf{y}(t_0)$$

as you may verify. Note that the solution, for all $\mathbf{y}(t_0)$, is bounded as $t \to \infty$. Since $\operatorname{tr} \hat{\mathscr{A}}(t) = -2\epsilon^{-3t}$, we find $\int_{t_0}^{t} \operatorname{tr} \hat{\mathscr{A}}(t)\, dt = \tfrac{2}{3}(\epsilon^{-3t} - \epsilon^{-3t_0})$ and hence condition (72a) is satisfied.

Now examine the other conditions of (72). For $\mathscr{A}(t) - \hat{\mathscr{A}}(t)$ we find

$$\mathscr{A}(t) - \hat{\mathscr{A}}(t) = \begin{bmatrix} t \sin 2t\epsilon^{-3t} & -\epsilon^{-t} \\ \epsilon^{-t} & 2t\epsilon^{-3t} \end{bmatrix}$$

and as a consequence

$$\|\mathscr{A}(t) - \hat{\mathscr{A}}(t)\|_1 = \max\ \{\epsilon^{-t} + t\,|\sin 2t\,|\ \epsilon^{-3t},\ \epsilon^{-t} + 2t\epsilon^{-3t}\}$$

$$= \epsilon^{-t} + 2t\epsilon^{-3t}.$$

Thus condition (72b) is satisfied, since

$$\int_{t_0}^{\infty} \|\mathscr{A}(t) - \hat{\mathscr{A}}(t)\|_1\ dt = \epsilon^{-t_0} + \tfrac{2}{9}(1 + 3t_0)\epsilon^{-3t_0} < \infty.$$

By this theorem, we may now make the following statement: For all excitation voltages $e(t)$ for which condition (72c) is satisfied, the state $[q_2(t)\ \lambda_5(t)]'$ of the network of Fig. 8 is bounded as $t \to \infty$.

10.4 FORMULATION OF STATE EQUATION FOR NONLINEAR NETWORKS

Although solving linear, time-varying equations is more difficult than solving linear, time-invariant ones, nevertheless the basic condition of linearity permits the application to time-varying networks of many procedures that were originally developed for the time-invariant case. This is no longer true for nonlinear networks, to which we now turn our attention. We assume at the very outset that the networks may be time-varying, as well as nonlinear. For such networks, in fact, it is necessary to return to the fundamental laws in order to formulate the appropriate equations.

TOPOLOGICAL FORMULATION

The formulation of the state equations must combine the topological relationships expressed by Kirchhoff's equations for a normal tree (or forest)—which are valid independent of the state of linearity or time-variation—with the nonlinear and time-varying expressions that relate the variables describing the branch relationships. The steps will parallel those in Chapter 4 that were used in the derivation of the state equations for linear, time-invariant networks. The topological equations given in (113) of Chapter 4 are repeated here for convenience. You should look back there for a review of the notation.

$$\mathbf{i}_E = -\mathbf{Q}_{EC}\mathbf{i}_{Cl} - \mathbf{Q}_{ER}\mathbf{i}_{Rl} - \mathbf{Q}_{EL}\mathbf{i}_{Ll} - \mathbf{Q}_{EJ}\mathbf{i}_{J} \qquad (73a)$$

$$\mathbf{i}_{Ct} = -\mathbf{Q}_{CC}\mathbf{i}_{Cl} - \mathbf{Q}_{CR}\mathbf{i}_{Rl} - \mathbf{Q}_{CL}\mathbf{i}_{Ll} - \mathbf{Q}_{CJ}\mathbf{i}_{J} \qquad (73b)$$

$$\mathbf{i}_{Rt} = \qquad\quad -\mathbf{Q}_{RR}\mathbf{i}_{Rl} - \mathbf{Q}_{RL}\mathbf{i}_{Ll} - \mathbf{Q}_{RJ}\mathbf{i}_{J} \qquad (73c)$$

$$\mathbf{i}_{Lt} = \qquad\qquad\qquad\quad -\mathbf{Q}_{LL}\mathbf{i}_{Ll} - \mathbf{Q}_{LJ}\mathbf{i}_{J} \qquad (73d)$$

$$\mathbf{v}_{Cl} = \mathbf{Q}'_{EC}\mathbf{v}_{E} + \mathbf{Q}'_{CC}\mathbf{v}_{Ct} \qquad (73e)$$

$$\mathbf{v}_{Rl} = \mathbf{Q}'_{ER}\mathbf{v}_{E} + \mathbf{Q}'_{CR}\mathbf{v}_{Ct} + \mathbf{Q}'_{RR}\mathbf{v}_{Rt} \qquad (73f)$$

$$\mathbf{v}_{Ll} = \mathbf{Q}'_{EL}\mathbf{v}_{E} + \mathbf{Q}'_{CL}\mathbf{v}_{Ct} + \mathbf{Q}'_{RL}\mathbf{v}_{Rt} + \mathbf{Q}'_{LL}\mathbf{v}_{Lt} \qquad (73g)$$

$$\mathbf{v}_{J} = \mathbf{Q}'_{EJ}\mathbf{v}_{E} + \mathbf{Q}'_{CJ}\mathbf{v}_{Ct} + \mathbf{Q}'_{RJ}\mathbf{v}_{Rt} + \mathbf{Q}'_{LJ}\mathbf{v}_{Lt}. \qquad (73h)$$

The details of the formulation will be different depending on the specific variables which are selected as components of the state vector. It is possible for these to be voltage and current, or charge and flux linkage. We shall formulate the branch relationships so as to make linear combinations of capacitor charges and linear combinations of inductor flux linkages the elements of the state vector. We know that

$$\mathbf{i}_{Ct} = \frac{d}{dt}\,\mathbf{q}_{Ct} \qquad (74a)$$

$$\mathbf{i}_{Cl} = \frac{d}{dt}\,\mathbf{q}_{Cl}, \qquad (74b)$$

where the elements of $\mathbf{q}_{Ct}$ and $\mathbf{q}_{Cl}$ are the charges on the twig capacitors and link capacitors, respectively. We shall suppose that the charges are nonlinear functions of the capacitor voltages. Thus

$$\mathbf{q}_{Ct} = \mathbf{f}_{Ct}(\mathbf{v}_{Ct}) \qquad (75a)$$

$$\mathbf{q}_{Cl} = \mathbf{f}_{Cl}(\mathbf{v}_{Cl}), \qquad (75b)$$

where $\mathbf{f}_{Ct}$ and $\mathbf{f}_{Cl}$ are vector-valued functions of $\mathbf{v}_{Ct}$ and $\mathbf{v}_{Cl}$. In addition, they may be functions of time. After substituting (74a) and (74b) in (73b) and rearranging terms, we obtain

$$\frac{d}{dt}\,(\mathbf{q}_{Ct} + \mathbf{Q}_{CC}\mathbf{q}_{Cl}) = -\mathbf{Q}_{CR}\mathbf{i}_{Rl} - \mathbf{Q}_{CL}\mathbf{i}_{Ll} - \mathbf{Q}_{CJ}\mathbf{i}_{J}. \qquad (76)$$

Next, let us express the linear combination of charges $\mathbf{q}_{Ct} + \mathbf{Q}_{CC}\mathbf{q}_{Cl}$ in terms of the relations in (75). Thus

$$\mathbf{q}_{Ct} + \mathbf{Q}_{CC}\mathbf{q}_{Cl} = \mathbf{f}_{Ct}(\mathbf{v}_{Ct}) + \mathbf{Q}_{CC}\mathbf{f}_{Cl}(\mathbf{v}_{Cl}). \tag{77}$$

Substituting $\mathbf{v}_{Cl}$ from the Kirchhoff equation 73e leads to

$$\mathbf{q}_{Ct} + \mathbf{Q}_{CC}\mathbf{q}_{Cl} = \mathbf{f}_{Ct}(\mathbf{v}_{Ct}) + \mathbf{Q}_{CC}\mathbf{f}_{Cl}(\mathbf{Q}'_{EC}\mathbf{v}_E + \mathbf{Q}'_{CC}\mathbf{v}_{Ct}). \tag{78}$$

We *assume* that this equation possesses a unique solution for $\mathbf{v}_{Ct}$, which we write as

$$\mathbf{v}_{Ct} = \mathbf{g}_{Ct}(\mathbf{q}_{Ct} + \mathbf{Q}_{CC}\mathbf{q}_{Cl}, \mathbf{v}_E), \tag{79}$$

where $\mathbf{g}_{Ct}$ is a vector-valued function of its arguments $\mathbf{q}_{Ct} + \mathbf{Q}_{CC}\mathbf{q}_{Cl}$ and $\mathbf{v}_E$.

Before considering (76) and (79) further, we turn our attention to the branch relationships for the inductors. We know that

$$\frac{d}{dt}\boldsymbol{\lambda}_{Lt} = \mathbf{v}_{Lt} \tag{80a}$$

$$\frac{d}{dt}\boldsymbol{\lambda}_{Ll} = \mathbf{v}_{Ll}, \tag{80b}$$

where the elements of $\boldsymbol{\lambda}_{Lt}$ and $\boldsymbol{\lambda}_{Ll}$ are the flux linkages through the twig inductors and link inductors, respectively. We shall suppose that the flux linkages are nonlinear functions of the inductor currents. Thus

$$\boldsymbol{\lambda}_{Lt} = \mathbf{f}_{Lt}(\mathbf{i}_{Lt}, \mathbf{i}_{Ll}) \tag{81a}$$

$$\boldsymbol{\lambda}_{Ll} = \mathbf{f}_{Ll}(\mathbf{i}_{Lt}, \mathbf{i}_{Ll}), \tag{81b}$$

where $\mathbf{f}_{Lt}$ and $\mathbf{f}_{Ll}$ are vector valued functions of $\mathbf{i}_{Lt}$ and $\mathbf{i}_{Ll}$. Also they may be functions of time. Next we substitute (80a) and (80b) into the topological expression (73g) and rearrange terms to establish the following:

$$\frac{d}{dt}(\boldsymbol{\lambda}_{Ll} - \mathbf{Q}'_{LL}\boldsymbol{\lambda}_{Lt}) = \mathbf{Q}'_{CL}\mathbf{v}_{Ct} + \mathbf{Q}'_{RL}\mathbf{v}_{Rt} + \mathbf{Q}'_{EL}\mathbf{v}_L. \tag{82}$$

The linear combination of flux linkages $\boldsymbol{\lambda}_{Ll} - \mathbf{Q}'_{LL}\boldsymbol{\lambda}_{Lt}$ may be expressed in terms of the relations in (81) as follows

$$\boldsymbol{\lambda}_{Ll} - \mathbf{Q}'_{LL}\boldsymbol{\lambda}_{Lt} = \mathbf{f}_{Ll}(\mathbf{i}_{Lt}, \mathbf{i}_{Ll}) - \mathbf{Q}'_{LL}\mathbf{f}_{Lt}(\mathbf{i}_{Lt}, \mathbf{i}_{Ll}). \tag{83}$$

Substituting $\mathbf{i}_{Lt}$ from the Kirchhoff equation 73d establishes

$$\boldsymbol{\lambda}_{Ll} - \mathbf{Q}'_{LL}\boldsymbol{\lambda}_{Lt} = \mathbf{f}_{Ll}(-\mathbf{Q}_{LL}\mathbf{i}_{Ll} - \mathbf{Q}_{LJ}\mathbf{i}_J, \mathbf{i}_{Ll})$$
$$- \mathbf{Q}'_{LL}\mathbf{f}_{Lt}(-\mathbf{Q}_{LL}\mathbf{i}_{Ll} - \mathbf{Q}_{LJ}\mathbf{i}_J, \mathbf{i}_{Ll}). \quad (84)$$

We shall *assume* this equation possesses a unique solution for $\mathbf{i}_{Ll}$ in terms of $\boldsymbol{\lambda}_{Ll} - \mathbf{Q}'_{LL}\boldsymbol{\lambda}_{Lt}$ and $\mathbf{i}_J$, which we may express as

$$\mathbf{i}_{Ll} = \mathbf{g}_{Ll}(\boldsymbol{\lambda}_{Ll} - \mathbf{Q}'_{LL}\boldsymbol{\lambda}_{Lt}, \mathbf{i}_J), \quad (85)$$

where $\mathbf{g}_{Ll}$ is a vector-valued function of its arguments.

Let us now substitute (85) into (76) and (79) into (82). Thus

$$\frac{d}{dt}(\mathbf{q}_{Ct} + \mathbf{Q}_{CC}\mathbf{q}_{Cl}) = -\mathbf{Q}_{CR}\mathbf{i}_{Rl} - \mathbf{Q}_{CL}\mathbf{g}_{Ll}(\boldsymbol{\lambda}_{Ll} - \mathbf{Q}'_{LL}\boldsymbol{\lambda}_{Lt}, \mathbf{i}_J) - \mathbf{Q}_{CJ}\mathbf{i}_J$$
$$(86a)$$

$$\frac{d}{dt}(\boldsymbol{\lambda}_{Ll} - \mathbf{Q}'_{LL}\boldsymbol{\lambda}_{Lt}) = \mathbf{Q}'_{CL}\mathbf{g}_{Ct}(\mathbf{q}_{Ct} + \mathbf{Q}_{CC}\mathbf{q}_{Cl}, \mathbf{v}_E) + \mathbf{Q}'_{RL}\mathbf{v}_{Rt} + \mathbf{Q}'_{EL}\mathbf{v}_E.$$
$$(86b)$$

Were it not for the presence of the resistor branch variables $\mathbf{i}_{Rl}$ and $\mathbf{v}_{Rt}$, we would have the desired pair of first-order vector differential equations for $\mathbf{q}_{Ct} + \mathbf{Q}_{CC}\mathbf{q}_{Cl}$ and $\boldsymbol{\lambda}_{Ll} - \mathbf{Q}'_{LL}\boldsymbol{\lambda}_{Lt}$. Therefore we turn our attention next to the branch relations for the resistors.

Suppose the relationship between the resistor voltages and resistor currents may be expressed by the implicit vector equation

$$\mathbf{f}_R(\mathbf{i}_{Rt}, \mathbf{v}_{Rt}, \mathbf{i}_{Rl}, \mathbf{v}_{Rl}) = \mathbf{0}. \quad (87)$$

This is the counterpart of (125) in Chapter 4. Substituting the Kirchhoff equations 73c and 73f into this implicit equation leads to

$$\mathbf{f}_R(-\mathbf{Q}_{RR}\mathbf{i}_{Rl} - \mathbf{Q}_{RL}\mathbf{i}_{Ll} - \mathbf{Q}_{RJ}\mathbf{i}_J, \mathbf{v}_{Rt}, \mathbf{i}_{Rl},$$
$$\mathbf{Q}'_{ER}\mathbf{v}_E + \mathbf{Q}'_{CR}\mathbf{v}_{Ct} + \mathbf{Q}'_{RR}\mathbf{v}_{Rt}) = \mathbf{0}. \quad (88)$$

Next, substituting (79) and (85) into this expression, we obtain

$$\mathbf{f}_R(-\mathbf{Q}_{RR}\mathbf{i}_{Rl} - \mathbf{Q}_{RL}\mathbf{g}_{Ll}(\boldsymbol{\lambda}_{Ll} - \mathbf{Q}'_{LL}\boldsymbol{\lambda}_{Lt}, \mathbf{i}_J) - \mathbf{Q}_{RJ}\mathbf{i}_J, \mathbf{v}_{Rt}, \mathbf{i}_{Rl},$$
$$\mathbf{Q}'_{ER}\mathbf{v}_E + \mathbf{Q}'_{CR}\mathbf{g}_{Ct}(\mathbf{q}_{Ct} + \mathbf{Q}_{CC}\mathbf{q}_{Cl}, \mathbf{v}_E) + \mathbf{Q}'_{RR}\mathbf{v}_{Rt}) = \mathbf{0}. \quad (89)$$

This is the counterpart of (126) in Chapter 4. Recall that there we simply assumed that certain matrices were nonsingular so that those equations could be inverted. Similarly, we *assume* that (89) may be solved for $\mathbf{v}_{Rt}$ and $\mathbf{i}_{Rl}$ in terms of $\mathbf{q}_{Ct} + \mathbf{Q}_{CC}\mathbf{q}_{Cl}, \boldsymbol{\lambda}_{Ll} - \mathbf{Q}'_{LL}\boldsymbol{\lambda}_{Lt}, \mathbf{v}_E$, and $\mathbf{i}_J$. The solution will be expressed as

$$\mathbf{v}_{Rt} = \mathbf{g}_{Rt}(\mathbf{q}_{Ct} + \mathbf{Q}_{CC}\mathbf{q}_{Cl}, \boldsymbol{\lambda}_{Ll} - \mathbf{Q}'_{LL}\boldsymbol{\lambda}_{Lt}, \mathbf{v}_E, \mathbf{i}_J) \tag{90a}$$

$$\mathbf{i}_{Rl} = \mathbf{g}_{Rl}(\mathbf{q}_{Ct} + \mathbf{Q}_{CC}\mathbf{q}_{Cl}, \boldsymbol{\lambda}_{Ll} - \mathbf{Q}'_{LL}\boldsymbol{\lambda}_{Lt}, \mathbf{v}_E, \mathbf{i}_J). \tag{90b}$$

Substitution of these two expressions for $\mathbf{v}_{Rt}$ and $\mathbf{i}_{Rl}$ into (86) gives the desired differential equations

$$\frac{d}{dt}(\mathbf{q}_{Ct} + \mathbf{Q}_{CC}\mathbf{q}_{Cl}) = -\mathbf{Q}_{CR}\mathbf{g}_{Ri}(\mathbf{q}_{Ct} + \mathbf{Q}_{CC}\mathbf{q}_{Cl}, \boldsymbol{\lambda}_{Ll} - \mathbf{Q}'_{LL}\boldsymbol{\lambda}_{Lt}, \mathbf{v}_E, \mathbf{i}_J)$$
$$-\mathbf{Q}_{CL}\mathbf{g}_{Ll}(\boldsymbol{\lambda}_{Ll} - \mathbf{Q}'_{LL}\boldsymbol{\lambda}_{Lt}, \mathbf{i}_J) - \mathbf{Q}_{CJ}\mathbf{i}_J \tag{91a}$$

$$\frac{d}{dt}(\boldsymbol{\lambda}_{Ll} - \mathbf{Q}'_{LL}\boldsymbol{\lambda}_{Lt}) = \mathbf{Q}'_{RL}\mathbf{g}_{Rt}(\mathbf{q}_{Ct} + \mathbf{Q}_{CC}\mathbf{q}_{Cl}, \boldsymbol{\lambda}_{Ll} - \mathbf{Q}'_{LL}\boldsymbol{\lambda}_{Lt}, \mathbf{v}_E, \mathbf{i}_J)$$
$$+\mathbf{Q}'_{CL}\mathbf{g}_{Ct}(\mathbf{q}_{Ct} + \mathbf{Q}_{CC}\mathbf{q}_{Cl}, \mathbf{v}_E) + \mathbf{Q}'_{EL}\mathbf{v}_E. \tag{91b}$$

We take the elements of the state vector to be linear combinations of capacitor charges and inductor flux linkages. Thus,

$$\mathbf{x} = \begin{bmatrix} \mathbf{q}_{Ct} + \mathbf{Q}_{CC}\mathbf{q}_{Cl} \\ \boldsymbol{\lambda}_{Ll} - \mathbf{Q}'_{LL}\boldsymbol{\lambda}_{Lt} \end{bmatrix}. \tag{92}$$

The excitation vector is defined as

$$\mathbf{e} = \begin{bmatrix} \mathbf{v}_E \\ \mathbf{i}_J \end{bmatrix}, \tag{93}$$

as before. Then the two differential equations in (91) may be combined and expressed as the one vector differential equation

$$\frac{d}{dt}\mathbf{x} = \mathbf{h}(\mathbf{x}, \mathbf{e}), \tag{94}$$

where $\mathbf{h}$ is a vector-valued function of $\mathbf{x}$ and $\mathbf{e}$, which is determined by the right-hand sides of the differential equations (91) and by the defining relations (92) and (93) for **x** and **e**. You should determine the actual expression for **h**. Furthermore, **h** may also be an explicit function of time; when it is necessary to make this fact evident, we shall write (94) as

$$\frac{d}{dt}\mathbf{x} = \mathbf{h}(\mathbf{x}, \mathbf{e}, t). \tag{95}$$

Since **x** is a *state vector* for the network, either of these last two differential equations for **x** will be called the *state equation*.

We have shown a method for developing the state equation for a non-linear network in which the branch-variable relationships can be expressed in the form specified in (74), (80), and (87) and for which the algebraic equations (78), (84), and (89) have solutions as assumed. For any specific problem, you should not take the final expression and substitute the appropriate numerical values. Rather you should repeat the steps in the procedure for the particular problem at hand. The results given are general; in a specific case linearity of some components may permit simplifications that come to light only when the detailed steps of the formulation procedure are followed.

Let us now apply this formulation procedure to the network shown in Fig. 9a. With the usual low-frequency model for the triode, the network may be redrawn as in Fig 9b. There is a nonlinear inductor and a non-linear controlled source. A network graph and normal tree are shown in Fig. 9c. There are no degeneracies, so the fundamental cut-set matrix is easily determined and used to express the Kirchhoff equations as in (73). Thus

$$[i_1] = \begin{bmatrix} -1 & 0 & -1 & 0 & 0 \end{bmatrix} \begin{bmatrix} i_6 \\ i_7 \\ i_8 \\ i_9 \\ i_{10} \end{bmatrix} + [0][i_{11}]$$

$$\begin{bmatrix} i_2 \\ i_3 \\ i_4 \end{bmatrix} = \begin{bmatrix} 1 & 0 & 1 & 0 & 0 \\ 1 & -1 & 0 & -1 & 0 \\ 0 & 0 & 0 & 1 & -1 \end{bmatrix} \begin{bmatrix} i_6 \\ i_7 \\ i_8 \\ i_9 \\ i_{10} \end{bmatrix} + \begin{bmatrix} 0 \\ 0 \\ -1 \end{bmatrix} [i_{11}]$$

$$[i_5] = [0 \quad 0 \quad 0 \quad -1 \quad 0] \begin{bmatrix} i_6 \\ i_7 \\ i_8 \\ i_9 \\ i_{10} \end{bmatrix} + [0][i_{11}]$$

$$\begin{bmatrix} v_6 \\ v_7 \\ v_8 \\ v_9 \\ v_{10} \end{bmatrix} = \begin{bmatrix} 1 \\ 0 \\ 1 \\ 0 \\ 0 \end{bmatrix} [v_1] + \begin{bmatrix} -1 & -1 & 0 \\ 0 & 1 & 0 \\ -1 & 0 & 0 \\ 0 & 1 & -1 \\ 0 & 0 & 1 \end{bmatrix} \begin{bmatrix} v_2 \\ v_3 \\ v_4 \end{bmatrix} + \begin{bmatrix} 0 \\ 0 \\ 0 \\ 1 \\ 0 \end{bmatrix} [v_5]$$

$$[v_{11}] = [0][v_1] + [0 \quad 0 \quad 1] \begin{bmatrix} v_2 \\ v_3 \\ v_4 \end{bmatrix} + [0][v_5].$$

For the branch relations we have, corresponding to (75),

$$\begin{bmatrix} q_2 \\ q_3 \\ q_4 \end{bmatrix} = \begin{bmatrix} 1 & 0 & 0 \\ 0 & 1 & 0 \\ 0 & 0 & \frac{1}{10} \end{bmatrix} \begin{bmatrix} v_2 \\ v_3 \\ v_4 \end{bmatrix}.$$

In this case the functional relation is linear. Of course, this is easy to solve for $[v_2 \quad v_3 \quad v_4]'$ in terms of $[q_2 \quad q_3 \quad q_4]'$; we get

$$\begin{bmatrix} v_2 \\ v_3 \\ v_4 \end{bmatrix} = \begin{bmatrix} 1 & 0 & 0 \\ 0 & 1 & 0 \\ 0 & 0 & 10 \end{bmatrix} \begin{bmatrix} q_2 \\ q_3 \\ q_4 \end{bmatrix}$$

as the particular case of (79). Similarly, corresponding to (81), we have

$$[\lambda_{11}] = [\tfrac{1}{10} \tanh i_{11}],$$

from which

$$[i_{11}] = [\tanh^{-1} 10\lambda_{11}].$$

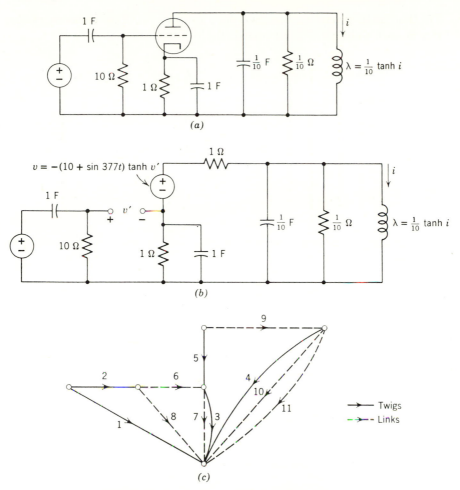

Fig. 9.   Nonlinear, time-varying amplifier circuit.

which is the particular case of (85).

Corresponding to (87) for the resistive branches, we have

$$
\begin{bmatrix}
v_5 - (10 + \sin 377t)\tanh v_6 \\[4pt]
i_6 \\[4pt]
i_7 - v_7 \\[4pt]
i_8 - \tfrac{1}{10} v_8 \\[4pt]
i_9 - v_9 \\[4pt]
i_{10} - 10 v_{10}
\end{bmatrix}
=
\begin{bmatrix}
0 \\[4pt]
0 \\[4pt]
0 \\[4pt]
0 \\[4pt]
0 \\[4pt]
0
\end{bmatrix}.
$$

The Kirchhoff equation that gives $[v_6 \quad v_7 \quad v_8 \quad v_9 \quad v_{10}]'$ in terms of the tree branch voltages is substituted; the result is

$$
\begin{bmatrix} 0 \\ 0 \\ 0 \\ 0 \\ 0 \\ 0 \end{bmatrix} = \begin{bmatrix} v_5 \\ i_6 \\ i_7 \\ i_8 \\ i_9 \\ i_{10} \end{bmatrix} - \begin{bmatrix} -(10 + \sin 377t) \tanh (v_1 - v_2 - v_3) \\ 0 \\ v_3 \\ \frac{1}{10}(v_1 - v_2) \\ v_3 - v_4 + v_5 \\ 10v_4 \end{bmatrix}.
$$

Upon substituting the previous expression for $[v_2 \quad v_3 \quad v_4]'$, the equation becomes

$$
\begin{bmatrix} 0 \\ 0 \\ 0 \\ 0 \\ 0 \\ 0 \end{bmatrix} = \begin{bmatrix} v_5 \\ i_6 \\ i_7 \\ i_8 \\ i_9 \\ i_{10} \end{bmatrix} - \begin{bmatrix} -(10 + \sin 377t) \tanh (v_1 - q_2 - q_3) \\ 0 \\ q_3 \\ \frac{1}{10} v_1 - \frac{1}{10} q_2 \\ q_3 - 10q_4 + v_5 \\ 100q_4 \end{bmatrix}.
$$

The solution for $[v_5]$ and $[i_6 \quad i_7 \quad i_8 \quad i_9 \quad i_{10}]'$ is obviously

$$
[v_5] = [-(10 + \sin 377t) \tanh (v_1 - q_2 - q_3)]
$$

$$
\begin{bmatrix} i_6 \\ i_7 \\ i_8 \\ i_9 \\ i_{10} \end{bmatrix} = \begin{bmatrix} 0 \\ q_3 \\ \frac{1}{10} v_1 - \frac{1}{10} q_2 \\ q_3 - 10q_4 - (10 + \sin 377t) \tanh (v_1 - q_2 - q_3) \\ 100q_4 \end{bmatrix}.
$$

These equations are the particular cases of (90).

The dynamical relation associated with the capacitors and corresponding to (74) is

$$
\frac{d}{dt}\begin{bmatrix} q_2 \\ q_3 \\ q_4 \end{bmatrix} = \begin{bmatrix} i_2 \\ i_3 \\ i_4 \end{bmatrix} = \begin{bmatrix} 1 & 0 & 1 & 0 & 0 \\ 1 & -1 & 0 & -1 & 0 \\ 0 & 0 & 0 & 1 & -1 \end{bmatrix} \begin{bmatrix} i_6 \\ i_7 \\ i_8 \\ i_9 \\ i_{10} \end{bmatrix} + \begin{bmatrix} 0 \\ 0 \\ -1 \end{bmatrix} [i_{11}].
$$

The second equality results from using the Kirchhoff equation giving $[i_2 \quad i_3 \quad i_4]'$ in terms of the link currents. Next, we substitute the previously established expressions for $[i_6 \quad i_7 \quad i_8 \quad i_9 \quad i_{10}]'$ and $[i_{11}]$ to obtain

$$
\frac{d}{dt}\begin{bmatrix} q_2 \\ q_3 \\ q_4 \end{bmatrix} = \begin{bmatrix} \tfrac{1}{10} v_1 - \tfrac{1}{10} q_2 \\ -2q_3 + 10q_4 + (10 + \sin 377t)\tanh{(v_1 - q_2 - q_3)} \\ q_3 - 110q_4 - (10 + \sin 377t)\tanh{(v_1 - q_2 - q_3)} \end{bmatrix}
$$

$$
+ \begin{bmatrix} 0 \\ 0 \\ -\tanh^{-1} 10\lambda_{11} \end{bmatrix},
$$

which is the counterpart of (91a) in this example.

Now turn to the inductor. The particular case of the dynamical inductor relation (80) is

$$
\frac{d}{dt}[\lambda_{11}] = [v_{11}] = [0][v_1] + [0 \quad 0 \quad 1]\begin{bmatrix} v_2 \\ v_3 \\ v_4 \end{bmatrix} + [0][v_5].
$$

The second equality reflects the Kirchhoff equation for $[v_{11}]$. Substitution of the previously derived expression for $[v_2 \quad v_3 \quad v_4]'$ yields

$$
\frac{d}{dt}[\lambda_{11}], = [10q_4]
$$

which is the particular case of (91b).

After combining these two first-order differential equations, we obtain the state equation

$$
\frac{d}{dt}
\begin{bmatrix}
q_2 \\
q_3 \\
q_4 \\
\lambda_{11}
\end{bmatrix}
=
\begin{bmatrix}
\frac{1}{10} v_1 - \frac{1}{10} q_2 \\
-2q_3 + 10q_4 + (10 + \sin 377t) \tanh (v_1 - q_2 - q_3) \\
q_3 - 110q_4 - (10 + \sin 377t) \tanh (v_1 - q_2 - q_3) - \tanh^{-1} 10\lambda_{11} \\
10q_4
\end{bmatrix}.
$$

With $\mathbf{x} = [q_2 \quad q_3 \quad q_4 \quad \lambda_{11}]'$ and $\mathbf{e} = [v_1]$, the right-hand side is the vector-valued function $\mathbf{h}(\mathbf{x}, \mathbf{e}, t)$ of the state equation 95. It may not be evident that this right-hand side is indeed a function of $\mathbf{x}$ and $\mathbf{e}$, since only dependence on the elements of $\mathbf{x}$ and $\mathbf{e}$ is evident. Therefore we shall show that the second row on the right-hand side, as a typical row, is a function of $\mathbf{x}$ and $\mathbf{e}$. It is obvious that

$$
[-2q_3 + 10q_4 + (10 + \sin 377t) \tanh(v_1 - q_2 - q_3)]
$$

$$
= [0 \; -2 \quad 10 \quad 0]
\begin{bmatrix}
q_2 \\
q_3 \\
q_4 \\
\lambda_{11}
\end{bmatrix}
$$

$$
+ (10 + \sin 377t) \tanh \left\{ [v_1] + [-1 \quad -1 \quad 0 \quad 0]
\begin{bmatrix}
q_2 \\
q_3 \\
q_4 \\
\lambda_{11}
\end{bmatrix} \right\}.
$$

Thus the second row is a function of $\mathbf{x}$ and $\mathbf{e}$; equivalently, we have found $h_2(\mathbf{x}, \mathbf{e}, t)$, the second element of $\mathbf{h}(\mathbf{x}, \mathbf{e}, t)$.

OUTPUT EQUATION

We turn now to the problem of specifying the output variables in terms of the state variables, the excitation, and the derivative of the excitation. Let us start with the inductors. The branch variables of primary interest are the branch fluxes, elements of $\boldsymbol{\lambda}_{Ll}$ and $\boldsymbol{\lambda}_{Lt}$; and the branch currents, elements of $\mathbf{i}_{Ll}$ and $\mathbf{i}_{Lt}$. By (85), $\mathbf{i}_{Ll}$ is given in terms of $\boldsymbol{\lambda}_{Ll} - \mathbf{Q}'_{LL}\boldsymbol{\lambda}_{Lt}$ (part of $\mathbf{x}$) and $\mathbf{i}_J$ (part of $\mathbf{e}$). Thus

$$\mathbf{i}_{Ll} = \mathbf{g}_{Ll}(\boldsymbol{\lambda}_{Ll} - \mathbf{Q}'_{LL}\boldsymbol{\lambda}_{Lt}, \mathbf{i}_J).$$

The Kirchhoff equation 73d, combined with this expression for $\mathbf{i}_{Ll}$, yields

$$\mathbf{i}_{Lt} = -\mathbf{Q}_{LL}\mathbf{g}_{Ll}(\boldsymbol{\lambda}_{Ll} - \mathbf{Q}'_{LL}\boldsymbol{\lambda}_{Lt}), \mathbf{i}_J) - \mathbf{Q}_{LJ}\mathbf{i}_J.$$

Substitution of these two expressions for $\mathbf{i}_{Ll}$ and $\mathbf{i}_{Lt}$ into the branch equations

$$\boldsymbol{\lambda}_{Ll} = \mathbf{f}_{Ll}(\mathbf{i}_{Lt}, \mathbf{i}_{Ll})$$

$$\boldsymbol{\lambda}_{Lt} = \mathbf{f}_{Lt}(\mathbf{i}_{Lt}, \mathbf{i}_{Ll})$$

establishes $\boldsymbol{\lambda}_{Ll}$ and $\boldsymbol{\lambda}_{Lt}$ as functions of $\boldsymbol{\lambda}_{Ll} - \mathbf{Q}'_{LL}\boldsymbol{\lambda}_{Lt}$ and $\mathbf{i}_J$.

The branch variables of secondary interest are the branch voltages, elements of $\mathbf{v}_{Ll}$ and $\mathbf{v}_{Lt}$. It is left to you as an exercise to show that $\mathbf{f}_{Lt}(\mathbf{i}_{Lt}, \mathbf{i}_{Ll})$ must be identically zero—no inductive twigs—or that $\mathbf{f}_{Lt}(\mathbf{i}_{Lt}, \mathbf{i}_{Ll})$ and $\mathbf{g}_{Ll}(\boldsymbol{\lambda}_{Ll} - \mathbf{Q}'_{LL}\boldsymbol{\lambda}_{Lt}, \mathbf{i}_J)$ must be differentiable functions in order to establish $\mathbf{v}_{Ll}$ and $\mathbf{v}_{Lt}$ as functions of $\mathbf{x}$, $\mathbf{e}$, and $d\mathbf{e}/dt$.

Next, consider the capacitors, The variables of primary interest are the branch charges, elements of $\mathbf{q}_{Ct}$ and $\mathbf{q}_{Cl}$; and voltages, elements of $\mathbf{v}_{Ct}$ and $\mathbf{V}_{Cl}$. Equation 79 gives $\mathbf{v}_{Ct}$ as

$$\mathbf{v}_{Ct} = \mathbf{g}_{Ct}(\mathbf{q}_{Ct} + \mathbf{Q}_{CC}\mathbf{q}_{Cl}, \mathbf{v}_E).$$

Combined with the Kirchhoff equation 73e we find

$$\mathbf{v}_{Cl} = \mathbf{Q}'_{CC}\mathbf{g}_{Ct}(\mathbf{q}_{Ct} + \mathbf{Q}_{CC}\mathbf{q}_{Cl}, \mathbf{v}_E) + \mathbf{Q}'_{EC}\mathbf{v}_E.$$

Substitution of the above two relations in the branch equations

$$\mathbf{q}_{Ct} = \mathbf{f}_{Ct}(\mathbf{v}_{Ct})$$

$$\mathbf{q}_{Cl} = \mathbf{f}_{Cl}(\mathbf{v}_{Cl})$$

yields $\mathbf{q}_{Ct}$ and $\mathbf{q}_{Cl}$ as functions of $\mathbf{q}_{Ct} + \mathbf{Q}_{CC}\mathbf{q}_{Cl}$ and $\mathbf{v}_E$.

The branch variables of secondary interest are the currents, elements of $\mathbf{i}_{Ct}$ and $\mathbf{i}_{Cl}$. You may show that $\mathbf{i}_{Ct}$ and $\mathbf{i}_{Cl}$ are functions of $\mathbf{x}$, $\mathbf{e}$, and

de/dt if $\mathbf{f}_{Cl}(\mathbf{v}_{Cl})$ is identically zero—no capacitive links—or if $\mathbf{f}_{Cl}(\mathbf{v}_{Cl})$ and $\mathbf{g}_{Ct}(\mathbf{q}_{Ct} + \mathbf{Q}_{CC}\,\mathbf{q}_{Cl}, \mathbf{v}_E)$ are differentiable functions.

Last, consider the resistive components. The elements of $\mathbf{v}_{Rt}$, $\mathbf{v}_{Rl}$, $\mathbf{i}_{Rt}$, and $\mathbf{i}_{Rl}$ are of interest. Equation 90 gives $\mathbf{v}_{Rt}$ and $\mathbf{i}_{Rl}$ as

$$\mathbf{v}_{Rt} = \mathbf{g}_{Rt}(\mathbf{q}_{Ct} + \mathbf{Q}_{CC}\,\mathbf{q}_{Cl}, \boldsymbol{\lambda}_{Ll} - \mathbf{Q}'_{LL}\boldsymbol{\lambda}_{Lt}, \mathbf{v}_E, \mathbf{i}_J)$$

$$\mathbf{i}_{Rl} = \mathbf{g}_{Rl}(\mathbf{q}_{Ct} + \mathbf{Q}_{CC}\,\mathbf{q}_{Cl}, \boldsymbol{\lambda}_{Ll} - \mathbf{Q}'_{LL}\boldsymbol{\lambda}_{Lt}, \mathbf{v}_E, \mathbf{i}_J).$$

The other resistive variables are obtained by substituting these relations along with the expressions for $\mathbf{i}_{Ll}$ and $\mathbf{v}_{Ct}$ in the Kirchhoff equations 73c and 73f:

$$\mathbf{i}_{Rt} = -\,\mathbf{Q}_{RR}\mathbf{i}_{Rl} - \mathbf{Q}_{RL}\mathbf{i}_{Ll} - \mathbf{Q}_{RJ}\mathbf{i}_J$$

$$\mathbf{v}_{Rl} = \mathbf{Q}'_{ER}\mathbf{v}_E + \mathbf{Q}'_{CR}\mathbf{v}_{Ct} + \mathbf{Q}'_{RR}\mathbf{v}_{Rt}.$$

We have shown the way in which the various network variables may be expressed in terms of $\mathbf{x}$, $\mathbf{e}$, and de/dt. Therefore, if $\mathbf{w}$ is a vector whose elements are a set of response variables for the network, then $\mathbf{w}$ may be expressed as

$$\mathbf{w} = \hat{\mathbf{h}}\left(\mathbf{x}, \mathbf{e}, \frac{de}{dt}\right) \tag{96}$$

or

$$\mathbf{w} = \hat{\mathbf{h}}\left(\mathbf{x}, \mathbf{e}, \frac{de}{dt}, t\right), \tag{97}$$

where, in the latter case, the output is explicitly dependent on time.

10.5 SOLUTION OF STATE EQUATION FOR NONLINEAR NETWORKS

Once a state equation has been written for a nonlinear network, the next task is to solve this equation. For convenience, in the equation $d\mathbf{x}/dt = \mathbf{h}(\mathbf{x}, \mathbf{e}, t)$, we shall set $\mathbf{h}(\mathbf{x}, \mathbf{e}, t) = \mathbf{f}(\mathbf{x}, t)$. This is proper, since, when $\mathbf{e}(t)$ is given, $\mathbf{h}[\mathbf{x}, \mathbf{e}(t), t]$ is an explicit function of $\mathbf{x}$ and t only. Thus the vector differential equation that will be the center of attention in this section is

$$\frac{d}{dt}\mathbf{x} = \mathbf{f}(\mathbf{x}, t). \tag{98}$$

There is no known way by which the solution of an arbitrary nonlinear differential equation can be obtained. In fact, closed-form analytic solutions are known for only a few, restricted classes of such equations. Therefore the effort devoted to the study of such equations is concentrated on the conditions for the existence and uniqueness of a solution, properties of a solution, and approximation of a solution. We shall give some consideration to the first two; the third is outside the scope of this text.*

EXISTENCE AND UNIQUENESS

The search for conditions under which (98) is known to have a solution, possibly a unique solution, is a very important task; because, amongst other reasons, only if a solution, usually a unique solution, is known to exist does it become meaningful to seek an approximate solution. Furthermore, the existence or uniqueness, if one exists, is by no means certain. We discovered this for time-varying networks through an example in Section 10.3.

As an additional example, in this case only of nonuniqueness, consider the network in Fig. 10. You should find it easy to verify that the state equation is

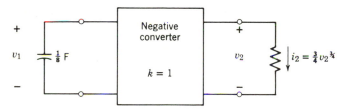

Fig. 10.   Nonlinear network with nonunique response.

$$\frac{dq_1}{dt} = 3q_1^{2/3}.$$

Suppose $q_1(t_0) = 0$. Then it is a simple matter to show that

$$q_1(t) = 0 \qquad (t_0 \leq t < \alpha)$$
$$= (t - \alpha)^3 \qquad (\alpha \leq t)$$

is a solution for all $\alpha \geq t_0$. Therefore this network does not have a unique solution when $q_1(t_0) = 0$.

---

* Almost any text on nonlinear analysis will provide considerable information on this subject; for example, Nicholas Minorsky, *Nonlinear Oscillations*, D. Van Nostrand Co., Princeton, N.J., 1962.

The conditions for a solution to exist for all $t \geq t_0$, the time interval of interest to us, are usually established in two parts. First, given $t_1 \geq t_0$, conditions are established whereby a solution exists in an interval $t_2 \geq t \geq t_1$, where t_2 is determined by properties of the function $\mathbf{f}(\mathbf{x}, t)$ in the interval $t_2 \geq t \geq t_1$. Then conditions are established by which a solution can be extended from one interval of time to the next until the solution for all $t \geq t_0$ is insured.

A solution of (98) *in the ordinary sense* requires that $\mathbf{x}$ satisfy the differential equation for all finite $t \geq t_0$. It is often possible to find a function that satisfies (98) for almost all t, although $d\mathbf{x}/dt$ fails to exist for some discrete values of t. To admit this type of function as a solution, we consider the integral equation

$$\mathbf{x}(t) = \mathbf{x}(t_1) + \int_{t_0}^{t} \mathbf{f}[\mathbf{x}(\tau), \tau] d\tau \tag{99}$$

associated with (98). We shall call any solution of (99) a solution of (98) *in the extended sense*. You should convince yourself that (99) can possess a solution for $t \geq t_0$ and yet (98) will not have a solution in the ordinary sense for $t \geq t_0$.

Apropos of existence, we state the following theorem without proof:*

Theorem 12. *Given any* $t_1 \geq t_0$ *and* $\mathbf{x}(t_1)$, *the differential equation* 98 *possesses at least one continuous solution in the extended sense for* $t_2 > t \geq t_1$, *equal to* $\mathbf{x}(t_1)$ *at time* t_1, *if* $\mathbf{f}(\mathbf{x}, t)$ *is continuous in* $\mathbf{x}$ *for fixed* $t \geq t_0$ *and locally integrable in* t *for* $t \geq t_0$ *and fixed* $\mathbf{x}$, *and if* $\|\mathbf{f}(\mathbf{x}, t)\|$ *is bounded in any bounded neighborhood of the origin* $\mathbf{0}$ *by a locally integrable function of* t *for* $t \geq t_0$.

The conditions of this theorem are more extensive than needed to establish existence of a solution for $t_2 > t \geq t_1$; however, they are appropriate to the statement of the next theorem on the extension of the solution to the entire half-line $t \geq t_0$.

Theorem 13. *Suppose the conditions of Theorem 12 on existence apply, then any solution of* (98) *in the extended sense, equal to* $\mathbf{x}(t_0)$ *at time* t_0, *may be extended to yield a defined solution for all* $t \geq t_0$, *if* $\|\mathbf{f}(\mathbf{x}, t)\|_1 \leq \alpha(t)\phi(\|\mathbf{x}\|_1)$, *for* $t \geq t_0$, *where* $\alpha(t)$ *is non-negative and locally integrable and where* $\phi(v)$ *is positive and continuous for* $v > 0$ *and satisfies*

$$\lim_{u \to \infty} \int_{u_0}^{u} \frac{dv}{\phi(v)} = +\infty \tag{100}$$

for some $u_0 > 0$.

---

* See the footnote on page 757.

Other extension theorems exist* and may be more useful in a given situation. One of these other theorems, because it is extremely simple, is given in the problem set as a theorem to be proved.

Before continuing, let us indicate under what additional restrictions the solution in the extended sense is also a solution in the ordinary sense. Applying the definition of a derivative to the integral equation (99) yields

$$\frac{\mathbf{x}(t + \Delta t) - \mathbf{x}(t)}{\Delta t} = \frac{1}{\Delta t} \int_t^{t + \Delta t} \mathbf{f}[\mathbf{x}(\tau), \tau] d\tau.$$

The solution in the extended sense is continuous. Therefore, if $\mathbf{f}(\mathbf{x}, t)$ is continuous in t, as well as being continuous in $\mathbf{x}$, then $\mathbf{f}[\mathbf{x}(\tau), \tau]$ will be a continuous function of τ. Then the first mean-value theorem of integral calculus may be used and, as a consequence, the above relation yields

$$\frac{\mathbf{x}(t + \Delta t) - \mathbf{x}(t)}{\Delta t} = \mathbf{f}[\mathbf{x}(\theta), \theta] \quad \cdot$$

for $t \leq \theta \leq t + \Delta t$, if $\Delta t > 0$, or $t + \Delta t \leq \theta \leq t$, if $\Delta t < 0$. Because $\mathbf{f}[\mathbf{x}(\theta), \theta]$ is a continuous function of θ, the limit, on the right-hand side of the equation, as Δt tends to zero exists and equals $\mathbf{f}[\mathbf{x}(t), t]$. Thus

$$\frac{d}{dt} \mathbf{x} = \mathbf{f}(\mathbf{x}, t)$$

since $\lim_{\Delta t \to 0} [\mathbf{x}(t + \Delta t) - \mathbf{x}(t)]/\Delta t = d\mathbf{x}/dt$. Hence when $\mathbf{f}(\mathbf{x}, t)$ is continuous in t, as well as $\mathbf{x}$, the solution of the integral equation is differentiable and also satisfies the differential equation.

Now that we know under what conditions a continuous solution exists on the half-line $t \geq t_0$, it is significant to inquire as to what additional restrictions are needed to guarantee a unique solution. This brings us to the next theorem.†

Theorem 14. *Suppose the conditions of Theorems 12 and 13 on existence apply, then (98) possesses a continuous solution in the extended sense for $t \geq t_0$ which is unique, if $\|\mathbf{f}(\mathbf{x}_1, t) - \mathbf{f}(\mathbf{x}_2, t)\| \leq \gamma(t) \|\mathbf{x}_1 - \mathbf{x}_2\|$ for $t \geq t_0$ and for all $\mathbf{x}_1$ and $\mathbf{x}_2$ in some neighborhood of every point, where $\gamma(t)$ is a non-negative, locally integrable function.*

The added condition, $\|\mathbf{f}(\mathbf{x}_1, t) - \mathbf{f}(\mathbf{x}_2, t)\| \leq \gamma(t) \|\mathbf{x}_1 - \mathbf{x}_2\|$ for $t \geq t_0$ and

---

* As an initial reference, see G. Sansone and R. Conte, *Non-Linear Differential Equations*, The Macmillan Co., New York, 1964.

† See the footnote on page 757.

all x_1 and x_2, that is needed to insure uniqueness is known as a *Lipschitz condition*.

Let us pause for a moment and reexamine the example shown in Fig. 10 in the light of these theorems and side comments. The scalar q_1 establishes the state of the network. For compatibility of notation with the theorems, we shall replace q_1 by x. Then $\mathbf{f}(x, t)$ is the scalar function $3x^{2/3}$. By the existence theorem, a solution $\mathbf{x}(t)$ starting from $x = 0$ at $t = 0$ exists for t in some positive time interval since $3x^{2/3}$ is continuous in x and bounded in any bounded neighborhood of the origin; for example, if for all x in some neighborhood of the origin $|x| \leq \rho$, then $|3x^{2/3}| \leq 3\rho^{2/3} < \infty$. By the extension theorem, a solution can be extended to the entire half-line $t \geq t_0$ if $|3x^{2/3}| \leq \phi(|x|)$ for some continuous $\phi(v)$ with the property (100). The condition is satisfied by taking $\phi(v) = 3v^{2/3}$. By the uniqueness theorem, the solution will be unique if $3x^{2/3}$ satisfies a Lipschitz condition; that is, if $|3x_1^{2/3} - 3x_2^{2/3}| \leq \gamma |x_1 - x_2|$ for all x_1 and x_2 in some neighborhood of every point, and for some finite γ. This condition is *not* satisfied. Take $x_2 = 0$; then for any finite γ the inequality is violated by taking $|x_1|$ sufficiently small. The application of these theorems gives results that conform with our prior knowledge of the solution as given in the example.

It is not necessarily an easy task to verify that the vector function $\mathbf{f}(\mathbf{x}, t)$ satisfies a Lipschitz condition. Therefore it is of value to realize that if the partial derivatives $\partial f_i(\mathbf{x}, t)/dx_i$ exist and are continuous in $\mathbf{x}$ for almost all $t \geq t_0$, and are bounded in magnitude in some neighborhood of every point by non-negative locally integrable functions of $t \geq t_0$, then the Lipschitz condition is satisfied. The proof of this fact will not be given, but is suggested to you as a problem.

As an illustration of this means of showing that the Lipschitz condition is satisfied, consider the network in Fig. 11 for $t \geq -1$. The state equation is

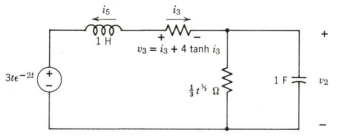

Fig. 11.   Nonlinear time-varying network.

$$\frac{d}{dt} \begin{bmatrix} q_2 \\ \lambda_5 \end{bmatrix} = \begin{bmatrix} -\dfrac{3}{t^{1/3}}\, q_2 - \lambda_5 \\[2ex] q_2 - \lambda_5 - 4 \tanh \lambda_5 - 3t\epsilon^{-2t} \end{bmatrix}.$$

You should show that the conditions of the existence and extension theorems are satisfied. Let us then turn our attention to the Lipschitz condition. We shall set $q_2 = x_1$ and $\lambda_5 = x_2$; then

$$\mathbf{f}(\mathbf{x}, t) = \begin{bmatrix} -\dfrac{3}{t^{1/3}} x_1 - x_2 \\ x_1 - x_2 - 4 \tanh x_2 - 3t\epsilon^{-2t} \end{bmatrix}.$$

The partial derivatives of $f_i(\mathbf{x}, t)$ with respect to x_j are

$$\frac{\partial f_1}{\partial x_1} = -\frac{3}{t^{1/3}}, \qquad \frac{\partial f_1}{\partial x_2} = -1$$

$$\frac{\partial f_2}{\partial x_1} = 1, \qquad \frac{\partial f_2}{\partial x_2} = -1 - 4 \operatorname{sech}^2 x_2.$$

Obviously, the derivatives exist and are continuous in x_1 and x_2 for almost all t; $\partial f_1/\partial x_1$ does not exist for $t = 0$. Further, $|\partial f_1/\partial x_1| \leq 3|t|^{-1/3}$, $|\partial f_1/\partial x_2| \leq 1$, $|\partial f_2/\partial x_1| \leq 1$, and $|\partial f_2/\partial x_2| \leq 5$; therefore, for all $\mathbf{x}$ and hence in some neighborhood of every $\mathbf{x}$, the derivatives are bounded by non-negative locally integrable functions of t for $t \geq -1$. Thus the Lipschitz condition is satisfied, and, by the uniqueness theorem, the solution will be unique.

It can in fact be shown, as you should verify, that, when the sum-magnitude vector norm is used;

$$\|\mathbf{f}(x_1, t) - \mathbf{f}(x_2, t)\|_1 \leq \max \{3|t|^{-1/3} + 1, 6\} \|\mathbf{x}_1 - \mathbf{x}_2\|_1$$

$$\leq (3|t|^{-1/3} + 6) \|\mathbf{x}_1 - \mathbf{x}_2\|_1.$$

The same result would be obtained if the max-magnitude vector norm were used.

PROPERTIES OF THE SOLUTION

In the case of time-varying networks we discovered various properties of the solution by relating the state equation to a reference homogeneous equation. A similar approach provides answers in the nonlinear case also.

Suppose the right-hand side of the nonlinear equation

$$\frac{d}{dt} \mathbf{x} = \mathbf{f}(\mathbf{x}, t)$$

is close, in a sense we shall define in subsequent theorems, to the right-hand side of

$$\frac{d}{dt}\mathbf{x} = \mathbf{A}(t)\mathbf{x}.$$

Then we might anticipate that, if all solutions of the latter equation are bounded or approach zero as $t \to \infty$, the solutions of the former equation will do likewise. It turns out that this inference is true if $\mathbf{A}(t)$ is a constant, or periodic, matrix. It will be left as a task for you to show that a theorem similar to those to be given does not exist when $\mathbf{A}(t)$ is an arbitrary time-varying matrix.

For networks that are close to being described by a time-invariant, homogeneous differential equation (namely, $\mathbf{A}(t) = \mathbf{A}$) we shall present a sequence of useful theorems. The first theorem describes conditions for a bounded response.

Theorem 15. *Suppose all solutions of the reference equation*

$$\frac{d}{dt}\mathbf{y} = \mathbf{A}\mathbf{y}, \tag{101}$$

where $\mathbf{A}$ is a constant matrix, are bounded as t *tends to infinity. Further, suppose* $\mathbf{f}(\mathbf{x}, t) = \mathbf{A}\mathbf{x} + \hat{\mathbf{f}}(\mathbf{x}, t) + \mathbf{g}(t)$. *Then all solutions of*

$$\frac{d}{dt}\mathbf{x} = \mathbf{f}(\mathbf{x}, t),$$

with an initial vector $\mathbf{x}(t_0)$ *such that* $\|\mathbf{x}(t_0)\| \leq \delta$, *where δ is a constant that depends on* $\mathbf{f}(\mathbf{x}, t)$, *are bounded as* t *tends to infinity if*

$$\|\hat{\mathbf{f}}(\mathbf{x}, t)\| \leq \beta(t)\|\mathbf{x}\| \quad \text{for} \quad \|\mathbf{x}\| \leq \zeta \tag{102a}$$

$$\int_{t_0}^{\infty} \beta(t)\,dt < \infty \tag{102b}$$

$$\int_{t_0}^{\infty} \|\mathbf{g}(t)\|\,dt \leq \gamma\delta, \tag{102c}$$

where γ is a suitably chosen positive constant.

The proof makes use of Gronwall's lemma. The state equation

$$\frac{d}{dt}\mathbf{x} = \mathbf{A}\mathbf{x} + \hat{\mathbf{f}}(\mathbf{x}, t) + \mathbf{g}(t)$$

is equivalent to the integral equation

$$\mathbf{x}(t) = \epsilon^{\mathbf{A}(t-t_0)}\mathbf{x}(t_0) + \int_{t_0}^{t} \epsilon^{\mathbf{A}(t-\tau_0)}\{\hat{\mathbf{f}}[\mathbf{x}(\tau),\,\tau] + \mathbf{g}(\tau)\}\,d\tau,$$

where $\epsilon^{\mathbf{A}(t-\tau)}$ is the transition matrix associated with (101). Upon taking the norm of both sides of this equation and applying the usual norm inequalities, it is found that

$$\|\mathbf{x}(t)\| \leq \|\epsilon^{\mathbf{A}(t-t_0)}\|\,\|\mathbf{x}(t_0)\| + \int_{t_0}^{t} \|\epsilon^{\mathbf{A}(t-\tau)}\|\,\{\|\mathbf{f}[\mathbf{x}(\tau),\,\tau]\| + \|\mathbf{g}(\tau)\|\}d\tau.$$

Since all solutions of (101) are bounded, a positive constant α exists such that $\|\epsilon^{\mathbf{A}(t-\tau)}\| \leq \alpha$ for $t_0 \leq \tau \leq t$ and for all $t \geq t_0$. Using this bound, that of (102a), and $\|\mathbf{x}(t_0)\| \leq \delta$, we get

$$\|\mathbf{x}(t)\| \leq \alpha\delta + \int_{t_0}^{t} \alpha[\beta(\tau)\|\mathbf{x}(\tau)\| + \|\mathbf{g}(\tau)\|]d\tau$$

provided $\|\mathbf{x}(\tau)\| \leq \zeta$ for $t_0 \leq \tau \leq t$. Gronwall's lemma may be applied to this relation by setting $\varphi(t) = \|\mathbf{x}(t)\|$, $\psi(t) = \alpha\beta(t)$, and $\theta(t) = \alpha\|\mathbf{g}(t)\|$. The resulting inequality is

$$\|\mathbf{x}(t)\| \leq \alpha\delta \exp\left\{\int_{t_0}^{t}[\alpha\beta(\tau) + \alpha\|\mathbf{g}(\tau)\|/\delta]d\tau\right\}$$

$$\leq \alpha\delta \exp\left[\int_{t_0}^{\infty}\alpha\beta(\tau)d\tau + \alpha\gamma\right].$$

The second inequality, obtained by first letting t approach infinity and then invoking (102c), shows that $\|\mathbf{x}(t)\|$ is bounded as t tends to infinity, because $\int_{t_0}^{\infty}\alpha\beta(\tau)d\tau$ is bounded by virtue of (102b). It further shows that $\|\mathbf{x}(t)\|$ is uniformly bounded for all t. Hence $\|\mathbf{x}(t)\| \leq \zeta$ for $t_0 \leq \tau \leq t$ and all $t \geq t_0$, as required to satisfy a preceding condition if

$$\delta \leq (\zeta/\alpha)\exp\left[-\int_{t_0}^{\infty}\alpha\beta(\tau)d\tau - \alpha\gamma\right].$$

Thus, unlike the case of the time-varying linear networks considered previously, the boundedness depends on the initial state; that is, $\|\mathbf{x}(t_0)\|$ must be sufficiently small. Furthermore, the function $\mathbf{g}(t)$ is restricted as indicated in (102c). (*Note*: the conclusion of the theorem becomes valid for all initial states only if $\zeta = +\infty$.

As the next theorem shows, the conclusions of the previous theorem can be strengthened, if all solutions of (101) approach zero as t tends to infinity.

Theorem 16. *Suppose all solutions of the reference equation 101 in which* **A** *is a constant matrix, approach zero as* t *tends to infinity. Furthermore, suppose* $\mathbf{f}(\mathbf{x}, t) = \mathbf{A}\mathbf{x} + \mathbf{B}(t)\mathbf{x} + \hat{\mathbf{f}}(\mathbf{x}, t) + \mathbf{g}(t)$. *Then all solutions of the nonlinear equation 98 with initial vector* $\mathbf{x}(t_0)$ *such that* $\|\mathbf{x}(t_0)\| \le \delta$, *where* δ *is a constant that depends on* $\mathbf{f}(\mathbf{x}, t)$, *approach zero as* t *tends to infinity if*

$$\int_{t_0}^{\infty} \|\mathbf{B}(t)\| \, dt < \infty \tag{103a}$$

$$\epsilon^{\beta(t-t_0)} \|\mathbf{g}(t)\| \le \gamma\delta \tag{103b}$$

$$\lim_{\mathbf{x} \to 0} \frac{\|\hat{\mathbf{f}}(\mathbf{x}, t)\|}{\|\mathbf{x}\|} \to 0 \quad \text{for} \quad t \ge t_0, \tag{103c}$$

where β *and* γ *are suitably chosen positive constants.*

The proof of this theorem is somewhat different from previous proofs. The integral equation equivalent to (98) is

$$\mathbf{x}(t) = \epsilon^{\mathbf{A}(t-t_0)}\mathbf{x}(t_0) + \int_{t_0}^{t} \epsilon^{\mathbf{A}(t-\tau)}\{\mathbf{B}(\tau)\,\mathbf{x}(\tau) + \hat{\mathbf{f}}[\mathbf{x}(\tau), \tau] + \mathbf{g}(\tau)\}d\tau.$$

Taking the norm of both sides and applying the usual norm inequalities yields

$$\|\mathbf{x}(t)\| \le \|\epsilon^{\mathbf{A}(t-t_0)}\| \, \|\mathbf{x}(t_0)\| + \int_{t_0}^{t} \|\epsilon^{\mathbf{A}(t-\tau)}\|\{\|\mathbf{B}(\tau)\| \, \|\mathbf{x}(\tau)\|$$
$$+ \|\hat{\mathbf{f}}[\mathbf{x}(\tau), \tau]\| + \|\mathbf{g}(\tau)\|\}d\tau.$$

Since all solutions of (101) approach zero as t tends to infinity, positive constants α and ν exist such that

$$\|\epsilon^{\mathbf{A}(t-\tau)}\| \le \alpha\epsilon^{-\nu(t-\tau)} \quad \text{for} \quad t_0 \le \tau \le t \quad \text{and all} \quad t \ge t_0.$$

Further, by (103c) there exists a positive constant ζ such that

$$\|\hat{\mathbf{f}}(\mathbf{x}, t)\| \le \left(\frac{\nu}{2\alpha}\right)\|\mathbf{x}\| \quad \text{for} \quad t \ge t_0 \quad \text{if} \quad \|\mathbf{x}\| \le \zeta.$$

Using these bounds and $\|\mathbf{x}(t_0)\| \leq \delta$, we find

$$\|\mathbf{x}(t)\| \leq \alpha\delta\epsilon^{-\nu(t-t_0)} + \int_{t_0}^{t} \alpha\epsilon^{-\nu(t-\tau)}\left\{\left[\|\mathbf{B}(\tau)\| + \left(\frac{\nu}{2\alpha}\right)\right]\|\mathbf{x}(\tau)\| + \|\mathbf{g}(\tau)\|\right\}d\tau,$$

or, equivalently,

$$\|\mathbf{x}(t)\|\epsilon^{\nu t} \leq \alpha\epsilon^{\nu t_0}\delta + \int_{t_0}^{t}\left\{\left[\alpha\|\mathbf{B}(\tau)\| + \left(\frac{\nu}{2}\right)\right]\|\mathbf{x}(\tau)\|\epsilon^{\nu\tau} + \alpha\epsilon^{\nu t}\|\mathbf{g}(\tau)\|\right\}d\tau,$$

provided

$$\|\mathbf{x}(\tau)\| \leq \zeta \quad \text{for} \quad t_0 \leq \tau \leq t.$$

Gronwall's lemma is applicable, with $\varphi(t) = \|\mathbf{x}(t)\|\epsilon^{\nu t}$, $\quad \psi(t) = \alpha\|\mathbf{B}(t)\| + \dfrac{\nu}{2}$,
$\theta(t) = \alpha\epsilon^{\nu t}\|\mathbf{g}(\tau)\|$. The result obtained is

$$\|\mathbf{x}(t)\|\,\epsilon^{\nu t} \leq \alpha\delta\epsilon^{\nu t_0}\exp\left[\int_{t_0}^{t}\alpha\|\mathbf{B}(\tau)\|d\tau\right]\epsilon^{\nu(t-t_0)/2}\exp\left\{\frac{\int_{t_0}^{t}\alpha\|\mathbf{g}(\tau)\|\,\epsilon^{\nu\tau}d\tau}{\alpha\delta\epsilon^{\nu t_0}}\right\}$$

$$\leq \alpha\delta\epsilon^{\nu t_0}\exp\left[\int_{t_0}^{\infty}\alpha\|\mathbf{B}(\tau)\|d\tau\right]\epsilon^{\nu(t-t_0)/2}\epsilon^{\gamma(t-t_0)}$$

The second inequality follows from the first upon letting t approach infinity in the first integral and using (103b) with $\beta = \nu$ in the second integral. Multiplication by $\epsilon^{-\nu t}$ yields

$$\|\mathbf{x}(t)\| \leq \alpha\delta\exp\left[\int_{t_0}^{\infty}\alpha\|\mathbf{B}(\tau)\|d\tau\right]\exp\left[-\left(\frac{\nu}{2}-\gamma\right)(t-t_0)\right].$$

Now set $\gamma < \nu/2$; then $\|\mathbf{x}(t)\|$ is uniformly bounded, by virtue of (103a), and less than

$$\alpha\delta\exp\left[\int_{t_0}^{\infty}\alpha\|\mathbf{B}(\tau)\|d\tau\right]$$

for all $t \geq t_0$. Hence, to satisfy the condition that $\|\mathbf{x}(\tau)\| \leq \zeta$ for $t_0 \leq \tau \leq t$ and all $t \geq t_0$, it is merely necessary to select δ such that

$$\delta \leq \left(\frac{\zeta}{\alpha}\right)\exp\left[-\int_{t_0}^{\infty}\alpha\|\mathbf{B}(\tau)\|d\tau\right].$$

Further, with reference to the bound on $\|\mathbf{x}(t)\|$, we see that $\mathbf{x}(t)$ must approach zero as t tends to infinity since $\|\mathbf{x}(t)\|$ is bounded by a decaying exponential.

Other theorems are possible. The two we have given establish the type of conditions that are usually needed to get a proof. You will find it to be a relatively easy task to vary these conditions and still get a valid proof. For example, in the last theorem, condition (103c) can be replaced by $\|\hat{\mathbf{f}}(\mathbf{x}, t)\| \leq \mu \|\mathbf{x}\|$ for $\|\mathbf{x}\| \leq \zeta$ and μ sufficiently small. This variation is significant because it permits a small linear term to be part of $\hat{\mathbf{f}}(\mathbf{x}, t)$.

In the two theorems given, if we replaced $\mathbf{A}$ by $\mathbf{A}(t)$, where $\mathbf{A}(t)$ is periodic, we would still have true theorems. The proofs vary only slightly from those given. You should determine what changes are needed.

As an illustration of the second theorem, consider the network in Fig. 12 for $t \geq 0$. With $q_2 = x_1$ and $\lambda_5 = x_2$, the state equation is

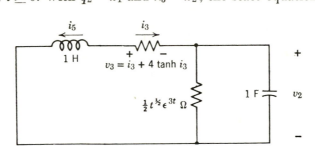

Fig. 12.   Nonlinear time-varying network with no excitation.

$$\frac{d}{dt}\begin{bmatrix} x_1 \\ x_2 \end{bmatrix} = \begin{bmatrix} -\dfrac{2\epsilon^{-3t}}{t^{1/2}}\, x_1 - x_2 \\ x_1 - x_2 - 4\tanh x_2 \end{bmatrix}.$$

First of all, we shall express this equation in the form to which the theorem applies. Thus

$$\frac{d}{dt}\begin{bmatrix} x_1 \\ x_2 \end{bmatrix} = \begin{bmatrix} 0 & -1 \\ 1 & -5 \end{bmatrix}\begin{bmatrix} x_1 \\ x_2 \end{bmatrix} + \begin{bmatrix} -\dfrac{2\epsilon^{-3t}}{t^{1/2}} & 0 \\ 0 & 0 \end{bmatrix}\begin{bmatrix} x_1 \\ x_2 \end{bmatrix} + \begin{bmatrix} 0 \\ 4x_2 - 4\tanh x_2 \end{bmatrix}$$

$$= \mathbf{A}\mathbf{x} + \mathbf{B}(t)\mathbf{x} + \hat{\mathbf{f}}(\mathbf{x}, t).$$

The characteristic polynomial of $\mathbf{A}$ is $d(s) = s^2 + 5s + 1$. The eigenvalues are clearly negative real. (If the characteristic equation had been of higher degree, then the Routh criterion or one of the other similar criteria could have been used to show whether or not all the roots had negative real part.) Thus all the solutions of (101) approach zero as $t \to \infty$. By using the sum-magnitude vector norm, we get

$$\|\mathbf{B}(t)\|_1 = \left\| \begin{bmatrix} -\dfrac{2\epsilon^{-3t}}{t^{1/2}} & 0 \\[2mm] 0 & 0 \end{bmatrix} \right\|_1 = \dfrac{2\epsilon^{-3t}}{|t^{1/2}|}$$

and

$$\int_0^\infty \|\mathbf{B}(t)\|_1 \, dt = 2\sqrt{\dfrac{\pi}{3}}.$$

Thus condition (103a) is satisfied. Condition (103b) is met since $\mathbf{g}(t) \equiv 0$. Now let us consider

$$\hat{\mathbf{f}}(\mathbf{x}, t) = \begin{bmatrix} 0 \\ 4x_2 - 4\tanh x_2 \end{bmatrix}.$$

Now

$$\dfrac{\|\hat{\mathbf{f}}(\mathbf{x}, t)\|_1}{\|\mathbf{x}\|_1} = \dfrac{4|x_2 - \tanh x_2|}{|x_1| + |x_2|} \le 4\,\dfrac{|x_2 - \tanh x_2|}{|x_2|}.$$

The inequality follows from the fact that $|x_2| \le |x_1| + |x_2|$. Now, as $\|\mathbf{x}\|_1$ and hence $|x_2|$ approach zero, $4|x_2 - \tanh x_2|/|x_2|$ and hence $\|\hat{\mathbf{f}}(\mathbf{x}, t)\|_1/\|\mathbf{x}\|_1$ approach zero. This follows from the fact that

$$\dfrac{|x_2 - \tanh x_2|}{|x_2|} = \dfrac{\left| \frac{1}{3}x_2^3 - \frac{2}{12}x_2^5 + \frac{17}{315}x_2^7 - \cdots \right|}{|x_2|}$$

$$= \left| \tfrac{1}{3}x_2^2 - \tfrac{2}{12}x_2^4 + \tfrac{17}{315}x_2^6 - \cdots \right| \to 0 \quad \text{as} \quad |x_2| \to 0.$$

Thus condition (103c) is satisfied.

All the conditions of the theorem are satisfied; therefore, by Theorem 16, all solutions approach zero as $t \to \infty$ for $\|\mathbf{x}(0)\|_1 \le \delta$. The value of δ is not known, but we know that it does exist. You should take it upon yourself to find a value of δ.

The assignment of the terms of $\hat{\mathbf{f}}(\mathbf{x}, t)$ to $\mathbf{A}\mathbf{x}$ and to $\hat{\mathbf{f}}(\mathbf{x}, t)$ is not unique. In fact, any linear term can be assigned to $\mathbf{A}\mathbf{x}$ if its negative is assigned to $\hat{\mathbf{f}}$. This flexibility can be utilized to help satisfy the condition (103c). This is indeed the reason for expressing the term $-x_2$ in the example as $-5x_2 + 4x_2$.

10.6 NUMERICAL SOLUTION

The several situations in which numerical solution of the state equation is called for are (1) when the exact analytic solution cannot be determined, (2) when an approximate analytic solution of sufficient accuracy can be determined only by an inordinate amount of work, and (3) when a family of solutions for only a limited number of parameter value variations is sought. In this section we shall turn our attention to the numerical solution of the state equation as expressed in the form

$$\frac{d}{dt}\mathbf{x} = \mathbf{f}(\mathbf{x}, t). \tag{104}$$

We shall develop some of the elementary methods of numerical solution of (104) and state without proof one sophisticated method.

NEWTON'S BACKWARD-DIFFERENCE FORMULA

Several methods of numerical solution of the state equation become easy to establish if one starts from an expression for the value of a function at some point in time in terms of its values at previous points. We shall therefore treat this subject first.

To provide a basis in familiar terms for the formula to be discussed, consider the vector function $\mathbf{y}(t)$ expressed as a truncated power series with a remainder; that is

$$\mathbf{y}(t) = \mathbf{y}(t_i) + \mathbf{y}^{(1)}(t_i)(t - t_i) + \frac{1}{2!}\mathbf{y}^{(2)}(t_i)(t - t_i)^2 + \cdots$$

$$\cdots + \frac{1}{j!}\mathbf{y}^{(j)}(t_i)(t - t_i)^j + \mathbf{r}(t), \quad (105)$$

where $\mathbf{y}^{(k)}(t_i) = d^k\mathbf{y}(t)/dt^k|_{t=t_i}$. Often $\mathbf{y}(t)$ can be approximated by the polynomial obtained when $\mathbf{r}(t)$ is neglected. Of course, if $\mathbf{y}(t)$ is a polynomial of degree no greater than j, then there is no error in such an approximation. The major fault with using (105) is that it requires knowledge of the derivatives of $\mathbf{y}(t)$, and such knowledge may not be available.

To avoid the problem of evaluating derivatives of $\mathbf{y}(t)$, we shall search for an alternative, but equivalent, representation of $\mathbf{y}(t)$. Let t_{i-1}, t_{i-2}, $\cdots$, t_{i-j} be a set of j distinct values of time. If we substitute each of these values into (105), we obtain j equations with the derivatives as unknowns. These may be solved for $\mathbf{y}^{(k)}(t_i)$ in terms of $\mathbf{y}(t_i)$, $\mathbf{y}(t_{i-k})$, and the remainder $\mathbf{r}(t_{i-k})$, where $k = 1, \cdots, j$. If these solutions for the $\mathbf{y}^{(k)}(t_i)$ are substituted into (105), the resulting expression for $\mathbf{y}(t)$ may be put into the following form:

$$\mathbf{y}(t) = a_0(t)\, \mathbf{y}(t_i) + a_1(t)\, \mathbf{y}(t_{i-1}) + a_2(t)\, \mathbf{y}(t_{i-2}) + \cdots$$

$$\cdots + a_j(t)\, \mathbf{y}(t_{i-j}) + \hat{\mathbf{r}}(t), \qquad (106)$$

where $\hat{\mathbf{r}}(t)$ is a remainder that is, of course, different from $\mathbf{r}(t)$; the $a_k(t)$ are polynomials in t of degree no greater than j. By neglecting $\hat{\mathbf{r}}(t)$, a polynomial that approximates $\mathbf{y}(t)$ is obtained.

The coefficients $a_k(t)$ are not as easy to evaluate as might be desired. However, if the terms in (106) are rearranged to express $\mathbf{y}(t)$ in terms of sums of and differences between, the various $\mathbf{y}(t_{i-k})$, the new coefficients are easy to evaluate. This will become evident as the representation evolves.

Let us first define a set of functions, which are sums and differences of the values of $\mathbf{y}(t)$ at different points in time, as follows:

$$\boldsymbol{\delta}[t_i] = \mathbf{y}(t_i) \qquad (107a)$$

$$\boldsymbol{\delta}[t_{i-1}, t_i] = \frac{\boldsymbol{\delta}[t_i] - \boldsymbol{\delta}[t_{i-1}]}{t_i - t_{i-1}} \qquad (107b)$$

. . . .

$$\boldsymbol{\delta}[t_{i-k}, \ldots, t_{i-1}, t_i] = \frac{\boldsymbol{\delta}[t_{i-k+1}, \cdots, t_i] - \boldsymbol{\delta}[t_{i-k}, \cdots, t_{i-1}]}{t_i - t_{i-k}}. \qquad (107c)$$

Each function, after the first one, is the difference between the preceding function at two successive instants in time, divided by a time difference. Thus these functions are known as *divided differences*. You should note that, since the first divided difference is just $\mathbf{y}(t_i)$ by definition, all successive divided differences are sums and differences of $\mathbf{y}(t)$ at various points in time, divided by time intervals.

Our next task is to express $\mathbf{y}(t)$ in terms of sums and differences of its values at t_i, $t_{i-1}, \cdots$; that is, in terms of the divided differences. We shall first express $\mathbf{y}(t)$ in terms of $\delta[t_i, t]$. Then, by expressing $\delta[t_i, t]$ in terms of $\delta[t_{i-1}, t_i]$ and $\delta[t_{i-1}, t_i, t]$, we shall establish $\mathbf{y}(t)$ in terms of $\delta[t_{i-1}, t_i]$ and $\delta[t_{i-1}, t_i, t]$. This substitution process will be continued to get $\mathbf{y}(t)$ in terms of the divided differences of all orders.

Let us start with the divided differences $\delta[t_i, t]$. By (107b) we know that

$$\delta[t] = \delta[t_i] + (t - t_i)\delta[t_i, t].$$

By substituting (107a), we get

$$\mathbf{y}(t) = \mathbf{y}(t_i) + (t - t_i)\delta[t_i, t]. \tag{108a}$$

Next, consider the divided difference $\delta[t_{i-1}, t_i, t]$. By (107c) we obtain an expression for $\delta[t_{i-1}, t_i, t]$ which, when terms are rearranged, may be written as

$$\delta[t_i, t] = \delta[t_{i-1}, t_i] + (t - t_{i-1})\delta[t_{i-1}, t_i, t].$$

By substituting in (108a), we obtain

$$\mathbf{y}(t) = \mathbf{y}(t_i) + (t - t_i)\delta[t_{i-1}, t_i] + (t - t_i)(t - t_{i-1})\delta[t_{i-1}, t_i, t]. \tag{108b}$$

From (107c) we know that

$$\delta[t_{i-k}, \cdots, t_i, t] = \delta[t_{i-k-1}, \cdots, t_i] - (t - t_{i-k-1})\delta[t_{i-k-1}, \cdots, t_i, t].$$

By repeated use of this relation, (108b) becomes

$$\begin{aligned}
\mathbf{y}(t) = \mathbf{y}(t_i) &+ (t - t_i)\delta[t_{i-1}, t_i] \\
&+ (t - t_i)(t - t_{i-1})\delta[t_{i-2}, t_{i-1}, t_i] \\
&+ \cdots\cdots\cdots\cdots\cdots\cdots\cdots\cdots\cdots\cdots\cdots\cdots\cdots \\
&+ (t - t_i)(t - t_{i-1}) \cdots (t - t_{i-j+1})\delta[t_{i-j}, \cdots, t_{i-1}, t_i] \\
&+ e(t),
\end{aligned} \tag{108c}$$

where

$$e(t) = (t - t_i) \cdots (t - t_{i-j})\delta[t_{i-j}, \cdots, t_i, t] \tag{109}$$

is viewed as the error in approximating $\mathbf{y}(t)$ by the polynomial

$$\mathbf{y}(t_i) + (t - t_i)\delta[t_{i-1}, t_i] + \cdots + (t - t_i) \cdots (t - t_{i-j+1})\delta[t_{i-j}, \cdots, t_i]. \tag{110}$$

Based on previous comments relevant to (106), we know that $e(t) \equiv 0$ if $y(t)$ is a polynomial of degree no greater than j. By examining (109) you will also observe that $e(t) = 0$ for $t = t_i, t_{i-1}, \cdots, t_{i-j}$ whether or not $y(t)$ is a polynomial.

Since divided differences are not as easy to calculate as simple differences, we shall now reformulate (110) in terms of the simple differences taken backwards, as follows.

$$\nabla y(t_l) = y(t_l) - y(t_{l-1}) \tag{111a}$$

$$\nabla^2 y(t_l) = [\nabla y(t_l) - \nabla y(t_{l-1})] \tag{111b}$$

$$\cdots\cdots\cdots\cdots\cdots\cdots\cdots\cdots\cdots$$

$$\nabla^k y(t_l) = [\nabla^{k-1} y(t_l) - \nabla^{k-1} y(t_{l-1})]. \tag{111c}$$

These are defined as the *backward differences*. Let us assume in all that follows that differences between adjacent values of time are equal; that is, $t_k - t_{k-1} = h$ for $k = i - j + 1, \cdots, i$. Then, starting with (111a), we obtain

$$\nabla y(t_i) = (t_i - t_{i-1}) \frac{y(t_i) - y(t_{i-1})}{t_i - t_{i-1}}$$

$$= h \, \delta[t_{i-1}, t_i].$$

Similarly, starting with (111b), we get

$$\nabla^2 y(t_i) = (t_i - t_{i-2}) \frac{\nabla y(t_i) - \nabla y(t_{i-1})}{t_i - t_{i-2}}$$

$$= 2h^2 \frac{\delta[t_{i-1}, t_i] - \delta[t_{i-2}, t_{i-1}]}{t_i - t_{i-2}}$$

$$= 2h^2 \, \delta[t_{i-2}, t_{i-1}, t_i].$$

The second line follows from the first by using the previously derived expression for $\nabla y(t_i)$, and the third line follows from the second by using (107c). Continuing in this manner, it is found that

$$\nabla y(t_i) = h \, \delta[t_{i-1}, t_i] \tag{112a}$$

$$\nabla^2 y(t_i) = 2h^2 \, \delta[t_{i-2}, t_{i-1}, t_i] \tag{112b}$$

$$\cdots\cdots\cdots\cdots\cdots\cdots\cdots\cdots$$

$$\nabla^k y(t_i) = k! \, h^k \, \delta]t_{i-k}, \cdots, t_i], \tag{112c}$$

or, equivalently,

$$\delta[t_{i-k}, \cdots, t_i] = \frac{1}{k!h^k} \nabla^k \mathbf{y}(t_i). \tag{113}$$

If this relationship between the divided differences and backward differences is now inserted into (108c), we get

$$\mathbf{y}(t) = \mathbf{y}(t_i) + \frac{(t - t_i)}{h} \nabla \mathbf{y}(t_i) + \frac{(t - t_i)(t - t_{i-1})}{2!h^2} \nabla^2 \mathbf{y}(t_i) + \cdots$$

$$\cdots + \frac{(t - t_i) \cdots (t - t_{i-j+1})}{j!h^j} \nabla^j \mathbf{y}(t_i) + \mathbf{e}(t). \tag{114}$$

Similarly, the approximating polynomial of (110), obtained by neglecting $\mathbf{e}(t)$, is

$$\mathbf{y}(t_i) + \frac{(t - t_i)}{h} \nabla \mathbf{y}(t_i) + \cdots + \frac{(t - t_i) \cdots (t - t_{i-j+1})}{j! h^j} \nabla^j \mathbf{y}(t_i). \tag{115}$$

Equation 114 is *Newton's backward-difference formula*; (115) is that formula truncated after $j + 1$ terms.

OPEN FORMULAS

Using Newton's backward difference formula, we shall construct some quite general formulas for the solution of the state equation at time t_{i+1} in terms of the backward differences of its derivatives at preceding points in time. These formulas, known as *open formulas*, are the basis for many of the specific numerical methods to be presented subsequently.

Denoting d/dt by a dot and setting $\mathbf{y}(t) = \dot{\mathbf{x}}(t)$, we get from Newton's backward-difference formula (115), truncated after $j + 1$ terms,

$$\dot{\mathbf{x}}(t) = \dot{\mathbf{x}}(t_i) + \frac{(t - t_i)}{h} \nabla \dot{\mathbf{x}}(t_i) + \cdots + \frac{(t - t_i) \cdots (t - t_{i-j+1})}{j! h^j} \nabla^j \dot{\mathbf{x}}(t_i). \tag{116}$$

If $\dot{\mathbf{x}}(t)$ is integrated from t_i to $t_{i+1} = t_i + h$, the result is $\mathbf{x}(t_{i+1}) - \mathbf{x}(t_i)$; equivalently,

$$\mathbf{x}(t_{i+1}) = \mathbf{x}(t_i) + \int_{t_i}^{t_i+h} \dot{\mathbf{x}}(t) \, dt. \tag{117}$$

By substituting (116) into (117), we get

$$\mathbf{x}(t_{i+1}) = \mathbf{x}(t_i) + h \sum_{k=0}^{j} b_k \nabla^k \dot{\mathbf{x}}(t_i), \tag{118}$$

where $b_0 = 1$, and, for $k > 0$,

$$b_k = \int_{t_i}^{t_i+h} \frac{(t - t_i) \cdots (t - t_{i-k+1})}{k! \, h^{k+1}} \, dt$$

$$= \int_0^1 \frac{\tau(\tau + 1) \cdots (\tau + k - 1)}{k!} \, d\tau. \tag{119}$$

The second integral in (119) is obtained from the first by the change of variable $\tau = (t - t_i)/h$ and the fact that $t_{i-l} = t_i - lh$. Upon evaluating this integral for several values of k and substituting the result into (118), we find that

$$\mathbf{x}(t_{i+1}) = \mathbf{x}(t_i) + h[\dot{\mathbf{x}}(t_i) + \tfrac{1}{2}\nabla\dot{\mathbf{x}}(t_i) + \tfrac{5}{12}\nabla^2\dot{\mathbf{x}}(t_i)$$

$$+ \tfrac{3}{8}\nabla^3\dot{\mathbf{x}}(t_i) + \tfrac{251}{720}\nabla^4\dot{\mathbf{x}}(t_i) + \cdots]. \tag{120}$$

We have, in a sense, achieved the goal. If we know $\mathbf{x}(t_0)$, $\mathbf{x}(t_1)$, $\cdots$, and $\mathbf{x}(t_j)$, then we can use the state equation to determine $\dot{\mathbf{x}}(t_0)$, $\dot{\mathbf{x}}(t_1)$, $\cdots$, and $\dot{\mathbf{x}}(t_j)$. The backward differences $\nabla\dot{\mathbf{x}}(t_j)$, $\nabla^2\dot{\mathbf{x}}(t_j)$, $\cdots$, and $\nabla^j\dot{\mathbf{x}}(t_j)$ are computed next, and (118) is used to evaluate $\mathbf{x}(t_{j+1})$. The steps are then repeated to evaluate $\mathbf{x}(t_{j+2})$ starting with $\mathbf{x}(t_1)$, $x(t_2)$, $\cdots$, and $x(t_{j+1})$. Continuing in this way, the value of $x(t)$ at $t = t_{j+1}, t_{j+2}, \cdots$ is established. We shall give this subject greater attention later.

Equation 118 states the dependency of $\mathbf{x}(t_{i+1})$ on the immediately preceding value of $\mathbf{x}$ (namely, $\mathbf{x}(t_1)$) and on $\dot{\mathbf{x}}(t_i)$, $\nabla\dot{\mathbf{x}}(t_i)$, $\cdots$, and $\nabla^j\dot{\mathbf{x}}(t_i)$. We can just as easily establish the dependency of $\mathbf{x}(t_{i+1})$ on $\mathbf{x}(t_{i-l})$ for $l \leq 0$ and on $\dot{\mathbf{x}}(t_i)$, $\nabla\dot{\mathbf{x}}(t_i)$, $\cdots$, and $\nabla^j\dot{\mathbf{x}}(t_i)$. This is accomplished by integrating $\dot{\mathbf{x}}(t)$ from $t_{i-l} = t_i - lh$ to $t_{i+1} = t_i + h$. The integral relation obtained thereby is

$$\mathbf{x}(t_{i+1}) = \mathbf{x}(t_{i-l}) + \int_{t_i-lh}^{t_i+h} \dot{\mathbf{x}}(t) \, dt. \tag{121}$$

Upon substituting (116), we obtain

$$\mathbf{x}(t_{i+1}) = \mathbf{x}(t_{i-l}) + h \sum_{k=0}^{j} b_k(l) \, \nabla^k \dot{\mathbf{x}}(t_i), \tag{122}$$

where $b_0(l) = l + 1$ and, for $k > 0$,

$$b_k(l) = \int_{-l}^{1} \frac{\tau(\tau + 1) \cdots (\tau + k - 1)}{k!} \, d\tau. \tag{123}$$

For subsequent use, values of the coefficients $b_k(l)$ that appear in the very general relation (122), are given in Table 1.

Table 1.   Values of $b_k(l)$

k	l					
	0	1	2	3	4	5
0	1	2	3	4	5	6
1	$\frac{1}{2}$	0	$-\frac{3}{2}$	-4	$-\frac{17}{2}$	-12
2	$\frac{5}{12}$	$\frac{1}{3}$	$\frac{3}{4}$	$\frac{8}{3}$	$\frac{85}{12}$	15
3	$\frac{3}{8}$	$\frac{1}{3}$	$\frac{3}{8}$	0	$-\frac{55}{24}$	-9
4	$\frac{251}{720}$	$\frac{29}{90}$	$\frac{27}{80}$	$\frac{14}{45}$	$\frac{83}{144}$	$\frac{33}{10}$
5	$\frac{95}{299}$	$\frac{14}{45}$	$\frac{51}{160}$	$\frac{14}{45}$	$\frac{95}{288}$	0

Nothing has been said about the error arising from using the truncated version of Newton's backward-difference formula. A good treatment of truncation errors is to be found in most texts on numerical analysis.* However, it should be noted that the error in (122) is proportional to h^{j+2}. Thus, in an intuitive sense, if h is sufficiently small and j is sufficiently large, the error should be quite small.

CLOSED FORMULAS

The relations obtained from (122) for the various values of l are known as *open formulas* because $\mathbf{x}(t_{i+1})$ depends not on $\dot{\mathbf{x}}$ at t_{i+1} but only on $\dot{\mathbf{x}}$ at previous points in time. On the other hand, the *closed formulas* to be derived next exhibit a dependency relationship between $\mathbf{x}(t_{i+1})$ and $\dot{\mathbf{x}}$ at t_{i+1} as well as preceding points in time.

Set $\mathbf{y}(t) = \dot{\mathbf{x}}(t)$ and then replace t_i by t_{i+1} in Newton's backward-

* See, for example, R. W. Hamming, *Numerical Methods for Scientists and Engineers,* McGraw-Hill Book Co., New York, 1962.

difference formula, truncated after $j+1$ terms. The resulting equation is

$$\dot{\mathbf{x}}(t) = \dot{\mathbf{x}}(t_{i+1}) + \frac{(t-t_{i+1})}{h} \nabla \dot{\mathbf{x}}(t_{i+1}) + \cdots + \frac{(t-t_{i+1})\cdots(t_{i-j+2})}{j!h^j} \nabla^j \dot{\mathbf{x}}(t_{i+1}).$$

$$(124)$$

The integration of $\dot{\mathbf{x}}(t)$ between $t_{i-l} = t_i - lh$ and $t_{i+1} = t_i + h$ leads to the relation

$$\mathbf{x}(t_{i+1}) = \mathbf{x}(t_{i-l}) + \int_{t_i - lh}^{t_i + h} \dot{\mathbf{x}}(t)\, dt. \qquad (125)$$

Upon substituting (124), it is found that

$$\mathbf{x}(t_{i+1}) = \mathbf{x}(t_{i-l}) + h \sum_{k=0}^{j} c_k(l) \nabla^k \dot{\mathbf{x}}(t_{i+1}), \qquad (126)$$

where $c_0(l) = l+1$ and, for $k > 0$,

$$c_k(l) = \int_{-l}^{1} \frac{(\tau-1)\tau(\tau+1)\cdots(\tau+k-2)}{k!}\, d\tau. \qquad (127)$$

Values of $c_k(l)$, for the completely general closed formula (126), have been computed and are given in Table 2.

Table 2.   Values of $c_k(l)$

				l		
k	0	1	2	3	4	5
0	1	2	3	4	5	6
1	$-\frac{1}{2}$	-2	$-\frac{9}{2}$	-8	$-\frac{25}{2}$	-18
2	$-\frac{1}{12}$	$\frac{1}{3}$	$\frac{27}{12}$	$\frac{20}{3}$	$\frac{175}{12}$	27
3	$-\frac{1}{24}$	0	$-\frac{3}{8}$	$-\frac{8}{3}$	$-\frac{69}{8}$	-24
4	$-\frac{19}{720}$	$-\frac{1}{90}$	$-\frac{3}{80}$	$\frac{14}{45}$	$\frac{425}{84}$	$\frac{123}{10}$
5	$-\frac{3}{160}$	$-\frac{1}{90}$	$-\frac{3}{160}$	0	$-\frac{95}{288}$	$-\frac{33}{10}$

As in the case of the open formulas, the error in (126) due to truncating Newton's backward difference formula for $\dot{\mathbf{x}}(t)$ is proportional to h^{j+2}.

At first glance it might appear as though closed formulas are of little value, since the state equation cannot be used to compute $\dot{\mathbf{x}}(t_{i+1})$ until $\mathbf{x}(t_{i+1})$ is known, and (126) cannot be used to determine $\mathbf{x}(t_{i+1})$ until $\dot{\mathbf{x}}(t_{i+1})$ is known. However, the closed formulas are useful in numerical solution of the state equation by what are known as predictor-corrector methods, which we shall examine shortly.

EULER's METHOD

Consider, simultaneously, the open formula (118) with $j = 0$ and the state equation evaluated at t_i. Thus

$$\mathbf{x}(t_{i+1}) = \mathbf{x}(t_i) + h\,\dot{\mathbf{x}}(t_i) \tag{128a}$$

$$\dot{\mathbf{x}}(t_i) = \mathbf{f}[\mathbf{x}(t_i), t_i]. \tag{128b}$$

The value of $\mathbf{x}$ at $t = t_i$ inserted into the right-hand side of the second equation gives the value of the derivative $\dot{\mathbf{x}}(t_i)$. When this is inserted into the right-hand side of the first equation, the result is the value of $\mathbf{x}$ at t_{i+1}. Alternate use of these two expressions leads to the values of x at $t_i + kh$ for all values of k. This numerical procedure is called *Euler's method*.

We shall not devote much more attention to this elementary method because the error is significantly larger than that associated with other methods. There is the further undesirable feature that the error may grow significantly as time progresses. This is best illustrated by example. Consider the scalar equation

$$\frac{dx}{dt} = x,$$

with $x(0) = 1$. The exact solution ϵ^t is concave upwards, as shown in Fig. 13. In Euler's method the value of $x(h)$ is computed by using $x(0)$ and the slope of the solution passing through the point $[x(0), 0]$. As illustrated in Fig. 13. $x(h)$ is less than the exact solution value. You will find it easy to convince yourself that the numerical solution, as it is computed for subsequent points in time, will depart by ever increasing amounts from the exact solution—a most undesirable situation. This is illustrated in Fig. 13, with $h = 0.5$, which is an unusually large value, so as to make this phenomenon more clearly evident.

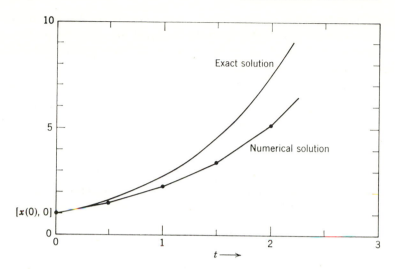

Fig. 13.   Numerical solution of $dx/dt = x$ by the Euler method, with $h = 0.5$.

THE MODIFIED EULER METHOD

A modification of Euler's method is designed to overcome one of its shortcomings by making $\mathbf{x}(t_{i+1})$ dependent on $\dot{\mathbf{x}}$ at both t_i and t_{i+1} rather than at t_i alone.

We start with the closed formula (126), with $j = 1$ and $l = 1$; this gives

$$\mathbf{x}(t_{i+1}) = \mathbf{x}(t_i) + h[\dot{\mathbf{x}}(t_{i+1}) - \tfrac{1}{2}\nabla \dot{\mathbf{x}}(t_{i+1})], \qquad (129)$$

which, if $\nabla \dot{\mathbf{x}}(t_{i+1}) = \dot{\mathbf{x}}(t_{i+1}) - \dot{\mathbf{x}}(t_i)$ is substituted, becomes

$$\mathbf{x}(t_{i+1}) = \mathbf{x}(t_i) + h\,\frac{\dot{\mathbf{x}}(t_{i+1}) + \dot{\mathbf{x}}(t_i)}{2}. \qquad (130)$$

Thus, if (129) is used, $\mathbf{x}(t_{i+1})$ is determined by $\mathbf{x}(t_i)$ and the average of $\dot{\mathbf{x}}$ at t_i and t_{i+1}. At first glance it might appear that this relation is useless, since $\dot{\mathbf{x}}(t_{i+1})$ can be determined from the state equation only when $\mathbf{x}(t_{i+1})$, which is the quantity being evaluated, is known. This difficulty is overcome in the following manner: Starting with $\mathbf{x}(t_i)$, Euler's method is used to predict the value of $\mathbf{x}(t_{i+1})$. With this *predicted* value of $\mathbf{x}(t_{i+1})$, equation 129, together with the state equation, is used to compute a corrected value of $\mathbf{x}(t_{i+1})$. This last step is then repeated until successive corrected values of $\mathbf{x}(t_{i+1})$ are equivalent, to the desired numerical accuracy; for example, equivalent to four significant decimal digits.

This is the *modified Euler method*; as the words used were meant to imply, it is a *predictor-corrector* method. It is just one method, perhaps the most elementary one, from a large class of such predictor-corrector methods, some others of which will be considered.

To illustrate the calculations required in the modified Euler method, suppose that

$$\frac{d}{dt}\begin{bmatrix} x_1 \\ x_2 \end{bmatrix} = \begin{bmatrix} -x_1{}^3 \\ x_1 - x_2 \end{bmatrix}$$

is the state equation and $\mathbf{x}(0) = [1.000 \quad -2.000]'$. We shall arbitrarily set $h = 0.1$. Now, $\mathbf{x}(0)$ and the state equation together yield $\dot{\mathbf{x}}(0) = [-1 \quad 3]'$. Equation 128$a$ in the Euler method gives

$$\mathbf{x}(h) = [0.900 \quad -1.700]'$$

as the predicted value of $\mathbf{x}(h)$. The state equation then gives the corresponding value of $\dot{\mathbf{x}}(h)$ as $\dot{\mathbf{x}}(h) = [-0.729 \quad 2.600]'$. When (130) of the modified Euler method is used, the first corrected value of $\mathbf{x}(h)$ becomes $[0.914 \quad -1.740]'$. The state equation now gives a new value of $\dot{\mathbf{x}}(h)$, and (130) yields the second corrected value of $\mathbf{x}(h)$. These are

$$\dot{\mathbf{x}}(h) = \begin{bmatrix} -0.764 \\ 2.654 \end{bmatrix}, \qquad \mathbf{x}(h) = \begin{bmatrix} 0.912 \\ -1.717 \end{bmatrix}.$$

Repetition of this step yields a new $\dot{\mathbf{x}}(h)$ and a third corrected value of $\mathbf{x}(h)$; they are

$$\dot{\mathbf{x}}(h) = \begin{bmatrix} -0.758 \\ 2.629 \end{bmatrix}, \qquad \mathbf{x}(h) = \begin{bmatrix} 0.912 \\ -1.719 \end{bmatrix}.$$

Assuming that three-digit accuracy is adequate, the calculation of $\mathbf{x}(h)$ is terminated, since the last two calculated values of $x(h)$ are equivalent to that accuracy. The evaluation of $\mathbf{x}(2h)$, $x(3h)$, $\cdots$ is carried out in the same manner, with the next calculations beginning from the value of $\mathbf{x}(h)$ just established.

THE ADAMS METHOD

The open formula (118) for some j, such as the particular case of $j = 3$ given here,

$$\mathbf{x}(t_{i+1}) = \mathbf{x}(t_i) + h[\dot{\mathbf{x}}(t_i) + \tfrac{1}{2}\nabla\dot{\mathbf{x}}(t_i) + \tfrac{5}{12}\nabla^2\dot{\mathbf{x}}(t_i) + \tfrac{3}{8}\nabla^3\dot{\mathbf{x}}(t_i)] \qquad (131)$$

forms the basis for the *Adams method*. The state equation is used, of course, to evaluate $\dot{\mathbf{x}}(t_i)$ from $\mathbf{x}(t_i)$.

Like other methods that use the open and closed formulas with $j \geq 1$, the Adam's method is *not self-starting*; that is, it is not sufficient to know just the state equation and $\mathbf{x}(t_0)$; the values of $\mathbf{x}(t_1), \cdots$, and $x(t_j)$ must also be known, Only with this added information can the first complete set of backward differences be calculated, at time t_j, and (118) used to evaluate $\mathbf{x}(t_{j+1})$.

The additional values of $\mathbf{x}$ needed to start the Adams method can be generated by some other method. Alternatively, a truncated Taylor series expansion at t_0 for $\mathbf{x}(t)$ may be used to evaluate $\mathbf{x}(t_1), \cdots$, and $x(t_j)$. To keep the errors in the Adams method and the series evaluation of the starting values equivalent, the Taylor series should be truncated after $j + 2$ terms; that is, the following truncated series should be used:

$$\mathbf{x}(t) = \mathbf{x}(t_0) + \mathbf{x}^{(1)}(t_0)(t - t_0) + \tfrac{1}{2}\,\mathbf{x}^{(2)}(t_0)(t - t_0)^2$$

$$+ \cdots + \frac{1}{(j + 1)!}\,\mathbf{x}^{(j+1)}(t_0)(t - t_0)^{j+1}. \quad (132)$$

This series method clearly requires that $\mathbf{f}(\mathbf{x}, t)$ be sufficiently differentiable (j times in $\mathbf{x}$ and t) at t_0 ; only then can $\mathbf{x}^{(k)}(t_0)$ be evaluated; for example, $\ddot{\mathbf{x}}(t_0) = \mathbf{f}_\mathbf{x}[\mathbf{x}(t_0), t_0]\dot{\mathbf{x}}(t_0) + \mathbf{f}_t[\mathbf{x}(t_0), t_0] = \mathbf{f}_\mathbf{x}[\mathbf{x}(t_0), t_0]\mathbf{f}[\mathbf{x}(t_0), t_0] + \mathbf{f}_t[\mathbf{x}(t_0), t_0]$.

To illustrate the Adams method, let us consider the very simple state equation

$$\frac{d}{dt}x = -x + \epsilon^{-t},$$

with $x(0) = 0$. Set $h = 0.1$ and take $j = 3$. Thus (131) will be the particular case of (118) used here.

The starting values will be obtained using a truncated Taylor series. First,

$$\dot{x}(0) = [-x(t) + \epsilon^{-t}]\big|_{t=0} = 1$$

$$\ddot{x}(0) = [-\dot{x}(t) - \epsilon^{-t}]\big|_{t=0} = -2$$

$$\dddot{x}(0) = [-\ddot{x}(t) + \epsilon^{-t}]\big|_{t=0} = 3$$

$$x^{(4)}(0) = [-\dddot{x}(t) - \epsilon^{-t}]\big|_{t=0} = -4.$$

Hence, the truncated series (132) is, in this case,

$$x(t) = t - t^2 + \tfrac{1}{2}t^3 - \tfrac{1}{6}t^4.$$

By using this series, $x(h)$, $x(2h)$, and $x(3h)$ are evaluated; thus $x(h) = 0.0905$, $x(2h) = 0.1637$, $x(3h) = 0.2222$. With these values (131) can be used to compute subsequent values of $x(ih)$. These values along with those of the several backward differences are shown in Table 3 for values of i up to 10. For comparison, the exact values, $(ih)\epsilon^{-(ih)}$, of x at $t = ih$ are also tabulated.

Table 3

i	Exact $x(ih)$	Numerical $x(ih)$	$\dot{x}(ih)$	$\nabla\,\dot{x}(ih)$	$\nabla^2\,\dot{x}(ih)$	$\nabla^3\,\dot{x}(ih)$
0	0.0000	0.0000	1.0000			
1	0.0905	0.0905	0.8143	-0.1857		
2	0.1637	0.1637	0.6550	-0.1593	0.0264	
3	0.2222	0.2222	0.5186	-0.1364	0.0229	-0.0035
4	0.2681	0.2681	0.4022	-0.1164	0.0200	-0.0029
5	0.3032	0.3032	0.3033	-0.0989	0.0175	-0.0025
6	0.3293	0.3292	0.2196	-0.0837	0.0152	-0.0023
7	0.3476	0.3475	0.1491	-0.0705	0.0132	-0.0020
8	0.3595	0.3594	0.0899	-0.0592	0.0113	-0.0019
9	0.3659	0.3658	0.0408	-0.0491	0.0101	-0.0012
10	0.3679	0.3678				

MODIFIED ADAMS METHOD

A modification of the Adams method can be made, similar to the modification of the Euler method. The Adams method is used to predict a value for $\mathbf{x}(t_{i+1})$. The closed formula (126) for some j and $l = 0$ is used repetitively until, to the desired numerical accuracy, successive corrected values of $\mathbf{x}(t_{i+1})$ are equivalent. As a typical case, (131) [i.e., (118) with $j = 3$] is used to predict the value of $\mathbf{x}(t_{i+1})$; then

$$\mathbf{x}(t_{i+1}) = \mathbf{x}(t_i) + h[\dot{\mathbf{x}}(t_{i+1}) - \tfrac{1}{2}\nabla\,\dot{\mathbf{x}}(t_{i+1}) - \tfrac{1}{12}\nabla^2\,\dot{\mathbf{x}}(t_{i+1}) - \tfrac{1}{24}\nabla^3\,\dot{\mathbf{x}}(t_{i+1})],$$

$$(133)$$

which is (126) with $j = 3$ and $l = 0$, is used to compute corrected values of $\mathbf{x}(t_{i+1})$. This method is called the *modified Adams method*; like the modified Euler method, it is a predictor-corrector method.

MILNE METHOD

The *Milne method* is another predictor-corrector method; in particular it is a method that makes good use of zeros that appear in Tables 1 and 2. Equation 122 with $j = 3$ and $l = 3$ is used to predict the value of $\mathbf{x}(t_{i+1})$. Because $b_3(3) = 0$, the term $\nabla^3 \dot{\mathbf{x}}(t_i)$ need never be included in the equation. Thus

$$\mathbf{x}(t_{i+1}) = \mathbf{x}(t_{i-3}) + h[4\dot{\mathbf{x}}(t_i) - 4\nabla \dot{\mathbf{x}}(t_i) + \tfrac{8}{3} \nabla^2 \dot{\mathbf{x}}(t_i)]. \tag{134}$$

Note that only two backward differences are computed, as when truncating the open formula (122) at $j = 2$. However, the accuracy is the same as that achieved when truncating the open formula at $j = 3$. The closed formula (126) with $j = 3$ and $l = 1$ is used to compute the corrected values of $\mathbf{x}(t_{i+1})$. Since $c_3(1) = 0$, only two, rather than three, backward differences must be computed. Thus the equation may be written without the $\nabla^3 \dot{\mathbf{x}}(t_{i+1})$ term as

$$\mathbf{x}(t_{i+1}) = \mathbf{x}(t_{i-1}) + h[2\dot{\mathbf{x}}(t_{i+1}) - 2\nabla \dot{\mathbf{x}}(t_{i+1}) + \tfrac{1}{3} \nabla^2 \dot{\mathbf{x}}(t_{i+1})]. \tag{135}$$

A second Milne method uses the open formula (122) with $j = 5$ and $l = 5$ to predict $\mathbf{x}(t_{i+1})$, and the closed formula (126) with $j = 5$ and $l = 3$ to correct $\mathbf{x}(t_{i+1})$. The fact that $b_5(5)$ and $c_5(3)$ are zero reduces the computing effort to that expended when (122) and (126) are truncated at $j = 4$.

PREDICTOR-CORRECTOR METHODS

Several of the methods we have examined are predictor-corrector methods. They belong to a large class of such methods that use the open formula (122), for some j and l, to predict $\mathbf{x}(t_{i+1})$; and the closed formula (126), for some j and l, to correct $\mathbf{x}(t_{i+1})$. The indices j and l may be different in the two equations. As a matter of convention, (122) is called a *predictor* and (126) is called a *corrector*.

To illustrate how these equations provide a particular method, let us see how the modified Euler method can be improved without adding to the amount of computing involved. The corrector

$$\mathbf{x}(t_{i+1}) = \mathbf{x}(t_i) + h \frac{\dot{\mathbf{x}}(t_i) + \dot{\mathbf{x}}(t_{i+1})}{2}$$

is (126) with $j = 1$ and $l = 0$. The predictor

$$\mathbf{x}(t_{i+1}) = \mathbf{x}(t_i) + h \dot{\mathbf{x}}(t_i)$$

is (122) with $j = 0$ and $l = 0$. Thus the predictor is not as accurate as the corrector. This means that the corrector, most likely, will have to be used more times to arrive at successive values of $\mathbf{x}(t_{i+1})$ that are equivalent, than would be necessary if the predictor and corrector had the same accuracy. If you examine Table 1, you will find that $b_1(1) = 0$. Thus (122) with $j = 1$ and $l = 1$ yields a predictor

$$\mathbf{x}(t_{i+1}) = \mathbf{x}(t_{i-1}) + 2h\,\dot{\mathbf{x}}(t_i),$$

with the same accuracy as the corrector; and the amount of computing needed to predict $\mathbf{x}(t_{i+1})$ is the same as with the original predictor.

You should make particular note of the fact that most of the predictor-corrector methods based on (122) and (126) are not self-starting. The exceptions are (122) with $j = 0$ and $l = 0$ and (126) with $j = 0$ or 1 and $l = 0$.

RUNGE-KUTTA METHOD

The fourth-order *Runge-Kutta method* is widely known and used for obtaining a numerical solution of the state equation. It is self-starting—a distinct advantage—and quite accurate.

The fourth-order Runge-Kutta method is expressed by the equation

$$\mathbf{x}(t_{i+1}) = \mathbf{x}(t_i) + \tfrac{1}{6}[\mathbf{w}_0 + 2\mathbf{w}_1 + 2\mathbf{w}_2 + \mathbf{w}_3], \tag{136}$$

where

$$\mathbf{w}_0 = h\,\mathbf{f}[\mathbf{x}(t_i), t_i]$$

$$\mathbf{w}_1 = h\,\mathbf{f}\left[\mathbf{x}(t_i) + \frac{\mathbf{w}_0}{2}, t_i + \frac{h}{2}\right]$$

$$\mathbf{w}_2 = h\,\mathbf{f}\left[\mathbf{x}(t_i) + \frac{\mathbf{w}_1}{2}, t_i + \frac{h}{2}\right]$$

$$\mathbf{w}_3 = h\,\mathbf{f}[\mathbf{x}(t_i) + \mathbf{w}_2, t_i + h].$$

The validation of this equation will not be attempted, as it is quite long and provides no useful information for development of other methods. It is known that the error term is proportional to h^5. Thus, if h is sufficiently small, we may well expect the error to be negligible.

Observe that no corrector is employed in the method. Compared with other methods that have similar accuracy and no corrector, more computational effort is required. For each time increment $\dot{x}$ must be evaluated four times compared with only one evaluation in the other methods. Furthermore, in a predictor-corrector method employing a sufficiently small step size h, $\dot{x}$ will seldom be evaluated more than twice. Thus, the Runge-Kutta method also compares poorly with a predictor-corrector method.

Let us observe, however, that the advantages and disadvantages of the Runge-Kutta method and the predictor-corrector methods complement each other. The Runge-Kutta method is self-starting and the predictor-corrector methods require less computation. Thus the Runge-Kutta method is best used to start one of the predictor-corrector methods.

ERRORS

We have said very little about errors in the numerical solution of the state equation except to point out the dependence on h of the error due to truncation of Newton's backward-difference formula. There are other errors, and you should be aware of how they occur.

There are errors that occur because arithmetic is done with numbers having a limited number of significant digits. This type of error is known as roundoff error, since the word "roundoff" denotes eliminating insignificant digits and retaining significant digits.

The truncation and roundoff errors occurring at each step in the calculation affect not only the error of the numerical solution at that step but also the error at subsequent steps; that is, the error propagates.

Another source of error is properly viewed as a dynamic error and occurs in the following manner. The equations used to obtain a numerical solution of the state equation may exhibit more dynamically independent modes than the state equation. If any of the additional modes are unstable then the numerical solution may depart radically from the actual solution.

We shall not consider errors any further but refer you to books on numerical analysis. (See Bibliography.)

10.7 LIAPUNOV STABILITY

In the case of linear networks, for which general analytic solutions of the state equation exist and can be determined, it is possible to examine the solution and study its properties. In particular, it is possible to make observations about the stability properties of the solution—whether it remains bounded or even approaches zero as $t \to \infty$. For nonlinear networks general analytic solutions do not exist, and so quantitative observations cannot be made about the solution. It is important therefore to have some idea of the qualitative behavior of a solution of the state equation, in particular its behavior as $t \to \infty$, which describes the degree of stability or instability of the solution.

STABILITY DEFINITIONS

The nonlinear state equation is repeated here:

$$\frac{d\mathbf{x}}{dt} = \mathbf{f}(\mathbf{x}, t). \tag{137}$$

It is observed that at those points at which $\mathbf{f}(\mathbf{x}, t) = \mathbf{0}$ for all $t \geq t_0$, the time rate of change of $\mathbf{x}$ is identically zero. This implies of course that if the state starts at or reaches one of these points, it will remain there. Clearly these points are distinctive; they are therefore, given the special name of *singular points*.

The subject of stability is concerned with the behavior of the state equation solution relative to a singular point. It is a matter of convenience in definitions and theorems to make the singular point in question the origin. To see that this is always possible, let $\mathbf{x}_s$ be a singular point. Next, set $\mathbf{y}(t) = \mathbf{x}(t) - \mathbf{x}_s$. Then $\mathbf{x}(t) = \mathbf{x}_s$ corresponds to $\mathbf{y}(t) = \mathbf{0}$. The equivalent identification $\mathbf{x}(t) = \mathbf{y}(t) + x_s$ substituted in the state equation 137 gives

$$\frac{d}{dt}\mathbf{y} = \mathbf{f}(\mathbf{y} + \mathbf{x}_s, t) = \hat{\mathbf{f}}(\mathbf{y}, t). \tag{138}$$

Since $\mathbf{y}$ and $\mathbf{x}$ differ only by the constant vector $\mathbf{x}_s$, either determines the state of the network. If we view $\mathbf{y}$ as the state vector, then (138) is the state equation and the origin is a singular point. Thus, without loss of generality, we shall assume that (137) establishes the origin as a singular point.

In what follows we shall be using norms without reference to a particular norm; however, the illustrations will pertain to the Euclidean norm. We shall let $S\rho$ denote the spherical region $\|\mathbf{x}\| < \rho$ in vector space and B_ρ denote the boundary of the spherical region S_ρ. Thus B_ρ is the sphere $\|\mathbf{x}\| = \rho$.

It is assumed that a unique, continuous solution of the state equation exists in the spherical region S_E. The locus of points $\mathbf{x}(t)$ in vector space for $t \geq t_0$ is called a *positive half-trajectory* or, for brevity, *trajectory* and is denoted by $\mathbf{x}^+$.

Each of the following three definitions is illustrated in Fig. 14 to give a visual as well as a verbal interpretation of stability.

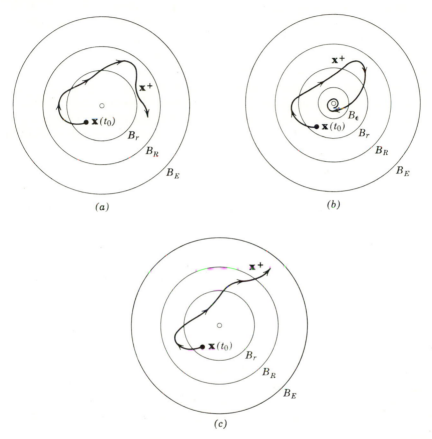

Fig. 14. Illustrations of stability definitions: (*a*) Stable origin: (*b*) asymptotically stable origin; (*c*) unstable origin.

Definition. *The origin is stable, if for each* $\mathrm{R} < \mathrm{E}$ *there is an* $\mathrm{r} \leq \mathrm{R}$ *such that any trajectory* $\mathbf{x}^+$ *originating in* $\mathrm{S_r}$ *remains in* $\mathrm{S_R}$. [*The point at which the trajectory originates is* $\mathbf{x}(t_0)$.]

The asymptotic behavior of the solution as $t \to \infty$ is determined by the set of points that the trajectory approaches at $t \to \infty$. If the trajectory tends to the origin, the particular type of stability is given a name.

More precisely,

Definition.   *The origin is asymptotically stable if it is stable and if for any* $\epsilon > 0$ *there exists a* t_ϵ *such that the trajectory remains in* S_ϵ *for* $t > t_\epsilon$.

This definition states in a precise way that $\mathbf{x}(t)$ approaches zero as t tends to infinity, by requiring the existence of a value of time after which the norm of the solution remains less than any abritrarily small number. Observe that both stability and asymptotic stability are *local*, or *in-the-small*, properties, in that their definitions permit $r > 0$ to be as small as necessary to satisfy the defintion. On the other hand, if, when $R = +\infty$, the origin is asymptotically stable for $r = +\infty$, then the origin is said to be *asymptotically stable in-the-large*, or *globally asymptotically stable*. In other words, $\mathbf{x}(t)$ approaches zero as t tends to infinity for all $\mathbf{x}(t_0)$.

Since not all networks are stable, the concept of instability must be made precise; this is done by the next definition.

Definition.   *The origin is unstable, if for some* $R < E$ *and any* $r \leq R$ *there exists at least one trajectory originating in* S_r *that crosses* B_R .

The conditions under which the origin is stable or asymptotically stable are stated in terms of the existence of certain classes of functions. We shall now define these functions.

Definition.   *The scalar function* $V(\mathbf{x})$ *is said to be* positive definite, *if* (1) $V(\mathbf{x})$ *and its first partial derivatives are continuous in an open region* D containing the origin,* (2) $V(\mathbf{0}) = 0$, *and* (3) $V(\mathbf{x}) > 0$ *for* $\mathbf{x} \neq \mathbf{0}$ *in D.*

Because the scalar function V may sometimes be an explicit function of t as well as of $\mathbf{x}$, the notion of positive definiteness must be extended to such cases.

Definition.   *The scalar function* $V(\mathbf{x}, t)$ *is said to be* positive definite, *if* (1) $V(\mathbf{x}, t)$ *and its first partial derivatives are continuous for* $t \geq t_0$ *and in an open region D containing the origin,* (2) $V(\mathbf{0}, t) = 0$ *for* $t \geq t_0$, *and* (3) $V(\mathbf{x}, t) \geq W(\mathbf{x})$ *for* $t \geq t_0$, *where* $W(\mathbf{x})$ *is a positive definite function of* $\mathbf{x}$ *alone.*

The continuity of the first partial derivatives guarantees the existence of $\boldsymbol{\nabla} V(\mathbf{x}, t)$, the gradient of $V(\mathbf{x}, t)$. Therefore we may write

$$\frac{d}{dt} V(\mathbf{x}, t) = \frac{\partial V}{\partial t} (\mathbf{x}, t) + [\boldsymbol{\nabla} V(\mathbf{x}, t)]' \frac{d}{dt} \mathbf{x}$$

$$= \frac{\partial V}{\partial t} (\mathbf{x}, t) + [\boldsymbol{\nabla} V(\mathbf{x}\ t)]' \mathbf{f}(\mathbf{x}, t). \tag{139}$$

---

* A region is said to be open if it contains none of its boundary points.

The last form follows since $dx/dt = \mathbf{f}(\mathbf{x}, t)$ along a trajectory of the network. Thus it is meaningful to talk about the time rate of change of V along a trajectory of the network. A very important class of functions is defined on the basis of the sign of this rate of change.

Definition. *A positive definite function* $V(\mathbf{x}, t)$ *is called a* Liapunov function *if*—$dV/dt \geq 0$ *along trajectories in* D.

On the basis of these definitions we can now discuss the question of stability and the conditions under which a singular point is stable.

STABILITY THEOREMS

In order to make the theorems relating to stability more meaningful we shall observe the behavior of a particularly simple Liapunov function $V(\mathbf{x})$, which is time invariant and involves a state vector with only two elements, x_1 and x_2.

Since $V(\mathbf{x}) > 0$ for $\mathbf{x} \neq \mathbf{0}$ and $V(\mathbf{x}) = 0$ for $\mathbf{x} = 0$, $v = V(\mathbf{x})$ may be depicted as a bowl-shaped surface tangent to the $x_1 - x_2$ plane at the origin, as shown in Fig. 15. The intersection of this surface with the horizontal planes $v = C_i$, where $C_1 < C_2 < \cdots$, will be closed curves. If these

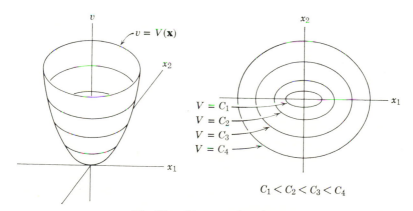

Fig. 15.   Liapunov function.

curves are projected vertically into the $x_1 - x_2$ plane, they will form a set of concentric constant-V contours with the value of V decreasing toward zero as the contours are crossed in approaching the origin. This is illustrated in Fig. 15. Since V is a Liapunov function, V must be nonincreasing along a trajectory. Hence a trajectory originating inside the contour $V(\mathbf{x}) = C_i$ can never cross that contour. The trajectory is thus constrained to a neighborhood of the origin. This is very nearly the condition laid down for stability. Therefore, we may properly anticipate that stability follows from the existence of a time-invariant Liapunov function. Although this illustration used a two-element state vector, it is not difficult to imagine the generalization to a state vector with n elements.

The theorems to be given next will make these notions more precise and will permit the Liapunov function to vary with time.

Theorem 17.   *The origin is stable if in an open region* D *containing the origin there exists a Liapunov function such that* $V(\mathbf{x}, t) \leq U(\mathbf{x})$ *for all* $t \geq t_0$ *where* $U(\mathbf{x})$ *is a positive definite function.*

This theorem is proved as follows. Given a number R, let C be the minimum value* of $W(\mathbf{x})$ for all $\mathbf{x}$ such that $\|\mathbf{x}\| = R$. Let $\mathbf{x}_r$ be a vector $\mathbf{x}$ having the least norm for which $U(\mathbf{x}) = C$. This vector exists and is not the zero vector, since $U(\mathbf{x}) = 0$ if and only if $\mathbf{x} = 0$. Further, because $U(\mathbf{x}) \geq W(\mathbf{x})$ for all $t \geq t_0$, $\|\mathbf{x}_r\| \leq R$. Let $r = \|\mathbf{x}_r\|$. Then any trajectory originating in S_r does not leave S_R. This is verified as follows: By continuity, if $\mathbf{x}(t_0)$ is contained in S_r, then $\mathbf{x}(t)$ must be contained in S_R for small values of $t - t_0$. Suppose, at t_1, that $\|\mathbf{x}(t_1)\| = R$. Then

$$V[\mathbf{x}(t_1), t_1] \geq \mathbf{W}[\mathbf{x}(t_1)] \geq C. \tag{140}$$

Because $\mathbf{x}_r$ was a vector $\mathbf{x}$ of least norm for which $U(\mathbf{x}) = C$ and because $V(\mathbf{x}, t) \leq U(\mathbf{x})$ for all $t \geq t_0$, $V(\mathbf{x}, t_0) < C$ for all $\mathbf{x}$ such that $\|\mathbf{x}\| < \|\mathbf{x}_r\| = r$. Thus for all $\mathbf{x}(t_0)$ in S_r,

$$V[\mathbf{x}(t_0), t_0] < C. \tag{141}$$

Next, because $-dV/dt \geq 0$,

$$V[\mathbf{x}(t_1), t_1] \leq V[\mathbf{x}(t_0), t_0]. \tag{142}$$

Clearly, (140) is in contradiction with (141) and (142). Thus t_1 does not exist and $\mathbf{x}^+$ is contained in S_R.

This proof can be given the following geometrical interpretation when $\mathbf{x}$ is a 2-vector, as illustrated in Fig. 16. The relation $W(\mathbf{x}) = C$ defines a closed contour K_1 contained in S_R plus its boundary B_R. The relation $U(\mathbf{x}) = C$ defines a closed contour K_2 contained in the closed region†

---

* Recall from the definition for a time-varying positive definite function $V(\mathbf{x}, t)$ that $W(\mathbf{x})$ is a time-invariant positive definite function such that $V(\mathbf{x}, t) \geq W(\mathbf{x})$ for all $t \geq t_0$.

† A region is said to be closed if it includes all the points of its boundary.

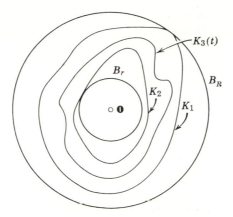

Fig. 16. Liapunov-function contour $K_3(t)$ for stable origin.

bounded by K_1. The spherical region S_r is the largest in the open region
bounded by K_2. Since $W(\mathbf{x}) \le V(\mathbf{x}, t) \le U(\mathbf{x})$ for all $t \ge t_0$, $V(\mathbf{x}, t) = C$
defines a closed contour $K_3(t)$ contained in the annular region bounded
between K_1 and K_2. Now, since $V[\mathbf{x}(t_0), t_0] < C$ for all $\mathbf{x}(t_0)$ in S_r and
$-dV/dt \ge 0$, a trajectory originating in S_r cannot cross $K_3(t)$ for $t \ge t_0$.
This means that $\mathbf{x}^+$ cannot cross K_1 either and must remain in S_R.

Now let us turn to asymptotic stability. To establish asymptotic
stability, an additional hypothesis must be introduced.

Theorem 18. *The origin is asymptotically stable if in an open region* D
containing the origin there exists a Liapunov function such that $V(\mathbf{x}, t) \le U(\mathbf{x})$
for all $t \ge t_0$, *where* $U(\mathbf{x})$ *is a positive definite function, and such that*
$-dV/dt$ *is positive definite.*

The additional requirement for asymptotic stability is that $-dV/dt$ be
positive definite. The proof of this theorem begins where the proof of the
previous theorem stopped. Select any ϵ that satisfies $0 < \epsilon \le r$. We must
show that there exists a t_ϵ such that x^+ is contained in S_ϵ for all $t \ge t_\epsilon$.
In the proof of the previous theorem, the value of r depended only on R
and *not* on t_0. Therefore we may infer the existence of a δ dependent
only on ϵ such that a trajectory, passing through a point of S_δ at time t_ϵ,
remains in S_ϵ for $t > t_\epsilon$. To complete the proof, we must show that $\mathbf{x}^+$
originating in S_r at time t_0, passes through a point of S_δ. Let w be the
least value of $W(\mathbf{x})$ for $\mathbf{x}$ such that $\delta \le \|\mathbf{x}\| \le R$. Then, since $V(\mathbf{x}, t)$
$\ge W(\mathbf{x})$ for $t \ge t_0$, $V[\mathbf{x}(t), t] < w$ only if $\mathbf{x}(t)$, which remains in S_R for all
$t \ge t_0$, is also in S_δ. We shall use this fact to show by contradiction that
$\mathbf{x}(t)$ is contained in S_δ for some $t \ge t_0$. Because $-dV/dt$ is positive definite,
a positive definite function $\hat{W}(\mathbf{x})$ exists such that $-dV/dt \ge \hat{W}$ for all

$t \geq t_0$. Let $\hat{w}$ be the least value of $\hat{W}(\mathbf{x})$ for $\mathbf{x}$ such that $\delta/2 \leq \|\mathbf{x}\| \leq R$. Then, under the restriction $\delta/2 \leq \|\mathbf{x}\| \leq R$,

$$V[\mathbf{x}(t), t] = V[\mathbf{x}(t_0), t_0] + \int_{t_0}^{t} \frac{dV[\mathbf{x}(\tau), \tau]}{d\tau} \, d\tau.$$

Since $dV/dt \leq -\hat{w}$ this relation yields the inequality

$$V[\mathbf{x}(t), t] \leq V[\mathbf{x}(t_0), t_0] - (t - t_0)\hat{w}, \tag{143}$$

valid for $\delta/2 \leq \|\mathbf{x}\| \leq R$. Now suppose $\mathbf{x}(t)$ is not in S_δ for any $t \geq t_0$; that is $\delta \leq \|\mathbf{x}(t)\| \leq R$ for all $t \geq t_0$. Then, by (143),

$$V[\mathbf{x}(t), t] < w \quad \text{for} \quad t > t_0 + \{V[\mathbf{x}(t_0), t_0] - w\}/\hat{w}.$$

This is a contradiction, since $V[\mathbf{x}(t), t] < w$ only if $\mathbf{x}(t)$ is in S. The proof is now complete.

There remains the question of asymptotic stability in-the-large, which we shall address by means of a theorem. The additional condition in this theorem requires the use of a particular type of positive definite function, which we shall now define. Loosely speaking, a function is *radially unbounded* if $V(\mathbf{x}, t)$ grows without bound, independent of the value of $t \geq t_0$, as $\mathbf{x}$ moves away from the origin or, alternatively, as $\|\mathbf{x}\|$ increases. In more precise terms, $V(\mathbf{x}, t)$ is said to be *radially unbounded* if, given any $M > 0$, there exists an m such that $V(\mathbf{x}, t) > M$ for all $t \geq t_0$ whenever $\|x\| > m$. With this background, we may state the theorem.

Theorem 19.   *The origin is asymptotically stable in-the-large if there exists a Liapunov function defined everywhere (the state equation has a solution) such that* $\mathrm{V}(\mathbf{x}, t) \leq \mathrm{U}(\mathbf{x})$ *for all* $t \geq t_0$, *where* $\mathrm{U}(\mathbf{x})$ *is a positive definite function, such that* $\mathrm{W}(\mathbf{x})$ *is radially unbounded* and such that* $-\mathrm{dV}/\mathrm{dt}$ *is positive definite.*

The proof does not depart very much from those given for the previous two theorems. Therefore it is left for you as an exercise.

Examples

Let us now illustrate some of the preceding theory with examples. Consider the network shown in Fig. 17. It is described by the state equation

$$\frac{d}{dt}\begin{bmatrix} x_1 \\ x_2 \end{bmatrix} = \begin{bmatrix} -x_1 - x_2 \\ x_1 - 4x_2 - \tanh x_2 \end{bmatrix}.$$

---

* See the footnote on page 757.

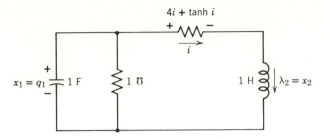

Fig. 17.   Nonlinear network.

Though we shall not do so, it can be shown that the solution of this state equation exists and is unique for all initial-state vectors $\mathbf{x}(t_0)$; that is, the positive scalar E that appears in the stability definitions is infinite.

A decision as to stability requires the discovery of a Liapunov function. There is no algorithm that can be followed to arrive at one. Experience is a guide; so, to gain some experience, let us see if ·the positive-definite function

$$V(\mathbf{x}) = \tfrac{1}{2}\,x_1{}^2 + \tfrac{1}{2}x_2{}^2$$

is a Liapunov function. To make this determination we must examine dV/dt; the result is

$$\frac{d}{dt}\,V(\mathbf{x}) = x_1\,\frac{dx_1}{dt} + x_2\,\frac{dx_2}{dt}$$

$$= -\,x_1{}^2 - 4x_2{}^2 - x_2\tanh x_2\,.$$

The second line is obtained by substituting dx_1/dt and dx_2/dt from the state equation. Clearly $-dV/dt$ is positive definite. Thus V is a Liapunov function that, furthermore, satisfies the additional restrictions of the theorem on asymptotic stability in-the-large. You may feel we forgot to consider the problem of selecting $U(\mathbf{x})$; however, a moment's reflection will convince you that, when V is time invariant, the supplemental condition calling for the existence of a $U(\mathbf{x})$ with the properties indicated is trivially satisfied by setting $U(\mathbf{x}) = V(\mathbf{x})$. We now know that the origin is asymptotically stable in-the-large.

It is not always possible to establish asymptotic stability in-the-large. Often only local stability properties can be certified. As an example,

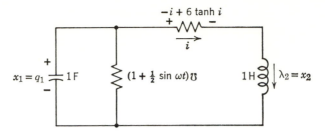

Fig. 18.   Nonlinear time-varying network.

consider the network illustrated in Fig. 18. It has the state equation

$$\frac{d}{dt}\begin{bmatrix} x_1 \\ x_2 \end{bmatrix} = \begin{bmatrix} -(1 + \frac{1}{2}\sin \omega t)x_1 - x_2 \\ x_1 - 6 \tanh x_2 + x_2 \end{bmatrix},$$

for which a unique solution may be shown to exist for all initial-state vectors. We shall use the same positive-definite function as before to be a trial Liapunov function; that is,

$$V(\mathbf{x}) = \tfrac{1}{2}x_1{}^2 + \tfrac{1}{2}x_2{}^2.$$

To verify that this function is a Liapunov function, we must evaluate dV/dt; the result is

$$\frac{d}{dt} V(\mathbf{x}) = x_1 \frac{d}{dt} x_1 + x_2 \frac{d}{dt} x_2$$

$$= -(1 + \tfrac{1}{2}\sin \omega t)x_1{}^2 - 6x_2 \tanh x_2 + x_2{}^2.$$

The second line follows by substitution of dx_1/dt and dx_2/dt from the state equation. Observe that $-dV/dt$ is positive definite if $6x_2 \tanh x_2 - x_2{}^2$ is positive for $x_2 \neq 0$. This is approximately equivalent to $|x_2| < 5.99999$. Now, in the open region $|x_2| < 5.99999$ (all values of x_1 are allowed) $-dV/dt$ is bounded from below by the positive-definite function

$$\tfrac{1}{2}x_1{}^2 + 6x_2 \tanh x_2 - x_2{}^2,$$

which we may take to be $\hat{W}(\mathbf{x})$. This is just one $\hat{W}(\mathbf{x})$ function; other positive-definite functions may also be found and used for $\hat{W}(\mathbf{x})$.

All the conditions of the theorem on asymptotic stability have been satisfied; therefore the origin is locally asymptotically stable. In fact, it may be shown, though we shall not do so, that all trajectories originating inside the disk of radius 5.99999 in the $x_1 - x_2$ plane approach the origin asymptotically.

INSTABILITY THEOREM

The stability theorems only specify sufficient conditions for stability, *not necessary conditions*. Therefore, if the conditions of the theorem are not satisfied, it is still possible for the origin, or other singular point being investigated, to be stable in one of the three senses considered. Another way in which the question of stability can be answered with certainty is to find sufficient conditions for *instability*. Thus, if it can be shown that the sufficiency conditions of some instability theorem are satisfied, then the origin cannot be stable. The following instability theorem, due to Chetaev, embraces two instability theorems originally formulated by Liapunov:

Theorem 20. *Let* D *be an open region containing the origin and let* D̃ *be an open region in* D *such that* (1) *the origin is a boundary point of* D̃; (2) *for all* $t \geq t_0$ *the function* V(x, t) *is positive and, along with its first partial derivatives, is continuous in* D̃; (3) $dV(x, t)/dt \geq \hat{W}(x)$ *for* $t \geq t_0$, *where* $\hat{W}(x)$ *is positive and continuous in* D̃; (4) $V(x, t) \leq U(x)$ *for* $t \geq t_0$, *where* U(x) *is continuous in* D̃; *and* (5) U(x) = 0 *at the boundary points of* D̃ *in* D. *Then the origin is unstable.*

The proof is quite short. The illustrations in Fig. 19 are helpful in the proof. Pick R such that D̃ does not lie entirely within S_R. For any arbitrarily small, positive $r \leq R$, it is possible to find a point in both S_r and D̃. Let $x(t_0)$ be such a point. By conditions 4 and 5 there exists a region

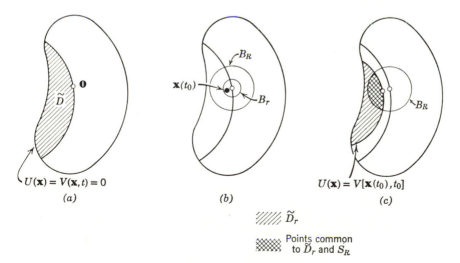

$U(x) = V(x, t) = 0$ $U(x) = V[x(t_0), t_0]$

(a)   (b)   (c)

     D̃$_r$

     Points common to D̃$_r$ and S_R

Fig. 19. Unstable origin.

contained entirely within $\tilde{D}$ such that $U(\mathbf{x})$ is greater than $V[\mathbf{x}(t_0), t_0] > 0$; denote it by $\tilde{D}_r$. Let $\hat{w} > 0$ denote the greatest lower bound assumed by $\hat{W}(\mathbf{x})$ at the points common to both $\tilde{D}_r$ and S_R. Now

$$V[\mathbf{x}(t), t] = V[\mathbf{x}(t_0), t_0] + \int_{t_0}^{t} \frac{d}{d\tau} V[\mathbf{x}(\tau), \tau] d\tau$$

yields the inequality

$$V[\mathbf{x}(t), t] \geq V[\mathbf{x}(t_0), t_0] + (t - t_0)\hat{w},$$

since $V(\mathbf{x}, t) \geq V[x(t_0), t_0]$ and $dV/dt \geq \hat{w}$ at the same points. By condition 4, $V(\mathbf{x}, t)$ is bounded from above at those points that are common to both $\tilde{D}_r$ and S_R. Therefore the above inequality indicates that the trajectory originating at $\mathbf{x}(t_0)$ must reach the boundary of those points in both $\tilde{D}_r$ and S_R. By the way we constructed $\tilde{D}_r$, the trajectory must, in fact, reach a point of B_R, the boundary of S_R.

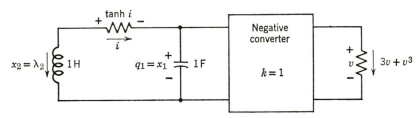

Fig. 20.   Unstable nonlinear network.

To illustrate this instability theorem, consider the network shown in Fig. 20. It possesses the state equation

$$\frac{d}{dt} \begin{bmatrix} x_1 \\ x_2 \end{bmatrix} = \begin{bmatrix} 3x_1 + x_1{}^3 - x_2 \\ x_1 - \tanh x_2 \end{bmatrix}.$$

It is possible to show, though we shall not do so, that a unique solution exists for all $\mathbf{x}(t_0)$. The function

$$V(\mathbf{x}) = \tfrac{1}{2} x_1{}^2 - \tfrac{1}{2} x_2{}^2$$

is positive for $|x_2| < |x_1|$. If the conditions of the instability theorem are to be satisfied, then we must show that $dV/dt > 0$ for $|x_2| < |x_1|$ in some neighborhood of the origin, To start with,

$$\frac{d}{dt} V(\mathbf{x}) = x_1 \frac{d}{dt} x_1 - x_2 \frac{d}{dt} x_2.$$

Upon substituting dx_1/dt and dx_2/dt, we get

$$\frac{d}{dt} V(\mathbf{x}) = 3x_1{}^2 + x_1{}^4 - 2x_1 x_2 + x_2 \tanh x_2$$

$$= 2x_1{}^2 + x_1{}^4 + (x_1 - x_2)^2 + x_2(\tanh x_2 - x_2).$$

The second line was obtained from the first by adding and subtracting $x_2{}^2$. Clearly,

$$\frac{d}{dt} V(\mathbf{x}) \geq 2x_1{}^2 + x_1{}^4 + (x_1 - x_2)^2 - x_2{}^2$$

$$\geq x_1{}^2 + x_1{}^4 + (x_1 - x_2)^2 + (x_1{}^2 - x_2{}^2).$$

The latter expression is clearly positive for $|x_2| < |x_1|$ in every neighborhood of the origin. Thus the origin must be unstable.

LIAPUNOV FUNCTION CONSTRUCTION

Observe the nature of the stability theorems. We are *not* given a definite prescription to follow; that is, there is no defined set of steps at the end of which we can reach an unambiguous conclusion as to whether or not the network is stable. Rather, the theorems provide a "hunting license." They ask that we seek a Liapunov function whose value remains bounded by a time-invariant positive-definite function. Finding an appropriate Liapunov function is a creative, inductive act, not a deductive one.

The functional form of a Liapunov function is not rigidly fixed. On the one hand, this is an advantage, because it affords a greater opportunity to establish stability by trying numerous potential Liapunov functions. On the other hand, it is a disadvantage, because there are no guidelines in picking a potential Liapunov function from the countless positive-definite functions one can think of. We shall now discuss a particular Liapunov function for time-invariant networks and establish, as a consequence, an alternative set of conditions for stability. Following this, we shall discuss a method of generating Liapunov functions.

Suppose that the network under consideration is described by the time-invariant state equation

$$\frac{d}{dt} \mathbf{x} = \mathbf{f}(\mathbf{x}). \tag{144}$$

Let us make explicit note of the fact that $\mathbf{f}(\mathbf{x})$ is a real-valued vector function of $\mathbf{x}$. In searching for a possible Liapunov function, consider

$$V(\mathbf{x}) = \mathbf{f}'(\mathbf{x}) \mathbf{f}(\mathbf{x}). \tag{145}$$

This is first of all a positive-definite function. We must therefore examine dV/dt. The time derivative of $\mathbf{f(x)}$ is

$$\frac{d}{dt}\,\mathbf{f(x)} = \mathbf{F(x)}\,\frac{d}{dt}\,\mathbf{x} = \mathbf{F(x)}\,\mathbf{f(x)}, \tag{146}$$

where $\mathbf{F(x)}$ is the Jacobian matrix of $\mathbf{f(x)}$; that is,

$$\mathbf{F(x)} = \begin{bmatrix} \dfrac{\partial f_1}{\partial x_1} & \dfrac{\partial f_1}{\partial x_2} \cdots \dfrac{\partial f_1}{\partial x_n} \\[2ex] \dfrac{\partial f_2}{\partial x_1} & \dfrac{\partial f_2}{\partial x_2} \cdots \dfrac{\partial f_2}{\partial x_n} \\[1ex] \vdots & \vdots \qquad \vdots \\[1ex] \dfrac{\partial f_n}{\partial x_1} & \dfrac{\partial f_n}{\partial x_2} \cdots \dfrac{\partial f_n}{\partial x_n} \end{bmatrix}. \tag{147}$$

Now, differentiation of (145) and subsequent substitution of (146) leads to

$$\begin{aligned} \frac{d}{dt}\,V(\mathbf{x}) &= \left[\frac{d}{dt}\,\mathbf{f'(x)}\right]\mathbf{f(x)} + \mathbf{f'(x)}\left[\frac{d}{dt}\,\mathbf{f(x)}\right] \\ &= \mathbf{f'(x)}\,\mathbf{F'(x)}\,\mathbf{f(x)} + \mathbf{f'(x)}\,\mathbf{F(x)}\,\mathbf{f(x)} \\ &= \mathbf{f'(x)}\,[\mathbf{F'(x)} + \mathbf{F(x)}]\,\mathbf{f(x)}. \end{aligned} \tag{148}$$

The matrix $-[\mathbf{F'(x)} + \mathbf{F(x)}]$ is symmetric and real; if it is also positive semidefinite in some neighborhood of the origin, then $-dV/dt \geq 0$ and the theorem on stability is verified. Other fairly obvious conditions give either local or global asymptotic stability. These results, due to Krasovskii, are stated precisely in the following theorem:

Theorem 21. *Let $\mathbf{f(x)}$ be differentiable with respect to $\mathbf{x}$ and let $\mathbf{f(0)} = \mathbf{0}$; then the origin is (1) stable if $-[\mathbf{F'(x)} + \mathbf{F(x)}]$ is positive semidefinite in some neighborhood of the origin, (2) asymptotically stable if $-[\mathbf{F'(x)} + \mathbf{F(x)}]$ is positive definite in some neighborhood of the origin, or (3) asymptotically stable in-the-large if $-[\mathbf{F'(x)} + \mathbf{F(x)}]$ is positive definite for all $\mathbf{x}$ and $\mathbf{f'(x)}\,\mathbf{f(x)}$ is radially unbounded.*

To illustrate this theorem, consider the network in Fig. 21. The state equation for this network

$$\frac{d}{dt}\begin{bmatrix} x_1 \\ x_2 \\ x_3 \end{bmatrix} = \begin{bmatrix} -x_1 + x_2 \\ x_1 - 3x_2 + x_2^3 - x_3 \\ x_2 - 4x_3 - \tanh x_3 \end{bmatrix}$$

has a unique solution for all $\mathbf{x}(t_0)$. The Jacobian matrix is

$$\mathbf{F}(\mathbf{x}) = \begin{bmatrix} -1 & 1 & 0 \\ 1 & -3 + 3x_2^2 & -1 \\ 0 & 1 & -4 - \operatorname{sech}^2 x_3 \end{bmatrix}.$$

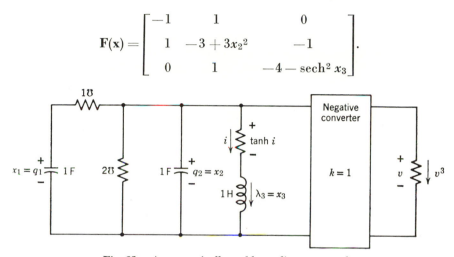

Fig. 21.   Asymptotically stable nonlinear network.

Upon adding $\mathbf{F}'(\mathbf{x})$ to $\mathbf{F}(\mathbf{x})$ and taking the negative of the resulting matrix, we find

$$-[\mathbf{F}'(\mathbf{x}) + \mathbf{F}(\mathbf{x})] = \begin{bmatrix} 2 & -2 & 0 \\ -2 & 6 - 6x_2^2 & 0 \\ 0 & 0 & 8 + 2 \operatorname{sech}^2 x_3 \end{bmatrix}.$$

The ascending principal cofactors are

$$2, \qquad 8 - 12x_2^2, \qquad \text{and} \qquad (8 - 12x_2^2)(8 + 2 \operatorname{sech}^2 x_3).$$

Clearly, each of these cofactors is positive and, hence, $-[\mathbf{F}'(\mathbf{x}) + \mathbf{F}(\mathbf{x})]$ is positive definite for all x_1, $x_2^2 < \frac{2}{3}$, and all x_3.* Thus the origin is

---

* See Section 7.2 for a discussion of the conditions for positive definiteness of a matrix. Then see Problem 18 in Chapter 7 for the particular criterion for positive definiteness used here.

asymptotically stable. The asymptotic stability is not global because the x_2 needed to make $-[\mathbf{F}'(\mathbf{x}) + \mathbf{F}(\mathbf{x})]$ positive definite is bounded from above and below.

This last theorem, though useful, is restrictive in that the Liapunov function is completely specified once the state equation is given. To make full use of the stability theorems, one must have some freedom to try different potential Liapunov functions. Obviously, however, something more than trial and error would be valuable. What is needed is some guideline for generating a Liapunov function. We shall now discuss such a procedure.

In what follows, the state equation is permitted to be time-varying; however, the Liapunov function is required to be time-invariant. The time derivative of V along a trajectory of the state equation is

$$\frac{d}{dt} V(\mathbf{x}) = [\boldsymbol{\nabla} V(\mathbf{x})]' \mathbf{f}(\mathbf{x}, t) \tag{149}$$

where

$$\boldsymbol{\nabla} V = \begin{bmatrix} \dfrac{\partial V}{\partial x_1} \\[4pt] \dfrac{\partial V}{\partial x_2} \\[4pt] \vdots \\[4pt] \dfrac{\partial V}{\partial x_n} \end{bmatrix} = \begin{bmatrix} \boldsymbol{\nabla} V_1 \\[4pt] \boldsymbol{\nabla} V_2 \\[4pt] \vdots \\[4pt] \boldsymbol{\nabla} V_n \end{bmatrix}$$

is the gradient of V. From (149) it is clear that the sign of dV/dt is determined by the sign of the gradient of V, since $\mathbf{f}(\mathbf{x}, t)$ is known. Hence, instead of looking for a Liapunov function V the sign of whose derivative will be suitable, we can look for a gradient function $\boldsymbol{\nabla} V$ that, in (149), makes dV/dt have the appropriate sign. Then the Liapunov function V itself can be determined by the line integral of the gradient from $\mathbf{0}$ to $\mathbf{x}$:

$$V(\mathbf{x}) = \int_{\mathbf{0}}^{\mathbf{x}} [\boldsymbol{\nabla} V(\mathbf{y})]' \, d\mathbf{y}. \tag{150}$$

If the scalar function V is to be uniquely determined by this line integral of its gradient, then the Jacobian matrix of ∇V with respect to $\mathbf{x}$

$$
\mathbf{D}(\mathbf{x}) = \begin{bmatrix}
\dfrac{\partial \nabla V_1}{\partial x_1} & \dfrac{\partial \nabla V_1}{\partial x_2} & \cdots & \dfrac{\partial \nabla V_1}{\partial x_n} \\[2ex]
\dfrac{\partial \nabla V_2}{\partial x_1} & \dfrac{\partial \nabla V_2}{\partial x_2} & \cdots & \dfrac{\partial \nabla V_2}{\partial x_n} \\[1ex]
\vdots & \vdots & & \vdots \\[1ex]
\dfrac{\partial \nabla V_n}{\partial x_1} & \dfrac{\partial \nabla V_n}{\partial x_2} & \cdots & \dfrac{\partial \nabla V_n}{\partial x_n}
\end{bmatrix}
\tag{151}
$$

must be symmetric.* Assume $\mathbf{D}(\mathbf{x})$ is symmetric and hence $V(\mathbf{x})$ is unique. This implies the integral is independent of the path of integration. Thus we may use the most convenient path. Such a path would lie along coordinate axes or parallel to coordinate axes, as in the following expanded form of (150):

$$
\begin{aligned}
V(\mathbf{x}) = &\int_0^{x_1} \nabla V(\mathbf{y})_1 \, dy_1 \bigg|_{y_2=0,\, y_3=0,\, \cdots,\, y_n=0} \\
&+ \int_0^{x_2} \nabla V(\mathbf{y})_2 \, dy_2 \bigg|_{y_1=x_1,\, y_3=0,\, \cdots,\, y_n=0} \\
&+ \cdots\cdots \\
&+ \int_0^{x_n} \nabla V(\mathbf{y})_n \, dy_n \bigg|_{y_1=x_1,\, y_2=x_2,\, \cdots,\, y_{n-1}=x_{n-1}}.
\end{aligned}
\tag{152}
$$

As you see, the problem of finding a Liapunov function by picking an arbitrary positive-definite function $V(\mathbf{x})$, and then ascertaining whether or not $-dV/dt$ is non-negative or positive definite, has been replaced by picking a function $\nabla V(\mathbf{x})$ such that $-dV/dt$ is non-negative or positive definite, as determined from (149), and $\mathbf{D}(\mathbf{x})$ is symmetric, and then ascertaining whether or not $V(\mathbf{x})$ is positive definite. Usually there is less guesswork involved in using this latter method, which is known as the *variable-gradient method*. However, this method of finding a Liapunov function is not really decisive; if the $V(\mathbf{x})$ found from the selected gradient function does not turn out to be positive definite, it cannot be concluded that the

---

* The requirement that $\mathbf{D}(\mathbf{x})$ be symmetric is equivalent to requiring that the curl of $\nabla V(\mathbf{x})$ be zero. It is a known theorem of vector analysis that, when the curl of a vector is zero, the vector is the gradient of a scalar function. For further discussion see H. Lass, *Vector and Tensor Analysis*, McGraw-Hill Book Co., New York, 1950, p. 297.

origin is not stable in some sense. It only means that a suitable Liapunov function has not yet been found.

It is usual practice to start by selecting the gradient of V to be of the form

$$\nabla V(\mathbf{x}) = \begin{bmatrix} \alpha_{11}x_1 + \alpha_{12}x_2 + \cdots + \alpha_{1n}x_n \\ \alpha_{21}x_1 + \alpha_{22}x_2 + \cdots + \alpha_{2n}x_n \\ \vdots \qquad \vdots \qquad\qquad \vdots \\ \alpha_{n1}x_1 + \alpha_{n2}x_2 + \cdots + \alpha_{nn}x_n \end{bmatrix}. \tag{153}$$

The scalars α_{ij} may be functions of $\mathbf{x}$, though for subsequent ease in evaluating the line integral (152), it is desirable for them to be constant.

The variable-gradient method may be summarized by a simple set of rules: (1) determine dV/dt as specified in (149) by using $\nabla V(\mathbf{x})$ of (153); (2) select the coefficients α_{ij} such that $-dV/dt$ is non-negative or positive definite and $\mathbf{D}(\mathbf{x})$ is symmetric; (3) evaluate $V(\mathbf{x})$ by using the line integral of (152); and (4) determine whether or not $V(\mathbf{x})$ is positive definite.

To illustrate the variable-gradient method, consider the network illustrated in Fig. 22. The state equation is

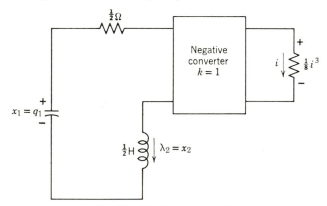

Fig. 22.   Nonlinear network.

$$\frac{d}{dt}\begin{bmatrix} x_1 \\ x_2 \end{bmatrix} = \begin{bmatrix} -2x_2 \\ x_1 - x_2 + x_2{}^3 \end{bmatrix}.$$

We first compute dV/dt from (149) as follows:

$$\frac{d}{dt}V(\mathbf{x}) = \begin{bmatrix} \alpha_{11}x_1 + \alpha_{12}x_2 \\ \alpha_{21}x_1 + \alpha_{22}x_2 \end{bmatrix}'\begin{bmatrix} -2x_2 \\ x_1 - x_2 + x_2{}^3 \end{bmatrix}$$

$$= \alpha_{21}x_1{}^2 + [-2\alpha_{11} - \alpha_{21}(1 - x_2{}^2) + \alpha_{22}]x_1 x_2$$

$$+ [-2\alpha_{12} - \alpha_{22}(1 - x_2{}^2)]x_2{}^2.$$

Let us pick α_{11} such that the coefficient of $x_1 x_2$ is zero; that is,

$$\alpha_{11} = \tfrac{1}{2}\left[\alpha_{22} - \alpha_{21}(1 - x_2{}^2)\right].$$

Then, if $-dV/dt$ is to be positive definite, we must have

$$-\alpha_{21} > 0$$
$$2\alpha_{12} + \alpha_{22}(1 - x_2{}^2) > 0.$$

Let $\alpha_{21} = -b$ and $\alpha_{22} = a$, where a and b are positive constants. Then

$$\boldsymbol{\nabla} V(\mathbf{x}) = \begin{bmatrix} \dfrac{a}{2}\,x_1 + \dfrac{b}{2}\,(1 - x_2{}^2)x_1 + \alpha_{12}\,x_2 \\[2mm] -bx_1 + ax_2 \end{bmatrix}$$

and

$$\mathbf{D}(\mathbf{x}) = \begin{bmatrix} \dfrac{a}{2} + \dfrac{b}{2}(1 - x_2{}^2) + \dfrac{\partial\alpha_{12}}{\partial x_1}\,x_2 & -bx_2 x_1 + \alpha_{12} + \dfrac{\partial\alpha_{12}}{\partial x_2}\,x_2 \\[2mm] -b & a \end{bmatrix}.$$

If $\mathbf{D}(\mathbf{x})$ is to be symmetric, then α_{12} must satisfy the partial differential equation

$$x_2\,\frac{\partial\alpha_{12}}{\partial x_2} + \alpha_{12} = -b + bx_1 x_2.$$

As you may easily verify, the solution of this equation is

$$\alpha_{12} = -b + \frac{b}{2}\,x_1 x_2.$$

Hence substituting this expression for α_{12} into the relation for $\boldsymbol{\nabla} V(\mathbf{x})$ yields

$$\boldsymbol{\nabla} V(\mathbf{x}) = \begin{bmatrix} \dfrac{a + b}{2}\,x_1 - bx_2 \\[2mm] -bx_1 + ax_2 \end{bmatrix}.$$

Before taking the line integral of $\nabla V(\mathbf{x})$, we must verify that $-dV/dt$ is positive definite for the chosen α_{21} and α_{22}, and the resultant α_{12} derived by the symmetry condition of $\mathbf{D(x)}$. To verify that $-dV/dt$ is positive definite, we must show that the two previously stated inequalities hold. First $\alpha_{21} = -b$, so $-\alpha_{21}$ is positive. Second,

$$2\alpha_{12} + \alpha_{22}(1 - x_2{}^2) = (a - 2b) + (bx_1 - ax_2)x_2 \,;$$

so, for $a > 2b$ and x_1 and x_2 sufficiently small, $2\alpha_{12} + \alpha_{22}(1 - x_2{}^2)$ is positive. The line integral of $\nabla V(\mathbf{x})$ according to (152) is

$$V(\mathbf{x}) = \int_0^{x_1} \frac{a+b}{2}\, y_1 dy_1 + \int_0^{x_2} (-bx_1 + ay_2)\, dy_2$$

$$= \frac{a+b}{4}\, x_1{}^2 - bx_1 x_2 + \frac{a}{2}\, x_2{}^2 \,.$$

This function may be expressed as the quadratic form

$$V(\mathbf{x}) = [x_1 \quad x_2] \begin{bmatrix} \dfrac{a+b}{4} & -\dfrac{b}{2} \\ -\dfrac{b}{2} & \dfrac{a}{2} \end{bmatrix} \begin{bmatrix} x_1 \\ x_2 \end{bmatrix} ,$$

which you may easily verify is positive definite provided $a > 0$, a condition already imposed, and $a^2 + ab > 2b^2$. This latter inequality is satisfied when the previous requirement $a > 2b$ is satisfied. Thus, for $a > 2b$, a Liapunov function has been constructed such that $-dV/dt$ is positive definite in a suitably small neighborhood of the origin. This implies that the origin is asymptotically stable.

In this section we have introduced the basic concepts of Liapunov stability relative to a singular point, have proven some basic theorems on stability, and have given two methods by which to guide a search for a Liapunov function. Much more is known. The concept of stability can be extended in a useful way to stability relative to a set of points. More sophisticated theorems on stability and instability exist. Also, other guidelines for selecting a Liapunov function for specific classes of problems are known. These advanced topics are treated in the books on stability listed in the bibliography.

PROBLEMS

1. Derive the state equations for the time-varying networks shown in Fig. P1 by using the state vector of (5). Repeat with the state vector of (10) Assume the element parameters are in Farads, Ohms, or Henrys.

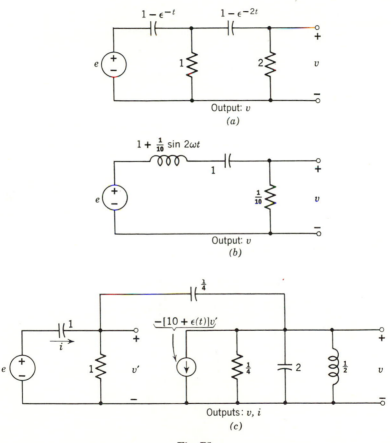

Fig. P1

2. Suppose

$$\mathscr{A}(t)\,\mathscr{A}(\hat{t}) = \mathscr{A}(\hat{t})\,\mathscr{A}(t)$$

for all t, $\hat{t} \geq t_0$. Prove that

$$\mathscr{A}(t)\left(\int_{t_0}^{t} \mathscr{A}(\tau)\, d\tau\right) = \left(\int_{t_0}^{t} \mathscr{A}(\tau)\, d\tau\right)\mathscr{A}(t)$$

for all $t \geq t_0$.

3. Use the condition in Problem 2 to determine which of the following matrices $\mathscr{A}(t)$ commutes with its integral $\int_{t_0}^{t} \mathscr{A}(\tau)\, d\tau$ for all $t \geq t_0$:

(a) $\mathscr{A}(t) = \begin{bmatrix} 1+t & -2 \\ 2 & 3t^2 \end{bmatrix}$,   (b) $\mathscr{A}(t) = \begin{bmatrix} -(2+t) & -3 \\ 3 & -(2+t) \end{bmatrix}$

(c) $\mathscr{A}(t) = \begin{bmatrix} -(1+2t) & -4t \\ 4t & -(1+2t) \end{bmatrix}$,   (d) $\mathscr{A}(t) = \begin{bmatrix} 2+\sin t & -t \\ t & 3+\sin t \end{bmatrix}$.

4. For those matrices $\mathscr{A}(t)$ of Problem 3 that commute with their integral $\int_{t_0}^{t} \mathscr{A}(\tau)\, d\tau$ for all $t \geq t_0$, express $\exp\left[\int_{t_0}^{t} \mathscr{A}(\tau)\, d\tau\right]$, the solution of (21), as a matrix.

5. Which of the following state equations, with $t_0 = 0$, have a solution in the extended sense? Of those that do, which have a solution in the ordinary sense? Of those that have a solution only in the extended sense, indicate whether or not the associated homogeneous equation has a solution in the ordinary sense.

(a) $\dfrac{d}{dt}\begin{bmatrix} x_1 \\ x_2 \\ x_3 \end{bmatrix} = \begin{bmatrix} -2\tanh t & 2 & -4 \\ -2 & -u(t) & 2 \\ 4 & 2 & -6 \end{bmatrix}\begin{bmatrix} x_1 \\ x_2 \\ x_3 \end{bmatrix}$

$$+ \begin{bmatrix} 1 & -1 \\ 2 & 0 \\ -1 & 2 \end{bmatrix}\begin{bmatrix} u(t) - u(t-1) \\ 2te^{-t} \end{bmatrix}$$

(b) $\dfrac{d}{dt}\begin{bmatrix} x_1 \\ x_2 \end{bmatrix} = \begin{bmatrix} \dfrac{-5}{t^{1/2}} & -1 \\ 1 & \dfrac{-3}{(1-t)^{1/3}} \end{bmatrix}\begin{bmatrix} x_1 \\ x_2 \end{bmatrix} + \begin{bmatrix} -2 \\ 1 \end{bmatrix}[tu(t) - 2tu(t-1)]$

(c) $\dfrac{d}{dt}\begin{bmatrix} x_1 \\ x_2 \end{bmatrix} = \begin{bmatrix} -(1+2t) & -1 \\ 2 & -2+t\epsilon^{-t} \end{bmatrix}\begin{bmatrix} x_1 \\ x_2 \end{bmatrix}$

$$+ \begin{bmatrix} 1 & 0 & -1 \\ 2 & 1 & 0 \end{bmatrix}\begin{bmatrix} \sin t \\ \cos t \\ -(t-1)u(t-1) \end{bmatrix}$$

(d) $\dfrac{d}{dt}\begin{bmatrix} x_1 \\ x_2 \\ x_3 \end{bmatrix} = \begin{bmatrix} \dfrac{1-2t^2}{1+t^2} & -1 & -2t \\ 1 & -3 & -1+\epsilon^{-t} \\ 2t & -1+\epsilon^{-t} & -2 \end{bmatrix}\begin{bmatrix} x_1 \\ x_2 \\ x_3 \end{bmatrix}$

$$+ \begin{bmatrix} 1 \\ -2 \\ 1 \end{bmatrix}\begin{bmatrix} -2 & (2-t)^{1/3} \end{bmatrix}$$

(e) $\dfrac{d}{dt}\begin{bmatrix} x_1 \\ x_2 \end{bmatrix} = \begin{bmatrix} -2 & -4 \\ 4 & -8+t\epsilon^{-t} \end{bmatrix}\begin{bmatrix} x_1 \\ x_2 \end{bmatrix} + \begin{bmatrix} 1 & -1 \\ -1 & 1 \end{bmatrix}\begin{bmatrix} \dfrac{2}{(t-1)^2} \\ \sin t \end{bmatrix}.$

6. Determine $\mathbf{Y}(t)$ and then $\mathbf{Q}(t)$ and $\mathbf{P}$ in $\mathbf{Y}(t) = \mathbf{Q}(t)\epsilon^{\mathbf{P}t}$ when

(a) $\mathscr{A}(t) = \begin{bmatrix} -2+\sin t & +1 \\ -1 & -4+\sin t \end{bmatrix}$

(b) $\mathscr{A}(t) = \begin{bmatrix} -3+\dfrac{\cos t}{[2+\sin t]^{1/2}} & +2 \\ -2 & -2+\dfrac{\cos t}{(2+\sin t)^{1/2}} \end{bmatrix}$

(c) $\mathscr{A}(t) = \begin{bmatrix} -6+\sin t\,(\epsilon^{\cos t}) & -1 \\ +1 & -2-\sin t\,(\epsilon^{\cos t}) \end{bmatrix}.$

(*Hint:* In each case $\mathscr{A}(t)$ commutes with $\displaystyle\int_{t_0}^{t}\mathscr{A}(\tau)\,d\tau$; therefore (28) is valid.)

7. By Theorem 5, with $t_0 = 0$, for which of the following state equations are all solutions bounded as $t \to \infty$?

(a)
$$\frac{d}{dt}\begin{bmatrix} x_1 \\ x_2 \\ x_3 \end{bmatrix} = \begin{bmatrix} te^{-t} & 1 & 0 \\ -1 & 0 & 0 \\ 1 & 1 & -2-\epsilon^{-t} \end{bmatrix}\begin{bmatrix} x_1 \\ x_2 \\ x_3 \end{bmatrix} + \begin{bmatrix} 1 & 0 \\ 1 & 1 \\ -1 & 2 \end{bmatrix}\begin{bmatrix} \epsilon^{-3t}\cos t \\ \epsilon^{-3t}\sin t \end{bmatrix}$$

(b)
$$\frac{d}{dt}\begin{bmatrix} x_1 \\ x_2 \\ x_3 \end{bmatrix} = \begin{bmatrix} -2-\epsilon^{-t} & -1 & 1 \\ -2 & -1 & 2+2t\epsilon^{-t} \\ 6+\epsilon^{-2t} & 5 & -3 \end{bmatrix}\begin{bmatrix} x_1 \\ x_2 \\ x_3 \end{bmatrix} + \begin{bmatrix} -1 \\ 2 \\ -1 \end{bmatrix}[u(t)-u(t-1)]$$

(c)
$$\frac{d}{dt}\begin{bmatrix} x_1 \\ x_2 \\ x_3 \\ x_4 \end{bmatrix} = \begin{bmatrix} \dfrac{\cos t}{1+t^2} & 1 & -1 & 2 \\ -1 & \dfrac{\epsilon^{-2t}}{t^{1/2}} & 1 & -1 \\ 0 & 0 & -2 & \epsilon^{-t} \\ 0 & 0 & \epsilon^{-t} & -3 \end{bmatrix}\begin{bmatrix} x_1 \\ x_2 \\ x_3 \\ x_4 \end{bmatrix} + \begin{bmatrix} -1 \\ 0 \\ 0 \\ 1 \end{bmatrix}[t\epsilon^{-t^2}]$$

(d)
$$\frac{d}{dt}\begin{bmatrix} x_1 \\ x_2 \end{bmatrix} = \begin{bmatrix} -1+\dfrac{1}{t^{1/3}} & 1 \\ 2 & -3 \end{bmatrix}\begin{bmatrix} x_1 \\ x_2 \end{bmatrix} + \begin{bmatrix} 1 & 0 \\ 2 & -1 \end{bmatrix}\begin{bmatrix} \epsilon^{-t^2} \\ (1-t)\epsilon^{-2t} \end{bmatrix}$$

8. If all eigenvalues of $\mathscr{A}$ have negative real part, select an α and δ such that

$$\|\epsilon^{\mathscr{A}t}\| \le \delta\epsilon^{-\alpha t}.$$

(*Hint:* Start with $\epsilon^{\mathscr{A}t}$ expressed in terms of the constituent matrices and eigenvalues of $\mathscr{A}$.)

9. By Theorem 6, with $t_0 = 0$, for which of the following state equations are all solutions bounded as $t \to \infty$?

(a)
$$\frac{d}{dt}\begin{bmatrix} x_1 \\ x_2 \end{bmatrix} = \begin{bmatrix} -5+te^{-t} & 4 \\ -3 & 2-\dfrac{1}{1+t^2} \end{bmatrix}\begin{bmatrix} x_1 \\ x_2 \end{bmatrix} + \begin{bmatrix} 0 \\ 2 \end{bmatrix}[\sin 2t]$$

$$(b) \ \frac{d}{dt} \begin{bmatrix} x_1 \\ x_2 \\ x_3 \end{bmatrix} = \begin{bmatrix} \dfrac{t^2}{1+t^4} & 2 & 0 \\ -2 & 0 & 0 \\ -1 & 0 & -3+t^2\epsilon^{-2t} \end{bmatrix} \begin{bmatrix} x_1 \\ x_2 \\ x_3 \end{bmatrix} + \begin{bmatrix} 1 & 0 \\ 0 & 1 \\ -1 & -1 \end{bmatrix} \begin{bmatrix} \epsilon^{-t} \\ u(t) \end{bmatrix}$$

$$(c) \ \frac{d}{dt} \begin{bmatrix} x_1 \\ x_2 \end{bmatrix} = \begin{bmatrix} -6+\dfrac{t}{1+t^2} & -8 \\ 2 & 2 \end{bmatrix} \begin{bmatrix} x_1 \\ x_2 \end{bmatrix} + \begin{bmatrix} -2 \\ -3 \end{bmatrix} [t^3\epsilon^{-t}]$$

$$(d) \ \frac{d}{dt} \begin{bmatrix} x_1 \\ x_2 \end{bmatrix} = \begin{bmatrix} -4+\epsilon^{-t}-\epsilon^{-2t} & 3 \\ -2 & 1+t\epsilon^{-3t} \end{bmatrix} \begin{bmatrix} x_1 \\ x_2 \end{bmatrix}.$$

10. Prove Theorem 7.

11. Prove Theorem 9.

12. Prove Theorem 10.

13. Consider the following state equations:

$$(a) \ \frac{d}{dt} \begin{bmatrix} x_1 \\ x_2 \end{bmatrix} = \begin{bmatrix} -3 & 1+\mu t\epsilon^{-t} \\ -1 & -1+\mu \end{bmatrix} \begin{bmatrix} x_1 \\ x_2 \end{bmatrix}$$

$$(b) \ \frac{d}{dt} \begin{bmatrix} x_1 \\ x_2 \\ x_3 \end{bmatrix} = \begin{bmatrix} -2+\mu\epsilon^{-t} & 1 & 2 \\ 0 & -1 & \mu(1-\epsilon^{-t}) \\ \mu & -1 & -3+\mu\dfrac{t}{1+t} \end{bmatrix} \begin{bmatrix} x_1 \\ x_2 \\ x_3 \end{bmatrix}$$

$$(c) \ \frac{d}{dt} \begin{bmatrix} x_1 \\ x_2 \end{bmatrix} = \begin{bmatrix} -7 & -5+\mu(2-\epsilon^{-t}) \\ -3+\mu & 1 \end{bmatrix} \begin{bmatrix} x_1 \\ x_2 \end{bmatrix}.$$

Take $t_0 = 0$. By Theorem 7, for which of these state equations do all solutions approach zero as $t \to \infty$, if μ is sufficiently small? In those cases to which Theorem 7 applies, determine an upper bound on μ.

14. For each of the following state equations, indicate which of Theorems 8, 9, and 10 have their conditions satisfied. If the state equation involves a parameter μ, indicate permissible values that μ may assume. Take $t_0 = 1$.

$$(a) \ \frac{d}{dt} \begin{bmatrix} x_1 \\ x_2 \end{bmatrix} = \begin{bmatrix} -1+\cos t & -1-t\epsilon^{-t} \\ 1-t\epsilon^{-2t} & -1+\cos t \end{bmatrix} \begin{bmatrix} x_1 \\ x_2 \end{bmatrix} + \begin{bmatrix} 2 \\ -1 \end{bmatrix} [1-t^3\epsilon^{-t}]$$

(b) $\dfrac{d}{dt}\begin{bmatrix} x_1 \\ x_2 \\ x_2 \end{bmatrix} = \begin{bmatrix} \cos^2 t & -t^2\epsilon^{-2t} & -1+t^{-3} \\ t^2\epsilon^{-2t} & \sin^2 t & -1+t^{-3} \\ t^{-3} & t^{-3} & -2 \end{bmatrix}\begin{bmatrix} x_1 \\ x_2 \\ x_3 \end{bmatrix} + \begin{bmatrix} -1 \\ 0 \\ 1 \end{bmatrix}\left[\dfrac{t^2}{1+t^4}\right]$

(c) $\dfrac{d}{dt}\begin{bmatrix} x_1 \\ x_2 \end{bmatrix} = \begin{bmatrix} -2+\sin^2 t+\dfrac{2}{t^3} & -1+\epsilon^{-t} \\ 1 & -2+\sin^2 t \end{bmatrix}\begin{bmatrix} x_1 \\ x_2 \end{bmatrix}$

$\qquad\qquad\qquad + \begin{bmatrix} 1 & 1 \\ 0 & -1 \end{bmatrix}\begin{bmatrix} 4-\epsilon^{-t} \\ \epsilon^{-t}\cos t \end{bmatrix}$

(d) $\dfrac{d}{dt}\begin{bmatrix} x_1 \\ x_2 \end{bmatrix} = \begin{bmatrix} -2+\sin t\cos t & -1+\mu\sin t \\ 1+\mu\sin t & -2+\sin t\cos t \end{bmatrix}\begin{bmatrix} x_1 \\ x_2 \end{bmatrix}$

(e) $\dfrac{d}{dt}\begin{bmatrix} x_1 \\ x_2 \\ x_3 \end{bmatrix} = \begin{bmatrix} 1-2\epsilon^{-t}\sin t & \mu(1-\epsilon^{-t}) & 1 \\ \mu & -3+\cos t & 3 \\ -\mu & -3+\mu\epsilon^{-t} & -3+\cos t \end{bmatrix}\begin{bmatrix} x_1 \\ x_2 \\ x_3 \end{bmatrix}$

(f) $\dfrac{d}{dt}\begin{bmatrix} x_1 \\ x_2 \end{bmatrix} = \begin{bmatrix} \dfrac{1}{t^3}+\cos 3t & -3\cos t+t\epsilon^{-t} \\ 3\cos t-t\epsilon^{-t} & \dfrac{t^2}{1+t^4}+\cos 3t \end{bmatrix}\begin{bmatrix} x_1 \\ x_2 \end{bmatrix} + \begin{bmatrix} 0 \\ -1 \end{bmatrix}[\epsilon^{-t}-2\epsilon^{-2t}]$

15. Verify that the network illustrated in Fig. 8 has the state equation given on p. 743.

16. Consider the following state equations:

(a) $\dfrac{d}{dt}\begin{bmatrix} x_1 \\ x_2 \end{bmatrix} = \begin{bmatrix} t\epsilon^{-t} & (1+t) \\ -(1+t) & t^2\epsilon^{-t} \end{bmatrix}\begin{bmatrix} x_1 \\ x_2 \end{bmatrix} + \begin{bmatrix} 1 & 1 \\ -1 & 0 \end{bmatrix}\begin{bmatrix} \epsilon^{-t}-2\epsilon^{-2t} \\ -u(t)+u(t-1) \end{bmatrix}$

(b) $\dfrac{d}{dt}\begin{bmatrix} x_1 \\ x_2 \end{bmatrix} = \begin{bmatrix} \cos t & t\sin t \\ -t\sin t & \cos t \end{bmatrix}\begin{bmatrix} x_1 \\ x_2 \end{bmatrix} + \begin{bmatrix} 2 \\ 3 \end{bmatrix}[t^2\epsilon^{-t}].$

Take $t_0 = 0$. By Theorem 11, for which of these state equations do all solutions approach zero at $t \to \infty$?

17. Let $\mathbf{Y}(t)\ \mathbf{Y}(\tau)^{-1}$ be the state-transition matrix associated with (71). Prove the theorem obtained after replacing condition (72a) in Theorem 11 by

$$\|\mathbf{Y}(t)\ \mathbf{Y}(\tau)^{-1}\| \leq \delta < \infty \quad \text{with} \quad t_0 \leq \tau \leq t \quad \text{for} \quad t \geq t_0.$$

Does this new theorem apply to the example following Theorem 11? If so, find a value for δ. This new theorem is less restrictive than Theorem 11, but it may be more difficult to apply. Explain why.

18. Show that the state equations in the examples of Section 10.7 are those for the networks illustrated in Figs. 17, 18, 20, 21, and 22.

19. Derive the state equations for each of the networks in Fig. P19.

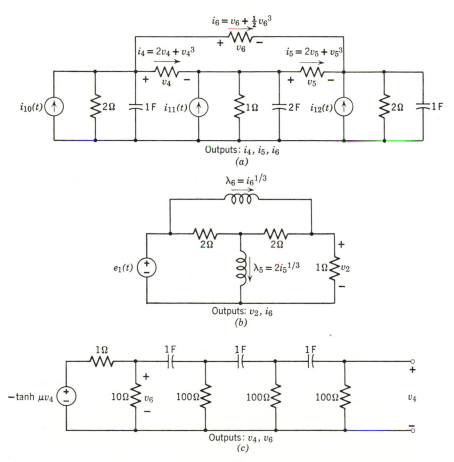

Fig. P19

20. Derive the state equations for the amplifier depicted in Fig. P20a. Use the transistor model shown in Fig. P20b.

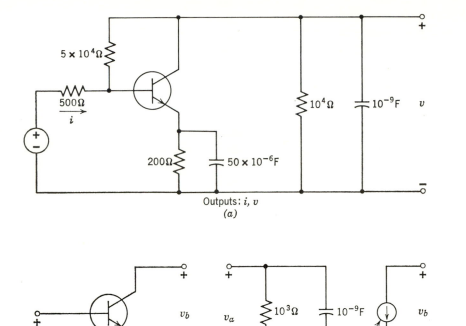

Outputs: i, v

(a)

(b)

Fig. P20

21. Derive the state equations for the amplifier depicted in Fig. P21. Use the transistor model shown in Fig. P20b.

22. Derive the state equations for the amplifier shown in Fig. P22. Use the transistor model depicted in Fig. P20b.

23. Derive the state equations for the amplifier depicted in Fig. P23. Use the transistor model shown in Fig. P20b.

24. (a) Show that $\mathbf{f}_{Lt}$ in (81a) must be identically zero or that $\mathbf{f}_{Lt}$ and $\mathbf{g}_{Ll}$ in (85) must be differentiable functions in order to express $\mathbf{v}_{Ll}$ and $\mathbf{v}_{Lt}$ as functions of $\mathbf{x}$, $\mathbf{e}$, and $d\mathbf{e}/dt$.

(b) Show that $\mathbf{f}_{Cl}$, in (75b), must be identically zero or that $\mathbf{f}_{Cl}$ and $\mathbf{g}_{Ct}$ in (79) must be differentiable functions in order to express $\mathbf{i}_{Ct}$ and $\mathbf{i}_{Cl}$ as functions of $\mathbf{x}$, $\mathbf{e}$, and $d\mathbf{e}/dt$.

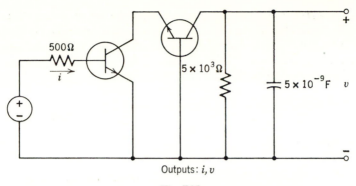

Outputs: i, v

Fig. P21

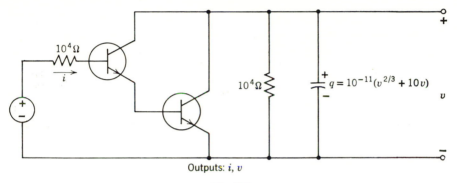

Outputs: i, v

Fig. P22

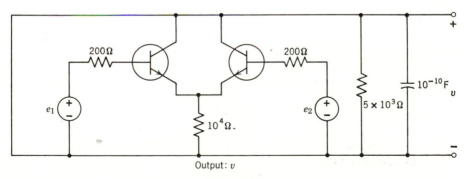

Output: v

Fig. P23

25. Consider the networks that can be represented as the interconnection of a capacitor subnetwork, an inductor subnetwork, and a resistor and independent source subnetwork as shown in Fig. P25a. Formulate the state equation in terms of the port parameters of each subnetwork when the state vector is the concatenation of a set of linearly independent port charge variables, $\mathbf{q}_C$, for the capacitor subnetwork and a set of linearly independent port flux variables, $\boldsymbol{\lambda}_L$, for the inductor subnetwork. Apply the result to establish a state equation for the network in Fig. P25*b*.

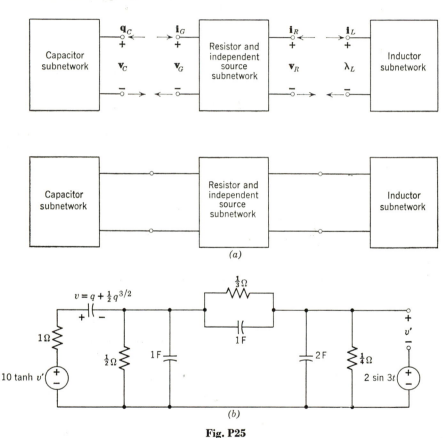

(a)

(b)

Fig. P25

26. Using Theorem 13, determine which of the following state equations has a solution for all $t \geq 2$.

$$(a) \quad \frac{d}{dt} \begin{bmatrix} x_1 \\ x_2 \end{bmatrix} = \begin{bmatrix} -x_1 + x_2 - t\epsilon^{-t} \\ -x_1 - \dfrac{x_2{}^3}{1 + x_2{}^2} \end{bmatrix}$$

$$(b) \quad \frac{d}{dt} \begin{bmatrix} x_1 \\ x_2 \\ x_3 \end{bmatrix} = \begin{bmatrix} -2x_1 + (x_2 - x_1) \tanh^2 (x_2 - x_1) + x_3 \\ (x_2 - x_1) \tanh^2 (x_2 - x_1) - \dfrac{x_2}{1 + x_2^2} \\ -x_1 - 3x_3 \end{bmatrix}$$

$$(c) \quad \frac{d}{dt} \begin{bmatrix} x_1 \\ x_2 \\ x_3 \end{bmatrix} = \begin{bmatrix} -\dfrac{t^2}{1 + t^4} x_1 + (x_2 - x_1)^3 + x_3 - 2 \sin t \\ (x_2 - x_1)^3 - 2(x_3 - x_2)^3 + 2 \cos t \\ x_1 - 2(x_3 - x_2)^3 - 8 \dfrac{x_3 - x_3^3}{1 + x_3^2} \end{bmatrix}$$

$$(d) \quad \frac{d}{dt} \begin{bmatrix} x_1 \\ x_2 \end{bmatrix} = \begin{bmatrix} -(2 - \epsilon^{-t})x_1 \tanh^2 x_1 + \dfrac{x_2^3}{1 + x_2^2} \\ -x_1 + 2 \tanh x_2 + (1 - \epsilon^{-t}) \end{bmatrix}$$

$$(e) \quad \frac{d}{dt} \begin{bmatrix} x_1 \\ x_2 \end{bmatrix} = \begin{bmatrix} -\dfrac{x_1^3}{1 + x_1^2} + x_2^2 \tanh x_2 \\ -(1 - t\epsilon^{-t})x_1 - x_2 \end{bmatrix}.$$

27. Prove the following theorem: *Suppose the conditions of Theorem 12 on existence apply, then any solution of* (98) *in the extended sense, equal to* $\mathbf{x}(t_0)$ *at time* t_0, *may be extended to yield a defined solution for all* $t \geq t_0$, *if* $\mathbf{x}'\mathbf{f}(\mathbf{x}, t) \leq 0$ *for* $t \geq t_0$. (*Hint:* Start by writing the differential equation that is satisfied by the scalar $\mathbf{x}'\mathbf{x}$.)

28. Using the theorem proved in Problem 27, determine which of the following state equations has a solution defined for all $t \geq 1$:

$$(a) \quad \frac{d}{dt} \begin{bmatrix} x_1 \\ x_2 \end{bmatrix} = \begin{bmatrix} -x_1 + x_2 \\ -x_1 - x_2^3 \end{bmatrix}, \qquad (b) \quad \frac{d}{dt} \begin{bmatrix} x_1 \\ x_2 \\ x_3 \end{bmatrix} = \begin{bmatrix} x_2 \\ x_3 \\ -6x_1 - 11x_2 - 6x_3^5 \end{bmatrix}$$

$$(c) \quad \frac{d}{dt} \begin{bmatrix} x_1 \\ x_2 \\ x_3 \end{bmatrix} = \begin{bmatrix} -5x_1 - 1250x_1^3 - 10(x_2 + 2x_3)^3 \\ -12x_2 + x_3 \\ 2x_2 - 11x_3 \end{bmatrix}$$

(d) $\dfrac{d}{dt}\begin{bmatrix} x_1 \\ x_2 \end{bmatrix} = \begin{bmatrix} -(3 - \epsilon^{-t})x_1 - x_1{}^3 - x_2 \\ -x_1 - x_2 \end{bmatrix}$

(e) $\dfrac{d}{dt}\begin{bmatrix} x_1 \\ x_2 \\ x_3 \end{bmatrix} = \begin{bmatrix} x_1{}^3 + x_2 - x_3 \\ -x_1 - 3x_2 \\ x_1 - 2(2 + \sin t)x_3{}^3 \end{bmatrix}.$

Would Theorem 13 also have established the existence of those solutions?

29. (a) Prove that the theorem stated in Problem 27 remains valid if the condition $\mathbf{x}'\,\mathbf{f}(\mathbf{x}, t) \le 0$ is replaced by $\mathbf{x}'\mathbf{P}\,\mathbf{f}(\mathbf{x}, t) \le 0$, where $\mathbf{P}$ is a symmetric positive-definite matrix.
(b) If possible, apply the result of (a) to the following state equations:

$$\frac{d}{dt}\begin{bmatrix} x_1 \\ x_2 \\ x_3 \\ x_4 \end{bmatrix} = \begin{bmatrix} x_2 \\ x_3 \\ x_4 \\ -x_1 - 4x_2 - 6x_3 - 4x_4{}^3 \end{bmatrix}$$

$$\frac{d}{dt}\begin{bmatrix} x_1 \\ x_2 \\ x_3 \end{bmatrix} = \begin{bmatrix} -x_1{}^5 - x_2 + x_3 \\ 2x_1 - 4x_2 + 2x_3 \\ -3x_1 - 3x_2 - 6x_3 - x_3{}^5 \end{bmatrix}.$$

30. Prove that $\mathbf{f}(\mathbf{x}, t)$ satisfies a Lipschitz condition if the partial derivatives $\partial f_i\,(\mathbf{x}, t)/\partial x_j$ exist, are continuous in $\mathbf{x}$ for almost all $t \ge t_0$, and are bounded in magnitude in some neighborhood of every point by non-negative locally integrable functions of $t \ge t_0$. Show by counterexample that these conditions are not necessary and therefore are only sufficient.

31. Verify that each of the functions $\mathbf{f}(\mathbf{x}, t)$ in Problems 28 and 29 satisfies a Lipschitz condition in the neighborhood $\|\mathbf{x} - \hat{\mathbf{x}}\| < 1$ of the arbitrary point $\hat{\mathbf{x}}$ by finding a $\gamma(t)$ that satisfies the conditions set forth in Theorem 14.

32. For each of the following state equations, verify whether the conditions of Theorems 15 and/or 16 are satisfied. Set $t_0 = -1$.

(a) $\dfrac{d}{dt}\begin{bmatrix} x_1 \\ x_2 \end{bmatrix} = \begin{bmatrix} -x_1 + x_2 + t\epsilon^{-t} \\ -x_1 - x_2{}^3 \end{bmatrix}$

(b) $\dfrac{d}{dt}\begin{bmatrix} x_1 \\ x_2 \end{bmatrix} = \begin{bmatrix} -(3-\epsilon^{-t})x_1 - \dfrac{x_1{}^3}{1+x_1{}^2} - x_2 \\[4mm] -x_1 - x_2 + \epsilon^{-t/2} \end{bmatrix}$

(c) $\dfrac{d}{dt}\begin{bmatrix} x_1 \\ x_2 \\ x_3 \end{bmatrix} = \begin{bmatrix} -5x_1 - 1250x_1{}^3 - 10(x_2 + 2x_3)^3 \\[2mm] -12x_2 + x_3 \\[2mm] 2x_2 - 11x_3 \end{bmatrix}$

(d) $\dfrac{d}{dt}\begin{bmatrix} x_1 \\ x_2 \\ x_3 \end{bmatrix} = \begin{bmatrix} -t\epsilon^{-t}x_1 + x_2 \\[2mm] -x_1 - \dfrac{x_2{}^3}{1+2x_2{}^2} - 2t\epsilon^{-3t} \\[2mm] x_1 - 2x_3 \end{bmatrix}$

(e) $\dfrac{d}{dt}\begin{bmatrix} x_1 \\ x_2 \end{bmatrix} = \begin{bmatrix} -x_1 + \dfrac{2x_1{}^5}{2+x_1{}^4} - x_2 \\[4mm] -x_1 - 2x_2 + \dfrac{t^2}{1+t^4} \end{bmatrix}.$

In each case, specify an upper bound on $\|\mathbf{x}(t_0)\|$.

33. Prove Theorems 15 and 16 after replacing $\mathbf{A}$ by $\mathbf{A}(t)$, where $\mathbf{A}(t)$ is a periodic matrix.

34. Prove Theorem 16 after replacing (103c) by the condition $\|\hat{\mathbf{f}}(\mathbf{x}, t)\| \leq \mu \|\mathbf{x}\|$ for $\|\mathbf{x}\| \leq \zeta$ and μ sufficiently small.

35. Prove the following theorem: *Suppose all solutions of the reference equation*

$$\frac{d}{dt}\mathbf{y} = \mathbf{A}(t)\mathbf{y}$$

are bounded as t *tends to infinity. Further, suppose* $\mathbf{f}(\mathbf{x}, t) = \mathbf{A}(t) + \hat{\mathbf{f}}(\mathbf{x}, t) + \mathbf{g}(t)$. *Then all solutions of*

$$\frac{d}{dt}\mathbf{x} = \mathbf{f}(\mathbf{x}, t)$$

with an initial vector $\mathbf{x}(t_0)$ *such that* $\|\mathbf{x}(t_0)\| \leq \delta$, *which is a constant depending on* $\mathbf{f}(\mathbf{x}, t)$, *are bounded as* t *tends to infinity, if*

$$\|\mathbf{Y}(t)\,\mathbf{Y}(\tau)^{-1}\| \leq \alpha \quad \text{with} \quad t_0 \leq \tau \leq t \quad \text{for} \quad \text{all } t \geq t_0,$$

where $\mathbf{Y}(t) \, \mathbf{Y}(\tau)^{-1}$ *is the state-transition matrix associated with the reference equation, and if*

$$\|\mathbf{f}(\mathbf{x}, t)\| \le \beta(t) \, \|\mathbf{x}\| \quad \text{for} \quad \|\mathbf{x}\| \le \zeta$$

$$\int_{t_0}^{\infty} \beta(t) \, dt < \infty$$

$$\int_{t_0}^{\infty} \|\mathbf{g}(t)\| \, dt \le \gamma\delta,$$

where γ *is a suitably chosen positive constant.*

36. Consider the second-order system

$$\frac{d\mathbf{x}}{dt} = \mathbf{f}(\mathbf{x}).$$

Let $\mathbf{x}_s$ be a singular point. Suppose $\mathbf{f_x} \, (\mathbf{x}_s) = \mathbf{A} \ne 0$. If the eigenvalues of $\mathbf{A}$ are not imaginary, it is known* that the behavior of the solution in the neighborhood of $\mathbf{x}_s$ is the same as that of $\mathbf{y} + \mathbf{x}_s$, where $\mathbf{y}$ satisfies the equation of first approximation at $\mathbf{x}_s$; that is,

$$\frac{d\mathbf{y}}{dt} = \mathbf{A}\mathbf{y}.$$

Let $\mathbf{y} = \mathbf{P}\mathbf{z}$, where $\mathbf{P}$ is a nonsingular 2×2 matrix. Then $\mathbf{z}$ is the solution of the equation

$$\frac{d\mathbf{z}}{dt} = \mathbf{P}^{-1}\mathbf{A}\mathbf{P}\mathbf{z} = \mathbf{B}\mathbf{z}.$$

(*a*) When $\mathbf{A}$ has distinct real eigenvalues and, in some cases, equal real eigenvalues, there exists a $\mathbf{P}$ such that

$$\mathbf{B} = \begin{bmatrix} \lambda_1 & 0 \\ 0 & \lambda_2 \end{bmatrix}.$$

Solve for the elements of $\mathbf{P}$ in terms of the elements of $\mathbf{A}$ and the eigenvalues λ_1 and λ_2.

---

* Solomon Lefschetz, *Differential Equations: Geometric Theory*, 2nd ed., Interscience, New York, pp. 188–195.

(b) Solve for **z** and sketch a typical set of trajectories in the $z_1 - z_2$ plane, when (i) both eigenvalues are positive, (ii) both eigenvalues are negative, and (iii) the eigenvalues have opposite sign.

(c) In case (iii) above, the trajectories corresponding to $z_1(t_0) = 0$ or $z_2(t_0) = 0$ terminate at the origin and are known as *separatrices*. Since the word separatrix means something that separates, why do you think the word is applied to each of these particular trajectories?

37. With reference to Problem 36 and for each of the following **A** matrices, (i) determine the eigenvalues λ_1 and λ_2, (ii) determine the transformation **P**, (iii) sketch a typical set of trajectories in the $z_1 - z_2$ plane, and (iv), using the results from (iii) and the transformation $\mathbf{y} = \mathbf{Pz}$, sketch a typical set of trajectories in the $y_1 - y_2$ plane.

(a) $\mathbf{A} = \begin{bmatrix} -5 & -4 \\ 3 & 2 \end{bmatrix}$, (b) $\mathbf{A} = \begin{bmatrix} -4 & 7 \\ -2 & 5 \end{bmatrix}$

(c) $\mathbf{A} = \begin{bmatrix} 2 & 0 \\ 0 & 2 \end{bmatrix}$, (d) $\mathbf{A} = \begin{bmatrix} 1 & -1 \\ 2 & 4 \end{bmatrix}$.

38. With reference to Problem 36:

(a) When **A** has equal real eigenvalues ($\lambda_1 = \lambda_2 = \lambda$) and is not diagonal, there exists a **P** such that

$$\mathbf{B} = \begin{bmatrix} \lambda & 1 \\ 0 & \lambda \end{bmatrix}.$$

Solve for the elements of **P** in terms of the elements of **A** and the eigenvalue λ.

(b) Solve for **z** and sketch a typical set of trajectories in the $z_1 - z_2$ plane, when (i) λ is positive and (ii) λ is negative.

(c) For

$$\mathbf{A} = \begin{bmatrix} 3 & 1 \\ -1 & 1 \end{bmatrix}$$

(i) determine λ, (ii) determine **P**, (iii) sketch a typical set of trajectories in the $z_1 - z_2$ plane, and (iv), using the result from (iii) and the transformation $\mathbf{y} = \mathbf{Pz}$, sketch a typical set of trajectories in the $y_1 - y_2$ plane.

39. With reference to Problem 36:

(a) When $\mathbf{A}$ has conjugate complex eigenvalues $(\lambda_1 = \bar{\lambda}_2 = \sigma + j\omega)$, there exists a $\mathbf{P}$ such that

$$\mathbf{B} = \begin{bmatrix} \sigma & \omega \\ -\omega & \sigma \end{bmatrix}.$$

Solve for the elements of $\mathbf{P}$ in terms of the elements of $\mathbf{A}$ and the real and imaginary parts, σ and ω, of the eigenvalues.

(b) Polar coordinates in the $z_1 - z_2$ plane are r and θ, where $r^2 = z_1^2 + z_2^2$ and $\tan \theta = z_2/z_1$. Verify that

$$\frac{dr}{dt} = \sigma r, \qquad \frac{d\theta}{dt} = \omega.$$

(c) Solve for r and θ and sketch a typical set of trajectories in the $z_1 - z_2$ plane, when (i) σ is positive and (ii) σ is negative.

(d) For

$$\mathbf{A} = \begin{bmatrix} 3 & 2 \\ 13 & -7 \end{bmatrix}$$

(i) determine α and β, (ii) determine $\mathbf{P}$, (iii) sketch a typical set of trajectories in the $z_1 - z_2$ plane, and (iv), using the result from (iii) and the transformation $\mathbf{y} = \mathbf{P}\mathbf{z}$, sketch a typical set of trajectories in the $y_1 - y_2$ plane.

40. With reference to Problem 36: The singular point $\mathbf{x}_s$ is called (i) a *node* if $\mathbf{A}$ has distinct, real eigenvalues of the same sign, (ii) a *log-node* if $\mathbf{A}$ has equal, real eigenvalues, (iii) a *saddle* if $\mathbf{A}$ has real eigenvalues of opposite sign, and (iv) a *focus* if $\mathbf{A}$ has conjugate complex eigenvalues with nonzero real part. The singular point is said to be *stable*, if both eigenvalues are negative real or have negative real part; otherwise, the singular point is said to be *unstable*. Let $\delta_1 = \det \mathbf{A}$ and $\delta_2 = \operatorname{tr} \mathbf{A}$. Divide the $\delta_1 - \delta_2$ plane into regions corresponding to the different classifications for a singular point such as stable node.

41. With reference to Problem 36: The $x_1 - x_2$ plane is known as the *phase plane*, and a typical set of trajectories in the phase plane is called a *phase portrait*. For each of the following state equations find the singular points and, using the results of Problems 36 through 40, sketch a phase

portrait in the neighborhood of each singular point and classify the
singular point:

(a) $\dfrac{d}{dt}\begin{bmatrix} x_1 \\ x_2 \end{bmatrix} = \begin{bmatrix} x_2 \\ -2x_2 + x_1 - x_1^3 \end{bmatrix}$,

(b) $\dfrac{d}{dt}\begin{bmatrix} x_1 \\ x_2 \end{bmatrix} = \begin{bmatrix} x_1 + x_2 \\ x_1 - x_1^5 - 3x_2 \end{bmatrix}$

(c) $\dfrac{d}{dt}\begin{bmatrix} x_1 \\ x_2 \end{bmatrix} = \begin{bmatrix} -2x_1 - 3x_2 + 2x_2^3 \\ 5x_1 + 6x_2 - 2x_2^3 \end{bmatrix}$

(d) $\dfrac{d}{dt}\begin{bmatrix} x_1 \\ x_2 \end{bmatrix} = \begin{bmatrix} x_2 \\ 1 - 3x_2 - 2\sin x_1 \end{bmatrix}$

(e) $\dfrac{d}{dt}\begin{bmatrix} x_1 \\ x_2 \end{bmatrix} = \begin{bmatrix} x_1 - 2x_1 x_2 \\ 2x_2 - 3x_1 x_2 \end{bmatrix}$.

Sketch a phase portrait valid throughout the phase plane using the
local phase portraits just completed. You will find it useful to keep in
mind that x_i, for $i = 1$ and 2, is increasing along a trajectory in that
region of the phase plane where $f_i(\mathbf{x}) > 0$ and decreasing along a trajectory in that region of the phase plane where $f_i(\mathbf{x}) < 0$. Furthermore,
the trajectory is vertical [horizontal] where it crosses the curve
$f_1(\mathbf{x}) = 0\,[f_2(\mathbf{x}) = 0]$.

42. Obtain a numerical solution of each of the following state equations by
using Euler's method. Set $h = 0.1$ and compute $\mathbf{x}(ih)$ for $i = 1, 2, \cdots, 10$.

(a) $\dfrac{d}{dt} x = -\dfrac{x^3}{1 + x^2} - 1,$ $\qquad\qquad\qquad\qquad\qquad x(0) = 2$

(b) $\dfrac{d}{dt} x = -x(1 + 2\tanh^2 x) + te^{-t},$ $\qquad\qquad\qquad x(0) = 0$

(c) $\dfrac{d}{dt}\begin{bmatrix} x_1 \\ x_2 \end{bmatrix} = \begin{bmatrix} 0 & 2 \\ -2 & 0 \end{bmatrix}\begin{bmatrix} x_1 \\ x_2 \end{bmatrix},$ $\qquad\qquad \begin{bmatrix} x_1(0) \\ x_2(0) \end{bmatrix} = \begin{bmatrix} 0 \\ 1 \end{bmatrix}$

(d) $\dfrac{d}{dt}\begin{bmatrix} x_1 \\ x_2 \end{bmatrix} = \begin{bmatrix} -1 & 1 \\ 0 & -1 \end{bmatrix}\begin{bmatrix} x_1 \\ x_2 \end{bmatrix} + \begin{bmatrix} 1 \\ -1 \end{bmatrix}[e^{-t}],$ $\quad \begin{bmatrix} x_1(0) \\ x_2(0) \end{bmatrix} = \begin{bmatrix} 0 \\ -1 \end{bmatrix}$

(e) $\dfrac{d}{dt}\begin{bmatrix} x_1 \\ x_2 \end{bmatrix} = \begin{bmatrix} x_2 \\ -2x_2 + x_1 - x_1^3 \end{bmatrix},$ $\qquad \begin{bmatrix} x_1(0) \\ x_2(0) \end{bmatrix} = \begin{bmatrix} 1 \\ -1 \end{bmatrix}$

(f) $\dfrac{d}{dt}\begin{bmatrix} x_1 \\ x_2 \\ x_3 \end{bmatrix} = \begin{bmatrix} x_2 \\ x_3 \\ -x_1^3 - 3x_2 - 3x_3 + (1 - e^{-t}) \end{bmatrix},$ $\begin{bmatrix} x_1(0) \\ x_2(0) \\ x_3(0) \end{bmatrix} = \begin{bmatrix} 0 \\ 1 \\ 0 \end{bmatrix}$

Next, obtain the solutions by using the modified Euler method. Assume three-digit accuracy is adequate. For the state equations in (c) and (d) compare the two numerical solutions and the exact solutions evaluated at ih.

43. Repeat Problem 42 by using the Adams method with $j = 3$ and the modified Adams method with $j = 3$ in place of Euler's method and the modified Euler method, respectively. Obtain starting values by using a truncated Taylor series.

44. Obtain a numerical solution of the following state equation by using the Adams method with $j = 1, 2, 3, 4, 5$. Set $h = 0.1$ and compute $\mathbf{x}(ih)$ for $i = 1, 2, \cdots, 15$. Obtain the starting values by using a truncated Taylor series. Compare the numerical solutions and the exact solution evaluated at ih.

$$\frac{d}{dt}x = -x(1 + x^2), \qquad x(0) = -5.$$

45. Obtain a numerical solution of each of the state equations in Problem 42 by using the Milne method. Set $h = 0.1$ and compute $\mathbf{x}(ih)$ for $i = 1, 2, \cdots, 10$. Assume three-digit accuracy is adequate. Obtain the starting values by using the Runge-Kutta method.

46. Obtain a numerical solution of the following state equation by using the modified Euler method. Set $h = 0.1$ and compute $\mathbf{x}(ih)$ for $i = 1, 2, \cdots, 10$. Assume three-digit accuracy is adequate.

$$\frac{d}{dt}x = -x(1 + x^2), \qquad x(0) = -5.$$

Next, obtain the solution by using the corrector in the modified Euler method and (122), with $j = 1$ and $l = 1$ as the predictor. Obtain the starting values by using the preceding numerical solution. Compare the number of times the corrector had to be applied at each time step in each of the two methods.

47. Create three predictor-corrector pairs from (122) and (126) and use them to obtain a numerical solution of the following state equations. Set $h = 0.1$ and compute $\mathbf{x}(ih)$ for $i = 1, 2, \cdots, 10$. Assume three-digit accuracy is adequate. Obtain starting values by using a truncated Taylor series.

(a) $\dfrac{d}{dt}x = -x(1 - x^2), \qquad x(0) = -5$

(b) $\dfrac{d}{dt}\begin{bmatrix} x_1 \\ x_2 \end{bmatrix} = \begin{bmatrix} -2t & -1 \\ 1 & -2t \end{bmatrix}\begin{bmatrix} x_1 \\ x_2 \end{bmatrix} + \begin{bmatrix} 0 \\ -1 \end{bmatrix}\begin{bmatrix} t \\ 1+t^2 \end{bmatrix}, \qquad \begin{bmatrix} x_1(0) \\ x_2(0) \end{bmatrix} = \begin{bmatrix} 0 \\ 1 \end{bmatrix}.$

48. Let $\mathcal{M}$ be a set of m non-negative integers and $\mathcal{N}$ be a set of n non-negative integers. Let

$$\mathbf{x}(t) = \mathbf{C}_1 + \mathbf{C}_2(t - t_i) + \mathbf{C}_3(t - t_i)^2 + \cdots + \mathbf{C}_{m+n}(t - t_i)^{m+n-1};$$

then

$$\dot{\mathbf{x}}(t) = \mathbf{C}_2 + 2\mathbf{C}_3(t - t_i) + \cdots + (m + n - 1)\mathbf{C}_{m+n}(t - t_i)^{m+n-2}.$$

The $\mathbf{C}_l$ may be solved for in terms of $\mathbf{x}(t_{i-j})$ for each j in $\mathcal{M}$, and $\dot{\mathbf{x}}(t_{i-h})$ for each k in $\mathcal{N}$. When the results are substituted in the expression for $\mathbf{x}(t)$ and t is set equal to t_{i+1}, a predictor of the following general form is obtained:

$$\mathbf{x}(t_{i+1}) = \sum_{j \text{ in } \mathcal{M}} a_j \mathbf{x}(t_{i-j}) + \sum_{k \text{ in } \mathcal{N}} b_k \dot{\mathbf{x}}(t_{i-k}).$$

Let $t_{i+1} = t_i + h$ for each value of i. Then determine the a_j and b_k when:

(a) $\mathcal{M} = \{0, 1\}$; $\mathcal{N} = \{0, 1\}$
(b) $\mathcal{M} = \{0, 1, 2\}$; $\mathcal{N} = \{0, 1\}$
(c) $\mathcal{M} = \{, 1\}$; $\mathcal{N} = \{0, 1, 2\}$
(d) $\mathcal{M} = \{0, 1, 2\}$; $\mathcal{N} = \{0, 1, 2\}$
(e) $\mathcal{M} = \{0, 2\}$; $\mathcal{N} = \{0, 1\}$
(f) $\mathcal{M} = \{0, 1\}$; $\mathcal{N} = \{0, 2\}$
(g) $\mathcal{M} = \{0, 2\}$; $\mathcal{N} = \{1\}$
(h) $\mathcal{M} = \{1, 2\}$; $\mathcal{N} = \{0, 1, 2\}$

49. By using each of the predictors in Problem 48, obtain a numerical solution of the following state equations. Set $h = 0.2$ and compute $\mathbf{x}(ih)$ for $i = 1, 2, \cdots, 10$. Assume three-digit accuracy is adequate. Obtain starting values using a truncated Taylor series.

(a) $\dfrac{d}{dt} x = -x(1 - x^2)$, $x(0) = -5$

(b) $\dfrac{d}{dt} \begin{bmatrix} x_1 \\ x_2 \end{bmatrix} = \begin{bmatrix} -2t & -1 \\ -1 & -2t \end{bmatrix} \begin{bmatrix} x_1 \\ x_2 \end{bmatrix} + \begin{bmatrix} 0 \\ -1 \end{bmatrix} \begin{bmatrix} \dfrac{t}{1 + t^2} \end{bmatrix}$, $\begin{bmatrix} x_1(0) \\ x_2(0) \end{bmatrix} = \begin{bmatrix} 0 \\ 1 \end{bmatrix}$.

50. Let $\hat{\mathcal{M}}$ be a set of $\hat{m}$ non-negative integers and $\hat{\mathcal{N}}$ be a set of $\hat{n}$ integers, one of which equals -1 and the rest are non-negative. Let

$$\mathbf{x}(t) = \hat{\mathbf{C}}_1 + \hat{\mathbf{C}}_2(t - t_i) + \cdots + \hat{\mathbf{C}}_{\hat{m}+\hat{n}}(t - t_i)^{\hat{m}+\hat{n}-1};$$

then

$$\dot{\mathbf{x}}(t) = \hat{\mathbf{C}}_2 + 2\hat{\mathbf{C}}_3(t - t_i) + \cdots + (\hat{m} + \hat{n} - 1)\hat{\mathbf{C}}_{\hat{m}+\hat{n}}(t - t_i)^{\hat{m}+\hat{n}-2}.$$

The $\hat{C}_l$ may be solved in terms of $x(t_{i-j})$ for each j in $\hat{\mathcal{M}}$, and $\hat{b}(t_{i-k})$ for each k in $\hat{\mathcal{N}}$. When the results are substituted in the expression for $x(t)$ and t is set equal to t_{i+1}, a corrector of the following general form is obtained:

$$x(t_{i+1}) = \sum_{j \text{ in } \hat{\mathcal{M}}} \hat{a}_j x(t_{i-j}) + \sum_{k \text{ in } \hat{\mathcal{N}}} \hat{b}_k \dot{x}(t_{i-k}).$$

Let $t_{i+1} = t_i + h$ for each value of i. Then determine the $\hat{a}_j$ and $\hat{b}_k$ when

(a) $\hat{\mathcal{M}} = \{0, 1\}$; $\hat{\mathcal{N}} = \{-1, 0\}$

(b) $\hat{\mathcal{M}} = \{0, 1\}$; $\hat{\mathcal{N}} = \{-1, 1\}$

(c) $\hat{\mathcal{M}} = \{0, 1, 2\}$; $\hat{\mathcal{N}} = \{-1, 0\}$

(d) $\hat{\mathcal{M}} = \{0, 1, 2\}$; $\hat{\mathcal{N}} = \{-1, 1\}$

(e) $\hat{\mathcal{M}} = \{0, 1, 2\}$; $\hat{\mathcal{N}} = \{-1, 0, 1\}$

(f) $\hat{\mathcal{M}} = \{0, 2\}$ $\hat{\mathcal{N}} = \{-1, 0, 1\}$

(g) $\hat{\mathcal{M}} = \{0, 2\}$; $\hat{\mathcal{N}} = \{-1, 1\}$

(h) $\hat{\mathcal{M}} = \{1\}$; $\hat{\mathcal{N}} = \{-1, 1\}$

51. Create three predictor-corrector pairs from the results of Problems 48 and 50. In each case select $\hat{m} + \hat{n} = m + n$ so that the predictor and corrector have the same accuracy. By using each of these predictor corrector pairs, obtain the numerical solutions called for in Problem 49.

52. The correctors derived from (126) for specific j and l when combined with the state equation (104) yield implicit equations of the form

$$x(t_{i+1}) = g[x(t_{i+1})].$$

It was proposed in Section 10.6 that solution of a corrector should be accomplished by the iteration

$$x(t_{i+1})^{(n)} = g(x(t_{i+1})^{(n-1)}),$$

where $x(t_{i+1})^{(n)}$ denotes the nth iterate; the zero-th iterate $x(t_{i+1})^{(0)}$ is the solution of a predictor. An alternate method for solving the corrector uses the well known Newton-Raphson iteration; in this case the iterates are given by

$$x(t_{i+1})^{(n)} = x(t_{i+1})^{(n-1)} - \{U - g_x[x(t_{i+1})^{(n-1)}]\}^{-1}[x(t_{i+1})^{(n-1)} - g(x(t_{i+1})^{(n-1)})]$$

where $g_x(x(t_{i+1})^{(n-1)})$ is the Jacobian matrix of $g(x)$ evaluated at the $(n-1)$-st iterate for $x(t_{i+1})$.[*] This latter method for solving the

---

[*] This approach to numerical solution of a differential equation was proposed in I. W. Sandberg, "Numerical Integration of Systems of Stiff Nonlinear Differential Equations," *Bell System Technical Journal*, Vol. 47, No. 4, April 1968, pp. 511–528.

corrector usually requires fewer iterations to achieve a specified level of accuracy; however, it does require evaluation of the Jacobian derivative $g_x(x)$.

(a) Consider the correctors derived from (126) with:

$$\text{(i) } j = 0 \text{ and } l = 1, \qquad \text{(ii) } j = 1 \text{ and } l = 1,$$

$$\text{(iii) } j = 2 \text{ and } l = 1, \qquad \text{(iv) } j = 2 \text{ and } l = 2.$$

Let $x(t_{i+1})^{(0)} = x(t_i)^{(1)}$ and determine $x(t_{i+1})^{(1)}$ in terms of $x(t_i)^{(1)}$ using the Newton-Raphson iteration. If the Newton-Raphson iteration is terminated at this point and $x(t_{i+1})^{(1)}$ is viewed as the (approximate) solution of the corrector, then the resulting equations are explicit expressions for $x(t_{i+1})^{(1)}$ in terms of $x(t_i)^{(1)}$; they are predictors derived by truncating iterations used in solving correctors.

(b) Repeat (a) using a linear projection from the two preceding solution points to get $x(t_{i+1})^{(0)}$; that is

$$x(t_{i+1})^{(0)} = x(t_i)^{(1)} + h \frac{x(t_i)^{(1)} - x(t_{i-1})^{(1)}}{h} = 2x(t_i)^{(1)} - x(t_{i-1})^{(1)}$$

[Note: Higher order polynomial fitting of preceding solution points could be employed to give a more refined projection used to determine $x(t_{i+1})^{(0)}$.]

(c) Repeat (a) using the solution of the Euler predictor for $x(t_{i+1})^{(0)}$; that is

$$x(t_{i+1})^{(0)} = x(t_i)^{(1)} + h\dot{x}(t_i)^{(1)}.$$

[Note: Other predictors could be used to determine $x(t_{i+1})^{(0)}$.]

(d) Obtain a numerical solution of each of the following state equations using the predictors created in (a) through (c). Set $h = 0.2$ and compute $x(ih)$ for $i = 1, 2, \cdots, 10$.

(i) $\dfrac{dx}{dt} = -x(1 + x^2), \; x(0) = -5$

(ii) $\dfrac{d}{dt} \begin{bmatrix} x_1 \\ x_2 \end{bmatrix} = \begin{bmatrix} -1 & 1 \\ 0 & -10 \end{bmatrix} \begin{bmatrix} x_1 \\ x_2 \end{bmatrix}, \; \begin{bmatrix} x_1(0) \\ x_2(0) \end{bmatrix} = \begin{bmatrix} 0 \\ 2 \end{bmatrix}$

(iii) $\dfrac{d}{dt} \begin{bmatrix} x_1 \\ x_2 \end{bmatrix} = \begin{bmatrix} -2t & -1 \\ 1 & -2t \end{bmatrix} \begin{bmatrix} x_1 \\ x_2 \end{bmatrix}, \; \begin{bmatrix} x_1(0) \\ x_2(0) \end{bmatrix} = \begin{bmatrix} 0 \\ 1 \end{bmatrix}$

(iv) $\dfrac{d}{dt} \begin{bmatrix} x_1 \\ x_2 \end{bmatrix} = \begin{bmatrix} x_2 \\ -2x_2 + x_1 - x_1^3 \end{bmatrix}, \; \begin{bmatrix} x_1(0) \\ x_2(0) \end{bmatrix} = \begin{bmatrix} 1 \\ -1 \end{bmatrix}.$

Obtain starting values by using a truncated Taylor series. For (i) through (iii) compare the three different predictors by considering the accuracy of the numerical solutions relative to the exact solutions.

53. For many networks the state vector varies rapidly in some intervals of time and slowly in others. In the former the time interval between solution points must be smaller than that permitted in the latter if the accuracy is to be comparable in both instances. Therefore, in the interest of computational efficiency it is desirable to have a criterion by which to adjust the integration interval size as the numerical evaluation of the solution progresses. An integration formula used in conjunction with such a criterion should possess the property that for infinite precision arithmetic the difference between the numerical solution and the actual solution (at the points in time when the numerical solution is evaluated) vanishes with increasing time for all possible values of the integration interval. A study of integration formulas relative to such a property is beyond the scope of this text. Therefore, we will consider the particular integration formula established in Part (a) (i) of Problem 52, which possesses this property for stable, time-invariant, linear state equations and all values of the integration interval. The corrector error is $-\frac{1}{2}h_{i+1}^2\,\ddot{\mathbf{x}}(\theta)$ for some θ such that $t_i \leq \theta \leq t_{i+1} = t_i + h_{i+1}$. Now, if the error is too large, h_{i+1} should be made smaller and the calculation of $\mathbf{x}(t_{i+1})$ should be repeated; if the error is sufficiently small, the value of the integration interval at the next step, h_{i+2}, should be made larger than h_{i+1}. This serves as justification for the following criterion:

Set $h_{i+2} = 2h_{i+1}$ if	$\frac{1}{2}h_{i+1}^2\,\lVert\ddot{\mathbf{x}}(t_{i+1})\rVert_\infty \leq \frac{1}{4} \times 10^{-3}$
Set $h_{i+2} = h_{i+1}$ if	$\frac{1}{4} \times 10^{-3} < \frac{1}{2}h_{i+1}^2\,\lVert\ddot{\mathbf{x}}(t_{i+1})\rVert_\infty < 10^{-3}$
Repeat calculation of $\mathbf{x}(t_{i+1})$ with h_{i+1} half as large if	$10^{-3} \leq \frac{1}{2}h_{i+1}^2\,\lVert\ddot{\mathbf{x}}(t_{i+1})\rVert_\infty$

Apply the integration formula indicated and this criterion for determining h to obtain the numerical solution of

$$\frac{d}{dt}\begin{bmatrix} x_1 \\ x_2 \end{bmatrix} = \begin{bmatrix} -1 & 1 \\ 0 & -10 \end{bmatrix}\begin{bmatrix} x_1 \\ x_2 \end{bmatrix}, \quad \begin{bmatrix} x_1(0) \\ x_2(0) \end{bmatrix} = \begin{bmatrix} 1 \\ 1 \end{bmatrix}$$

at twenty-five points in time. Take $\frac{1}{2} \times 10^{-2}$ as the initial value for h_{i+1}.

Indicate the range of values h_{i+1} takes on. To determine $\ddot{\mathbf{x}}(t_{i+1})$ you will find the following fact useful:

$$\ddot{\mathbf{x}}(t_{i+1}) = \mathbf{f_x}(\mathbf{x}(t_{i+1}), t_{i+1})\dot{\mathbf{x}}(t_i) = \mathbf{f_x}(\mathbf{x}(t_{i+1}), t_{i+1})\mathbf{f}(\mathbf{x}(t_{i+1}), t_{i+1}).$$

54. Which of the following scalar functions are positive definite for all $\mathbf{x}$? Let $t_0 = 1$.

(a) $V(\mathbf{x}) = 6x_1^4 + 6x_2^2 - 8x^3 + 3x^4$

(b) $V(\mathbf{x}) = \dfrac{2x_1^2}{1 + x_1^2} + \dfrac{x_2^4}{1 + x_2^2} + x_1^2 x_3^4 + \dfrac{x_3^2}{1 + x_1^2}$

(c) $V(\mathbf{x}) = x_1^2 + x_2^2 - 2x_1^4 - 2x_2^4 + x_1^6 + x_2^6$

(d) $V(\mathbf{x}) = x_1^2 + 2\epsilon^{-t}x_2^4 + (2 + \cos t)x_3^2$

(e) $V(\mathbf{x}) = \dfrac{2 + t}{1 + t}\, x_1^6 + (x_1 - x_2)^2 + x_3^4 + (x_1 - x_3 - x_4)^2$

(f) $V(\mathbf{x}) = (x_1 - x_2)^2 + (x_1 - x_2 - 2x_3)^2 + x_3^4$

55. Consider a linear, time-varying RLC network with no external excitation. The state equation (not in normal form) is

$$\frac{d}{dt}\begin{bmatrix} \mathscr{C} & \mathbf{0} \\ \mathbf{0} & \mathscr{L} \end{bmatrix}\begin{bmatrix} \mathbf{v}_{Ct} \\ \mathbf{i}_{Ll} \end{bmatrix} = \begin{bmatrix} -\mathscr{Y} & \mathscr{H} \\ \mathscr{G} & -\mathscr{Z} \end{bmatrix}\begin{bmatrix} \mathbf{v}_{Ct} \\ \mathbf{i}_{Ll} \end{bmatrix}$$

and the stored energy is

$$V = [\mathbf{v}'_{Ct} \quad \mathbf{i}'_{Ll}]\begin{bmatrix} \mathscr{C} & \mathbf{0} \\ \mathbf{0} & \mathscr{L} \end{bmatrix}\begin{bmatrix} \mathbf{v}_{Ct} \\ \mathbf{i}_{Ll} \end{bmatrix}.$$

Under what condition will V be a Liapunov function?

56. Determine an open region D, containing the origin, within which the following scalar functions are positive definite. Let $t_0 = 0$.

(a) $V(\mathbf{x}) = (x_1 - x_2)^2 - (x_1 - x_2)^4 + x_2^2 - x_2^3$

(b) $V(\mathbf{x}) = (x_1 - x_2)^2 - (x_1 - x_2)^4 + 4(x_1 + x_2)^2 - (x_1 + x_2)^4$

(c) $V(\mathbf{x}) = 2\,\dfrac{2 + t}{1 + t}\, \tanh^2 x_1 - x_1^2 + 2\,\dfrac{x_2^2 - x_2^4}{1 + x_1^2}$

57. Prove Theorem 19.

58. Each of the following state equations is paired with a positive-definite function. Determine whether the positive-definite function is a Liapunov function for the state equation and, if so, which of the Theorems 17, 18, and/or 19 is satisfied. Set $t_0 = 1$.

(a)
$$\frac{d}{dt}\begin{bmatrix} x_1 \\ x_2 \\ x_3 \end{bmatrix} = \begin{bmatrix} -x_1 + x_2 - x_3 \\ x_1 - 2x_2 - x_2{}^3 \\ x_1 - \tanh x_3 \end{bmatrix}$$

$$V(\mathbf{x}) = \tfrac{1}{2}[x_1{}^2 + x_2{}^2 + x_3{}^2]$$

(b)
$$\frac{d}{dt}\begin{bmatrix} x_1 \\ x_2 \end{bmatrix} = \begin{bmatrix} -\dfrac{(x_1 - x_2)}{1 + (x_1 - x_2)^2} \\ -\dfrac{(x_2 - x_1)}{1 + (x_1 - x_2)^2} - x_2 - x_2{}^3 \end{bmatrix}$$

$$V(\mathbf{x}) = 2x_1{}^2 - 6x_1x_2 + 5x_2{}^2$$

(c)
$$\frac{d}{dt}\begin{bmatrix} x_1 \\ x_2 \end{bmatrix} = \begin{bmatrix} -\epsilon^{-t}x_1 + x_2 \\ -x_1 - x_2 \end{bmatrix}$$

$$V(\mathbf{x}) = (x_1 + x_2)^2 + (\epsilon^{-t}x_1 + x_2)^2$$

(d)
$$\frac{d}{dt}\begin{bmatrix} x_1 \\ x_2 \end{bmatrix} = \begin{bmatrix} -t\epsilon^{-t}x_1 - x_2 \\ -2x_1 + \dfrac{x_2}{1 + x_2{}^2} \end{bmatrix}$$

$$V(\mathbf{x}) = \tfrac{1}{2}x_1{}^2 + \tfrac{1}{4}x_2{}^4$$

59. Describe how the stability theorems 17, 18, and 19 establish that a solution of the state equation 98, equal to $\mathbf{x}(t_0)$ at time t_0, may be extended to yield a defined solution for all $t \geq t_0$. Indicate any restrictions on $\mathbf{x}(t_0)$. Find a state equation, of order greater than 1, to which these results apply and to which Theorem 13 does not apply.

60. Each of the following state equations is paired with a scalar function of the state vector. Let the scalar function be $V(\mathbf{x}, t)$ in Theorem 20; then, by this theorem, determine whether the origin is unstable. Set $t_0 = 1$.

(a) $\dfrac{d}{dt}\begin{bmatrix} x_1 \\ x_2 \end{bmatrix} = \begin{bmatrix} -\epsilon^{-t}x_1 - x_2 \\ -x_1 + \tanh x_2 \end{bmatrix}$

$V(\mathbf{x}) = -x_1 x_2$

(b) $\dfrac{d}{dt}\begin{bmatrix} x_1 \\ x_2 \\ x_3 \end{bmatrix} = \begin{bmatrix} -2x_1 + x_2 \\ x_1 - 3x_2 + x_2{}^5 - x_3 \\ x_2 - 5x_3 - x_3 \tanh^2 x_3 \end{bmatrix}$

$V(\mathbf{x}) = (x_1{}^2 - x_2{}^2)x_3$

(c) $\dfrac{d}{dt}\begin{bmatrix} x_1 \\ x_2 \end{bmatrix} = \begin{bmatrix} -2\,\dfrac{x_1}{1 + 2x_1{}^2} + 5x_2 \\ -x_1 + 4\,\dfrac{x_2 + 2x_2{}^3}{1 + x_2{}^2} \end{bmatrix}$

$V(\mathbf{x}) = x_1(2x_2 - x_1)$

61. Apply Theorem 21 to each of the state equations in Problem 58. How do the results, here obtained with Theorem 21, compare with those obtained in Problem 58?

62. Use Theorem 21 to show that the origin is asymptotically stable in-the-large when

$$f(\mathbf{x}) = \mathbf{A}\mathbf{x},$$

where $\mathbf{A}$ is a constant $n \times n$ matrix with negative eigenvalues.

63. By using the variable-gradient method, seek a Liapunov function for each of the following state equations:

(a) $\dfrac{d}{dt}\begin{bmatrix} x_1 \\ x_2 \\ x_3 \end{bmatrix} = \begin{bmatrix} -x_1 + x_2 - x_3 \\ x_1 - 2x_2 - x_2{}^3 \\ x_1 - \tanh x_3 \end{bmatrix}$

(b) $\dfrac{d}{dt}\begin{bmatrix} x_1 \\ x_2 \end{bmatrix} = \begin{bmatrix} x_2{}^3 \\ -x_2 + \tfrac{1}{3}x_2{}^3 - x_1 \end{bmatrix}$

(c) $\dfrac{d}{dt}\begin{bmatrix} x_1 \\ x_2 \end{bmatrix} = \begin{bmatrix} -\dfrac{x_1}{1+x_1^2} + x_2 \\ x_1 - 2x_2 - x_2^3 \end{bmatrix}$

64.* Prepare a program flow chart and a set of program instructions, in some user language such as FORTRAN IV, for a digital computer to obtain a numerical solution of the state equation 104 by the

(a) Euler method
(b) Modified Euler method
(c) Adams method $(j=4)$
(d) Modified Adams method $(j=4)$
(e) Milne method
(f) Predictor-corrector pairs of Problem 47
(g) Predictors of Problem 48
(h) Predictor-corrector pairs of Problem 50

The Runge-Kutta method should be used to generate starting values. Assume that $\mathbf{x}(t_0)$, $\mathbf{f}(\mathbf{x}, t)$, h, and the total number of time steps will be specified by the user of the program. Include a set of user instructions for each program.

Appendix 1

GENERALIZED FUNCTIONS

Analysis of physical systems is often facilitated by employing the impulse function and/or its derivatives. Now, the impulse function is not a function in the usual sense; therefore we are, strictly speaking, in violation of rigorous mathematics in applying theorems developed for ordinary point functions to relations involving the impulse function. Thus, from 1927, when Dirac popularized the impulse function as a tool of mathematical physics,* till 1950, when Schwartz published a complete and rigorous basis for it, the impulse function stood in mathematical disrepute but was nevertheless used by physicists and engineers.

The *theory of distributions* developed by Schwartz† provides the basis for using the impulse function in mathematical analysis, the impulse function being a *distribution* within this theory. In addition to refinements to and to more lucid presentations of distribution theory published since 1950,‡ work has progressed along other lines,§ as well. A particularly

---

* P. A. M. Dirac, "The Physical Interpretation of the Quantum Mechanics," *Proc. Roy. Soc.*, Ser. A., Vol. 113, 1926–1927, pp. 621–641.

P. A. M. Dirac, *The Principles of Quantum Mechanics*, Oxford University Press, London, 1930.

† L. Schwartz, *Théorie des distributions*, Vols. I and II, Herman, Paris, 1950 and 1951.

‡ I. M. Gel'fand and G. E. Shilov, *Generalized Functions*, Vol. 1, "Properties and Operations," Academic Press, New York, 1964.

V. Dolezal, *Dynamics of Linear Systems*, Publishing House of the Czechoslovak Academy of Sciences, Prague, 1964.

A. H. Zemanian, *Distribution Theory and Transform Analysis*, McGraw-Hill, New York, 1965.

§ J. Mikusiński, *Operational Calculus*, 5th ed. (in English) Macmillan, New York, 1959. The first edition (in Polish) was published in 1953.

G. Temple, "Theories and Applications of Generalized Functions," *J. London Math. Soc.*, Vol. 28, 1953, pp. 134–148.

M. J. Lighthill, *Introduction to Fourier Analysis and Generalized Functions*, Cambridge University Press, Cambridge, 1958.

829

understandable alternative to distribution theory was introduced by Mikusiński in this period.

Convolution in the set of continuous functions is similar to multiplication in the set of integers. When division—inverse operation to multiplication—of two integers is defined, the resulting set of rational numbers contains the integers as a proper subset. Perhaps a suitably defined convolution division—inverse operation to convolution—of two continuous functions will establish, similarly, a set of convolution quotients or generalized functions, containing the continuous functions as a proper subset. As Mikusiński discovered, this does happen. In particular, the impulse function and all its derivatives are included in this set of generalized functions.

In this appendix we shall give a brief description of the theory of generalized functions based on convolution division. For simplicity, we shall treat only scalar-valued, and not vector-valued, functions. The extension of these concepts to vector-valued functions will be left to you. The discussion here is not complete, but it is adequate for putting the use of impulse functions in Chapter 5 on a firm basis.

To preclude notational confusion—a possibility when new concepts are presented—we shall adhere rather rigidly to the following conventions. A lower case Greek letter, with the exception of δ, will denote a scalar. A function of time, in its entirety for non-negative time, will be denoted by a lower case italic letter, with the exception of s and t. If f is an arbitrary function of time, $f(t)$ will denote the value of f at time t. Note that $f(t)$ is a scalar. There are situations in which a function will be given explicitly; for example, $t\epsilon^{-t}$. To denote this function in its entirety, we shall use the notation $\{t\epsilon^{-t}\}$. Observe that $\{\alpha\}$ is not a scalar but the function that assumes the constant value α for $t \geq 0$. The product of two scalars, of a scalar and a function, and of two functions will be denoted in the usual manner; for example, $\alpha\beta$, αf or $\{\alpha f(t)\}$, and fg or $\{f(t)\,g(t)\}$. The usual exponent convention for a repeated product will apply; for example, $\alpha\alpha\alpha\alpha = \alpha^4$ and $fff = f^3$. As in the main body of the text, convolution will be denoted by

$$f * g = \left\{ \int_0^t f(t - \tau)\, g(\tau)\, d\tau \right\}.$$

J. D. Weston, "Operational Calculus and Generalized Functions," *Proc. Roy. Soc.*, Ser. A, Vol. 250, 1959, pp. 460–471.

J. D. Weston, "Characterization of Laplace Transforms and Perfect Operators," *Arch. Rat. Mech. and Anal.*, Vol. 3, 1959, pp. 348–354.

A. Erdélyi, *Operational Calculus and Generalized Functions*, Holt, Rinehart & Winston, New York, 1962. This is a lucid and succinct description of Mikusiński's convolution quotients as the basis for a theory of generalized functions.

To denote repeated convolution we shall use a positive, boldface exponent; for example, $f * f * f = f^3$.

A1.1 CONVOLUTION QUOTIENTS AND GENERALIZED FUNCTIONS

In solving a scalar equation, such as

$$\alpha \zeta = \beta, \tag{1}$$

where $\alpha \neq 0$ and β are known scalars, we express the solution as $\zeta = \beta/\alpha$; that is, the quotient of β by α. This is possible because scalar multiplication has a uniquely defined inverse—scalar division.

Let us now consider the convolution equation

$$a * z = b, \tag{2}$$

where $a \neq \{0\}$ and b are known functions. Since the convolution of functions has the same algebraic properties as scalar multiplication—associativity, commutativity, and so on—it is tempting to suppose that there exists a uniquely defined inverse operation to convolution, which permits us to express the solution as $z = b//a$; that is, the *convolution quotient* of b by a. (Observe the double slant line used as the symbol.) We shall show that the notion of a convolution quotient is meaningful.

To set the background for defining convolution quotients, let us turn again to the algebraic equation (1), when α and β are known integers, and consider its solution. It is the uniqueness of that solution that gives meaning to the quotient, β/α, by which the solution is expressed. Suppose (1) has more than one solution; then let ζ_1 and ζ_2 denote two distinct solutions. Since $\alpha\zeta_1$ and $\alpha\zeta_2$ both equal β, then $\alpha(\zeta_1 - \zeta_2) = 0$. But α is assumed to be a nonzero integer; hence $\zeta_1 - \zeta_2 = 0$ and, by contradiction, the solution of (1) must be unique. If the solution ζ is an integer, then that is the value assigned to β/α. Of course, (1) may not have an integer solution. However, because the solution is unique, we could say that β/α is the quantity, called a quotient, that denotes the unique solution; in this way β/α is made meaningful. In fact, history has given the name *rational number* to the quotient of two integers. Now, β/α is a rational-number solution of $\alpha\zeta = \beta$ that is unique to within the equivalence $\beta/\alpha = \gamma\beta/\gamma\alpha$, where γ is any nonzero integer.

Now let us give similar consideration to the convolution equation (2), $a * z = b$, when a and b are known continuous functions and $a \neq \{0\}$.

If we can show that it has a unique solution, then that solution ascribes meaning to the convolution quotient, $b/\!/a$, that expresses the solution. Suppose (2) has more than one solution; then let z_1 and z_2 be two distinct solutions. Since $a * z_1$ and $a * z_2$ both equal b, then $a * (z_1 - z_2) = \{0\}$. There is a theorem due to Titchmarsh* that states the following:

Theorem 1.   *If* f *and* g *are continuous and* f $*$ g $= \{0\}$, *then either* f $= \{0\}$ *or* g $= \{0\}$ *or both.*

Since the function $a \neq \{0\}$ in the previous result $a * (z_1 - z_2) = \{0\}$, then $z_1 - z_2$ must be the zero function $\{0\}$. But this is a contradiction of the assumption that z_1 and z_2 are distinct. Hence the solution of (2) must be unique.

If the solution of (2) is a continuous function, then $b/\!/a$ is identified with that function. Of course, there may be no continuous function that solves (2), just as $\alpha \zeta = \beta$ may not have an integer solution; for example, if $b(0) \neq 0$, then the solution cannot be a continuous function. (Why?) However, because the solution is unique, we could extend the meaning of the term " function " by saying $b/\!/a$ is the quantity, called a *convolution quotient*, that denotes the unique solution. In this way $b/\!/a$ becomes meaningful.

Now suppose the function $c \neq \{0\}$ and consider the convolution equation

$$c * a * z = c * b, \tag{3}$$

or, equivalently,

$$c * (a * z - b) = \{0\}.$$

By Theorem 1, $a * z - b = \{0\}$. Thus (3) has a solution if and only if (2) has a solution; furthermore, a solution of (2) must also be a solution of (3). Since each has a unique solution and since $b/\!/a$ denotes the solution of (2) and $c * b/\!/c * a$ denotes the solution of (3), we must have $b/\!/a = c * b/\!/c * a$. Hence $b/\!/a$ is the unique solution of (2) to within the equivalence $c * b/\!/c * a$, where $c \neq \{0\}$.

We shall say that two convolution quotients $b/\!/a$ and $d/\!/c$ are *equivalent* —that is, $b/\!/a = d/\!/c$—if $b * c = a * d$. The set of all convolution quotients that are equivalent to any one convolution quotient is called an *equivalence class*. Obviously, any one convolution quotient is in one and only one equivalence class; that is, the convolution quotients are sorted into equivalence classes in a unique manner. Thus each equivalence class is identi-

---

* This theorem is a special case of Theorem 152 in E. C. Titchmarsh, *Introduction to the Theory of Fourier Integrals*, 2nd ed., Oxford University Press, London, 1948.

fied by any one of its convolution quotients. Therefore it is the equivalence class containing $b/\!/a$ that should be viewed as the unique solution of $a * z = b$ and referred to as a *generalized function*.

A1.2   ALGEBRA OF GENERALIZED FUNCTIONS

Let us temporarily denote a generalized function by brackets around any convolution quotient in its equivalence class; for example, $[b/\!/a]$. The first relation we must establish is that of equality. This we do by noting that

$$\left[\frac{b}{a}\right] = \left[\frac{d}{c}\right] \quad \text{if and only if} \quad b * c = a * d. \tag{4}$$

This is an immediate consequence of the following: (1) If $b * c = a * d$, then $b/\!/a$ and $d/\!/c$ are equivalent and, hence, determine the same equivalence class. (2) If $[b/\!/a] = [d/\!/c]$, then $b/\!/a$ and $d/\!/c$ are in the same equivalence class and, hence, $b * c = a * d$.

We now define addition of two generalized functions, convolution of two generalized functions, and the product of a scalar and a generalized function as follows:

$$\left[\frac{b}{a}\right] + \left[\frac{d}{c}\right] = \left[\frac{b * c + a * d}{a * c}\right] \tag{5a}$$

$$\left[\frac{b}{a}\right] * \left[\frac{d}{c}\right] = \left[\frac{b * d}{a * c}\right] \tag{5b}$$

$$\alpha\left[\frac{b}{a}\right] = \left[\frac{\alpha b}{a}\right]. \tag{5c}$$

You should verify that the generalized functions on the right are independent of the specific convolution quotients that characterize each of the generalized functions on the left; for example, suppose $b/\!/a = b'/\!/a'$ and $d/\!/c = d'/\!/c'$; then

$$\left[\frac{b * d}{a * c}\right] = \left[\frac{b}{a}\right] * \left[\frac{d}{c}\right]$$

and

$$\left[\frac{b' * d'}{a' * c'}\right] = \left[\frac{b'}{a'}\right] * \left[\frac{d'}{c'}\right]$$

will be equal if $b * d * a' * c' = a * c * b' * d'$. This equality is an obvious consequence of $b * a' = a * b'$, $d * c' = c * d'$, and the commutative property of convolution. Because the algebraic operations defined in (5) are independent of the specific convolution quotient that characterizes the corresponding generalized function, the brackets need not be used in denoting a generalized function. We shall therefore dispense with the brackets and refer to $b/\!/a$ as a generalized function. This should cause no confusion.

With the equality relation established in (4) and the basic operations defined in (5), it is easily shown that the usual laws of algebra given in (6) below are valid. The verification is left to you.

$$\frac{b}{a} + \frac{d}{c} = \frac{d}{c} + \frac{b}{a} \tag{6a}$$

$$\frac{b}{a} + \left(\frac{d}{c} + \frac{f}{e}\right) = \left(\frac{b}{a} + \frac{d}{c}\right) + \frac{f}{e} \tag{6b}$$

$$\alpha\left(\frac{b}{a} + \frac{d}{c}\right) = \alpha\frac{b}{a} + \alpha\frac{d}{c} \tag{6c}$$

$$(\alpha + \beta)\frac{b}{a} = \alpha\frac{b}{a} + \beta\frac{b}{a} \tag{6d}$$

$$\alpha\left(\frac{b}{a} * \frac{d}{c}\right) = \left(\alpha\frac{b}{a}\right) * \frac{d}{c} = \frac{b}{a} * \left(\alpha\frac{d}{c}\right) \tag{6e}$$

$$(\alpha\beta)\frac{b}{a} = \alpha\left(\beta\frac{b}{a}\right) \tag{6f}$$

$$\frac{b}{a} * \frac{d}{c} = \frac{d}{c} * \frac{b}{a} \tag{6g}$$

$$\frac{b}{a} * \left(\frac{d}{c} * \frac{f}{e}\right) = \left(\frac{b}{a} * \frac{d}{c}\right) * \frac{f}{e} \tag{6h}$$

$$\left(\frac{b}{a} + \frac{d}{c}\right) * \frac{f}{e} = \frac{b}{a} * \frac{f}{e} + \frac{d}{c} * \frac{f}{e}. \tag{6i}$$

The cancellation law is also valid; that is, if $e \neq \{0\}$ and $f \neq \{0\}$, then

$$\frac{b}{a} * \frac{f}{e} = \frac{d}{c} * \frac{f}{e} \quad \text{if and only if} \quad \frac{b}{a} = \frac{d}{c}. \tag{7}$$

Theorem 1 is required in the proof of the cancellation law.

Next, we note that if $a \neq \{0\}$, then $\{0\}/\!/a$ is the " zero " generalized function and that $a/\!/a$ is the " unit " generalized function. This follows by showing that both are unique and verifying that

$$\frac{\{0\}}{a} + \frac{d}{c} = \frac{d}{c} \tag{8a}$$

$$\frac{\{0\}}{a} * \frac{d}{c} = \frac{\{0\}}{a} \tag{8b}$$

$$\frac{a}{a} * \frac{d}{c} = \frac{d}{c}. \tag{8c}$$

This task is left to you.

CONVOLUTION QUOTIENT OF GENERALIZED FUNCTIONS

The convolution division of two ordinary functions was defined as the inverse operation of convolution. It is only natural to inquire whether such an inverse operation, convolution division, can be defined for generalized functions. The answer is given by proceeding in the same manner as we did in introducing the convolution quotient of ordinary functions. This was done by showing that the convolution equation (2) has a unique solution. Similarly, we must show that the following convolution equation of generalized functions,

$$\frac{b}{a} * \frac{x}{y} = \frac{d}{c}, \tag{9}$$

where $b/\!/a \neq \{0\}/\!/a$, has a unique, generalized function solution. Clearly, by substitution in (9) and application of (8c), $x/\!/y = a * d/\!/b * c$ is a solution of (9); hence we need to show only that it is the unique solution. Suppose $x'/\!/y'$ is another solution. Then $(b/\!/a) * (x'/\!/y') = d/\!/c$ combined with (9) yields $(b/\!/a) * (x/\!/y - x'/\!/y') = \{0\}/\!/a$ or $x/\!/y = x'/\!/y'$, since $b/\!/a \neq \{0\}/\!/a$. Thus $a * d/\!/b * c$ is the unique, generalized-function solution of (9).

The solution of (9) indicates that the convolution quotient of the generalized function $d/\!/c$ by the generalized function $b/\!/a$, that is, $(d/\!/c)/\!/(b/\!/a)$, has the following meaning:

$$\frac{(d/\!/c)}{(b/\!/a)} = \frac{a * d}{b * c}. \tag{10}$$

Recall that repeated convolution is denoted by a positive, boldface exponent. Let us arbitrarily set $(b/\!/a)^{\mathbf{0}} = a/\!/a$, the unit generalized function. Next, let us denote the convolution quotient of $a/\!/a$ by $(b/\!/a)^{\mathbf{n}}$ as

$$\frac{(a/\!/a)}{(b/\!/a)^{\mathbf{n}}} = (b/\!/a)^{-\mathbf{n}}. \tag{11}$$

By using (10) it is easily shown that $(b/\!/a)^{-\mathbf{n}} = (a/\!/b)^{\mathbf{n}}$.

With this background you should have no difficulty verifying the usual operations with exponents; that is,

$$\left(\frac{b}{a}\right)^{\mathbf{m}}\left(\frac{b}{a}\right)^{\mathbf{n}} = \left(\frac{b}{a}\right)^{\mathbf{m+n}} \tag{12a}$$

$$\left(\left(\frac{b}{a}\right)^{\mathbf{m}}\right)^{\mathbf{n}} = \left(\frac{b}{a}\right)^{\mathbf{mn}} \tag{12b}$$

$$\left(\frac{b}{a} * \frac{d}{c}\right)^{\mathbf{n}} = \left(\left(\frac{b}{a}\right)^{\mathbf{n}}\right) * \left(\left(\frac{d}{c}\right)^{\mathbf{n}}\right), \tag{12c}$$

where $\mathbf{m}$ and $\mathbf{n}$ are integers: positive, zero, or negative.

A1.3   PARTICULAR GENERALIZED FUNCTIONS

Equations for most dynamic systems are formulated in terms of scalars; in terms of ordinary point functions—usually continuous, piecewise continuous, and locally integrable; and in terms of operators—usually differential, integral, and delay. In this and the next section we shall show that all of these scalars, functions, and operators may be identified as generalized functions. In this way the equation for a dynamic system will be meaningful as an equation on the set of generalized functions.

Let us begin with the set of scalars. This set can be embedded in the set of generalized functions if the following two conditions are satisfied:

1. There is a one-to-one correspondence between the set of scalars and a subset of the set of generalized functions.

2. Every algebraic operation that is defined on the set of scalars has a counterpart that is defined on the set of generalized functions.

If a is *any* continuous function other than $\{0\}$, then the generalized function $\alpha a /\!/ a$ stands in a one-to-one relation with the scalar α; thus condition 1 is satisfied. This relation is symbolically denoted as

$$\alpha \Leftrightarrow \frac{\alpha a}{a}. \tag{13}$$

The verification of condition 2 is left to you; however, to illustrate how this is done, we shall verify that addition of scalars has its counterpart. If $\alpha \Leftrightarrow \alpha a /\!/ a$ and $\beta \Leftrightarrow \beta a /\!/ a$, then

$$\alpha + \beta \Leftrightarrow \frac{(\alpha + \beta)a}{a} = (\alpha + \beta)\frac{a}{a} = \frac{\alpha a}{a} + \frac{\beta a}{a}.$$

The several steps make use of (5c) and (8d).

Let us next consider the set of continuous functions. This set can also be embedded in the set of generalized functions if the same two conditions as for scalars are satisfied, with the words "continuous function" replacing the word "scalar." To establish condition 1, let a be *any* continuous function other than $\{0\}$; then the continuous function c stands in a one-to-one relation with the generalized function $c * a /\!/ a$; this relation is symbolically denoted by

$$c \Leftrightarrow \frac{c * a}{a}. \tag{14}$$

We leave the task of verifying condition 2 to you; however, we shall illustrate how this is done. If $b \Leftrightarrow b * a /\!/ a$ and $c \Leftrightarrow c * a /\!/ a$, then

$$b * c \Leftrightarrow \frac{(b * c) * a}{a} = \frac{b * a * c * a}{a * a} = \frac{b * a}{a} * \frac{c * a}{a}.$$

The several steps make use of (5b) and (8c).

These embeddings give justification to viewing α and $\alpha a /\!/ a$ as being interchangeable and c and $c * a /\!/ a$ as being interchangeable. In addition, the identification of c with $c * a /\!/ a$ permits us to assign a value to $c * a /\!/ a$ at time t. In general, this is not possible; a generalized function is an entity for which the value at time t may have no meaning. We shall later show that some generalized functions other than those identified with continuous functions may be assigned a value at time t.

CERTAIN CONTINUOUS FUNCTIONS

A function that will appear often is the constant function $\{1\}$, which we shall denote henceforth by u; that is,

$$u = \{1\}. \tag{15}$$

This is the well-known unit step function with its value specified as 1 at $t = 0$ as well as at $t > 0$.

Repeated convolution of u with itself yields an often-encountered set of functions. Thus

$$u^2 = u * u = \left\{ \frac{t}{1} \right\}$$

$$u^3 = u * u^2 = \left\{ \frac{t^2}{1 \cdot 2} \right\}$$

$$u^4 = u * u^3 = \left\{ \frac{t^3}{1 \cdot 2 \cdot 3} \right\} \tag{16}$$

$$\cdots\cdots\cdots\cdots\cdots$$

$$u^{n+1} = u * u^n = \left\{ \frac{t^n}{n!} \right\},$$

as you should verify by carrying out the indicated convolutions. The generalized function that is identified with the continuous function u^n may be expressed as $u^n * a /\!/ a$, where $a \neq \{0\}$. Setting $a = u$, we get the interesting result

$$u^n \Leftrightarrow \frac{u^{n+1}}{u}. \tag{17}$$

Many other continuous functions that are commonly encountered in analysis can be identified with generalized functions, each expressed as a convolution quotient of polynomials in u; for example,

$$\{\epsilon^{-\alpha t}\} \Leftrightarrow \frac{\{\epsilon^{-\alpha t}\} * u}{u} = \frac{u^2}{u + \alpha u^2} \tag{18a}$$

$$\{\sin \omega t\} \Leftrightarrow \frac{\{\sin \omega t\} * u}{u} = \frac{\omega u^3}{u + \omega^2 u^3} \tag{18b}$$

$$\{\cos \omega t\} \Leftrightarrow \frac{\{\cos \omega t\} * u}{u} = \frac{u^2}{u + \omega^2 u^3}. \tag{18c}$$

Each of these equalities can be checked according to the general equality relation (4). As an illustration, consider (18b). The two expressions are equal if and only if

$$\{\sin \omega t\} * (u^2 + \omega^2 u^4) = \omega u^4.$$

By Theorem 1, u may be " cancelled " on both sides. Thus we only need to verify

$$\{\sin \omega t\} * (u + \omega^2 u^3) = \omega u^3,$$

or, equivalently,

$$\int_0^t \sin \omega(t - \tau)\left[1 + \frac{\omega^2}{2}\tau^2\right] d\tau = \frac{\omega}{2}t^2. \tag{19}$$

You should verify this integral relation, thus verifying (18b).

The one-to-one relationships between continuous functions and generalized functions, such as given in (18), are useful in solving some convolution equations. As an example, consider

$$(u + \omega^2 u^3) * z = u^2.$$

The generalized-function solution is

$$z = \frac{u^2}{u + \omega^2 u^3}.$$

Now, by (18c) we know that this is the same as the continuous function $\{\cos \omega t\}$.

LOCALLY INTEGRABLE FUNCTIONS

A class of functions that is somewhat less " well behaved " than continuous functions is the class of *locally integrable* functions. By a locally integrable function we mean a function (1) that is continuous except at a finite number of points in any finite interval and (2) that possesses a proper or absolutely convergent improper Riemann integral on every finite interval over the interior of which the function is continuous. Under these conditions the integral on the interval $0 \leq t \leq \hat{t}$ for each $\hat{t} > 0$ is defined as the sum of the Riemann integrals over the continuous subintervals, over the interior of which the function is continuous. It is of interest to note that the convolution of two locally integrable functions is a locally integrable function and that the convolution of a locally integrable function with a continuous function is a continuous function.

It is clear from the definition of a locally integrable function that two such functions will have the same integral on the interval $0 \leq t \leq \hat{t}$ for all $\hat{t} > 0$ if they differ in value only at a finite number of points in any finite interval or, equivalently, if they have the same value at points where both are continuous. We shall follow the usual practice of regarding two such functions as being equivalent.

Let us see if locally integrable functions can be identified with some generalized functions. This set can be embedded in the set of generalized functions if the same two conditions as before are satisfied; the word " scalar " must, of course, be replaced by the words " locally integrable." To see whether they are satisfied, note that, if a is any continuous function other than $\{0\}$ and b is a locally integrable function, then the generalized function $b * a /\!/ a$ stands in a one-to-one relation with b; thus condition 1 is satisfied. This relation is symbolically expressed as

$$b \Leftrightarrow \frac{b * a}{a}, \tag{20}$$

which is the same type of relationship as given in (14) for continuous functions. We leave the verification of condition 2 to you.

The set of locally integrable functions obviously contains the set of continuous functions. More significantly, however, it also contains the set of *piecewise continuous* functions. Note that any piecewise continuous function can be expressed as the sum of a continuous function and of a weighted sum of displaced step functions.* The displaced step function, to

---

* This statement is easy to prove provided the equivalence among the locally integrable functions is kept in mind.

be denoted by u_α, is defined as

$$u_\alpha = \{u_\alpha(t)\}, \tag{21}$$

where

$$u_\alpha(t) = 0 \qquad (0 \leq t < \alpha)$$
$$= 1 \qquad (\alpha \leq t).$$

The generalized function identified with u_α is, of course $u_\alpha * a /\!/ a$. We shall consider this generalized function again, in the next section.

Up to this point a lower case italic letter has designated a function—in general, a locally integrable function. Now there is a one-to-one correspondence between such functions and a subset of generalized functions. Hence it is possible to let the italic letter itself stand for the generalized function. Thus if b is a locally integrable function, the corresponding generalized function has heretofore been expressed as $b * a /\!/ a$. For convenience, we shall henceforth express this generalized function simply as b. This will cause no notational difficulties, because each operation on ordinary functions, including convolution division, has its counterpart on generalized functions.

Besides being a notational simplification, the preceding plan may lead to more meaningful things. Thus the u^n on the left side in (17) was an ordinary function. With the new notation it can be regarded as a generalized function, and (17) becomes

$$u^n = \frac{u^{n+1}}{u}. \tag{22}$$

Although when first introduced in (17), n was a positive integer, we can now set $n = 0$ and find

$$u^0 = \frac{u}{u}. \tag{23}$$

The significance of this quantity will be discussed in the next section.

In view of the above comments the relations in (18) may now be replaced by

$$\{\epsilon^{-\alpha t}\} \Leftrightarrow \frac{u}{u^0 + \alpha u} \tag{24a}$$

$$\{\sin \omega t\} \Leftrightarrow \frac{\omega u^2}{u^0 + \omega^2 u^2} \tag{24b}$$

$$\{\cos \omega t\} \Leftrightarrow \frac{u}{u^0 + \omega^2 u^2}. \tag{24c}$$

We shall discuss one of these and leave to you the examination of the others. Focus attention on the right side of (24c). It is the generalized-function solution of the convolution equation

$$(u^0 + \omega^2 u^2) * z = u. \tag{25}$$

There is a term in this equation unlike any we have previously encountered, $u^0 * z$. From (23) we see that $u^0 * z$, in more precise terms, is $u * z /\!/ u$. Now, if z is a continuous or locally integrable function, $u * z /\!/ u$ is, by (14) or (20), just z.* With this in mind, the convolution equation, defined on the set of continuous or locally integrable functions, corresponding to (25) is

$$z(t) + \omega^2 \int_0^t (t - \tau) z(\tau) \, d\tau = 1. \tag{26}$$

By (24c) we know $z(t) = \cos \omega t$ is the solution of (26). The other relations in (24) may be examined in a similar manner.

A1.4 GENERALIZED FUNCTIONS AS OPERATORS

Up to this point we have discussed the relationship of functions to generalized functions. We shall now look at integral and differential operators and discuss their relationship to generalized functions.

Consider again the function u. The convolution of u with the function a is

$$u * a = \left\{ \int_0^t a(\tau) \, d\tau \right\}. \tag{27}$$

Thus, in addition to being regarded as the continuous function $\{1\}$, u may be viewed as an integral operator. Similarly, u^{n+1} for n positive, which was

---

* Note that this is the same result as that obtained by convolving an impulse function with z. We shall discuss this identification of u^0 with the impulse function in the next section.

seen earlier to be the function $\{t^n/n!\}$, may be thought of as an $(n+1)$ fold integral operator. As an illustration of this operator interpretation of u^{n+1}, observe that the operator view indicates that the integral equation corresponding to (25), on the space of continuous or locally integrable functions, is

$$z(t) + \omega^2 \int_0^t \int_0^\tau z(\nu) \, d\nu \, d\tau = 1. \tag{28}$$

Note that (28) has the same solution, $z(t) = \cos \omega t$, as the convolution equation (26) that corresponds to (25).

Before going on, let us use this idea of u^n being an n-fold integration to help assign a value to a generalized function at time t. Let a denote an arbitrary generalized function. Suppose $u^n * a$ stands in a one-to-one relation with the ordinary function b, which possesses an nth derivative for $\tau_1 < t < \tau_2$. Then, for $\tau_1 < t < \tau_2$, we shall assign the value $b^{(n)}(t)$ to a at time t. The value assigned to a in this manner is unique.*

As an example, take $a = u^0$ and determine a value to be assigned to it. Now $u * u^0 = u^2//u$ is by (14) the same as the continuous function u, which is differentiable for $0 < t$. Since $du(t)/dt = 0$ for $t > 0$, we assign the value zero to u^0 for $t > 0$.

Now we turn to an interpretation of u^n for negative n. Alternatively, we shall examine u^{-n} for positive n. For convenience, let $p^n = u^{-n}$. Next suppose that a is a function that is $n-1$ times continuously differentiable and that possesses a locally integrable nth derivative. Then

$$p^n * a = a^{(n)} + a^{(n-1)}(0) \, p^0 + a^{(n-2)}(0) \, p + \cdots + a(0) \, p^{n-1}, \tag{29}$$

where $a^{(k)}$ denotes the kth derivative of a with respect to t. This relation is easily established by induction. We shall start the proof and leave its completion to you. Let $n = 1$; then (29) becomes

$$p * a = a^{(1)} + a(0) \, p^0,$$

which in more precise notation is

$$\frac{u}{u * u} * \frac{a * u}{u} = \frac{a^{(1)} * u}{u} + \frac{a(0)u}{u}.$$

---

* The uniqueness is established by Theorem 6 in Arthur Erdélyi, *Operational Calculus and Generalized Functions*, Holt, Rinehart & Winston, New York, 1962.

By the relations in (5), (6), and (8), this equation becomes

$$\frac{\overline{u * a}}{u * u} = \frac{\overline{a^{(1)} * u + a(0)u}}{u}.$$

If this is a valid relation, then, by (4),

$$u * u * a = u * u * u * a^{(1)} + a(0)u * u * u,$$

or, equivalently,

$$u * u * u * a^{(1)} = u * u * (a - a(0)u).$$

By Theorem 1, this becomes

$$u * a^{(1)} = a - a(0)u,$$

or

$$\int_0^t a^{(1)}(\tau) \, d\tau = a(t) - a(0),$$

which is true if $a^{(1)}$ is locally integrable. Thus (29) is valid for $n = 1$; the proof is completed by induction.

If a is sufficiently differentiable and $a(0) = a^{(1)}(0) = \cdots = a^{(n-1)}(0) = 0$, (29) shows that the generalized function $p^n * a$ stands in a one-to-one relation with the function $a^{(n)}$. Hence the generalized function p^n should be viewed as a differential operator. If a is not sufficiently differentiable or if one or more of the $a^{(k)}(0)$, $k = 1, \cdots, n-1$, are not zero, then $p^n * a$ does not stand in a one-to-one relation with an ordinary function. In this case $p^n * a$ exists only as a generalized function; we shall refer to $p^n * a$ as the generalized nth derivative of a.

Let us now apply some of these results to an example. The ordinary differential equation

$$\frac{d}{dt} z(t) + \alpha z(t) = 0 \tag{30}$$

has the generalized-function counterpart

$$p * z - z(0)p^0 + \alpha p^0 * z = 0.$$

For $z(0) = 2$, the generalized-function solution is

$$z = 2\, \frac{p^0}{p + \alpha p^0}. \tag{31}$$

Recalling that $p = u^{-1}$ and that $p^0 = u^0$, we get

$$z = 2\, \frac{u}{u^0 + \alpha u},$$

which, by (24a) is the same as $2\{\epsilon^{-\alpha t}\}$. This agrees with our knowledge that the solution of the differential equation (30) is $2\{\epsilon^{-\alpha t}\}$ for $z(0) = 2$.

In finding the ordinary function standing in a one-to-one relation with the generalized function in (31), it was useful to have had the relation in (24a). In the solution of other differential equations, which will be considered in the next section, it would be helpful to have relationships between generalized functions expressed in terms of p and the corresponding ordinary functions. Such relationships are shown in Table 1.

Table 1.   Ordinary Function—Generalized Function Pairs

Ordinary Function	Generalized Function	Ordinary Function	Generalized Function
$\{1\}$	$\dfrac{p^0}{p}$	$\{\sin \omega t\}$	$\dfrac{\omega p^0}{p^2 + \omega^2 p^0}$
$\{t\}$	$\dfrac{p^0}{p^2}$	$\{\cos \omega t\}$	$\dfrac{p}{p^2 + \omega^2 p^0}$
$\left\{\dfrac{t^2}{2}\right\}$	$\dfrac{p^0}{p^3}$	$\{\epsilon^{-\alpha t} \sin \omega t\}$	$\dfrac{\omega p^0}{(p + \alpha p^0)^2 + \omega^2 p^0}$
$\left\{\dfrac{t^n}{n!}\right\}$	$\dfrac{p^0}{p^{n+1}}$	$\{\epsilon^{-\alpha t} \cos \omega t\}$	$\dfrac{p + \alpha p^0}{(p + \alpha p^0)^2 + \omega^2 p^0}$
$\{\epsilon^{-\alpha t}\}$	$\dfrac{p^0}{p + \alpha p^0}$	$\{t\epsilon^{-\alpha t} \sin \omega t\}$	$\dfrac{2\omega(p + \alpha p^0)}{[(p + \alpha p^0)^2 + \omega^2 p^0]^2}$
$\{t\epsilon^{-\alpha t}\}$	$\dfrac{p^0}{(p + \alpha p^0)^2}$	$\{t\epsilon^{-\alpha t} \cos \omega t\}$	$\dfrac{(p + \alpha p^0)^2 - \omega^2 p^0}{[(p + \alpha p^0)^2 + \omega^2 p^0]^2}$
$\left\{\dfrac{t^n \epsilon^{-\alpha t}}{n!}\right\}$	$\dfrac{p^0}{(p + \alpha p^0)^{n+1}}$		

THE IMPULSE FUNCTION

Most of the preceding effort has been devoted to considering the generalized functions that have a one-to-one relationship with ordinary functions or the properties of operators on ordinary functions. We shall now turn to a consideration of the relationship of generalized functions to the impulse function and its derivatives. We shall find that $p^0 = u^0$ is properly interpreted as the impulse function. To verify this, let a be a continuous function; then

$$a = \frac{a * u}{u} = a * p^0. \tag{32}$$

If, following custom, we let δ be a symbolic function corresponding to p^0, then (32) is equivalent, in a formal sense, to the following:

$$a(t) = \int_0^t a(\tau)\, \delta(t - \tau)\, d\tau. \tag{33}$$

This is what is called the *sifting property* associated with the impulse function δ.

The generalized function p is properly interpreted as the first derivative of the impulse function. To show this, let the function a possess a continuous first derivative. Then by using (29) we have

$$p * a = a^{(1)} + a(0)p^0. \tag{34}$$

If $a(0) \neq 0$, $p * a$ is a generalized function that does not stand in a one-to-one relation with an ordinary function. However, as previously shown, p^0 can be assigned the value 0 for all $t > 0$. This is a useful fact. Since p is the generalized first derivative of p^0, which is interpreted as the impulse function, let $\delta^{(1)}$ denote a symbolic function corresponding to p; then for $t > 0$ (34) is equivalent in a formal sense to

$$a^{(1)}(t) = \int_0^t a(\tau)\, \delta^{(1)}(t - \tau)\, d\tau. \tag{35}$$

We can continue in this manner to interpret p^n as the nth derivative of the impulse function. The detailed justification of this statement is left to you.

Consider now the generalized function $p * u_\alpha$ as an operator, where u_α

is the displaced step given in (21). Suppose the generalized function a is also a continuous function; then

$$p * u_\alpha * a = p * \left\{ \int_0^t u_\alpha(t-\tau)\, a(\tau)\, d\tau \right\}$$

$$= p * \left\{ \begin{array}{ll} 0 & (0 \leq t < \alpha) \\ \int_0^{t-\alpha} a(\tau)\, d\tau & (\alpha \leq t) \end{array} \right\}.$$

By applying (34), we obtain from this relation

$$p * u_\alpha * a = \left\{ \begin{array}{ll} 0 & (0 \leq t < \alpha) \\ a(t-\alpha) & (\alpha < t) \end{array} \right\}. \tag{36}$$

Thus the generalized function $p * u_\alpha$ has the property of a delay, or shift, operator. This property makes $p * u_\alpha$ useful in the solution of difference equations. Since difference equations are not encountered in this book, we shall not give further consideration to $p * u_\alpha$.

A1.5 INTEGRODIFFERENTIAL EQUATIONS

In the study of linear, time-invariant, lumped-parameter networks, the subject of the major portion of this book, the dynamical equations encountered are integrodifferential equations. (In the state formulation they are purely differential equations.) These equations, considered as equations on the set of generalized functions, have generalized-function solutions. It would be desirable, if possible, to identify these generalized functions with ordinary functions. We shall show how to accomplish this by extending the concept of partial-fraction expansion to generalized functions.

The generalized-function solution of a linear integrodifferential equation can be expressed as the sum of (1) a convolution quotient of polynomials in p and (2) a convolution quotient of polynomials in p convolved with another generalized function. As an example, consider the equation

$$\frac{dz(t)}{dt} + 3z(t) + 2 \int_0^t z(\tau)\, d\tau + 1 = 2\epsilon^{-t}, \tag{37}$$

with $z(0) = -1$. The corresponding equation on the set of generalized functions is

$$p * z + p^0 + 3p^0 * z + 2p^{-1} * z + p^0 = \{2\epsilon^{-t}\} * p^0.$$

The generalized-function solution for z is

$$z = \frac{-2p^0}{p + 3p^0 + 2p^{-1}} + \frac{p^0}{p + 3p^0 + 2p^{-1}} * \{2\epsilon^{-t}\},$$

or, equivalently,

$$z = \frac{-2p}{p^2 + 3p + 2p^0} + \frac{p}{p^2 + 3p + 2p^0} * \{2\epsilon^{-t}\}. \tag{38}$$

Clearly, if we can identify $p /\!/ (p^2 + 3p + 2p^0)$ with an ordinary function, then z will also be identifiable as an ordinary function.

In the general case, which we leave to you to verify, we have the solution of a linear integrodifferential equation expressed as follows:

$$z = \frac{\alpha_m p^m + \alpha_{m-1} p^{m-1} + \cdots + \alpha_0 p^0}{p^n + \gamma_{n-1} p^{n-1} + \cdots + \gamma_0 p^0} + \frac{\beta_l p^l + \beta_{l-1} p^{l-1} + \cdots + \beta_0 p^0}{p^n + \gamma_{n-1} p^{n-1} + \cdots + \gamma_0 p^0} * f. \tag{39}$$

If each of these convolution quotients can be identified with ordinary functions and if the generalized function f is also an ordinary function, then the generalized function z that constitutes the solution will be an ordinary function.

Since both convolution quotients in (39) are of the same form, we shall examine only the second:

$$\frac{\beta_l p^l + \beta_{l-1} p^{l-1} + \cdots + \beta_0 p^0}{p^n + \gamma_{n-1} p^{n-1} + \cdots + \gamma_0 p^0}.$$

In the event that $l \geq n$, it is easy to verify that scalars $\xi_0, \cdots, \xi_{l-n}$ and $\nu_0, \cdots, \nu_{n-1}$ exist such that

$$\frac{\beta_l p^l + \cdots + \beta_0 p^0}{p^n + \cdots + \gamma_0 p^0} = \xi_{l-n} p^{l-n} + \cdots + \xi_0 p^0 + \frac{\nu_{n-1} p^{n-1} + \cdots + \nu_0 p^0}{p^n + \cdots + \gamma_0 p^0}. \tag{40}$$

The denominator $p^n + \cdots + \gamma_0 p^0$ is factorable as

$$p^n + \cdots + \gamma_0 p^0 = (p + \lambda_1 p^0) \cdots (p + \lambda_n p^0), \tag{41}$$

where the λ_i are the zeros of $\lambda^n + \gamma_{n-1}\lambda^{n-1} + \cdots + \gamma_0\lambda^0$. This may be verified by (4), after setting $p = u/\!/u^2$ and $p^0 = u/\!/u$. Thus

$$\frac{\nu_{n-1}p^{n-1} + \cdots + \nu_0 p^0}{p^n + \cdots + \gamma_0 p^0} = \frac{\nu_{n-1}p^{n-1} + \cdots + \nu_0 p^0}{(p + \lambda_1 p^0) \cdots (p + \lambda_n p^0)}. \qquad (42)$$

Assume that the λ_i are distinct; the right side may then be expressed as a sum of convolution quotients. Thus

$$\frac{\nu_{n-1}p^{n-1} + \cdots + \nu_0 p^0}{(p + \lambda_1 p^0) \cdots (p + \lambda_n p^0)} = \frac{\mu_1 p^0}{p + \lambda_1 p^0} + \cdots + \frac{\mu_n p^0}{p + \lambda_n p^0}, \qquad (43)$$

where

$$\mu_i = \left[(\lambda - \lambda_i) \frac{\nu_{n-1}\lambda^{n-1} + \cdots + \nu_0}{\lambda^n + \cdots + \gamma_0} \right]_{\lambda=\lambda_i}. \qquad (44)$$

The right side of (43) is the partial-fraction expansion of the left side, when the λ_i are distinct. You should verify (44). If the λ_i are not distinct, the partial-fraction expansion is more complicated and cumbersome in notation. However it is quite similar to the ordinary variable case. We shall not consider this case in detail. To complete the case of distinct λ_i, substitution of (43) into (42) and of that result into (40) establishes the following:

$$\frac{\beta_l p^l + \cdots + \beta_0 p^0}{p^n + \cdots + \gamma_0 p^0} = \xi_{l-n} p^{l-n} + \cdots + \xi_0 p^0$$

$$+ \frac{\mu_1 p^0}{p + \lambda_1 p^0} + \cdots + \frac{\mu_n p^0}{p + \lambda_n p^0}. \qquad (45)$$

We shall call the convolution quotient on the left a *rational convolution quotient* in p and say that it is proper, if $l < n$ or, equivalently, if $\xi_0 = \cdots = \xi_{l-n} = 0$.

Each of the terms $\mu_i p^0 /\!/ (p + \lambda_i p^0)$ in (45) stands in a one-to-one relation with a continuous function, as seen from Table 1: $\mu_i p^0 /\!/ (p + \lambda_i p^0) \Leftrightarrow \{\mu_i \epsilon^{-\lambda_i t}\}$ and each of the terms $\xi_i p^i$ can be assigned the value zero for $t > 0$. Thus we can make the following statements about the generalized function z in (39):

1. If $m < n$, $l < n$, and f is a locally integrable function. then z is a continuous function.

2. If $m \geq n$, $l < n$, and f is a locally integrable function, then z can be assigned a value for $t > 0$.

3. If $m < n$, $l \geq n$, and f has a continuous $(l - n)$th derivative, then z can be assigned a value for $t > 0$.

4. If $m \geq n$, $l \geq n$, and f has a continuous $(l - n)$th derivative, then z can be assigned a value for $t > 0$.

To illustrate (45), consider the particular z given in (38). Now

$$\frac{p}{p^2 + 3p + 2p^0} = \frac{p}{(p + p^0)(p + 2p^0)} = \frac{-p^0}{p + p^0} + \frac{2p^0}{p + 2p^0}$$
$$= \{-\epsilon^{-t}\} + \{2\epsilon^{-2t}\}$$

The last line is obtained from Table 1. By substituting this result in (38), we get

$$z(t) = 2\epsilon^{-t} - 4\epsilon^{-2t} + \int_0^t [-\epsilon^{-(t-\tau)} + 2\epsilon^{-2(t-\tau)}]2\epsilon^{-\tau}\,d\tau$$

as the solution of (38).

A1.6   LAPLACE TRANSFORM OF A GENERALIZED FUNCTION

In the analysis of linear, time-invariant networks, it is common practice to use Laplace-transform techniques and, in doing so, to introduce the transform of the impulse function and its derivatives. To justify this latter act, we must extend the definition of the Laplace transform to include the generalized functions, or, at the very least, the specific generalized functions p^n, $n \geq 0$, which are associated with the impulse function and its derivatives.

The Laplace transform of a large subset of generalized functions can be defined by a theorem due to Krabbe.* Before stating this very useful theorem we must define several sets of functions. Let $\mathcal{M}$ denote the set of all functions of the complex variable s that are regular, except at a finite number of poles, in some right half-plane [Re $(s) >$ Re $(\hat{s})$ for some $\hat{s}$]. Let $\hat{\mathcal{M}}$ denote the subset of functions in $\mathcal{M}$, each of which is equal in

---

* G. Krabbe, "Ratios of Laplace Transforms, Mikusiński Operational Calculus," *Math. Annalen*, Vol. 162, 1966, pp. 237–245. This paper is reprinted in *Contributions to Functional Analysis*, Springer-Verlag, New York, 1966.

some right half-plane to the ratio of two functions that are bounded and regular in that right half-plane; for example, $1/(s+2)$ and $(s+1)/(s+3)$ are bounded, regular functions of s for Re $(s) > -2$; their ratio $[1/(s+2)]/[(s+1)/(s+3)] = (s+3)/(s+1)(s+2)$ is regular, except at the pole $s = -1$, for Re $(s) > -2$. Thus $(s+3)/(s+1)(s+2)$ is a function in $\hat{\mathscr{M}}$. Next, let $\hat{\mathscr{C}}$ denote the set of all continuous functions c such that

$$\mathfrak{L}\{c\} = \left\{\int_0^\infty c(t)\epsilon^{-st}\, dt\right\} \tag{46}$$

exists for some $s = \hat{s}$. Then $\mathfrak{L}\{c\}$ is regular in the right half-plane Re $(s) >$ Re $(\hat{s})$. Finally, let $\hat{\mathscr{G}}$ be the subset of generalized functions $b/\!/a$ such that a and b are functions in $\hat{\mathscr{C}}$ and $\mathfrak{L}\{b\}/\mathfrak{L}\{a\}$ is a function in $\hat{\mathscr{M}}$. With these definitions, Krabbe's theorem can be stated as follows:

Theorem 2.   *There exists a one-to-one relation, denoted $\hat{\mathfrak{L}}^{-1}$, from the set of functions $\hat{\mathscr{M}}$ onto the set of generalized functions $\hat{\mathscr{G}}$. The relation $\hat{\mathfrak{L}}^{-1}$ possesses an inverse $\hat{\mathfrak{L}}$ that is linear and satisfies*

$$\hat{\mathfrak{L}}\{g_1 * g_2\} = \hat{\mathfrak{L}}\{g_1\}\hat{\mathfrak{L}}\{g_2\}, \tag{47}$$

where g_1 and g_2 are two generalized functions in $\hat{\mathscr{G}}$. Furthermore if $a/\!/b$ is a generalized function in $\hat{\mathscr{G}}$, then

$$\hat{\mathfrak{L}}\{b/\!/a\} = \mathfrak{L}\{b\}/\mathfrak{L}\{a\}. \tag{48}$$

Now let $\hat{\mathscr{I}}$ denote the set of all locally integrable functions b such that

$$\mathfrak{L}\{b\} = \left\{\int_0^\infty b(t)\epsilon^{-st}\, dt\right\}$$

exists for some $s = \hat{s}$. Next, consider the generalized function $u * b/\!/u$ identified with b. The function

$$u * b = \left\{\int_0^t b(\tau)\, d\tau\right\}$$

is continuous, and its Laplace transform exists for $s = \hat{s}$, since b is in $\hat{\mathscr{I}}$. Hence $u * b$ is in $\hat{\mathscr{C}}$. Furthermore, $\mathfrak{L}\{u * b\} = \mathfrak{L}\{b\}/s$ is bounded and regular in some right half-plane, and $\mathfrak{L}\{u\} = 1/s$ is bounded and regular in the right half-plane Re $(s) > \epsilon$ for any $\varepsilon > 0$. Thus $u * b // u$ is in $\hat{\mathscr{G}}$. Therefore

$$\hat{\mathcal{L}} \left\{ \frac{u * b}{u} \right\} = \frac{\mathcal{L}\{u * b\}}{\mathcal{L}\{u\}} = \mathcal{L}\{b\}. \tag{49}$$

This shows us that the transformation $\hat{\mathcal{L}}$, of a generalized function into a function of the complex variable s, coincides with the Laplace transformation, if that generalized function is also a Laplace transformable, locally integrable function. Thus the transformation $\hat{\mathcal{L}}$ is an extension of the Laplace transformation $\mathcal{L}$. This being the case, no confusion will arise if we henceforth drop the circumflex from $\hat{\mathcal{L}}$.

Let us now turn our attention to the generalized functions $p^n = u // u^{n+1}$. Since u and u^{n+1} are both in $\hat{\mathscr{C}}$ and since $\mathcal{L}\{u\} = 1/s$ and $\mathcal{L}\{u^{n+1}\} = 1/s^{n+1}$ are both in $\hat{\mathscr{M}}$, $\mathcal{L}\{p^n\}$ exists. By (48),

$$\mathcal{L}\{p^n\} = s^n. \tag{50}$$

Recall that p^n is the generalized function associated with $\delta^{(n)}$, the nth derivative of the impulse function. The Laplace transform of $\delta^{(n)}$, in books on applications of Laplace-transform theory, is given as s^n; (50) is in agreement with the notion that $\mathcal{L}\{\delta^{(n)}\} = s^n$.

We shall not give further consideration to the Laplace transform of a generalized function. It is sufficient that it is an extension of the usual Laplace transform and preserves the heuristic Laplace transform of the impulse function and its derivatives.

Appendix 2

THEORY OF FUNCTIONS
OF A COMPLEX VARIABLE

The purpose of this appendix on functions of a complex variable is twofold. First, it will serve as a reference for those who are familiar with the subject through an earlier encounter but would like to refresh their memories on specific points. Secondly, it will provide a skeleton that an instructor can augment by supplying proofs, examples, etc. The material is almost entirely in summary form. There is no attempt to provide motivation, and few proofs are given. Nevertheless, results are stated precisely.

A2.1 ANALYTIC FUNCTIONS

We assume familiarity with the algebra of complex numbers (addition, subtraction, multiplication, and division) and the representation of complex numbers as points on a plane. We also assume familiarity with the elements of the theory of functions of a real variable.

Let $s = \sigma + j\omega$ denote a complex variable. We say that another complex variable $F = U + jX$ is a *function* of the complex variable s, if to each value of s (in some set), there corresponds a value of F or a set of values of F. We write $F = F(s)$, where $F(\,\cdot\,)$ is the rule that associates the values of F with values of s. If to each value of s (in the set) there is only one value of F, we say that F is a *single-valued* function of s; otherwise it is *multivalued*.

Continuity for a function of a complex variable is formally defined in the same way as for functions of a real variable; namely, $F(s)$ is *continuous* at s_0 if it is defined in a neighborhood of s_0 and if

$$\lim_{s \to s_0} F(s) = F(s_0) = F_0 . \tag{1}$$

We may interpret this statement in the complex plane as follows. Let $\varepsilon > 0$ be a given number. We consider a circular neighborhood of $F(s_0)$ as in Fig. 1, where all the points within the circle of radius ε around

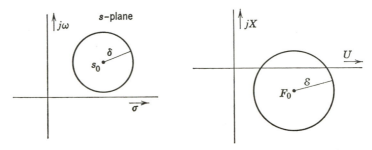

Fig. 1.   Neighborhoods in the s-plane and F-plane.

$F(s_0)$ belong to this neighborhood. Now (1) is equivalent to the following claim. We can find a small enough neighborhood of s_0, of radius $\delta > 0$, such that the values of $F(s)$ at all points in this neighborhood fall within the circle of radius ε about $F(s_0)$.

Differentiability in the complex plane is also defined by the same formal relation as on the real line, but is conceptually of much greater significance.

$F(s)$ is *differentiable* at s_0, with the derivative $F'(s_0)$, provided

$$F'(s_0) = \lim_{s \to s_0} \frac{F(s) - F(s_0)}{s - s_0} \tag{2}$$

exists and is finite.

Implicit in this definition is the assumption that s may approach s_0 in any direction, or may spiral into it, or follow any other path. The limit in (2) must exist (and be unique) independently of how s approaches s_0. It is this fact that makes differentiability in the complex plane a very strong requirement. In consequence, differentiable functions of a complex variable are extremely " well-behaved," as contrasted with real functions, which can be " pathological."

It can be shown (this is only one of many " it can be shown's " that we shall meet in this appendix) that the usual rules for derivatives of sums, products, quotients, etc., carry over from the real case, with no changes.

So does the chain rule for the function of a function; and all the familiar functions have the same derivatives as on the real line, except that the variable is now complex. We summarize these results below.

Let $F_1(s)$ and $F_2(s)$ be two differentiable functions. Then

$$\frac{d}{ds}[F_1(s) + F_2(s)] = \frac{d}{ds}F_1(s) + \frac{d}{ds}F_2(s) \tag{3}$$

$$\frac{d}{ds}[F_1(s)F_2(s)] = F_1(s)\frac{d}{ds}F_2(s) + \left[\frac{d}{ds}F_1(s)\right]F_2(s) \tag{4}$$

$$\frac{d}{ds}\frac{F_1(s)}{F_2(s)} = \frac{F_2(s)\,dF_1(s)/ds - F_1(s)\,dF_2(s)/ds}{F_2{}^2(s)}, \qquad F_2(s) \neq 0 \tag{5}$$

$$\frac{d}{ds}F_1[F_2(s)] = \frac{dF_1}{dF_2}\frac{dF_2(s)}{ds} \tag{6}$$

$$\frac{d}{ds}(s^n) = ns^{n-1}. \tag{7}$$

If a function F of a complex variable is differentiable at the point s_0 *and at all points in a neighborhood of s_0*, we say that $F(s)$ is *regular* at s_0.

Notice that the statement " $F(s)$ is regular at s_0 " is a very much stronger statement than " $F(s)$ is differentiable at s_0." A function $F(s)$ that has *at least one regular point* (i.e., a point at which the function is regular) in the complex plane is called an *analytic function*. A point s_0 at which the analytic function $F(s)$ is *not regular* is a *singular point* of the function. $F(s)$ is said to have a singularity at s_0. In particular, a point at which the derivative does not exist is a singular point.

Although the requirement of regularity is a very strong condition and therefore the class of analytic functions is a " very small" subset of the set of all functions, almost all functions that we meet in physical applications are analytic functions. An example of a nonanalytic function is $|s|^2$. This function has a derivative at $s = 0$ and nowhere else. Hence it has no regular points. The function $\bar{s}(= \sigma - j\omega)$ is another simple example of a nonanalytic function. The function $F(s) = 1/(s-1)$ is a simple example of an analytic function. Its *region of regularity* consists of the whole plane exclusive of the point $s = 1$. The point $s = 1$ is a singular point of this function

The singularities of an analytic function are extremely important, as we shall see. For the present we can only distinguish between two kinds of singularities. The point s_0 is an *isolated singularity* of $F(s)$, if s_0 is a singular point, but there is a neighborhood of s_0 in which all other points (except s_0) are regular points. If no such neighborhood exists, s_0 is a *nonisolated essential singularity*. Thus in every neighborhood of a non-isolated singularity there is at least one other singular point of the function. Hence a nonisolated singularity is a *limit point* (or *point of accumulation*) of singularities and conversely.

Rational functions (quotients of polynomials) are examples of functions that have only isolated singularities. To give an example of a function that has nonisolated singularities, we have to use trigonometric functions that we have not defined yet. Nevertheless, an example of a nonisolated singularity is the point $s = 0$ for the function

$$F(s) = \frac{1}{\sin 1/s}. \tag{8}$$

The denominator becomes zero whenever

$$s = \frac{1}{k\pi}, \tag{9}$$

and so these points are singular points of $F(s)$. The origin is a limit point of these singularities.

The famous French mathematician Augustin Cauchy (who originated about half of complex function theory) gave the following necessary and sufficient condition for the differentiability of a function of a complex variable: The function

$$F(s) = U(\sigma, \omega) + jX(\sigma, \omega)$$

is differentiable at s_0 *if and only if the partial derivatives* $\partial U/\partial \sigma$, $\partial U/\partial \omega$, $\partial X/\partial \sigma$, *and* $\partial X/\partial \omega$ *exist and are continuous at* (σ_0, ω_0) *and satisfy the equations*

$$\partial U/\partial \sigma = \partial X/\partial \omega \tag{10a}$$

$$\partial X/\partial \sigma = -\partial U/\partial \omega \tag{10b}$$

at this point.

The necessity is proved by letting s approach s_0 in (2) by first letting σ approach σ_0 and then letting ω approach ω_0 for one computation, and reversing the order for another computation. Equating the two derivatives so obtained leads to (10). The sufficiency is proved by using the concept of the total differential of a function of two variables and the definition of the derivative.

The equations in (10) are known as the *Cauchy–Reimann equations* in honor of the German mathematician Bernhard Reimann (who made these equations fundamental to the theory of analytic functions) in addition to Cauchy. We can use the Cauchy-Riemann equations as a test for the regularity of a function as follows.

If the four partial derivatives are continuous in a region of the complex plane and if they satisfy the Cauchy–Riemann equations at every point of this region, then $F(s)$ is regular in the region.

Notice that this condition involves the neighborhood about s_0 just as the definition of the regularity of a function does. The proof of the result again depends on the concept of a total differential for a function of two variables.

By differentiating one of the two equations in (10) with respect to σ and the other with respect to ω, and combining, we may observe the important fact that the real and imaginary parts of an analytic function satisfy Laplace's equation in two dimensions, within the region of regularity; that is,

$$\partial^2 U/\partial\sigma^2 + \partial^2 U/\partial\omega^2 = 0 \qquad (11a)$$

$$\partial^2 X/\partial\sigma^2 + \partial^2 X/\partial\omega^2 = 0. \qquad (11b)$$

Thus the real and imaginary parts of an analytic function are *harmonic functions*. The converse of this statement is also true. Every harmonic function (in two dimensions) is the real part of an analytic function, and the imaginary part of another analytic function. This fact makes analytic functions of considerable interest in two-dimensional potential theory.

A2.2 MAPPING

A function of a real variable can be represented geometrically as a graph. However, for a function of a complex variable, a " graph " would require four dimensions, two for the variable and two for the function. Hence it is impossible to draw a graph for an analytic function. Nevertheless, the concept of a geometrical representation can still be used for analytic functions to provide a better understanding of these functions. We use two planes, an s-plane for the variable and a F-plane for the function, as in Fig. 1, and thus get four coordinate axes.

To draw a complete picture, showing what the value of the function is at each point in the s-plane, is futile, since this merely results in a smear. Therefore we choose certain representative lines in the s-plane and show in the F-plane the functional values of $F(s)$ at points on these lines. Single-valued functions $F(s)$ will give us smooth lines in the F-plane, as a result. As an example, we have a representative sketch of the function $F(s) = s^2$ in Fig. 2. Here we have taken some lines along which either σ

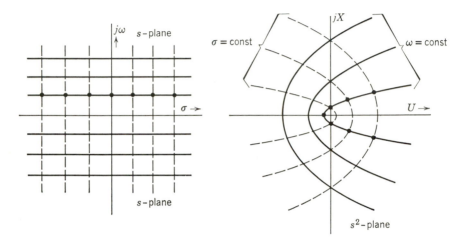

Fig. 2.   Representation of the mapping $F(s) = s^2$.

or ω is constant as representative lines. The corresponding lines in the s^2-plane are all parabolas. The two sets of parabolas, corresponding to $\sigma = $ const. and $\omega = $ const., are orthogonal families. If we had chosen other representative lines in the s-plane, we would have obtained other types of curves in the s^2-plane.

We refer to this graphical concept as a *mapping*. The s-plane is said to be *mapped into* the F-plane; the F-plane is a map of the s-plane. The lines in the F-plane are *images* of the lines in the s-plane, under the function $F(s)$. We also refer to $F(s)$ as a *transformation*. The function $F(s)$ transforms points in the s-plane into points in the F-plane. The concept of a mapping by an analytic function is a very useful one.

The fact that the parabolas of Fig. 2 constitute an orthogonal family is no accident. The reason is that the original lines in the s-plane intersect at right angles, and an analytic function preserves angles, except when the derivative does not exist or is zero. Let us make a definition before establishing this fact. A *conformal transformation* F is one in which the angle of intersection of two image curves in the F-plane is the same in (both

magnitude and in sense) as the angle of intersection of the two corresponding curves in the s-plane.

*The mapping by an analytic function is conformal at all points at which
the function is regular and the derivative is nonzero.*

To prove this result we take two smooth curves C_1 and C_2 in the
s-plane that intersect at s_0. Let s be an arbitrary point on C_1. Let us
introduce polar coordinates about s_0, by defining

$$s - s_0 = r\epsilon^{j\theta_1}. \tag{12}$$

Then as s approaches s_0, the angle θ_1 approaches the angle α_1, which is
the angle of the tangent to C_1 at s_0. By the definition of the derivative,

$$\lim_{s \to s_0} \frac{F(s) - F(s_0)}{s - s_0} = F'(s_0). \tag{13}$$

Since this derivative exists, we may take the limit along C_1, and since the
derivative is nonzero, we may write

$$F'(s_0) = \rho\epsilon^{j\beta}. \tag{14}$$

Then from (13),

$$\lim_{s \to s_0} \left| \frac{F(s) - F(s_0)}{s - s_0} \right| = \rho \tag{15a}$$

and

$$\lim_{s \to s_0} \{\arg [F(s) - F(s_0)] - \arg (s - s_0)\} = \beta. \tag{15b}$$

Equation 15b can be rewritten

$$\lim_{s \to s_0} \arg [F(s) - F(s_0)] = \beta + \lim_{s \to s_0} \arg (s - s_0) = \beta + \alpha_1. \tag{16}$$

The point $F(s)$ is on the curve C_1', which is the image of C_1 under the
mapping $F(s)$. Thus the left side of (16) is the angle of the tangent to
C_1' at $F(s_0)$. Thus from (16), the curve C_1' has a definite tangent at $F(s_0)$,
making an angle $\beta + \alpha_1$ with the positive real axis. An identical argument
gives the angle of the tangent to C_2' at $F(s_0)$ to be $\beta + \alpha_2$. Thus the angle
between the two tangents, taken from C_1' to C_2', is $(\alpha_2 - \alpha_1)$, which is the

same (in magnitude and sign) as the angle between the curves C_1 and C_2 at s_0 measured from C_1 to C_2.

Incidentally, we see from (15a) that the *local magnification* (i.e., the increase in linear distance near s_0) is independent of direction and is given by the magnitude of the derivative. Thus, locally, the mapping by an analytic function [when $F'(s_0) \neq 0$] produces a linear magnification $|F'(s_0)|$ and a rotation arg $F'(s_0)$, thus preserving shapes of small figures.

An auxiliary consequence is that the images of smooth curves are also smooth curves; that is, they cannot have " corners."

We have not yet defined some point-set-topological concepts about regions and curves that are really needed to clarify the earlier discussions. Let us proceed to rectify this omission, although we cannot be completely precise without introducing very complex ideas, which we do not propose to do. Therefore we shall take a few concepts such as path, continuous curve, etc., to be intuitively obvious.

A *simple arc* is a continuous path in the complex plane that has no crossover or multiple points. A *simple closed curve* is a path in the complex plane that, if cut at any one point, becomes a simple arc. If the end points of a simple arc are joined, we form a simple closed curve.

An *open region* is a set of points in the complex plane each of which has a neighborhood all of whose points belong to the set. The region " inside " a simple closed curve, not counting the curve itself, is an example. If we add the points on the boundary of an open set to the open set itself, the combined region is called a *closed region*. An open or closed region is said to be *connected* if any two points in the region can be connected by a line all points on which are in the region.

In the preceding paragraph the word " inside " was put in quotation marks. Although we have a strong intuitive feeling that the inside of a closed curve is well defined, nevertheless this requires a proof. The *Jordan curve theorem* gives the desired result. It states that *every simple closed curve divides the complex plane into two regions, an " inside " and an " outside," the curve itself being the boundary of these two regions*. If we start at some point on the curve and traverse it in a counterclockwise sense, the region to the left of the curve will be called the inside; that to the right, the outside.

If we do not permit a closed curve to pass through infinity, then the " inside " region, as just defined, will be *bounded*; that is, all points in the region will satisfy the condition $|s| \leq M$, where M is a fixed positive number. On the other hand, if the closed curve goes through infinity, then neither the inside nor the outside is bounded.

The question arises as to what is meant by a closed curve passing through infinity. The path consisting of the imaginary axis, for example, is such a curve. But this may appear to be a simple arc rather than a closed curve. The *Reimann* sphere will serve to clarify this point.

Consider a sphere placed on the complex plane with its "south pole" at the origin, as illustrated in Fig. 3. Now consider joining by a straight line

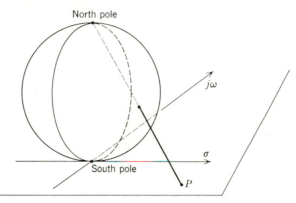

Fig. 3. The Riemann sphere.

each point in the plane to the "north pole" of the sphere. These lines will all intersect the sphere, thus setting up a one-to-one correspondence between the points in the plane and those on the sphere. Each point in the finite plane will have its counterpart on the sphere. As we go further and further away from the origin of the plane in any direction, the point of intersection of the lines with the sphere will approach closer and closer to the north pole. Thus the north pole corresponds to infinity. On the sphere infinity appears to be a unique point. Both the real and the imaginary axes become great circles on the sphere, and a great circle appears like a simple closed curve.

The concept of the Reimann sphere serves another purpose; it permits us to look upon "infinity" as a single point, whenever this is convenient. We refer to infinity as the *point at infinity*.

Very often we wish to talk about the behavior of a function at the point infinity. A convention in mathematics is that no statement containing the word "infinity" is to be considered meaningful unless the whole statement can be defined without using this word. This convention is introduced to avoid many inconsistencies that would otherwise arise. The behavior of a function at the point infinity is defined as follows.

That behavior is assigned to the function $F(s)$ at $s = \infty$, as is exhibited by the function

$$G(s) = F(1/s)$$

at $s = 0$; for example, the function $F(s) = 1/s$ is regular at $s = \infty$ since $G(s) = F(1/s) = s$ is regular at $s = 0$. Similarly, the function $F(s) = as^2 + bs$ is not regular at infinity since $G(s) = a/s^2 + b/s$ has a singularity at $s = 0$.

By a similar artifice we can also talk about the value of a function at a point in the complex plane being ∞, if we are careful. By this statement we mean that the reciprocal of the function is zero at this point.

A2.3   INTEGRATION

The definite integral of a function of a complex variable is defined in a manner similar to the definition of real integration. In the case of real variables the definite integral can be interpreted as an area. For complex variables such a geometrical interpretation is not possible. In Fig. 4

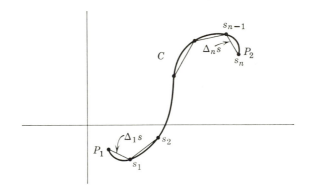

Fig. 4.   The definite integral.

two points P_1 and P_2 are connected by a simple arc C. The path is divided into intervals by the points s_k; the chords* joining these points are labeled $\Delta_k s$. Suppose we multiply each of the chords by the value of a function $F(s)$ evaluated at some point s_k^* of the interval and then add all these products. Now we let the number of intervals increase with a simultaneous decrease in the lengths of the chords. We define the *definite integral* of $F(s)$ as the limit of this sum as the number of intervals goes to infinity while the length of each chord goes to zero. More precisely,

$$\int_{C\ P_1}^{P_2} F(s)\, ds = \lim_{\substack{n \to \infty \\ \max|\Delta_k s| \to 0}} \sum_{k=1}^{n} F(s_k^*)\Delta_k s \tag{17}$$

provided the limit on the right exists.

* Here the chords are taken to be expressed as complex numbers. Thus $\Delta_k s = s_k - s_{k-1}$.

Note that in addition to the lower and upper limits P_1 and P_2, we have indicated that in going from P_1 to P_2 we shall follow the path C. It is conceivable that a different answer will be obtained if a different path is followed. It would not be necessary to write the limits on the integration symbol if we were to always show the path of integration on a suitable diagram together with the direction along the path. Because the path, or contour, is inseparable from the definition of an integral, we refer to it as a *contour integral*.

To determine the conditions under which the definite integral in (17) exists, we must first express this integral as a combination of real integrals. With $F(s) = U + jX$, and after some manipulation, (17) becomes

$$\int_C F(s)\, ds = \int_C U\, d\sigma - \int_C X\, d\omega + j\left[\int_C U\, d\omega + \int_C X\, d\sigma\right]. \tag{18}$$

Each of the integrals on the right is a real line integral; if these integrals exist, then the contour integral will exist. From our knoweldge of real integrals we know that continuity of the integrand is a sufficient condition for the existence of a real line integral. It follows that *the contour integral of a function $F(s)$ along a curve C exists if $F(s)$ is continuous on the curve.*

CAUCHY'S INTEGRAL THEOREM

The question still remains as to the conditions under which the integral between two points is independent of the path joining those points. Consider Fig. 5, which shows two points P_1 and P_2 joined by two simple

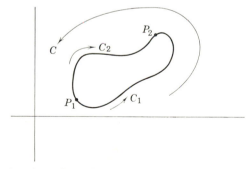

Fig. 5. Conditions for the value of an integral to be independent of the path of integration.

paths C_1 and C_2. Note that the directions of these paths are both from P_1 to P_2. The combined path formed by C_1 and the negative of C_2

forms a simple closed curve, which we shall label $C = C_1 - C_2$. If the integral of a function $F(s)$ along path C_1 is to equal the integral along path C_2, then the integral along the combined path C must be equal to zero, and conversely. The inquiry into conditions under which an integral is independent of path is now reduced to an inquiry into conditions under which a contour integral along a simple closed curve is equal to zero. The question is answered by the following theorem, which is known as *Cauchy's integral theorem.*

Let $F(s)$ be a function that is regular everywhere on a simple closed curve C and inside the curve. Then

$$\int_C F(s)\, ds = 0. \tag{19}$$

This is a very powerful and important theorem, but we shall omit its proof.

A word is in order about the *connectivity* of a region in the complex plane. Suppose we connect any two arbitrary points P_1 and P_2 that lie in a region by two arbitrary simple arcs C_1 and C_2 also lying in the region. The region is said to be *simply connected* if it is possible to slide one of these arcs along (distortion of the arc is permitted in this process) until it coincides with the other, without ever passing out of the region. Cauchy's theorem is proved *ab initio* for just such a region. The hatched region between the two closed curves in Fig. 6 is called *doubly connected.*

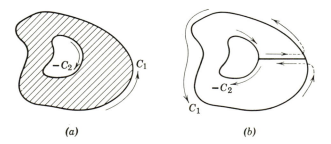

(a)   (b)

Fig. 6.   A doubly-connected region.

Such a region can be reduced to a simply connected region by the artifice of " digging a canal " between the two closed curves. The region now is bounded by the composite curve whose outline is shown by the arrows in Fig. 6*b*.

Suppose that a function $F(s)$ is regular in the hatched region shown in Fig. 6*a* including the boundaries. Cauchy's theorem can be applied here

to the composite curve consisting of the inner and outer curves and the
" canal." The canal is traversed twice, but in opposite directions, so that
its contribution to the complete contour integral is zero. If we denote the
outside and inside curves by C_1 and C_2, respectively, both in the counterclockwise direction, then Cauchy's theorem will lead to the result that

$$\int_{C_1} F(s) \, ds = \int_{C_2} F(s) \, ds. \tag{20}$$

As a matter of fact, if we choose any other closed path between the inner
and outer ones in Fig. 6, the same reasoning will tell us that the integral
around this path in the counterclockwise direction will be equal to each
of the integrals in (20).

This reasoning leads us to conclude that the value of a contour integral
around a simple closed curve will not change if the contour is distorted,
so long as it always stays inside a region of regularity.

Turn again to Fig. 5. The points P_1 and P_2 are in a simply connected
region R throughout which a function $F(s)$ is single valued and regular.
Let P_1 be a fixed point that we shall label s_0, and P_2 a variable point
that we shall label s. We have stated that the integral from s_0 to s is
independent of the path of integration so long as the paths remain in the
region of regularity. Hence we can define the function $G(s)$ as

$$G(s) = \int_{s_0}^{s} F(z) \, dz, \tag{21}$$

where z is a dummy variable of integration. This function is a single-
valued function of the upper limit s for all paths in the region of regularity.
It is easy to show that $G(s)$ is regular in R and that its derivative is $F(s)$.
We call it the *antiderivative* of $F(s)$. (For each s_0 we get a different antiderivative.)

Actually it is not necessary to assume that $F(s)$ is regular in the region.
Instead it is sufficient to assume that $F(s)$ is *continuous* in R and that its
closed-contour integral for all possible simple closed curves in R is zero.
However, Morera's theorem, which we shall discuss later, states that a
function satisfying these conditions is regular.

In evaluating a definite integral in real variables we often look for an
antiderivative of the integrand. The same procedure is valid for complex
variables; that is, if an antiderivative of $F(s)$ is $G(s)$, then

$$\int_{s_1}^{s_2} F(s) \, ds = G(s_2) - G(s_1). \tag{22}$$

CAUCHY'S INTEGRAL FORMULA

Let us now consider a simple closed curve C within and on the boundary of which a single-valued function $F(s)$ is regular. It is possible to express the value of the function at any point s_0 inside the curve in terms of its values along the contour C. This expression is

$$F(s_0) = \frac{1}{2\pi j} \int_C \frac{F(s)}{s - s_0} \, ds. \tag{23}$$

It is referred to as *Cauchy's integral formula* (as distinct from Cauchy's theorem). This result can be proved by noting that in the integral involved, the contour C can be replaced by a circular contour C' around the point s_0 without changing its value, according to the discussion centering around (20). The purely algebraic step of adding and subtracting $F(s_0)$ in the integrand then permits us to write

$$\int_{C'} \frac{F(s)}{s - s_0} \, ds = \int_{C'} \frac{F(s_0)}{s - s_0} \, ds + \int_{C'} \frac{F(s) - F(s_0)}{s - s_0} \, ds. \tag{24}$$

The last integral on the right can be shown to be zero. It remains to evaluate the first integral on the right.

Let us write $s - s_0 = re^{j\theta}$; then $ds = jre^{j\theta}\, d\theta$, since the contour C' is a circular one and only θ varies. Then

$$\int_{C'} \frac{ds}{s - s_0} = \int_0^{2\pi} j d\theta = 2\pi j. \tag{25}$$

The desired expression now follows immediately upon substituting this result into (24).

Cauchy's integral formula sheds much light on the properties of analytic functions. We see that the value of an analytic function that is regular in a region is determined at any point in the region by its values on the boundary. Note that the point s_0 is *any point whatsoever* inside the region of regularity. We should really label it with the general variable s, which would then require that in (23) we relabel the variable s (which merely represents points on the boundary and is thus a dummy variable) with some other symbol. For clarity, we shall rewrite (23) as

$$F(s) = \frac{1}{2\pi j} \int_C \frac{F(z)}{z - s} \, dz. \tag{26}$$

Here s represents any point inside a contour C in which $F(s)$ is regular, and z refers to points on the contour.

Another very important fact about analytic functions can be determined from Cauchy's integral formula. Let us try to find the nth order derivative of an analytic function $F(s)$. For the first and second derivatives we can use the definition of a derivative directly on (26), without getting bogged down in a great mass of algebra. The result will be

$$F'(s) = \frac{1}{2\pi j} \int_C \frac{F(z)}{(z-s)^2}\, dz \qquad (27a)$$

$$F''(s) = \frac{2}{2\pi j} \int_C \frac{F(z)}{(z-s)^3}\, dz. \qquad (27b)$$

The form of these expressions, which seems to indicate that we simply differentiate with respect to s under the integral sign, suggests the following expression for the nth derivative.

$$F^{(n)}(s) = \frac{n!}{2\pi j} \int_C \frac{F(z)}{(z-s)^{n+1}}\, dz. \qquad (28a)$$

This result can be corroborated by the use of mathematical induction.

An extremely important implication of the points we have just been discussing is the following. *If a single-valued function* F(s) *is regular at a point, it follows that the function will have derivatives of all orders at that point.* This same statement cannot be made for a function of a real variable.

Having seen that the derivative of an analytic function is itself analytic and has the same region of regularity, we can now make a statement that appears to be the converse of Cauchy's theorem *Let* F(s) *be a function that is continuous in a region R and whose closed contour integral around all possible paths in the region is zero.* These conditions ensure that $F(s)$ has an antiderivative $G(s)$ that is regular in the region R. But the derivative of $G(s)$ is $F(s)$; consequently $F(s)$ is also regular in R. This result is known as *Morera's theorem.*

MAXIMUM MODULUS THEOREM AND SCHWARTZ'S LEMMA

Cauchy's formula leads to some other very interesting results. However, we shall demonstrate these same results from the viewpoint of mapping. Let $F = F(s)$ be an analytic function that is regular within and on a curve

C in the s-plane; let this region, including the curve C, be R. The map of the the curve C may take one of the forms shown in Fig. 7. Note that the maps

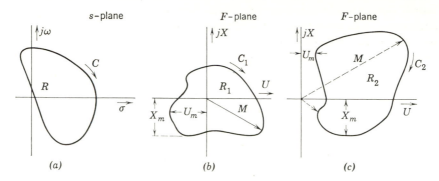

Fig. 7. Demonstration of the principles of maximum and minimum.

of the region R cannot extend to infinity, since infinity in the F-plane corresponds to a singular point in the s-plane, and there are no singular points in R. Both maps of the curve C have been shown as simple closed curves for simplicity; they need not, and usually will not, be. In the map shown in part (b) the origin of the F-plane is inside the region. This corresponds to the possibility that $F(s)$ has a zero in the region R. The origin is not included inside the map of region R shown in part (c).

In either of these cases it is clear from the figures that the point in R_1 or R_2 that lies farthest from the origin of the F-plane lies on the boundary of the region, which is the map of curve C. Similarly, if $F(s)$ does not have a zero in region R, then the point in R_2 which lies closest to the origin of the F-plane lies on the boundary, as illustrated in part (c) of the figure. It is also clear from the figure that the minimum values in region R_2 of the real part of F, and the imaginary part, lie on the boundary. The last statement is also true when $F(s)$ has a zero in the region, as part (b) of the figure illustrates. But in this case the smallest value of the magnitude, which is zero, lies inside the region and not on the boundary. We shall summarize these results as follows.

Let a closed curve C and its interior constitute a region R in the s-plane and let $F = F(s) = U + jX$ be regular in R. The largest value reached by the magnitude $|F(s)|$, the real part U and the imaginary part X in region R occurs for some point or points on the boundary. Likewise, the minimum values reached by the real part and the imaginary part in R occur on the boundary. The last is also true for the magnitude if $F(s)$ has no zero in region R. The statements concerning the magnitude are referred to as the *maximum modulus theorem* and the *minimum modulus theorem*. Similar designations can be applied to the other cases by replacing "modulus" by "real part" and "imaginary part."

A somewhat stronger statement than the maximum modulus theorem can be made if, $F(s)$ satisfies the additional condition that $F(0) = 0$, and if the region R is a circle. More specifically, suppose $F(s)$ is regular within a circular region of radius r and has a zero at $s = 0$. Let its maximum magnitude on the circle be M. Then $F(s)/s$ is also regular within the circle and satisfies the conditions of the maximum modulus theorem. Hence, $|F(s)/s| \leq M/r$. That is, for all points within the circle,

$$|F(s)| \leq \frac{|s|}{r} M. \tag{28b}$$

The equality holds only at $s = 0$ or if $F(s) \equiv Ms/r$. This result is called *Schwartz's lemma*.

A2.4   INFINITE SERIES

Let $f_1(s), f_2(s), \cdots$ be an infinite sequence of functions and consider the sum of the first n of these:

$$S_n(s) = \sum_{j=1}^{n} f_j(s). \tag{29}$$

This is called a *partial sum* of the corresponding infinite series. Now consider the sequence of partial sums, $S_1, S_2, \cdots, S_n$. We say that this sequence *converges* in a region of the complex plane if there is a function $F(s)$ from whose value at a given point the value of the partial sum S_n differs as little as we please, provided that we take n large enough. The function $F(s)$ is called the *limit function* of the sequence. More precisely, we say that the sequence converges in a region R if, given any positive number ε, there exists an integer N_j and a function $F(s)$ such that at any point s_j in the region

$$|S_n(s_j) - F(s_j)| < \varepsilon \tag{30}$$

for all values of n greater than N_j. The value of the integer N_j will depend on the number ε and on the point s_j.

We say that the sequence is *uniformly convergent* in a closed region if the same integer N can be used in the role of N_j for all points in the region instead of having this integer depend on the point in question. (N still depends on ε.)

The infinite series is said to converge (or converge uniformly) to the function $F(s)$ if the sequence of partial sums converges (or converges uniformly). An infinite series is said to *converge absolutely* if the series formed by taking the absolute value of each term itself converges. Absolute convergence is a stronger kind of convergence. It can be shown that *if a series converges absolutely in a region* R *it also converges in the region.*

We shall now state a number of theorems about sequences of functions without giving proofs.*

Theorem 1. *If a sequence of continuous functions* $S_n(s)$ *is uniformly convergent in a region* R, *then the limit function of the sequence is continuous in the same region* R.

Theorem 2. *If a sequence of continuous functions* $S_n(s)$ *converges uniformly to a limit function* $F(s)$ *in a region* R, *then the integral of* $F(s)$ *along any simple arc* C *in the region* R *can be obtained by first finding the integral along* C *of a member* $S_n(s)$ *of the sequence and then taking the limit as* $n \to \infty$; *that is,*

$$\int_C F(s)\ ds = \int_C [\lim_{n \to \infty} S_n(s)]\ ds = \lim_{n \to \infty} \int_C S_n(s)\ ds. \tag{31}$$

Theorem 3. *If a sequence of analytic functions* $S_n(s)$ *is regular in a region* R *and if they converge uniformly in* R *to a limit function* $F(s)$, *then* $F(s)$ *is regular in the region* R.

Theorem 4. *If the members of a sequence of analytic functions* $S_n(s)$ *are regular in a region* R *and if the sequence converges uniformly in* R *to a limit function* $F(s)$, *then the sequence of derivatives* $S_n'(s)$ *converges uniformly to the derivative of* $F(s)$ *for all interior points in* R.

Repeated applications of the theorem shows that the sequence of kth order derivatives $S_n{}^{(k)}(s)$ converges uniformly to $F^{(k)}(s)$.

---

* All of these theorems have to do with conditions under which two limit operations can be interchanged. They are of the general character

$$\lim_{x \to a} \lim_{y \to b} f(x, y) = \lim_{y \to b} \lim_{x \to a} f(x, y).$$

This interchange is permissible if both limits (separately) exist and one of them (say $x \to a$) exists uniformly with respect to the other variable.

TAYLOR SERIES

These theorems can be used to establish many important properties of infinite series by letting the sequence of functions $S_n(s)$ represent the partial sums of a series. Let us consider an important special case of infinite series.

We shall define a *power series* as follows:

$$F(s) = \sum_{n=0}^{\infty} a_n(s - s_0)^n. \tag{32}$$

The partial sums of a power series are polynomials in $(s - s_0)$; hence they are regular in the entire finite complex plane (this implies that they are continuous as well). If we can now determine the region of uniform convergence, we can use Theorems 1 through 4 to deduce properties of the limit function.

Suppose that a power series converges for some point $s = s_1$. It is easy to show that the series will converge absolutely (and hence it will also converge) at any point inside the circle with center at s_0 and radius $|s_1 - s_0|$. The *largest* circle with center at s_0 within which the series converges is called the *circle of convergence*, the radius of the circle being the *radius of convergence*. It follows that a power series *diverges* (does not converge) at any point outside its circle of convergence, because if it does converge at such a point s_2, it must converge everywhere inside the circle of radius $|s_2 - s_0|$, which means that the original circle was not its circle of convergence.

Let R_0 be the radius of convergence of a power series and suppose that R_1 is strictly less than R_0. Then it can be shown that the given series is uniformly convergent in the closed region bounded by the circle of radius $R_1 < R_0$ with center at s_0.

Suppose now that a power series converges to a function $F(s)$ in a circle of radius R_0. This means that the sequence of partial sums $S_n(s)$ will have $F(s)$ as a limit function. Since $S_n(s)$ is a continuous function, it follows from Theorem 1 that $F(s)$ is also continuous everywhere inside the circle. Furthermore, since the partial sums are regular in the region of uniform convergence, it follows from Theorem 3 that $F(s)$ is regular in the region. Thus *a power series represents an analytic function that is regular inside its circle of convergence.*

Two other important conclusions about power series follow from Theorems 2 and 4. According to Theorem 2, since the partial sums of a power series satisfy the conditions of the theorem, a *power series that converges to* F(s) *can be integrated term by term and the resulting series will converge to the integral of* $F(s)$ *for every path inside the circle of convergence.* Similarly, according to Theorem 4, *a power series may be differentiated term by term, and the resulting series will converge to the derivative of* $F(s)$ *everywhere inside the circle of convergence.* The circles of convergence of both the integrated series and the differentiated series are the same as that of the original series.

We saw that a power series converges to an analytic function that is regular within the circle of convergence. The converse of this statement, which is more interesting, is also true. Every analytic function can be represented as a power series about any regular point s_0. The desired result is *Taylor's theorem*, which states: *Let* $F(s)$ *be regular everywhere in a circle of radius* R_0 *about a regular point* s_0. *Then* $F(s)$ *can be represented as*

$$F(s) = \sum_{n=0}^{\infty} a_n (s - s_0)^n, \tag{33}$$

where the coefficients are given by

$$a_n = \frac{1}{n!} F^{(n)}(s_0). \tag{34}$$

The circle of convergence of the power series is the largest circle about s_0 *in which* $F(s)$ *is defined or is definable as a regular function.*

This series is referred to as a *Taylor series*. The theorem is proved by starting with Cauchy's integral formula given in (23) and expanding $(z - s)^{-1}$ as a finite number of terms in inverse powers of $(z - s_0)$ (after adding and subtracting s_0 to the denominator of the integrand), together with a remainder term. Use of the integral formulas for the derivatives of an analytic function given in (28) leads to a polynomial in $(s - s_0)$ plus a remainder term. The proof is completed by noting that the remainder term vanishes as the order of the polynomial in $(s - s_0)$ approaches infinity.

An important consequence of Taylor's theorem is that the circle of convergence of any power series passes through a singular point of the analytic function represented by it, because by Taylor's theorem, the radius of convergence is the distance from the point s_0 to the nearest singular point.

To find the power-series representation of a function, it is not necessary to use the formulas given in Taylor's theorem. But independent of the method used to find the power series representation, we shall end up with Taylor's series, with the coefficients satisfying Taylor's formula. This fact is established through the following *identity theorem for power series. If the two power series*

$$\sum_{n=0}^{\infty} a_n(s - s_0)^n \quad \text{and} \quad \sum_{n=0}^{\infty} b_n(s - s_0)^n$$

have positive radii of convergence and if their sums coincide for an infinite number of distinct points having the limit point s_0, *then* $a_n = b_n$ *for all* n; *that is, they are identical.*

In particular, the conditions of the theorem are satisfied if the two series agree in a neighborhood of s_0 or along a line segment (no matter how small) that contains s_0. This result is proved by induction on n. Thus the representation of an analytic function by a power series about a given regular point s_0 is unique.

LAURENT SERIES

We have seen that a power-series representation can be found for an analytic function in the neighborhood of a regular point with a region of convergence which extends to the nearest singular point of the function. The question arises whether it is possible to find other infinite-series representations for an analytic function that converge in other regions. Consider the annular region between the two concentric circles C_1 and C_2 with center at s_0 shown in Fig. 8. A function $F(s)$ is regular on C_1, C_2, and

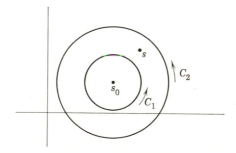

Fig. 8.   Region of convergence of a Laurent series.

the region between them. The point s_0 may be a regular point or a singular point of $F(s)$. Also there may be other singular points of $F(s)$ inside the inner circle. The annular region can be made simply connected by the device of "digging a canal" discussed in a preceding section. If we now apply Cauchy's integral formula, we get

$$F(s) = \frac{1}{2\pi j} \int_{C_2} \frac{F(z)}{z - s} \, dz - \frac{1}{2\pi j} \int_{C_1} \frac{F(z)}{z - s} \, dz, \tag{35}$$

where s is a point in the interior of the annular region and z represents points on the contours of the two circles. For the quantity $(z - s)^{-1}$ we can write

$$\frac{1}{z - s} = \frac{1}{z - s_0} \sum_{k=0}^{n} \left(\frac{s - s_0}{z - s_0}\right)^k + \frac{1}{z - s} \left(\frac{s - s_0}{z - s_0}\right)^{n+1} \tag{36}$$

$$\frac{1}{z - s} = -\frac{1}{s - s_0} \sum_{k=0}^{n} \left(\frac{z - s_0}{s - s_0}\right)^k + \frac{1}{z - s} \left(\frac{z - s_0}{s - s_0}\right)^{n+1}. \tag{37}$$

These can be checked by noting that the expression

$$\frac{1}{1 - w} = \sum_{k=0}^{n} w^k + \frac{w^{n+1}}{1 - w} \tag{38}$$

is an identity for all values of w except $w = 1$. Equation 36 is obtained by adding and subtracting s_0 in the denominator on the left and then writing it in the form of (38) with

$$w = \frac{s - s_0}{z - s_0}. \tag{39}$$

A similar case obtains for (37), except that w is now

$$w = \frac{z - s_0}{s - s_0}. \tag{40}$$

Now let us use (36) in the first integral in (35) and (37) in the second integral. Each integral will give a finite number of terms plus a remainder term. It can be shown, as in the proof of Taylor's theorem, that the remainder terms vanish as $n \to \infty$. The final result is

$$F(s) = \sum_{k=0}^{\infty} a_k(s - s_0)^k + \sum_{k=1}^{\infty} a_{-k}(s - s_0)^{-k} \tag{41}$$

or

$$F(s) = \sum_{k=-\infty}^{\infty} a_k(s - s_0)^k, \tag{42}$$

where a_k in the last expression is given by

$$a_k = \frac{1}{2\pi j} \int_C \frac{F(z)}{(z - s_0)^{k+1}} \, dz. \tag{43}$$

The contour C is any closed contour in the annular region between C_1 and C_2.

The series just obtained is called a *Laurent series*. It is characterized by having negative as well as positive powers. Its region of convergence is an annular region, as contrasted with the region of convergence of a Taylor series, which is a circle.* For a given function $F(s)$ and a point of expansion s_0 there can be more than one Laurent series with different regions of convergence. The point of expansion can be a regular point or a singular point. As in the case of a Taylor series, it is not necessary to use the formula in order to determine the coefficients in any particular case. But the identity theorem for Laurent series, which follows the statement of the residue theorem in the next section, tells us that no matter how the Laurent series of a function may be obtained, it must be unique, for a given region of convergence.

Let us now consider the particular case of a Laurent expansion of a function $F(s)$ about a point s_0, which is a singular point. The inner circle in Fig. 8 is to enclose no other singularities (this implies that the singularity is isolated). Hence we should expect the Laurent series to tell us something about the nature of the singularity at s_0. Remember that the Laurent series consists of two parts, the positive powers and the negative powers. Let us define the *regular part* $F_r(s)$ of the Laurent expansion as the series of positive powers and the constant, and the *principal part* $F_p(s)$ as the series of negative powers. If there were no principal part, the Laurent series would reduce to a Taylor series and s_0 would be a regular point. Thus the principal part of the Laurent series contains the clue regarding the nature of the singularity at s_0.

To describe the singularity at s_0 we make the following definitions. We say $F(s)$ has a *pole of order* n at s_0 if the highest negative power in the principal part is n. (A pole of order 1 is also called a *simple* pole.) On the other hand, if the principal part has an infinite number of terms, the singularity at s_0 is called an *isolated essential singularity*. (The word "isolated" is often omitted.)

FUNCTIONS DEFINED BY SERIES

One of the results that we noted previously is that a power series defines an analytic function that is regular inside its circle of convergence. We shall now use this fact to define some specific functions. Up until now

---

* This property of Laurent series can be interpreted as saying that the series of positive powers in $(s - s_0)$ converges everywhere inside C_2 of Fig. 8 and the series of negative powers converges everywhere outside of C_1, the two converging simultaneously in the annular region between C_1 and C_2.

we have explicitly mentioned rational functions. In the case of real variables we know the importance of such functions as exponentials, trigonometric and hyperbolic functions, and others. However, we have no basis for taking over the definitions of such functions from real variables. The tangent of a complex variable, for instance, cannot be defined as the ratio of two sides of a triangle.

We use the above-quoted property of power series to *define* an exponential function as follows:

$$\epsilon^s = 1 + s + \frac{s^2}{2!} + \frac{s^3}{3!} + \cdots = \sum_{k=0}^{\infty} \frac{s^n}{n!}$$

$$= \epsilon^{\sigma}(\cos \omega + j \sin \omega). \tag{44}$$

The last form is obtained by inserting $s = \sigma + j\omega$ in the series; expanding the powers of s; collecting terms; and finally identifying the real power series representing ϵ^{σ}, $\cos \omega$, and $\sin \omega$. We are not completely free in choosing a defining series for ϵ^s, because it must reduce to the correct series when s is real.

To determine the radius of convergence of the defining series we can resort to various tests for the convergence of series (which we have not discussed). Alternatively, since the series represents an analytic function, we can use the Cauchy–Riemann equations. In the latter case we find that there are no singular points in the entire finite plane, since the Cauchy–Riemann equations are satisfied everywhere. Hence the series converges everywhere. (The same result is, of course, obtained by testing the series for convergence.)

We can now follow the same procedure and define other transcendental functions in terms of series. However, it is simpler to define the trigonometric and hyperbolic functions in terms of the exponential. By definition, then,

$$\sin s = \frac{\epsilon^{js} - \epsilon^{-js}}{2j}, \qquad \cos s = \frac{\epsilon^{js} + \epsilon^{-js}}{2}, \qquad \tan s = \frac{\sin s}{\cos s} \tag{45}$$

$$\sinh s = \frac{\epsilon^s - \epsilon^{-s}}{2}, \qquad \cosh s = \frac{\epsilon^s + \epsilon^{-s}}{2}, \qquad \tanh s = \frac{\sinh s}{\cosh s}. \tag{46}$$

From the behavior of the exponential we see that the sines and cosines, both trigonometric and hyperbolic, are regular for all finite values of s. The singular points of $\tan s$ occur when $\cos s = 0$; namely, for an infinite number of real values of s at the points $s = (2k-1)\pi/2$ for all integral values of k. Similarly, the singular points of $\tanh s$ occur when $\cosh s = 0$; namely, at an infinite number of imaginary values of s at the points $s = j(2k-1)\pi/2$ for all integral values of k.

The trigonometric and hyperbolic functions of a complex variable satisfy practically all of the identities satisfied by the corresponding real functions.

A2.5 MULTIVALUED FUNCTIONS

In real function theory we define a number of "inverse" functions. These functions can be extended into the complex plane as analytic functions. As we know, most of these functions (the nth root, inverse sine, etc.) are multivalued on the real line. We may therefore expect similar behavior in the complex plane.

THE LOGARITHM FUNCTION

Let us begin by extending the concept of the logarithm. We define

$$G(s) = \log F(s) \tag{47a}$$

if and only if

$$F(s) = \epsilon^{G(s)}. \tag{47b}$$

(In this appendix we shall conform to the mathematical convention of writing log for the logarithm to the base ϵ.) Since we know the meaning of (47b), we also know the meaning of (47a). Let us first observe that if $G(s)$ satisfies (47b), so does $G(s) + j2k\pi$, since

$$\epsilon^{G(s)+j2k\pi} = \epsilon^{G(s)}\,\epsilon^{j2k\pi} = \epsilon^{G(s)}. \tag{48}$$

(We are using several results for the exponential function which we have not established in the complex plane, but which can be proved very easily.) Thus (47a) does not define a unique functional value for $G(s)$. However, we can show that any two values satisfying (47b) can differ by at most $j2k\pi$. (Do this.) Thus, although the function $\log F(s)$ is multivalued, its values are related by the simple additive constants $j2k\pi$. We shall find a formula for one of these multiple values by writing

$$F(s) = |F(s)|\epsilon^{j\,\arg F(s)} \tag{49}$$

where arg $F(s)$ is the principal value of the argument defined by

$$-\pi < \arg F(s) \leq \pi. \qquad (50)$$

Expressing $|F(s)|$ as $\exp[\log|F(s)|]$, (49) becomes

$$
\begin{aligned}
F(s) &= \epsilon^{\log|F(s)|}\, \epsilon^{j\,\arg\,F(s)} \\
&= \epsilon^{[\log|F(s)|+j\,\arg\,F(s)]}.
\end{aligned}
\qquad (51)
$$

Therefore from the definition of the logarithm *one of the values* of this function is

$$\log F(s) = \log|F(s)| + j \arg F(s). \qquad (52)$$

This particular value, which is *unique* by virtue of (50) is known as the *principal value* of the logarithm function. We signify this conventionally by writing a capital " L " in log $F(s)$; similarly, arg $F(s)$ always means the principal value given in (50). Thus we can write, for all values of the log function,

$$
\begin{aligned}
\log F(s) &= \text{Log } F(s) + j2k\pi \\
&= \log |F(s)| + j[\arg F(s) + 2k\pi],
\end{aligned}
\qquad (53)
$$

where k is an integer—positive, negative, or zero.

Thus there are an infinite number of values for the logarithm function, one for each value of k. Because of this difficulty, we might try to simplify by using only the principal value, Log $F(s)$. Before considering Log $F(s)$, let us first consider the behavior of the function Log s in the complex plane. Log s is given by

$$\text{Log } s = \text{Log } r + j\theta, \qquad (54)$$

where

$$s = r\,\epsilon^{j\theta} \quad \text{with} - \pi < \theta \leq \pi. \qquad (55)$$

We notice that the angle θ is undefined at $s = 0$. Therefore this equation does not define Log s at $s = 0$. But no matter how we define Log 0, Log s will not be continuous at $s = 0$, since the imaginary part of Log s takes on all values from $-\pi$ to π in any neighborhood of $s = 0$. Therefore $s = 0$ is a singular point of Log s. Even though we restrict ourselves to the principal value, Log s is discontinuous at any point on the negative real axis; for, the imaginary part of Log s here is π, but there are points arbitrarily close to it at which the imaginary part is very nearly $-\pi$. Thus Log s is not regular at any point on the negative real axis, including $s = 0, \infty$. (The behavior at ∞ is identical to the behavior at 0, since Log $1/s = -$Log s, as you can verify.)

BRANCH POINTS, CUTS, AND RIEMANN SURFACES

However, if we consider the complex plane to be "cut" along the negative real axis, as illustrated in Fig. 9, preventing us from going from

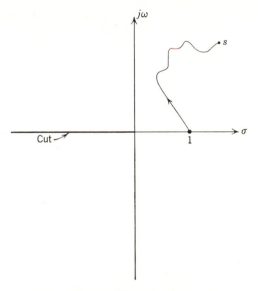

Fig. 9.   The cut s-plane.

one side of it to the other, Log s is regular in the rest of the complex plane. In fact, we have

$$\frac{d}{ds}\,\text{Log } s = \frac{1}{s} \tag{56}$$

at all other points of this "cut" plane. Thus Log s is an antiderivative of $1/s$. We can show that

$$\text{Log } s = \int_1^s \frac{dz}{z} \tag{57}$$

provided the path of integration does not go through the " cut " negative real axis.

Similar remarks apply to the other values of the logarithm function. The restriction to principal values is unnecessary. The only thing we need to do is to restrict the imaginary part of log s to some 2π range. For (57) to apply, we have to add a suitable multiple of $j2\pi$ to the right side. It is not even necessary that the cut be along the negative real axis. We may cut the plane along any radius vector by defining

$$\theta_1 < \arg F(s) \leq 2\pi + \theta_1 .\tag{58}$$

Even this is unnecessary. Any simple path from $s = 0$ to $s = \infty$ will do.

Thus by suitable restrictions we can make the function log s single-valued and regular in any neighborhood. The only exceptional points are $s = 0$, ∞. No matter what artifice we employ, we cannot make log s regular and single valued in a deleted neighborhood of $s = 0$, ∞. (Since these points are singular points, we have to delete them from the neighborhood if we hope to make the function regular.) Thus these two singular points are different in character from the ones we have met so far. Therefore we give them a different name. They are called *branch points*. Precisely, a branch point is defined as follows.

The point s_0 is a *branch point* of the function $F(s)$ if s_0 is an isolated singular point and there is no deleted neighborhood of s_0 in which $F(s)$ is defined or is definable as a single-valued regular function.

We now see that the plane has to be cut along a simple path from one branch point of log s to the other branch point. Each value of log s so obtained is called a *branch of the function*. Thus Log s is a branch of log s.

Riemann introduced an artifice that allows us to consider the complete log function and treat it as a single-valued function. This important concept is known as the *Riemann surface*. It is quite difficult to define this term precisely, and we shall not attempt it. Instead let us describe a few Riemann surfaces. For the function log s the Riemann surface has the following structure. We consider the s-plane to consist of an infinite number of identical planes. One of these is the plane in which arg s is restricted to its principal value. There are an infinite number of *sheets* above this and another infinity below. All of these planes are cut along the negative real axis. All of these have the same origin and ∞, so that the sheets are all joined together at these points. Each sheet is also joined to the ones immediately above and below, along the negative real axis. The upper edge of the negative real axis of each sheet is joined to the lower edge of the negative real axis of the sheet immediately above it. The whole Riemann surface looks somewhat like an endless spiral ramp.

Let us consider $\log s$ on such a surface. On each sheet.

$$\log s = \log |s| + j(\arg s + 2k\pi), \tag{59}$$

where k is a *fixed integer*. The integer k increases by 1 as we go to the sheet immediately above and decreases by 1 as we go to the sheet immediately below. On this Riemann surface, therefore, $\log s$ is a single-valued regular function with two singular points, $s = 0, \infty$.

We can now return to the function $\log F(s)$. We are considering $\log F(s)$ as a function of s. In the F-plane the branch cut goes from $F(s) = 0$ to $F(s) = \infty$. Let us consider only the simplest case where $F(s)$ is rational. The other cases are somewhat more complicated. The branch points in the s-plane are the zeros and poles of $F(s)$. Each branch cut goes from a zero to a pole. The number of branch cuts at a zero or a pole is equal to the multiplicity. The branch cuts are chosen so as not to intersect except at branch points.

As another example of the concept of the Riemann surface, let us consider the inverse of the function

$$F(s) = s^2. \tag{60}$$

The inverse of this function is called the *square root*, written

$$G(s) = s^{1/2} = \sqrt{s}. \tag{61}$$

(Formally we define powers of s other than integral powers as

$$s^\alpha = \epsilon^{\alpha \log s}, \tag{62}$$

where α may be any complex number.) As in the real case, the square root is a double-valued function. The two values G_1 and G_2 are related by

$$G_1(s) = -G_2(s) = G_2(s)\epsilon^{j\pi}. \tag{63}$$

We may make this function single-valued by restricting the angle of s as before; that is,

$$-\pi < \arg s \leq \pi, \tag{64}$$

and defining the " positive square root " as

$$G_1(s) = \sqrt{|s|}\, \epsilon^{1/2 j \arg s}, \tag{65}$$

where $\sqrt{|s|}$ is a real positive number.

Again we find that $G_1(s)$ is not continuous on the negative real axis, including $s = 0$, ∞. The points $s = 0$, ∞ are seen to be branch points of this function $G(s)$. The Riemann-surface concept may be introduced as follows. We need two sheets of the Riemann surface, both cut along the negative real axis. To make $G(s)$ continuous and regular on this surface, we " cross-connect " the two sheets along the negative real axis. The upper edge of the negative real axis of *each sheet* is connected to the lower edge of the negative real axis of the *other sheet*. (Obviously, it is useless to attempt to draw a picture of this in three dimensions.) On this Riemann surface, $G(s)$ is regular and single valued except at $s = 0$, ∞.

We see that the branch points of the function $\log s$ are somewhat different from the branch points of $s^{1/2}$. In one case we have an infinite number of branches, and in the other case we have only a finite number. Therefore we sometimes distinguish between these by calling the former a *logarithmic singularity* (or a logarithmic branch point) and the other an *algebraic singularity* (or an algebraic branch point).

For example, we can extend this discussion to other algebraic irrational functions,

$$F(s) = \left\{ \frac{as^2 + bs + c}{ds + e} \right\}^{1/3} \tag{66}$$

in an obvious way.

CLASSIFICATION OF MULTIVALUED FUNCTIONS

We have seen that the singularities of an analytic function are extremely important. In fact, we can *classify* analytic functions according to the type and locations of its singular points. This we shall do in the following brief discussion.

The simplest case is that of an analytic function that possesses no singularities at all, either in the finite plane or at ∞. In this case a theorem known as *Liouville's theorem* tells us that the function is simply a constant. The next case we might consider is that of a function that has no finite singularities, that is, the only possible singularity is at $s = \infty$. The exponential function is an example of this class. A function that has no singularities in the finite s-plane is known as an *entire* (or *integral*) function. If the singularity at ∞ is a pole, we see from the Laurent expansion about ∞ that this function is a *polynomial* (also called *entire rational* or *integral rational*). If the singularity at ∞ is an essential singularity the function is an *entire transcendental* function. The functions ϵ^s, $\sin s$, $\cos s$, etc., belong to this category.

The quotient of two entire functions is a *meromorphic function*. The only singularities of a meromorphic function in the finite plane are the points at which the entire function in the denominator goes to zero. Thus a meromorphic function can have only poles in the finite part of the *s*-plane. Again the behavior at infinity divides this class into two subclasses. If the point ∞ is either a regular point or a pole, then it can be shown that the function has only a finite number of poles (by using the theorem known as the Bolzano-Weierstrass theorem). Then by using the partial-fraction expansion, to be given in Section A2.7, we can show that this function is a *rational function*—that is, a quotient of two polynomials. Conversely, every rational function is a meromorphic function with at most a pole at $s = \infty$. An example of a nonrational meromorphic function is tan *s* or cosec *s*.

All of these functions are single-valued functions. The multivalued functions can be classified according to the number of branch points and the number of branches at each branch point. A function with a finite number of branch points and a finite number of branches is an *algebraic irrational* function. We saw examples of these. The logarithm function can be used to construct examples for infinite number of branches. The function log *s* has a finite number of branch points but an infinite number of branches. The function log sin *s* has an infinite number of branch points and an infinite number of branches, whereas the function $\sqrt{\sin s}$ has an infinite number of branch points with a finite number of branches at each branch point. These three classes have no special names associated with them.

A2.6 THE RESIDUE THEOREM

Cauchy's theorem tells us about the value of a closed-contour integral of a function when the function is regular inside the contour. We now have the information required to determine the value of a closed-contour-integral when the contour includes one or more singular points of the function. For this purpose turn to the formula for the coefficients of a Laurent series given in (43) and consider the coefficient of the first inverse power term, $k = -1$. This is

$$a_{-1} = \frac{1}{2\pi j} \int_C F(z)\, dz. \tag{67}$$

This is an extremely important result. It states that if a function is integrated around a closed contour inside which the function has one singular point, the value of the integral will be $2\pi j$ times the coefficient of the first negative power term in the Laurent series. None of the other terms in the series contributes anything; they all "wash out." We call this coefficient the *residue*. Note that the function is regular on the contour.

If the contour in question encloses more than one singular point (but a finite number), we can enclose each singular point in a smaller contour of its own within the boundaries of the main contour. By " digging canals " in the usual way, we find the value of the integral around the original contour to be equal to the sum of the integrals around the smaller contours, all taken counterclockwise. Now we consider a Laurent series about each of the singular points such that no other singular points are enclosed. According to the preceding paragraph, the value of the integral about each small contour is equal to $2\pi j$ times the corresponding residue. Hence the integral around the original contour is equal to $2\pi j$ times the sum of the residues at all of the singular points inside the contour ; that is,

$$\int_C F(s)\, ds = 2\pi j \sum \text{ residues at enclosed singularities.} \qquad (68)$$

This statement is referred to as the *residue theorem*. To find the value of a closed-contour integral, then, all we need to do is to calculate the residues at all of the singular points in a manner independent of the formula for the coefficients of the Laurent series.

Consider a function $F(s)$ that has a pole of order n at s_0. If the Laurent series about s_0 is multiplied by $(s - s_0)^n$, the result will be

$$(s - s_0)^n F(s) = a_{-n} + a_{-n+1}(s - s_0) + \cdots + a_{-1}(s - s_0)^{n-1}$$
$$+ a_0(s - s_0)^n + \cdots. \qquad (69)$$

The function on the left is regular at s_0, and the series on the right is the Taylor series representing it in the neighborhood of s_0. Hence, by using the formula for the Taylor coefficients, we get

$$a_{-1} = \frac{1}{(n-1)!} \frac{d^{n-1}}{ds^{n-1}} [(s - s_0)^n F(s)]|_{s=s_0}. \qquad (70)$$

For a simple pole this reduces to the following simple form:

$$a_{-1} = (s - s_0) F(s)|_{s=s_0}. \qquad (71)$$

In the case of poles, at least, we now have an independent way of finding residues.

There are alternative ways of expressing the residue at a simple pole, which are useful in computations. If the given function is expressed as

$$F(s) = G(s)/H(s), \tag{72}$$

where s_0 is a simple pole of $F(s)$, in the nontrivial case $H(s)$ has a simple zero at s_0 and $G(s)$ is regular and nonzero at s_0. In this case we may write

$$\text{residue of } F(s) \text{ at } s_0 = \lim_{s \to s_0} (s - s_0) F(s) = \frac{\lim_{s \to s_0} G(s)}{\lim_{s \to s_0} H(s)/(s - s_0)} \tag{73}$$

since $G(s)$ is regular at a_0. Thus the limit in the numerator is simply $G(\dot{s}_0)$. For the limit in the denominator we subtract $H(s_0)$ from $H(s)$, which is permissible since $H(s_0) = 0$, getting

$$\text{residue of } F(s) \text{ at } s_0 = \frac{G(s_0)}{\lim_{s \to s_0} [H(s) - H(s_0)]/(s - s_0)} = \frac{G(s_0)}{H'(s_0)} \tag{74}$$

since the limit of the difference quotient is by definition the derivative.

If, on the other hand, we write

$$F(s) = \frac{1}{F^{-1}(s)} \tag{75}$$

and follow through the same argument, we conclude that

$$\text{residue of } F(s) \text{ at } s_0 = \frac{1}{\left. \dfrac{d}{ds}\left(\dfrac{1}{F(s)}\right)\right|_{s=s_0}}. \tag{76}$$

Thus the residue at a simple pole is the reciprocal of the derivative of the reciprocal function.

One of the important applications of the residue theorem is the following identity theorem for Laurent series:

If the two Laurent series

$$\sum_{n=-\infty}^{\infty} a_n(s - s_0)^n \quad \text{and} \quad \sum_{n=-\infty}^{\infty} b_n(s - s_0)^n$$

have a common region of convergence $R_1 < |s - s_0| < R_2$ *and represent the same function in this region, then*

$$a_n = b_n \qquad (\text{all } n, \; -\infty < n < \infty).$$

Since the two series represent the same function,

$$\sum_{n=-\infty}^{\infty} a_n(s - s_0)^n = \sum_{n=-\infty}^{\infty} b_n(s - s_0)^n \qquad (R_1 < |s - s_0| < R_2). \qquad (77)$$

Since the positive and negative series are power series, they converge uniformly for $|s - s_0| \leq R_2 - \varepsilon, (\varepsilon > 0)$ and $|s - s_0| \geq R_1 + \varepsilon$, respectively. Therefore in the annular region $R_1 + \varepsilon \leq |s - s_0| \leq R_2 - \varepsilon$ the Laurent series are uniformly convergent. We now multiply both sides of (77) by $(s - s_0)^{k-1}$, where k is an integer—positive, negative, or zero—and integrate along a circular path C lying in the region of uniform convergence and enclosing s_0. By the residue theorem we get

$$a_{-k} = b_{-k} \qquad (\text{all } k, \; -\infty < k < \infty),$$

which proves the result.

EVALUATING DEFINITE INTEGRALS

The residue theorem (which, incidentally, includes Cauchy's theorem) provides a means for evaluating many real definite integrals that cannot be evaluated by other means. We choose a function of s that reduces to the given real integrand when s is real, and we choose a closed contour that includes as part of it the desired interval in the definite integral. Now if we can find the residues at the singularities of the integrand that might lie inside the chosen contour, and if we can independently calculate the contribution to the closed-contour integral of the parts of the path other than the desired interval, the value of the desired integral can be found.

In evaluating such integrals two circumstances often arise. In the first place it may happen that the integrand has a simple pole on the path of integration. In order to apply the residue theorem the function must be regular on the closed contour. This situation is remedied by distorting, or indenting, the contour by a small semicircular arc, as shown in Fig. 10. The new contour is, of course, different from the old one. However, we eventually let the radius of the semicircle approach zero. It remains to calculate the contribution of the semicircle to the closed-contour integral.

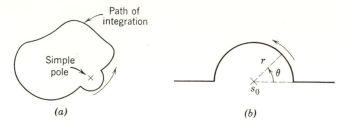

Fig. 10. Distortion of contour of integration around pole.

Consider the semicircular path shown in Fig. 10b around a simple pole at s_0. The Laurent expansion of $F(s)$ about s_0 has the form

$$F(s) = \frac{a_{-1}}{s - s_0} + \sum_{n=0}^{\infty} a_n(s - s_0)^n. \tag{78}$$

Note that the direction of the path is counterclockwise around the pole when we are to indent the contour in such a way that the pole is inside. We can also indent the contour to exclude the pole. Then the value obtained will be the negative of that obtained here. The series in this equation can be integrated term by term; let $(s - s_0) = re^{j\theta}$ and let C represent the semicircle. On the semicircle θ varies from 0 to π. The integral of $F(s)$ on the semicircle becomes

$$\int_C F(s)\,ds = \int_0^{\pi} \frac{a_{-1}}{re^{j\theta}}\, jre^{j\theta}\,d\theta + \sum_n a_n \int_0^{\pi} r^n e^{jn\theta}\, jre^{j\theta}\,d\theta$$

$$= j\pi a_{-1} + \sum_n \frac{a_n r^{n+1}}{n+1}\left(e^{j(n+1)\pi} - 1\right). \tag{79}$$

The first term is seen to be independent of the radius r of the semicircle. As we let r approach zero, each term in the summation will vanish. Hence

$$\int_{\text{semicircle}} F(s)\,ds = j\pi a_{-1}; \tag{80}$$

that is, the integral around half a circle about a simple pole will have one half the value of an integral around a complete circle. In fact, by the same reasoning, if the contour is a fraction k of a circular arc, the contribution will be $k(2\pi j a_{-1})$.

JORDAN'S LEMMA

The second circumstance that often arises in definite integrals is the need to evaluate an integral with infinite limits, such as

$$I = \int_{-\infty}^{\infty} f(\omega) \, d\omega. \tag{81}$$

Such an integral is called an *improper integral*. The notation means

$$I = \lim_{R_0 \to \infty} \int_{-R_0}^{R_0} f(\omega) \, d\omega. \tag{82}$$

The value obtained by going to negative and positive infinity in a symmetrical fashion is called the *principal value* of the integral.

This type of integral can be evaluated by choosing a contour consisting of the imaginary axis from $-R_0$ to R_0 and a large semicircle in the right or left half-plane, such as the one shown in Fig. 11. The integrand must

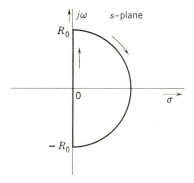

Fig. 11.   Contour for evaluating infinite integrals.

be a function $F(s)$ that reduces to the given integrand on the imaginary axis. Use of the residue theorem will now permit the evaluation of the desired integral, provided that the integral along the semicircular arc tends to a limit as $R_0 \to \infty$ and that this limit can be found. It would be best if there were no contribution from this arc. Let $F(s)$ be the integrand of the contour integral. It can be shown that if $sF(s)$ on the arc approaches zero uniformly* as the radius of the circle approaches infinity, then there

---

* That is, the limit is approached at the same rate for all angles of s within this range. The range is $|\arg s| \leq \pi/2$ for a semicircle in the right half-plane and $|\arg s| \geq \pi/2$ for a semicircle in the left half-plane. In the ϵ-δ language, the magnitude of the difference between $sF(s)$ and the limit (in this case 0), can be made less than ϵ, so long as $|s| > N(\epsilon)$, where $N(\epsilon)$ is independent of arg s in the appropriate range.

will be no contribution from the infinite arc; for example, if $F(s)$ is a ratio of two polynomials, the degree of the denominator must exceed that of the numerator by 2 or more.

Let t be a real variable and suppose the integrand has the form

$$F(s) = G(s)\epsilon^{st}. \tag{83}$$

Then it can be shown that for $t > 0$ the infinite arc in the left half-plane will not contribute to the integral, nor will the arc to the right for $t < 0$, provided that $G(s)$ vanishes uniformly as the radius of the semicircle approaches infinity. This result is called *Jordan's lemma*. The presence of the exponential loosens the restriction on the remaining part of the integrand. Thus, if $G(s)$ is a ratio of two polynomials, it is enough that the degree of the denominator exceed that of the numerator by 1 (or more).

As an example of the evaluation of integrals consider

$$I = \int_0^\infty \frac{\sin \omega t}{\omega} \, d\omega. \tag{84}$$

By substituting the definition of a sine function in terms of exponentials, we get,

$$I = \int_0^\infty \frac{\epsilon^{j\omega t}}{2j\omega} \, d\omega + \int_0^\infty \frac{-\epsilon^{-j\omega t}}{2j\omega} \, d\omega. \tag{85}$$

In the second integral, if we replace ω by $-\omega$, the integrand will become identical with that of the first integral, whereas the limits will become $-\infty$ to zero. The two integrals can then be combined to yield,

$$I = \frac{1}{2j} \int_{-\infty}^\infty \frac{\epsilon^{j\omega t}}{\omega} \, d\omega. \tag{86}$$

Now consider the integral

$$\int_C \frac{\epsilon^{st}}{s} \, ds, \tag{87}$$

where the contour C is the closed contour shown in Fig. 12. The integrand has a simple pole on the original contour so that we indent the contour around the pole as shown. The complete contour consists of two portions of the $j\omega$-axis and two semicircles, the radius of one of which will

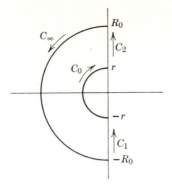

Fig. 12.   Path of integration for the evaluation of an integral.

approach zero while the other will approach infinity. Since the integrand is regular everywhere inside the contour, the closed-contour integral will vanish. We can write

$$\int_C \frac{\epsilon^{st}}{s}\, ds = 0 = \int_{-R_0}^{-r} \frac{\epsilon^{j\omega t}}{\omega}\, d\omega + \int_{C_0} \frac{\epsilon^{st}}{s}\, ds + \int_r^{R_0} \frac{\epsilon^{j\omega t}}{\omega}\, d\omega + \int_{C_\infty} \frac{\epsilon^{st}}{s}\, ds.$$

$$(88)$$

The integrand satisfies Jordan's lemma, so that the last integral in this equation will vanish. The value of the integral on C_0 is $-j\pi$ times the residue of the integrand at $s = 0$, according to (80). To calculate the residue we use (71) and find it to be unity. Hence

$$\int_{C_0} \frac{\epsilon^{st}}{s}\, ds = -j\pi.$$

$$(89)$$

We can now write (88) as

$$\lim_{\substack{r \to 0 \\ R_0 \to \infty}} \left\{ \int_{-R_0}^{-r} \frac{\epsilon^{j\omega t}}{\omega}\, d\omega + \int_r^{R_0} \frac{\epsilon^{j\omega t}}{\omega}\, d\omega \right\} = j\pi.$$

$$(90)$$

But by the improper integral in (86) we mean precisely the left side of the last equation. Hence, finally,

$$I = \int_0^\infty \frac{\sin \omega t}{\omega}\, d\omega = \frac{1}{2j}\, j\pi = \frac{\pi}{2}.$$

$$(91)$$

PRINCIPLE OF THE ARGUMENT

As another application of the residue theorem we shall now prove a very useful theorem called the "argument principle." Let $F(s)$ be an analytic function that is regular in a region R except possibly for poles. We would like to evaluate the integral.

$$\int_C \frac{F'(s)}{F(s)} \, ds \tag{92}$$

around a closed contour C in region R in the counterclockwise direction, where the prime denotes differentiation. There should be no poles or zeros of $F(s)$ on the contour C.

Suppose $F(s)$ has a zero of order n at a point s_1 in R. Then we can write

$$F(s) = (s - s_1)^n \, F_1(s)$$

$$F'(s) = n(s - s_1)^{n-1} \, F_1(s) + (s - s_1)^n \, F_1'(s)$$

$$\frac{F'(s)}{F(s)} = \frac{n}{s - s_1} + \frac{F_1'(s)}{F_1(s)}. \tag{93}$$

We see that this function has a simple pole at the zero of $F(s)$ with a residue n. The function $F_1(s)$ can now be treated in the same way and the process repeated until all the zeros of the original function $F(s)$ have been put into evidence. Each zero will lead to a term like the first one on the right side of (93).

Now suppose that $F(s)$ has a pole of order m at a point s_2 in R. Then we can write

$$F(s) = \frac{F_2(s)}{(s - s_2)^m}$$

$$F'(s) = \frac{(s - s_2) F_2'(s) - m F_2(s)}{(s - s_2)^{m+1}}$$

$$\frac{F'(s)}{F(s)} = -\frac{m}{s - s_2} + \frac{F_2'(s)}{F_2(s)}. \tag{94}$$

The desired function is seen to have a simple pole at the pole of $F(s)$, with a residue that is the negative of its order. Again the same process can be repeated and each pole of $F(s)$ will lead to a term like the first one on the right side of the last equation. The only singularities of $F'(s)/F(s)$ in the region R will lie at the zeros and the poles of $F(s)$. Hence, by the residue theorem, the value of the desired contour integral will be

$$\int_C \frac{F'(s)}{F(s)} \, ds = 2\pi j \left[\sum n_j - \sum m_j \right], \tag{95}$$

where the n_j are the orders of the zeros of $F(s)$ in R and the m_j are the orders of the poles.

Note, however, that

$$\frac{F'(s)}{F(s)} = \frac{d}{ds} \log F(s). \tag{96}$$

Hence we can evaluate the contour integral by means of the antiderivative of $F'(s)/F(s)$, which is $\log F(s)$. In going around the contour C we mean to start at a point and return to the same point. Note that the multivalued function $\log F(s)$ will have the same real part after returning to the starting point. Hence the value of the integral will be j times the increase in angle of $F(s)$ as s traverses the contour C in the counterclockwise direction. This should equal the right side of (95). If we now divide by 2π, the result should be the number of times the locus of the contour C in the F-plane goes around its origin counterclockwise (increase in angle divided by 2π is the number of counterclockwise encirclements of the origin).

Let us now state the principle of the argument. *If a function* F(s) *has no singular points within a contour* C *except for poles, and it has neither zeros nor poles on* C, *then the number of times the locus of the curve* C *in the* F-plane *encircles its origin in the counterclockwise direction is equal to the number of zeros minus the number of poles of* F(s) *inside* C. *Each zero and pole is to be counted according to its multiplicity.*

Before concluding this section, let us consider another contour integration problem. This is the problem of integrating a function partway around a logarithmic singularity.

Let us therefore consider the integral

$$\int_P \log F(s) \, ds,$$

where the path P is an arc of a circle around a zero or a pole of $F(s)$, as shown in Fig. 10b. Let $F(s)$ have a zero (or a pole) of order k at s_0. Then we may write

$$F(s) = (s - s_0)^k \, F_1(s) \qquad (97a)$$

$$\log F(s) = k \log (s - s_0) + \log F_1(s). \qquad (97b)$$

If s_0 is a pole, we let k be a negative integer in these expressions, thus including a zero of order k and a pole of order $-k$ simultaneously in the discussion. As we let the radius of the circle approach zero, $\log F_1(s)$ will not contribute anything to the integral, since it is regular at s_0. Thus it is sufficient to consider

$$\int_P \log(s - s_0) \, ds$$

if we wish to take the limit, as we do.

On the arc of radius r, we may estimate

$$\left| \int_P \log (s - s_0) \, ds \right| \leq r\theta |\log r| + r\theta^2 = \theta |r \log r| + r\theta^2, \qquad (98)$$

where θ is the angle subtended by the arc at the center. Now it is a well-known result that

$$\lim_{r \to 0} r \log r = 0. \qquad (99)$$

Hence

$$\lim_{r \to 0} \int_P \log F(s) \, ds = 0, \qquad (100)$$

which is the result we wish to establish. We have shown that a logarithmic singularity lying on a path of integration does not contribute anything to the integral.

A2.7 PARTIAL-FRACTION EXPANSIONS

The Laurent expansion of a function about a singular point describes a function in an annular region about that singular point. The fact that the function may have other singular points is completely submerged and there is no evidence as to any other singular points. It would be useful to have a representation of the function that would put into evidence all of its singular points.

Suppose a function $F(s)$ has isolated singularities at a finite number n of points in the finite plane. It may also have a singularity at infinity. Let us consider expanding $F(s)$ in a Laurent expansion about one of the singular points say s_1. The result will be

$$F(s) = F_{p1}(s) + F_{r1}(s), \tag{101}$$

where the subscripts refer to the principal part and the regular part.

Now consider $F_{r1}(s)$, which is simply the original function $F(s)$ from which has been subtracted the principal part of the Laurent series about one of its singularities. This function is regular at s_1 but has all the other singularities of $F(s)$. Let us expand it in a Laurent series about one of the other singularities, s_2:

$$F_{r1}(s) = F(s) - F_{p1}(s) = F_{p2}(s) + F_{r2}(s). \tag{102}$$

The function $F_{p1}(s)$ is regular at the singularity s_2; hence it will not contribute anything to the principal part $F_{p2}(s)$. This means that the principal part $F_{p2}(s)$ will be the same whether we expand $F_{r1}(s)$ or the original function $F(s)$.

We now repeat this process with $F_{r2}(s)$ and keep repeating with each singularity. At each step we subtract the principal part of the Laurent expansion until all the singular points are exhausted. The regular part of the last Laurent expansion will have no other singularities in the finite plane. Hence it must be an entire function. In this fashion we have succeeded in obtaining a representation of $F(s)$ that has the form

$$F(s) = \sum_{k=1}^{n} F_{pk}(s) + F_r(s). \tag{103}$$

Each of the terms in the summation is the principal part of the Laurent series of $F(s)$ expanded about one of its singularities. The last term is an entire function. If $F(s)$ is regular at infinity, this term will be a constant. If $F(s)$ has a pole of order n at infinity, this term will be a polynomial of degree n. Finally, if $F(s)$ has an essential singularity at infinity, this term will be an infinite power series. The representation of an analytic function given in (103) is called a *partial-fraction expansion*.

Suppose a function has an infinite number of poles and no essential singularities in the finite plane (this makes it a meromorphic function).

In such cases also a partial-fraction expansion can be found. However, the summation of principal parts in (104) will be an infinite series and may not converge in general. Nevertheless, it is always possible to so modify the terms that the series converges. But now the form of the expansion is changed. Of course, in some cases such a modification is not necessary, but the statement of the conditions when this is true is not a simple one, and we will not pursue the subject any further. (This expansion is known as the *Mittag–Leffler* expansion.)

A2.8 ANALYTIC CONTINUATION

Near the beginning of this appendix we defined an analytic function as one that is differentiable everywhere in a neighborhood, however small, of a point. Later, from the Taylor expansion of an analytic function about a point s_0 at which the function is regular, we saw that knowledge of all the derivatives of an analytic function at a point permits us to represent the function everywhere in a circle about the point, a circle that extends up to the closest singularity of the function. We stated that once a power-series representation of a function about a point is obtained, no matter by what procedure, this series is unique. We can state this result in a different way as follows. *If two functions are regular in a region* R *and if they coincide in some neighborhood, no matter how small, of a point* s_0 *in* R, *then the two functions are equal everywhere in* R. This theorem is called the *identity theorem for analytic functions.* (In fact, the two functions need coincide only on a segment of path no matter how small; or even only on an infinite number of distinct points with a limit point at s_0.)

Now let us consider two functions $F_1(s)$ and $F_2(s)$, which are respectively regular in overlapping regions R_1 and R_2, the common region being R_0, as shown in Fig. 13. (The regions need not be circular as shown here.) The two functions $F_1(s)$ and $F_2(s)$ determine each other uniquely. This follows from the identity theorem since only one function can be regular in R_1 (or R_2) and have the same values in R_0.

Suppose we were starting with the function $F_1(s)$ in R_1 and could find a function $F_2(s)$ in R_2 with the property just described. We would say that $F_1(s)$ has been *analytically coninued* beyond its original region into region R_2. But we might just as well consider $F_2(s)$ to be the original one and $F_1(s)$ its *analytic continuation* into region R_1. For this reason we say that each of them is but a *partial representation*, or an *element*, of a single function $F(s)$ that is regular in both R_1 and R_2.

Consider now the problem of starting with one element $F_1(s)$ of a

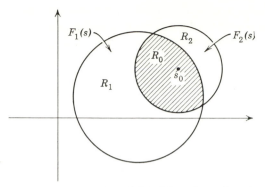

Fig. 13. Common region of definition of two functions.

function, which is in the form of a power series, and determining its analytic continuation outside its original circle of convergence. Figure 13 can again be used. Suppose we choose a point s_0 in region R_1. From the given element $F_1(s)$ we can evaluate all the derivatives at s_0 and form a new power series about s_0. This series will certainly converge in R_1, the original region of convergence of $F_1(s)$, and may also converge in a circle that extends beyond the original circle, as the illustration in Fig. 13 shows. The series then defines another element $F_2(s)$ of the function of which $F_1(s)$ is also an element. We can now choose another point within the new region R_2, but not common with R_1, and again calculate a new series that may converge in a circle extending beyond the boundaries of R_2.

This procedure can now be repeated. The only circumstance that will prevent any one circle from extending beyond the preceding one is the existence of a singular point on the circumference of the first circle that lies on the radius of the first circle drawn through the center chosen for the second one. But this can be rectified by choosing a different point for the center of the second circle, *unless every point on the first circle happens to be a singular point.* This is a possible occurrence, but it is not common. If such is the case, the original function is said to have a *natural boundary* beyond which it cannot be analytically continued.

Barring a natural boundary, then, an element can be analytically continued into the whole plane by this process of overlapping circles. The only points that will be excluded from the interiors of any of the circles will be the singular points. The sequence of functions defined in the circles will all be elements of a single function. It is now clear why an analytic function was defined as it was.

The process we have described has very little practical value, since we would not ever contemplate the actual construction of all the elements of a function in this manner. However, it has very great significance in providing insight into the fundamental behavior of functions.

In the process of constructing (in imagination, at least) the overlapping circles, suppose one of them overlaps one of the earlier ones (thus forming a closed chain). The question will arise whether the functional values given by the latest function will be the same as those given by the previous one in the common region of the two circles. If these values are not the same, then the function defined by this set of elements will be *multivalued*.

Let us now consider another aspect of analytic continuation. Suppose a function is defined along a simple arc which lies in a region R. It may be possible to find a function that is regular in R and coincides with this one on the simple arc. This function is also called the analytic continuation of the original one. The simple arc may be, for example, part or all of the $j\omega$-axis. If we define, as an example, a function to have the value $1 + j\omega$ for the interval $1 \leq \omega \leq 2$, its analytic continuation is $1 + s$. There is no other function that is regular in the region containing the given interval on the $j\omega$-axis and that coincides with the given function in that interval.

THEORY OF LAPLACE TRANSFORMS

As for the appendix on functions of a complex variable, this appendix on Laplace transforms will serve as a reference for those already familiar with the subject. It will also provide an instructor with an outline that he can augment by filling in discussions, illustrations, etc.

The concept of transforming a function can be approached from the idea of making a change of variable in order to simplify the solution of a problem. Thus if we have a problem involving the variable x, we substitute some other expression for x in terms of a new variable (e.g., $x = \sin \theta$), with the anticipation that the problem has a simpler formulation and solution in terms of the new variable θ. After obtaining the solution in terms of the new variable, we use the opposite of the previous change and thus have the solution of the original problem.

A more complicated "change of variable," or transformation, is often necessary. If we have a function $f(t)$ of the variable t, we define an *integral transform of $f(t)$* as

$$\text{integral transform of } f(t) = \int_a^b f(t)\, K(t,s)\, dt. \tag{1}$$

The function $K(t, s)$, which is a function of two variables, is called the *kernal* of the transformation. Note that the integral transform no longer depends on t; it is a function of the variable s on which the kernel depends.

A3.1 LAPLACE TRANSFORMS: DEFINITION AND CONVERGENCE PROPERTIES

The type of transform that is obtained and the types of problem in which it is useful depend on two things: the kernel and the limits of integration. For the particular choice of the kernel $K(s, t) = \epsilon^{-st}$ and the limits 0 and infinity, the transform is called a *Laplace transform* and is denoted by $\mathcal{L}\{f(t)\}$. Thus

$$\mathcal{L}\{f(t)\} = \int_0^\infty f(t)\epsilon^{-st}\,dt. \tag{2}$$

The Laplace transform of $f(t)$ is thus a function of the complex variable s. We denote the Laplace transform of $f(t)$ by $F(s)$.

Because it is defined as an integral, the Laplace transform is a *linear functional*; that is, if $f_1(t)$ and $f_2(t)$ have Laplace transforms $F_1(s)$ and $F_2(s)$, and k_1, k_2 are constants,

$$\mathcal{L}\{k_1 f_1(t) + k_2 f_2(t)\} = k_1\,F_1(s) + k_2\,F_2(s). \tag{3}$$

Since the defining equation contains an integral with infinite limits, one of the first questions to be answered concerns the existence of Laplace transforms. A simple example of a function that does not have a Laplace transform is ϵ^{ϵ^t}. Let us therefore state a few theorems (a few of which we shall also prove) concerning the convergence of the Laplace integral. Since s appears as a significant parameter in (2) we may expect the convergence to depend on the particular value of s. In general, the integral converges for some values of s and diverges for others.

In all of the theorems to follow we shall consider only integrable functions $f(t)$ without specifically saying so each time. As a first theorem, consider the following:

If the function f(t) *is bounded for all* $t \geq 0$, *then the Laplace integral converges absolutely for* $\mathrm{Re}(s) > 0$.

To prove the theorem, note that the condition on $f(t)$ means $|f(t)| < M$ for all $t \geq 0$, where M is a positive number. Then for $\sigma > 0$ we shall get

$$\int_0^T \left|\epsilon^{-st}f(t)\right|\,dt < M \int_0^T \epsilon^{-st}\,dt = \left(\frac{M}{\sigma}\right)(1 - \epsilon^{-\sigma T}). \tag{4}$$

In the limit, as T approaches infinity, the right-hand side approaches M/σ. Hence

$$\int_0^\infty \left|\epsilon^{-st}f(t)\right|\,dt < \frac{M}{\sigma} \qquad (\sigma > 0). \tag{5}$$

The familiar sine and cosine functions, and other periodic functions such as the square wave, satisfy the conditions of the theorem. Before commenting on this theorem, let us consider one more theorem.

If the Laplace integral converges for some $s_0 = \sigma_0 + j\omega_0$, *then it converges for all* s *with* $\sigma > \sigma_0$.

Let

$$\int_0^\infty \epsilon^{-s_0 t} f(t)\, dt = k_0,\tag{6}$$

where k_0 is a constant, since s_0 is a fixed complex number. Let us define the auxiliary function

$$\int_0^\tau \epsilon^{-s_0 t} f(t)\, dt = g(\tau).\tag{7}$$

Then $g(\tau)$ has a limit as τ goes to ∞; namely, k_0. Hence $g(\tau)$ is bounded for all τ. Next, we shall write the Laplace integral as below and integrate by parts to get

$$\int_0^T \epsilon^{-st} f(t)\, dt = \int_0^T \epsilon^{-(s-s_0)t} \epsilon^{-s_0 t} f(t)\, dt$$

$$= \epsilon^{-(s-s_0)t} g(t) \Big|_0^T - \int_0^T -(s-s_0)\epsilon^{-(s-s_0)t} g(t)\, dt,\tag{8}$$

or

$$\int_0^T \epsilon^{-st} f(t)\, dt = \epsilon^{-(s-s_0)T} g(T) - g(0) + (s-s_0)\int_0^T \epsilon^{-(s-s_0)t} g(t)\, dt.\tag{9}$$

Now $g(0) = 0$, $g(\infty) = k_0$, and if $\sigma > \sigma_0$, $\epsilon^{-(s-s_0)T} g(T)$ approaches 0 as T approaches ∞. Also, by the preceding theroem, the last integral in (9) converges absolutely for $\sigma > \sigma_0$ as T approaches ∞. Thus the result is proved. In fact

$$\int_0^\infty \epsilon^{-st} f(t)\, dt = (s-s_0)\int_0^\infty \epsilon^{-(s-s_0)t} g(t)\, dt \qquad (\sigma > \sigma_0).\tag{10}$$

This result can be strengthened to show that the Laplace integral converges *absolutely* for $\sigma > \sigma_0$ if it converges for σ_0. However, we shall not need this result in the general case. For functions of exponential order (to be defined shortly) we can prove this result with greater ease. Thus the *region of convergence* of the Laplace integral is a *half-plane*, since by this theorem, whenever the integral converges for some point in the s-plane, it converges at all points to the right. Thus we can define an *abscissa of convergence* σ_c such that the Laplace integral converges for all s with $\sigma > \sigma_c$ and diverges for all s with $\sigma < \sigma_c$. The stronger result, which we have not proved, is that the region of convergence is also the region of absolute convergence. The behavior of the Laplace integral is thus somewhat analogous to the behavior of power series. The function $f(t)$ plays the role of the coefficients of the power series, and the function ϵ^{-st} plays the part of $(s - s_0)^n$. Just as a power series may have any behavior on the circle of convergence, the Laplace integral may also have any behavior on the abscissa of convergence. The only difference concerns the existence of a singular point on the circle of convergence, which we shall examine a little later.

With infinite series, we have many tests for convergence. All of these have analogues in Laplace transforms. We shall be content to state just two of these. The analogue of the ratio test is the following:

If $|f(t)| \leq M\epsilon^{ct}$ for some constant M and some number c, for all t (or only for t greater than some T_0), then the Laplace integral converges absolutely for $\sigma > c$.

We see this result immediately since

$$\int_0^\infty |\epsilon^{-st} f(t)| \, dt \leq M \int_0^\infty \epsilon^{-\sigma t} \epsilon^{ct} \, dt = \frac{M}{\sigma - c} \qquad (\sigma > c). \qquad (11)$$

We thus have a sufficient criterion for the existence of the Laplace integral. Functions satisfying the inequality

$$|f(t)| \leq M\epsilon^{ct} \quad \text{for some } M \text{ and for } t > T \qquad (12)$$

are called functions of *exponential order*. The *order* of the function is the smallest number σ_0 such that the inequality (12) is satisfied by any

$$c = \sigma_0 + \delta \qquad (\delta > 0) \qquad (13)$$

and by no $c = \sigma_0 - \delta$. In this case we have established that the Laplace integral converges absolutely for $\sigma > \sigma_0$ and diverges for $\sigma < \sigma_0$.

Many functions that are not of exponential order have Laplace transforms. However, we can state the following necessary and sufficient condition, which shows that the integral of a transformable function is of exponential order.

The function $f(t)$ *is transformable, with the abscissa of convergence* $\sigma_0 > 0$ *if and only if the function.*

$$g(t) = \int_0^t f(x)\, dx \tag{14}$$

satisfies

$$|g(t)| \leq M_\epsilon e^{ct} \tag{15}$$

for any $c = \sigma_0 + \delta$.

The proof of this result depends on the Stieltjes integral, and so we cannot give it here. We can use this theorem to get an analogue for the Cauchy root test for power series.

Let $g(t)$ *be the function defined in* (14). *If*

$$\lim_{t \to \infty} \frac{\ln |g(t)|}{t} = c \neq 0, \tag{16}$$

then the abscissa of convergence of the Laplace integral of $f(t)$ *is* c. *The integral converges for* $\sigma > c$ *and diverges for* $\sigma < c$. *If* $c = 0$, *the test is inconclusive.*

In the case of power series the regions of convergence, absolute convergence, and uniform convergence coincide. We have stated that in the case of the Laplace integral the regions of convergence and absolute convergence coincide, both of them being half-planes. Therefore we may ask whether the region of uniform convergence also coincides with the region of convergence. The answer to this question is in the negative in the general case. The region of uniform convergence is described in the following theorem, which we shall not prove:

If the Laplace integral converges for $s = \sigma_0$, *then it converges uniformly in the sector.*

$$\left| \arg(s - \sigma_0) \right| \leq \left(\frac{\pi}{2} \right) - \delta \tag{17}$$

for every $\delta > 0$.

This region is shown in Fig. 1. We may take σ_0 to be the abscissa of convergence σ_c if the integral converges at this point. Otherwise, σ_0 is a point arbitrarily close to σ_c and to the right.

In the case of functions of exponential order, however, the region of uniform convergence coincides with the region of convergence; that is, we may take $\delta = 0$ in the theorem above.

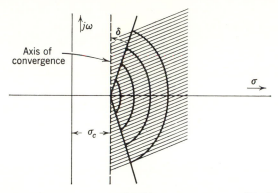

Fig. 1.   Regions of convergence and uniform convergence of Laplace integral.

For functions of exponential order the region of uniform convergence is the half-plane

$$\sigma \geq \sigma_c + \delta \qquad (\delta > 0),$$

where σ_c is the abscissa of convergence.

The proof of this result is quite similar to the proof given earlier for absolute convergence and so is omitted.

Thus the convergence behavior of the Laplace integral for functions of exponential order is identical with the behavior of power series.

A3.2   ANALYTIC PROPERTIES OF THE LAPLACE TRANSFORM

Using the power-series analogy again, a power series defines an analytic function within the circle of convergence. We may therefore wonder whether the analogy extends this far. The answer to this question is in the affirmative, as stated by the following theorem:

If the integral

$$F(s) = \int_0^\infty \epsilon^{-st} f(t)\, dt \qquad (18)$$

converges for $\sigma > \sigma_c$, then the function $F(s)$ defined by the integral is regular in the half-plane $\sigma > \sigma_c$. In fact, the derivative of $F(s)$ is given by

$$\frac{dF(s)}{ds} = \int_0^\infty \epsilon^{-st}(-t) f(t)\, dt \qquad (19a)$$

and in general

$$F^{(n)}(s) = \int_0^\infty \epsilon^{-st}(-t)^n f(t)\ dt. \tag{19b}$$

Given any point s with $\sigma > \sigma_c$, we can surround this point with a circle that is entirely within the region of uniform convergence, since ε in (17) is arbitrary. Now, because of the uniform convergence, the limit operations of integration and differentiation can be interchanged. Hence

$$\frac{d}{ds}\ F(s) = \frac{d}{ds}\int_0^\infty \epsilon^{-st} f(t)\ dt$$

$$= \int_0^\infty \frac{\partial}{\partial s}\ \epsilon^{-st} f(t)\ dt. \tag{20}$$

This leads to (19a). The convergence of (19a) is easily established for functions of exponential order. For the general case we integrate by parts.

Thus the Laplace integral defines a regular function within the half-plane of convergence. However, although the function $F(s)$ is *defined* by the integral only in the half-plane of convergence, we can use the technique of *analytic continuation* to extend the function across the abscissa of convergence whenever it may be continuable. [In practice this is merely a formality, the "analytic continuation" being merely an extension of the formula for $F(s)$.] It is this more general analytic function that is referred to as the *Laplace transform.* If $F(s)$ is the Laplace transform of $f(t)$, we refer to $f(t)$ as the *determining function* and $F(s)$ as the *generating function.*

In this more general concept of a Laplace transform the generating funtion will, in general, have singularities. They will have to lie in the half-plane $\sigma \leq \sigma_c$ or at ∞. Here we may revert to the power-series analogy again. The function defined by a power series always has a singular point on the circle of convergence. We may ask whether $F(s)$ has a finite singular point on the abscissa of convergence σ_c. Here the analogy breaks down. In general, there may be no singular point on $\sigma = \sigma_c$. The following example has been given by Doetsch:

$$f(t) = -\pi\epsilon^t \sin{(\pi\epsilon^t)}. \tag{21}$$

For this function the abscissa of convergence is zero. However, its transform satisfies the difference equation

$$F(s) = 1 - \frac{s(s+1)}{2}\ F(s+2) \tag{22}$$

so that $F(s)$ is an entire function.

However, in certain special cases, the transform has a singular point on $s = \sigma_c + j\omega$. For instance, if $f(t)$ is ultimately non-negative, then it can be shown that the real point on the abscissa of convergence is a singular point. This result is too specialized to be of interest to us and so we omit its proof. The important result as far as we are concerned is that the Laplace transform is an analytic function that is regular in the half-plane of convergence of the defining integral. The general Laplace transform is the function obtained by analytically continuing the original function.

One of the important analytic properties of the Laplace transform is its behavior at ∞. Concerning this we have the following theorem:

If the determining function f(t) *is a real or complex valued function of* t *and the Laplace integral converges at* s_0, *then as* s *approaches* ∞ *from within the sector*

$$|\arg(s - s_0)| \leq \left(\frac{\pi}{2}\right) - \delta \quad \text{with} \quad \delta > 0,$$

the generating function F(s) approaches 0.

The proof of this result proceeds as follows. We begin with a given $\varepsilon > 0$. Since $f(t)$ is an integrable function and therefore bounded for all t, we can find a T_1 so small that for $\sigma > 0$

$$\left| \int_0^{T_1} \epsilon^{-st} f(t) \, dt \right| < \int_0^{T_1} |f(t)| \, dt < \frac{\varepsilon}{3}. \tag{23}$$

Since the Laplace integral is uniformly convergent in this sector, we can find T_2 so large that $T_2 > T_1$ and

$$\left| \int_{T_1}^{\infty} \epsilon^{-st} f(t) \, dt \right| < \frac{\varepsilon}{3} \tag{24}$$

for all s in this sector. These two conditions fix T_1 and T_2 and therefore the value of the integral

$$\int_{T_1}^{T_2} |f(t)| \, dt = M. \tag{25}$$

Finally, we find σ_1 so large that

$$\epsilon^{-\sigma_1 T_2} < \frac{\varepsilon}{3M}$$

and then the product of the two is again a power series:

$$F_1(s)\ F_2(s) = \sum_{n=0}^{\infty} c_n\, s^n, \tag{32a}$$

where

$$c_n = \sum_{k=0}^{n} a_{n-k}\, b_k = \sum_{k=0}^{n} a_k\, b_{n-k}. \tag{32b}$$

The product series converges in the common region of convergence of the two individual series. The sums in $(32b)$ are known as *convolution sums*. We get a similar result in Laplace transforms.

If

$$F_1(s) = \int_0^{\infty} \epsilon^{-st} f_1(t)\ dt$$

and $\tag{33}$

$$F_2(s) = \int_0^{\infty} \epsilon^{-st} f_2(t)\ dt$$

have finite abscissae of convergence σ_1 and σ_2, then the product $F_1(s)\ F_2(s)$ *is also a Laplace transform*

$$F_1(s)\ F_2(s) = \int_0^{\infty} \epsilon^{-st} g(t)\ dt, \tag{34a}$$

where

$$g(t) = \int_0^{t} f_1(x)\, f_2(t-x)\ dx = \int_0^{t} f_1(t-x)\, f_2(x)\ dx, \tag{34b}$$

with an abscissa of convergence equal to the larger of σ_1, σ_2.

If $F_1(s)$ and $F_2(s)$ are Laplace transforms of $f_1(t)$ and $f_2(t)$, with abscissae of convergence σ_1 and σ_2, the Laplace transform of the product $f_1(t)f_2(t)$ is given by

$$\mathcal{L}\{f_1(t) f_2(t)\} = \frac{1}{2\pi j} \int_{c-j\infty}^{c+j\infty} F_1(z)\, F_2(s-z)\ dz, \tag{35}$$

where $\sigma_1 < c < \sigma - \sigma_2$ and $\sigma = \text{Re}(s)$ is greater than the abscissa of convergence $\sigma_1 + \sigma_2$.

The first of these two results is of considerable interest in network theory and is proved in Chapter 5. The second result is not of particular interest to us; we shall omit its proof. The integrals in (34b) and (35) are known as *convolution integrals*, the first being a real convolution and the second a complex convolution.

DIFFERENTIATION AND INTEGRATION

Next we shall consider the analytic operations of differentiation and integration in both domains. These correspond, as we shall see, to multiplication or division by s or t. Differentiation in the s-domain has already been considered; let us repeat the result here.

If

$$\mathcal{L}\{f(t)\} = F(s), \tag{36a}$$

then

$$\mathcal{L}\{t^n f(t)\} = (-1)^n F^{(n)}(s), \tag{36b}$$

the abscissae of convergence being the same.

As might be expected, the inverse operations, division by t and integration in s, correspond. The negative sign is missing, however.

If

$$\mathcal{L}\{f(t)\} = F(s), \tag{37a}$$

then

$$\mathcal{L}\left\{\frac{f(t)}{t}\right\} = \int_s^\infty F(z)\,dz, \tag{37b}$$

where the abscissae of convergence are the same and the path of integration is restricted to the sector of uniform convergence.

This result is proved by integrating by parts in the s-domain, noting that $F(s)$ approaches 0 as s approaches ∞. More important operations than the proceding ones, as far as the application to network theory is concerned, are differentiation and integration in the t-domain. These operations are found to resemble duals of the ones above.

Let f(t) *be differentiable (and therefore continuous) for* t > 0, *and let the derivative* f′(t) *be transformable. Then* f(t) *is also transformable and with the same abscissa of convergence. Further*

$$\mathfrak{L}\{f'(t)\} = sF(s) - f(0+), \qquad (38a)$$

where

$$F(s) = \mathfrak{L}\{f(t)\} \qquad (38b)$$

and

$$f(0+) = \lim_{\substack{t \to 0 \\ t > 0}} f(t). \qquad (38c)$$

Since $f'(t)$ is transformable, it follows that $f(t)$ is of exponential order and therefore transformable. The rest follows on integrating

$$\int_0^T \epsilon^{-st} f'(t) \, dt$$

by parts and taking the limit as T goes to ∞.

Let f(t) *be an integrable and transformable function. Let*

$$g(t) = \int_0^t f(x) \, dx. \qquad (39)$$

Then g(t) *is also transformable and with the same abscissa of convergence. Further*

$$G(s) = \frac{1}{s} F(s), \qquad (40)$$

where G(s) *and* F(s) *are Laplace transforms of* g(t) *and* f(t) *respectively.*

The first part follows as before. Equation 40 follows from (38b) on observing that $g(0+) = 0$ by (39).

These results can easily be extended to higher order derivatives and integrals of $f(t)$ by repeated applications of these theorems.

INITIAL-VALUE AND FINAL-VALUE THEOREMS

Two other limit operations are of considerable interest in estimating the behavior of the transient response of linear systems. In the first one we seek to relate the value of $f(t)$ at $t = 0$ to a specific value of $F(s)$. The definition of the Laplace transform gives us only a relationship between the values of $f(t)$ on the whole of the real positive t-axis and the behavior of $F(s)$ in a complex half-plane. The desired relationship is the following:

If $\mathcal{L}\{f(t)\} = F(s)$ *with a finite abscissa of convergence* $\sigma_c < \sigma_0$, *and if* $f'(t)$ *is transformable, then*

$$f(0+) = \lim_{s \to \infty} sF(s), \tag{41}$$

where the limit on the right is to be taken in the sector

$$|\arg(s - \sigma_0)| \le \left(\frac{\pi}{2}\right) - \delta.$$

This is called the *initial-value theorem.*

To prove this result we start with the derivative formula

$$\mathcal{L}\{f'(t)\} = sF(s) - f(0+) \tag{42}$$

and take the limit as s goes to ∞ in the sector specified. Since $\mathcal{L}\{f'(t)\}$ is a Laplace transform, it goes to zero as s goes to ∞ in this sector. The result is (41).

We might analogously expect to get the final value $f(\infty)$ by taking the limit of (42) as s approaches 0. We run into difficulties here, since

$$\lim_{s \to 0} sF(s) = f(0+) + \lim_{s \to 0} \int_0^\infty f'(t)\epsilon^{-st}\, dt, \tag{43}$$

where the limits are to be taken with $|\arg s| \le (\pi/2) - \delta$.

It is first of all not clear that the last limit exists. If it does, we do not see what it might be. If we can interchange the limit and the integral, however, we shall get

$$\int_0^\infty \lim_{s \to 0} f'(t)\epsilon^{-st}\, dt = \int_0^\infty f'(t)\, dt = f(\infty) - f(0+). \tag{44}$$

If we assume uniform convergence of the Laplace integral for $f'(t)$ in a region including $s = 0$, then this interchange can be made. In such a case, however, the abscissae of convergence of both $f'(t)$ and $f(t)$ must be negative. This is possible only if $f(t)$ approaches 0 as t approaches ∞; that is,

$$f(\infty) = 0. \tag{45}$$

But in this instance the whole theorem will be devoid of content. Hence, in order to establish the theorem, the interchange of the limit and the integral must be justified by finer criteria than uniform convergence. This need takes the proof of the theorem outside the scope of this text. The desired theorem can be stated as follows:

If f(t) *and* f′(t) *are Laplace transformable, and if* sF(s) *is regular on the* jω-*axis and in the right half-plane, then*

$$\lim_{t \to \infty} f(t) = \lim_{s \to 0+} sF(s), \tag{46}$$

where the limit on the right is to be taken along the positive real axis. This result is known as the *final-value theorem*.

SHIFTING

The last two operations that we shall consider are multiplication of $f(t)$ or $F(s)$ by an exponential function. Let us first consider multiplication of $F(s)$ by ϵ^{-as}, where a is a real number. We have

$$\epsilon^{-as} F(s) = \epsilon^{-as} \int_0^\infty \epsilon^{-st} f(t) \, dt$$

$$= \int_0^\infty \epsilon^{-s(t+a)} f(t) \, dt. \tag{47}$$

If we make the substitution $x = t + a$, and then change the dummy variable of integration back to t, we shall get

$$\epsilon^{-as} F(s) = \int_a^\infty \epsilon^{-st} f(t - a) \, dt. \tag{48}$$

If we assume that $f(t)$ vanishes for $t < 0$, then $f(t - a)$ will vanish for $t < a$ and the lower limit of the integral can be replaced by zero. To indicate that $f(t - a)$ is zero for $t < a$ we can write it in the form $f(t - a)u(t - a)$. The function $u(x)$ is the unit step, defined as zero for negative x and unity for positive x. This leads to the following result:

If $\mathcal{L}[f(t)] = F(s)$ *and* a *is nonnegative real, then*

$$\mathcal{L}[f(t - a)u(t - a)] = \epsilon^{-as} F(s), \tag{49}$$

with the same abscissa of convergence.

This result is called the *real shifting* or *translation theorem*, since $f(t - a)$ is obtained by shifting $f(t)$ to the right by a units.

The operation of multiplying $f(t)$ by ϵ^{at} leads to a similar result. This is called the *complex shifting theorem*.

If $\mathcal{L}[f(t)] = F(s)$ with abscissa of convergence σ_c; then

$$\mathcal{L}[\epsilon^{at}f(t)] = F(s + a), \tag{50}$$

with the abscissa of convergence $\sigma_c + Re(a)$.

This theorem follows directly from the definition of the Laplace transform.

A3.4   THE COMPLEX INVERSION INTEGRAL

We now consider the problem of finding the determining function $f(t)$ from a knowledge of the generating function $F(s)$. Since the uniqueness theorem tells us that two essentially different functions $f(t)$ cannot lead to the same function $F(s)$, we can expect to find an inverse transformation that will give us $f(t)$. We might intuitively expect that the inverse transformation would also be an integral, this time a complex integral in the s-domain. It must involve some kernel function of s and t, since we must end up with a function of t. Such is indeed the case, as stated by the following theorem:

Let $\mathcal{L}\{f(t)\} = F(s)$, with an abscissa of convergence σ_c. Then

$$\frac{1}{2\pi j}\int_{c-j\infty}^{c+j\infty} F(s)\epsilon^{st}\,ds = \begin{cases} 0 & t < 0 \\ \frac{1}{2}f(0+) & t = 0 \\ \frac{1}{2}[f(t+) + f(t-)] & t > 0 \end{cases} \tag{51}$$

where $c \geq 0$, $c > \sigma_c$. This is known as the inversion integral.

The proof of this important theorem involves a knowledge of the Fourier integral theorem and several results from the theory of Lebesgue integration. Usually we understand the normalization implied and write

$$\frac{1}{2\pi j}\int_{c-j\infty}^{c+j\infty} F(s)\epsilon^{st}\,ds = f(t)u(t) \tag{52}$$

or simply

$$\frac{1}{2\pi j}\int_{c-j\infty}^{c+j\infty} F(s)\epsilon^{st}\,ds = f(t), \tag{53}$$

the assumption $f(t) = 0$ for $t < 0$ being understood.

When the function $F(s)$ alone is given we do not generally know σ_c.

However we do know that $F(s)$ is regular for $\sigma > \sigma_c$. Hence, in such a case, we take the path of integration to be a vertical line to the right of all the singular points of $F(s)$. Such a path is known as a *Bromwich path* after the famous mathematician T. J. I'A. Bromwich, who made many significant contributions to the theory of Laplace transformation. The abbreviation " Br " is sometimes used on the integral sign, instead of the limits, to signify this contour.

We saw in Section 6 of Appendix 2 that the residue theorem can often be used to evaluate integrals of this type. In order to use the residue theorem we have to close the contour. Let us consider the two closed paths shown in Fig. 2. If the integrand $F(s)\epsilon^{st}$ satisfies Jordan's Lemma on either of the semicircular arcs, we can evaluate the integral by the residue theorem. If Jordan's lemma is satisfied on the arc to the right; that is, if

$$\lim_{s \to \infty} s F(s)\epsilon^{st} = 0, \qquad |\arg (s - \sigma_0)| \leq \pi/2 - \delta$$

(which will be true, for instance, if $t < 0$), the integral on C_1 of Fig. 2 is

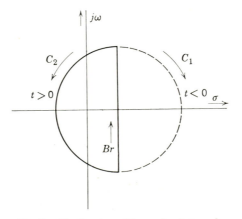

Fig. 2.   Evaluation of inversion integral.

zero. Since, in addition, the closed-contour integral is zero because no singularities are enclosed, the inversion integral yields zero.

If Jordan's lemma is satisfied on the semicircular arc to the left, which is much more often the case, the integral on the closed contour C_2 is $2\pi j$ times the sum of the residues of $F(s)\epsilon^{st}$ at the enclosed singular points. Including also the $1/2\pi j$ in (52), we get the following result:

If $F(s) \to 0$ *as* $s \to \infty$, *uniformly in the sector* $|arg\ (s - \sigma_0)| \geq \pi/2 - \delta$, *then*

$$f(t) = \sum \text{residues of } F(s)\epsilon^{st} \text{ at finite singularities of } F(s).$$

This is an extremely useful result. For simple functions, for example, rational functions, we can evaluate $f(t)$ very easily by this theorem. For a rational function to be a Laplace transform, the degree of the denominator polynomial must exceed the degree of the numerator polynomial by the condition of (28). Thus this inversion by residues is always applicable to rational functions.

In our brief discussion of the Laplace transform we have had to omit many of the proofs and several results that are of considerable importance. However, we have at least stated all the results that have been used in the main text. Those who would like a more thorough treatment are referred to the standard texts listed in the bibliography. We shall conclude our discussion with a very short table of transform pairs, which will be adequate for the present application.

Table of Transform Pairs

$f(t)$	$F(s) = \mathcal{L}\{f(t)\}$
df/dt	$sF(s) - f(0+)$
d^2f/dt^2	$s^2 F(s) - sf(0+) - f^{(1)}(0+)$
$d^n f/dt^n$	$s^n F(s) - s^{n-1}f(0+) - \cdots - f^{(n-1)}(0+)$
$\displaystyle\int_0^t f(\tau)\,d\tau$	$\dfrac{1}{s}F(s)$
$\displaystyle\int_0^t \int_0^\tau f(\lambda)\,d\lambda\,d\tau$	$\dfrac{1}{s^2}F(s)$
$t^n f(t)$	$(-1)^n \dfrac{d^n}{ds^n}F(s)$
$\dfrac{1}{t}f(t)$	$\displaystyle\int_s^\infty F(\sigma)\,d\sigma$
$\displaystyle\int_0^t f_1(t-\tau)f_2(\tau)\,d\tau$	$F_1(s)F_2(s)$
$f(t)\epsilon^{-\alpha t}$ $f(t/\alpha)$	$F(s+\alpha)$ $\alpha F(\alpha s)$
$f(t-\alpha)u(t-\alpha)$	$\epsilon^{-\alpha s}F(s)$
$\delta^{(n)}(t)$	s^n
$\delta^{(1)}(t)$	s

Table of Transform Pairs—Continued

$f(t)$	$F(s) = \mathcal{L}\{f(t)\}$
$\delta(t)$	1
$u(t)$	$\dfrac{1}{s}$
t	$\dfrac{1}{s^2}$
t^n	$\dfrac{n!}{s^{n+1}}$
$\epsilon^{-\alpha t}$	$\dfrac{1}{s+\alpha}$
$t\epsilon^{-\alpha t}$	$\dfrac{1}{(s+\alpha)^2}$
$t^n\epsilon^{-\alpha t}$	$\dfrac{n!}{(s+\alpha)^{n+1}}$
$\sin \omega t$	$\dfrac{\omega}{s^2+\omega^2}$
$\cos \omega t$	$\dfrac{s}{s^2+\omega^2}$
$\epsilon^{-\alpha t}\sin \omega t$	$\dfrac{\omega}{(s+\alpha)^2+\omega^2}$
$\epsilon^{-\alpha t}\cos \omega t$	$\dfrac{s+\alpha}{(s+\alpha)^2+\omega^2}$
$t\sin \omega t$	$\dfrac{2\omega s}{(s^2+\omega^2)^2}$
$t\cos \omega t$	$\dfrac{s^2-\omega^2}{(s^2+\omega^2)^2}$
$t\epsilon^{-\alpha t}\sin \omega t$	$\dfrac{2\omega(s+\alpha)}{[(s+\alpha)^2+\omega^2]^2}$
$t\epsilon^{-\alpha t}\cos \omega t$	$\dfrac{(s+\alpha)^2-\omega^2}{[(s+\alpha)^2+\omega^2]^2}$

BIBLIOGRAPHY

This bibliography is not intended to be exhaustive; the entries listed constitute some major and alternate references for the subjects treated in this text

1. MATHEMATICAL BACKGROUND

Complex Variable Theory

Churchill, R. V., *Introduction to Complex Variables and Applications*, McGraw-Hill Book Co., New York, 1948.

Hille, E., *Analytic Function Theory*, Ginn and Co., New York, vol. I, 1959.

Knopp, K., *Theory of Functions*, Dover Publications, New York, vol. I, 1945, vol. II, 1947.

LePage, W. R., *Complex Variables and the Laplace Transform for Engineers*, McGraw-Hill Book Co., New York, 1961.

Spiegel, M. R., *Theory and Problems of Complex Variables*, McGraw-Hill Book Co., New York, 1964.

Computer Programming

Healy, J. J., and Debruzzi, D. J., *Basic FORTRAN IV Programming*, Addison-Wesley Publishing Co., Reading, Mass., 1968.

McCalla, T. R., *Introduction to Numerical Methods and FORTRAN Programming*, John Wiley & Sons, New York, 1967.

McCracken, D. D., *A Guide to ALGOL Programming*, John Wiley & Sons, New York, 1962.

McCracken, D. D., *A Guide to FORTRAN IV Programming*, John Wiley & Sons, New York, 1965.

McCracken, D. D., *FORTRAN with Engineering Applications*, John Wiley & Sons, New York, 1967.

McCracken, D. D., and Dorn, W. S., *Numerical Methods and FORTRAN Programming*, John Wiley & Sons, New York, 1964.

Organick, E. I., *A FORTRAN IV Primer*, Addison-Wesley Publishing Co., Reading, Mass., 1966.

Differential Equations

Ayres, F., Jr., *Theory and Problems of Differential Equations*, McGraw-Hill Book Co., New York, 1952.

Bellman, R., *Stability Theory of Differential Equations*. McGraw-Hill Book Co., New York, 1953.

Coddington, E. A., and Levinson, N., *Theory of Ordinary Differential Equations*, McGraw-Hill Book Co., New York, 1955.

Coppel, W. A., *Stability and Asymptotic Behavior of Differential Equations*, D. C. Heath and Co., Boston, 1965.

Hartman, P., *Ordinary Differential Equations*, John Wiley & Sons, New York, 1964.

Lefschetz, S., *Differential Equations: Geometrical Theory*, 2nd edition, John Wiley & Sons, New York.

Sansone, G., and Conti, R., *Nonlinear Differential Equations*, revised edition, The Macmillan Co., New York, 1964.

Struble, R. A., *Nonlinear Differential Equations*, McGraw-Hill Book Co., New York, 1962.

Laplace Transform Theory

Churchill, R. V., *Operational Mathematics*, McGraw-Hill Book Co., New York, 1958.

Doetsch, G., *Handbuch der Laplace Transformation*, Birkhauser, Basel, vol. 1, 1950.

LePage, W. R., *Complex Variables and the Laplace Transform for Engineers*, McGraw-Hill Book Co., New York, 1961.

Spiegel, M. R., *Theory and Problems of Laplace Transforms*, McGraw-Hill Book Co., New York, 1965.

Matrix Algebra

Ayres, R., Jr., *Theory and Problems of Matrices*, McGraw-Hill Book Co., New York, 1962.

Bellman, R., *Introduction to Matrix Analysis*, McGraw-Hill Book Co., New York, 1960.

Gantmacher, F. R., *The Theory of Matrices*, Chelsea Publishing Co., New York, vol. I, 1959, vol. II, 1959.

Hohn, F. E., *Elementary Matrix Algebra*, The Macmillan Co., New York, 1958.

Pease, M. C., III, *Methods of Matrix Algebra*, Academic Press, New York, 1965.

Perlis, S., *Theory of Matrices*, Addison-Wesley Publishing Co., Cambridge, Mass., 1952.

Numerical Analysis

Beckett, R., and Hurt, J., *Numerical Calculations and Algorithms*, McGraw-Hill Book Co., New York, 1967.

Hamming, R. W., *Numerical Methods for Scientists and Engineers*, McGraw-Hill Book Co., New York, 1962.

Henrici, P., *Discrete Variable Methods in Ordinary Differential Equations*, John Wiley & Sons, New York, 1962

Henrici, P., *Elements of Numerical Analysis*, John Wiley & Sons, New York, 1964.

Householder, A. S., *Principles of Numerical Analysis*, McGraw-Hill Book Co., New York, 1953.

Kelly, L. G., *Handbook of Numerical Methods and Applications*, Addison-Wesley Publishing Co., Reading, Mass., 1967.

Macon, N., *Numerical Analysis*, John Wiley & Sons, New York, 1963.

Scheid, F., *Theory and Problems of Numerical Analysis*, McGraw-Hill Book Co., New York, 1968.

Wilkinson, J. H., *Rounding Errors in Algebraic Processes*, Prentice-Hall, Englewood Cliffs, N.J., 1963.

2. NETWORK TOPOLOGY AND TOPOLOGICAL FORMULAS

Kim, W. H., and Chien, R. T-W., *Topological Analysis and Synthesis of Communication Networks*, Columbia University Press, New York, 1962.

Seshu, S. and Balabanian, N., *Linear Network Analysis*, John Wiley & Sons, New York, 1959.

Seshu, S., and Reed, M. B., *Linear Graphs amd Electrical Networks*, Addison-Wesley Publishing Co., Reading, Mass., 1961.

3. LOOP, NODE-PAIR, MIXED-VARIABLE EQUATIONS

Desoer, C. A., and Kuh, E. S., *Basic Circuit Theory*, McGraw-Hill Book Co., New York, 1966.

Guillemin, E. A., *Theory of Linear Physical Systems*, John Wiley & Sons, New York, 1963.

Seshu, S., and Balabanian, N., *Linear Network Analysis*, John Wiley & Sons, New York, 1959.

4. NETWORK FUNCTIONS AND THEIR PROPERTIES

Balabanian, N., *Network Synthesis*, Prentice-Hall, Englewood Cliffs, N.J., 1958.

Bode, H. W., *Network Analysis and Feedback Amplifier Design*, D. Van Nostrand Co., New York, 1945.

Cauer, W., *Synthesis of Communications Networks*, McGraw-Hill Book Co., New York, translation second German edition, 1958.

Guillemin, E. A., *Synthesis of Passive Networks*, John Wiley & Sons, New York, 1957.

Kuh, E. S. and Rohrer, R. A., *Theory of Linear Active Networks*, Holden-Day, San Francisco, 1967.

Newcomb, R. W., *Linear Multiport Synthesis*, McGraw-Hill Book Co., New York, 1966.

Seshu, S., and Balabanian, N., *Linear Network Analysis*, John Wiley & Sons, New York, 1959.

Van Valkenburg, M. E., *Introduction to Modern Network Synthesis*, John Wiley& Sons, New York, 1960.

5. STATE EQUATIONS

Calahan, D. A., *Computer-Aided Network Design*, McGraw-Hill Book Co., New York, 1968.

DeRusso, P. M., Roy, R. J., and Close, C. M., *State Variables for Engineers*, John Wiley & Sons, New York, 1965.

Desoer, C. A., and Kuh, E. E., *Basic Circuit Theory*, McGraw-Hill Book Co., New York, 1966.

Gupta, S. C., *Transform and State Variable Methods in Linear Systems*, John Wiley & Sons, New York, 1966.

Koenig, H. E., Tokad, Y., and Kesavan, H. K., *Analysis of Discrete Physical Systems*, McGraw-Hill Book Co., New York, 1967.

Roe, P. H. O'N., *Networks and Systems*, Addison-Wesley Publishing Co., Reading, Mass., 1966.

Zadeh, L. A. and Desoer, C. A., *Linear System Theory*, McGraw-Hill Hill Book Co., New York, 1963.

6. NETWORK RESPONSE AND TIME-FREQUENCY RELATIONSHIPS

Cheng, D. K., *Analysis of Linear Systems*, Addison-Wesley Publishing Co., Reading, Mass., 1959.

Guillemin, E. A., *The Mathematics of Circuit Analysis*, John Wiley & Sons, New York, 1949.

Kuo, F. F., *Network Analysis and Synthesis*, 2nd edition, John Wiley & Sons, New York, 1966.

7. NETWORK SYNTHESIS

Balabanian, N., *Network Synthesis*, Prentice-Hall, Englewood Cliffs, N.J., 1958.

Chirlian, P. M., *Integrated and Active Network Analysis and Synthesis*, Prentice-Hall, Englewood Cliffs, N.J., 1967

Ghausi, M. S., and Kelly, J. J., *Introduction to Distributed-Parameter Networks*, Holt, Rinehart and Winston, New York, 1968.

Guillemin, E. A., *Synthesis of Passive Networks*, John Wiley & Sons, New York, 1957.

Karni, S., *Network Theory: Analysis and Synthesis*, Allyn and Bacon, Boston, 1966.

Newcomb, Robert W., *Active Integrated Circuit Synthesis*, Prentice-Hall, Englewood Cliffs, N.J., 1968.

Su, K. L., *Active Network Synthesis*, McGraw-Hill Book Co., New York, 1965.

Van Valkenburg, M. E., *Introduction to Modern Network Synthesis*, John Wiley & Sons, New York, 1960.

Weinberg, L., *Network Analysis and Synthesis*, McGraw-Hill Book Co., New York, 1962.

8. SCATTERING PARAMETERS

Carlin, H. J., and Giordano, A. B., *Network Theory*, Prentice-Hall, Englewood Cliffs, N.J., 1964.

Kuh, E. S. and Rohrer, R. A., *Theory of Linear Active Networks*, Holden-Day, San Francisco, 1967.

9. SIGNAL-FLOW GRAPHS

Horowitz, I. M., *Synthesis of Feedback Systems*, Academic Press, New York, 1963.

Huggins, W. H., and Entwisle, D. R., *Introductory Systems and Design*, Blaisdell Publishing Co., Waltham, Mass., 1968.

Lorens, C. S., *Flowgraphs for the Modeling and Analysis of Linear Systems*, McGraw-Hill Book Co., New York, 1964.

Mason, S. J., and Zimmerman, H. J., *Electronic Circuits, Signals, and Systems*, John Wiley & Sons, New York, 1960.

Robichaud, L. P. A., Boisvert, M., and Robert, J., *Signal Flow Graphs and Applications*, Prentice-Hall, Englewood Cliffs, N.J., 1962.

Truxal, J. G., *Control System Synthesis*, McGraw-Hill Book Co., New York, 1955.

Ward, J. R. and Strum, R. D., *The Signal Flow Graph in Linear Systems Analysis*, Prentice-Hall, Englewood Cliffs, N.J., 1968.

10. SENSITIVITY

Bode, H. W., *Network Analysis and Feedback Amplifier Design*, D. Van Nostrand Co., New York, 1945.

Horowitz, I. M.,*Synthesis of Feedback Systems*, Academic Press, New York, 1963.

Kuh, E. S., and Rohrer, R. A., *Theory of Linear Active Networks*, Holden-Day, San Francisco, 1967.

Truxal, J. G., *Automatic Feedback Control System Synthesis*, McGraw-Hill Book, Co., New York, 1955.

11. STABILITY

Routh, Hurwitz, Lienard-Chipart, and Nyquist Criteria

Ogata, K., *State Space Analysis of Control Systems*, Prentice Hall, Englewood Cliffs, N.J., 1967.

Schwarz, R. J., and Friedland, B., *Linear Systems*, McGraw-Hill Book Co., New York, 1965.

Zadeh, L. A., and Desoer, C. A., *Linear System Theory*, McGraw-Hill Book Co., New York, 1963.

Liapunov's Method

Hahn, W., *Theory and Application of Liapunov's Direct Method*, Prentice-Hall, Englewood Cliffs, N.J., 1963.

LaSalle, J., and Lefschetz, S., *Stability by Liapunov's Direct Method*, Academic Press, New York, 1961.

Liapunov, A. M., *Stability of Motion*, Academic Press, New York, 1966.

Ogata, K., *State Space Analysis of Control Systems*, Prentice-Hall, Englewood Cliffs, N.J., 1967.

Zubov, V. I., *Methods of A. M. Liapunov and Their Application*, P. Noordhoff, Groningen, The Netherlands, 1964.

12. TIME-VARYING AND NONLINEAR NETWORK ANALYSIS

Blaquiere, A., *Nonlinear System Analysis*, Academic Press, New York, 1966.

Butenin, N. V., *Elements of the Theory of Nonlinear Oscillations*, Blaisdell Publishing Co., New York, 1965.

Cunningham, W. J., *Introduction to Nonlinear Analysis*, McGraw-Hill Book Co., New York, 1958.

d'Angelo, H., *Time-varying Networks*, Allyn and Bacon, Boston, 1969.

Desoer, C. A. and Kuh, E. S., *Basic Circuit Theory*, McGraw-Hill Book Co., New York, 1966.

Minorsky, N., *Nonlinear Oscillations*, D. Van Nostrand Co., Princeton, N.J., 1962.

Stern, T. E., *Theory of Nonlinear Networks and Systems*, Addison-Wesley Publishing Co., Reading, Mass., 1965.

INDEX